The Biology and Ecology of Streams and Rivers

The Biology and Ecology of Streams and Rivers

Alan Hildrew

School of Biological and Behavioural Sciences, Queen Mary University of London, UK
Freshwater Biological Association, Ambleside, Cumbria, UK

Paul Giller

School of Biological, Earth and Environmental Sciences, University College Cork,
Distillary Fields, North Mall, Cork, Ireland

OXFORD
UNIVERSITY PRESS

Great Clarendon Street, Oxford, OX2 6DP,
United Kingdom

Oxford University Press is a department of the University of Oxford.
It furthers the University's objective of excellence in research, scholarship,
and education by publishing worldwide. Oxford is a registered trade mark of
Oxford University Press in the UK and in certain other countries

Published in the United States of America by Oxford University Press
198 Madison Avenue, New York, NY 10016, United States of America

British Library Cataloguing in Publication Data
Data available

Library of Congress Control Number: 2022948852

ISBN 978–0–19–851610–1
ISBN 978–0–19–851611–8 (pbk.)

DOI: 10.1093/oso/9780198516101.001.0001

Printed and bound by
CPI Group (UK) Ltd, Croydon, CR0 4YY

Preface

While we started this book several years ago, the bulk of the work took place during the 2020–22 COVID pandemic. In fact, it might be true to say that it might not have been finished even by now in more ordinary times—when other tasks could have interfered. It is remarkable that Oxford University Press, mainly in the person of Ian Sherman, showed such patience with us—for which we are extremely grateful. The book transformed as we wrote it. This was mainly because we perhaps had not fully appreciated at the outset quite how much the field had changed and expanded over the past few decades—we had been too close to it to take an overview. Noel Hynes wrote, in the preface to his (1970) classic *The Ecology of Running Waters*, that of the 1,500 references in his text (intended to be 'a comprehensive and critical review of the literature on the biology of rivers and streams') over half had been published since 1955. In other words, the literature on running waters was already growing very quickly. The biology and ecology of streams and rivers is not the fastest-moving area of science but the literature is nevertheless expanding rapidly, and progress is very encouraging, particularly in the light of the many challenges running-water systems are facing.

Although scientific progress may be good, the state of rivers and streams worldwide is not, even though this component of our environment is absolutely crucial to us humans and our future. It is our view that a real understanding of the natural history and ecology of running waters should be brought even more prominently into river management. If rivers can ever be anything like sustainable, ecology must take its place as an equal to the physical sciences such as hydrology and geomorphology. It is our purpose here, therefore, to provide an up-to-date and internationally focused view of the biology and ecology of running waters. It is aimed at advanced undergraduates of the aquatic, ecological and environmental sciences, graduate students studying or researching running waters, and at more established practitioners, managers and conservationists, as well as researchers in both lotic (running-water)

and general ecology who want a modern account of a particular new or developing area. While we look forward to new developments, at the same time we have tried to increase understanding by retaining something of the intellectual thread leading from the past.

Some 25 years ago, one of us (PSG) co-wrote a book, *The Biology of Streams and Rivers*, with the late Bjorn Malmqvist in the Biology of Habitats series published by Oxford University Press. The present book is not a revision of that title. If we ever thought it would be possible to produce a revised version, within a similar length, we clearly realised that was not what we wanted to do, even if it had been feasible. That early book ran to around 250 pages of text, in a small format. All but less than 30 pages of that book were about the habitat itself, the makeup of the biota, their adaptations, communities and a little about applications. The single chapter on energy and nutrients was less than 20 pages in length. Of course, that was the purpose of the series, aimed at undergraduates and with a more-or-less exclusive emphasis on natural history and biology. While we certainly wanted to reflect advances in those aspects, some of the most spectacular progress over the last 25 years has been in ecosystem ecology, in which the particular identity of organisms is not the main focus but rather the processes in which they are involved—of energy flow and the cycling of materials. This area has been very greatly expanded in the present book. We have attempted this by taking a much more ecological approach, building on the physico-chemical foundations of the habitat templet, progressing from the population and community ecology and diversity of organisms and linking them to ecosystem process via the complexities of food webs. We also stress interactions with rivers as ecosystems within the wider biosphere—another strong thread in modern lotic ecology. While not a book about application *per se*, each chapter refers to how humans affect rivers and are affected by them, and examples are embedded throughout, while a final chapter seeks to point the way forward in some key strategic areas and to adopting knowledge of natural history and ecology into policy and management (including how science itself is informed by lessons from practice). These aspects, after all, are what the modern student demands (having become aware of the basics!).

Of course, writing a book during a pandemic did have the advantage that we had less to distract us than normal. However, working in different countries, the limited travel opportunities also meant that we had little chance to interact face to face, and this had to be accommodated as the true scale of the task we had undertaken became clear. Nevertheless, and perhaps at the cost of a greater degree of overlap between the original draft chapters than we anticipated (and which we have worked hard to disentangle and reconcile), we have avoided major 'injury' and emerge with our long-standing friendship intact. Speaking of friendships, and whilst this is definitely not a revision of the Giller and Malmquist book, we both want to acknowledge Bjorn's untimely passing and the loss not just of a friend and colleague but also of a great stream ecologist and natural historian. His work keeps his memory alive.

We have many people to thank. We have already mentioned Ian Sherman at OUP, but it is worth repeating, and various others from OUP who have been involved from time to time (these include Lucy Nash, Bethany Kershaw, Charles Bath, and Giulia Lipparini) as well as Shanmugapriyan Gopathy who oversaw the production activities. We do thank them all for their patience and understanding. We thank Mike Winterbourn (University of Canterbury, New Zealand) for his usual direct and extremely useful reading of much of this book—he brought us down to earth on many occasions—thanks Mike! We are grateful to Jack Webster (Virginia Polytechnic Institute and State University, USA) for his expert (and speedy) help with Chapter 9. The inclusion of the Topic Boxes from a wide variety of top river scientists from around the world on a number of issues of contemporary importance adds a lot to this book and we thank them

all: Jon Benstead, Nuria Bonáda, Eric Chauvet, Alan Covich, Wyatt Cross, Russell Death, Sylvain Dolédec, Deb Finn, Mark Gessner, Steve Ormerod, Vince Resh, Tenna Riis, Belinda Robson, Emma Rosi, Dave Strayer, Bruce Wallis and Christina Zarfl. We are also grateful to the many publishers, societies and individual running-water biologists and ecologists for their permission to include the figures and photographs that illustrate the book and that we acknowledge in the figure legends. Of course, any mistakes are our own.

As we wrote this book, across the world we were experiencing both extreme floods and droughts, largely laid at the door of climate change. For example, much of western Europe, including southern and eastern Britain, in 2022 experienced the driest year since 1976, an extremely rare event. In parts of the UK there were bans on using hosepipes in what people see, quite wrongly, as a very wet country. In the UK, water use currently stands at >150 L per person per day and rising. The media are full of demands to build more reservoirs, halt the leaks and the 'wasteful' discharge of rivers into the sea and to restrain sewage overflows and pollution—in fact, anything so that consumers and farmers can continue to use water as they wish and to pay very little for it. No doubt there are similar demands elsewhere. The problem with water is there is either too much or too little or it is too polluted and when an immediate crisis has passed, people forget about it—until the next time. As societies and on a global basis, we really do need to have a serious conversation about the sustainability and management of fresh water and to learn more about it—we hope this book helps.

On finishing, we thank our families and above all Pam and Janet—without your forbearance and support this wouldn't have happened at all!

Alan Hildrew
London

Paul Giller
Cork

Contents

CHAPTER 1

Rivers as ecological systems

There is something inherently fascinating about running water, not least from a cultural or aesthetic point of view. Fresh water is essential for life on earth. Beyond that freshwater habitats, including running waters, are key ecological systems in their own right, home to an enormous variety and abundance of animals, plants and microbes and responsible for many important ecological processes. Of course, streams and rivers are also of enormous practical importance, bringing many benefits that have underpinned the development of civilisations around the world—though at the same time they are often a great natural hazard.

Running waters are different from other fresh waters, essentially because of their ribbon-like channels, and the inexorable flow downstream as water carves its way through the landscape. They come in a huge range of sizes, from the small trickles originating from precipitation or as groundwater springs (which can sometimes be substantial) at the upmost 'fingertips' of river networks, coalescing into shallow streams, often tumbling down an open mountainside or flowing through a forest, then joining with other streams to form bigger waterways. Channel width and the flow of water normally increase as we move down the river network. At some point, a stream becomes a river. There is really no formal definition of this change, and terminology is largely a point of view and is culturally determined. Ecologists sometimes refer to a stream as a channel that can at least normally be studied by wading and is of restricted width. What is clear though is that biologically and ecologically small streams and large rivers are different, and their study brings contrasting challenges, requiring methods and techniques of different scale and cost (Figure 1.1).

The world's largest rivers can be kilometres wide and tens of metres deep at some points in their long journey to the oceans. The first civilisations and most historical empires were founded around major floodplain rivers, such as the monsoon-controlled Huang He (Yellow River), the Indus and the Nile (which all have summer season floods lasting three or four months

followed by much-reduced flow), and the mid-latitude Tigris and Euphrates Rivers (with prominent springtime floods). The floods brought nutrient-rich sediments that supported floodplain agriculture. Rivers have thus shaped cultures and economies while their valleys became areas for settlements, infrastructure and food production. They provided drinking water and were early routes for transport. They were used for waste disposal, as sources of energy and offered cultural and aesthetic value (Macklin & Lewin 2015; Wantzen et al. 2016; Böck et al. 2018). Archaeological and later written evidence has shown that rivers have been used in these ways since ancient times. For example, the Sadd-el-Kafara Dam on the Nile was built about 4,510 years ago and is one of the oldest constructions of its kind anywhere (Hassan 2011). River channels and floodplains have fluctuated in their extent over timescales from decades to millennia, influencing the sustainability of the very civilisations they supported. Human settlements have been abandoned due to prolonged drought and reduced river flows or, conversely, due to an increase in the intensity or frequency of destructive floods. Even now, the average annual impact of water-related hazards (inadequate water and sanitation, drought and floods) affects more people and leads to more deaths and greater economic losses than are caused by earthquakes, epidemics and conflict (UNESCO 2019). It is estimated that around 2.2 billion people live in river basins stressed by water shortages; these basins account for 27% of global food crop production and 28% of global GDP and are losing freshwater storage—that is, they are drying (Huggins et al. 2022). About 4 billion people experience severe water scarcity during at least one month of the year (UNESCO 2021).

Although fresh water is arguably the most essential of natural resources and provides many benefits (which we now often refer to as *ecosystem services*), freshwater systems are directly threatened by a range of human activities. The growth and development of human societies is often based on the unsustainable

The Biology and Ecology of Streams and Rivers. Alan Hildrew and Paul Giller, Oxford University Press. © Alan Hildrew and Paul Giller (2023).
DOI: 10.1093/oso/9780198516101.003.0001

Figure 1.1 Contrasting running waters. (a) A wadable stream (a tributary of the Araglin River, Cork, Ireland); (b) the large Columbia River (Washington State, USA).
Source: (a) photo by Paul Giller; (b) licensed under the Creative Commons Attribution 2.0 Generic licence.

consumption of the 'natural capital' of rivers (Wantzen et al. 2016). A glance at the titles of the recent UNESCO World Water Development reports provides a picture of the challenges faced by mankind; these include 'Water in a changing world', 'Water for a sustainable world', 'Leaving no one behind', 'Water and climate change' and 'Valuing water'.

While covering less than 1% of the Earth's surface, fresh waters hold at least 100,000 identified species (nearly 6% of the global total) (Abell et al. 2008) with many more to come. Vorosmarty et al. (2010) estimated that the biodiversity sustained by around 65% of global river discharge and its associated aquatic habitat was under moderate to high threat. More recently, Maasri et al. (2022) predicted an 84% decline in the abundances of almost a thousand freshwater vertebrate species within less than 50 years. In addition, climate change (including, but not exclusively, an increase in temperature) is an ongoing threat to freshwater ecosystems and their biodiversity. Partly due to climate change and overexploitation of water, there have been declines identified in the actual area of freshwater ecosystems (largely wetlands of various kinds) in Europe of about 50% from 1970 to 2008, while globally there has been a decline of around 64% between 1997 and 2011 (Vári et al. 2022), and it is thought that an increasing fraction of rivers and streams now undergo periodic drying (IPCC 2022).

Our book, written against this background, explores the key aspects of the biology and ecology of streams and rivers and their unique physicochemical environment. It illustrates their wonderful biodiversity that drives important ecosystem processes. We came to the ecology of running waters through an interest in and curiosity about their natural history, and this science provides the surest information required to conserve and restore them. Rapid growth in understanding, and advances in the techniques now available, offers significant hope for the future of streams and rivers in our ever-changing world. We cannot afford otherwise.

1.1 Rivers in context

Fresh water represents only a small proportion of the total global water, almost all of which (97.5%) is salty, while some is frozen. A global map based on Landsat remote-sensing imagery data (Figure 1.2) gives a clear impression of the extent and distribution of the largest rivers on earth while, at a somewhat smaller scale (Figure 1.3), river networks become even more apparent. Rivers and streams have an estimated total global surface area (at mean annual discharge) of 773,000 +/− 79,000 km^2; equivalent to 0.58 +/− 0.06% of the earth's non-glaciated surface (Allen & Pavelsky 2018). This estimate is greater than earlier calculations by Raymond et al. (2013) at 536,000 km^2 and Downing et al. (2012) at between 485,000 and 662,000 km^2, and was developed from an extensive Landsat database containing over 58 million river width measurements (for rivers > 90 m wide), validated by *in situ* river measurements from North America, and a statistical analysis for stream and smaller (< 90 m wide) river surface area. Larger rivers (> approximately 50 m wide) do cover a greater surface area globally (Downing et al. 2012; Allen & Pavelsky 2018) while the total length of running waters is dominated by small stream channels.

Effectively the entire land surface of the earth (excepting Antarctica and a few of the driest deserts) is

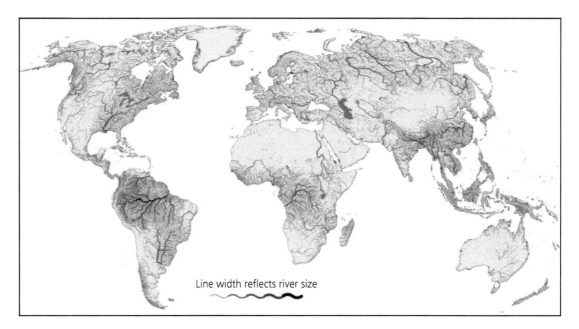

Line width reflects river size

Figure 1.2 The global river network, derived from HydroSHEDS version 1 database (https://www.hydrosheds.org), an applied global hydrographic mapping tool and a database using satellite imagery that provides catchment boundaries, river networks and lakes. Only large rivers can be shown, the thickness of the line denoting the size (width) of the river.
Source: Lehner et al. 2022, with permission from Bernhard Lehner and Eos: Science News by AGU (American Geophysical Union).

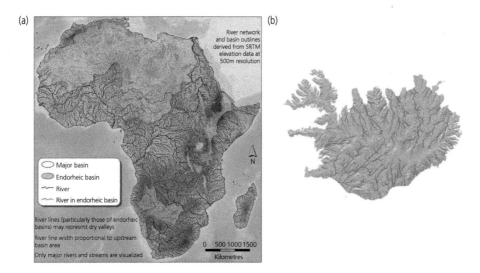

Figure 1.3 Higher-resolution digital maps based on the HydroSHEDS database showing river networks in: (a) Africa (showing the major river basins and endorheic areas). Endorheic drainage basins retain water and allow no outflow to external bodies of water, such as large rivers or oceans. Drainage converges instead into lakes or swamps and is lost by evaporation/evapotranspiration); (b) Iceland (only medium to large rivers are depicted).
Source: (a) from Lehner et al. 2022, with permission from Bernhard Lehner and Eos: Science News by AGU (American Geophysical Union); (b) from Lehner et al. 2008 with permission from the American Geophysical Union and John Wiley and Sons).

made up of river catchments (approximately 150 million km²), and the mean instantaneous water volume carried in river channels is just over 2,000 km³ (Giller & Malmqvist 1998). This represents a tiny fraction of fresh water on earth (approximately 0.006%), most of which is stored in the polar ice caps (69.56%; the majority in Antarctica), with the rest held in groundwater and soil (30.1%), the atmosphere (0.04%) and lakes and marshes (0.29%) (Keller 1984). There is, however, a rapid turnover of water in rivers and streams. The mean residence time of water in a river channel (i.e. the time between water entering a river and its reaching a downstream monitoring point) across 323 UK catchments was estimated at around 30 hours (Worrall et al. 2014). The longest time of those assessed was that for the highly regulated River Thames at 6.3 days but, for most systems, it was less than a day. At the scale of the whole river basin, residence time is of course much longer. The runoff maximum for the Amazon lags behind the maximum rainfall by some 2 to 3 months (Dai & Trenberth 2002) and, based on various river basins in the USA, residence times ranged from 2 to 20 years (Michel 1992), although most of these systems had water storage infrastructure such as reservoirs. Residence times are therefore important as they indicate not only the dynamic nature of streams and rivers but also the timescale in which the basin might respond to anthropogenic inputs.

Flowing water usually ends up in the sea, with 37,000–38,000 km³ being discharged from rivers to the world's oceans every year (Dai & Trenberth 2002). The 50 largest rivers account for about 57% of this global runoff. The eastern coasts of the Americas account for about 40% of global discharge, draining into the Atlantic Ocean (in large part from the Amazon at 17% of global runoff—6,600 km³ yr^{-1}); 15% comes from eastern Asia into the North Pacific, and 10% from the western coast of South America, also into the Pacific. The impacts of this runoff can be far reaching; for example, recent studies have shown a significant impact of the Amazon runoff on the northern hemispheric climate (Jahfer et al. 2017).

1.2 The hydrological cycle

The rapid turnover of water in rivers and stream channels is a dynamic part of the hydrological cycle, which involves the continuous recycling of water among the various storage compartments in the biosphere (Figure 1.4). Essentially, water evaporates from the oceans and land surface, and transpires from terrestrial vegetation, processes driven by solar energy. Clouds form when the atmosphere becomes saturated and

water vapour condenses into droplets or ice crystals around nuclei of dust, smoke particles or salt. Precipitation as rain, snow, sleet and hail deposits atmospheric water onto the land or directly back into the oceans.

At any one time, the atmosphere holds water equivalent to the annual rainfall of the Amazon basin (Keller 1984), though water in the atmosphere 'turns over' every nine days. Very little rain falls directly into streams and rivers. Some precipitation is intercepted by vegetation before reaching the ground and evaporates. Some water passes through the vegetation and into the soil, where it is taken up by plants, and again returns to the atmosphere via evapotranspiration. Drying of soils by evaporation also returns water to the atmosphere. Still more water percolates down through the soil into the water table, from where it recharges the groundwater. Alternatively, precipitation may be held over winter as snow or ice. The remainder finds its way downhill to streams and rivers (Figure 1.4).

Depending on the permeability and infiltration capacity of soil, rainfall can either travel as surface runoff directly and rapidly into stream and river channels (a major component of flooding), though this overland flow is relatively rare, or percolate into soil and flow just below the surface (*subsurface flow*) if there is a relatively impermeable layer between the surface and the water table. Alternatively, rainfall that has infiltrated deeper into the soil can be released more slowly from groundwater stores by displacement to enter stream channels from below the surface (*groundwater flow*) (see Chapter 2, section 2.2.4 and Figure 2.7 for more details). In addition, many headwater streams originate from springs where groundwater emerges directly onto the surface of the land. The hydrological cycle is completed by the downstream flow of water in streams and rivers, most of which eventually discharges into the sea.

The mean annual global rainfall for 2021 was estimated at 970.9 mm and the 40-year average at 981.85 mm (NCEI, 2021), with more falling over the oceans (1,036 mm y^{-1}) than over the land (832.2 mm y^{-1}). The proportion of this that ends up as streamflow depends on the weather, soil type and development, vegetation, slope of the land, properties of aquifers (groundwaters), the nature and extent of human alterations to the landscape and other local factors. Runoff is a small proportion of precipitation in dry deserts but much higher in moister areas. It has been estimated that, on average, around 35% of global precipitation runs off via rivers (Leopold 1962; Dai & Trenberth 2002). The rest is evaporated or transpired or will be stored in aquifers (for periods that vary between weeks and months in

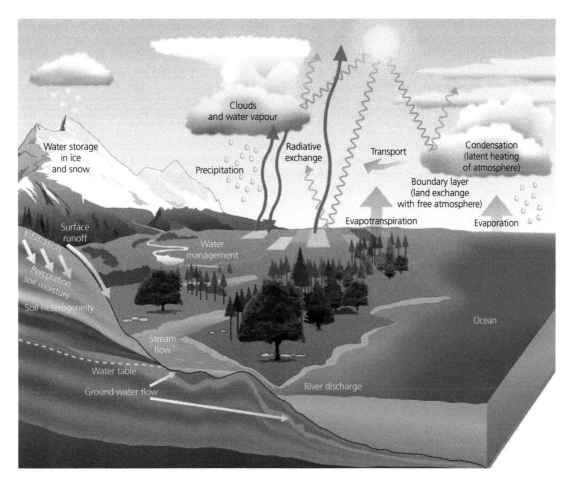

Figure 1.4 The hydrological cycle illustrating the various sources, storage compartments and flows within the biosphere (see text for details). *Source:* from UK Meteorological Office, with permission—© Crown copyright, Met Office (http://www.nationalarchives.gov.uk/doc/open-government-licence/version/3/).

karst—limestone—aquifers to thousand or even millions of years in deeper aquifers). Ongoing global warming is likely to lead to a more active hydrological cycle, fuelled by greater total cloud cover, wind speeds and global precipitation (Dodds & Whiles 2019).

From an ecological point of view, an important aspect of the hydrological cycle is that the water in streams and rivers has had intimate contact with the atmosphere and can take different routes through catchments, variously contacting vegetation, soil and rocks before entering the freshwater habitat. This has profound implications for the nature of the stream and river habitats—as described in the next chapter.

1.3 Rivers in space and time

The river *catchment* or *drainage basin* is the natural unit of landscape, combining the linked terrestrial and aquatic systems, and it encompasses the entire area of

land drained by the various tributaries and the main river. The size of the catchment obviously increases with distance from headwaters as more and more tributary streams merge, carrying the influence of their separate catchments with them. Movements of water and elements through the catchment link various components of the system; terrestrial and aquatic, plants and soils, atmosphere and vegetation, and soils, geology and water. The landscape governs the direction that water takes, with higher ground (ridges and hilltops) marking divisions between catchments. These boundaries are known as *watersheds*—that is, they dictate whether water flows into one catchment or another. This is the original terminology, and is used in this book, though 'watershed' is often used synonymously with catchment, with the boundaries between catchments then known as 'divides'.

Among the most influential papers on stream and river ecology is H.B.N. Hynes's (1975) famous essay

'The stream and its valley'. He asserted that in every respect 'the valley rules the stream'. The geology and morphology of the valley determines the soil (and availability of ions) and the slope of the land. Soil and climate together determine the vegetation. The vegetation determines the supply of organic matter and modifies the light reaching the stream and, together with the soil, influences water chemistry and water inputs to the stream. Human activity in catchments greatly affects streams and rivers, sometimes directly and obviously, sometimes subtly.

Clearly, streams and rivers can be characterised as extremely 'open' ecological systems, with exchanges of materials and energy with neighbouring systems. This in turn has a dramatic effect on their biology and ecology and how we study them—in no sense can they be described as 'microcosms' of wider

ecosystems. To understand the biology of streams and rivers holistically, it is therefore necessary to consider the entire drainage basin, incorporating both the aquatic system and its surrounding catchment. This means that the management of running waters should ideally be at the landscape scale rather than simply associated with the channel itself. A number of pioneering research programmes have adopted this 'whole-ecosystem' approach, such as the famous Hubbard Brook catchment (Likens & Bormann 1974). A further example that has also made an enormous contribution to lotic (running-water) ecology in particular is the long-term research carried out at the Coweeta Hydrologic Laboratory in the Little Tennessee River basin in the USA. In Topic Box 1.1, Bruce Wallace outlines the types of research conducted and the impressive range of results achieved over the years in one of the best-studied sets of

---

Topic Box 1.1 The Coweeta Hydrologic Laboratory

J. Bruce Wallace

The Coweeta Hydrologic Laboratory is in the Blue Ridge Mountains in the Southern Appalachians of Western North Carolina, USA, one of the world's oldest continually functioning hydrologic laboratories. Located in the headwaters of the Little Tennessee basin (latitude 35° 03' N, longitude 83° 25' W), the laboratory was established in 1934 by the United

States Forest Service (USFS) in response to severe downstream flooding of large rivers, including Tennessee, Ohio and Mississippi. The objective was to assess the influence of headwater forestry and farming practices on downstream water yield and quality. Thus, early experiments primarily tested the effects of land use on water yield and sediments. An example of the kind of infrastructure installed in Coweeta is shown in Box Figure 1.1a.

Box Figure 1.1a An example of an experimental weir for hydrological and hydrochemical monitoring in an experimental catchment at Coweeta.
Source: photo provided by Bruce Wallace.

Topic Box 1.1 *Continued*

These studies included effects of farming on steep gradients, logging practices, conversion of hardwoods to conifers and conversion of forests to grazed grasslands. Swank & Crossley (1988) describe many of these early studies at Coweeta. The laboratory consists of two main basins, Coweeta Creek (1,626 ha) and Dryman's Fork (559 ha) and includes first- to fourth-order streams (Box Figure 1.1b). Elevation ranges from 670 to 1,610 m asl. Mean precipitation, max to min (cm y^{-1}) is around 240 (range = 151 to 361) at high altitude and 181 (range = 107 to 272) at lower altitude over an 86-year record. For a more extensive history of the laboratory, maps, site descriptions, individual catchment histories, underlying geology and long-term records, see Swank & Crossley (1988) and Miniat et al. (2021). Within the Coweeta Creek Basin, there are 56 km of streams based on a 1:7,200 scale map; however, only 0.8 km appear on a 1:500,000 scale map, and only 24.4 km show on a 1:28,000 scale map. Thus, at least in North America, small streams are under-represented on most maps (Meyer & Wallace 2001). A thick canopy of

evergreen riparian *Rhododendron* borders most Coweeta streams draining the deciduous oak–hickory forest. These evergreens provide significant year-round shading of the stream bed and contribute to low instream primary production, as do low stream nutrient concentrations as they drain underlying crystalline rock formations.

In contrast, following clearcutting of Watershed (= Catchment) 7 in 1976, combined with removal of riparian *Rhododendron*, there were short-term increases in stream nutrients, light, and increases in primary production and corresponding changes in invertebrate community structure. These changes were short-lived, as within two decades of forest succession the invertebrate community reverted to that of before clearcutting. Swank & Webster (2014) summarised these and many other changes in response to clearcut logging.

The numerous small streams and individual catchments within the Coweeta Laboratory (Box Figure 1.1b) have enabled many experimental studies of streams in addition to

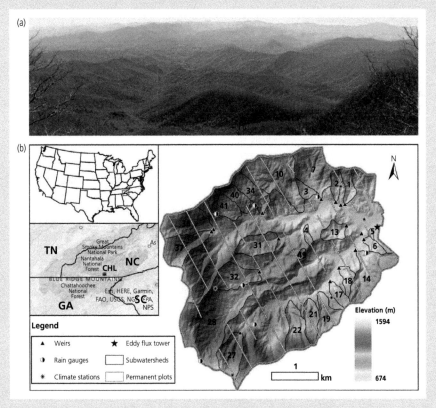

Box Figure 1.1b (a) Panoramic photo from Albert Mountain, ~ 1,609 m asl, looking down on the headwaters of the Coweeta Creek Basin, at the Coweeta Hydrologic Laboratory in Western North Carolina in early spring. The dark green catchments are former deciduous forests that have been converted to white pine. (Photo with permission of Jason Love). (b) Measurement networks and the various numbered catchments (watersheds) at the Coweeta Hydrologic Lab, in western North Carolina.
Source: Miniat et al. 2021, with Permission from John Wiley and Sons.

Topic Box 1.1 *Continued*

the clearcutting and others mentioned above. These include experimental studies examing the role of wood in headwater streams and examining the relationship between instream woody debris, nutrient availability, basal detritus resources and aquatic populations, including secondary production. Other studies examined the effects of whole-stream manipulations of invertebrate populations with insecticides to evaluate their role in coarse particulate organic matter (CPOM) processing and the subsequent export of organic and inorganic matter to downstream reaches. A 15-year study of litter exclusion from a headwater stream demonstrated the importance of the surrounding forest as a source of allochthonous organic material that supported the secondary production of invertebrate consumers and their invertebrate and salamander predators (see also section 7.2.2 and Figure 7.23). The exclusion of terrestrial litter also influenced nutrient uptake and showed that a few taxa could use existing instream wood as a food resource until it was removed and replaced by PVC piping and tubing. Rosemond and colleagues manipulated nitrogen and phosphorus concentrations and effects on organic matter breakdown and secondary production of invertebrates and salamanders (see Rosemond et al. 2021 for details of these and other manipulations). Higher nutrient concentrations enhance breakdown rates of CPOM and increased generation of FPOM and subsequent export to downstream reaches. Nutrient additions reduce leaf litter standing crop, especially in summer and early autumn. Grossman and colleagues have examined fish over stream gradients illustrating how individuals and species respond in patchy environments to optimise energetics and the effects of environmental variations on fish population dynamics.

The website http://www.Coweeta.UGA.edu provides many more details about the Coweeta Laboratory's site description, history, aquatic and terrestrial studies, and a list of publications for aquatic and terrestrial sites. Under 'publications' at the top of the page, these are listed either as signature publications; by author and year, year and author, and those cross-site comparisons of several Long-Term Ecolgical Research (LTER) sites. Box Table 1.1 lists

various authors of publications in stream ecology found on the website.

Box Table 1.1 A list of past and current Principal and Co-Principal investigators involved in stream projects at Coweeta and surrounding area. Post-doctoral associates are listed at the bottom of this table. Coweeta papers published by these authors can be found at the website http://www.Coweeta.UGA.edu.

Name of Investigator	Location
Benfield, E.F.	Virginia Tech (VT)
Benstead, J.P.	University of Alabama (UA)
Grossman, G.D.	University of Georgia (UGA)
Helfman, G.S.	UGA
Jackson, C.R.	UGA
Leigh, D.S.	UGA
Maerz, J.	UGA
Meyer, J.L.	UGA
Pringle, C.M.	UGA
Rosemond, A.D.	UGA
Suberkropp, K.	UA
Valett, H.M.	VT
Wallace, J.B.	UGA
Webster, J.R.	VT

Post-doctoral Associates: Cuffney, T.F. (UGA); Eggert, S.L. (UGA); Georgian, T. (UGA); Gulis, V. (UA); Harding, J. (VT); Huryn, A.D. (UGA); Kominoski, J. (UGA); Tank, J.L. (VT); Thomas, S. (VT).

The publications include those from the 1970s to the present and include signature publications for the following: Catchment 7 clearcut existing forest and natural regeneration; forest ecology; stream ecology; hill slope–stream interactions; social and land-use drivers of ecosystem change; and the most recent publications. Many of these publications are available as direct pdf downloads from the US Forest Service, Southern Research Station from the following US Department of Agriculture website: http://www.SRS.fs.usda.gov. This website also includes additional information and maps to the Coweeta Laboratory.

Professor J. Bruce Wallace is an Emeritus faculty member at the Department of Entomology and Odum School of Ecology, University of Georgia, Athens, GA, USA.

---

stream catchments in the world. All these studies addressed mainly relatively small streams in the 'headwaters' of river catchments. The study of much larger rivers, with their proportionately larger-scale catchments, is even more challenging and requires even more resources.

Running waters also link land and ocean, transporting downstream enormous quantities of sediment particles and solutes and providing a potential route for the upstream movement of marine-derived nutrients and organic matter, along with the migrations of animals, as we will see in later chapters.

1.3.1 Hierarchical networks and scale

Scale is important in biology. The way we look at the stream and river habitat, from the size of our sampling unit to the frequency and duration of our observations, will influence how we identify biological responses to the environment and environmental change, how we perceive the various patterns in biotic and abiotic environmental factors, and what ecological processes appear to be important in the functioning of lotic systems. One of the things ecologists have learnt is that different processes operate on different timescales. In a stream, highly localised surges of the current occur over seconds, while large-scale variation in, for example, water level and temperature occur over days, to decades to hundreds or thousands of years. When we carry out research in running-water systems most biological samples collected and environmental measurements taken are at the spatial scale of the sampling point, the size of the sampling device, or the transect or stream reach. As Hynes (1975) clearly illustrated, if the spatial scale is extended further, it rapidly leads the researcher out onto the surrounding landscape and into groundwaters. In effect, stream and river ecology at larger spatial scales merges with terrestrial ecology and landscape ecology (Hildrew & Giller 1994).

Different scientific questions, levels of generalisation and types of investigations therefore apply, depending on the spatial and temporal scale at which we are working. Lotic ecologists face spatial and temporal scales extending over approximately 16 orders of magnitude (Minshall 1988), easily visualised at the extremes from the single substratum particle to the drainage basins of the largest rivers (Hildrew & Giller 1994). Conceptually, these can be collapsed to six general categories, from the microhabitat to the river basin (Figure 1.5). The larger the scale, the slower the evident processes and rates of change. The 'microhabitat' system, at a scale of centimetres, consists of patches of substratum, particulate organic matter, local current velocity or vegetation persisting for hours to days, weeks or even to years. The habitat scale, for instance of the 'riffle–pool' system we describe in Chapter 2 (Figure 2.4), includes the substratum surface and the hyporheic zone (below the substratum) and regions of exchange between them. Riffles are areas of broken, turbulent flow while pools are less turbulent and deeper. The hyporheic subsystem is biologically active, with water upwelling and downwelling to the surface or into the sediments via the wider forces of flow (Grimm 1994). The *reach* system includes one to several sub-reaches of upwelling and downwelling zones and pool-and-riffle sequences, covering metres to tens or even hundreds of metres and, excepting really major

disturbance events, persisting in their general location for years, decades or longer. The whole stream or river sitting within its broader catchment, covering tens to hundreds of square kilometres, is likely to have existed for hundreds to thousands of years, although the exact position of the channel may have changed. It is at this scale that the openness of the lotic system and the interactions with the landscape become apparent. Entire drainage basins covering thousands of square kilometres probably have a long geological history. Moving up these scales changes the nature of the ecological study from habitat preferences and behaviour through species interactions to population dynamics and community ecology and on to macroecology, biogeography and evolution (Figure 1.5).

A further feature of river systems is that they seem to be 'hierarchically organized' (Frissell et al. 1986; Hildrew & Giller 1994). The levels in the hierarchy are 'spatially nested'—that is, the higher-scale systems impose constraints on features of the systems below them. For instance, features of the riffle–pool sequence are determined by higher-order characteristics such as the slope of the whole reach, water runoff and sediment inputs from the catchment. The algal distribution on a single stone is mainly affected by the flow and hydraulic forces acting on it—compared with others around it. At the larger scale of the stream reach, algal biomass and community structure may depend on nutrient limitation or perhaps shading from the riparian vegetation. In longer stretches of the stream, local zones of high nutrient supply may alleviate this limitation, allowing a shift of control of algae to grazing invertebrates or fish where they are important (Fisher 1994). Over long periods of time, algal biomass in large stream reaches may become closely associated with the flood disturbance regime governed by catchment-scale features.

Thus, external physical processes act on different spatio-temporal scales, and each level in the hierarchy has its own characteristic persistence time, disturbance regime and spatial extent. The biota respond in turn to this set of interlinked physical subsystems— the *habitat templet* of Southwood (1977)—which we explore in Chapter 2. Adaptations to the templet are described in Chapter 4 and determine which species from those 'available' locally can persist and what their species traits are (see Chapter 6). Further, different organisms perceive (or experience) the habitat templet in different ways, depending on their size, life cycle and ecology. If we wish to understand the match between organism and habitat, therefore, we need to assess the environment with an 'organism's eye view' (Hildrew & Giller 1994). A large, long-lived

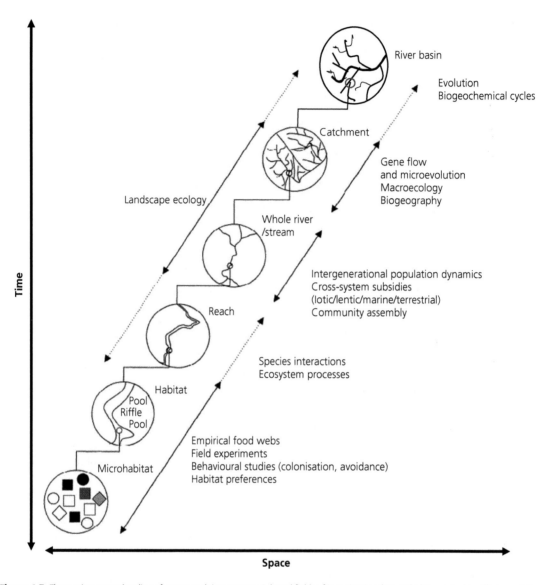

Figure 1.5 The spatio-temporal scaling of stream and river systems. Selected fields of investigation that might be expected to affect population dynamics, community structure and ecosystem processes at the different scales are highlighted. Solid double-headed arrows indicate the typical spatio-temporal limits of these investigations; the dashed arrows indicate rarer instances, where these limits are exceeded.
Source: modified from Woodward & Hildrew 2002; with permission from John Wiley & Sons.

hippo 'sees' the physical environment of a stream or river in a completely different way to a small, relatively short-lived insect. To a hippo, a small stream can look like a relatively simple two-dimensional expanse, but to an insect larva it is a hugely complicated world, where small-scale variations in the current and substratum provide a highly structured, three-dimensional habitat. The hippo might 'perceive' an environment close to the size of a catchment during its lifetime; an insect might 'perceive' a stream reach whilst an attached diatom may spend its entire (brief) lifespan in a few square micrometres of space. Over the long life of the hippo it will be subject to many seasons and potentially large-scale environmental changes, which would exceed the life-time and tolerance of a single insect.

There are four dimensions to the hierarchically organised river systems (three spatial and one of time; Figure 1.6) which all coincide with clear biological and ecological patterns.

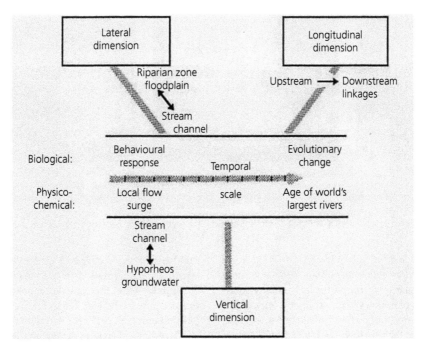

Figure 1.6 Ward's (1989) concept of the four-dimensional nature of stream and river ecosystems.
Source: from Giller & Malmqvist 1998.

i. *Longitudinal dimension*—upstream and downstream—along which there are major changes in physicochemical conditions and, as a consequence, longitudinal shifts in biotic variables.

ii. *Lateral dimension*—there are clear biotic patterns along the lateral spatial gradient between the stream channel, the riparian zone and the surrounding catchment (i.e. incorporating the land–water interface). These involve shifts in species abundance and composition, as well as key processes (exchanges of organic matter and nutrients). There are also changes laterally, but below the substratum surface, where extensive water-filled, interstitial, systems have sometimes been documented.

iii. *Vertical dimension*—this primarily involves the interaction between river waters and the contiguous groundwaters, with a *hyporheic* zone between the two and in which river water and deeper groundwater may be exchanged. Along the vertical dimension, there is also the air–water interface, in which a number of specialist organisms live and across which important exchanges of gases occur.

iv. *Temporal dimension*—rivers are temporally highly dynamic across different timescales, from short-term biological and physicochemical responses, to brief environmental fluctuations, through more predictable seasonal changes, to unpredictable and more prolonged variations in the environment and biota, to long-term changes in the structure and function of stream and river ecosystems relating to profound climatic, geomorphic and evolutionary events.

The importance of these various spatio-temporal dimensions will become clear as we develop these themes later in the book.

1.4 The variety of streams and rivers

Running waters range in size from small trickles to enormous rivers draining major parts of the continents. The largest of all, in terms of discharge, is the Amazon, at over 6,400 km in length and more than 11 km wide in places. The nature of lotic habitats is characterised by the flow of water within the channel. It ranges from swift or torrential and cascading in headwaters arising in upland/mountainous areas, to the sluggish, quiet downstream margins or poorly connected floodplain waterbodies of larger rivers (where, in many respects, conditions resemble those of still waters). We can also group streams and rivers loosely based on the continuity of their flow over the year (Figure 1.7). *Perennial streams* and rivers flow all year

Figure 1.7 Examples of (a) *perennial* (Araglin River, Cork, Ireland); (b) *intermittent* (the Albarine River, France); and (c) *ephemeral* (unnamed river in Southern Australia) rivers.
Source: (a) photo Paul Giller; (b) from Datry et al. 2014; with permission from Oxford University Press; (c) from Datry et al. 2016, CC BY-NC 3.0, Credit A. J. Boulton.

round and are fed by groundwater, with the local water table always above the stream bed. Flow near the source of many rivers, particularly in dry landscapes, is often temporary or intermittent. *Intermittent streams* and rivers maintain flow only in certain seasons when the water table is high, as for example during the rainy/wet season. At other times the channel is dry or surface water is restricted to disconnected pools. Lastly, *ephemeral streams* and rivers receive no contribution from groundwaters and flow only after heavy rains, as seen in arid zones. These groups do overlap somewhat, however.

We usually think of streams and rivers sitting within a sizeable landscape, but streams do occur even on many small and remote oceanic islands. The evolution of their flora and fauna is interesting. In Topic Box 1.2, Alan Covich explains how these streams originate, how they differ from mainland systems and what this means for their species diversity, composition and food webs.

There have been a number of attempts to classify streams and rivers more precisely into 'types', aiming to group them in terms of the geomorphological, hydrological, physicochemical and biological features common to members of each type. In reality, there is more or less continuous variation among systems and not all can fit into 'clean' categories. Nevertheless, such classifications do help us to 'organise' information, explore ecological patterns, and develop management and restoration tools and approaches. Essentially, they are a practical attempt to reduce masses of detail. For instance, the present legislation for water management in Europe, the *Water Framework Directive* (WFD), requires EU Member States to develop 'typologies' for freshwater ecosystems based on a set of environmental variables or type descriptors. However, variation in the

number of types recognised in the various European countries raises questions about the way such systems are applied. Thus, there are six types in Denmark, 40 in Germany, 68 in the UK, 145 in France and 367 in Italy (see Solheim et al. 2019), the variation not being entirely explained by differences in land area and environmental heterogeneity (such as geology, relief and climate).

Solheim et al. (2019) recognised the problem and attempted a new general typology linking national water-body types with high similarity to a few broad European types. Basing the classification on altitude, region (Mediterranean and rest of Europe), geology (related to alkalinity, calcium, colour, and bedrock or deposits) and three categories of catchment size, they arrived at a set of 20 broad river types (described in the paper) to which they assigned 65,840 (77%) of the 85,500 water bodies from the countries that could be included in the analysis.

On a global scale, Keith et al. (2020, 2022) have produced a typology for all ecosystems (both terrestrial and aquatic), moving beyond just the physicochemical descriptors to include ecological drivers and dependencies and convergent biotic traits in the derivation of 'ecosystem functional groups'. The typology is hierarchical with its upper levels defining ecosystems by their convergent ecological functions and the lower levels distinguishing ecosystems with contrasting species assemblages engaged in these functions. The potential value of this classification approach is well described by Keith et al. in the above publications. For running waters, they identified seven such ecosystem groups together with two other groups that can also be considered as associated lotic habitats (Figure 1.8). These are listed in Table 1.1 along with some of the generalised physical attributes and ecological traits associated with them.

Topic Box 1.2 Freshwater streams on tropical islands

Alan P. Covich

Many tropical oceanic islands have relatively small, linear river networks connected to coastal estuaries and relatively low species richness compared to larger, reticulate mainland river networks. The remoteness of many oceanic islands restricts colonisation by species that can only fly or swim relatively short distances. The lack of any river connectivity between islands and the mainland prevents any 'true' freshwater fish and other freshwater and amphibious vertebrates from reaching them. Consequently, diversity often depends on species derived from marine invertebrates (e.g. shrimps, crabs and neritid gastropods) and vertebrates that have adapted to freshwater (Resh & De Szalay 1995) and that complete all or part of their life cycle in coastal rivers on tropical islands. In addition, some widely dispersed aquatic insects (e.g. simulid blackflies) can dominate specific habitats such as waterfalls, where they avoid direct competition with or predation by abundant decapods and fish that occur below the waterfalls (Craig 2003). In general, altitude is a key variable that influences orographic precipitation and stream-flow regimes that affect seasonal species distributions (Jenkins & Jupiter 2011). Insular river basins are especially affected by extreme disturbances such as hurricanes and prolonged severe droughts that affect species abundance and ecosystem processes (Covich et al. 2006; Gutiérrez-Fonseca et al. 2020).

Biotic diversity of island streams, much like that in continental streams, depends on island age and origin as well as size and isolation. Islands have three main origins. Volcanic islands, such as the Hawaiian chain, begin when steep volcanic cones emerge above sea level. These slowly erode as rainfall weathers the rock and drainage networks with pool, riffle and reach habitats form along altitudinal gradients that in turn influence species distributions. Calcareous islands originate as coral reefs build atolls that rise above sea level and are often found at great distances from mainland sources of freshwater species. Rainfall and weak carbonic acids erode steep drainage channels and waterfalls in the carbonate rocks (Box Figure 1.2a). Some shallow groundwater storage can lead to spring-fed streams. These streams can support dense populations of submerged plants and shell-bearing molluscan species relative to volcanically originated islands. The third type of island is 'continental' in origin because some insular locations, such as Trinidad, Madagascar and Borneo, were once part of the mainland during their geological history. Their tropical streams have higher diversity because they contain many species that evolved in continental inland waters as well as those that adapted to variable salinities in coastal rivers

Box Figure 1.2a Steep waterfalls function as species-specific barriers to upstream movement of migratory fish (Rio Espiritu Santo, Luquillo Mountains, Puerto Rico).
Source: photo by Freshwaters Illustrated.

Topic Box 1.2 *Continued*

(Longo & Blanco 2014). They are also often relatively close to their original continental sources, so recolonisation can occur following disturbances.

In general, island biotas often result from colonisation by species highly adapted for dispersal as well as from genetic diversification among endemic species that occupy distinct ecological niches (Schubart & Santl 2014; Gomard et al. 2018). As an example, the well-studied Central Sulawesi Islands of Indonesia contain endemic species of freshwater diatoms, hydrobiid snails, limpets, sphaeriid clams, sponges, crabs and fish (see e.g. Klaus et al. 2013; Tweedley et al. 2013; Albrecht et al. 2020). The importance of the rivers (Larona, Petea, Tominaga) that connect this series of ancient deep lakes is illustrated by the different subsets of the total species of shrimps that occur in each river (von Rintelen et al. 2010). A succession of river–lake reconnections and recolonisation events that occurred in response to long periods of changing sea level, extreme floods and droughts resulted in 21 endemic shrimp species. The age of tropical islands is important in determining the extent to which species have adapted to coexist or compete with or avoid predation from other species. Some widely distributed species, such as amphidromous gobiid fish and decapod crustaceans, are well adapted to climb steep waterfalls (Kinzie 1988; Bauer 2013; Lagarde et al. 2021) which can sever upstream connections while also providing spatial refugia for such species from most fish predators (Baker et al. 2017; Box Figure 1.2a). These species have a life history that includes larval dispersal by marine currents that allows them to occupy many insular and continental rivers. For example, the filter-feeding shrimp *Atya scabra* occupies fast-flowing rivers and has a wide Atlantic distribution from Mexico to Brazil, including the large Caribbean islands and the Cape Verde islands in the Gulf of Guinea off Africa. In contrast, highly restricted shrimp species and fish are affected by large-scale ecohydrological changes that can first connect and then isolate insular river and lake habitats.

The species assemblages on different tropical islands are often distinct from each other and from continental streams. A continuum of increased numbers of species is based initially on their isolation and geomorphology. The most remote and most recently formed oceanic islands, such as emergent coral atolls, are species poor with the least complex food webs dominated by marine-derived species of decapods, gastropods and fish (Resh & De Szalay 1995; Benstead et al. 2009). Higher species richness occurs on volcanic islands, where the downstream reaches have numerous species of marine origin—in contrast to the headwaters with fewer species and less complex food webs (Hein et al. 2011). Continental islands generally reflect the high species richness characteristic of the larger, more complex reticulate mainland stream networks.

Some species of consumers such as detritivorous insects are often under-represented in insular food webs where processing of organic matter is dominated by shrimps and crabs (Crowl et al. 2001). Shrimps occupy similar ecological niches to aquatic insects, which are often relatively scarce but still functionally important in island food webs (Cross et al. 2008; Rosas et al. 2020). High abundance of atyid shrimps can affect growth of periphytic diatoms and the abundance of grazing insects (Sousa & Moulton 2005; Macías et al. 2014). Atyid shrimps scrape periphyton and decomposing leaf litter but also filter suspended organic matter at base flow (Box Figure 1.2b). Other shrimps are effective predators on small insects and shredders of leaf litter (Box Figure 1.2b). Smaller shrimp species detect chemicals produced by large, predatory species (*Macrobrachium carcinus*) and change their movement, thus reducing risk (Crowl & Covich 1994). In the presence of predatory fishes, some shrimp species (e.g. Box Figure 1.2b) grow extended rostral peaks that lower their risk of predation. Populations upstream and above steep waterfalls have short rostrums (Covich et al. 2009; Ocasio-Torres et al. 2021).

Box Figure 1.2b (a) Atyid shrimp, such as *Atya lanipes* (basket shrimp), can filter suspended organic matter and scrape rock or leaf surfaces using the chelae which have setae in the form of brushes with microscopic teeth. (b) Some species of shrimp, such as *Xiphocaris elongata* ('yellow-nose shrimp'), can shred leaf litter and other detritus that falls into headwater streams as well as feed on small insect larvae.

Source: (a) photo by Freshwaters Illustrated; (b) photo by Alan Covich.

Topic Box 1.2 *Continued*

The absence or scarcity of species in some major functional feeding groups (see Chapter 4) among island streams seems to make these ecosystems especially vulnerable to introduced species. For example, the accidental and intentional introductions of the widespread and omnivorous freshwater crayfish (*Procambarus clarkii*) affects detrital processing (Larned et al. 2003). Many other species are small and parthenogenetic. For example, snails such as *Melanoides tuberculata* and *Thiara granifera* are found in both tropical insular and continental streams where they compete with native neritid snails (Myers et al. 2000). The formation of these 'novel' assemblages is increasing as non-native species are introduced by releases from home aquaria into many tropical island streams.

In summary, the total insular freshwater species diversity includes species from five sources: (1) continents; (2) other islands; (3) marine-derived species that slowly adapt to freshwater; (4) speciation resulting in the *in situ* evolution of endemics; and (5) those species introduced intentionally or accidentally by humans. This total varies with island location, size, age, altitude and climate, especially rainfall. Concentrations of endemic freshwater species generally occur in older remote islands. The distributions of insular endemic species appear to reflect the same dynamic patterns often associated with the diversity of insular terrestrial species and continental inland waters. Specific connections among the various abiotic (e.g. spatial isolation, drought frequencies or hurricanes) and biotic (e.g. inter-specific competition, mutualism, parasitism and predation) variables that affect ecological processes and food webs are still unclear. However, the number of tropical insular streams under study is rapidly increasing and new species and their relationships continue to be discovered.

Professor Alan Covich is in the Odum School of Ecology, University of Georgia, Athens, GA, USA.

Figure 1.8 The seven river and stream functional types and two associated lotic habitats identified under the IUCN global ecosystem typology (Keith et al. 2020, 2022). (a) permanent upland streams—Episodic Eyre Creek, Queensland, Austral; (b) permanent lowland streams—Rio Carrao, Venezuela; (c) freeze–thaw rivers and streams—Volga River, Zubstov, Russia.; (d) seasonal upland streams—Yamuna River Mussoorie, India; (e) seasonal lowland rivers—Patalon Chaung, Myanmar; (f) episodic arid rivers—Cooper Creek in central Australia; (g) large lowland rivers—Amazon River near Iquitos, Peru; (h) irrigation canal—irrigation canal and valve, California, USA; (i) artesian spring—Washington Oaks State Gardens, Florida. The general physicochemical attributes and ecological traits are summarised in Table 1.1.
Source: (a)–(h) photos from Keith et al. 2020, copyright 2020 IUCN, International Union for Conservation of Nature and Natural Resources; (i) from Wikipedia Creative Commons.

Table 1.1 Running-water ecosystem functional groups identified under the new IUCN Global Ecosystem Typology, together with two additional groups that include running-water canals, ditches and storm drains (classified under Artificial Wetlands) and Artesian springs (classified under the Lakes). Key attributes are shown (summarised from Keith et al 2020).

Running-water 'ecosystem functional groups'	Generalised physical attributes and ecological traits
Permanent upland streams	1st–3rd-order systems, generally with steep gradients, fast flows, coarse substrata, often with a riffle–pool sequence of habitats, and periodic (usually seasonal) high-flow events. Many organisms have specialised morphological and behavioural adaptations to high-flow-velocity environments. Where in forested catchments, deciduous riparian trees produce copious leaf fall that provide energy subsidies to the systems.
Permanent lowland rivers	Small–medium lowland rivers (stream orders 4–7). Locally or temporally important erosional processes redistribute sediment and produce geomorphically dynamic depositional features. Nutrient concentrations depend on riparian/floodplain inputs and vary with catchment geochemistry. Productive depositional ecosystems with simpler and less diverse trophic webs than large lowland rivers. Intermittently connected oxbow lakes or billabongs increase the complexity of associated habitats, providing more lentic waters for a range of aquatic fauna and flora. Aquatic biota have physiological, morphological and behavioural adaptations to lower oxygen concentrations, which may vary seasonally and diurnally
Freeze–thaw rivers and streams	In seasonally cold montane and boreal environments, under low winter temperatures and seasonal freeze–thaw regimes where surfaces of both small streams and large rivers freeze in winter. These systems have relatively simple trophic networks with low functional and taxonomic diversity. In the larger rivers, fish, and particularly migratory salmonids returning to their natal streams and rivers for breeding, can provide significant nutrient and energy inputs to the rivers and associated terrestrial fauna.
Seasonal upland streams	Upland streams (orders 1–4) tend to be shallow with highly seasonal flows, and highly variable flood regimes between marked wet and dry seasons. Associated changes in water quality occur as solute concentration varies with volume. They may be perennial, with flows much reduced in the dry season, or seasonally intermittent with flows ceasing and water persisting in isolated stagnant pools. They have low to moderate productivity and a simpler trophic structure than lowland rivers. Taxonomic diversity varies between streams, but can be lower than permanent streams and relatively high in endemism. Compared to lowland rivers there tend to be low numbers of larger predators.
Seasonal lowland rivers	Large riverine systems (stream orders 5–9) shaped by seasonal hydrology and linkages to floodplain wetlands with cyclical, seasonal flow regimes. High-volume flows and floods occur during summer in the tropics or winter–spring at temperate latitudes. These systems tend to possess significant biophysical heterogeneity, which, together with this temporal variability, promotes functional diversity in the biota.
Episodic arid rivers	High temporal variability in flows and resource availability. Low elevational gradients and shallow channels result in low turbulence and low to moderate flow velocity. Lowland stream channels are broad, flat and often anastomosing, with mostly soft sandy sediments. These systems have a low-diversity biota with periodically high abundance of some organisms. Productivity is episodically high and punctuated by longer periods of low productivity (i.e. boom–bust dynamics). The trophic structure can be complex. Episodic rivers are hotspots of biodiversity and ecological activity in arid landscapes.
Large lowland rivers	Typically stream orders 8–12 with shallow gradients with low turbulence, low to moderate flow velocity and very high flow volumes (>10,000 m^3 s^{-1}), which are continuous but may vary seasonally depending on catchment area and precipitation. These are highly productive environments with complex trophic webs. Primary production is mostly from phytoplankton and riparian macrophytes. The fauna includes a significant diversity of pelagic organisms. Floodplain zones vary in complexity from forested banks to productive oxbow lakes and extensive and complex flooded areas where emergent and floodplain vegetation grows.
Canals, ditches and storm drains	These are artificial but function as rivers or streams with low heterogeneity. Flows in some ditches may approach lentic regimes. Irrigation, transport or recreation canals usually have steady perennial flows but may be seasonal for irrigation or intermittent where the water source is small. Substrata and banks vary from earthen material to hard surfaces. They often have simplified habitat structure and trophic networks, although some older ditches have fringing vegetation, which contributes to structural complexity. While earthen banks and linings may support macrophytes and a rich associated fauna, sealed or otherwise uniform substrata limit the diversity and abundance of benthic biota.
Artesian springs	Groundwater-dependent systems with little surface inflow, permanently disconnected from surface-stream networks though some have outflow streams. Hydrological variability is low and discharge waters tend to be warm and enriched with minerals reflecting their geological origins. Most biota are poorly dispersed and have continuous life cycles and other traits specialised for persistence in hydrologically stable, warm or hot mineral-rich water. Artesian springs and oases tend to have simple trophic structures

There are some more specialised lotic habitats that are also worth mentioning, such as madicolous habitats, where a thin flowing film of water constantly seeps over rock faces or other types of substratum; torrential habitats and waterfalls; and lake and reservoir outlets. However, it should be recognised that the designation of river types is purely descriptive, and is a practical measure that does not necessarily capture real discontinuities among lotic habitats overall. What these classifications illustrate is that there is a great variety of running-water habitats which, together with the variability in geological, geomorphic, biogeographical and climatic settings, underly the enormous diversity of habitats within and between stream and river systems. This in turn provides the backdrop to the diversity of life and ecological patterns and processes explored in this book.

1.5 The naming of parts—living assemblages in rivers and streams

One early approach to studying the biota of streams and rivers—now somewhat outdated—seeks to distinguish, describe and classify particular assemblages of lotic organisms—groupings of species that tend to be found in similar kinds of habitats. While itself having little explanatory power as to the processes by which the assemblages come about, and at some risk of drowning in a sea of specialist names, this approach does force us to consider the wide range of habitats and kinds of assemblages that are represented in them. We begin by briefly doing so here then deal with assembly processes for lotic communities in Chapter 6.

A basic division of aquatic assemblages in general is between organisms living in contact with substratum and those in the water column (though some do alternate between the two). With the exception of fish, the substratum-dwellers (those living within and on bed sediments) dominate in streams and rivers, particularly in erosive and turbulent waters.

Organisms of the water column that are capable of a good deal of active swimming—and are not largely at the mercy of water currents—make up the *nekton*, a term which distinguishes them from those merely floating, that is, the *plankton*. In contrast to common preconceptions, however, planktonic organisms are often abundant in rivers, particularly in the lowland course of larger rivers, in areas where there is limited water flow (hydraulic retention zones, side-arms and the like, see e.g. Bergfeld et al. 2009; see Chapter 5). The plankton includes both prokaryotes and eukaryotes,

ranging in size from viruses to metazoan zooplankton and fish larvae. Larger fish are the most abundant and important members of the nekton, although in places with low current speed, some large, actively swimming, fully aquatic insects, like many beetles and 'true' bugs (hemipterans) are found. There is also a much less well-known assemblage, the *neuston*, living in or on the water surface of many running waters, and the organisms that comprise this again range from some quite large arthropods to viruses and bacteria. The neuston lives at the interface between the water and the atmosphere, the most important site of gas exchange between the two, and one that must be passed through by emerging insects at the end of their aquatic lives, and by a good deal of the terrestrial material (mineral, organic and some living) entering river systems.

Even among the fish, those in very erosive, fast-flowing, usually headwater channels often are adapted to live close to or on the substratum. There they join the *benthos*, a plethora of living organisms that has delighted so many of us who have wielded a simple net in the current and caught animals and plants disturbed by kicking over stones on the bed of a stream. Benthic organisms are diverse, abundant and productive and can tell us a great deal about the nature of rivers and streams (the reason why they are so often used in bioassessments of ecological 'quality'—see Chapter 10). Benthic organisms include small photosynthetic species (algae and blue-green bacteria or cyanobacteria), protozoans, fungi, bacteria and viruses and a very wide range of invertebrate animals. Many microorganisms live embedded within a slimy *biofilm* growing on underwater surfaces. This layer is a crucially important assemblage in terms of ecosystem processes (see Chapters 8 and 9), the 'slime' (chemically a polysaccharide) being produced by bacteria and some algae. Organisms living attached to surfaces are sometimes divided into subcategories depending on the surface concerned. These groups include the *epilithon* (the biofilm on stones and rocks), *epipelon* (algae growing on mud), *epipsammon* (on sand grains) and *epiphyton* (on the leaves or stem of a larger plant). However, attached assemblages of small photosynthetic organisms growing on any surface may simply be called the *periphyton*. While some algae are characteristic of the particular surfaces on which they grow (and thus justify the different names), many are much less 'fussy' and grow on more or less any firm surface underwater.

Some further terminology relating to stream and river communities may occasionally be useful, and often relates to the spatial complexity of river systems referred to throughout this book. Recalling Ward's

(1989) four dimensions (one of which is time) of river ecosystems (Figure 1.6), notably different assemblages are arrayed along the three spatial dimensions (longitudinal, lateral, vertical). We return in more detail to longitudinal patterns in river communities in Chapter 6 (section 6.3), but it is worth mentioning here that ecologists have distinguished three general assemblages along that dimension. In spring-fed rivers, we may find a *crenal* community living close to the springhead (the *crenon*). The spring habitat generally offers a stable, moderate temperature (cool in summer, above air temperature in winter, except in hot springs of course) and a rather stable flow and substratum. The water may be rich in carbon dioxide and minerals, and plant growth, particularly mosses and liverworts, is often prolific. Particular benthic animals are sometimes restricted to living very close (a few metres) to the springhead and may include cool stenothermic (narrow temperature tolerance) groundwater species that require a low and stable temperature, as well as sedentary and 'attached' invertebrates. Hot spring specialists are also found. River systems normally include a so-called *rhithral* community (of the *rhithron*—essentially, 'streams') living in relatively steep, turbulent, hard-bottomed upstream reaches, followed by a *potamal* assemblage (of the *potamon*, the 'river') living in the lowland reaches, which are deeper, generally less turbulent and may have a bed of fine sediments, including mud and sand. This latter assemblage may be dominated by a burrowing 'infauna', a prominent plankton, and rooted higher plants if the channel is not too deep. These terms denote only the most general characteristics of river assemblages, and they need not apply to all running waters, but one may well encounter them in the literature.

Other assemblages have been distinguished along vertical and lateral gradients below and away from the channel itself and tend to be particularly important in the geologically unconstrained alluvial lower reaches of rivers. These assemblages may seem to be less obviously 'riverine' in their species composition, but nevertheless are a definite part of the overall river community. This is a good place to remind the reader that the limits of a river can be difficult to define, as discussed in Chapter 2, and that rivers are not isolated 'microcosms' but well-connected components of the biosphere. In the language of ecologists interested in the vertical dimension, invertebrates of the substratum surface and superficial bed sediments are described as *epigean*, adapted to life in the light and in flowing water at the surface, whereas those living deeper beneath the bed, or in groundwater, are

hypogean. Epigean and hypogean communities may not be sharply distinguishable in practice as there is usually a gradual transition from one to the other. Epigean animals include the familiar benthic organisms, although some may also be preadapted to interstitial life by having, for instance, elongate, flexible bodies and a tolerance of occasionally low oxygen concentration (see Chapter 4). In the right circumstances of flow and substratum type, such epigean organisms can penetrate tens of metres into an alluvial bed, especially where surface water downwells into the bed sediments. In contrast, normally interstitial forms (that live in spaces between substratum particles) may be found at the surface where water upwells from deep in the bed. Such species provide links between the food webs of the more superficial river bed and the deeper *hyporheic* zone, whose characteristic assemblage is termed the *hyporheos* (see Stanford & Ward 1993; Ward et al. 1998).

Lateral to the river channel is the *riparian* zone, essentially the fauna and flora of the bankside. Its biotic assemblages interact with those of the river itself and link it to the adjacent terrestrial system. Riparian species are notably adapted to, or tolerant of, disturbances caused by periodic flooding. As we move away from the river channel and out over the broader floodplain, partially empty channels may be encountered, some of which may be occasionally, or perhaps more regularly, connected to active channels in very wet conditions, while others are effectively totally isolated. Sometimes permanent or temporary ponds and lakes may also be present in a floodplain, which can be many kilometres wide. Typically, the aquatic assemblages of the floodplain freshwater habitats become progressively more like those of ponds and lakes (see Brönmark & Hanson 2017), although the river channel and the entire floodplain system can sometimes be considered as a single ecosystem in terms of energy flow, nutrient transformations and the carbon cycle at an appropriate spatial and temporal scale (see Chapters 8 and 9).

Finally, we deal briefly with a terminology that has been adopted for fractions of the overall river-animal community based on body size, and which relates to some extent to the assemblages identified above as inhabiting particular parts of the river habitat templet. Benthic organisms (excluding benthic fish) range in size from less than 0.0001 mm for viruses to more than about 200 mm for a few freshwater mussels, crabs and crayfish (Chapter 3). It is usual to categorise the size range of benthic animals into the *microfauna* (essentially protists, passing through a mesh of 42 μm), the *meiofauna* (essentially a very wide range of small meta-zoans, retained on a 42 μm sieve but passing through a mesh of 500 μm (or by some definitions, 1,000 μm)), and *macroinvertebrates* (larger animals retained on a mesh of 500 μm), including the later instars of benthic insects, 'macro'-crustaceans, larger oligochaetes, molluscs and others. These assemblages have to be studied in different ways, are taxonomically rather distinct, and our extent of knowledge of them differs greatly (see Chapter 3). The vast majority of studies of benthic organisms in streams and rivers concentrate on the macroinvertebrates, which are also one of the assemblages of choice for bioassessment. Despite the best efforts of a band of enthusiasts (e.g. Rundle et al. 2002; Schmid-Araya et al. 2002a; Majdi et al. 2017) the meiofauna are still not usually assessed sufficiently, and this remains a major limitation in the study of lotic ecosystems overall, but one that perhaps is slowly being addressed with the introduction of new technologies (see Chapter 10). The smaller animals (the micro- and meiofauna) are certainly everywhere throughout the river system, and are increasingly dominant in interstitial habitats, and the hyporheic zone.

1.6 The biology and ecology of streams and rivers

Rivers and streams are indeed fascinating ecologically, but are highly variable both within and between systems and in space and time. At one extreme, all rivers appear to be identical (bearing water down the slope), but at the other extreme, all are different, as no two streams have exactly the same complement of species at the same relative abundance or the same physico-chemical conditions. Referring to the inexorable flow of water downstream, the Greek philosopher Heraclitus quite rightly pointed out that 'one cannot step into the same river twice'.

So, what distinguishes running waters from other aquatic habitats? A number of major characteristics are apparent that help to summarise the uniqueness of running water ecosystems (Giller & Malmqvist 1998).

1. *A unidirectional, although far from uniform, flow.* This means that downstream reaches are influenced to a greater or lesser extent by upstream ones, whereas the reverse is less true.
2. *Linear form.* Rivers and streams are uniquely long, thin systems. Headwater streams are widely scattered in the landscape, and river systems are fairly isolated from each other, at least via a continuous aquatic route, and together occupy a small fraction

of the landscape, thus resembling aquatic 'islands' in a 'sea' of land.
3. *A dynamic channel and bed morphology.* The shearing action of flowing water transports and deposits material from the bank and substratum and thus continually changes the physical environment, sometimes on a small scale, sometimes quite catastrophically.
4. *Open ecosystems.* There is predominantly downstream transport of dissolved and particulate organic matter and nutrients in the flow from source to mouth, though in some systems there is also upstream transport from the sea into rivers via migratory organisms. There are also cross-system interactions between the stream and the surrounding terrestrial ecosystem. Materials and nutrients move from land to water via the flow of water or simply gravity, and biotic linkages (e.g. terrestrial predators taking aquatic prey or emerging aquatic organisms) may move some energy and materials back from water to the land. In the lower reaches of larger rivers, there are often two-way movements of sediment, nutrients and detritus, driven by the filling and draining of floodplains, as water rises and recedes.
5. *High within-system spatial and temporal heterogeneity at all scales.* Spatially, this varies from small-scale variations in particle size of the substratum, patchy instream vegetation and, more importantly, extreme local differences in current velocity and hydraulic stress, to larger-scale longitudinal gradients in discharge, riparian vegetation and water chemistry. These influence both the diversity and the nature of the biota. Over time, there are relatively short-term fluctuations in current velocity, discharge, temperature and water chemistry, and seasonal changes in the inputs of organic matter (such as leaf litter) and light are frequent. Rivers are affected by climatic extremes and weather events. The occurrence of droughts and/or major floods is typical of nearly all lotic systems given sufficient time. Over historical/geological timescales, entire drainage patterns may be altered and river flows reversed by geological upheavals. Few other ecosystems possess either the frequency or intensity of such environmental changes (Power et al. 1988; Hildrew & Giller 1994).
6. *A hierarchical organisation of the physical system.* The characteristics of progressively smaller sections of streams or rivers are determined by features and processes of the system above them in size.
7. *High variability among streams and rivers.* Streams and rivers vary chemically and physically. These basic

characteristics are determined by the geological setting, soil type and relief of the catchment, latitude and altitude and, more locally, the land cover and use, and any human impacts.

8. *A special biota*. The animals and plants of streams and rivers are distinctive, many being specialised for life in the flow and in an environment that is frequently disturbed. There are hotspots of very high productivity and diversity. The flow replenishes mineral nutrients and oxygen, while floodplains and deltas are focal points for the accumulation of enormous quantities of detrital carbon, much of it terrestrial in origin. Riverine wetlands are amongst the most productive ecosystems in the world.

1.7 The structure of the book

We start from the view that running waters are distinct and important, and that the challenges to their study require methods and approaches sufficiently different to warrant a more or less specialised treatment. Nevertheless, the biology and ecology of rivers and streams are part of the wider field of ecological biology, and share the same principles. We talk of biology *and* ecology, because we deal prominently with the store of biological diversity to be found in the world's running waters, in the form of the biota, their natural history and particular features, while we also deal with running waters as ecosystems. Although much of ecology is indeed part of biology, the terms remain distinct. As in any science, we endeavour to make generalisations and predictions from the plethora of information available that will test our understanding of underlying governing processes. For rivers and streams this requires a multidisciplinary approach, including biology and ecology. A better understanding leads on to the improved management and rehabilitation of systems that have been profoundly affected by humans, yet are of ongoing importance to human well-being.

We begin (Chapter 2) by exploring the *habitat templet* of streams and rivers, covering their geomorphological setting, and the physicochemical backdrop including the fundamentally important nature of flow and hydraulics, the substratum, water chemistry, temperature and oxygen. We then discuss the impressive diversity of organisms in Chapter 3 and, in Chapter 4, concentrate on their adaptations to life in running waters and the ecological concept of species traits. In Chapter 5 we explore what is so special about stream and river populations and discuss species distributions, abundance and refugia across a range of scales, as well as population dynamics, migration and mobility in running-water habitats. Chapter 6 moves on to the community, where we address phenomena and patterns that occur at the multispecies level of organisation. This includes community diversity, multispecies patterns in space and time and the underlying processes that govern them. In Chapter 7 we delve into the various species interactions, both positive and negative, and investigate food webs in streams and rivers. Chapters 8 and 9 shift to the ecosystem level, where we consider energy flow through river ecosystems, and both how this energy flow drives the flux or cycles of nutrients and how nutrient cycling affects energy flow. Finally, in Chapter 10 we address the burgeoning topic of the future of running waters in a progressively populated and changing world, and the more applied side of stream and river biology and ecology. This includes ways in which humans affect rivers but, increasingly, feedbacks from rivers on human well-being, for rivers bring many benefits that are not so widely appreciated. We also explore some of the new and emerging techniques and approaches to studying streams and rivers and give some pointers as to what kinds of skills the running-water biologist and ecologist might need into the future. The literature on running water biology and ecology has exploded over recent years and we have tried to reflect this in our extensive bibliography. We have also supplemented our text with a number of 'Topic Boxes' from invited specialists, who bring their expertise and insights to bear on a range of topic areas.

The book reflects many years (about 90 between us!) of experience studying and writing about running-water biology and ecology, driven by our fascination for streams and rivers. Over this time, we have seen the subject change and understanding grow, linked in part to the introduction of new techniques and massive expansion in the amount of information available. There is a growing realisation of the importance of natural ('healthy') streams and rivers. We hope that some of our enthusiasm for the subject is apparent, and that the book leads on from the heritage of former 'giants' of the subject (like Thienemann, Margalef, Macan, Hynes, Likens and many others) to a better future for this important fraction of the natural world.

CHAPTER 2

The habitat templet

2.1 Introduction

The physical, chemical and biological environment of streams and rivers—the habitat templet—poses distinctive challenges to their inhabitants. As for any habitat, in the shorter-term the environment ultimately determines which species from a wider pool are able to colonise and persist in a particular location—a process we can call *community assembly* (see Chapter 6). The wider species pool is in turn determined both by longer-term evolutionary processes, working on genetic variation, and also by 'accidents' of biogeographical history, such as continental drift, past climate changes and mountain and other barrier formation. To understand how the biota meets the environmental 'challenge' of its habitat, we must understand the key features of running waters that influence the wide variety of animals and plants that live in them. This may seem straightforward, but animals and plants of differing size and longevity 'perceive' their environment in different ways—think of an individual algal cell suspended briefly in the water column of a river compared with a freshwater pearl mussel that may live for over 100 years in the bed of the same river. As individuals, they are subjected to quite different opportunities and pressures from their environment.

Further complications arise from the interdependence of factors operating in the ecosystem. At a small scale, and in a stream more or less unperturbed by human activities, water movements, temperature and the substratum seem to be key physical variables. Oxygen supply is also extremely important to the biota and largely depends on the first two of these physical factors. The prevailing climate and the nature of the catchment influence flow (through slope and runoff) and temperature (through altitude and the riparian vegetation). At a larger scale, the geological and biogeographical setting and surrounding land use influence water chemistry (particularly the supply of nutrients and other ions) while, at least for streams and smaller rivers, the riparian vegetation influences the nature of

energy inputs (via the light supply and allochthonous organic matter). The flow of the river down the slope, with associated changes in discharge, channel size and form, superimposes more or less predictable longitudinal variation in physicochemical factors, such as coarseness of the substratum and hydraulic forces in the water column and on the riverbed. It is the combination of all these natural features that sets the habitat templet.

In this chapter we deal first with the geomorphological setting of rivers, encompassing the catchment landscape, river drainage patterns and larger-scale patterns in flow and discharge. We then consider flow and hydraulics and their variability, so important to the ecology of lotic biota. Flow interacts with the stream and river substratum, on and in which most of the lotic community resides, while the geology of the catchment and its land use play a pivotal role in relation to water chemistry. We then consider two fundamental factors—temperature and oxygen—that affect the basic physiology of lotic organisms.

2.2 The geographical and evolutionary setting and river morphology

Rivers and streams are key features of the landscape and are major 'engineers' of our world. Landforms are built by erosion (by ice, wind and water), sedimentation and geological activity driven by volcanism, and continental drift. These complex processes together determine the outline of the catchments within which precipitation falls, flows downhill and forms the network of channels that are running waters. The physical pattern of a river within the landscape reflects a complex set of factors, driven essentially by runoff, the nature of the underlying geology and the supply of sediment.

Rivers are highly dynamic systems that are created, age and move within the landscape. The ageing process involves continuous erosion of the stream channel

The Biology and Ecology of Streams and Rivers. Alan Hildrew and Paul Giller, Oxford University Press. © Alan Hildrew and Paul Giller (2023).
DOI: 10.1093/oso/9780198516101.003.0002

Table 2.1 Size and age of some of the largest and oldest rivers on earth (various sources).

River	Length (km)	Drainage area (km^2)	Estimated age (million years, Ma)
Amazon, S. America	6,992	7,050,000	200
Colorado, USA	2,334	640,000	75
Finke, Australia	600	115,000	350–400
Indus, Asia	3,610	960,000	45
Meuse, Europe	925	34,548	320–380
Mississippi–Missouri, USA	6,275	2, 980,000	100–300
Nile, Africa	6,800	3,254,555	65–75
Rhine, Europe	1,233	185,000	240
Danube, Europe	2,888	817,000	4
Yangtse, China	6,300	1, 800,000	23–36.5
Congo, Central Africa	4,700	3,680,000	1.5–2
Yellow River (Huang He), China	5,464	745,000	1.2–1.6
Mekong, South East Asia	4,350	810,000	17

back towards its source or across its floodplain. The rate of erosion differs greatly from place to place. It is particularly high in parts of China and North America, as is evident from particularly heavy silt loads, and hence colour, of many of the rivers there. The Grand Canyon in Arizona, USA, at up to 16 km wide and 1.6 km deep, is an extreme example of this erosional power, here delivered by the Colorado River. Typically, river channels have an extended lifespan compared to lakes (Table 2.1) but may change radically in form and position within the landscape. Whereas rivers flowing on consolidated bedrock may be described as 'constrained', in valleys consisting of more or less unconsolidated sediment (*alluvium*) the channel can 'migrate' laterally over long distances, meandering over the floodplain. Lateral movements of 5 to 90 km over several thousands of years have been identified in extant rivers in for example the USA (see e.g. Osborn & du Toit 1990), and repeated rapid change on a vast scale is evident over the last 2,500 years in the Yellow River (Hwang Ho) in north-east China as a result of large-scale floods (Twidale 2004).

At times of large-scale climate change, such as during glacial periods, rivers can be 'consumed' by the ice sheet, and then reform in old channels or be created in new ones as the climate ameliorates. We can get a glimpse of the development of these new riverine habitats today where glaciers are retreating, as in Glacier Bay, Alaska (Milner & Petts 1994; Milner et al. 2000). These cold streams show an interesting rapid succession of colonisers, macrospcopic animal life being led by a small assemblage of chironomid midges that is gradually replaced by a more complex community of including mayflies, stoneflies and caddisflies, and

eventually joined by salmonids as stream temperature increases (see Chapter 6, pp. 212–213).

2.2.1 Patterns in drainage basins and river classification

Despite the extreme physical variability in streams and rivers, some general patterns in drainage basins can be distinguished, reflecting their origins and development. Rivers in their drainage basins have been described as *hierarchical dendritic networks*; 'hierarchical' because there is a hierarchy of branches as the smaller tributaries coalesce into larger and larger channels, contributing water and materials downstream; and 'dendritic' because the whole river pattern across the landscape resembles a branching network of plant roots (see Chapter 1, Figure 1.3, p. 3). There would appear to be 'physical rules' that govern fluvial drainage patterns, and landforms resembling those on earth have been discovered on all of the inner planets and some of their satellites in our solar system [see Baker et al. (2015) for some stunning images]. These patterns have been variously caused by fluid lava flows (on Mercury, Venus, the Moon), liquid methane (on Titan, a satellite of Saturn) or possibly water (on Mars). On earth natural patterns include (i) *dendritic* (the most common, with many tributaries merging into the main river and in which water flow is most efficient), (ii) *parallel* (where rivers flow in the same direction, typically following natural faults or erosion, and have very few tributaries), (iii) *radial* (where streams flow away from a common high-point source such as a mountain peak or volcano) and (iv)

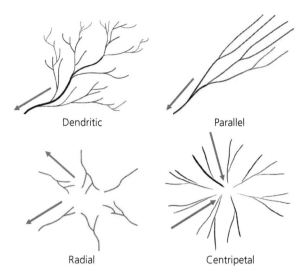

Dendritic   Parallel

Radial   Centripetal

Figure 2.1 The major types of drainage basin pattern (arrows indicate the direction of river flow).

centripetal (where streams converge on a common low point or depression, as in arid regions) (Figure 2.1). Other more minor patterns do occur under different circumstances, including tectonic, glacial and volcanic activity and man-made diversions (see Twidale 2004 for detailed descriptions of river drainage patterns). By establishing such a classification, knowledge about geomorphological, hydrological and eventually biological processes in rivers can be generalised and is potentially of value in management (Tadaki et al. 2014).

2.2.2 Stream size and stream order

Taking an aerial view, for most running-water systems a hierarchy of tributaries is apparent (e.g. Figure 2.2) which can be labelled using the stream-order scheme of Strahler (1952).

Stream order is a fairly straightforward system of describing the position of different channels within the network. First-order streams are single, unbranched headwater channels—the 'fingertips' of the network. Second-order streams are formed when two first-orders meet, third-order streams are formed when two second-orders combine, and so on. Stream order only increases when two streams of equivalent rank merge (Figure 2.3). Large rivers, such as the Mississippi are 10th order, the Nile is 11th order and the Amazon 12th order (Downing et al. 2012). Classifying streams and rivers in this way is useful for organising information of a spatial nature and helps the biologist when

analysing longitudinal changes in stream characteristics within a single catchment. This seems beguilingly simple but can be problemetic 'on the ground' in certain circumstances. One has to choose a map scale to decide what constitutes a 'first-order stream' and, in some areas, much of the drainage network includes channels in which flow is not permanent but intermittent. Such systems may be an increasing feature in a regime of a changing climate and greater water abstraction.

Drainage networks vary enormously in size depending on the geomorphology and tectonic history of the landscape. Compare the short networks of the coastal river catchments of much of eastern Australia with the enormous Amazon river basin. The 'branches' offer very different environments from the mainstem in terms of flow velocity and discharge, influence and nature of riparian zones, water chemistry, substratum and temperature. Thus, there is usually a sequential downstream pattern in conditions from the small tributaries to the main river.

Clearly, the total number of streams of each order decreases as stream order increases. For Great Britain, for example, Smith & Lyle (1978) estimated that there are over 146,000 first-order streams, over 36,000 second-order streams, but only 66 sixth-order and four seventh-order rivers. If one plots the log number of streams of each order against stream order, a straight line results. In fact, similar plots of 'log mean length of stream' of each order or 'average drainage area of streams' in each order against stream order also give straight-line relationships (Leopold et al. 1964). The river length (*L*) actually increases with drainage area (*A*) according to the following relationship:

$$L = 1.4A^{0.6}.$$

Bankfull discharge (the amount of water that fills the entire stream channel to the top of the banks) also increases log-linearly with drainage basin area. Thus, size and depth of a channel increase downstream as more water is discharged from the increasing catchment area.

2.2.3 Stream morphometry patterns

The basic spatial unit recognised in rivers and streams is the *reach*—technically, a section or stretch of river and floodplain along which conditions are sufficient to maintain a near-consistent internal set of geomorphological process interactions between the channel and floodplain—which normally should not be shorter than 20 times the mean channel width (Rinaldi et al.

Figure 2.2 Landsat imagery of the Salmon River, Central Idaho, the longest free-flowing river in the United States, showing a number of sub-catchments joining the main branch from multiple mountain ranges. Direction of river flow is indicated by the yellow arrow.
Source: courtesy of NASA Goddard Space Flight Center and US Geological Survey.

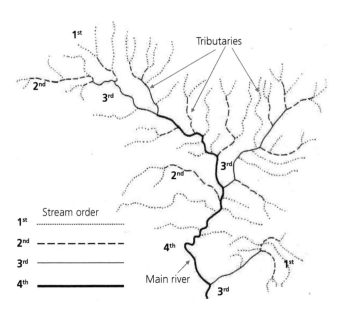

Figure 2.3 The Strahler system in which individual streams and rivers are assigned to *stream orders*, based on how the tributaries flow into each other in the drainage basin.
Source: modified from Giller & Malmqvist 1998.

THE GEOGRAPHICAL AND EVOLUTIONARY SETTING AND RIVER MORPHOLOGY

2016). In practice, the reach may be more arbitrarily defined as a stretch of river (of tens to perhaps hundreds of metres) along which similar overall hydrologic conditions prevail, such as discharge, depth, slope and area. Streams do not usually flow far in straight lines but have an inherent tendency to meander with gentle or sharper bends, unless constrained by bedrock (or human engineering), or sometimes divide into a series of branches—called *braids*. The latter are most common in the middle to lower reaches of rivers in areas with high sediment transport and often draining mountainous areas like the Himalayas or in desert regions. In both straight and meandering segments, water velocity varies longitudinally. In areas where the river or stream channel curves, the maximum water velocity occurs at the outside of the bend, where erosion takes place, and it is slower on the inside of the bend, allowing deposition of sediment (Dodds & Whiles 2019). Thus, bed sediments are eroded from some areas and deposited in others, leading to alternating sequences of shallow *riffles*, with turbulent broken water (where flow velocity at any point fluctuates irregularly and there is continual mixing of water) and coarse substrata, and deeper *pools* with more steady or laminar flow and a fine substratum (Figure 2.4). In steep mountain streams, the pool–riffle sequence is replaced by a pool–step sequence, where water plunges over short waterfalls into small scour pools. Due to the hydrodynamic features of running water, riffles tend to be spaced five to seven stream widths apart and so there are typically two riffles per 'wavelength' of a meandering channel reach (Leopold et al. 1964). The distance between meanders is related to

several factors, and the average meander wavelength is 10–14 times channel width (with a worldwide median estimated at 12.67; Frasson et al. 2019). Aerial photos of a large river and of the meanders of a small stream show a similar pattern. More recent work based around Landsat images for all rivers wider than 90 m between 60°N and 56°S and modelling confirms such 'classical' patterns even at global scales (Frasson et al. 2019).

Numerous schemes have been developed for grouping rivers based on river morphology and on flow regimes (see next subsection). As in the case for stream order, these schemes can provide a general sense of the nature of the physical environment to which the biota is exposed and in which ecological interactions and ecosystem processes play out. Seven basic river channel types are often recognised (Figure 2.5), though they intergrade between each other and, in any one river catchment, a variety of types may be found. Highly modified reaches (such as urban and channelised rivers) are considered separately, and more recent work has, at least in the European context, extended the number of river channel types quite a bit further (Rinaldi et al. 2016).

2.2.4 Runoff and flow regimes

In addition to spatial patterns in river morphology, the discharge of natural running waters varies over a range of temporal scales (Figure 2.6). *Discharge* is measured as the volume of water flowing past a point in the channel in a given time (usually expressed as $m^3 s^{-1}$). This discharge is derived from a combination of surface water, soil water and ground water and is effectively

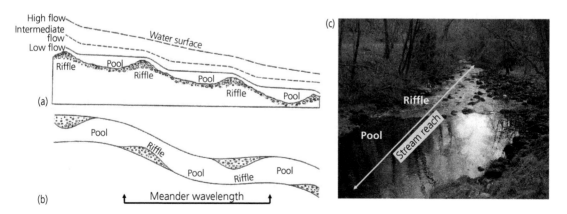

Figure 2.4 The pool–riffle sequence in typical streams or small rivers ((a) longitudinal section, and (b) top-down view). The meander wavelength (typically including two riffles) and water depth at different flows are also shown. (c) Photo of a stream reach showing a pool and riffle. *Source:* (a) and (b) redrawn after Dunne & Leopold 1978.

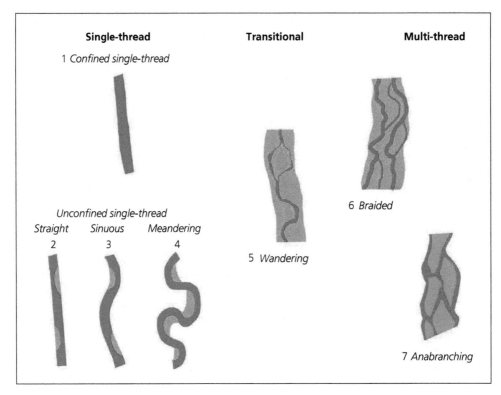

Figure 2.5 The seven basic types of river channel morphology.
Source: Rinaldi et al. 2016 with permission from Springer Nature.

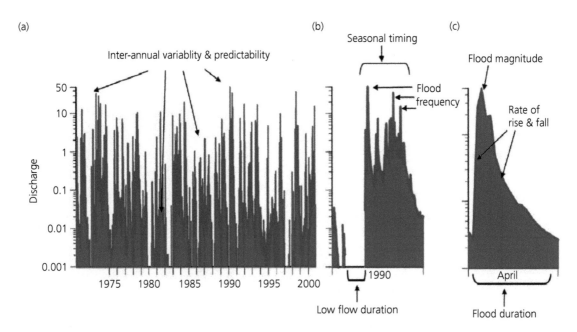

Figure 2.6 Different temporal scales of discharge variation which form different components of the flow regime: (a) interannual flow regime, (b) annual flow variability, (c) an individual flood hydrograph in April. Discharge is shown on a log scale (m^3 s^{-1}).
Source: Olden et al. 2012, with permission from John Wiley & Sons.

the portion of precipitation that eventually finds its way into streams and rivers and leaves the catchment as surface runoff (Poff et al. 1997). With measurements of discharge over relatively long periods, characteristic seasonal/annual discharge patterns, known as *flow regimes*, are apparent (Figure 2.6a and Figure 2.8). All lotic systems, including large rivers, increase in discharge following individual rainfall events in their catchments (Figure 2.6b). A simple peak in discharge is often called a *spate*. Beyond that, streams and rivers can flood (properly reserved for events in which the water leaves the channel—above *bankfull discharge*), on average about once every 1.5 years (Leopold et al. 1964). Note that the terms spate and flood are not used consistently, and almost any substantial peak in discharge is often called a 'flood'.

Most rivers continue to flow during periods of no rainfall due to inputs from groundwater or sometimes lakes (ephemeral streams excepted). This sustained low discharge is known as *base flow*. A *hydrograph* is a plot of discharge over time. Single storms in headwaters usually result in a quite rapid rise to peak discharge and a more gradual fall to base flow once the rainfall event ceases (see Figure 2.6c). The response time in such a flood hydrograph is indicated by the time to peak discharge following the peak in rainfall and

relates to catchment size, shape, gradient, soils and land cover, and basically depends on how much and how quickly water runs off the catchment and into the stream channel. Small streams show considerable short-term variation in discharge but the larger catchments and greater volume of water in lowland rivers tend to dampen the effects of individual local storms.

Discharge is extremely sensitive to land use, and catchments of similar size and subject to similar precipitation can respond very differently to a single rainfall event depending on their *infiltration capacity*—the capacity for rainfall to penetrate the soil surface. Any compaction of the soil surface caused by agricultural practices or trampling by livestock for instance, or an increase in hard, impervious surfaces following urbanisation, is associated with a much greater fraction of water from precipitation immediately leaving the catchment via surface flow. This potentially leads to high storm flows. In contrast, in catchments with high infiltration capacity, excess overland flow is much less and subsurface flows become important until, in excessively wet weather, soils can become saturated leading to saturated overland flow. These two extremes are illustrated in the schematic in Figure 2.7. These differences in infiltration capacity generate different hydrographs, generally with 'quick-flow' components (e.g.

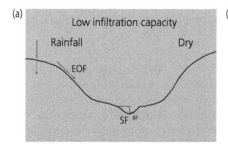

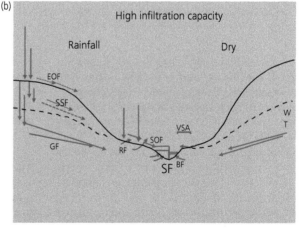

Figure 2.7 A sketch of hydrological pathways in two stream catchments, (a) of very low infiltration capacity, (b) of high infiltration capacity, in wet (left-hand side) or dry (right-hand side) weather. In (a) rainfall exceeds infiltration capacity so excess overland flow (EOF) predominates, leading to rapid and high storm flows (SF) that are not sustained in prolonged dry weather, when base flow (BF) is extremely low or absent. In (b) rainfall normally infiltrates, so EOF is minimal or absent. Rain penetrates to the soil and a further fraction then to the groundwater. Where there are cracks or 'pipes' (animal burrows, tree roots), some water may move laterally (down the slope) in the soil, as subsurface stormflow (SSF). Near the channel, the water table reaches the surface, creating a variable saturated area (VSA, shown only on the right-hand side of this diagram) which expands in wet weather and contracts in dry. In wet weather some soil water issues back to the surface as return flow (RF). Rainfall directly on the saturated area cannot infiltrate but runs directly to the channel as saturated overland flow (SOF). In dry weather base flow (BF) in the channel is sustained by groundwater flow (GF), whereas EOF, SOF and SSF are 'quick' components of the hydrograph.
Source: based on Burt 1996 and others.

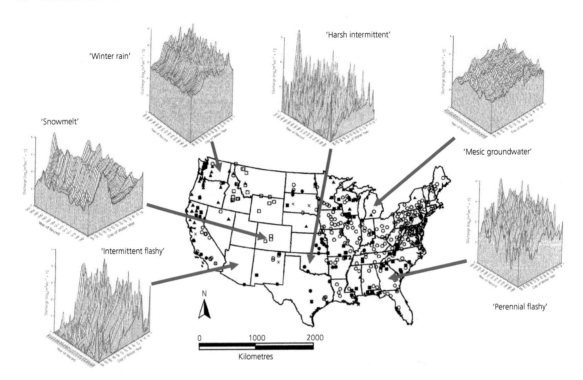

Figure 2.8 Examples from the flow regime classification based on long-term daily mean discharge records from across the contiguous United States. The graph axes are: y-axis: discharge ($\log_e[m^3\ sec^{-1}+1]$); x-axis: year of record; z-axis: day of water year. The general shape of the flow regime classes is more important than the detailed discharge values. The symbols on the map represent locations for stream groups based on 420 gauged streams (Olden & Poff 2003) and the classification scheme of Poff (1996): black triangle—winter rain, ×—harsh intermittent, black square—mesic/stable groundwater, open circle—perennial flashy or runoff, black circle—intermittent flashy, open square—snowmelt.
Source: images compiled from Olden & Poff 2003; Poff & Ward 1989, 1990; Poff 1996; Poff et al. 1997.

excess overland flow) dominating in a low-infiltration catchment and the delayed 'slow-flow' component (groundwater flow) dominating in a high-infiltration catchment. The latter can sustain surface stream-flow for long periods, even in the absence of further precipitation, while the former cannot. Water in the channel thus often consists of different components of various residence times in the catchment and whose chemistry is modified along these different hydrological pathways.

The pathways by which precipitation moves through catchments have great ecological consequences. Surface runoff causes erosion and the dumping of soil into running waters (sedimentation), while prolonged contact with soil and the underlying geology moderates temperature fluctuations and stream chemistry (it may buffer 'acid rain' for instance). Most importantly, the flow regimes of streams and rivers that are dominated by surface runoff show marked and short-term variations, with high but brief peak flows and prolonged low flows (or

no surface flow at all) in dry weather. These impose different *disturbance regimes* on river organisms and ecosystems—to which we return in later chapters. We should note that overall the proportion of precipitation that eventually runs off downstream tends to be relatively high in moist tropical rainforest areas in South-East Asia, West Africa and tropical South America, and also in some temperate areas including western Canada, southern Alaska, western Norway, southern Iceland, northern Scotland, the Alps, and south-western Chile. It is very low in the arid areas of the world, a few of which are 'arheic'—lacking any permanent streamflow at all—due to rapid evaporation after rainfall.

As well as such small-scale features of the catchment, large-scale factors naturally determine the shape and nature of flow regimes. These include climate, seasonality of precipitation, the natural biome (large-scale distribution of forests, grasslands, deserts etc), geographical relief, soils and underlying geology. By monitoring hydrologic metrics and their variability

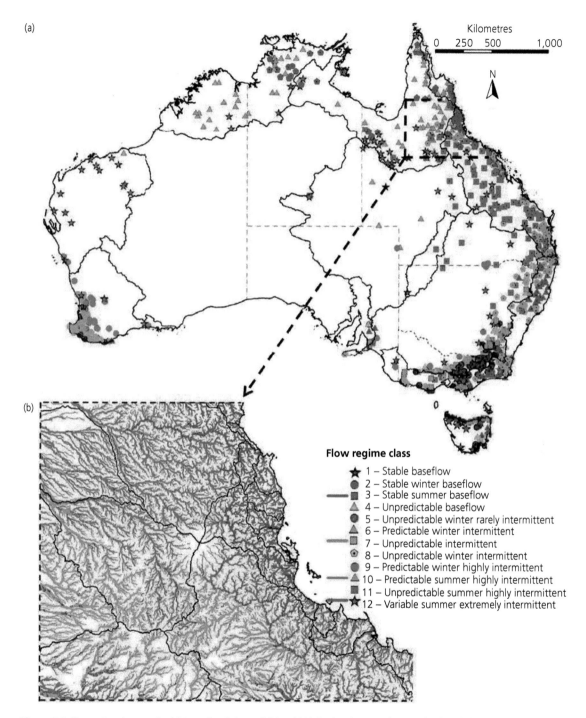

Figure 2.9 Flow regimes in Australia. (a) Australian drainage divisions (thick lines) and state and territory borders (dashed lines) and the distribution of the 12 flow regime classes (coded in the key by coloured symbols) are shown. The inset figure (b) shows predicted flow regime types of north-eastern Australian streams based on climate and catchment topographic characteristics (flow regime classes are indicated in the key). Note that large areas of the dry central and southern continent lack river basins. This figure incorporates data that are copyrighted by the Commonwealth of Australia (GeoSciences Australia 2006).

Source: from Olden et al. 2012, based on classification of regimes by Kennard et al. 2010. With permission from John Wiley & Sons.

at the fine scale of daily records, and over several years (including seasonal patterns of flows; timing of extreme flows; the frequency, predictability and duration of floods, droughts and intermittent flows), we can categorise flow variation and predictability in a system. This has led to the development of river streamflow classifications in a range of different regions, focused on different elements of the flow regime (Olden et al. 2012). Poff & Ward (1989, 1990) recognised three types of intermittent streams and six types of perennial streams based on flow regimes at the scale of the contiguous USA (i.e. excluding Alaska and Hawaii). The same geographical region may include several different flow regimes (Figure 2.8 shows some examples). Alaska itself accounts for about one third of all runoff in the USA, with pronounced climatic and topographical gradients, and has six hydrologic regions (Oswood et al. 1995). In Australia, Kennard et al. (2010) developed a classification scheme of 12 flow regime classes. This was based on 830 stream gauges and 15–30 years of mean daily discharge data, differing in the seasonal pattern of discharge, degree of flow permanence, variation in flood magnitude and flow predictability and variability (Figure 2.9). Global flow regime patterns are also apparent. One of the earliest attempts at a global classification was based on nearly 1,000 stations across 66 countries, with data over 33 years on average at each station. Haines et al. (1988) identified 15 groups of river systems largely based on monthly peak flows, ranging from uniform (with little seasonality in stream flows) to, for example, late spring–early summer, extreme late summer, moderate autumn and extreme winter peak flows. More recently, Poff et al. (2006) used 460+ daily streamflow gauges from five continents to compare intercontinental flow regimes. There was considerable similarity in overall flow regimes (between Australia and the USA, for instance), whereas New Zealand was the most regionally distinct but with some similarities to Europe.

There have also been major and increasing modifications of natural flow regimes through a range of anthropogenic impacts (such as dams, urbanisation, flood protection etc). This can be clearly seen in the establishment of a dam on the Green River, Utah which truncated the extremes of floods and droughts (Lytle & Poff 2004) (Figure 2.10). There are already impoundments on most of the world's larger rivers (and myriad schemes for small-scale hydropower generation on small but steep streams), and plans for many more. Such developments are a profound threat to biodiversity and involve many 'conflicts of interest'. Christine Zarfl discusses this in more detail in Topic Box 2.1 (and see also Chapter 10).

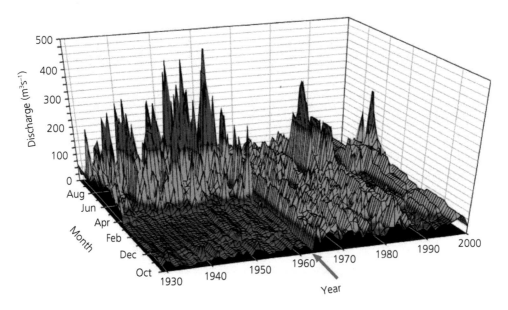

Figure 2.10 A long-term hydrograph of the Green River, Utah, USA (1929–2000) showing monthly discharge patterns before and after the Flaming Gorge Dam was completed in 1963 (arrow). The natural flow regime involved floods from spring snowmelt and prolonged low flows during autumn and winter. Damming truncated the flow extremes; peak discharge is smaller and very high peaks are less frequent while very low flows are now rare.

Source: from Lytle & Poff, 2004, with permission from Elsevier.

Topic Box 2.1 Dammed rivers—are they a blessing or are they damned?

Christiane Zarfl

The power of rivers has been exploited by humans for centuries, rivers and streams having been 'tamed' by dams of greatly differing size. Dams can serve a single or a variety of purposes (Box Figure 2.1); for instance, they are useful for flood control (by reducing hydrological extremes) and, conversely, to store water for irrigation, public supply or industry. In valleys of sufficient slope, dams are used to harness the energy of the river via devices ranging from ancient waterwheels to modern hydroelectric plants. Hydroelectricity is presently the biggest source of renewable electricity and is widely seen as 'green energy'—a means to satisfy an increasing demand for energy while mitigating climate change. However, hydroelectricity is not climate neutral (Hermoso 2017). Why not?

Rivers are complex systems connecting 'components' (water, sediment, nutrients, organisms etc), highly dynamic in space and time, and we can only understand them by combining several disciplines and their interactions. Dams have many environmental consequences. First, they lead to the longitudinal, lateral and vertical fragmentation of rivers (Grill et al. 2019), change the flow of water and energy and interrupt the transport of sediments downstream. Delta areas are then lost due to a combination of missing sediments and rising sea level (Dunn et al. 2018). River fragmentation also reduces nutrient transport (Maavara et al. 2020) and impedes the longitudinal migration of organisms (most obviously but not only fish) to their spawning or feeding grounds. Dam operation can be crucial. Artificial water releases cause hydropeaking and abrupt changes in temperature downstream (when cold water from deep in the impoundment is released) and erode the riverbed. Turning a river into a reservoir leads to upstream flooding and downstream loss of floodplains. This often has social and economic impacts, when human resettlements are required, and cultural sites or agricultural land are lost. Ecologically, lotic and floodplain habitats are lost and their freshwater biodiversity is reduced.

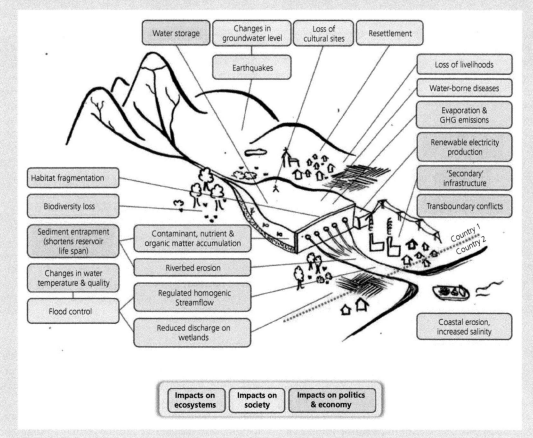

Box Figure 2.1 Overview of the ecological, social, political and economic impacts of (hydropower) dams on river catchments (GHG = greenhouse gases).
Source: adapted from Peters et al. 2021.

Topic Box 2.1 *Continued*

At the same time, new lentic habitats are created and can foster species invasions and often an increase in waterborne diseases. Changes in the level of groundwater may even trigger earthquakes.

Reservoirs usually stratify and evaporative losses increase—that is, there is a change in the water cycle of the catchment. Sediments from upstream are deposited in the reservoir, reducing its lifespan, and contaminants and organic matter accumulate along with those sediments. In a stratified reservoir anoxic conditions can develop, leading to the production and release of greenhouse gases, especially methane. Emissions are particularly high in the first 30 years after reservoir flooding as the original terrestrial vegetation and organic matter decomposes. Large dams also have enormous economic and political dimensions. Their creation leads to a 'secondary' infrastructure, including industrial and urban development. Furthermore, such projects involve not only local people and interest groups but often an international portfolio of investors and companies, while time and cost overruns in such 'megaprojects' are common. In addition, the potential for political conflict is particularly high where dams are built on transboundary rivers and upstream and downstream countries compete for water. In summary, the costs and benefits of dams are complex and depend on competing interests.

We know that many new impoundments are proposed and under construction, but how many dams are there and where? Recently, global data have been compiled by a team of research groups, but mostly include only those dams greater than 10 to 15 m in height and visible by satellite imagery (http://www.globaldamwatch.org). Using these records for about 20,000 medium- to larger-sized dams, the degree of connectivity of rivers worldwide has also been quantified. It was estimated that 63% of formerly free-flowing rivers longer that 1,000 km have been lost globally (Grill et al. 2019). Even this number is still conservative and may only be the 'tip of the iceberg', since the study did not include smaller barriers. Thus, a pan-European atlas of river barriers, including some even smaller than 2 m, has been compiled for 36 countries (Belletti et al. 2020). This includes no less than 630,000 unique barriers, recorded in local, regional and national databases. Based on an additional field survey of 147 rivers (2,715 km of river length in total), an estimate of at least 1.2 million river barriers in these 36 European countries has been made, almost 70% of them < 2 m in height. A similarly expanded worldwide database on existing barriers is clearly required to provide a basis for river management.

What lies ahead? Spatially explicit data have also been compiled for projected dam construction on a global scale (Zarfl et al. 2015). This is restricted so far to hydropower dams with a capacity ≥ 1 MW. This is fundamental for informing the discussion on sustainable hydropower development, taking the river catchment as a whole into account and allowing economic, social and ecological aspects to be weighed against each other (Peters et al. 2021). Nevertheless, more data and information will be required on future dams, many of them small, to be built for alternative purposes, such as flood control and irrigation. These will also expand in line with expected increases in climatic extremes.

Finally, while we are seeing a renewed boom in dam construction, especially in countries of the Global South, and also in the Balkan region of Europe, dams are being removed in countries like the USA, Spain and Norway, partly due to an increased awareness of the ecological (and economic) impacts of dams, and partly due to the declining safety of dilapidated dams and the high costs of their repair. Dam removal, however, does not mean that there will be a return to a natural (pre-impoundment) status. Depending on the 'type' of dam, removal might even lead to the destruction of downstream ecosystems (e.g. by the sudden flush of contaminated sediments) before a new stable ecosystem state is reached. The better the river ecosystem is understood, the better the consequences of dam removal, and the greater the likelihood that an appropriate river restoration and management programme can be implemented.

Professor Dr Christiane Zarfl is based at the Centre for Applied Geoscience, Environmental Systems Analysis, Eberhardt Karls University of Tübingen, Germany.

Five key components of the flow regime are recognised as influencing the ecology of the lotic systems (Poff & Ward 1989; Poff et al. 1997): (i) *magnitude* of discharge during any given time interval; (ii) *frequency of occurrence* of specific flow events above a given magnitude (usually above baseflow); (iii) *duration* of a specific flow condition; (iv) the *timing* or *predictability* of a certain flow magnitude (i.e. how regularly they occur); and lastly (v) *rate of change* or 'flashiness', relating to how quickly flow changes from one magnitude to another. These components have particular relevance to flow disturbances in relation to the ecology and management of streams and rivers, as we will see in Chapters 5, 6 and 10. Further refinement of the characterisation of river flow regimes through the inclusion of the additional factors of sediment input, flow and storage has been suggested as an important contribution to future river management (Wohl et al. 2015). What is also now attracting the attention of freshwater biologists is the potential impact of climate change on flow regimes and on extreme events (Schneider et al. 2013; Woodward et al. 2016).

2.2.5 Groundwater and up- and downwelling zones

While the water flowing above the substratum is what one normally considers as the river or stream, much of the water in the system actually often lies beneath the surface, in what is called the *hyporheic zone* (Figure 2.11). This is essentially an interface or mixing zone between surface water and groundwater, and can extend into permeable sediments beneath the channel and also below the banks and out into the alluvial floodplain.

This extends the river as an ecosystem well beyond the physical confines of the channel itself. Where such sediments are sufficiently coarse, and there is some below-ground active flow, metazoan (multicellular) invertebrates have been found in groundwater up to several kilometres lateral to the river channel itself (Ward et al. 1998). Thus, large numbers of riverine invertebrates were collected within a grid of shallow (10 m) wells located on the floodplain up to 2 km from the channel of the Flathead River, Montana, USA (Stanford & Ward 1988). This phenomenon is most clearly seen in gravel-bed rivers, and commonly in the valley bottoms of glaciated mountain systems, such as those found in the Rocky Mountains of North America, the Alps of Europe, the Andes of Patagonia, the Southern Alps of New Zealand, and the high Himalayas of Asia (Hauer et al. 2016; Figure 2.12).

Some of these invertebrates are subterranean specialists, rarely found at the surface, but others are species common in the stream benthos where the hyporheic zone offers both habitat and refuge, particularly for the early stages of macroinvertebrates and

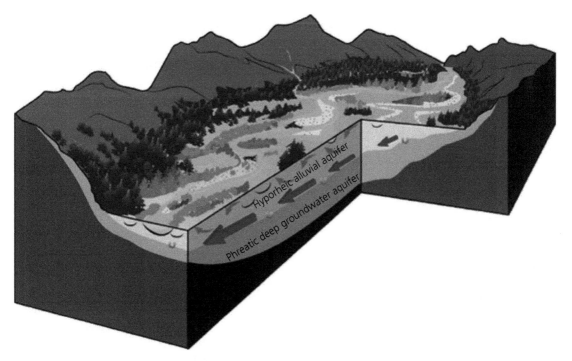

Figure 2.11 The three-dimensional structure of a gravel-bed river. The larger blue arrows signify the hyporheic zone that develops at the upper end of the floodplain. Hyporheic water flows through the gravel substratum to discharge back to the surface at the lower end of the floodplain. The smaller arrows near the surface illustrate small-scale water exchange between surface waters and the upper hyporheic waters (upwelling and downwelling) in the shallow bed sediments. The deeper, phreatic groundwaters are stored for longer periods of time.
Source: from Hauer et al. 2016, under Creative Commons Attribution Non-Commercial Licence 4.0.

Figure 2.12 A segment of the Flathead River located in south-eastern British Columbia lies in the valley bottom of a heavily glaciated mountain system. The white arrow illustrates the width of the gravel-bed river floodplain system in this river segment.
Source: from Hauer et al. 2016, under Creative Commons Attribution Non-Commercial Licence 4.0.

fish, as well as for meiofauna (smaller metazoans, see Chapter 3) and microorganisms (particularly heterotrophic bacteria). The hyporheic zone may be critical for nutrient cycling and ecosystem metabolism, and is a site of the exchange of carbon and nutrients between surface and subsurface waters (e.g. in desert streams, see Grimm & Fisher 1984 and Chapters 8 and 9). Hence, hyporheic zones can have a strong influence on both stream biota and hydrochemistry.

Hydrological connections between the groundwater, the hyporheic zone and the stream channel occur in up- and downwelling zones, where water issues up into the channel from beneath or down into the hyporheic zone from the surface water above, respectively (Figure 2.11). At the scale of the stream riffle, there may a classic flow pattern in which surface water downwells into the hyporheic zone at the head of the riffle while hyporheic water returns (upwells) to the stream surface at the tail of the riffle (Franken et al. 2001). This is caused by decreasing stream depth creating a high-pressure zone at the head of a riffle, leading to downwelling into the sediments. This water flows interstitially beneath the bed for some distance. At the downstream end of the riffle and start of the pool, increasing stream depth produces a low-pressure zone, causing an upwelling of hyporheic water. These hydrologic patterns generate further heterogeneity in biological and physicochemical patterns in the hyporheic zone, and may be much more complex than this simple pattern described, or the nature of the substratum may inhibit hydrological exchange altogether.

2.3 Flow and hydraulics

Both terrestrial and aquatic habitats consist of a 'fluid' medium (air or water) overlying a solid. Whilst the lives of terrestrial animals and plants are relatively decoupled from the motion of the air (except for dispersal of some propagules, spores and pollen), for aquatic organisms the medium influences all aspects of their existence. Water is denser than air, offering more mechanical support but making it harder to move through. In lakes, much of the biota floats or swims freely above the 'ground' (the lake bottom), whilst in lotic systems, most of the biota is much more closely associated with the riverbed. Flow forces are undoubtedly the major architects of the physical habitat in streams and rivers, through their influence on the particle size and nature of the substratum and channel morphology, the supply of dissolved oxygen, the distribution and turnover of food and other resources, and through direct physical forces within the water column and on the substratum.

2.3.1 The nature of flow

Flow is the directional ('advective') movement of water, predominantly downstream. It is also complex in both space and time, and at any instant each fluid particle can move longitudinally, laterally or vertically. As flow is so important to the biology and ecology of running-water organisms, we need to understand a little about its nature. At the same velocity, flow can be either *turbulent* or *laminar* which, as Statzner et al. (1988) point out, presents lotic organisms with two distinct physical worlds. Laminar, hydraulically smooth flow conditions can exist over similarly smooth solid surfaces (mud bottoms, flat bedrock) or through dense strands of macrophytes and, on a smaller scale, over the flat blades of some macrophyte leaves. Here, the fluid moves in parallel layers which may slide past each other at differing speeds but move in the same direction. Pure laminar flow is rare, however, and usually occurs at low velocity. Turbulent, hydraulically rough flow occurs in areas of coarser substratum and at higher velocity, and involves chaotic eddies and swirls in every direction. This disrupts orderly laminar flow and has important mixing effects on heat and water chemistry, and carries oxygen and other dissolved gases close to and into the substratum or to the water surface (where they may equilibrate with the atmosphere).

Two other physical properties of a fluid influence the nature of flow: *viscocity* and *inertia*. Viscosity is related to how rapidly a fluid can be deformed, the resistance being due to the coherence of its molecules. Cold water is more viscous (thick and 'syrupy') than warmer water. Inertia reflects the resistance of fluid particles to forces that, when applied, cause them to accelerate or decelerate. High inertial forces promote turbulence; high viscous forces promote laminar flow. The ratio of inertial to viscous forces within a fluid produces the dimensionless *Reynolds number* (*Re*), and can describe both the transport properties of a fluid or the forces experienced by a particle or organism moving in the fluid. It can be estimated for the stream channel as a whole or for an individual organism (see Gordon et al. 2006 for more details).

For a fluid, a large *Re*, where inertial forces dominate, indicates turbulent flow. For example, a small babbling brook with a current of 0.1 m s^{-1} has an *Re* of about 10,000 (Reynolds 1994). In contrast, a small *Re*, where viscous forces dominate, indicates laminar flow and, for any given depth, flow becomes laminar when velocity decreases such that *Re* drops well below 2,000. *Re* also indicates the forces experienced by an animal. Both the movement of the fluid and the movement of the animal will govern the Reynolds number. In general, small organisms close to the stream bed (where velocity is low) have low *Re* values and will be more

subject to viscous forces (e.g. a bacterium has a *Re* of $< 10^{-4}$ and a mayfly larva in the region of 10^2 (Vogel 1994). The advantage for an organism living at a low Reynolds number is that it is essentially protected from turbulence as it is surrounded by a 'coat' of viscous fluid, although the supply of nutrients and gases relies on slow diffusion. For many aquatic invertebrates life may start at low Reynolds numbers (1–10) and end up in conditions of *Re* of 1,000 or higher as they grow (Statzner et al. 1988). Large organisms in higher water velocity have higher *Re* values (e.g. salmon have an $Re > 10^7$), where viscous forces are less important and the organisms will be subject to more inertial forces. The importance of the physics of flow will become clearer when we consider how current changes with depth and how animal size influences life in flowing water.

2.3.2 Discharge and current

River or stream discharge (see Section 2.2.4) is related to stream width, depth, current velocity and roughness of the substratum. Here we need to be careful with terminology. Hydrologists use 'flow' to mean the discharge of a river—as in 'flow regimes' dealt with earlier. Ecologists often mean water velocity— as measured by a 'flow meter'. These are not quite the same thing. Discharge normally increases downstream, due to the addition of tributaries, along with an increase in depth and change to a smoother substratum. This is clearly seen from the huge data set from the USA examined by McManamay and DeRolph (2019; Figure 2.13) which shows a strong relationship between discharge and stream order. The actual discharge reached depends on the length of river, size of drainage basin, amount of runoff and climate (as well as the withdrawal of water), which explains the level of variation within each stream order seen in Figure 2.13. The pattern is not always so clear, however. For instance, in large rivers flowing through arid zones without permanent tributaries, as often found in the dry tropics, discharge can actually decline along their length, as water is lost via evapotranspiration and commonly through water withdrawals for irrigation agriculture. A classic example is the African River Nile, which is of a similar length to the Amazon (6,400 km compared to 6,300 km, respectively) but has an annual discharge to the sea of just 2,700 m^3 s^{-1} compared to the Amazon's 175,000 m^3 s^{-1}.

While organisms are likely to be affected by the mean and range of discharge (especially the extremes), discharge itself is of less biological interest than the *current*

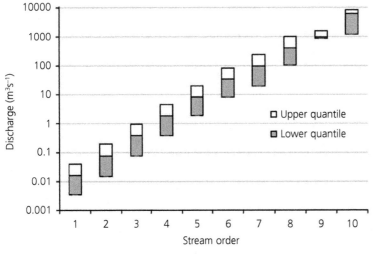

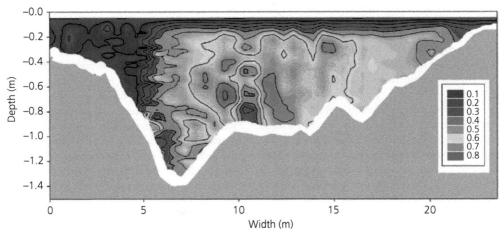

Figure 2.13 Extensive data from the 48 contiguous states of the USA based on a spatial framework of over 2.6 million stream reaches illustrate the relationship between discharge and stream order. Box plots (with upper and lower quantiles) show that discharge values ranged widely within stream orders, but this does not disguise the clear positive relationship.
Source: from McManamay & DeRolph 2019, under Creative Commons Attribution 4.0 International Licence.

Figure 2.14 A flow velocity profile (a transverse section across the river channel) of the Kansas River, measured with acoustic Doppler velocity equipment. This type of method relies on reflection of sound and light waves, respectively, from small particles in the flowing water and allow very short-duration measurements (many per second) on very fine spatial scales (sub-mm). The colours from blue through to red indicate increasing current velocity (m s^{-1}) while the white area is too close to the substratum for this equipment to measure velocity.
Source: from Dodds & Whiles 2019, with permission from Elsevier.

(water velocity) in which the organism actually lives, and its associated flow forces such as drag, lift and shear. New techniques have allowed us to study water velocity at the scale of the individual invertebrate, as we will see later (Chapter 4, Section 4.2.1, p. 109). Water velocity is usually measured in metres per second and mean velocity in natural channels rarely exceeds 3 m s^{-1}. River discharge, the gradient (slope) of the stream bed, the nature of the substratum and water depth all affect mean reach water velocity. An important feature of streams and rivers is that water velocity varies enormously at small scales. At a given discharge, the current decreases exponentially with depth, giving

a vertical velocity profile or gradient (Figure 2.14). This results from a thin layer of fluid in contact with the substratum being slowed to zero by frictional drag. The influence of friction diminishes with distance from the bed into the water body and as the current speed increases. Similarly, friction from the channel sides and from surface tension will cause transverse and vertical velocity gradients. Thus, the highest current speeds at any cross section of a river will normally occur just below the surface at the deepest point of the channel. Average velocity at any cross section usually occurs at around 0.6 of the depth from the surface and is about 80–90% of surface velocity.

2.3.3 Boundary layers and dead zones

The steepest part of the vertical velocity gradient is close to the stream bed, where the flow is reduced and becomes laminar. Here current speeds approach zero, forming a so-called *boundary layer*. The overall thickness of this layer shrinks with increasing current velocity in the water column above. Within the boundary layer is a viscous laminar sublayer, which varies in thickness depending on the nature of the stream bed. Where the substratum is relatively smooth (i.e. individual substratum particles are small in comparison to channel depth), hydraulically smooth conditions occur, leading to a relatively thick laminar sublayer (as found over bedrock in small streams, or in deep lowland rivers with a substratum of fine particles) (Figure 2.15a). However, most stream substrata have large surface irregularities and are said to be 'hydraulically rough', and the viscous sublayer is much thinner (Figure 2.15b). Near-bed patterns then become extremely complex and patches of high velocities can occur relatively close to the stream bed. Where flow hits an object protruding from the bottom (like a cobble or boulder), the front edge is fully exposed to the flow, while on the top and at the sides the current is fast with a thin laminar boundary layer. The outer edge of the boundary layer forms a transitional

zone between fully turbulent flow and laminar flow and is characterised by patches of turbulence (Dodds & Whiles, 2019). The boundary layer increases in thickness as current speed reduces, and progressively thickens as one passes along the object in a downstream direction (Statzner et al. 1988). If current speed increases, as in a spate, the boundary layers narrow at both the front edge and downstream end of the object. At the downstream edge, flow 'separates' and the current becomes less or non-existent directly behind the object (in its 'flow shadow' so to speak), forming a so called *dead zone* which leads to a depositional microhabitat (Figure 2.15c). Many macroinvertebrates are too large to exist fully within the boundary layer but can avoid the full forces of the current in various ways (there are examples in Chapter 4). Indeed, some animals can even control the nature of the boundary layer over their surface by altering their profile, such as the Water Penny (*Psephenus herricki*, Coleoptera; McShaffrey & McCafferty 1987).

2.3.4 Shear stress

Steep vertical gradients in water velocity produce shearing forces close to the bottom that are measured as *shear stress*—a force (per unit area) acting parallel

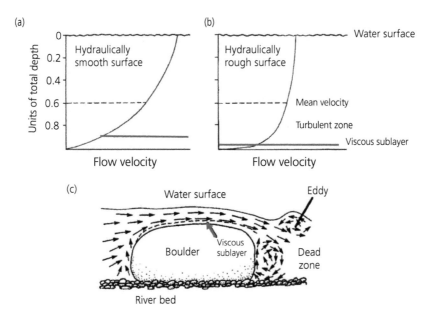

Figure 2.15 Velocity gradients over surfaces in a stream. Vertical gradients over hydraulically smooth (a) and rough (b) surfaces illustrate the difference in thickness of the viscous (boundary) layer. (c) shows the distribution of currents around a boulder, illustrating the dead zone
Source: modified from Giller & Malmqvist 1998.

to the stream bed. This determines both the potential to dislodge mineral and biotic particles and the inertia or energy required to withstand such forces. Thus, as water flows over a solid surface, be it a substratum particle or an organism, it generates lift and drag forces and, if the shear stress is sufficient, that particle or organism is set in motion. Particles may roll or bound along the stream bed (known as *bed load*) or be carried in suspension (entrained) by turbulent eddies (known as *suspension load*). Shear stress increases with water depth and is also correlated with the square of current velocity, so high flows exert a large force on the stream bed and influence the particle size and the quantity of bed load (Section 2.4). Fine material will be entrained at current speeds above about 0.2 m s^{-1}. Sandy beds, which form in flows between 0.2 and 0.4 m s^{-1}, will tend to be scoured at higher discharge, when depth and current velocity increase, and may refill as they decline. Shear stress will clearly act on organisms in the same way. In disturbing bed materials, high shear stress can also expose buried benthic detritus to oxygenated water, 'winnow away' that detritus, and turn over rocks whose upper surfaces bear photosynthetic organisms (plunging them into the relative darkness). In this way, flow fluctuations influence stream metabolism (the balance between ecosystem primary production and respiration, see Chapter 8). As we will see, many stream organisms have morphological features and/or behaviour that limits their passive entrainment into the flow to some extent (discussed in Chapter 4).

2.3.5 Water velocity and shear stress vary spatially

Within a stream reach, changes in gradient and stream depth lead to sequences of turbulent riffles (high shear stress) and hydraulically smooth pools (low shear stress) and, as the stream meanders, flow varies across the channel. Irregularities in the stream bed or the banks, woody debris and changing substrata also result in complex current patterns (eddies, reverse upstream flows, etc) even in low-gradient streams. Even at the scale of the individual cobble or boulder, the boundary layer varies spatially and the water velocity at a single point can fluctuate over a few seconds due to local surges in response to passing eddies and turbulent water movement. The stream and river reach is thus a mosaic of flow conditions, water travelling at different velocities and exerting different shear stresses on patches of the stream bed and on various organisms.

Larger-scale, longitudinal patterns in flow occur from headwaters to source. Rivers typically originate from springs and rivulets which in turn form turbulent streams in high-gradient uplands. Tributary streams then gradually coalesce, resulting in larger, smoothly flowing, deep but low-gradient rivers that wind their way towards the sea. We know discharge is greater downstream (with a few exceptions), but average current velocity (and Reynolds number) also usually increases downstream with increasing discharge (Statzner et al. 1988). This at first seems counterintuitive, as the relaxed 'lazy river' of the lowlands seems to be flowing much more slowly compared to the bustling streams of the uplands, but mean velocity is actually usually greater in the former. At a constant slope, current velocity will increase with size and depth of channel. Although stream slope declines downstream, channel width and especially depth increase, which 'overcompensates' for the decline in slope (Leopold & Maddock, 1953). In addition, the substrata become finer and offer less resistance to flow than the coarse substrata upstream, again offsetting the declining gradient. A 'less organised', turbulent, roughly flowing upland stream is probably travelling more slowly overall than the larger, deeper, smoother-flowing lowland river. Nevertheless, shear stress is highest in the turbulent shallower headwaters and decreases downstream.

2.3.6 Variability in discharge and shear stress over time

In addition to the spatial heterogeneity, most natural running waters show marked temporal variability in discharge and shear stress. Flow fluctuations, especially high and low extremes, can potentially act as quite severe disturbances to lotic communities (see Chapters 5 and 6). Floods and droughts will, of course, differ in their effect—the former are episodic and short-term, the latter build up slowly and are often longer lasting (see the flow regime examples in Figure 2.8). Surface flow can dry up in normally perennial systems during rare severe droughts, but droughts are a regular and 'normal' feature of intermittent streams (see Chapter 1). The smaller the stream and the drier the climate, the more likely flow will be intermittent, and perhaps around half of the world's river networks are intermittent (Datry et al. 2016b). Drying starts with shrinkage of the wetted channel and continues to a series of isolated pools and perhaps to complete loss of surface water, although subsurface flow may persist (see also Chapter 6, section 6.4.2.2). While regulation of rivers for irrigation, water supply and hydroelectric power can reduce variation in flow (see Figure 2.10 above), it can also lead to decreased flow and even zero flow below dams. Many countries now provide for a *stipulated minimum flow* or an *environmental flow* sufficient to sustain river ecosystems (see Chapter 10).

Contrastingly, floods, in which discharge overflows the banks, are just as much a feature of tropical, Mediterranean and desert streams as of those in wetter climates. The variability of discharge in Australia, for example, is far greater than elsewhere (Lake 2000) and extreme flash floods are common in the rainy season in many desert areas. At the other extreme, large floods occur following snow melt in upland alpine and subarctic regions, or during monsoon rains in Asia. These latter floods, which can also affect rivers in the lowlands, are more or less predictable, occurring in the same season every year, although their magnitude can vary. This has led to suggestions that species living in such places are somehow adapted to the conditions

(see Chapter 4). Unpredictable floods are common in other systems and can cause severe disturbances to the biota (see Chapter 5, Section 5.3.3, p. 166). Even though floods may be predictable, exceptionally severe floods can also cause major disturbances, and may be described as '1 in 50-year' or '1 in 100-year' events. Discharge may reach up to three times bankfull, often with devastating results to agricultural land and people (see Table 2.2).

In general, flooding inevitably leads to increased current velocity within the channel and in turn increased mean shear stress. However, this change in velocity and shear stress is not equal over all parts of the stream reach. For instance, Lancaster & Hildrew (1993a) recorded different patterns on small patches of the bed of a small English stream across a ten-fold

Table 2.2 Examples of flood disasters world-wide.

Date	Location	Disaster impact
September–October 1887	Huang He (Yellow River), China	900,000 deaths
January 1910	Seine, Paris	River rose 8.3 m above normal, causing the equivalent of $1.5billion damage
August 1931	Hueng He, Central China	110,000 km^2 completely or partially inundated, with estimates of 850,000 to 4 million flood-related deaths
June–July 1971	States of Bihar, Orissa, Uttar Pradesh and West Bengal, India	£300 million damage, 55 million people affected
August 1973	Central Mexico (Hurricane Brenda)	200,000 homeless
May–September 1993	Great flood of Mid-West USA by Mississippi and Missouri rivers	17 million acres of land inundated over three months. Destroyed > 10,000 homes and caused equivalent of $26 billion damage
January 1995	Rivers Waal, Maas and Rhine, the Netherlands	250, 000 persons evacuated
June 1997	Oder and Vistula drainage basins, Poland	650,000 ha inundated (1 in 1000-yr event) and material loss of several billion $
November 1997	Somalia (El Nino-related)	1,000 deaths, 250,000 homeless
December 1999	Vargas River, Venezuala	Death toll between 10,000 and 30,000
August 2002	Across Germany, Austria, Russia and Czech Republic	Overall €15 billion of damage at the time
May–August 2010	Yangtze, Yellow and Songhua Rivers across 28 provinces, municipalities and regions of China	> 230 million people affected, > 3,000 deaths and estimated damage amounting to the equivalent of over 275 billion Yuan ($4.1 billion)
August 2018	Kerala Province, India	> 1 million evacuated and an estimated cost of 400 billion rupees ($5.6 billion)
July 2021	Rhine and major tributaries, Germany	148 L m^2 of rain in 48 hours in a part of Germany that usually sees about 80 L in the whole of July. Estimated cost a minimum of €10 billion ($11.8 billion)
March 2022	New South Wales, Australia	Rainfall exceeding historic records for over 120 yrs with up to 900 mm in a week leading to up to 500,000 subject to evacuation notice
June 2022	Assam province, India	Unprecedented rainfall and flooding affected 32 of the 35 districts, killing at least 45 and displacing 4.5 million over one week

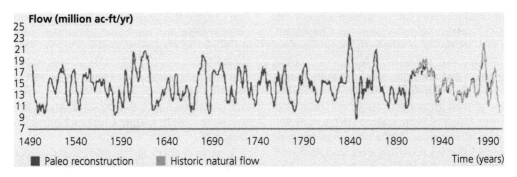

Figure 2.16 Palaeo-reconstruction of flow in the Colorado River from 1490 through to the 1990s, based on a combination of five-year running means of historic and palaeo-reconstructed stream flow at Lees Ferry, Arizona. One acre-foot (ac-ft) is equivalent to 1,233 m³.
Spirce: from WWAP (World Water Assessment Programme) 2012, which was based on Prairie et al. 2008.

increase in discharge. A few patches retained low shear stress and velocity, even at relatively high flows—attributable to the physical complexity of the channel. Such physical heterogeneity could add to the resilience of stream-dwelling populations by the presence of such refugia from high flows (see Chapter 5, Secion 5.3.3, pp. 166–167).

Long-term patterns of discharge are evident from rivers that have been monitored over considerable periods of time. For example, regular discharge measurements begun on the Rivers Thames and Lea in England in the 1880s, they have been collected for well over a century by the United States Geological Survey (Depetris 2021), while long-term river-gauging stations in Argentina have allowed computation of discharge time series from 1904–2003 for a number of rivers (Pasquini & Depetris 2007). With the help of palaeo-reconstruction and modelling it is also possible to take an even longer-term view, as shown for over 700 years of history of the Colorado River in the USA (Figure 2.16). The value of this kind of data lies in enhancing predictions of flood or drought risk. In the case of the Colorado, the very long-term data indicates a much higher frequency of dry spells and low flows relative to the streamflow in the twentieth century. For example, the prolonged drought on the Colorado River between 2000 to 2010 appears to be highly unusual based on the more recent observed data, but quite common based on the long paleo-reconstruction (Prairie et al. 2008). Such data can also point to long- and shorter-term discharge cycles, as seen in the River Nile over a period from AD622–1922 (Kondrashov et al. 2005) with a 256-year cycle and a quasi-quadriennial (4.2-year) and quasi-biennial (2.2-year) mode with connections to the El Nino/Southern Oscillation (ENSO) phenomenon in the Indo-Pacific.

North Atlantic influences have been attributed to a seven-year periodicity cycle in the Nile, with connotations for the biblical seven-year cycle of lean and fat years.

2.4 Substratum

Most lotic invertebrates are benthic (living in or on the channel bottom) and hence the nature of the substratum is of great importance. It is, in turn, the result of the interplay between sediment supply and the forces of flow. The substratum is the habitat of the benthos, providing, for instance, anchorage for rooted plants, shelter from high shear stresses on the bed, and a refuge for benthic prey from predators living in the water column. It also provides food directly, in the case of benthic organic matter and living plants and microbes in biofilms, or indirectly as a firm surface for the settlement of sedentary filter-feeders who capture food particles from the current.

2.4.1 Physical properties

The substratum itself comprises a wide variety of inorganic and organic particles. The inorganic material originates from the erosion of rocks associated with the river basin slopes, river channel and banks, and is modified by the current. The organic materials vary from particles, such as plant fragments and leaves, to fallen trees, and is derived ultimately from the surrounding catchment and upstream habitats, as well as from aquatic plants such as filamentous algae, moss and macrophytes.

Mineral (inorganic) substratum particles are most commonly classified according to size using the Wentworth Scale (Table 2.3) which is based on the diameter

Table 2.3 Wentworth classification of substratum particle size and current velocity necessary to move particles along, with more general descriptions of associated channel morphology and stability (modified from Giller & Malmqvist 1998, including information from Church 2006; van Rijn 2019; and Wohl & Merritt 2021).

Size category	Particle diameter (range in mm)	Approximate current velocity to move particle (ms^{-1})	General channel morphology	Likely channel stability
Bedrock	> 400		Uniform or variable bed gradient step-pool or plane-bed or channels with undulating walls	Permanent
Boulder	>256	3.0+	Step-pools or boulder cascades	Stable except under severe events (e.g. flash floods)
Cobble—Large	128–256	2.0–3.0	Single thread, step-pools,	Stable for long periods although subject to movement during catastrophic flow events
Cobble—Small	64–128	1.5–2.0	Cobble-gravel, highly structured sediment type	Relatively stable for extended periods, but subject to major floods causing lateral channel instability
Pebble—Large	32–64	1.25–1.75	Cobble-gravel highly structured substratum; single thread or wandering channel; relatively steep; and with low sinuosity	Relatively stable for extended periods, but subject to major floods causing lateral channel instability
Pebble—Small	16–32	1.0–1.25	Sandy-gravel to cobble- gravel sediment type	Subject to avulsion (tearing away of substratum) and frequent channel shifting; braid-form channels may be highly unstable, both laterally and vertically; deep scour possible at sharp bends
Gravel—Coarse	8–16	0.75	Single thread to braided channel. Limited, local bed structure and often complex bar development	Subject to avulsion (tearing away of substratum) and frequent channel shifting; braid-form channels may be highly unstable, both laterally and vertically; deep scour possible at sharp bends
Gravel—Medium	4–8	0.5		
Gravel—Fine	2–4		Sand to fine gravel sediment type	
Sand—Very coarse	1–2	0.25	Mainly single-thread, irregularly sinuous to meandering with lateral/point bar development by lateral and vertical accretion; levees present. Moderate gradient; sinuosity < 2; w/d< 40	Single-thread channels have irregular lateral instability or progressive meanders; braided channels are laterally unstable and degrading channels exhibit both scour and channel widening
Sand—Coarse	0.500–1		Single thread to braided; limited, local bed structure; complex bar development by lateral accretion	Subject to avulsion and frequent channel shifting; braid-form channels may be highly unstable, both laterally and vertically; single-thread channels subject to chute cutoffs at bends; deep scour possible at sharp bends
Sand—Medium	0.250–0.500		Sandy channel bed, fine-sand to silt banks sediment type	
Sand—Fine	0.125–0.250	0.1	Single thread or meandering with point bar development and significant levees. Low gradient channels sinuosity > 1.5 and w/d < 20; serpentine meanders with cutoffs	Single-thread, highly sinuous channel; loop progression and extension with cutoffs. Islands are protected by vegetation. There is vertical accretion in the floodplain and vertical degradation in the channel

continued

Table 2.3 *Continued*

Size category	Particle diameter (range in mm)	Approximate current velocity to move particle (ms⁻¹)	General channel morphology	Likely channel stability
Sand—Very fine	0.063–0.125			
Silt	0.0039–0.063	0.001–0.07	Silt to sandy channel bed; silty to clay-silt banks	Single-thread or anastomosed channels; common in deltas and inland basins. Vertical accretion of sediment in the floodplain; slow or no lateral movement of individual channels
Clay	< 0.0039		Single-thread or Anastomosed channels with prominent levees. Very low channel gradient and high sinuosity	

of each particle size fraction, each category of which is twice the preceding one. The substratum in a patch of stream bed or reach is then described according to the predominant size categories of the mineral particles (Figure 2.17).

Organic particles lie on, or are mixed in with, the mineral substratum and can also be classified on the basis of size (Table 2.4). The organic matter is derived from a number of sources (see section 2.5.5) but much of it is allochthonous (imported from the surrounding catchment or from upstream). Its importance lies in the fact that the organic matter plays such a major role in stream and river energetics (Chapter 8), and both coarse and fine particulate organic matter (both sedimented or suspended) provide a food source for a range of invertebrates (see Chapter 4, section

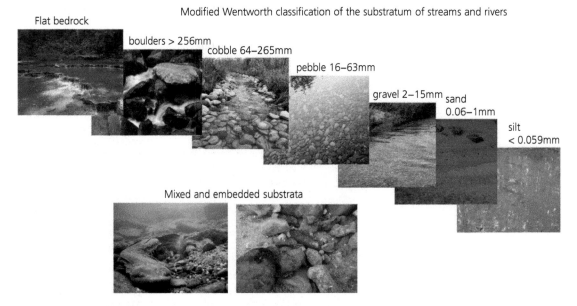

Modified Wentworth classification of the substratum of streams and rivers

Flat bedrock

boulders > 256mm

cobble 64–265mm

pebble 16–63mm

gravel 2–15mm

sand 0.06–1mm

silt < 0.059mm

Mixed and embedded substrata

Figure 2.17 The nature of the river bed can be described by the size (diameter) of the particles under the *Wentworth classification* of substratum types (see Table 2.3 for details). Usually, in headwater streams and smaller rivers, the substratum is a mixture of size classes with an embedded mixture of cobbles, pebbles, gravel and sand.

Table 2.4 Nature and size categories of non-living particulate organic matter (modified from Cummins 1974).

Organic matter and detritus: categories and subcategories	Approximate size ranges
Coarse particulate organic matter (CPOM)	> 1 mm
Large woody debris	> 64 mm
Terrestrial leaves forming leaf packs	> 16 to < 64 mm
Leaf, twig & bark fragments, needles, fruits, buds and flowers	> 4 to < 16 mm
Plant and animal detritus, faeces	> 1 to < 4 mm
Fine particulate organic matter (FPOM)	> 0.5 μm to < 1 mm
Ultrafine particulate organic matter (including microbes)	> 0.45 μm to < 75 μm
Dissolved organic matter (DOM)	< 0.45 μm

4.5.1). Dissolved organic matter contributes to microbial metabolism within the substratum and biofilm. The distribution of coarse organic particles is very patchy across the stream bed, driven to a large extent by variation in near-bed flow and the ability of the stream to retain it. This latter is also highly variable and depends on the nature of the substratum, presence of large woody debris and frequency of flow disturbances. The amount of organic matter is also temporally variable, particularly in seasonal climates associated with the riparian vegetation. We deal with the suspended fine particulate organic matter in more detail below (section 2.5.5) and with organic matter dynamics in Chapter 8 (section 8.5.1).

Whilst the size categories are subjective, general patterns in the nature of the substratum are evident. The greater the shear stress (and normally water velocity) the larger the particle size that can be moved, though note that the current velocity required to move a particle of any given size is greater than that required to keep the particle entrained in the flow. Thus current and substratum type are very strongly related. For example, pools are depositional zones, dominated by fine mineral substrata, and often deposited organic matter, while riffles are turbulent and shallow erosional zones with high shear stress and dominated by coarse substrata. There are also longitudinal gradients of grain (particle) size in rivers, the mean generally declining downstream. In headwaters, large particles predominate as a result of the proximity of bedrock and the erosional ability of turbulent flow, and high and patchy water velocity. In larger streams on shallow gradients, reduced turbulence and erosional ability (related to reduced shear stress) encourages sedimentation of smaller particles, thus leading to finer, more uniform substrata.

Larger substratum particles can 'protect' smaller ones from being washed away. In coarser substrata, finer sands and gravels can collect between or behind the larger elements (known as *embedded substrata*) and increase overall habitat heterogeneity (Figure 2.17). *Embeddedness* is an index of the degree to which these larger elements (boulders and cobbles) are surrounded or covered by finer sediments. A simple index relates to the percentage of surface area of the larger size particles covered by fine sediments (see e.g. Platts et al. 1983). The greater the embeddedness, the more homogeneous the substratum becomes, and embeddedness obviously increases in more depositional areas. The same properties also relate to collection of organic matter and detritus on the substratum—so-called *retentiveness* of the stream or river. Streams with many obstacles that can act as 'keys' to the accumulation of detritus particles, or with many depositional areas, will retain organic matter for longer than streams with a lower retentiveness. As we will see later (Chapters 5, 6 and 8), this has consequences for the diversity and abundance of animal life in the system and for ecosystem processes.

The *stability* of the substratum refers to its resistance to movement and is generally proportional to particle size. Consequently, there is also a relationship between channel morphology, substratum type and stability (Table 2.3). Redistribution of bed sediments and the movement of particles occurs at high discharge following rainstorms. The movement and small-scale redistribution of substratum particles can also lead to what are called *microform bed clusters*. These are 'organised' groups of stones on the stream bed that normally consist of a small or medium-sized boulder which acts as an anchor, against which other stones become stacked in a characteristic manner (Matthaei & Huber 2002). They are particularly resistant to entrainment during high-flow events and could be an important refugium for the benthos (see Chapters 4 and 5).

2.5 The chemical habitat and its dynamics

Streams, rivers and their estuaries are probably among the most chemically variable of environments on earth. Marine biologists work in a medium that varies relatively little in chemical composition and salinity (concentration of dissolved solids). However, one simply cannot generalise the actual chemical composition of stream and river water as it depends on the interplay of several variables that themselves vary among river catchments and even tributary sub-catchments. These include:

(i) the initial chemical composition, amount and distribution of rain and snowfall related to the proximity to the coast or industry and to climate;

(ii) the nature of the surrounding catchment and movement of water from the catchment to the river which modifies the rainfall chemistry in ways related to topography, geology, soils and riparian vegetation, and to the contribution of groundwater;

(iii) the distance from headwaters, and the timing of rainfall (e.g. seasonality and shorter-term fluctuataions);

(iv) the extent of exchange between the groundwater, the hyporheic zone and surface waters;

(v) the influence of human activity and land use and management in the catchment, such as agriculture, forestry and urbanisation.

These variables affect the *source* of the chemical constiuents of the river, the *mobilisation* of these constituents from their sources and their *delivery* to the receiving waters (Figure 2.18). A 'typical' river is essentially a dilute calcium bicarbonate solution dominated by a few cations and anions (Wetzel 1983); rivers have a global mean salinity of between 0.1 and 0.12 g L^{-1} (Berner & Berner 1987) and a range from 0.01 to 0.5 g L^{-1} (Ward 1992a). Other important variables are the acidity of the water (measured as the concentration of H^+ ions and its derivative pH), 'hardness' (which effectively measures the concentration of Ca^{2+} and Mg^{2+} ions), conductivity (which measures the total ionic content), alkalinity and 'acid neutralising capacity' (which mainly measures the concentration of carbonates and bicarbonates) and nutrients (notably forms of inorganic nitrogen and phosphorus).

Aside from any direct inputs of ions into the stream or river water from the channel substratum (largely related to its geological origins), the chemical composition is related to the hydrological cycle (discussed in Chapter 1), and particularly to the movement of rainfall through the surrounding catchment and via groundwaters to the channel. Additional inputs are related to land use and atmospheric processes and involve soils, sea spray (especially Na^+ and Cl^- ions), air pollution and volcanoes, all mediated through the action of rainfall and the hydrological cycle or through 'dry deposition'—atmospheric gases and particles. The amount of dissolved and particulate organic matter in rivers also varies, and can be high, ranging from 0.5 to 10 mg L^{-1} (Hynes 1970b). There is a vast body of data on stream and river water chemistry, particularly from northern temperate regions in Europe and N. America (e.g. Benke & Cushing 2005; Lyons et al. 2021; Tockner et al. 2022 and many major environmental agencies). Examples from elsewhere can be found from a wide range of locations including the Tibetan

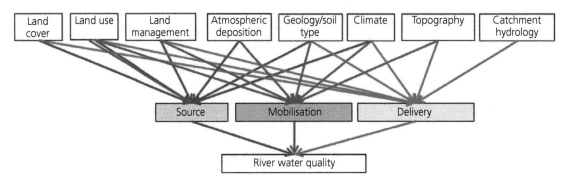

Figure 2.18 A schematic showing the influence of catchment landscape climatic characteristics on the source (the application or presence of constituent chemical sources within the catchment), mobilisation (of the constituent chemicals from their sources) and delivery of constituents to receiving streams and rivers (from source and point of mobilisation).

Source: from Linten et al. 2017, under Creative Commons Attribution Non-Commercial International Public Licence 4.0.

plateau (Fuquing et al. 2019), the Mayur River basin in Bangladesh (Roy et al. 2018), pre-glacial streams in the upper Santa River, Peru (Eddy et al. 2017), the Motueka River system in New Zealand (Young et al. 2005) and major rivers flowing into the Aegean Sea in Greece (Skoulikidis 1993).

2.5.1 Rainwater chemistry

Rainwater contains a wide variety of dissolved substances, although it is usually more dilute than stream water (salinity 0.020–0.040 g L^{-1} on average). The overall chemical composition depends on the atmospheric chemical composition. Natural rainwater is normally a weak solution of carbonic acid (CO_2 dissolved in atmospheric water droplets), with a pH of about 5.6. This is the threshold pH of the cloud water in equilibrium with atmospheric CO_2 (Charlson & Rodhe 1982), so values much below 5.5 can be considered 'acidic'. In the presence of atmospheric pollution from 'strong acid species', rain can be much more acidic. Oxides of sulphur and nitrogen dissolve to produce sulphuric and nitric acids and can be transported hundreds of kilometres before deposition (sulphur has a mean residence time in the atmosphere of two to four days). Changes in the direction of movement of air masses can lead to temporal fluctuations in rainwater acidity, as seen for example in precipitation chemistry over urban, rural and high-altitude Himalayan areas in eastern India (Roy et al. 2016). Thus on the east coast of India, the pH ranged from 4 to 7.5, with an average value of 5.5 over a three-year period (Das et al. 2010). Rainfall with a pH as low as 2.1 to 2.8 has formerly been recorded in the United States and Scandinavia (see Hildrew 2018). Following significant atmospheric pollution and acid rain up to the 1990s, acid precipitation in NW Europe and NE North America has greatly declined over the last 20 years or so, due to emissions control and reductions in the burning of sulphur-rich coal. Nitrogen deposition (mainly from road traffic) has also declined but much less markedly (e.g. RoTAP 2012).

Global spatial variability in the ionic composition of rainfall can be seen in the summary data shown in Figure 2.19 averaged over a three-year period (2005–07, Vet et al. 2014—this paper provides a detailed exposition of rainfall chemistry on a global scale). Major ions are shown in different colours, and it is clear where anthropogenic (e.g. with high sulphate concentration) or marine aerosols (sea salt—with abundant chloride (Cl$^-$) and base cations like sodium (Na$^+$)) influences are important. Sea salts (especially Na$^+$, Cl$^-$ and Mg^{2+} ions) reach the atmosphere in spray and

are transported over the land on the wind, values decreasing with distance from the sea. They vary geographically with the prevailing wind direction (e.g. Lyons et al. 2021). It is believed that the uncharacterised 'other' anions shown in white in Figure 2.19 are associated with weak organic acids (primarily formic and acetic) coming largely from biomass burning and photochemical oxidation of volatile organic compounds produced by plants. These are typically not measured and are unstable in unpreserved samples.

Thus, rainwater reaching the land already has the complex chemical composition of a dilute, weakly acidic seawater solution modified by dust (Moss 1988). Where rain reaches streams having had little contact with the soil, as in bogs or some tropical rainforest habitats, the streams will have low mineral content and often be slightly more acidic than the rain itself. More usually, however, the rain has been in contact with vegetation, the soil and often the deeper geology before reaching the stream or river, and has undergone considerable chemical alteration.

2.5.2 Geology, soils and stream pH

The chemical nature of precipitation is evidently modified before it enters streams and rivers. Rain is often intercepted by the vegetation (particularly trees) before dripping from the foliage, or running down trunks, as *throughfall* to the soil. Trees can trap atmospheric dust (dry deposited) or mist droplets, which in polluted areas are often a source of further acidifying pollutants. Soils differ with respect to their neutralising or buffering capacity of acidic precipitation and throughfall. This variability relates to the nature of the parent bedrock and to any glacial movements that occurred in the past and to land use. Table 2.5 illustrates the differences between rainwater and stream-water chemistry on three very different geologies and land uses in comparison to European and global average values.

The buffering capacity of various rock types varies substantially and relates to their content of base cations (principally calcium) and weatherable silicate minerals (Table 2.6). Residence time of water in soil also influences the rate and amount of buffering (Hornung et al. 1990). Hydrogen ions (H$^+$), produced by the dissociation of carbonic acid in rainwater, are the cause of acidity in clean areas and are neutralised by a solution of carbonate minerals and hydrolysis of silicate minerals as water percolates through rocks and soils. Buffered catchment water thus finds its way to streams and rivers, carrying carbonates and cations,

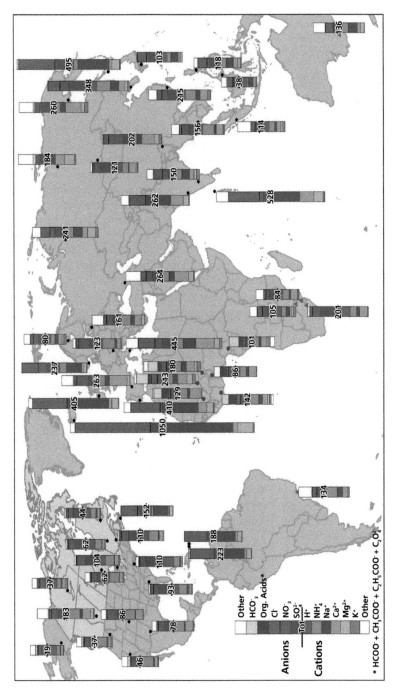

Figure 2.19 The contribution of a range of individual ions in precipitation to total ion composition globally for the period 2005–07 (the number at the centre of each bar in µeq L⁻¹ and the bar length is scaled) at selected regionally representative sites (i.e. each site is considered to be representative of most sites in the surrounding area). Sites labelled with black circles show the three-year average ion composition with no measured organic acid and bicarbonate data; pink circles show the values that include measured organic acid and bicarbonate data; red triangles show data from outside the 2005–07 period.

Source: from Vet et al. 2014; data from the GAW World Data Centre for Precipitation Chemistry (http://wdcpc.org/). Crown Copyright © 2013 Published by Elsevier Ltd: http://dx.doi.org/10.1016/j.atmosenv.2013.10.060. Open access under CC BY-NC-ND licence.

Table 2.5 The effect of geology and land use on streamwater chemistry compared with rainfall in the same general location from three different geographical areas. Note that the differences in streamwater quality are far greater than those for rainfall. European (unpolluted) and global mean values for rivers are also shown for comparison (modified from Burgis & Morris 1987).

Chemical parameters (mg L^{-1})	Igneous (insoluble) rocks, undisturbed forest: New Hampshire, USA		Chalk and glacial drift, lowland agriculture: Norfolk, UK		Volcanic/metamorphic, thornbush and rangeland: Rift Valley, Kenya		Global river mean**	European mean for unpolluted rivers**
	Rainfall	Stream	Rainfall	Stream	Rainfall	Malewa river		
Na$^+$	0.12	0.87	1.2	32.5	0.54	9.0	9.83	6.05
K$^+$	0.07	0.23	0.74	3.1	0.31	4.3	0.84	0.72
Mg^{2+}	0.04	0.38	0.21	6.9	0.23	3.0	3.48	8.97
Ca^{2+}	0.16	1.65	3.7	100.0	0.19	8.0	8.3	15.07
Cl$^-$	0.47	0.55	<1.0	47.0	0.41	4.3	4.6	3.72
HCO$_3^-$	0.006	0.92	0	288.0	1.2	70.0	13.98	21.52
SO$_4^{2-}$					0.72	6.2	0.905	1.63
pH	4.14	4.92	3.5	7.7		7.4–8*		

*Based on estimates from various published sources. ** Data converted from μM from Lyons et al. (2021).

Table 2.6 Buffering capacity of different rock types, the characteristic nature of low-order streams draining such geologies and the predicted impact of acidic precipitation on them (modified from Hornung et al. 1990).

Class	Buffering capacity	Major rock types	Characteristics of 1st- and 2nd-order stream	Impact of acidic precipitation on surface waters
1	Little or no buffering capacity	Granite and acid igneous rocks or metamorphic equivalents; granite gneisses, quartz sandstones and metamorphic equivalents; decalcified sandstones	Naturally acidic, low conductivity, poorly buffered	Widespread impact expected
2	Low to medium buffering capacity	Sandstones, shales, conglomerates and metamorphic equivalents; coal measures; intermediate igneous rocks	Weakly acidic, low conductivity, poorly buffered	Impact restricted to 1st- and 2nd-order streams and small lakes
3	Medium to high buffering capacity	Slightly calcareous rocks, (e.g. marlstones); basic and ultrabasic igneous rocks; Mesozoic mudstones; low-grade intermediate to mafic volcanic rocks	Circum-neutral. Well buffered	Impact improbable except for near-surface drainage in areas of acid soils
4	Infinite buffering capactiy	Limestones, chalk, highly fossiliferous sediments or metamorphic equivalents	Alkaline, high conductivity, highly buffered	No impact

and supplements those with any minerals released directly to the water from bed materials.

Catchments on hard, igneous rocks tend to be low in dissolved salts and have low buffering capacity and hence surface waters are soft (low alkalinity) in cleaner areas and acidic (pH around 3.5 to 5.5) in areas polluted by strong acids. Catchments on sedimentary, especially calcareous, rocks are rich in carbonates and streams are usually well-buffered, hard-water systems with higher pH, often between 7.5 and 8.5. Similarly, soils with free carbonates, a high content of weatherable silicates or high base saturation generally give rise to circumneutral (around pH 7.0), well-buffered waters. In catchments where soils are strongly leached and have lost their buffering capacity, as in most parts of Malaysia or the Amazon basin, acidic 'blackwaters' result (see Section 2.5.4), especially in forested areas (Dudgeon 1995). The freshwater systems in catchments with poorly buffered rocks and soils exhibit quite dramatic temporal change in pH over short periods associated with rainfall or snow-melt events (see Section 2.5.5).

In addition to the above, other factors can lead to low stream pH. Naturally acidic streams arise from the presence of sulphate-rich soils and/or ironstone ferrous carbonate in the catchment. This results in the oxidation of ferrous to ferric iron in aerobic surface waters that in turn generates H^+ ions (Townsend et al. 1983). Water issuing from abandoned and flooded coal mines is also often profoundly acidic and laden with metals (e.g. Hogsden & Harding 2012). Natural organic acids can arise from areas of *Sphagnum* bog in wet valley bottoms, swamps and peaty areas and contribute to acidic runoff from the catchment. Acidity also influences the solubility and speciation of metals in the soil and stream waters. Of particular interest is aluminium which becomes soluble below pH 4.5 and can be toxic to fish when in a particular form known as *labile monomeric aluminium* and above a concentration of about 0.2 mg L^{-1}. This has been one of the most problematic effects of acid rain in catchments of low buffering capacity, especially when planted with coniferous forests.

2.5.3 Vegetation effects

Riparian vegetation can influence stream-water chemistry through a number of processes, including directly through chemical uptake and indirectly through the supply of organic matter to soils and channels, modification of water movement, and stabilisation of riparian soil (Dosskey et al. 2010). Some processes are more strongly expressed under certain site conditions, such as denitrification where groundwater is near the surface, and by certain kinds of vegetation. Vegetation in general, and especially coniferous trees, scavenge ions (including sea salts and atmospheric pollutants) from rainfall and dry deposition from the air which are further increased in concentration following evaporation from the canopy surfaces and canopy exchanges. As discussed above, precipitation passing through the vegetation to the ground (*throughfall*) picks up these

additional ions which leads to a different chemical composition and higher concentrations of some ions than in the original rain. Once throughfall reaches the soil, root uptake for transpiration further increases concentrations of ions, and in-soil processes involving cation exchange, mineralisation of mineral and organic matter and uptake by organisms add to the changing chemical composition of what is now soil water that can potentially flow via subsurface pathways into the adjacent stream. This process is particularly important in upland regions near the headwaters of most streams (Gee & Stoner, 1989).

The other effect of vegetation will be on selective uptake of ions and nutrient fluxes in the soil. This is clearly illustrated by increases in nitrate and potassium concentration in stream waters following removal of riparian vegetation in stream stretches in the Plynlimon catchment, mid Wales (Hornung & Reynolds 1995). Similar increases in nitrate, potassium and phosphorus were found following clearcutting during the influential catchment-level Hubbard Brook studies (Likens & Bormann 1995), whereas removal of vegetation had much less effect on concentrations of calcium, magnesium, sodium and, especially, sulphur in this system. Removal of the canopy also reduces the 'scavenging' ability of the catchment vegetation and hence atmospheric inputs to the catchment, such that concentrations of ions like sodium, chloride and sulphates decrease in stream waters following clearfelling of trees. Subsequent revegetation of the catchment limits the loss of nutrients to the lotic system. In mature rainforests, moreover, streams are very dilute and most ions are released by weathering (Moss 1988).

2.5.4 Land use, nutrients, suspended solids and pollution

While sea spray and the weathering of rock still dominate the ionic composition of most of the world's fresh waters, human activities, through agricultural, industrial and urban pollution, probably have the greatest effect on nutrient concentrations. Nutrients can be important limiting factors for plant growth and productivity in aquatic habitats, just as they are in terrestrial ones (see Chapter 9). Rainfall has small to substantial amounts of nitrogenous compounds dissolved from the atmosphere (mainly nitric acid and ammonia). This can be an important source (of a limiting element) in desert streams (Grimm 1994), whereas nitrogen deposition is a source of pollution in more industrial areas. In wetter climates, as rainfall percolates through the vegetation and soil, the concentration increases and nitrogen, particularly nitrate, can find its way to stream waters. As we have seen, however, the nature and extent of vegetation and land use in the catchment regulates the flux to the lotic habitat. The concentration of nitrates also varies with the extent of arable land in the catchment (see Figure 9.5, Chapter 9), and large fluxes in total nitrogen (from 876 to 5,000 kg m^{-2} y^{-1}) can result from agricultural runoff (Billen et al. 1995) which can lead to over-enrichment (eutrophication) of fresh waters. Dry deposition of ammonia in areas of intensive animal production supplements nitrogen in the catchment.

Phosphorus (P) is very much less soluble than fixed nitrogen. It weathers from rocks and soils in small amounts but, due to its natural scarcity in the biosphere and the ability of plants to absorb and retain P, in more pristine areas it is generally found at extremely low concentrations in stream waters (largely as phosphate). Where there are inputs of treated sewage or agriculture in the catchment, phosphorus is much more available and is a noted pollutant. We return to this in Chapters 9 and 10 and the important interactions between phosphorus and the iron cycle are described in section 9.3.5.

Industrial effluents introduce new substances to fresh waters as well as increasing the concentration of natural ones. They enter river systems as 'point-source' effluents (such as factory effluents), 'diffuse pollutants' (via runoff from the landscape) or from the atmosphere. They comprise an enormous range of organic and inorganic substances, some of them now entirely synthetic (manufactured rather than naturally occurring) and physical pollutants (such as plastic particles of various sizes). Obviously, the concentrations and toxicity vary, but partly depend on the ratio of receiving water to effluent volume. Less obvious and insidious pollutants have been found to have unexpected effects—these include oestrogens, arising from sewage treatment outfalls, that are oestrogenic to fish and affect their reproduction (Harries et al. 2009). The latest concerns have arisen over microplastics, and whilst their longer-term biological impact is unknown at present, their ubiquity in freshwater systems is alarming. For example, microplastics were found in all sample locations along 820 km of the River Rhine (with almost 900,000 particles per km^{-2}; Mani et al. 2016), and were ubiquitous at all river mouths draining into Manilla Bay in the Philipines (with concentrations from 1,580 to 57,665 particles/m^3; Osorio et al. 2021).

Atmospheric pollution can also influence freshwater habitats far from their source, and this is clearly the case in relation to sulphur and nitrogen deposition causing acidification and eutrophication of rivers and lakes. We will revisit eutrophication and its effects in Chapter 9 (section 9.4.2 and Topic Box 9.2) and acidification and various other aspects of pollution in more detail in Chapter 10. Overall, pollution greatly modifies the chemical habitat in streams and rivers, almost always to its detriment.

Suspended solids consist of an inorganic fraction that includes small sediment particles (e.g. silts and clays) and an organic fraction (largely detrital particles, along with microbial cells etc). Here we concentrate on the former. Inorganic suspended solids originate from terrestrial sources, such as through soil disturbance followed by heavy rainfall, bank erosion etc. They are generally transported by overland flow pathways, although finer particles can be transported in subsurface flows (Figure 2.20). The significance of suspended solids, aside from their direct role in ecosystem energetics (discussed in Chapter 8), relates largely to the effects on subsurface light supply and indirectly to the nature of the substratum following sedimentation (see Section 2.4 above). Rivers need a sediment supply otherwise they erode downwards, and a balance between sediment and water supply largely creates the river habitat. The concentration of suspended solids can vary considerably; as an example, values from 5 to 540 mg L^{-1} have been recorded in rivers in northern South America (Lewis et al. 1995) and globally concentrations range from 10 to 1,700 mg L^{-1} (Lintern et al. 2017). The nature of suspended (and dissolved) materials conveys optical properties that can be used to classify rivers (Sioli 1975):

- *Blackwater* rivers are poor in dissolved inorganic and suspended solids, but dissolved organic matter produces a reddish-brown colour. These are typically of low pH with high tannin concentrations resulting from decay of wetland vegetation. Examples can be found predominatly in the Amazon basin and the southern United States but also occur in Africa (e.g. Congo) and several large rivers in Australia.
- *Whitewater* rivers have high concentrations of suspended solids with a muddy/silty appearance, as well as dissolved inorganic solids, tending to be alkaline with higher nutrient content than Blackwaters. Whitewaters are found in the Amazon basin and elsewhere in South America (including tributaries of the Orinoco and Parana which also have their sources in the Andes). Elsewhere whitewaters occur in parts of large African rivers such as Nile, Niger and Zambezi and in the Mekong in Asia and sections of the Danube in Europe.

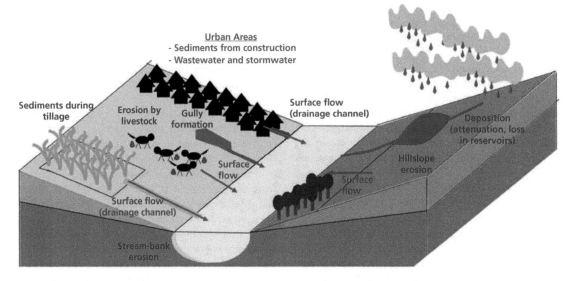

Figure 2.20 A schematic illustrating the sources (red text), mobilisation (purple text), and delivery (blue text, with arrows representing delivery processes) of sediments into rivers.
Source: from Linten et al. 2017, under Creative Commons Attribution Non-Commercial International Public Licence 4.0.

- *Clearwater* rivers vary in acidity and have little suspended material. Nutrients are usually scarce (similar to Blackwaters) or in moderate concentration. Again, the Amazon basin has many examples while, outside South America, examples can be found in Africa (e.g. upper Zambezi) and upland streams in major river basins in Southern Asia and northern Australia.

In addition to highly erodible soils, poor soil conservation (particularly associated with agriculture and forestry) can lead to extremely high sediment loads. Removal of vegetation in the catchment, either by clearcutting forests (especially on steep slopes) or through overgrazing of the uplands (especially by sheep and goats), can lead to large-scale soil erosion and influxes of sediments to waters draining the catchment. There is thus an inverse relationship between the amount of vegetation and erosion, which generally means drainage waters in arid areas have higher concentrations of suspended solids than in wetter areas (all else being equal!). Other anthropogenic activities influence the load of suspended solids, such as construction of dams, which reduce sediment transport downstream and thus influence the fertility of the downstream floodplain and perhaps even coastal marine ecosystems, as found with the Aswan Dam on the River Nile.

2.5.5 Variation in water chemistry over time

Water chemistry varies at different timescales. Changes in water chemistry have been documented under normal flow conditions where there are large quantities of macrophytes such as in chalk streams. Here, photosynthesis can cause diurnal fluctuations in pH from 7.4 (night) to 9.0 (day) in addition to significant changes in dissolved oxygen (supersaturated in the day to low concentration at night; see section 2.6.2). Normally, however, short-term reversible changes in chemistry follow the rise and fall of discharge associated with rainfall.

Base flows in streams usually carry higher concentrations of most ions than bank-full flows. Depending on the size of the stream, discharge rises to a peak some time after maximum rainfall (see example in Figure 2.6c), and much of this water has had only minimal contact time with soil and rock of the catchment. Dilution also plays a role. However, during such increases in discharge, the concentration of H^+ ions increases and thus pH falls (Figure 2.21a). This is known as an *acid pulse*. In conifer-afforested catchments on poorly buffered geologies, this decline in pH is even greater and may be accompanied by a rise in aluminium (see Section 2.5.2 above). At the same time, conductivity usually decreases (Figure 2.21b). These episodic events can have a marked effect on the biota of streams and rivers.

Typically, suspended solids increase during spates (Figure 2.21c), especially where riparian vegetation has been removed. These changes are even more marked with seasonal monsoons, as seen in the Mayur River in Bangladesh (Roy et al. 2018). If heavy rain follows a period of drought, accumulated solutes, which have increased in concentration through evaporation and oxidation in the dry soil, are flushed out. Runoff immediately after the drought will then contain large amounts of nitrates and other solutes (so called *first-flush* events; see Chapter 9, section 9.3.3 and Figure 9.12 for more details). The length of dry periods preceding runoff events impacts the amount of chemical constituents accumulated in the catchment and that can be flushed into the receiving waters. Similar accumulation of materials occurs in snow, and, again, thawing allows increased mineralisation and a flush of nitrification, picked up in runoff.

The effects of increased discharge are short lived, but those due to low-flow conditions last longer. Reduced discharge would normally lead to slightly raised conductivity. Thus, seasonal fluctuations in concentration of many ions are expected, related to seasonal changes in source, mobilisation and delivery (Lintern et al. 2017), although such changes can be dampened by the presence of natural or man-made lakes within the river system (as seen in the River Shannon in Ireland (Lyons et al. 2021)). Changes in wind direction and precipitation over the year will influence inputs of marine salts or pollutants, while seasonal changes in biological activity (particularly vegetation) both on land and in the water may also affect water chemistry. In intermittent streams and rivers longitundinal connectivity may be lost and pools contract during drying periods, resulting in increased concentrations of nutrients over time (Stubbington et al. 2017). Temperature and soil

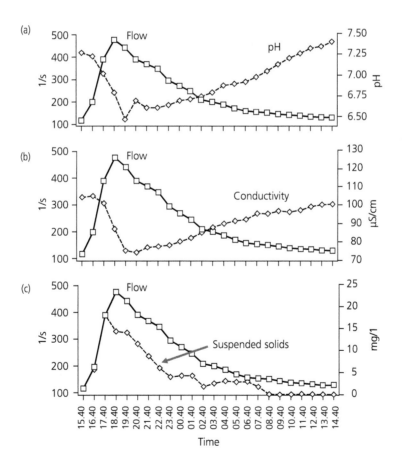

Figure 2.21 Short-term (24 h) variation in flow following a rainfall event and consequent changes in: (a) pH, (b) conductivity and (c) suspended solids in the conifer-afforested catchment of the River Douglas, southern Ireland.
Source: modified from Cleneghan et al. 1998, with permission from Springer Nature.

moisture (again seasonally variable) influence the connectivity of various hydrological pathways within the catchment and the extent and rate of vegetation growth and thereby its impact on rainfall chemistry. Sources of nutrients can also vary seasonally, related to land-use patterns such as seasonal fertiliser applications and tillage as well as seasonal shifts in vegetation growth (Lintern et al. 2017).

Long-term monitoring can detect the directional chemical changes in nutrients, salinity, suspended solids and oxygen that accompany the gradual eutrophication of rivers caused by pollution (see Section 9.4.2). Water chemistry also responds to land-use change in the catchment, such as afforestation or clear-cutting. Under such circumstances, conditions will not revert to the previous state unless pollution ceases or the original catchment vegetation is reinstated, or may do so only after a considerable time lag.

2.5.6 Spatial differences in water chemistry

Variation in water chemistry within a stream reach is not easy to detect because running waters are usually well mixed. However, high respiration and nitrogen mineralisation rates in the hyporheic zone can lead to increased nitrogen and reduced oxygen concentrations close to upwelling zones (Grimm 1994). This in turn can influence the distribution of benthic algae. Local hydrological processes, such as stream-bank erosion, groundwater influx and the addition of chemically contrasting tributaries flowing through sub-catchments on different geologies or with different land uses can lead to fairly abrupt longitudinal change in hydrochemistry between stream sections (see later in this section). The effect of tributary inputs on water chemistry in the main river depends on the relative discharges of the two, as well as the chemical difference between them. Of course, large 'point-source' pollutants and the inputs of land drains are also common causes of hydrochemical (and ecological) changes in rivers and streams.

There are also usually larger-scale, more gradual longitudinal hanges in hydrochemistry attributable to shifts in geology, soils, climate, vegetation and in human influence from up- to downstream. Most dissolved salts (thus conductivity, alkalinity and hardness), nutrient concentrations and pH therefore tend to increase in the downstream direction. In intermittent rivers and ephemeral streams, there is an interplay of physicochemical variables across a range of scales (Figure 2.22). Whilst most of these variables are similarly important in perennial rivers and streams,

in intermittent systems the critical wetting–drying cycle affects water quality in a way characteristic of each system (making generalisation problematic). We return to these longitudinal patterns and the effects on the biota in Chapters 5 and 6.

An example of within-catchment changes is seen in the 22.4 km-long Araglin river in Southern Ireland (Giller 2020). The headwaters drain peat bog (largely made up of *Sphagnum* mosses) and are joined by a number of tributaries. A significant increase in nutrients is associated with agriculture and fish farms. Upland streams are circumneutral and oligotrophic but with low acid-buffering capacity. At the sub-catchment scale, significant chemical changes in pH and nutrients can occur over fairly short distances. For instance, in one small tributary of the Araglin pH rises by about 1.7 units (from 5.1 to 6.8; about a 150 times decrease in hydrogen ion concentration) in just over 1 km (Clenaghan et al. 1998). In larger rivers water chemistry can also vary significantly; for example, in the 730 km^2 Jinshui river catchment in China, median total nitrogen (TN) and total dissolved phosphorus (TDP) concentrations ranged from 0.01 to 4.38 mg L^{-1} and 0.022 to 4.16 mg L^{-1}, respectively, across 12 sites (Bu et al. 2010). In the 26,219 km^2 catchment of the Han river of Korea, concentrations of total suspended solids (TSS), total phosphorus (TP) and TN varied from below detection to 36 mg L^{-1}, 1.6 mg L^{-1} and 35 mg L^{-1}, respectively (Chang 2008).

At a regional (between-river) scale, geology and soils are the major factors influencing differences in water chemistry, although climate (especially rainfall) and surrounding vegetation are also important. On a biogeographic scale, geology and climate are again important. Berner & Berner (1987) give information on average concentrations of a variety of parameters from different continents. Thus, in South America, with extensive rainforest and high annual precipitation, the mean concentration of total dissolved solids (TDS) is probably among the lowest in the world. The average for Africa is also low, with its predominately hard geology and weathered, ancient soils. European rivers seem to hold the highest concentrations. More of Europe has younger soils of more recent glacial orgin. Agricultural and urban runoff is also widespread. Bear in mind that these large-scale data are dominated by information from a few large rivers while regional variation within continents is likely to be greater than between-continent averages. The Santa river basin in Peru, for instance, drains two major mountain systems, one glaciated (the *Cordillera Blanca* with over 600 glaciers—the largest concentration of tropical glaciers

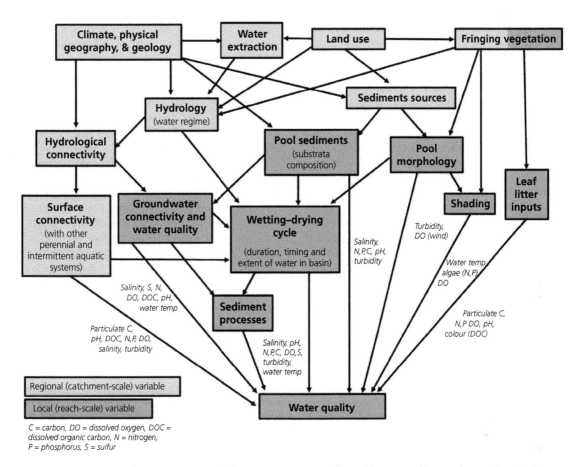

Figure 2.22 Water quality of intermittent rivers and ephemeral streams (IRES) is influenced by a variety of larger-scale regional and local reach-scale variables. These physicochemical variables are often interacting, many of them strongly influenced by the wetting–drying cycle, a key element shaping the water quality in IRES.
Source: from Gomez et al. 2017, with permission from Elsevier.

anywhere) and one not (the *Cordillera Negra*) (Eddy et al. 2017). Glacial meltwaters strongly reflect the geology over which they flow, such that high sulphate concentration leads to highly acidified waters below the glacial point sources.

2.6 Oxygen and temperature

Both temperature and the concentration of oxygen are fundamental, and closely related, aspects of the habitat templet for organisms in running water. Many of the organisms require a good supply of dissolved oxygen, and that in turn relates to temperature, the current and altitude. Variation in these factors creates some of the most important patterns and gradients in lotic ecology.

2.6.1 Temperature

Temperature is often regarded as a 'master factor' in ecology, and streams and rivers are no exception, with research on the factors affecting water temperature and on the effects of temperature on lotic organisms going back to the 1950s and earlier (e.g. Macan 1963; Caissie 2006). Temperature affects almost all aspects of life at a range of levels of organisation, from individuals (e.g. growth rates) to populations (growth rates and distribution) and from basic biochemical, metabolic and physiological process rates to large-scale ecosystem processes (such as productivity, decomposition and ecosystem metabolism) (Caissie 2006; Hildrew et al. 2017; Tiegs et al. 2019). Temperature also influences other aspects of the physical environment of streams and rivers, such as the density of water (greatest at

3.94°C), which in turn influences water flow and viscosity. Most solids dissolve in water more readily as temperature increases, whereas the solubility of gases in water (including oxygen) tends to decrease with increasing temperature. Water also has a high heat capacity, requiring a relatively large amount of energy to increase the temperature of liquid water; thus aquatic organisms are generally buffered from the rapid diel temperature changes that apply on land. The slow heating and cooling of water is evident in the five to seven day lag that is often found between air and stream water temperature in many regions (Dodds & Whiles 2019).

Ward (1985) suggested rivers could be grouped into equatorial, tropical and temperate, based on temperature range and their thermal maxima, but overall it has proved exceptionaly difficult to catagorise the thermal regimes of rivers (Caissie 2006), largely due to the variability among rivers, the complexity of the thermal processes at work and the range of contributing factors. These factors fall into four broad groups (Figure 2.23); (i) atmospheric conditions—responsible for the heat-exchange processes taking place at the water surface; (ii) topography or geographical setting—these influence atmospheric conditions but also aspect, altitude etc; (iii) stream discharge—mainly affects the heating

capacity (volume of water) and/or cooling through mixing of water from different sources (including the hyporheic zone); and (iv) streambed–water body interactions—conduction heat exchanges and ground-water input.

Based on the unique properities of water, meteorological and physical conditions and geographical setting, the temperature of streams and rivers fluctuates on a daily and seasonal basis. Daily temperature generally usually reaches a minimum around sunrise and a maximum in late afternoon to early evening but, due to the high heat capacity of water, the variation is much less than that in the surrounding air (Figure 2.24a). Diel variability often reaches a maximum in wide and shallow rivers (rivers generally wider than 50 m and < 1.5 m deep), and decreases downstream and in large, deep, lowland rivers (Caissie 2006). The lowland Amazon, for example, is always within a few degrees of 29°C (Lewis et al. 1995). At high altitudes, however, tropical system behave more like temperate streams and show diel and considerable longitudinal temperature variation (Covich 1988). Riparian canopy cover can also moderate daily fluctuations.

Seasonally, river temperature usually approximates a sinusoidal function, with temperature approaching freezing in winter (Figure 2.24b; Caissie 2006). An

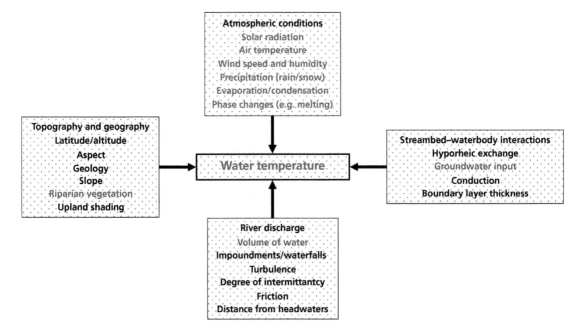

Figure 2.23 The major factors influencing the water temperature of streams and rivers. Factors strongly related to climatic conditions, and thus susceptible to the impact of climate change, are highlighted in blue.
Source: modified from Caissie 2006.

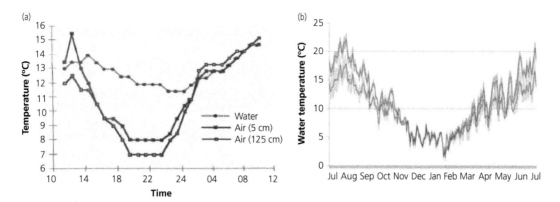

Figure 2.24 Diel and seasonal fluctuations in river temperature. (a) Diel temp variation in a wooded temperate shallow stream in March compared with air temperatures 5 cm and 125 cm above the stream surface (unpublished data, P. Giller). (b) An example of seasonal water temperature fluctuations in an upstream (20 km from source; orange) and downstream (70 km from source; turquoise) section of the River Pielach, in the Alpine foothills of Austria. Data are based on mean daily temperature (solid lines) and the min/max range (shaded).
Source: (b) from Pletterbauer et al. 2018, with permission under Creative Commons Attribution 4.0 International Licence.

annual range of over 20°C can be found in temperate and arctic streams, and up to 40°C in intermittent desert streams. In contrast, equatorial rivers and rainforest streams fluctuate much less (by only a few degrees; Lewis et al. 1995). In high-latitude systems, with an annual mean air temperature of −20°C or less, water is frozen for much of the year (Dodds et al. 2019), even if only at the surface, while shallow streams may freeze completely.

Over recent decades, stream and river temperatures seem generally to have increased as a result of climatic warming (Caissie 2006; Woodward et al. 2016; Hildrew et al. 2017; Pletterbauer et al. 2018; see also Chapter 10, section 10.6.5). The number of detailed examples is growing. From Wales increases of between 1.4°C and 1.7°C in mean winter temperatures have been described for forest and moorland headwater streams of Llyn Brianne over 25 years (Durance & Ormerod 2007), and mean daily maximum temperature increases from 0.49°C to 0.73°C have occurred in headwaters of the River Usk and 1.03°C in the main river over a 49-year period (Hildrew et al. 2017). In the upper River Danube, annual mean temperatures since 1901 have increased by 1.4°C–1.7°C, mostly in the decades since 1970 (Webb & Nobilis 2007), while an increase in annual mean temperature of 1.88°C over 40+ years (1969–2010) has been found in the Breitenbach, a spring-fed sandstone stream in central Germany (Baranov et al. 2020). Further, Soto (2016) reported a decadal rate of increase in mean water temperature in 11 rivers in northern Iberia of 0.16°C between 1986 and 2013.

Spatially, there are longitudinal patterns in water temperature, with the general trend being one of increasing temperature with increasing stream order (i.e. from source to mouth) although this is non-linear (Dodds & Whiles 2019). Mean daily water temperature is generally close to the groundwater temperature at the source and increases by about 0.6°C km^{-1} for small streams, 0.2°C km^{-1} for intermediate-sized rivers but only 0.09°C km^{-1} for larger rivers (Caissie 2006). This general, large-scale pattern is mediated more locally by small-scale variations at confluences with tributaries, in deep pools, downstream of dam releases, in intermittent rivers during periods of loss of connectivity, or in river sections with significant groundwater inflows.

2.6.2 Oxygen

Oxygen is required by all aerobic organisms for respiration, yet its concentration is about 33 times less in fresh water than in air, while its rate of diffusion is some 3×10^5 times lower (Verberk et al. 2011). Oxygen concentration in water is determined by its solubility, which decreases with increasing temperature; hence, pure water in equilibrium with air at standard atmospheric pressure has an oxygen concentration of 12.77 mg L^{-1} at 5°C but only 8.26 mg L^{-1} at 25°C (Wetzel 1983). In addition, oxygen concentration is affected by the partial pressure of oxygen (which decreases with altitude) (Jacobsen 2020). Due to this factor alone, oxygen concentration tends to decrease with altitude, although this is counteracted by a decline in temperatures, and hence increasing solubility. This pressure effect is really only important at very high altitudes (Jacobsen 2020). To complicate matters even further, oxygen diffusion rates increase with temperature such that, surprisingly, more oxygen is actually available to an organism in warmer equatorial habitats, even though oxygen concentrations would be lower than in polar waters (Verberk et al. 2011). The challenges to organsms then relate more to oxygen demand exceeding supply in warmer water than to the lower oxygen concentrations. To reflect oxygen supply to organisms more effectively, Verberk et al. (2011) developed an *oxygen supply index* (OSI), involving a combination of solubility, partial pressure and diffusivity. This relationship between oxygen and temperature therefore creates some interesting challenges for aquatic organisms. For example, increased oxygen demand at high temperature leads to a limit in oxygen exchange rates that affects maximum body size in equatorial waters, whereas the decreased oxygen demand in cold polar waters allows an increase in (maximal) body size, despite the lower oxygen availability (Verberk et al. 2011). These complex relationships will also pose interesting and differential challenges to stream animals under intensifying climate change (Jacobsen 2020 and see Chapter 10).

Oxygen availability also varies with current speed and turbulence; small, turbulent, unpolluted streams are usually saturated with oxygen. Oxygen availability declines downstream in larger rivers, which are less turbulent and warmer and have a lower surface-area-to-volume ratio for diffusion from the atmosphere. Pools and stagnant bays, especially with accumulations of dead organic matter, can be depleted in oxygen. The presence of macrophytes can also affect oxygen concentration. As oxygen is a by-product of photosynthesis, active plant growth (especially during summer) can lead to supersaturation during the day, although oxygen concentration declines significantly by night (often known as an *oxygen sag*) due to ecosystem respiration (e.g. Caraco & Cole 2002). Values of the order +/− 60% dissolved oxygen over the day have been recorded (Williams et al. 2000) but in some systems oxygen concentration can change to an even greater extent, such as from 36 to 164% saturation between night and day (Moss 1988). For example, extreme diel oxygen amplitudes (0 to 25 mg O_2 L^{-1}) have been recorded (Acuña et al. 2011) and the supersaturation can last for more than 12 hours, as in large macrophyte beds in the Hudson River (Findlay et al. 2006) (see also Chapter 8, section 8.2). Small-scale spatial variations in oxygen saturation have also been detected (of the order of

10–30% saturation), particularly in smaller, more productive rivers (see e.g. Williams et al. 2000). Upwelling groundwater sometimes has a low oxygen concentration and, similarly, any instream impoundments with releases of water from depth (*hypolimnial discharge*) will tend to reduce oxygen and increase carbon dioxide. Severe deoxygenation occurs in heavily organically polluted streams or through a combination of drought (low flow), high temperature and dense instream vegetation.

2.7 Concluding remarks and patterns in the habitat templet

Overall, streams and rivers are really dynamic ecosystems with great environmental heterogeneity. The hydrological cycle drives small-scale temporal changes and the climate generates larger-scale patterns in a range of environmental variables. As hierarchical dendritic networks rivers also vary spatially, with almost everything about them changing longitudinally as one travels from the headwaters downstream and tributaries coalesce into larger and larger channels flowing through a changing landscape. The typical lengthy river is often described as originating in mountainous areas from springs and rivulets, coalescing to turbulent and shallow streams. These in turn join with other tributaries to form a large, more smoothly flowing, deeper river that meanders through the lowlands to the sea. The role of lateral connections with the riparian zone and floodplain and vertical connections with the groundwaters is also important in contributing to the heterogeniety. Associated with the increase in stream size with distance from the source is a decrease in the direct influence of the surrounding landscape on the ecosystem. The boundary between the stream edge and the land is relatively sharp in headwaters but much less so as one progresses downstream, especially where there are seasonal changes in water levels and large floodplains. In a quite predictable longitudinal pattern downstream, the slope of the channel decreases, discharge and water temperature increase, variability and the nature of the dominant physical factor of flow and associated forces change, and, in unpolluted systems, so does water chemistry. Oxygen concentration also tends to decline, while its availability to organisms shows a more complex pattern along the river's course from its source.

This array of physicochemical features, and the high variability in time and space in environmental conditions across a wide range of scales, in turn provide the variety of microhabitats that can support the great biotic diversity found in natural streams and rivers, which we will explore in Chapter 3. Meeting the challenges posed by these physical and chemical characteristics involves adaptations and species traits that we discuss in Chapter 4. Physicochemical features also clearly have a major influence on large-scale distribution patterns of lotic species, as well as on their abundance and distribution at the finest of spatial scales (Chapter 5). The habitat templet creates the environment within which the ecological communities are established (Chapter 6) and biological interactions are played out (Chapter 7). The spatial and temporal variability and the hierarchical network of streams and rivers imposes significant consequential changes to, and sets limits on, ecosystem processes (such as decomposition, community respiration and primary and secondary production considered in Chapter 8) and patterns (such as standing stock of organic biomass, species richness and community structure, examined in Chapters 5, 6 and 8). Suffice to say at this point that these environmental changes should be visualised more as a complex patchwork of conditions than as smoothly continuous gradients of conditions so often seen on land. This is particularly so in the face of the impacts of human populations and their activities, often concentrated around rivers, which have a profound modifying effect on this natural background—by withdrawing water and flow regulation, through the discharge of pollutants and by land-use change as well as driving large-scale climate change (Chapters 9 and 10). In any event, these various and regular patterns in physical characteristics of river systems are dominated largely by the hydraulic processes associated with the nature of flowing water.

CHAPTER 3

River biota

The diversity of life in streams and rivers

3.1 Introduction

Running waters support a multitude of organisms, ranging in size from tiny viruses to large water plants and vertebrates. Nevertheless, the majority are small and fairly cryptic, so that a quick glance into a fast-flowing river or stream often reveals some larger plants and a few fish but little else. Relatively few organisms are able to persist out in the main water column unless the current is slow or they are relatively large, while most species are associated, in one way or another, with the substratum, living (or rooted) in the interstices between mineral particles of the sediment, on the substratum surface, in leaf packs, or between and underneath the stones, rocks and woody debris etc found on the river bed. On closer inspection, the most obvious fully aquatic inhabitants of a stream or small river, aside from perhaps fish, are filamentous (macro-) algae, mosses, and the larger macroinvertebrates including crustacea, larvae of many species of insects, and molluscs. Even closer examination usually reveals species belonging to many other taxonomic groups, largely invisible to the naked eye, including a vast array of microbes, microarthropods, and 'minor' phyla. The scale of this overall diversity is surprising. In the Breitenbach, an unremarkable first-order sandstone stream in central Germany, whose biodiversity has been particularly carefully assessed, some 2,000 animal species, including over 820 species of insects, have been identified over 42 years of study (Wagner et al. 2011). This may not be unusual, although streams have rarely been studied in this much taxonomic detail over such a prolonged period.

For perhaps a more immediately impressive display of biodiversity in and around rivers we need to look at an array of semi-aquatic organisms. These include the potentially enormous biomass of emergent plants on floodplains, the adapted trees of regularly flooded forests, semi-aquatic mammals from water shrews to hippos, and an enormous number of water birds (waders, ducks, geese etc), plus normally more or less terrestrial birds and mammals that find food in and around rivers at some time or another.

In the following sections we will give a brief overview of the more important groups of organisms living in streams and rivers, emphasising those more or less restricted to running waters and/or of particular importance to the ecological processes there. Our aim at the outset is to equip our readers with a broad appreciation of the living things in rivers—'life through a telescope' rather than, at this stage, 'life through a microscope'. Readers can find a huge amount of detailed and reliable information in a large range of other publications, and increasingly on the WWW. Much of this is relates to freshwater biodiversity in particular parts of the world and Table 3.1 highlights a few such sources which may be useful, some of them freely accessible. We acknowledge there is a bias towards the invertebrates. Aside perhaps from fish, the invertebrates as a whole, and the larger macroinvertebrates in particular, are without doubt the best-studied organisms in running-water systems. They are also, on the basis of present knowledge, highly diverse and play an extremely important role in ecosystem processes, as will become evident in later chapters.

3.2 Microorganisms

Organisms invisible to the naked eye, including viruses, prokaryotes (bacteria and archaea) and protists (single-celled eukaryotes), many fungi and microscopic metazoans (such as nematodes and rotifers)

The Biology and Ecology of Streams and Rivers. Alan Hildrew and Paul Giller, Oxford University Press. © Alan Hildrew and Paul Giller (2023).
DOI: 10.1093/oso/9780198516101.003.0003

Table 3.1 A list of publications highlighting texts that cover the taxonomy, identification and more general biology of running-water organisms.

Author(s)	Title	Notes
Dobson, M., Pawley, S., Fletcher, M. and Powell, A. (2012)	*Guide to British Freshwater Iinvertebrates* Freshwater Biological Association Scientific publications, volume 68	A guide to the identification of invertebrate families in British rivers, streams, lakes and wetlands
Yule, C.M. and Sen, Y.H. (2004)	*Freshwater Invertebrates of the Malaysian Region* Academy of Sciences Malaysia	Focuses on the ecology, distribution, identification and habitats of freshwater invertebrates
Dudgeon, D (1999)	*Tropical Asian Streams: Zoobenthos, Ecology and Conservation* Hong Kong: Hong Kong University Press	The bibliography provides a comprehensive list of the ecological and taxonomic literature published up to and including 1996. The book also provides a series of keys and guidelines for the identification of invertebrates in running waters of the region.
Hawking JH, Smith LM, LeBusque K, Davey C (eds) (2013)	Identification and Ecology of Australian Freshwater Invertebrates Murray–Darling Freshwater Research Centre, http://www.mdfrc.org.au/bugguide	Online resource is a compilation of taxonomic resources published by the various taxonomists working on invertebrates with aquatic life stages
Thorp, J.H. and Rogers, D.C. (2010)	*Field Guide to Freshwater Invertebrates of North America* Academic Press	Focuses on freshwater invertebrates that can be identified using at most an inexpensive magnifying glass
Thorp, J.H. and Rogers, D.C. (2014)	*Freshwater Invertebrates, 4th edition. Volume I: Ecology and General Biology* Academic Press	A revision and expansion of the classic *Thorp and Covich Ecology and Classification of North American Freshwater Invertebrates* and includes global coverage of freshwater invertebrate ecology
Thorp, J.H. and Rogers, D.C. (2016)	*Freshwater Invertebrates, Volume II: Keys to Nearctic Fauna* Elsevier	Presents a comprehensive revision and expansion of the earlier editions. Provides taxonomic coverage of inland water invertebrates of the Nearctic zoogeographic region with keys to insect families and to all other inland water invertebrates at the taxonomic level appropriate for the current scientific knowledge.
Merritt, R.W., Cummins, K.W and Berg, M.B. (2019)	*An Introduction to the Aquatic Insects of North America, 5th edition* Kendall Hunt	This text is intended to serve as a standard reference on the taxonomy, biology and ecology of aquatic insects, with updated keys to separate life stages of all major taxonomic groupings
Kriska, G. (2013)	*Freshwater Invertebrates in Central Europe: A Field Guide* Springer-Verlag.	Provides systematic information on freshwater macroinvertebrates of the central European region
Baldisserotto, B., Urbinati, E. and Cyrino, J. (2019)	*Biology and Physiology of Freshwater Neotropical Fish.* Academic Press	Provides updated systematics, classification, anatomical, behavioural, genetic and functioning systems information on freshwater neotropical fish species
Bellinger, E.G., and Sigee, D. C. (2010)	*Freshwater Algae: Identification and Use as Bioindicators.* Wiley-Blackwell	The key allows the user to identify the more frequently encountered algae to genus level: algae are separated on the basis of readily observable morphological features such as shape, motility, cell wall structure and colonial form
Haslam, S. M., (2014)	*River plants of Western Europe. The Macrophytic Vegetation of Watercourses of the European Economic Community* Cambridge University Press	Describes the vegetation of rivers and other watercourses in Europe with an emphasis on distributional, community and historical ecology
Collier, K.J. & Winterbourn, M.J. (Eds) (2000)	*New Zealand stream invertebrates: ecology and implications for management, New Zealand Limnological Society, Christchurch, 415 pp.*	Taxonomic guide to the stream invertebrates of New Zealand, plus notes of use to stream managers

are the most numerous component of any freshwater community. If we exclude viruses, about which very little is known in rivers and streams, the range in body mass across fully aquatic organisms in fresh waters spans around 16 orders of magnitude, from bacteria to large fish (Reiss 2018). Around one-third of this range is occupied by bacteria and Archaea (0.2–20 μm in length), while protists and microscopically small metazoans range from ~20 to 2,000 μm in length (and six orders of magnitude in terms of body mass). Due to their small size and the relatively specialised techniques required to study them, our understanding of the biology and ecology of the non-photosynthetic microorganisms in freshwater systems has, until fairly recently, been rather limited and a specialised pursuit. However, with the development of new molecular methods for identifying these microbes taxonomically, as well as studying their ecology and physiology, the role and enormous importance of the heterotrophic microbes (bacteria, Archaea and Fungi) in running-water ecosystems is becoming increasingly apparent (Findley 2010; see Chapters 8 and 9 and Topic Box 3.1 below). Okafor (2011) discusses the principles behind this as well as the detailed taxonomy of the bacteria, fungi, algae, protozoa and viruses (including bacteriophages), so we provide only a brief overview here.

3.2.1 Bacteria (and Archaea)

As in all other habitats, bacteria are often present in large numbers and are highly diverse, closely linked to environmental variation in streams and rivers (Zeglin 2015). One of the challenges in studying bacteria is that the species concept is difficult to apply, although modern DNA sequencing has revolutionised our understanding of bacterial ecology and diversity. Free-living bacteria are mainly associated with decomposing organic material, but also occur in the biofilm on the surfaces of rocks and vegetation, in the interstices of the substratum and suspended in the water. Most attention has been paid to the heterotrophic bacteria found in leaf packs and surface sediments (Findley 2010) but in the deeper, poorly oxygenated sediments anaerobic bacterial metabolism clearly occurs. In addition, bacteria occur as gut commensals and parasites in lotic animals.

Bacterial community composition (as identified from 16S rRNA gene sequence libraries) appears remarkably consistent within the main lotic substratum types (Zeglin 2015). Cyanobacteria ('blue-green algae') and Bacteroides (a group of Gram-negative, non-spore–forming, obligately anaerobic rod-shaped bacteria) dominate epilithic biofilms. Gammaproteobacteria (the largest and the most diverse group of Proteobacteria) and Betaproteobacteria (capable of living in low-nutrient environments; many are heterotrophic whilst others are important in denitrification) are the most common groups in the sediment. Betaproeobacteria, Bacteriodes and Actinobacteria (the latter a highly diverse bacterial phylum with an unrivalled metabolic versatility and a characteristic filamentous morphology) are the most common in the water column, Alpha-, Beta- and Gammaproteobacteria dominate CPOM and Acidobacteria (physiologically diverse, mostly aerobic heterotrophs and capable of thriving in strongly acidic conditions) dominate FPOM. Many of the bacteria suspended in the water column of running waters are involved in the decomposition of terrestrial organic matter, and their seasonal dynamics clearly reflect the seasonality of leaf-litter inputs. Concentrations of suspended bacteria in streams and rivers range from 5.2×10^4 to 2.5×10^7 to cells mL^{-1} (Lamberti & Resh 1987). Densities in sediments may exceed 10^7 cells mL^{-1}.

3.2.2 Fungi

While there are probably more than 3,000 species of fungi in streams and rivers, aquatic hyphomycetes are by far the most studied, being important in the decomposition of leaf litter and playing a pivotal role in stream food webs as highlighted in Topic Box 3.1 by Mark Gessner and Eric Chauvet. Other groups have been identified by molecular means but their contribution to decomposition remains relatively unknown (Krauss et al. 2011). Rather than being a phylogenetically distinct group, hyphomycetes are in fact the (often unknown) asexual stages of ascomycetes and basidiomycetes, the two major phyla of fungi that produce fruiting bodies (Findlay 2010). The reproductive cycle begins with rapid growth of the hyphae (the long, thread-like and branching structure of the fungus) over several days prior to spore production. The conidia of hyphomycetes (asexual spores cut off externally at the apex of specialised hyphae) are relatively large, being 10–40 times longer than the average terrestrial spores, and are released into the water column so their presence and abundance are relatively easy to track. The conidia typically have four diverging arms (they are 'tetraradiate') or, more rarely, a filiform and sigmoidal shape (see Box Figure 3.1a in Topic Box 3.1). These shapes and the sticky muscilage that covers the tips of the spores facilitate their attachment to leaves and other smooth surfaces and may reduce losses of conidia in the current (Webster & Descals 1981; Dang et al. 2007; Krauss et al. 2011).

Topic Box 3.1 Aquatic hyphomycetes: microbial multi-instrumentalists in streams

Mark Gessner and Eric Chauvet

When the British mycologist Clarence T. Ingold discovered some 80 years ago the stunning diversity of peculiar fungal spores in streams and rivers (Ingold 1942), he might have had a presentiment of the universe of freshwater fungi that would unfold in the coming decades. But it took another 30 years before the pivotal roles were recognised that these fungi play in both stream food webs and one of the most fundamental ecosystem processes, the decomposition of leaves that fall into streams from riparian canopies (Bärlocher 1974). Aquatic hyphomycetes, as Ingold called them, are sometimes referred to as Ingoldian fungi in recognition of the discoverer (Webster & Descals 1981). They are a polyphyletic group of asexual forms of filamentous fungi (Baschien et al. 2013) which stand out due to their hyaline (colourless) hyphae and the ability to sporulate under water, producing an abundance of spores with characteristic morphologies, a trait otherwise uncommon among fungi. More than 7 million spores may be released daily per gramme of leaf litter decomposing in streams (Gessner et al. 2007). At the stream scale, this release can result in spore concentrations per litre of flowing water exceeding 20,000 (Webster & Descals 1981; Gulis et al. 2019), equivalent to 17 billion transported downstream per day at a discharge of only 10 L s^{-1}.

A first synoptic account of progress made since Ingold's discovery 50 years earlier was published in a book dedicated to *The Ecology of Aquatic Hyphomycetes* (Bärlocher, 1992), and an update appeared in a special issue of *Fungal Ecology* (Bärlocher 2016). More than 300 species are known today, most of them ascomycetes and some basidiomycetes, with about 10% of them known to have a sexual stage (Shearer et al. 2007). Aquatic hyphomycetes occur worldwide, and some broad geographical patterns have been identified (Webster & Descals 1981; Seena et al. 2019). Contrary to intitial expectations, aquatic hyphomycetes are not restricted to small, swiftly flowing forest streams, but are also common in large rivers, polluted waters (Ferreira et al. 2014; Krauss et al. 2011), some extreme habitats such as immediately below glaciers in high-mountain streams above the treeline (Gessner & Robinson 2003), and even outside of water, including as root endophytes (Chauvet et al. 2016).

The distinct tetraradiate or sigmoidal spores of many species (Box Figure 3.1a) have been interpreted as an adaptation to flowing water, reflecting the need to attach effectively to suitable substrata in a turbulent environment exerting shear stress on surfaces. The spores that settle on submerged leaves also germinate quickly, the extending hyphae produce extracellular enzymes to macerate the leaf tissue, assimilate resources, grow pervasively in the colonised tissue, start reproducing early in the life cycle (Suberkropp 1992a; Gulis et al. 2019) and release sundry decomposition products, including dissolved and fine particulate organic matter, carbon dioxide (Gessner et al. 2007) and mineral nutrients. Drifting spores are also effectively

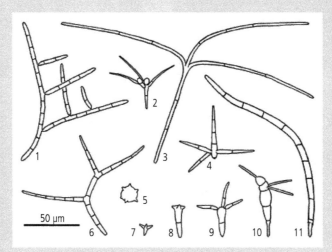

Box Figure 3.1a Selection of aquatic hyphomycete spores illustrating the diversity of size and morphology of these fungi proliferating in streams. (1) *Varicosporium elodeae*, (2) *Tetracladium marchalianum*, (3) *Tetrachaetum elegans*, (4) *Triscelophorus* sp., (5) *Goniopila monticola*, (6) *Tricladium kelleri*, (7) *Heliscella stellata*, (8) *Neonectria lugdunensis*, (9) *Heliscina antennata*, (10) *Culicidospora gravida*, (11) *Anguillospora longissima*.
Source: from Gulis et al. (2019) with permission.

Topic Box 3.1 *Continued*

trapped by foam forming in streams, which is conveniently sampled to determine the structure of aquatic hyphomycete communities and to isolate individual strains (Webster & Descals 1981; Bärlocher et al. 2020).

As a group, aquatic hyphomycetes have become a model for the study of the role of microbial species and communities in ecosystem processes (Bärlocher 2016; Box Figure 3.1b), prompted by their ubiquity and importance, ease of cultivation and often straightforward microscopic identification to species level (Box Figure 3.1a), which is virtually impossible with nearly all other heterotrophic microbes. The central role of aquatic hyphomycetes as decomposers of leaf litter in streams is now firmly established (Gulis et al. 2019; Swan et al. 2021). Evidence in support of this role includes their regular occurrence on submerged leaves (Seena et al. 2019); a rich enzymatic complement to degrade plant polymers, although their ligninolytic capabilities appear to be limited (Gessner et al. 2007; Krauss et al. 2011); the efficient degradation of leaves in pure culture (Suberkropp, 1992a); abundant sporulation on submerged leaves; substantial fungal biomass accumulating in decomposing leaf litter; and tight correlations between leaf decomposition rates in streams and both fungal biomass accrual and sporulation rate (Gessner et al. 2007; Gulis et al. 2019). These lines of evidence are further underscored by the effective decomposition of leaves in field situations where leaf-shredding macroinvertebrates are absent (Suberkropp & Wallace 1992) and by organic matter budgets attributing a large proportion of overall litter mass loss to fungi (Gessner et al., 2007; Gulis et al. 2019).

Resource use by aquatic hyphomycetes leading to litter decomposition also promotes their rapid growth in decomposing leaves, which translates to the whole-stream scale, where fungi have been estimated to produce large amounts of biomass (> 100 g C m^{-2} of stream bed per year; Suberkropp et al. 2010; Gulis et al. 2019), many times the production of macroinvertebrates. These numbers suggest a key role of aquatic hyphomycetes not only as decomposers in forest streams but equally as secondary producers. It is not surprising, therefore, that aquatic hyphomycetes assume a crucial role in stream food webs. In particular, they have long been recognised as mediators of energy flow from leaf litter to leaf-shredding detritivores (Bärlocher & Sridhar 2014). These detritivores show consistent feeding preferences for

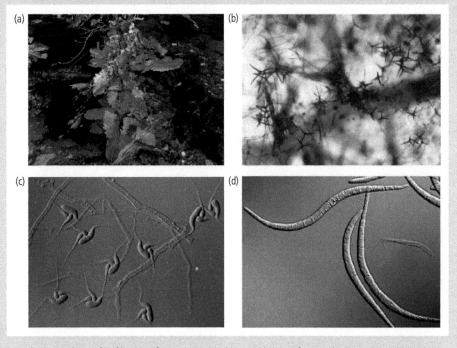

Box Figure 3.1b Accumulation of leaf litter in a forest stream (a), the typical habitat of aquatic hyphomycetes, which are sporulating on a decomposing alder leaf viewed by light microscopy (b). To increase contrast against the yellow-brownish background of the leaf, the spores, here mostly tetraradiate, were stained with a dark-blue dye. Differential interference contrast micrographs of spores of aquatic hyphomycete species: *Gyoerffyella gemellipara* (c) and *Anguillospora longissima* (d) in pure culture. *Source:* photo credits: (a) E. Chauvet, (b) M.O. Gessner, (c) E. Chauvet, (d) C.A. Shearer (with permission).

Topic Box 3.1 *Continued*

leaf litter colonised by fungi, a phenomenon referred to as leaf conditioning, and distinguish among individual species of aquatic hyphomycetes when selecting colonised leaf patches (Suberkropp 1992b).

Moreover, the release into the flowing water of large numbers of spores and mesophyll cells resulting from effective leaf tissue maceration by aquatic hyphomycetes also provides a resource to filter-feeding and particle-collecting invertebrates. This and other types of interactions of aquatic hyphomycetes in stream communities remain poorly investigated. Nevertheless, some evidence exists for both competitive and facilitative interactions with leaf-associated bacteria and microalgae in streams (Gulis et al. 2019).

The quantitative importance of single species in fungal communities has long been based on colony counts and the numbers of spores released into stream water. Now a range of molecular approaches facilitate refined species-level analyses to assess the diversity, abundance and activity of aquatic hyphomycetes, among others resolving the problem of capturing non-sporulating species and strains. As metabarcoding, quantitative real-time PCR, analyses of precursor rRNA, transcriptomics, proteomics, metabolomics and quantitative stable-isotope probing (see Cornut et al.

2019; Bärlocher et al. 2020) are systematically applied to aquatic hyphomycetes, biodiversity patterns at microscopic to global scales (Seena et al. 2019) will be greatly refined and the nature of interactions among fungal species and with other community members will be unravelled. Stream bioassessment and ecotoxicological studies will also benefit (Ferreira et al. 2014; Ittner et al. 2018). In addition, the array of molecular approaches now available will greatly enhance insights into a wide range of other, mostly poorly studied freshwater fungi, including aquatic ascomycetes, which are particularly prominent on decomposing wood in streams; so-called aero-aquatic hyphomycetes; yeasts; saprotrophic and parasitic chytrids; other parasitic fungi such as invertebrate-infecting trichomycetes; and even freshwater lichens and mycorrhizae of aquatic plants (Jones et al. 2014).

Professor Mark O. Gessner is at the Department of Plankton and Microbial Ecology, Leibniz Institute of Freshwater Ecology and Inland Fisheries, Germany.

Professor Eric Chauvet is at the Laboratoire écologie fonctionnelle et environnement, University of Toulouse, CNRS, Toulouse, France

---

Aquatic hyphomycetes colonise the dead leaves and woody debris that fall into streams, and their hyphae penetrate the leaf tissue (a process sometimes called *conditioning*) and begin to 'skeletonise' the leaves by macerating cells using pectinases. Conditioning makes the leaf more palatable and nutritious to detritivores (see Topic Box 3.1). In addition to the hyphae, fungal spores are consumed by detritus-feeding animals, a potentially substantial resource as described in Topic Box 3.1. Aquatic hyphomycetes are distributed from the tropics to the polar regions. Their activity in tropical forests is not well known, although more processing may take place before the leaf material enters streams than in temperate areas. Our knowledge from higher latitudes is also restricted, although large numbers of species, many of which tolerate freezing, are sometimes present (Marvanová & Müller-Haeckel 1980). In the first large-scale study of biodiversity of aquatic hyphomycetes, stream temperature rather than biogeography accounted for most variation in community composition (Seena et al. 2019) and a humped latitudinal diversity gradient is evident. Local distribution seems to be closely coupled to the kind of riparian

vegetation present, and to water chemistry, and fungal community composition varies in relation to water chemistry, temperature and other factors (including nutrient concentration) (Bärlocher 1992).

3.2.3 Protozoa

The protozoa are a heterogeneous group of microscopic unicellular or colonial eukaryotes (Figure 3.1). Corliss (1994) defines 34 separate protist phyla of which 16 are described as freshwater, and Finlay & Estaban (1998) provide a useful summary classifying and defining the characteristics and ecology of the free-living protozoa. They are of contrasting importance to humans, on the one hand being used in bioassessment and on the other as a significant health risk (e.g. the intestinal parasites *Giardia duodenalis* and *Cryptosporidium parvum*). The free-living protozoans are phagotrophic (bacterial-grazing) and, in quantitative terms, are the most important grazers of bacteria in aquatic environments (Finlay & Estaban 1998), while some species may also show some photosynthetic ability. Many protozoa are microaerobic, found in habitats with a low O_2

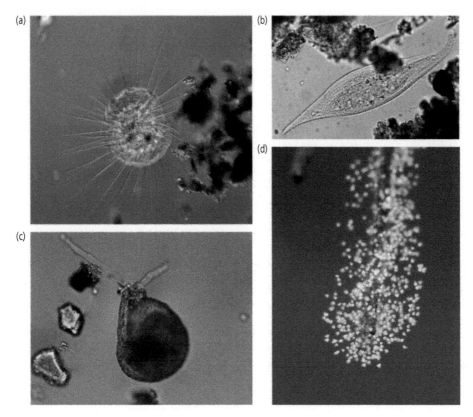

Figure 3.1 Protozoan meiofauna: (a) Heliozoa (*Actinosphaerium nucleofilum*) from the Calle-Calle River, Chile; (b) Ciliophora (*Amphileptus pleurosigma*) from the Calle-Calle River, Chile; (c) Testacea (*Lesquereusia spiralis*) from the Huicha River, Chiloé, Chile; (d) the ciliate *Vorticella* spp. *Source:* hotos (a)–(c) courtesy of Dr P.E. Schmid (PJSchmid.com), and Dr J.M. Schmid-Araya, Research Fellow University of Bournemouth UK; (d) with kind permission of Jan Hamrsky, Lifeinfreshwater.net.

tension but rich in microbial food, and are supported by nutritional symbionts such as sulphide-oxidising bacteria and endosymbiotic algae (Fenchel & Finlay 1989; Finlay & Esteban 1998).

In streams, protozoans mainly occur in biofilms (the organic layer that grows on wetted surfaces; Chapter 8), in interstitial habitats, associated with plants (e.g. ciliates and amoebae), and where organic matter settles. In larger channels they are also present in the water column although often vulnerable to being 'washed away'. Free-living protozoa have been classified into three broad functional groups (Finlay & Esteban 1998). The first, the ciliates, are relatively large (from < 20 μm to 2 mm), using simple cilia or ciliary organelles for movement, and grazing bacteria, unicellular algae and other protozoans either suspended or on sediments. Ciliates can reach huge numbers—maximum densities of over 900,000 to 6,000,000 per m^{-2}—and are abundant even in small

streams (Reiss & Schmid-Araya 2008) (Figure 3.1b, d). These forms constitute the majority of protozoan biomass in sediments and can match both the production and biomass of invertebrates in lake and river sediments (Finlay & Esteban 1998). The second functional group, the Sarcodina (including testate amoebae; Figure 3.1c), engulf bacteria or algae with pseudopodia and are generally associated with sediments and surfaces. The third group, the heterotrophic flagellates, move and feed using flagella and are important grazers of bacteria in sediments and on surfaces as well as feeding on suspended bacteria mainly by 'raptorial' (i.e. the bacterial cell is seized) or filter feeding.

Ciliate assemblages in biofilms from different streams and seasons differ considerably, with a greater diversity in nutrient-impacted sites than in pristine ones (Dopheide et al. 2009). That said, species composition and abundance of ciliates associated with biofilm did not differ with stream order within the

Rhine river network, apart from the largest river, which differed from the others (Ritz et al. 2017).

The diversity of freshwater protozoans has been difficult to assess and lists of morphospecies may be five to ten times lower than actual biological species (Dopheide et al. 2009). Due to their lack of hard parts, protozoans are often not recorded from predator guts, despite being eaten by many different species of scrapers, filterers and deposit feeders or by young stages of predatory macroinvertebrates, particularly within the biofilm food webs (see Chapter 8). For many small invertebrates, like naidid worms and orthoclad midge larvae, protozoans may indeed make up much of the diet.

3.3 Plants

Plants are a diverse and interesting group with a range of adaptations for life in and alongside running waters, a theme we return to in Chapter 4. They are also important in the ecology of streams and rivers. They produce oxygen, contribute to the physical habitat of animals and algal epiphytes and are of major importance in nutrient cycling and energy flow. It is also now recognised that the larger species are major 'engineers' of river channels (see Chapter 9).

3.3.1 Algae

Algae are the most significant primary producers in most streams and many rivers and primary production by benthic algae often makes up the major energy pathway in unshaded streams and in rivers with low turbidity (see Chapter 8, Section 8.2.1). Macroalgae are primarily filamentous, while most other algae are microscopic. Assemblages of algae (eukaryotes) and cyanobacteria (prokaryotes) form visible layers attached to the substratum and are collectively called *periphyton*. The main algal periphyton includes diatoms, green algae and phytoflagellates. Diatoms (e.g. Figure 3.2a) are a species-rich group which is often considered the most important food for many benthic invertebrate herbivores. Diatoms have also been used to establish biotic indices for acidification, eutrophication and assessment of general water quality (Kelly et al. 2008; Chapter 10) as alkalinity and nutrients are the major environmental factors influencing their broader distribution and community composition. At a more local level, flow and herbivory are important.

Epilithic algae grow on rocks and stones and either form thin crusts (e.g. *Hildenbrandia*, *Lithoderma*) or occur in 'meadows' or 'lawns' up to several mm thick

(e.g. many pennate diatoms) (Figure 3.2b). A few algae actually live within the surface of softer rocks (usually calcareous ones) and are called *endolithic* (e.g. *Schizothrix*). Although the instability of finer sediments makes life risky, populations of many species of algae and cyanobacteria can at least temporarily occur there in dense aggregations (e.g. *Oscillatoria*, *Phormidium*, *Microcoleus*, *Mastigocladus*, *Hydrodityon*, *Nitzschia*). Epiphytic algae live on higher aquatic plants and filamentous algae and a true phytoplankton suspended in the water column occurs chiefly in lowland, meandering rivers.

The larger (macrophytic) algae occur as threads or long filaments (e.g. *Cladophora*; Figure 3.2c), tufts (e.g. *Oedogonium*, *Ulothrix*, *Stigeoclonium*), or as branched structures with carbonate reinforcements (e.g. *Chara*, *Nitella*). Tubular algae may also be present (e.g. *Enteromorpha*, *Lemanea*). The amount of algal biomass reflects light, flow, availability of nutrients and grazing pressure and, because of spatial variability in these factors, large differences in algal growth may occur over small distances of stream bed and even over the surface of a single stone or boulder.

3.3.2 Bryophytes (mosses and liverworts)

In rivers and streams with a stable substratum, mosses can be abundant and can influence overall primary production, retention of fine particles and nutrient uptake. Occasionally, as in some boreal, forested catchments and acidic streams, they may contribute more to primary production than periphyton (Naiman 1983; Stream Bryophyte Group 1999). Estimates of biomass can vary from < 2 g m^{-2} for *Fontinalis squamosa* in Muskrat River Quebec to > 280 g m^{-2} for the same species in the high Pyrenees in France. Mosses sometimes cover entirely rocks, boulders and bedrock and are especially abundant where low light constrains periphyton growth. Moss growth makes up a three-dimensional habitat of much greater structural complexity than bare rocks or periphyton-covered substratum (Figure 3.3), which in turn provides protection from flow and increased living space for small animals, while trapped FPOM is a source of food (Englund 1991; Stream Bryophyte Group 1999).

The moss plant we see in a stream is the haploid gametophyte generation of the organism (as opposed to the diploid generation of higher plants). The large perennial species reproduce mainly vegetatively, while the smaller forms have sexual reproduction. Most species are identified on the basis of the shape, colour and texture of the leaves. Some moss species are truly

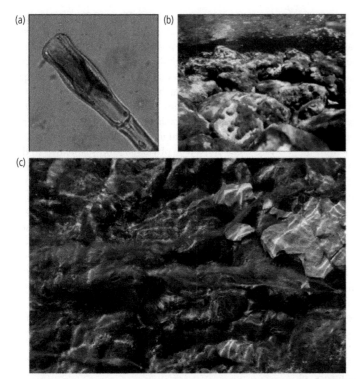

Figure 3.2 Examples of stream algae. (a) The benthic diatom *Didymosphenia geminata*,
(b) *D.geminata* mat formation in a nutrient-poor stream in the Canadian Rocky mountains
(c) Underwater photograph of long strands of filamentous green algae and algal 'lawns' growing on cobbles in the middle reaches of the Tukituki River, New Zealand.
Source: (a) and (b) modified from Bothwell & Taylor 2017, with permission from John Wiley & Sons; (c) photo John Quinn, NIWA, https://niwa.co.nz/file/41454.

Figure 3.3 The lotic moss *Hygroamblystegium tenax*, showing the dense clumping and branching from the main axis common in many stream mosses.
Source: from Hermann Schachner, Wikimedia Commons; (d, 143,553–482,324) 4252.

aquatic, while others are semi-aquatic (i.e. able to withstand prolonged periods above the water) and can often be found on exposed boulders along river banks. In addition to flow stability, water velocity is an obvious and important factor for mosses, with different species of moss differentially sensitive to flow variation. A comprehensive discussion of the reproduction,

physiology and ecology of stream mosses is provided by the Stream Bryophyte group (1999) review and by Glime (2020).

Freshwater liverworts are much less common and of much less importance than the mosses. They generally have two forms; an undifferentiated mass or thallus (thalloid form) and a leafy form with definite stem and with single-cell-thick leaves. Both forms have a definite dorsoventral organisation, in contrast to the mosses (Bowden et al. 2017).

3.3.3 Vascular plants

Mosses and liverworts lack specialist tissues for conducting water and minerals through the plant. Plants that do have such tissues, typically including true roots, are called 'vascular' plants and include most of the larger land and freshwater species. The major groups of vascular plants are the Pteridophyta—non-seed-bearing plants including the ferns and their allies—and the Spermatophyta—the seed-bearing plants. The latter are then divided into the flowering plants (Angiospermae, by far the most important group) and the non-flowering Gymnospermae (conifers and allies), though the latter have no aquatic species. A few important ferns (pteridophytes) do live in fresh water—often free-floating and in river systems limited to fringing flood plains. Of the several hundred thousand species of vascular plants, only about 1% of them (*c*.2,600 species) are aquatic. The familiar term for the aquatic species, *macrophyte*, just means 'large plants' and is a non-formal term technically including photosynthetic organisms big enough to see with the naked eye. It thus includes macroalgae and larger blue-green bacteria (Cyanobacteria) plus the mosses and liverworts, which we have already mentioned. The seed-angiosperms and ferns are by far the most familiar 'aquatic macrophytes' and the term is most commonly used to refer to them more or less exclusively (see review by Chambers et al. 2008).

The terrestrial vascular plants probably evolved from marine algae, with a small fraction of them returning to an aquatic life in fresh water (Chambers et al. 2008). The return to water took place many times in evolution, with 11 of the > 300 genera of ferns and allies (~ 3%) having freshwater species, and > 400 of the > 13,000 genera of angiosperms (~ 3%) also having freshwater species. Some families of aquatic angiosperms are entirely aquatic, presumably having diverged from a terrestrial ancestor very early in their evolution, while many others have only a few aquatic members of largely terrestrial groups. The move to a freshwater environment brings many environmental challenges, some of them specific to life in running waters as discussed in Chapter 4.

Freshwater vascular plants can also simply be grouped into a few different growth forms or alternatively based on the primary habitat in which they are found (Riis et al. 2001; Bowden et al. 2017).

In terms of growth form, there are basically four clear groups (Table 3.2). The first two include species whose leaves and stems are either fully submersed or float on the water surface. They have flexible stems and in running waters can bend with the current. Those whose leaves and stem remain submersed usually have roots anchored in the sediment. Those whose leaves float flat on the water are also usually rooted in the sediment, though a few float freely. The third group are free-floating forms, with no attachment to the substratum, although they do have free-hanging roots. These are often individually small plants but can form dense mats (e.g. *Lemna*, the duckweed) and are largely confined to sluggish waters. The well-known, larger floating species, the water hyacinth (*Pontederia* (formerly *Eichhornia*) *crassipes*) occurs in the Amazon basin and can grow up to a metre above the water surface. Its rapid growth rate and highly invasive nature has made it one of the most problematic of invasive species. The fourth group are 'emergent' species, rooted in the substratum and whose upper leaves and shoots grow above the water while the lower ones are able to grow whilst submerged. The leaves and stems have 'supporting' tissues that enable this upright growth form (see Chapter 4). Examples of all four groups can be seen in Figure 3.4.

Macrophytes can also be placed into three broad groups based on the primary habitat (Table 3.2). Obligate submerged plants live permanently submerged although may possess floating leaves. Amphibious plants can live on land or fully submerged, often with different morphological forms. As the name suggests, the third group of species, terrestrial plants, live on land, but can occur in streams although never as permanently submerged populations. Separately, it is also useful to recognise vascular plants that normally grow on or close to the banks of rivers and streams extending to the edge of the floodplain—'riparian vegetation'. This is a very important although very disparate group of plants, ranging from emergent aquatic species that can tolerate periods of low discharge to terrestrial plants that can survive occasional flooding.

In most running-water systems, rooted flowering plants, such as the examples in Figure 3.4, are generally restricted to those areas where the stream gradient and hence eroding power of the flow is low, allowing sufficient accumulation of fine materials to provide the necessary substratum for rooting. The most important additional environmental factors controlling macrophyte distribution and community composition include light availability (influenced by shading

Table 3.2 Groupings of stream macrophytes based on growth form and primary habitat (modified from Bowden et al. 2017).

Traditional grouping based on growth form	General characteristics	Examples
Submerged plants	Plants permanently submerged; produce floating, aerial, or submerged reproductive organs	*Heteranthera dubia* (water stargrass) native to North and Central America form rivers and lakes. Significant variability in leaf shapes and stem morphology when submerged compared to out of water. *Myriophyllum spicatum* (Eurasian watermilfoil) has wide geographic distribution across Europe, Asia and North Africa and introduced to N. America. Found in slow-flowing water including large rivers, as well as still water.
Floating-leaved plants	Plants permanently submerged and produce floating leaves that differ in morphology from submerged leaves in still and slow-flowing water; produce floating or aerial reproductive organs.	*Ranunculus aquatilis* (common water crowfoot) has perennial submerged leaves and palmately lobed (toothed) floating leaves and is found in ditches and streams. Native to Europe, the western USA and north-west Africa.
Free-floating plants	Plants not attached to the substratum; produce floating, aerial, or very rarely submerged reproductive organs	*Salvinia minima* (water spangles), a fern native to S. America and the West Indies, found in still waterways
Emergent plants	Plants normally erect and standing above the water surface, but some species tolerate submergence; all produce aerial reproductive organs	*Schoenoplectus lacustris* (common club-rush): widespread across Europe and extends east into Asia. Can grow up to 3.5 m tall.
Grouping based on primary habitat	**General characteristics**	**Examples**
Obligately submerged plants	Similar to submerged plants above but also includes the floating-leaved and free-floating plants	*Potamogeton perfoliatus* (perfoliate pondweed; characterised by the leaf encircling the stem). Occurs in most continents in both standing and flowing freshwater habitats. It does not tolerate drying out but is quite tolerant of flowing water.
Amphibious plants	Part of the emergent plants described above; able to live on land as well as emerged and submerged; some develop water forms	*Veronica anagallis-aquatica* (water speedwell): present on most continents along stream banks as well as other wetland habitats. *Sparganium emersum* (European bur-reed): grows in shallow water bodies like ponds and streams.
Terrestrial plants	Mainly present on land, occasionally in streams; do not live permanently submerged, nor do they ever develop water forms	*Epilobium hirsutum* (great willowherb): native to north Africa, most of Europe and parts of Asia, found on the banks of rivers and streams and ditches.

from the riparian zone, water depth and turbidity) and nutrient availability (Franklin et al. 2008). Biotic interactions, particularly competition, and increasingly river management, also influence diversity, abundance and distribution. Under certain circumstances, however, largely dependent on stable flow conditions, lush macrophyte communities can develop, as in the chalk streams of southern England (Figure 3.4), or in strongly regulated rivers where flow variation is moderated. Nutrient-enriched lowland rivers (e.g. within an agricultural landscape) can also develop thick macrophyte communities, especially during the summer.

While species diversity of aquatic flowering plants is, in most cases, quite low, the natural riparian zone of rivers often abounds with species, which are more or less dependent on the river water seasonally inundating this zone and depositing nutrients. In contrast, in regulated rivers, where the spring flood is reduced due to storage in large reservoirs, the development of river-margin vegetation is greatly impeded (Nilsson & Jansson 1995).

Not only is the distribution and abundance of higher plants affected by the river environment, but in turn macrophytes can influence the lotic systems in different ways through their role as biological or ecological

Figure 3.4 A lowland, macrophyte-rich, eutrophic river in southern England illustrating a variety of macrophytes. Species highlighted are A—*Lemna minor* (common duckweed), B—*Sparganium erectum* (branched bur-reed), C—*Nuphar lutea* (yellow water lily), D—*Potamogeton natans* (floating pondweed), E—*Schoenoplectus lacustris* (Common clubrush). *Source:* photo courtesy of Dr J.I. Jones, with kind permission.

engineers (Franklin et al. 2008). We discuss this in some detail in Chapter 10, along with the increasing threats to riverine higher plants from anthropogenically induced changes in flow regimes (e.g. damming and river regulation), changes in nutrient status of river systems through alterations in land use, and climate change. Macrophytes also provide structural habitat diversity and possible refuge from flow and predators as well as a substratum for an epiphytic microflora, which in turn attracts grazers. The extent of direct grazing on macrophytes has been debated over many years, as we discuss in Chapter 7 (section 7.2). Otherwise, these plants probably contribute to energy flow mainly via the decomposer food chain (see Chapter 8) and in their role in nutrient retention (see Topic Box 9.1).

3.4 Invertebrates

The invertebrates are multicellular animals (metazoans) including some of the best-known groups found in running waters, but conversely also include some of the least known. This dichotomy is largely determined by their body size, the methods available to sample and collect them, and the ease with which they can be identified.

3.4.1 Smaller metazoans—the meiofauna

Freshwater ecologists commonly divide metazoans into groups based on body size, with those passing through a sieve of mesh size 500 or 1,000 μm but retained on one of 20, 42 or 63 μm referred to as the *meiofauna* (Robertson et al. 2000; Reiss 2018; Majdi et al. 2020). This approach though does not clearly separate groups taxonomically. Larger protozoans (see Figure 3.1) may be retained on the smaller meshes while the smallest metazoans will pass through. Further, the small stages of what will grow into quite large invertebrates—the more familiar macroinvertebrates—easily pass through a mesh of 1,000 μm or even of 500 μm. This has the consequence that all the larval stages of many stream-dwelling invertebrates cannot be sampled quantitatively using common devices with too coarse a mesh.

Nevertheless, the meiofaunal fraction does contain many organisms that are often still completely ignored by stream ecologists, including more than 10 metazoan phyla (Roberston et al. 2000; Reiss 2018). Groups of meiofaunal organisms common in streams and rivers include the Microturbellaria, Rotifera and Gastrotricha, Nematoda, Hydrachnida, microcrustaceans (including Ostracoda; Branchiopoda, including Cladocera;

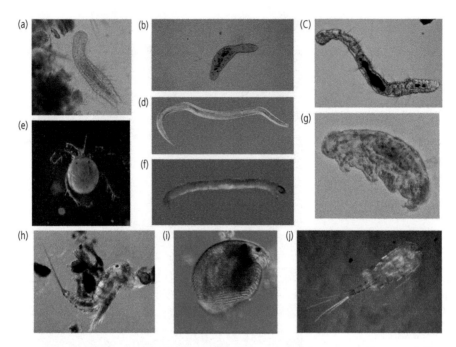

Figure 3.5 Some examples of freshwater meiofauna passing through 1,000 μm and retained on 20 μm mesh: (a) Gastrotricha (*Chaetonotus simili*) Lake Pratignano catchment, Italy (courtesy of M. Balsamo); (b) Microturbellaria (*Stenostomum grabbskogense*), from the Huicha River, Chiloé, Chile; (c) Oligochaete Aeolosomatidae (*Aelosoma hemprichi*) from Lone Oak stream, UK; (d) Nematoda (*Mononchus niddensis*) from the Lake Schösee catchment, Plön, North Germany (courtesy of W. Traunspurger); (e) Hydrachnidia (*Sperchon violaceus*) from the Adur River, UK; (f) Chironomidae (*Glyptotendipes pallens*) from the City Mills River and Lea Navigation, London, UK; (g) Tardigrada (*Macrobiotus* sp.) from the Afon Mynach, Wales, UK; (h) Harpacticoid (*Delachauxiella wiesseri*) from the Huicha River, Chiloé, Chile; (i) Cladoceran (*Alona* sp.) from Regents Canal, London, UK; (j) Cyclopoid copepod (*Acanthocyclops* sp.) from Huicha River, Chiloé, Chile.
Source: unless otherwise indicated, photos courtesy of Dr. P.E. Schmid (PJSchmid.com) and Dr. J.M. Schmid-Araya, Research Fellow, University of Bournemouth, UK.

Maxillopoda, including Copepoda) and Tardigrada (see examples in Figure 3.5). More details of the biology and ecology of these groups can be found in Kolasa (2000; microturbellarians), Ricci & Balsamo (2000; rotifers and gastrotrichs), Traunspurger (2000; nematodes), Di Sabatino et al. (2000; water mites, Hydrachnida), Dole-Olivier et al. (2000; microcrustaceans), and Nelson & Marley (2000, tardigrades).

The microcrustaceans are perhaps the best known and most studied of the meiofaunal groups. Ostracoda are small, usually between 0.4 and 3 mm long, and have a bivalve-like carapace. There are three superfamilies occurring in running waters. Most species have an interstitial lifestyle or are restricted to groundwater, and feed on fine sediments and fine particulate organic matter and the associated bacteria, algae and microfauna (Dole-Olivier et al. 2000). They are mainly parthenogenetic, and often only female ostracods are found. Cladocerans (Figure 3.5i) inhabit virtually any kind of freshwater habitat, including riverine and interstitial environments, and occur in the plankton of large rivers as well as in the benthos, but are not found in fast-flowing habitats. Species of the dominant family Chydoridae live in the interstices of stream gravels, although, like most of the cladocerans, they appear to lack particular morphological adaptations to this habitat. The benthic cladocerans are generally discus-shaped and < 2mm in length, and can crawl or swim. Whereas ostracods are largely restricted to approximately neutral waters, many chydorids are tolerant of low pH (between 3.8 and 5; Fryer 1993). They are generally parthenogenetic and periodically gametogenetic (sexual reproduction takes place mainly in the autumn). The diets vary, with most species either detrivores or algivores. Free-living copepods range in size from 0.2–5 mm and the body is normally divided into two parts (Figure 3.5j). Many Copepoda are planktonic, and as such may be found in large rivers and in lake and reservoir outlet streams. Small copepods of the suborder Harpacticoida (Figure 3.5h) (also to some extent members of Cyclopoida and Calanoida) occur on or in the stream bed and are one of the numerically dominant elements of the meiofauna (along with nematodes and rotifers). Most copepods reproduce sexually, with the males generally smaller than the females, the latter carrying one or two egg sacs. Locomotion varies and includes swimming, crawling and burrowing (mainly harpacticoid) species. Feeding modes vary from the predatory large-bodied cyclopoid copepods to omnivorous and herbivorous species (harpacticoids). Some species are parasitic and these have extreme morphology, making them very unlike other copepods.

Water mites have mainly been studied as part of the meiofauna, so it is appropriate to consider them here. They have evolved from several lineages of Acari that secondarily colonised fresh waters and are frequent members of both the benthos and, in early stages, the hyporheos (Smith et al 2010; Figure 3.5e). At least 10,000 species of water mites are known (Di Sabatino et al. 2008), many of them lotic. The abundance of water mites in stream riffles can reach up to 5,000 m^{-2} with up to 50 species in 30 genera (Goldschmidt 2016). *Sperchon* and *Aturus* are the most diverse genera in North American streams. The life cycle of water mites is complex, characterised by interactions with other macroinvertebrates. The eggs are laid into plant stems or into the tissues of sponges or mussels. Larvae developing from the egg are parasitic-phoretic (i.e. passengers) on the adult stage of an aquatic insect (predominantly chironomids). A succession of other stages follows, including the quiescent proto- and tritonymphs which are pupa-like resting stages, deutonymphs which are predators, the again dormant 'imagochrysalis' and, finally, the adult stage which is also predatory, feeding on crustaceans, insect eggs and larvae. Smith (1988) found the prevalence of insect infestation by larval water mites (i.e. the total number of parasitised hosts to the total number of potential hosts) frequently exceeded 20% for a variety of adult insects including Corixidae (water bugs), Dytiscidae (beetles), Libellulidae (dragonflies) and Culicidae and Chironomidae (mosquitoes and midges).

Progress has been made in integrating these small metazoans into lotic ecology, including their place in food webs and production (see Chapters 7 and 8). Recent evidence suggests that meiofauna are significant prey for many benthic macroinvertebrates (such as chironomids, shrimps and flatworms) as well as for juveniles of widespread bottom-feeding fish species (such as carp, gudgeon and catfish) (Ptatscheck et al. 2020). Students and researchers of lotic ecology must be aware of the meiofauna although their study remains largely a specialised task. The application of molecular methods in assessments of biodiversity and bioassessment may potentially bring them into greater focus (see Chapter 10).

3.4.2 Larger turbellarians

The Turbellaria (flatworms) are free-living members of the Platyhelminthes and are characterised by a ciliated epidermis. In lotic environments many are very small (the microturbellarians), and the 400 or so species known worldwide are members of the meiofauna (see section 3.4.1). Tricladida (planarians) are an important order of larger turbellarians and are common in streams and rivers. Of the 400 described species, most are located in the Palaearctic region, although it has been suggested that this number is at least an order of magnitude too low (Shockaert et al. 2008). Triclads are benthic, dorso-ventrally flattened animals (Figure 3.6a), 5–30 mm long, that glide on the substratum through the action of cilia on the ventral body surface beating in a thin layer of mucus that is secreted from special glands. They are predators and scavengers with a great capacity to detect food chemically and are quickly attracted to injured invertebrates. The pharynx is protrusible through the ventrally positioned mouth and digestion is often external to the body. Triclads are only rarely fed on by other predators. Many stream-living triclads are most abundant in cold, headwater streams. Reproduction may be sexual, resulting in the production of egg capsules fixed to the substratum on stalks (Figure 3.6b), or asexual (through fission), but most turbellarians are hermaphrodites (i.e. they possess both male and female sex organs). Turbellarians are well known for their remarkable regeneration capacity due to undifferentiated stem cells or neoblasts; if they are cut into pieces, each piece will regenerate into a whole flatworm.

3.4.3 Mollusca: Gastropoda

About 26 of the 409 currently recognised gastropod (snail) families are wholly or mostly restricted to fresh water, and a further four have significant representation in freshwater habitats (Strong et al. 2008). Freshwater gastropods are found on every continent except Antarctica and are widely distributed in streams and rivers. Many representatives from the subclasses Prosobranchia (with a horny operculum that closes the entrance to the shell) and Pulmonata (without an operculum) occur in flowing waters and small streams, springs and groundwaters. Common examples include

Figure 3.6 (a) The flatworm *Dugesia gonocephala*; note the flattened body shape and triangular head; (b) Flatworm (*Dugesia gonocephala*) egg capsule on a stalk.
Source: photos with kind permission of Jan Hamrsky, lifeinfreshwater.net.

Figure 3.7 (a) The freshwater limpet *Fellissia* spp. (Ancylini, Planorbidae); (b) *Viviparus* spp. (Viviparidae), the river snail; (c) a group of pearl mussels *Margaritifera margaritifera*; (d) the solid orb mussel *Sphaerium solidum* found in sandy substrata of large rivers.
Source: photos (a), (b) and (d) with kind permission of Jan Hamrsky, lifeinfreshwater.net; photo (c) Joel Berglund from Wikipedia commons.

members of the families Pleuroceridae and Viviparidae (Prosobranchia), and Ancylidae, Lymnaeidae, Physidae and Planorbidae (Pulmonata). Generally, snails favour streams with a high calcium concentration, since this is essential for the construction of their shell. A notable exception is the family Valvatidae, which is most common in waters with low carbonate concentration (Pennak 1989).

Species vary in their habitat preferences but most snail species are found in shallow waters. Ancylids (freshwater limpets; Figure 3.7a) occur on rock surfaces. Spiral-shelled snails include the pleurocerids, found on rocky or sandy sediments; physids and hydrobiids, on vegetation; and the lymnaeids, which

are quite unspecialised with regard to habitat. A detailed review of the taxonomic representation and distribution of gastropods is given by Strong et al. (2008).

Snails are almost entirely micro-herbivorous and/or micro-omnivorous, feeding on biofilms and periphyton which is scraped away with a special toothed or file-like rasping organ (the radula). Some physids and planorbids may, however, include large amounts of detritus in their diet. The Viviparidae (Figure 3.7b) can also filter feed on suspended organic microdebris. Practically all species reproduce sexually and most lay egg capsules from which emerge free-crawling juveniles. However, some species brood live young (e.g.

Viviparidae) while others, particularly associated with coastal streams, produce free-swimming veliger larvae that may develop at sea (e.g. Neritidae). The abundance and biomass of snails may be very high in streams. *Juga silicula*, a North American pleurocerid, can make up more than 90% of the invertebrate standing crop in some streams, leading to negative competitive relationships between the snails and other benthic invertebrates, the latter increasing in numbers after snail removal (Hawkins & Furnish 1987).

3.4.4 Mollusca: Bivalvia

Bivalve molluscs are often common in fresh waters (Figure 3.8), yet are also amongst the most threatened of invertebrate groups globally, with high extinction rates (Lopes-Lima et al. 2018). The most important freshwater bivalves are unionaceans (the freshwater mussels, including Unionidae and Margaritiferidae) with over 900 species identified, and sphaeriids (including *Sphaerium* and *Pisidium*) (Figure 3.7c, d). The smaller family Dreissenidae contains several highly invasive species (including the zebra and quagga mussels *Dreissena* spp.). Out of the 260 native and six introduced freshwater species in North America, which has

the most diverse bivalve fauna in the world, 227 are unionids. Europe has some 50 species.

Freshwater mussels of the Unionidae are an ancient group. Many species have very restricted distributions, making them especially vulnerable to extinction from a variety of sources. For example, the number of species in the Tennessee river dropped from 100 before construction of a large impoundment in 1936 to less than 45 subsequently (Pennak 1989). Throughout most of its range, the freshwater pearl mussel, *Margaritifera margaritifera* (Figure 3.7c), is threatened with extinction, while attempts at its conservation are widespread. Mussels in general have a long lifespan, some reaching more than 100 years. Topic Box 3.2 by David Strayer provides an overview of the biodiversity and the distinctive role of unionids in stream and river ecology as well as the threats to them posed by human activities [see also Graf & Cummings (2007) for a review of the systematics and diversity of freshwater mussels and Vaughn & Hoellein (2018) for a review of bivalve impacts on aquatic ecosystems].

In contrast to unionaceans, sphaeriids are more widely distributed, suggesting that their dispersal capacities may be fundamentally different (McMahon 1991). Unionaceans have unusual life cycles and show

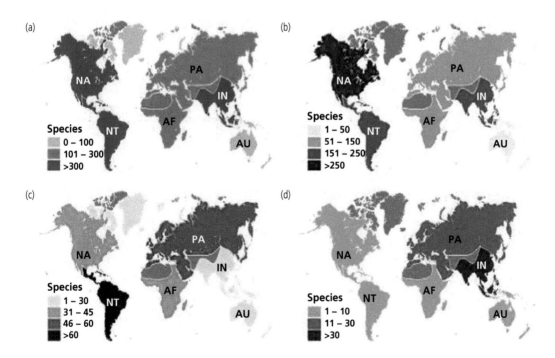

Figure 3.8 Diversity of bivalve molluscs by ecoregions. (a) All freshwater bivalves; (b) Unionida; (c) Sphaeriidae; (d) Cyrenidae plus remaining freshwater bivalve groups. NA Nearctic, NT Neotropical, PA Palaearctic, AF Afrotropical, IN Indo-tropical, AU Australasian. Glaciated and desert areas void of mussels in grey.
Source: from Lopes-Lima et al. 2018, with permission from Springer Nature.

Topic Box 3.2 The biodiversity and role of unionids in stream and river ecology

David Strayer

The Unionida, or freshwater mussels, includes six families and > 900 species of large bivalves in rivers, streams and lakes around the world (Graf & Cummings 2021). Their distinctive biology both gives them a unique place in freshwater ecology and makes them especially sensitive to human activities, so that many species are now imperilled or extinct.

In general, unionids are large-bodied and long-lived (Haag 2012). The sedentary adults typically are 2–20 cm long with lifespans of 5 to more than 100 years. Sexes are usually separate, sexual maturity is attained after 1–10 years, and adults reproduce every year or two thereafter. The distinctive larvae (glochidia or lasidia) are produced in prodigious numbers (often 10^4–10^6 larvae y^{-1}), and are obligate parasites of fish (or occasionally amphibians; Box Figure 3.2a). Some unionids are very host-specific, whereas others can use dozens of fish species as hosts (Strayer 2008; Haag 2012). As far as is known, early juveniles deposit-feed, and later juveniles and adults suspension-feed on phytoplankton, small zooplankton, detritus and bacteria (Strayer 2008). There are exceptions to nearly all of these generalisations. The substantial variation across species in traits such as body size, lifespan, fecundity, and number and identity of fish hosts correlates with habitat use and conservation status of individual species (Haag 2012).

Unionids are dispersed chiefly on fish, and their distributions often are dispersal-limited—that is, both unionids and their hosts may move rapidly through a river system that is free from barriers, but may not be able to surmount obstacles such as waterfalls, dams and drainage divides, even if given thousands of years (Strayer 2008). As a result, ancient, isolated drainage basins (e.g. the Alabama River basin of the south-eastern United States) often contain endemic species with small geographic ranges (Haag 2012). Furthermore, if a mussel population in an isolated site is eliminated, whether by natural causes or human activities, it may not be able to re-establish through natural dispersal. This problem is especially severe in modern river systems, many of which have been dismembered into small, isolated segments by dams and other human-made barriers (Strayer 2008; Fuller et al. 2015).

Several attributes of freshwater mussels give them distinctive roles in freshwater ecosystems (Box Figure 3.2b) and expose them to threats from humans. Unionids often live in more or less distinct beds, which can be large and dense. Although many beds cover just a few square metres at densities of 1–10 adults m^{-2}, mussel beds may cover many hectares at densities around 100 m^{-2}, especially in large rivers. For instance, before they were decimated by the dreissenid invasion,

(a) (b) (c) (d)

Box Figure 3.2a Examples of mussel lures and conglutinates, and interactions of these structures with fish. (a) Mantle lure of the pocketbook mussel *Lampsilis cardium*. (b) Redeye bass (*Micropterus coosae*) attacking mantle lure of *Lampsilis cardium*. (c) Conglutinates (packages of mussel larvae that look like fish food, in this case, larval fish) of the Ouachita kidneyshell *Ptychobranchus occidentalis*. (d) Orangethroat darter (*Etheostoma spectabile*) feeding on conglutinates of *Ptychobranchus occidentalis*.
Source: from Strayer et al. (2004), after original photographs by Wendell Haag and Chris Barnhart.

Topic Box 3.2 *Continued*

beds in the upper Hudson River estuary contained 720 million freshwater mussels over 9.6 km^2, with local densities > 400 m^{-2} (Strayer et al. 1994). Such large mussel populations filter huge volumes of water, which can reduce phytoplankton biomass, increase water clarity, and remove harmful human pathogens from the water (Vaughn 2018).

The material captured by freshwater mussels is released as undigested particles ('biodeposits'), accumulated in mussel tissue and shell, or released as soluble nutrients or CO_2. Mussel beds thus focus large amounts of plankton and other particles captured from the water column onto small areas of the streambed. This makes mussel beds hotspots for nutrient regeneration, microbial activity, periphyton growth, and invertebrate and fish populations (Vaughn 2018). Mussel shells may persist for decades, depending on local physical forces and water chemistry, and the combination of living mussels and accumulated spent shells provides habitat for other organisms, and probably affects sediment stability and dynamics.

The high density of sedentary freshwater mussels in beds makes them easy for humans to harvest for food, pearls and shells. These uses are ancient (doodles scratched by *Homo erectus* on a freshwater mussel shell date from 500,000 years ago), but is it probably only in the past few centuries that such harvests have had strong, widespread effects on mussel populations. Pearl fishers nearly eliminated local mussel populations at many sites in Europe and North America, and the button industry harvested immense numbers of mussels from American rivers. For instance, 13 million kg of shells from live mussels were taken from the state of Illinois in a single year, and 100 million mussels were taken from a single 73-ha mussel bed in the Mississippi River. Shell and pearl fisheries in the United States between 1857 and 1963 were worth ~ $10 billion (Strayer 2017). As staggering as these harvests were, they probably caused local to regional depletions of mussel populations rather than global extinctions.

Other human activities had more severe, long-lasting and widespread effects on freshwater mussel populations (Strayer 2008; Haag 2012). Dams changed riverine habitats; altered flow, sediment and temperature regimes; and blocked fish migration. Land-use change led to changes in water and sediment budgets and pollution by agricultural fertilisers and pesticides. Point-source pollution from

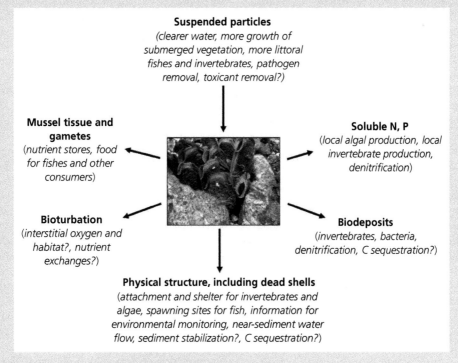

Suspended particles
(clearer water, more growth of submerged vegetation, more littoral fishes and invertebrates, pathogen removal, toxicant removal?)

Mussel tissue and gametes
(nutrient stores, food for fishes and other consumers)

Soluble N, P
(local algal production, local invertebrate production, denitrification)

Bioturbation
(interstitial oxygen and habitat?, nutrient exchanges?)

Biodeposits
(invertebrates, bacteria, denitrification, C sequestration?)

Physical structure, including dead shells
(attachment and shelter for invertebrates and algae, spawning sites for fish, information for environmental monitoring, near-sediment water flow, sediment stabilization?, C sequestration?)

Box Figure 3.2b Summary of ecosystem services that might be provided by freshwater mussels.
Source: modified from Strayer 2017), based on the ideas of Caryn Vaughn and others. Functions marked with a question mark probably occur but have not yet been definitively demonstrated.

Topic Box 3.2 *Continued*

factories and cities added sewage and toxic chemicals to rivers. Careless introductions of invasive species such as zebra mussels (Lucy et al. 2014) and now (perhaps) viral diseases (Richard et al. 2020) have decimated what remains of unionid populations in many lakes and rivers. The traits of freshwater mussels left them unable to escape when human activities degraded or destroyed a habitat; unable to recolonise a habitat if it was isolated by barriers; vulnerable to a very wide range of problems (e.g. dissolved pollutants, particle-bound pollutants, changes in sediment regimes, scouring or drying from altered flows); and vulnerable to losses of host fish populations due to human activities. As a result of these unique sensitivities, many unionid species disappeared, and many more are imperilled. The IUCN Red List has been able to assign a conservation status to only about half of the world's freshwater mussels; of these, 7% are globally extinct, and 38% more

are threatened with extinction. In contrast to other taxonomic groups, small-bodied mussels are at far higher risk of extinction than larger mussels, for unknown reasons (Strayer 2008). The large, widespread losses of mussel populations must have led to correspondingly large losses in the ecosystem processes and services described above, although the global extent of such losses has not been estimated.

On the brighter side, scientists recently learned to propagate freshwater mussels, including some very rare species, setting the stage for reintroduction of mussels to sites from which they were lost (Patterson et al. 2018). If the threats that destroyed the mussel populations in the first place are ameliorated, such reintroductions could restore mussel populations and their ecological functions to some of the world's streams and rivers.

Dr. David Strayer is at the Cary Institute of Ecosystem Studies, Millbrook, New York, USA.

parental care (brooding); spread primarily in their larval (glochidia) stage, which is ectoparasitic on the gills of fish; and are often dispersal-limited. In contrast, sphaeriids may disperse efficiently as passengers on birds, salamanders and insects and hence can more readily cross catchment boundaries. Their establishment is facilitated by the fact that only a single individual is needed to found a new population, as they are self-fertilising hermaphrodites. Many unionaceans inhabit large and fast-flowing rivers with a coarse sand or gravel substratum with little siltation. In contrast, members of the most species-rich genus of sphaeriids, *Pisidium*, prefer a substratum of finer particles.

Most mussels filter fine particles of organic matter suspended in the water. Cilia, and perhaps mucus, are instrumental in the capture of particles. Some species may feed from organic matter that is resuspended from the stream bed through 'pedal feeding', where benthic particles are collected by the foot and passed to the digestive tract (Brendelburger & Klauke, 2009). Bivalves may also be significant as consumers of riverine phytoplankton. Many of the larger vertebrate predators, including some fish, feed on bivalves.

3.4.5 Larger Crustacea—decapods, amphipods and isopods

Many kinds of crustaceans are associated with running waters and in many tropical and subtropical island

communities they commonly replace the ecological role of insects (see Topic Box 1.2 by Alan Covich). Here we emphasise the larger Malacostraca that includes the orders Decapoda, Amphipoda and Isopoda, primarily because these are very important detritivores and predators in many lotic communities.

Decapoda are among the largest running-water invertebrates. They may conveniently be divided into ('true') shrimps (Caridea), crayfish (Astacoidea and Parastacoidea) and crabs (Brachyura). Decapods inhabit all sorts of aquatic environments, but many species are confined to lotic environments. Crabs, atyid shrimps and prawns (Palaemonidae) have an important position in tropical Asian rivers and have penetrated fast-flowing upland streams (Dudgeon 2000).

Shrimps are common in the tropics and subtropics, especially in streams with lush vegetation and slow currents, where they are primarily filterers and grazers although some are scavangers/detritivores (Figure 3.9a). There are two main genera of palaemonid shrimps in sluggish streams. *Macrobrachium* is species rich, several of which are widely cultured for food. The second main genus, *Palaemonetes*, seems to be excluded by *Macrobrachium* from locations between 30° N and 30° S, except at higher altitudes.

Freshwater crabs are also restricted mainly to the warmer parts of the world, with about 1,400 known species. In Africa, for instance, the river crab genus *Potamonautes* contains more than 40 species. Southern

Europe is the home of three crab species (in the genus *Potamon*), while North America has only one (*Platychirograpsus typicus*), which occurs along the Mexican east coast and into Florida. Freshwater crabs are omnivorous (with some rather more herbivorous and some carnivorous) and include species that feed on fallen leaves, aquatic insects, gastropods, dead frogs or even snakes. The detritivorous species are important in nutrient-cycling in tropical fresh waters (Cumberlidge et al. 2009).

Of 640 extant species of freshwater crayfish, the Northern hemisphere superfamily Astacoidea (Figure 3.9b) includes nearly 460 species, the vast majority in the Nearctic, while the Southern hemisphere superfamily Parastacoidea has over 170 species, nearly all in the Australasian region and none in the Oriental realm (Crandall & Buhay 2008). A hotspot of diversity for crayfish is in the Appalachian mountains of the south-east United States and another is in southeast Australia. In contrast, Europe has less than half a dozen indigenous species. Most crayfish live in lakes and ponds, although some species are confined to flowing, sometimes even swift, streams. These forms typically have a large abdomen for swimming and do not tolerate a low oxygen concentration (Crandall & Buhay 2008). Shelter from the current is important and some burrow into the stream bed or into the stream banks or hide under boulders and cobbles. They typically live for about two years and are usually nocturnal. Most species are omnivorous and may be key processors of organic material, primarily allochthonous plant litter. They may also have strong effects as herbivores (see Chapter 7). In Europe and North America, non-indigenous crayfish have eliminated or reduced native crayfish, amphibians, invertebrates and aquatic vegetation in lakes and streams, sometimes displacing fish and invertebrates that use similar resources (Klose & Cooper, 2011). They also eat the eggs of some fish. Crayfish are fed on by a range of predators including herons, muskrat, mink, reptiles, amphibians, fish and dragonfly larvae. Humans eat crayfish in some parts of the world, rarely causing extinction but probably changing crayfish population structure. However, populations of European crayfish (*Astacus astacus* and, most significantly, the White-clawed crayfish *Austropotamobius pallipes*) have been severely reduced over most of their distributional range by the 'crayfish plague', a fungus brought in through the introduction of around 10 species of non-native North American species which have proved invasive and resistant to the disease (Olsson et al. 2009, see Chapter 7). This has led to the introduction of the European Union (Invasive Alien Species) (Freshwater Crayfish) regulations 2018 (SI 354/18). Ireland holds one of the largest

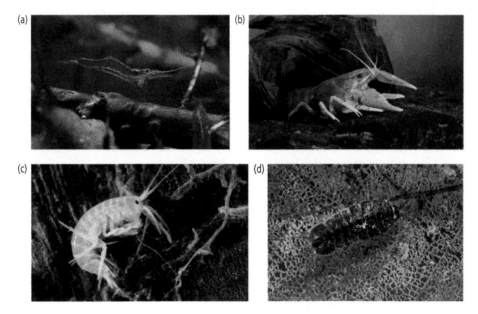

Figure 3.9 Lotic crustaceans. (a) The detrivorous yellow-nosed shrimp *Xiphocaris elongata* from Puerto Rico streams (photo courtesy of Alan Covich); (b) Noble or European crayfish (Decapoda—*Astacus astacus*); (c) freshwater shrimp (Gammaridae—*Gammarus* spp); (d) water hoglouse (Asellidae—*Asellus aquaticus*).
Source: photos (b), (c) and (d) by Jan Hamrsky, lifeinfreshwater.net.

surviving populations of the White-clawed crayfish and is currently (2022) the last European country without any invasive species. In contrast, crayfish are under significant threat in the USA with 20 species restricted to less than five localities (15 species are known only from a single locality), and over 210 species are considered endangered or threatened (Crandall & Buhay 2008).

While most of the *Amphipoda* (the second main order of the Malacostraca) are marine, they are also common in many types of fresh waters, including streams, rivers, springs and subterranean habitats. Amphipods (and also isopods) are must less abundant in Asian streams and rivers than in their north-temperate counterparts (Dudgeon 2000). Vainola et al. (2008) suggested there were 1,870 freshwater amphipod species (45% of them subterranean), but this may well be an underestimate. The global distribution suggests that they are cold stenothermic with 70% of species from the Palaearctic and relatively few in the tropics. The most important family is the Gammaridae (Figure 3.9c), with diversity centred in Europe. Although some gammarids appear to favour high calcium concentration, most amphipods appear to be restricted to streams with low or moderate concentrations.

In some headwater streams amphipods may be very abundant (with densities of several thousand to above ten thousand m^{-2}), especially where there is a stable substratum and ample food (e.g. Pennak 1989). Basically amphipods are omnivores, consuming dead organic material (particularly leaf litter), periphyton and some animal material. Their activity as detritivores, via shredding leaves, is highly significant (e.g. Woodward et al. 2012; see also Chapter 8).

The less diverse Asellidae (order *Isopoda*) has many species that are confined to groundwaters and subterranean rivers. *Asellus aquaticus* (Figure 3.9d) is perhaps the best-known lotic species in Europe. It is detritivorous and thrives in a wide range of habitats including quite heavily polluted rivers, although it is replaced by *A. meridianus* in acid waters. In North America, there are 12 species of *Asellus* in running waters, and six species of *Lirceus*, with the majority of both genera quite restricted regionally (Williams 1976). Again, isopods are much less abundant in Asian streams and rivers.

3.4.6 Other non-insect macroinvertebrates

Two frequently encountered groups in lotic environments are the Oligochaeta (true worms), and Hirudinaea (leeches), most species occupying slowly flowing and depositional marginal habitats.

Practically all freshwater *oligochaetes* are deposit-feeders and there are about 1,000 species. These are mainly small, very thin worms (apart from some larger species resembling earthworms) (Figure 3.10a, b). The Palaearctic region is the most diverse, with some 600 freshwater oligochaete species (Martin et al. 2008). Whilst most species are detritivorous, some naidid oligochaetes, such as *Chaetogaster limnaei*, are symbionts associated with molluscs and feed on trematode parasites (Hopkins et al. 2013; see Chapter 7). Oligochaetes may be very abundant in situations when other macroinvertebrates are absent, particularly in organically polluted systems with low oxygen concentrations. This makes them excellent indicators of organic enrichment, particularly the diverse Tubificidae (Figure 3.10b) with close to 600 freshwater species and a cosmopolitan distribution.

Most freshwater leeches (there are *c.*500 species) are predators, feeding on various invertebrates, although a few are blood-feeding ectoparasites of amphibians, fish, birds and mammals. The Glossiphoniidae (Figure 3.10c) is among the most species-rich of the leech families. They are dorsoventrally flattened and normally feed on the blood of turtles or amphibians, although some species of *Helobdella* and *Glossiphonia* feed on the haemolymph of snails and oligochaetes (Siddell et al. 2005). The Glossiphoniidae also show a remarkable parental care. The adults secrete a membranous bag covering the eggs, which are held on their underside and are fanned until the brood hatches. Young leeches then attach to the underside of the parent and are carried to their first blood meal on a host. European leech species appear highly divergent in terms of their ecological preferences (Koperski 2017): some are common and tolerant of water pollution (e.g. *Erpobdella octoculata*, *Glossiphonia complanata*) while others are more sensitive (*E. vilnensis*, *G. nebulosa*) or highly specialised (*Calliobdella mamillata*, *Trocheta bykowskii*).

3.4.7 Insects

Insects are among the most conspicuous inhabitants of streams and rivers all over the world and their study has been the focus of much stream research. In a few groups, both the larva and adult are aquatic but in most the larva is aquatic while the adult is terrestrial. Typically, the larval stage is prolonged while the the adult is short-lived. There are a number of orders of insects with species living in streams and rivers—some exclusively or predominantly aquatic—while in others most species are terrestrial. In all winged insects there may be a

Figure 3.10 Examples of annelid worms, Oligochaeta: (a) *Chaetogaster* sp. (note also the stalked protozoans in the bottom right), (b) the sludge (or 'red') worm *Tubifex tubifex* (Tubificidae)—red colour is haemoglobin; (c) the leech *Glossiphonia complanata*.
Source: photos (a) and (b) by Jan Hamrsky, lifeinfreshwater.net; (c) photo from MNHN & OFB [Ed]. 2003-2023. National inventory of natural heritage (INPN), Website: https://inpn.mnhn.fr/espece/cd_nom/236044/tab/fiche?lg=en Copyright J. -F.Cart

complete metamorphosis—termed a *holometabolous* life cycle—or an incomplete metamorphosis—termed a *hemimetabolous* life cycle. Holometabolous insects have a pupal stage, and the adult and larva bear no physical resemblance to each other, as the larval tissues are completely reorganised in the pupal stage, when the wings develop. Hemimetabolous insects lack a pupal stage, passing straight from a larva (or nymph) to an adult and the wings develop gradually and externally over the larval instars. Generally, there are more larval instars in hemimetabolous than in holometabolous insects.

The most important hemimetabolous insect groups in running waters are the Ephemeroptera (mayflies, Figure 3.11), Plecoptera (stoneflies, Figure 3.12), Hemiptera (bugs) and Odonata (dragonflies and damselflies, Figure 3.13). Holometabolous groups include the Trichoptera (caddisflies, Figure 3.14), Diptera (true flies, Figure 3.15), Coleoptera (beetles) and Megaloptera (alderflies and dobsonflies, Figure 3.16). A great deal of effort has gone into devising taxonomic keys for the identification of both adults and larvae, and into associating the larva with the 'correct adult'. Species names are given first to the adult stage. The different orders are usually easily distinguished and the identification of most lotic insects is relatively easy in parts of the world where the fauna is more limited and where there is a rich heritage of taxonomic work. In much of the world, however, the fauna is still

relatively poorly known and species identification, particularly of larvae, remains a significant problem. Some groups of insects, such as many Diptera, are particularly problematic—as is the case with some of the small-bodied groups of meiofaunal invertebrates described above. We should not underestimate the great contribution of taxonomy and taxonomists to the population and community ecology of rivers and streams. In addition, many methods of biological assessment of the 'health' of rivers and streams depend on an ability to identify stream invertebrates (see Chapter 10).

3.4.7.1 Ephemeroptera (mayflies)

Mayflies are the iconic stream insects (most species are restricted to running waters) and have a worldwide distribution, absent only from Antarctica and some remote oceanic islands. Most species are more or less restricted to running waters. Just under 3,700 species have been identified to date amongst 40 families and some 470 genera. The greatest diversity is found in the Neotropics (almost 900 species described), followed by the Palaearctic (830), Nearctic and Oriental (610 and 620, respectively), Afrotropical (440), Australasia (250) and the Pacific (48) (Jacobus et al. 2019). Three families are the most frequently encountered (Baetidae, Heptageniidae and Ephemerellidae; Figure 3.11). The larvae can be found on stony substrata, on higher plants or

Figure 3.11 Larval mayflies (Ephemeroptera) from various families: (a) Baetidae; (b) Siphlonuridae; (c) Heptageneidae; (d) Ephemerellidae. Note the three 'tails' and abdominal gills as well as the thoracic wing pads (most obvious in (b)).
Source: photos by Jan Hamrsky, lifeinfreshwater.net.

burrowing in the sediments, with various morphological adaptations associated with the specific lifestyles (see Chapter 4 and Jacobus et al. 2019).

Larval mayflies are characterised by the presence of three 'tails' (including two cerci, and between them (usually) a terminal filament. The variable body shape seen in Figure 3.11 is important in relation to the general microhabitat selection and activity, as discussed in Chapter 4. Mayflies go through many (often 20–30) postembryonic moults. The larvae feed mainly on algae and fine detritus, although a few genera are predatory (e.g. within the families Siphlonuridae, Metretopodidae and Behninigidae). Some members of the families Ephemeridae, Heptageniidae and Siphlonuridae have hair-fringed legs or maxillary palps that filter food particles from the water column. Feeding on leaf litter is rare among mayflies, but it does occur in the Ephemerellidae and Leptophlebiidae.

Respiration involves gills (paired organs borne on up to seven abdominal segments; see Figure 3.11a, b). In many species these gills beat, moving water over their surface. The tolerance to low oxygen concentration varies widely among species, making them useful

indicators of organic pollution. Most species are sensitive to acidity although a few (some Leptophlebiidae) can tolerate pH to values of around 4.

The adult lifespan is short, ranging from a few hours to a few days, rarely up to two weeks, and the adults do not feed. Mayflies are unique among insects in having two winged stages. The subimago, which is in effect the final larval stage, emerges from the water body, finds a resting place and moults to the adult stage (the imago). The emergence of adults is often synchronous (with 'mass emergences', as in some Ephemeridae), leading to large aggregations of mating flying adults. In extreme cases these aggregations may be enormous, with swarms emerging from large lakes and rivers sometimes extending over considerable areas ($\gg$ km^2), perhaps 'swamping' their predators and reducing *per capita* predation rate (Sweeney & Vannote 1982).

3.4.7.2 Plecoptera (stoneflies)

This is a widely distributed, ancient and rather homogeneous order with around 3,500 species belonging to 16 families, described from all continents except Antarctica (Fochetti & Tierno de Figueroa 2008).

Figure 3.12 Stonefly larvae (Plecoptera) from different families: (a) Perlodidae; (b) Nemouridae; (c) Leuctridae; (d) Taeniopterygidae.
Source: photos by Jan Hamrsky, lifeinfreshwater.net.

Almost 90% of species are found in the northern hemisphere, though there appears to be high endemicity in the tropics where the discovery of more species is to be expected. Stoneflies, like mayflies, have a hemimetabolous life cycle but are readily distinguishable by having two long cerci ('tails') (Figure 3.12). They are best represented in the temperate regions of both hemispheres or at higher altitudes in the tropics. Most stonefly larvae occur in stony streams, often in the interstices of the substratum or in leaf packs, and are characteristically inhabitants of cool, well-oxygenated, low-order streams, although several taxa occur in alpine and high-latitude lakes and a few in larger rivers. While they are sensitive to organic pollution and low oxygen concentrations, some (though by no means all) stoneflies tolerate acidic conditions. Many species lack external gills, presumably associated with their intolerance of low oxygen.

Their importance for ecosystem processes relates largely to their feeding activities. One group of (northern hemisphere) families, including Capniidae, Leuctridae, Nemouridae and Taeniopterygidae, are at least partially detritivorous, their life cycle matching autumnal leaf fall. Another group, again mainly from the northern hemisphere and including Chloroperlidae, Peltoperlidae, Perlidae, Perlodidae and Pteronarcyidae, consists chiefly of predators of smaller invertebrates. Many of these predatory species, some of which grow quite large (up to 3 cm length), may actually shift from herbivory on periphyton to carnivory in the course of their ontogeny (see also Chapter 7), whilst others may feed opportunistically on whatever is available. A detailed review of the plecopteran trophic ecology is presented by Tierno del Figueroa & Lopez-Rodriguez (2019).

Life cycles vary from several generations per year to one generation per two or even more years. Many stoneflies grow slowly in winter and some species have a larval diapause, though many grow well at temperatures near 0°C, an ability most common amongst the detritivores feeding on leaf litter. There is even an unusual winter species (*Mesocapnia arizonensis*), an intermittent stream specialist, whose nymphs are abundant within days of rewetting of the stream channel and adults start to emerge 42 days after flow resumes (Bogan 2017). Some stoneflies reach a considerable size; for example, the detritivorous nymphs of North American *Pteronarcys* may grow to 50 mm in length and the development takes several years. At the other end of the size spectrum, full-grown nymphs of several families are found that barely exceed 5 mm in length and the life cycles of these are completed within a year.

Adult stoneflies have a short lifespan of a few days to a few weeks and are poor fliers, probably explaining their virtual absence on oceanic islands. Often stoneflies have short wings (brachypterous) or are wingless (apterous)—this character increasing with altitude. Brachypterous stoneflies may use the wing flaps for 'sailing' on the water surface and for aerial gliding (Marden & Kramer 1995). Some stonefly adults feed on algae, leaves, buds and pollen and fungal spores. Interestingly, there appears to be quite complex mate-searching behaviour in some species, with adults attracting partners by drumming their abdomens against leaves or wood.

While some groups and species of Plecoptera are well known, this applies to a minority and, for the rest, much remains to be discovered of their life history, trophic interactions, growth, development, spatial distribution and behaviour (Fochetti & Tierno de Figueroa 2008). Of course, this applies to most groups of freshwater insects globally.

3.4.7.3 Hemiptera (bugs)

Water bugs belong to the large, mainly terrestrial (>90% of the species) order Hemiptera and all are found in the suborder Heteroptera. In total, 20 families, 326 genera and 4,656 species inhabit fresh water and occur on all continents except Antarctica, although they are most numerous in the tropics (Polhemus & Polhemus 2007). Few are lotic, but several are semi-aquatic in the sense that they live on the surface of pools, slowly flowing reaches or near the stream margins (Gerromorpha or 'water striders', Figure 3.13a). In slow-flowing streams some species are truly aquatic (e.g. 'backswimmers' (Notonectidae, genus *Anisops*) and 'lesser water boatmen' (family Corixidae); Figure 3.13b). Tropical streams and rivers often have many species, particularly in pools and backwaters.

Hemipterans usually have five nymphal stages, and most (except for the majority of corixids) are predatory on other insects. They pierce prey with specialised maxillary stylets and inject often highly toxic compounds in the saliva, paralysing the victim rapidly. Water striders prey on terrestrial or emerging insects caught in the surface film, locating the prey through vision and by vibrations on the water surface (see more in Chapter 4).

Figure 3.13 Water bugs and odonates: (a) water strider (Gerridae); (b) lesser waterboatman (*Corixa punctata*) scavenging on a *Baetis* nymph; (c) dragonfly nymph (*Cordulegaster boltonii*, Cordulegastridae) common in UK upland streams and rivers; (d) the lotic broad-winged damselfly nymph (*Calopteryx virgo*, Calopterygidae).
Source: photos by Jan Hamrsky, lifeinfreshwater.net.

3.4.7.4 Odonata (dragonflies and damselflies)

These are conspicuous, hemimetabolous insects with aquatic larvae and charismatic terrestrial adults. About 5,680 species are known worldwide, although the number could be closer to 7,000 according to Kalkman et al. (2008). There are two suborders with approximately similar numbers of recognised species: Anisoptera ('dragonflies') and Zygoptera ('damselflies') (Figure 3.13c, d). Anisoptera adults are stout, strong-flying insects; the fore- and hindwings are of different sizes and always held open. In contrast, Zygoptera adults are slender with similar-sized fore- and hindwings that are held closed above the body at rest. The larval body shape resembles that of the adults and is much larger in the dragonflies, whilst the Zygoptera possess 2–3 leaf-like caudal lamellae that aid respiration.

The highest diversity of lotic odonate species is found in flowing waters in tropical rainforests, particularly in the Oriental and Neotropical regions. The adults are often large, colourful and day active, and are efficient territorial predators that patrol around aquatic habitats. They are also one of the best studied of the insects in terms of reproductive behaviour because of the external secondary copulatory organs that not only transfer sperm to the female but also remove sperm depositied by previous males—a form of sperm competition. In the larval stages most dragonflies are found in lentic or slow-flowing habitats, although some are lotic specialists (e.g. *Zygonyx torrida* and *Cordulagaster boltonii*). Development involves 10–15 moults, which can involve a larval life of between a few weeks in some Zygoptera up to seven years in some Anisoptera. A detailed review of the life history and behaviour of the Odonata can be found in Corbett (1999).

Dragonfly larvae are voracious predators and they include various invertebrates as well as vertebrates (such as small fish and amphibians) in their diet. Prey are grasped with the characteristic long, hinged labium—the 'labial mask'—that is rapidly projected out as the prey are grasped between the labial claws. Prey are either stalked, captured using a sit-and-wait strategy (Zygoptera, Aeshnidae) in the vegetation or pursued by more active hunters (most Anisoptera) crawling among fine organic material on the substratum. The larvae (and indeed the adults) possess binocular vision which aids in prey capture. Due to their large size, odonates are often the favoured prey of fish. In fact, many species thrive best in the absence of fish.

3.4.7.5 Trichoptera (caddisflies)

The holometabolous Trichoptera is more speciose than all the other primarily aquatic insect orders combined (Morse et al. 2019), occupying the whole range of major freshwater environments and with a worldwide distribution (apart from Antarctica). Informally, caddisflies have been categorised broadly into species with 'free-living' (caseless and net-spinning) and those with 'case-building' larvae (Figure 3.14), and these differ from other holometabolus insect larvae by having segmented legs and anal prolegs terminating in claws (see section 4.2.3). The fixed retreats and nets of caddis larvae are made from silk (from modified salivary glands—see section 4.2.4 and Figure 4.11) and cases are usually combined with sand grains and other small mineral and/or organic, usually plant-derived, particles (Figure 3.14a and see Figure 4.12). Larval constructions often distinguish genera or families.

Trichoptera are taxonomically close to Lepidoptera and the adults are indeed moth-like, differing by the possession of hairs on the wings rather than the scales of moths and butterflies. Taxonomic work on the order has moved apace, with new species being discovered quite quickly; the latest data lists 16,266 extant species across 618 genera of 51 families (Morse et al. 2019). There is a fairly even number of trichopteran families distributed across the biogeographic regions, but Southern Asia (Oriental region) has by far the greatest number of species (close to 6,000) and the neighbouring East Palaearctic the lowest (1,200). The success of caddisflies is undoubtedly related to their capacity to spin silk (see Chapter 4) as well as the flexibility they show across the order in life history and diet. However, the Trichoptera have also been reported to be suffering a relatively high rate of species loss due to anthropogenic factors, especially in well-studied parts of the world (Europe and North America) (Sanchez-Bayo & Wyckhuys 2019).

Caddisflies are have five to eight larval instars and some caddis larvae attain considerable size (e.g. the cased Phryganeidae can reach up to 5 cm), while the Hydroptilidae (microcaddis) rarely exceed 6 mm. Similarly, adult size range extends from a wingspan of less than 3 mm to close to 100 mm. The adults will feed on nectar and are usually fairly short-lived (less than a month), though in larger species in the family Limnephilidae, the strongly flying adults may live for three months (Wiggins 1973).

There is often a single flight period per year, described as 'univoltine' (the individual lifespan being

Figure 3.14 Larvae from three families of larval Trichoptera (caddis): (a) case-building (Limnephilidae); (b) 'free-living' (Rhyacophilidae); (c) net-spinning (Polycentropodidae).
Source: photos by Jan Hamrsky, lifeinfreshwater.net.

about one year), though some species are bivoltine with two generations per year and others take longer than one year to complete the life cycle. Life-history plasticity can be found, as seen in one of the most detailed studies of caddis life history, growth patterns and adult flight periods for 14 species in a small Irish stream (Sangpradub et al. 1999). As discussed in Chapter 4 (section 4.4) this may be evidence of a 'spreading of risk' against environmental variation within life histories. Similar life-history variation has been found in trichopterans in high-altitude pategonian steams (Brand & Miserendino 2011).

Trichopteran species are found across a whole range of lotic habitats, from slow-flowing waters with debris accumulations to patches of vegetation, gravel bars and rocks. Some of the smaller cased caddis species are more or less sedentary on rock faces, and the net-spinning caddis construct nets between stones, moss stems or on hard surfaces. Their diets are equally varied. While many species ingest a wide variety of food, feeding may be highly specialised. The families Rhyacophilidae and Polycentropodidae for example are predatory; the cased families Limnephilidae, Lepidostomatidae and Sericostomatidae feed mainly on allochthonous leaf litter; Hydroptilidae pierce vegetation; the net-spinning Hydropsychidae

and Philopotamidae are filter feeders; and the cased Goeridae, Glossosomatidae and Helicopsychidae feed largely on periphyton. In addition, some taxa (e.g. the family Psychomyiidae) may actually garden and/or defend patches of biofilm around (or on) their (fixed) silken shelters. One striking feature of larval trichopterans is that their diet can change dramatically with age ('ontogenetically'). Limnephilid larvae, for example, are often considered to be detritivorous but penultimate and final instars have been shown to be voracious predators (Giller & Sangpradub 1993).

3.4.7.6 Diptera (true flies)

This is a very diverse and widespread order of insects. Most dipterans are terrestrial, although almost one-third of all fly species (roughly 46,000) have some developmental connection with an aquatic (almost exclusively freshwater) environment (Adler & Courtney 2019). As in most aquatic insects, the predominant aquatic phase of the life cycle is the larva, while the adults are terrestrial. The larvae of aquatic dipterans (Figure 3.15) are easily recognisable by their lack of jointed thoracic legs, although two types of 'false legs' may be present: ventral prolegs on the thoracic and anal segments (sometimes also on the abdominal segments) and 'creeping welts', transverse ridges

Figure 3.15 Larvae from the three most important lotic groups of Diptera: (a) Simuliidae (with a pupa on the right—note the spiracular filamentous gills); (b) Chironomidae (predatory midge, Tanypodinae); (c) Chironomidae (tube-building midge, *Rheotanytarsus* sp.); (d) Tipulidae (*Tipula* sp.).
Source: photos by Jan Hamrsky, lifeinfreshwater.net.

usually on the ventral margins of the first seven body segments. Dipterans have colonised freshwater habitats on every continent including Antarctica (some Chironomidae). They can inhabit some of the most inhospitable environments, from beneath anchor ice in Canadian rivers (Blethariceridae—the net-winged midges) to thermal springs over 70°C (some Chironomidae and Stratiomyidae). Some Chironomidae have even been found at a pH as low as 2 (Adler & Courtney 2019). Their numerical abundance, geographical range, tolerance to extreme environments and diversity of adaptations makes them one of the most significant groups in rivers and streams.

Many species of the suborder Nematocera (mostly aquatic) have a sclerotised head capsule with stout, toothed mandibles (e.g. Chironomidae). In contrast, members of a second suborder, Brachycera (mostly terrestrial), lack a head capsule, and their mandibles are claw-like and can be withdrawn into the head. Table 3.3 highlights the diversity and predominant feeding mode and habitat of the major dipteran families associated with lotic habitats. In light of their importance in running waters, it is useful to highlight three families that are all lotic (Simuliidae, Figure 3.15a) or have a substantial presence there (Chironomidae, Figure 3.15b, c; and Tipulidae, Figure 3.15d).

Simuliidae

While about 1,650 species of Simuliidae (blackflies) had been described by the late 1990s, there are now over 2,300 recognised (Table 3.3). The family is a morphologically homogeneous group and with very few exceptions the larvae are restricted to running water, filtering small particles from the water using 'head fans' originating from lateral brushes of hairs borne on the insect's 'upper lip' or labrum (Figure 3.15a; see also Figure 4.10). Larvae temporarily attach to the substratum, anchoring their abdominal proleg, which carries large numbers of hooklets, into a silken pad which they stick to the surface of suitable stones (see Figure 4.10).

Blackfly larvae are ecologically important and can occur in enormous densities (values of over 600,000 m^{-2} have been recorded in a lake outlet stream; Wotton et al. 1998). They collect large quantities of minute (< 5 μm) suspended particles which rapidly pass through the gut and are transformed into faecal pellets of > 50 μm diameter. Blackflies can have a major impact on particle transport and carbon cycling, or 'spiralling', in rivers through the retention of organic matter by increasing its particle size and sinking rate thus retarding its downstream transport (Malmqvist et al. 2001). Simuliids are important prey for predatory stoneflies and caddis larvae, for fish such as 'sculpins' or 'bullheads' (Cottidae) and for insectivorous birds

Table 3.3 The major dipteran families associated wholly or partly with lotic and related habitats, illustrating the total number of recognised species and the number of aquatic representatives. The major feeding mode is also indicated. Families are labelled as members of either the Nematocera (N) or the Brachycera (B) (modified from Aldler & Courtney 2019).

Family or superfamily	Total species	Aquatic species	Predominant feeding mode and habitat
Ceratopogonidae (N)	5,902	5,182	FPOM feeders and predators; diverse lentic and lotic
Chironomidae (N)	7,290	7,090	All trophic groups; all aquatic habitats
Simuliidae (N)	2,335	2,335	Filter feeders; small streams to large rivers
Blephariceridae (N)	330	330	Grazers; lotic (rocks)
Tipuloidea/Tipulidae (N)	15,803 4,413	11,062	All trophic groups; lentic and lotic
Athericidae (B)	133	133	Predators; lotic
Tabanidae (B)	4,434	4,434	Predators; lentic and lotic
Dolichopodidae (B)	7,358	3,182	Predators; lentic and lotic
Empididae (B)	3,142	671	Predators; lotic
Muscidae (B)	5,218	701	Predators; lentic and lotic

such as dippers and harlequin ducks. The terrestrial adult female blackflies feed on the blood of vertebrates and, in some tropical areas, transmit a nematode parasite that causes 'river blindness' (onchocerciasis) in humans and cattle, a potentially extremely serious, although not lethal, disease. They also transmit leucocytozoonosis, a protozoan blood parasite often fatal to birds.

Chironomidae

The Chironomidae (non-biting midges; Figure 3.15b, c) is recognised as one of the most speciose freshwater insect groups. Estimates can range up to 20,000 species for the family as a whole, although the exact number is uncertain for various reasons (Ferrington 2008). A recurrent problem is the difficulty of correctly identifying chironomids conventionally: keys are incomplete and usually only 'work' for final-larval-instar larvae or for certain regions. The Global Biodiversity Information Facility (https://www.gbif.org/) currently lists only 7,821 species. The Chironomidae is now subdivided into eleven subfamilies, of which the Tanypodinae, Diamesinae, Prodiamesinae, Orthocladiinae and Chironominae are the most important in relation to lotic environments.

Chironomids are the most widely distributed and often the most abundant group of insects in fresh water. They are found in cold, glacier-fed streams, springs and 'madicolous' habitats (thin layers of flowing water over rocky surfaces such as waterfalls), and from the smallest streams to the largest rivers. They

occur up to 5,600 m altitude in the Himalayas, and their latitudinal distribution spans continental Antarctica (68° S) and Ellesmere Island in the Northwest Territory of Arctic Canada (81° N). This is largely due to the fact that, as a group, they are the most tolerant of the freshwater insects in relation to temperature, pH and oxygen concentration. The life cycle varies between and within sub-families, although most species have four larval instars and appear to be univoltine to trivoltine in seasonal environments (Tokeshi 1994). Development rate varies greatly between species, ranging between egg to adult in < 7 days to a 7-year life cycle (Ferrington 2008).

The diet varies greatly between species, although most are detritivores (Berg 1995), including both free-living (Figure 3.15b) and tube-building (Figure 3.15c) forms. Chironomid tubes can be seen on rock surfaces, plant leaves and in muddy substrata and are often the most obvious indication of their presence. Often the tubes are regularly dispersed, indicating some degree of territoriality. In the subfamily Chironominae several genera, including *Chironomus* and *Rheotanytarsus* (Figure 3.15c), spin silken catch nets associated with the tubes to filter suspended particles. Many orthocladiines and diamesines are grazers and scrape algae (particularly diatoms) from rocks and higher plants. A number of Orthocladiinae feed on living vascular plants and macroalgae, whilst others shred leaf litter. Some orthoclads and chironomids also feed on wood. The predatory species are mainly found in the subfamily Tanypodinae (Figure 3.15b), which are amongst the largest of the chironomids and have the largest head capsules. Yet others live as commensals or parasites. For example, *Epoicocladius ephemerae* lives on the body of the burrowing mayfly *Ephemera danica*, and *Eukiefferiella ancyla* builds tubes on the inner rim of the shell of the limpet *Ancylus fluviatilis* (see Chapter 7). Chironomids are also important prey and most studies on predation and natural diets in running waters find that chironomids are victims of virtually all types of predators—often more than any other prey.

Tipulidae

The Tipuloidea (craneflies) is a very large superfamily, with over 15,200 species worldwide, and is among the most evolutionarily 'primitive' of the dipteran lineages (de Jong et al. 2008). Most have a larval stage restricted to terrestrial or semi-aquatic environments, but many species do occur in running waters, occupying a variety of habitats, from madicolous systems to small streams and large rivers. Our ecological knowledge of tipulids is, however, rudimentary for most species. Tipulids have four larval instars and the majority of species are univoltine or bivoltine (one or two generations per year), with most aquatic larvae requiring six–12 months for development. However, the life cycle can be as short as six weeks in some species and up to five years in Arctic species (Byers 1996).

The head capsule of larval tipulids is particularly characteristic in that it is incompletely sclerotised and can be retracted into the thorax (Figure 3.15d). Larvae of the subfamily Tipulinae are large (some reaching several centimetres in length) and have a tough cuticle (they are sometimes called 'leather jackets'), while most Limoniinae have an almost transparent cuticle. Aquatic tipulids are important in the decomposition of organic material, many feeding on decaying leaf detritus. *Tipula* may be quite dependent on a specialised flora of bacteria residing in the hind gut (Pritchard 1983). In the Limoniinae, there are also some important predaceous genera, notably *Dicranota*, *Hexatoma*, *Limnophila* and *Pedicia*. Cylindrotominae are mainly herbivorous (moss, higher plants) or feed on leaf detritus.

3.4.7.7 Coleoptera (beetles)

The beetles are an enormously species-rich order of mainly terrestrial insects, the 13,000 or so aquatic species representing less than 4% of the total number. Both lotic and lentic habitats are occupied, species in streams and rivers typically having smaller geographical ranges (Bilton et al. 2019). Usually both the larval and adult stages are aquatic.

The well-known riffle beetles are an arbitrary collection of taxa primarily including the entire family Elmidae (with over 1,300 described species, Figure 3.16a, b) and riffle-inhabiting members of Dryopidae, the genus *Lutrochus* (Lutrochidae) and the wonderfully adapted water penny (family Psephenidae *sensu lato*). Over 80 species of Elmidae have been recorded in North America, 46 in Europe, although the western and northern fringes of Europe have fewer species (12 species in Britain, four in Ireland and only three in Norway) (Elliott 2008). Most riffle beetles are small, generally slow-moving species clinging to the substratum, and have five to eight instars depending on the species. Virtually all the elmids have aquatic adults that respire via a plastron (a *plastron* is a thin film of atmospheric air trapped in dense hairs growing on the insect cuticle; see Figure 4.1), which allows them to remain underwater indefinitely (see Chapter 4, section 4.1.1). The water pennies have terrestrial adults, however. Both adults and larvae of many species feed on fine detritus with associated microorganisms and algae, scraped from the substratum. Other species eat wood (xylophagous e.g. *Lara*, Elmidae). Riffle beetles appear to be only very rarely eaten by the major invertebrate predators and seem to show some form of antipredator response (such as rolling up into a small ball or becoming immobile; Elliott 2008).

Other beetle taxa may also be important in running waters, particularly in slow-flowing sections. Among these, the Dytiscidae (diving beetles) is the most diverse family, with more than 2,500 freshwater species (White & Brigham 1996). Both larvae and adults are predators: larvae swallow their prey (they are 'engulfers') while adults suck body fluids from their prey via piercing mouthparts. Hydrophilid larvae are also predatory, while the adults are omnivores. Gyrinid (whirligig beetles) larvae are benthic predators, whereas the adults live on the water surface of slower-flowing streams and rivers (Figure 3.16c), attacking dead and living organisms trapped in the surface film. Dryopids are detritivores, and hydraenid adults are mainly grazers.

3.4.7.8 Megaloptera (alderflies, dobsonflies)

The Megaloptera is a medium-sized order with less than 5,000 species worldwide. Most species are terrestrial with only around 330 in freshwater. In running waters, members of the two megalopteran families Sialidae (alderflies) and Corydalidae (dobsonflies) are particularly important, as they are all voracious predators, having large mandibles. So far 131 species of Corydalinae, and 81 species of Sialidae are recognised

Figure 3.16 A riffle beetle (Elmidae) adult (a) and larva (b); (c) a whirligig beetle (Gyrinidae) showing the upper part of its divided eye above the water surface; and (d) a larval alderfly (Megaloptera, *Sialis lutaria*); (e) dobsonfly larva (hellgrammite) from a Tennessee stream.
Source: (a)–(d) photos by Jan Hamrsky, lifeinfreshwater.net; (e) photo by DellaRay923—own work, CC BY-SA 4.0, https://commons.wikimedia.org/w/index.php?curid=41276064.

(Cover & Resh 2008). Corydalid larvae (sometimes known as 'hellgrammites' in North America) are often found in fast-flowing riffles under gravel and cobbles and sialids are usually associated with soft sediments of slower-flowing streams as well as in lentic habitats. Sialids (Figure 3.16d) feed on a variety of aquatic invertebrates, in particular small insects such as chironomids, as well as oligochaetes and molluscs. Smaller larvae also include microcrustaceans in their diet, and the first instars consume microorganisms and detritus. The larger corydalids (Figure 3.16e), which can reach 65 mm in length, can feed on small fish as well as aquatic insects. The development of sialids can involve up to 10 moults over 1 to 2 years. In corydalids, the number of instars is 10–12, and the larval lifespan is 2 to 5 years. In both groups, pupation takes place in soil close to stream banks.

3.5 Vertebrates

Representatives from all vertebrate classes (most notably fish, of course) are found associated with lotic habitats. Because of their size, ease of identification, economic importance, conservation status and or general interest, freshwater vertebrates are generally very well studied and have been the subject of many specialist texts. Here we briefly introduce the groups and

provide some of the more interesting aspects of their biology and ecology relevant to life in streams and rivers.

3.5.1 Fish

These are without doubt the best-known and best-studied inhabitants in freshwater systems. This is related to the importance of commercial fisheries from the large rivers of the world as well as to their value for recreational angling and associated tourism, particularly in the northern hemisphere. Although fresh waters make up just 1% of the earth's surface, they support at least 15,750 species of fish, just under half (48%) of the global diversity of bony fishes and around 25% of total vertebrate diversity (Darwell & Freyhof 2016). This figure is likely to rise over time as new species are being found and described, particularly in areas like tropical South America. This remarkable diversity has led to a range of theories about their speciation and biogeography (see example, Seehausen & Wagner 2014) and provided many classic examples of adaptive radiation, driven in part by the fact that fresh waters are the equivalent of islands in a sea of land, which increases the likelihood of population isolation. Nearly 7,000 of the 13,400 species considered to be totally dependent on fresh water are endemic to a single (of the 426 described) ecoregion—a large area encompassing one or more freshwater systems with distinct assemblage of freshwater species that can be used as a conservation unit (Abell et al. 2008). The diversity is clearly concentrated in the neotropical realm of South and Central America, with over 4,000 freshwater fish species, whereas the Palaearctic (Eurasia, the Middle East and North Africa) supports just over 900 (Figure 3.17), with the European fauna particularly impoverished due largely to the relatively recent glaciations and east–west orientation of mountain ranges. This strong variation is mirrored to the same degree in the functional diversity, a measure of the range of ecosystem processes influenced by fish (in this case based on morphological characteristics, a topic we return to in the next chapter), with more than 75% of global functional diversity in the neotropical freshwater systems (Toussant et al. 2016). Large Asian river catchments (the Pearl River, Brahmaputra, Ganges and Mekong), the African Congo and areas of Guinea, and South America's Amazon, Orinoco and Parana, are all known biodiversity hotspots for freshwater fish. When ecoregion area is taken into account, Africa's Niger basin, river basins in the south-eastern USA, parts of Sumatra and Borneo, and even some remote islands like Fiji and Vanuatu, show particularly high species richness (Abell et al. 2008).

There are about 170 families of fish with freshwater species, but just a few groups are most speciose: the Characiformes (over 1,800 species including the characins such as pirhanas (Figure 3.18a) and tetras predominantly from the neotropics); Cypriniformes (over 3,700 including carp, minnows and loach (Figure 3.18b)) most abundant in the Palaearctic and oriental regions; Siluriformes (catfish, with about 3,000 species, predominantly neotropical); Gymnotiformes (a small group of some 130 species of neotropical knife fish, possessing electric organs); the Perciformes (over 3,400 species, notably the family Cichlidae); and the Cyprinodontiformes (around 1,100 species including tropical killifish, many species of which are live-bearers). In North American streams and rivers, darters (Percidae) (Figure 3.18b), minnows (Cyprinidae) and suckers (Catostomidae) are the most diverse groups. The order Salmoniformes has a single family, the Salmonidae (including salmon, trout, char and grayling (Figure 3.18c) with 66 species found across the northern hemisphere. This is one of the dominant and best-known families. Two groups of fish are especially important and diverse in South America: Characidae and Siluriformes. Families of cyprinoids, including cyprinids, loach (Cobitidae) and hillstream loach (Homalopteridae), contribute notably to the diversity of Asian rivers. Whilst most temperate

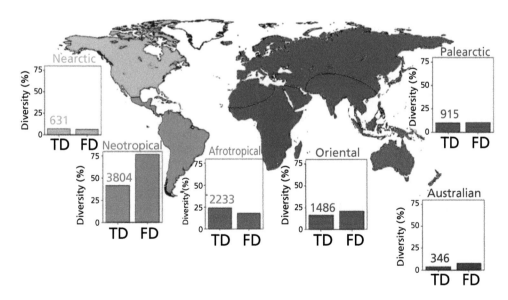

Figure 3.17 Taxonomic (species number; TD) and functional (based on 11 morphological measurements; FD) diversity of freshwater fish across the six biogeographical realms as a percentage of the world taxonomic or functional diversity.
Source: adapted from Figure 3.1 in Toussant et al. 2016 under Creative Commons CC BY licence.

Figure 3.18 (a) Red-bellied Piranha *(Pygocentrus nattereri)* a possible invasive in the Godavari River basin in Andhra Pradesh, India; (b) *Percina pantherine*, the leopard darter from the Little River system in Oklahoma and Arkansas, USA; (c) Brown trout *(Salmo trutta)*; (d) *Pomahaka galaxias* (Galaxias 'Pomahaka' as yet undescribed species in the Pomahaka River of Southern Otago).
Source: (a) photo Gregory Moine/ WikiMedia Commons under CC by 2.0, https://indianbiodiversitytalk.blogspot.com/2014/02/; (b) photo from https://alchetron.com/ Leopard-darter; (c) photo from Eric Engbretson for US Fish and Wildlife Service, public domain, via Wikimedia Commons; (d) photo from New Zealand Department of Conservation website at https://www.doc.govt.nz/nature.

riverine fish species are predators on macroinvertebrates, in tropical/subtropical rivers detritivorous and grazing fish species are also important.

Quite a number of fish species occupy both fresh and marine waters and some others migrate between the two at different stages of their life cycle (they are *diadromous*). Species that reproduce in fresh water but have their major growth phase in the sea are called *anadromous* species and include, for example, lampreys and salmonids. Species of salmon are important anadromous fish in northern-hemisphere rivers, spending at least one year in the river, migrating to the sea following significant physiological changes, then returning to reproduce in their natal river or stream. In the Atlantic, *Salmo salar* is the chief species while, in the Pacific, several species of the genus *Oncorhynchus* are similar ecologically. Salmonids have been particularly affected by habitat destruction and damming, as well as overfishing, and major restocking and management programmes are undertaken (e.g. in the Columbia river basin in the north-west USA). Eels are

catadromous—that is, they spawn in the sea and migrate as larvae to freshwater rivers where their main growth takes place before they return to their place of birth in the sea. Galaxiids in New Zealand (Figure 3.18d) reproduce in fresh water, and young fish spend a relatively short period at sea before returning to rivers as juveniles (a life cycle sometimes known as *amphidromous*; McDowell, 2010). The Hawaiian freshwater fish fauna consists only of *diadromous* gobioid species. In these, newly hatched larvae leave the streams in which they hatch to spend a period in the sea as part of the zooplankton (Radke & Kinzie 1996). 'Post-larvae' then return to streams and metamorphose. Juveniles and adults then live in fresh water, where they spawn.

Buried in a sandy substratum, larval lampreys (a primitive jawless fish-like vertebrate) filter the water both for organic particles to support their slow growth and for respiration. Larval lampreys are restricted to running waters whereas, after metamorphosis, the adults of many parasitic species migrate to lakes or the sea (Maitland 2003). After a period of parasitic feeding, predominantly on fish, they return to running waters to spawn and then die. Their adult life is restricted to one or two years, but their larval life may be extended over many years. Non-parasitic (small-bodied) species do not feed as adults; they spawn in the stream in the spring after metamorphosis.

The iconic fish species of European streams is arguably the brown trout, *Salmo trutta* (Figure 3.18c), which is particularly important in cool, headwater streams. Further downstream, they are progressively replaced by other species, a pattern that was recognised early in the subject and was used in attempts to 'classify' stream zones (see Chapter 6). Other salmonids occupy this niche in other continents, such as in North America (e.g. cut-throat *Oncorhynchus clarkia* and brook trout *Salvelinus fontinalis*). In parts of the world where trout are not native, one or another species has often been introduced (mainly for recreational angling) with significant impacts on native fauna (see Chapter 7). Trout feed on a wide range of aquatic invertebrates on the stream bed or drifting in the water column and, especially during the late summer and autumn, on terrestrial insects that have fallen onto to the water surface (Bridcut & Giller 1995). There is a strong relationship between individual small-scale habitat use and diet in these fish (Giller & Greenberg 2015) and there is evidence of density-dependent population regulation (Chapter 5, section 5.3.2.1). Other fish species (including sculpins (bullheads), darters and minnows) feed predominantly on invertebrates associated with the stream bed.

Extensive environmental changes to rivers and streams, including the effects of climate change, pollution, habitat destruction, damming, overfishing, agriculture and aquaculture, and invasive species, has severely affected natural populations of freshwater fish and estimates by the IUCN suggest that over 30% of the species are under significant threat of extinction (Darwell & Freyhof 2016). We will return to this important topic in Chapters 5 and 10.

3.5.2 Amphibians and reptiles

Amphibians and reptiles are widely distributed in fresh waters across the globe. The global diversity of amphibia was estimated at 8,120 species in 2020, an apparent increase of over 60% since 1985 (AmphibiaWeb 2020). This 'increase' is due entirely to efforts to collect and describe new species—they are actually extremely threatened. There are three orders, Urodela/Caudata (salamanders and newts), Anura (toads and frogs, by far the most diverse) and the

lesser known, exclusively tropical, Gymnophiona (a group of limbless snake or large worm-like amphibians known as caecilians). Over 70% species of amphibians are recognised as aquatic (living in fresh water for at least one of their life stages) (Vences & Kohler 2008), consisting mainly of anuran genera in the neotropical, afrotropical and oriental biogeographic regions (Figure 3.19a). The urodeles, in contrast, are more or less restricted to the Palaearctic and Nearctic. According to the IUCN, 64 of the 85 European species are endemic, and this endemicity is repeated in other areas as only six of the 348 aquatic genera occur in more than one biogeographic area (Vences & Kohler 2008).

Amphibians live at the interface between freshwater and terrestrial habitats, laying eggs and with larval development in water, and they are relatively short-lived. The larvae of the salamanders and caecilians largely resemble the adults but the larval tadpoles of the anurans are very different morphologically, physiologically and in their diet (larvae are largely omnivorous suspension feeders while the adults are carnivorous). Respiration in larvae is through gills and the skin, in adults through the skin and lungs. Few amphibians are totally restricted to habitats with flowing water. Of those that are, the most spectacular examples include two species of giant salamanders belonging to the genus *Andrias* (Cryptobranchidae) living in cold, highly oxygenated streams in China and Japan. They can reach a length of up to 1.5 m, are nocturnal and feed on crustaceans, fish and small amphibians. The North American 'hellbender' (*Cryptobranchus alleganiensis*, Figure 3.19b) belongs to the same family, and shows a similar affinity to swift rocky streams, where they feed on snails, crayfish and worms. Other salamanders spend time as larvae in stream pools. Nearly 41% of the freshwater species assessed by the IUCN are considered threatened by extinction through a range of factors including habitat loss, climate change, emerging pathogens, alien species and pollution (Ficetola et al. 2015). Amphibians are ecologically important in the control of pests, in bioturbation and via nutrient recycling by the tadpoles (Valencia-Aguilar et al. 2013).

Most reptiles (> 96% of the 9,546 species) are terrestrial, and those found in fresh water (turtles, 327 species; crocodiles, 25; some snakes) are air-breathing (Pincheiro-Donoso et al. 2013). Most of them lay eggs, but not in water. Aquatic reptiles are less diverse than freshwater amphibians, though they tend to be longer-lived. Some semi-aquatic lizards are known to use freshwater habitats (around 70 mainly tropical species, such as the Nile monitor lizard (*Varanus niloticus*)), though no aquatic lizard species are found in Europe,

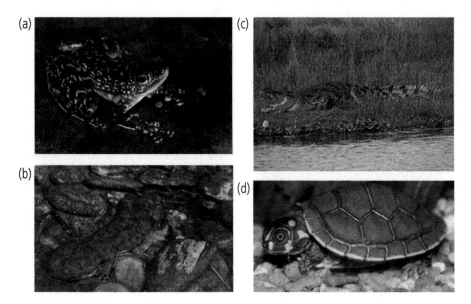

Figure 3.19 Amphibians and reptiles: (a) the rock skipper or Sabah splash frog (*Staurois latopalmatus*) found perched on vertical rock faces in or near rapids in swift rocky streams in Borneo; (b) the hellbender, a giant salamander endemic to fast-flowing streams of the eastern and central United States; (c) a Nile crocodile (*Crocodylus niloticus*); (d) The yellow-spotted Amazon river turtle (*Podocnemis unifilis*).
Source: (a) © 2012 Nathan Litjens from https://amphibiaweb.org; (b) photo by Todd Pearson 2010, Creative Commons Attribution Non-Commercial 3.0 (CC BY-NC 3.0); (c) photo from https://en.wikipedia.org/wiki/Crocodyloidea; (d) from the Reptile Database, https://reptile-database.reptarium.cz, photo with permission of Robert Wayne Van Devender.

North Africa or Palaearctic Asia. The 25 species of crocodiles and their relatives (caimans, alligators, gharials) are among the largest vertebrates of tropical and subtropical rivers (Figure 3.19c). They are opportunistic carnivores, usually the top predator in their habitat and significant predators of fish. Many of them occur in slow-flowing rivers, although the dwarf caymans found in South America do prefer fast-flowing streams (Martin 2008). Some species are used as food by indigenous peoples, including caimans in the Amazon basin (Valencia-Aguilar et al. 2013), and crododile farms have been established in a number of countries.

There are about 250 species of freshwater turtles (Order Chelonii, Figure 3.19d), predominantly found in the subtropics and tropics, with two general hotspots of diversity in south-east North America and the Indo-Malayan area of South-East Asia. Typical habitats include large rivers and lakes as well as marshes, bogs and sometimes estuaries (see Bour 2008 for further details). Freshwater turtles are largely carnivorous, feeding on invertebrates and small fish, but there are records of herbivory on algae and a more omnivorous diet. Turtles appear to be very sensitive to habitat disturbance and up to 60% of the extant species are more or less threatened.

3.5.3 Birds

About 5% of bird species (some 560; Dehorter & Guillemain, 2008) depend on freshwater habitats to some significant degree, yet a much greater proportion depend on the production of freshwater systems at some phase of their cycle (see Topic Box 3.3 by Steve Ormerod). Although many more species of birds forage in the margins of streams and rivers, only about 70 species from 16 families are recognised as running-water specialists (Buckton & Ormerod, 2002; see also Chapter 8). The greatest diversity (16 species) is in the eastern Himalayas (see Topic Box 3.3) and the overall diversity of river specialists peaks in fast-flowing rivers at mid latitudes (Buckton & Ormerod 2002; Sinha et al. 2019). Most of these specialists are relatively small species feeding on the adult and larval stages of aquatic insects. They are found in different parts of the world and include the European grey wagtail (*Motacilla cinerea*); North American black phoebe (*Sayornis nigricans*) and water thrushes (genus *Parkesia*); Asian water redstarts (genus *Phoenicurus*), forktails (genus *Enicurus*) and whistling thrushes (genus *Myophonus*); and the Central and South American buff-rumped warbler (*Myiothlypis fulvicauda*) and Andean torrent tyrannulet (*Serpophaga cinerea*), tyrant flycatchers (family Tyrannidae)

Figure 3.20 A selection of riverine birds from India: river lapwing (*Vanellus duvaucelii*), (b) Indian skimmer (*Rynchops albicollis*), (c) ibisbill (*Ibidorhyncha struthersii*) and (d) crested kingfisher (*Megaceryle lugubris*).
Source: photos with kind permission of Dr Nilanjan Chatterjee.

and wagtail tyrants (genus *Stigmatura*). A range of riverine birds is found across Asia (e.g. Figure 3.20 and Debata & Kal. 2021). Some are found on exposed sandbars along lowland rivers feeding on aquatic prey, such as the wading river lapwing *Vanellus duvaucelii*; Figure 3.20a. The Indian skimmer (*Rynchops albicollis*; Figire 3.20b) skims over the water surface with just its enlarged lower mandible in the water, while the ibisbill (*Ibidorhyncha struthersii*; Figure 3.20c) forages along shingle river and stream banks of the high plateau of the Himalayas, where they probe under rocks or amongst gravel for macroinvertebrates and small fish. The crested kingfisher (*Megaceryle lugubris*; Figure 3.20d) lives near fast-flowing streams in forested catchments and plunge-dives for fish and crustaceans. The larger, somewhat less specialised, species like herons and cormorants feed primarily on fish from rivers and standing waters. Finally, a small group of lotic species, the dippers (Cinclidae; five true species globally) and half a dozen species of ducks, feed in the water column or the benthic zone of streams and rivers.

Dippers (Figure 3.21a, b) are unique among passerines for their ability to swim underwater, and their adaptations and specialised ecology are explored in detail by Steve Ormerod in Topic Box 3.3. The dipper diet is affected by flow, with trichopteran larvae (particularly the large cased larvae of Limnephilidae) important at baseflow, whereas larval blackflies, mayflies (Baetidae) and terrestrial prey are more frequently taken during spates, when diving is more difficult (Taylor & O'Halloran 2001). Dippers are territorial, especially during the breeding season, when they can depress prey production, especially of caddisfly larvae. Studies in the USA (Harvey & Marti 1993) and in Wales (Ormerod & Tyler 1991) found that adult dippers can consume 8->18 g d^{-1} (dry mass) of Trichoptera alone during the breeding season, possibly exceeding the prey intake by trout (Harvey & Marti 1993).

Many species of waterfowl (ducks) occur in rivers, but only a limited number of species live and forage in fast-flowing streams and rivers for at least part of the year. These include the harlequin duck (*Histrionicus*

Figure 3.21 (a) Adult European dipper (*Cinclus cinclus*) with prey; (b) recently fledged dipper chick; (c) adult male torrent duck (*Merganetta armata*) from Alto Bío Bío, Lonquimay, Araucanía, Chile; (d) spectacled duck from Río Toltén, Villarrica, Araucanía, Chile.

Source: (a) and (b) photos: Darío Fernández-Bellon, University College Cork, under licence from NPWS, Ireland; (c) photo by Hederd Torres Garcia, with permission from eBird.org; (d) photo by Pio Marshall, with permission from eBird.org.

histrionicus) of Iceland, Greenland, Labrador, north-western and eastern North America, and north-eastern Siberia; the torrent duck (*Merganetta armata*) from the Andes; the blue duck (*Hymenolaimus malacorhynchos*) in New Zealand; and the Brazilian merganser (*Mergus octosetaceus*), which is found in inaccessible forest torrents in headwaters of the Paraná and Tocantins rivers in southern Brazil, eastern Paraguay and north-eastern Argentina. Like the dippers, many of these ducks exploit invertebrate prey which they capture in fast-flowing water, usually by diving, although some, such as the mergansers, also feed on fish. Although the harlequin ducks breed on torrential streams and rivers, they spend the non-breeding seasons on rocky coasts. In Iceland their diet consists mainly of blackfly larvae and pupae, the supply of which affects *per capita* reproduction (Einarsson et al. 2006). In Newfoundland, however, Chironomidae makes up the greatest proportion of the diet (Goudie 2008). In contrast, the Torrent duck (Figure 3.21c) remains on fast-flowing streams throughout the year, pairs holding territories of 1–2 km. Its range extends from the cold forests of Patagonia in Argentina, north to temperate Columbian forests and across a wide altitudinal range (sea level to 4,500 m). They have even been found to feed in hot springs (Ceron 2015). The Andean torrent duck feeds in a way similar to most (Eurasian) dippers. Curiously, the two Andean dipper species have a different mode of foraging, perhaps as a consequence of competitive interaction with the ducks (Tyler & Ormerod 1994; Topic Box 3.3). The omnivorous Salvadori's duck (*Salvadorina waigiuensis*) is endemic to New Guinea, favouring rushing mountain streams at altitudes up to 4,000 m, but it occasionally occurs in sluggish rivers and lakes. The African black duck (*Anas sparsa*) is also omnivorous and is found in less-torrential, wooded streams and rivers of upland parts of tropical Africa, while the omnivorous bronze-winged or spectacled duck (*Speculanas specularis*; Figure 3.21d) occurs in forested rivers and fast-flowing streams of the lower slopes of the Andes in southern South America.

Topic Box 3.3 The world's specialist river birds—and the case of the dippers (Cinclidae)

Steve Ormerod

While the contribution to freshwater biodiversity of aquatic vertebrates, such as fish and amphibians is well recognised, mammals and birds are more often linked with terrestrial realms. Yet globally, 10–25% of all bird species depend on production from freshwater ecosystems at some phase of their life cycle and for 5% the association is obligatory (Ormerod & Tyler 1993). Among this latter group, almost 70 bird species from 16 families are tied closely to fast-flowing streams and rivers where they breed and feed. This raises the possibility not only of bottom-up influences of river characteristics on their ecology and fitness, but also of their (top-down) influence on river ecosystem processes.

Among the specialist river birds, the world's five dipper species in the family Cinclidae are among the most remarkable. Present on upland and mountain streams across five continents, the dippers are unique among passerines in using their wings for swimming and diving to feed on benthic invertebrates and small fish which together form their entire diet. This potentially connects them more closely to aquatic production, for example, than salmonids. Adaptations to aquatic feeding include closable nostrils, specialised iris musculature and highly accommodating eye lenses for underwater vision, blood rich in haemoglobin with good oxygen-binding capacity, a high feather density with specialised water-repellent structure, a relatively larger preen ('uropygial', water-proofing) gland than almost any other bird, modifications to the wing bones and tail musculature linked to swimming, and even enlarged toe-pads on their un-webbed feet that facilitate grip on submerged or wet rocks. Their domed nest, created from aquatic or semi-aquatic mosses, is waterproofed and well insulated, while key events in their life cycle, such as breeding and moulting, are adapted to different hydrographic regimes around the world, such as annual snow melt, monsoonal rain or temperate flood seasonality (Tyler & Ormerod 1994; Smith et al. 2021).

There have been detailed assessments of the interactions of white-throated dippers, *Cinclus cinclus*, with other river organisms and ecosystem processes. By swimming, diving and probing the river bed, dippers feed highly selectively in ways that reflect the different phases of their annual cycle (Ormerod & Tyler 1991). Prior to breeding, increased energetic requirements of both males and females leads to a focus on larger Trichoptera and small fish, such as cottids and salmonid fry, while smaller invertebrates are taken opportunistically. Additionally, the demands for calcium in the egg-laying female are met by feeding on fish, benthic

molluscs and crustaceans such as gammarids. The provisioning of young at the nest in the centre of the pair's linear territory of 0.5–1.5 km requires particular energy efficiency. In the first few days after hatching, because of their small gape, nestlings are fed small invertebrates such as baetid or heptageniid mayflies. As the brood of four to six chicks grow, however, larger prey such as hydropsychid and limnephilid caddis are included, meeting their increased food requirements. Such prey are brought by both parents in bundles of four to eight larvae in around 300–350 daily visits to the nest. This requires deft handling to extract caddis from their cases or shelters as each item is added to the accumulating load in each bird's bill (Tyler & Ormerod 1994). Once successfully fledged, young dippers then concentrate on easily captured prey such as simuliid larvae or small mayflies (Yoerg 1994), and these same groups form a large numerical proportion of the dipper diet through the winter, when flooding can constrain foraging. Trichopteran larvae and small fish are taken when possible and dominate the winter diet by biomass. American dippers (*Cinclus mexicanus*) in the Pacific North West of their range take the eggs and fry of migratory salmon in large numbers during the winter and pre-breeding periods, linked to the life cycle and abundance of the various *Oncorhynchus* spp. (Morrissey et al. 2004).

This specific pattern of prey use in dippers has three possible consequences. First, the considerable energy requirements of a territorial pair (10.5–11.0 kg dry mass of fish and invertebrates, equivalent to 1–2.4 g m^{-2} of river bed annually) suggest that dippers could have significant effects on the biomass of various freshwater taxa, particularly Trichoptera and fish such as *Cottus gobio* (Ormerod & Tyler 1991). This possibility has been supported experimentally in American dippers (*Cinclus mexicanus*) (Harvey & Marti 1993). Second, dippers might potentially compete with other aquatic organisms, such as riverine fishes, but evidence for this is so far scarce (Nilsson et al. 2018). Third, the abundance and composition of available prey influence dipper distribution and productivity: their abundance, egg mass, clutch size, brood size and nestling growth all decline where the major prey types are scarce, for example along base-poor headwaters (Ormerod et al. 1991). This third point means that dippers provide effective indicators of stream quality, because their numbers decline where river quality is impaired but increase where there has been recovery from past pollution—for example the along formerly grossly polluted rivers in the UK. In both North America and Europe, dippers have also been used to indicate

Topic Box 3.3 *Continued*

biomagnifying pollutants, metals and, most recently, the food-web transfer of microplastics (Morrissey et al. 2004; D'Souza et al. 2020).

Among the other highly specialised and phylogenetically diverse river birds around the world—for example blue ducks (*Hymenolaimus malacorhynchos*) in New Zealand, torrent ducks (*Merganetta armata*) and torrent tyrannulets (*Serpophaga cinerea*) in South America, Louisiana waterthrush (*Parkesia motacilla*) in North America—ecological investigations do not yet match the details available for dippers, and species such as these are promising subjects for further research. Even basic surveys are still few and far between in Africa and South America (Politi et al. 2020). In the Himalayan mountains of Nepal and India, however, investigations have focused on patterns of niche use, resource partitioning and assembly among the most diverse community of specialist river birds on earth (Buckton & Ormerod 2002; Sinha et al. 2022). Here, around 16 specialist river species partition resources, through their

occurrence either at different altitudes or in sympatric groups with differing morphological traits or habitat use (Buckton & Ormerod 2008). Brown dippers (*Cinclus pallasii*), two wagtails (Motacillidae), several forktails, chats, the blue whistling Thrush (*Myophonus caeruleus*) (all Muscicapidae), two kingfishers (Alcedinidae) and the unique ibisbill (*Ibidoryncha struthersii*) are among the co-occurring species which divide their use of river production between benthic feeding in the margins or main channel, aquatic foraging in the splash zone of large boulders, fly-catching, ground gleaning in the riparian zone, or between piscivory and insectivory. In this Himalayan region, productivity and a complex habitat template appear to have worked alongside species radiation within and among different phylogenetic groups to produce a remarkable array of river organisms (Sinha et al. 2022).

Professor Steve Ormerod is at the Water Research Institute, Cardiff School of Biosciences, Cardiff University, Cardiff, UK.

Box Figure 3.3 Examples of Himalayan river birds: (a) spotted forktail, (b) little forktail (insectivores feeding respectively in the river margins and on the wetted perimeter of large boulders); (c) plumbeous water redstart, (d) brown dipper (feeding in the main channel by flycatching or benthic foraging).
Source: photos by Nilanjan Chatterjee.

3.5.4 Mammals

Rivers provide food and habitat for at least 220 'semi-aquatic' species from amongst most mammalian orders, but members of only two orders have fully aquatic species; the Cetacea (river dolphins) and Sirenia (dugongs and manatees). Freshwater mammals are represented among 70 genera and are found on every continent except Antarctica, with both geographically widespread and restricted species found. Many of these are considered endangered (Veron et al. 2008) and a few species have very recently become extinct. A range of other mammals do have a close relationship with rivers and streams whilst not being considered even as semi-aquatic, such as the many species of bat that feed extensively on flying insects along river channels and use bridges as roosting sites. Directly dependent on rivers are the beavers of northern boreal regions, otters, hippopotamus, tenrecs and desmans and the peculiar monotreme mammal, the platypus (*Ornithorhynchus anatinus*) from Australia.

The family Tenrecidae (tenrecs) includes a number of semiaquatic species, including otter shrews and the larger (80–100 g) web-footed tenrec (*Limnogale mergulus*), which lives along large, fast-flowing rivers in Madagascar. It establishes a permanent streamside burrow and is a nocturnal and tactile predator, finding its prey by sweeping the stream bed with its vibrissae, located on its snout (Figure 3.22a). This species feeds mainly on larval mayflies but also takes larvae of Odonata, Trichoptera and Lepidoptera and coleopteran adults, as well as decapod crustaceans and larval anurans (Benstead et al. 2001). Their diet is similar to those of the Pyrenean desman (*Galemys pyrenaicus*) and Tibetan water shrew (*Sorex thibetanus*) from western China. The Pyrenean desman (Figure 3.22b) is a small, semi-aquatic species endemic to the Pyrenees and the northern Iberian peninsula, and is a protected species in four countries (Spain, France, Portugal and Andorra). It lives in cold and well-oxygenated mountain streams between 15 and 2,700 m above sea level (Charbonnel et al. 2014). It is a dietary generalist, though stream invertebrates are widely taken, along with some terrestrial prey. Indeed, no less than 100 different invertebrate genera have been detected, using a variety of methods, in faecal pellets (Biffi et al. 2017; Esnaola et al. 2018). The ecology and natural history of this elusive little mammal remain relatively poorly known, however.

Beavers are important mammals associated with lower-order stream habitats in North America (*Castor canadensis*) and Europe (*Castor fiber*). They have been described as 'ecological engineers' since they can modify water flow and stream morphology by building dams across streams (see Chapter 10 for more details). Beavers are herbivorous rodents that feed to a large extent on the fresh bark of riparian trees (particularly willow and aspen), although in some areas aquatic vegetation may comprise 60–80% of the diet (Stringer & Gaywood 2016). Larger trees are felled with the aid of their powerful teeth and spectacular lodges are built in the river using branches and twigs, completed with mud and rocks. This can lead to the formation of braided channels, which were probably more common in times when beavers themselves were more abundant. After extensive hunting, beaver populations declined to a small fraction of their former size, but having been given some protection and through some reintroductions, numbers and range are increasing rapidly in North America and large parts of Europe (e.g. in Scotland; Stringer & Gaywood 2016). In the French River Loire, following reintroductions beginning in 1970, beavers had expanded their range over > 6,000 km of channel by the end of 2014 (Moatar et al. 2021). Other rodents, such as muskrats, coypu and voles, can also influence channel patterns through their burrowing into banks. The South American coypu (*Myocastor coypus*) was accidentally introduced into many countries in Europe as it escaped from fur farms and established free-living populations that have led to the destruction of macrophyte and reed beds to such an extent that in Britain, at least, this invasive species was actually proscribed and exterminated.

Otters (Mustelidae; Figure 3.22c) are streamlined mammals, up to about 1 m long for most species (genus *Lutra*) and 1.8 m for the south American Giant Otter (*Pteronura brasiliensis*). Otters are closely associated with aquatic habitats such as streams and rivers, lakes and the sea shore. Different species are found in Eurasia (two species), South-East Asia (four), Africa (four), North America (one) and Central and South America (four). The European otter (*L. lutra*) and the North American river otter (*L. canadensis*) have similar lifestyles, preying mainly on fish (particularly eels and salmonids) but seasonally including amphibians, birds and crayfish (Ottino & Giller 2004). They prefer habitats with a good riparian cover, especially deciduous trees with large rooting systems. Otters are territorial, marking boundaries with faecal deposits ('spraints'), and occupy underground 'holts' dug in or close to riverbanks. Individual territories range up to 15 km long (Ottino & Giller 2004). A smaller related mustelid, the mink, is found in North America (*Neovison vison*; also inadvertently introduced into Europe following

Figure 3.22 (a) Web-footed tenrec (*Limnogale mergulus*) from Ranomafana National Park in Madagascar, December 1996; (b) Pyrenean desman (*Galemys pyrenaicus*); (c) the European otter (*Lutra lutra*); (d) the platypus (*Ornithorhynchus anatinus*) from Australia and Tasmania.
Source: (a) from Veron et al. 2008, with permission from Springer Nature; (b) Esnaola et al. 2018, photo by Joxerra Aihartza. https://doi.org/10.1371/journal.pone. 0208986.g001, under Creative Commons Licence; (c) photo from Wikipedia Commons; (d) photo from Wikipedia Creative Commons Attribution-ShareAlike 2.0 Generic licence.

escapes from fur farms) and Europe (*Mustela lutreola*, the native Eurasian mink), and these have similar lifestyles to the otter, although the mink is more terrestrial, has a less specialised diet and can forage inland.

The hippopotamus (*Hippopotamus amphibius*) is a very large mammal that spends most of the daylight hours in rivers, lakes and marshes of sub-Saharan Africa. Formerly abundant, their numbers are widely in decline although they are still very important to river ecosystems as they graze on terrestrial vegetation at night but defaecate in the water, leading to the import of millions of tons of terrestrial organic matter into the African rivers (see Chapter 8). The related pygmy hippopotamus (*Hexaprotodon liberiensis*) of West Africa is much smaller and more terrestrial than its larger relative but is known to spend time in fresh waters during the day.

In great contrast to hippos, the Platypus (*Ornithorhynchus anatinus*) is a unique, small-bodied, egg-laying monotreme inhabiting a wide range of fresh waters in Eastern Australia and Tasmania (Figure 3.22d). It lives in large streams to rivers, as well as creeks, shallow lakes and wetlands and their riparian margins, in agricultural land and urban areas. It feeds exclusively in the water and rests in burrows, typically in the banks of water bodies. Foraging takes place underwater during short dives (30–145 seconds

per dive), with roughly 75 dives h^{-1}, over 8–16 h per day. Platypus diets are often dominated by relatively large aquatic macroinvertebrates including freshwater crayfish, although small chironomid species may also be consumed. Prey are located using its bill-shaped organ, which carries numerous electrosensors. Bino et al. (2019) provide an excellent review of all aspects of the biology of the platypus. Like many of the endemic specialist riverine species, the platypus is in decline and is now under the IUCN Red Listing as 'Near Threatened'.

The only fully aquatic mammals in fresh water are the river dolphins. There are (or were) two obligately freshwater species found in Asia: the Yangtze River dolphin or baiji (*Lipotes vexillifer*) and the blind South Asian species *Platanista gangetica* in the Indus, Ganges, Brahmaputra, Meghna and Karnaphuli river systems. Two other Asian species are referred to as facultative river species as they can also occur in estuaries and coastal waters; these are the Irrawaddy dolphin (*Orcaella brevirostris*) and finless porpoise (*Neophocaena phocaenoides*), found in major rivers including the Irrawaddy, Mekong and Yangtze. In South America the Amazon dolphin or boto (*Inia geoffrensis*) and the tucuxi (*Sotalia fluviatilis*) are found. There is also an estuarine and coastal marine form of the tucuxi (*S. guianensis*), which extends far up some major rivers such as

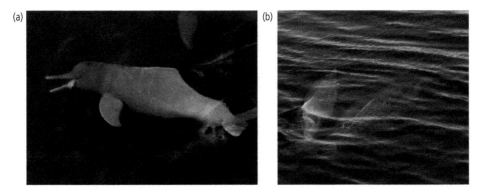

Figure 3.23 The newly described Amazonian Araguaian river dolphin (*Inia araguaiaensis*) (a) feeding on fish; (b) engaged in social activity.
Source: photos from DOI: 10.7717/peerj.6670/fig-2, Melo-Santos et al. 2019, under Creative Commons Attribution Licence.

the Orinoco (Veron et al. 2008). Surprisingly, a new Amazonian river dolphin species has recently been described (Figure 3.23a and b), the Araguaian river dolphin (*Inia araguaiaensis*) from Brazil (Hrbek et al. 2014).

'True' or obligate river dolphins share some superficial morphological traits, including reduced eyes, large paddle-like flippers, unfused cervical vertebrae (which allows considerable neck movement) and a long, narrow rostrum (beak) with numerous sharply pointed teeth (Smith & Reeves 2012). River dolphins have weak vision (the Indus and Ganges dolphins are virtually blind, the eye lacking a crystalline lens), and live in turbid rivers where vision is of little value. They prefer recirculating pools within the river, which provide shelter from downstream currents and where food sources may be concentrated (Smith & Reeves 2012). They feed mainly on fish, molluscs and crustaceans, locating prey by a sophisticated auditory sense, including echolocation, which also helps them to navigate and avoid obstacles. They also communicate through a wide range of acoustic signals (Melo-Santos et al. 2019). River dolphins are amongst the world's most threatened mammals, and populations and ranges of

these mammals, like many of the megafauna species of rivers, are small and decreasing, due to a range of factors (see Chapter 5, Section 5.3.1). Indeed the Yangtze River dolphin was declared functionally extinct in 2006.

3.6 End note

This chapter has briefly described the range and diversity of organisms found in rivers and streams. We have included just a few details of their natural history where appropriate and pointed to other aspects of their biology and ecology that we deal with in more detail in later chapters or that can be found in a range of other, more specialised, publications. We next consider how these species are 'equipped' for life in rivers and streams—do their biological features represent 'adaptations' to lotic life? These features can then be considered as part of what we can think of as 'species traits'—including body size and morphology, life cycle and mode of reproduction. This then helps us to look for any possible match between the occurrence of different species traits and the variable and varying nature of river and stream habitats.

CHAPTER 4

Matching the habitat templet

Adaptations and species traits

Terrestrial and aquatic habitats are similar in that they consist of a fluid (gaseous atmosphere or liquid water) overlying a solid (land surface or river- or lake-bed sediments). However, the greater density of water provides mechanical support that allows a vast array of aquatic organisms to live 'free from contact with the ground' suspended in the fluid medium, particularly in the oceans and lakes. It also means that any movement of the denser water is also of much greater and more immediate importance for aquatic organisms than air movements are for terrestrial ones. When the reduced availability of oxygen and the moderation to rates of daily and seasonal temperature change in aquatic habitats are also taken into account, the challenges to life in aquatic habitats are therefore clearly different from those in terrestrial systems.

The exploration of species adaptations to their environment has been a widespread approach to understanding the biology and ecology of organisms in their environment, and stream and river biology has been no exception, with studies going back over 100 years or more. The forces associated with flowing water, and the limited depths, distinguish most river habitats from lakes, although these differences decline as one considers the larger, higher-order channels of the world's major rivers, and certainly when we consider flood-plain rivers. As a consequence, we find unique adaptations among the lotic biota, covering physiological, morphological, life history and behavioural aspects of their biology. We can only touch on a relatively limited range of examples to illustrate the key features of animals and plants that seem to fit them for life in running waters, but we indicate other sources that delve further into this fascinating area.

4.1 Physiological adaptations

The physiological tolerances of different stream organisms influence their ability to cope in a specific situation or habitat and here we highlight some physiological systems of particular relevance to running waters.

4.1.1 Respiration

Two disadvantages of life in water are that oxygen availability is much lower than in air (by a factor of 20 or more depending on temperature) and that oxygen diffuses much more slowly in solution (approximately 300,000 times slower; Buchwalter et al. 2019), although in running water oxygen is often replenished at respiratory surfaces by water flow. As we have seen in Chapter 2, water temperature, the abundance of macrophytes, altitude, turbulence and organic pollution all influence oxygen concentration and will thereby interact with the respiratory physiology, morphology and behaviour of lotic organisms.

Gaseous exchange by single-celled or colonial periphytic algae and mosses is through simple diffusion from and to the surrounding water. In higher aquatic plants, the water provides mechanical support, so the epidermis can be thinner thus facilitating gaseous exchange. However, plant tissues with extensive intercellular gas-filled air spaces or lacunae (*aerenchyma*), found within plant stems and roots in many submerged and emergent macrophytes, facilitate the transport of gases internally, allowing oxygen to diffuse from shoots above the water to the submerged roots thus aerating them. Approximately 30% of the oxygen consumed in metabolism by the macrophytes *Egeria densa* and *Potamogeton crispus*, and more than 40% of that consumed by *Myriophylum triphyllum*, is supplied through this lacunar system (Sorrell & Dromgoole 1989). In some subtropical mangroves growing along the banks of slowly flowing upper tidal reaches of creeks and rivers (e.g. the river mangrove *Aegiceras corniculatum* (Myrsinacea) distributed throughout South East Asia and Australia), *pneumatophores* (upward growing roots projecting

The Biology and Ecology of Streams and Rivers. Alan Hildrew and Paul Giller, Oxford University Press. © Alan Hildrew and Paul Giller (2023).
DOI: 10.1093/oso/9780198516101.003.0004

from the substratum) exchange gases between the atmosphere and the internal tissues via openings (*lenticels*) in the otherwise impermeable root covering. By transporting air to the subsurface tissues, pneumatophores function as a highly specialised ventilation mechanism.

In lotic animals, gaseous exchange takes place across the body walls of most of the smaller, especially elongate, invertebrates (e.g. nematodes, annelids, small midge larvae) where the body surface can supply sufficient oxygen. However, as body size increases, tissue volume (and thus oxygen demand) increases more rapidly than its surface area, requiring specialist respiratory adaptations. An excellent review of aquatic insect respiration—Buchwalter et al. (2019)—allows us just to highlight the main features and some key examples.

Insects possess tracheal systems—networks of air-filled tubes that branch throughout the body, and eventually reach the cells with blind-ended tracheoles. Oxygen and carbon dioxide move through the smaller, air-filled tubes largely by diffusion. These tracheal systems mark the terrestrial origins of the insects. Most freshwater insects are only secondarily aquatic and, moreover, the adults are terrestrial with trachea open to the atmosphere via spiracles on the surface of the thorax and/or abdomen. In most aquatic larvae the spiracles are closed, and thus oxygen and carbon dioxide must diffuse across the cuticle. Like many lentic forms, lotic insect larvae often respire with the help of external gills, which increases the overall surface area of the respiratory surfaces, thus enhancing diffusion of oxygen into the body. Examples are common in the Ephemeroptera, Odonata and Trichoptera (see

for example Figures 3.11, 3.13 and 3.14). Since oxygen supply is only rarely a problem in fast-flowing and shallow streams, the gills of common lotic species are often relatively small or immobile. Under oxygen stress, however, ventilatory movements do occur (as described below). Taxa which use air stores underwater (including adult beetles and bugs) often have to visit the surface to replenish the oxygen. In a few groups specialised for living in faster-flowing, often quite deep water, such excursions would be dangerous or impossible. This is the case in some elmid beetles (see Figure 3.16), and in the Palaearctic water bug *Aphelocheiris aestivalis*, which has a *plastron* consisting of a very dense pile of hydrofuge (water-repellent) hairs that hold an air bubble into which oxygen diffuses, forming a permanent and incompressible 'physical gill' on the body surface. The density of the 6-μm-long hairs in *A. aestivalis* is as high as 2.5 million m^{-2} (Thorpe 1950; Figure 4.1a and b). The volume of air in the plastron is extremely small (0.14 mm^3), under slightly negative pressure and connected to the gas-filled tracheal system through spiracles on the cuticle (Seymour et al. 2015). The air film presents a large surface area for diffusion of oxygen and this respiratory system is therefore particularly efficient.

The spiracular gills of blackfly pupae (see Figure 3.15a) are filled with water, but the outer cuticle of the gill wall is hollow and forms an air film, providing another example of a plastron that is linked to the tracheal system of the pupa. In contrast, larval blackflies, like most stream invertebrates without gills, absorb oxygen through the body wall. Some other lotic aquatic insects (e.g. dytiscid beetles, hemipterans such as the tropical *Anisops* and other notonectids,

Figure 4.1 (a) Ventral side of the water bug *Aphelocheirus aestivalis* showing the broad, flat surface covered with the plastron and the spiracles (the light spots on the sternites, arrowed). The two bright spots on the second abdominal sternites are presumed to be sense organs. (b) Detailed morphology of the hydrofuge hairs comprising the plastron. *Source:* (a) from Seymour et al. 2015, with permission from the Company of Biologists Ltd; (b) from Buchwalter et al. 2019, with permission from Kendall Hunt Publishing Company.

(a)
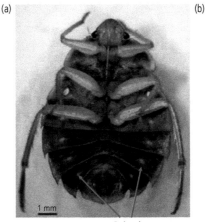
1 mm

Spiracles

(b)
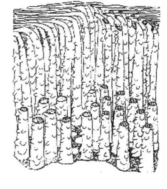

and some dipteran larvae) breathe atmospheric air from compressible air bubbles brought from the water surface, although they are largely confined to slower-flowing streams and rivers.

Gills are also found associated with the legs of amphipod and isopod crustaceans and in molluscs. Freshwater prosobranch snails, which have a horny operculum to shut the shell mouth, have a single gill (a *ctenidium*) located in the mantle cavity with leaf-like plates richly supplied with blood vessels. Oxygen-poor blood passes through the ctenidium in the opposite direction to oxygen-rich water currents created by cilia, a counter-current mechanism ensuring a positive diffusion of oxygen from the water to the blood (Brown 1991). In contrast, pulmonate snails, which have re-invaded fresh water from the land, have a 'lung', a richly vascularised pocket in the mantle. They either rely on coming to the surface to breathe or, in some species of Lymnaeidae or Physidae, the pocket is filled with water and acts as a gill.

For invertebrates that live in fast-flowing water, and that depend on simple diffusion of oxygen into respiratory tissues from the surrounding water, replenishment of oxygen occurs naturally through the stream flow and many do not survive for long in still water (e.g. if they become trapped in pools at very low flow or in floodplain ponds). However, in sluggish water or when oxygen concentration declines (e.g. as a result of organic pollution or high temperature) and/or oxygen demand (metabolism) increases, many species are able to undertake active ventilation. Ventilatory currents, which normally push water posteriorly over the body surface or gills, can be generated by body undulations, gill beating, abdominal contractions or a combination of these, as well as by active swimming. The frequency of these activities is usually related to oxygen concentration. Many mayfly larvae, for example, can beat their abdominal gills. Trichopterans, chironomids and some ephemeropterans 'pump' water across their abdomen or through their cases or tubes and burrows via body undulations. Anisoptera (dragonflies, Figure 3.13c) have a rectal pump which, through contraction and relaxation of dorsoventral abdominal muscles, produces a water flow over the internal gills located in a blind sac off the rectum. During periods of respiratory stress large stream-dwelling stoneflies also create water flows over the thoracic gills situated at the base of the limbs through unique ventilatory movements involving 'push-ups' (raising and lowering the body). Several forms of respiratory pigments are found in aquatic invertebrates. These range from haemoglobin commonly found in annelids such as tubifex worms (see Figure 3.10b) and in *Chironomus* midge spp. to haemocyanin in crustaceans and some stoneflies (Buchwalter et al. 2019).

Fish have a very efficient respiratory system involving gills and counter-current systems, while most amphibians can breathe underwater either through their body surfaces or via gills which are retained generally only in the juvenile stage. Other semi-aquatic vertebrates, including reptiles, birds and mammals, breathe air directly, and effectively 'hold their breath' while submerged, with associated changes in blood-flows to various parts of the body.

4.1.2 Osmoregulation

Organisms in fresh water are hypertonic (cell fluids or body fluids such as haemolymph and blood have higher ionic concentration than the medium). For example, in freshwater algae, osmotic concentration (as milliosmoles L^{-1}, where an 'osmole' is a measure of osmotic pressure exerted by solutes) can vary depending on species. In freshwater mussels, values range from 70–100 mOs mol L^{-1} and in aquatic insects from 200–400 mOs m L^{-1}, while fresh water itself typically ranges from 1–2 mOs m L^{-1}. This is a challenge, since the surrounding water therefore tends to enter the body, diluting the internal concentration of salts, while salts tend to diffuse into the surrounding water, compromising internal homeostasis. This is exacerbated by the permeability of the respiratory surfaces and, hence, respiration and osmoregulation are often linked. For example, the plastron described earlier effectively keeps water away from body surfaces. To counter the risk of diluting the body fluid, osmoregulatory processes involving several structures differentially retain and take up salts. In aquatic insects these ion-absorption sites include: (1) chloride cells on gill surfaces (Ephemeroptera, Plecoptera, Heteroptera); (2) chloride epithelia (abdominal—Trichoptera: Limnephilidae, Goeridae; anal—Diptera: Tabanidae, Stratiomyidae, Ephydridae, Muscidae; rectal—Odonata); (3) anal papillae (Diptera: Nematocera, Syrphidae; Trichoptera: Glossosomatidae, Philopotamidae); and (4) gut wall epithelium of drinking insects (Sialidae, Dytiscidae) (Ward 1992a). Many freshwater insects also excrete hypotonic (very dilute) urine and bivalves also eliminate excess water as urine via the kidney and recover lost ions via active transport over the gills and other epithelial surfaces (McMahon 1991). Freshwater fish have specialised membranes that retain salts and produce copious amounts of dilute urine, eliminating the large amounts of water that enter the body through the gills.

4.1.3 Drought resistance

Survival during adverse conditions often relies on various physiological adaptations, a clear example being the capacity to withstand periodic (sometimes seasonal) drought. The review by Lytle & Poff (2004) and the book on intermittent rivers and ephemeral streams edited by Datry et al. (2017a) provide some excellent and detailed examples.

Some of the clearest adaptations to drought conditions are found in aquatic plants (Sabater et al. 2017). Algae that can survive desiccation in intermittent rivers and ephemeral streams often produce extracellular mucilage (e.g. as in *Cymbella*) that increases cellular water retention, while intracellular osmoregulatory solutes also prevent water loss in drying sediments. Stalks or tubes in some algae 'wrap' cells in a protective filament, while migration to deeper and moister (particularly sandy) sediments, thick cell walls (e.g. Rhodophyta—red algae), reduced metabolism and rapid rehydration also aid survival. In addition, at the onset of drying, algae may also produce spores, cysts or zygotes.

Macrophytes demonstrate some additional mechanisms that overcome drought. Some are remarkably plastic in form and life cycle—differences associated with the prevailing conditions and, in particular, drought and stream drying. In *Ranunculus peltatus* (a plant often found in slow-flowing streams), individuals may be large and erect under fully aquatic conditions, reaching an approximate height of 27 cm, or small and prostrate under dry conditions, with height restricted to ~ 4 cm. The length of the flowering period also shifts and plants can set seed in droughts, even though fewer seeds are then produced (Volder et al. 1997). In seasonally drying rivers, sexual reproduction and seed production may lend *resilience* (the ability to recover after disturbances) to plant populations. Morpho-anatomical traits also contribute to survival amongst drought-tolerant plants. These include a decrease in plant size, water content and total dry mass, greater energy allocated to roots, and a decrease in leaf area and increase in leaf thickness (Sabater et al. 2017). Stomatal density and cuticle thickness also increase.

Most larval aquatic macroinvertebrates rapidly succumb in the absence of water, while some specialised taxa can tolerate long dry periods, but then usually only in a 'resting' dormant stage. In intermittent rivers and ephemeral streams, most invertebrates appear to be generalists. As Stubbington et al. (2017) point out, adaptations to drought include behaviour that leads to the use of refuges, resistance to desiccation and resilience traits. Refuges, which are discussed by Belinda Robson in Topic Box 4.1 (and also in relation to populations in Chapter 5, section 5.3.4), include upstream perennially flowing reaches, contracting pools (especially for taxa that can tolerate lower oxygen concentration), subsurface hyporheic refuges (with appropriate adaptations in body shape—see section 4.2.1) and life-history refugia in aquatic insects with flying adult stages. Resistance to drought can thus be related to morphology, physiology or life history. Body armouring (limiting water loss) and small size (enhancing access to the hyporheos) are two key structural adaptations. Physiological adaptations include anhydrobiosis—a state in which the organism's metabolism effectively ceases during a desiccation-tolerant state. Some of the meiofauna, such as nematodes and rotifers, can revive in as little as a few minutes even after a couple of decades (Wallace & Snell 2009). Desiccation-resistant life stages are common in intermittent rivers and ephemeral streams (as found in nemourid stoneflies and some crustacean and caddis eggs) as well as dormancy (e.g. *Hydrobaenus* chironomid larvae use protective tubes, and limnephilid caddis larvae seal their case) (Stubbington et al. 2017). Timing of life-history phases that avoid drought in intermittent streams, such as pupation within the substratum in the megalopteran *Neohermes filicornis* (Megaloptera: Corydalidae), can also be found (Figure 4.2).

Encystment is a key part of many protist life cycles, especially in lotic systems that have a propensity to dry out, or when conditions otherwise become unsuitable (Taylor & Sanders 2009). Sponges can undergo dormancy when active tissue transforms into dry gemmules during periods of environmental stress, and triclads encyst as entire animals or have eggs that tolerate desiccation, a feature of many other invertebrates able to 'resist' adverse conditions, especially among inhabitants of temporary streams. Most notable in this regard are the Crustacea. Ostracod eggs, for example, may remain viable for years and in some species the eggs even require a dry period in order to hatch. Some species of blackfly have eggs that survive dry seasons and droughts, such as the widely distributed *Simulium vernum* complex, whose females lay eggs in stream channels after the seasonal flow has ceased (Crosskey 1990). Life history adaptations are also important in surviving drought (see section 4.4.4), some features promoting 'resilience'—an enhanced ability to recover

Topic Box 4.1 Surviving drought in streams and rivers

Belinda Robson

Refuges are places in which species survive disturbances, such as stream drying, either on 'ecological timescales' (approximately seasonal or over a few seasons) or over 'evolutionary timescales' (many seasons to decades). Rivers and streams with intermittent flow normally dry up annually, while very prolonged ('supraseasonal') droughts over several seasons may be caused by climate change, increasing landscape aridity. In arid or drying landscapes, perennial, groundwater-fed waterbodies (e.g. subterranean aquifers, springs, riverine waterholes) can provide refuge for aquatic

organisms over both ecological and evolutionary timescales (Davis et al., 2013; Box Figure 4.1a). Species once common in wetter climates millennia ago are now found only in those refugia which have their own microclimate decoupled from the regional climate. Yet, because they provide fresh water during even the worst decadal-scale droughts, such places also provide short-term refuge for a range of more strongly dispersing, drought-adapted species capable of inhabiting a wide range of water bodies. Individual species may do exceptionally well in such refuges, as shown in the case of the mayfly *Bibulmena kadjina* (Carey et al. 2021; Box Figure 4.1b(d)).

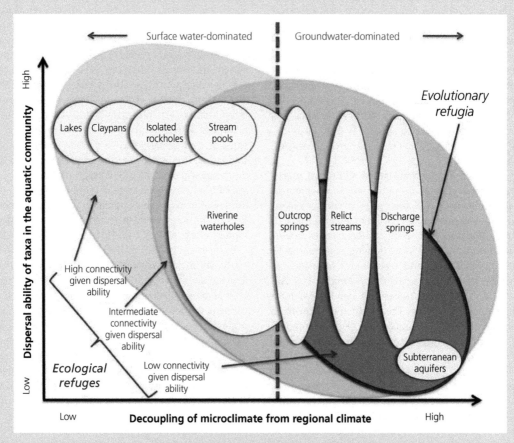

Box Figure 4.1a The source of water and its relationship to local climate determine refuge type and thus the dispersal capacity of species able to use those refuges. Evolutionary refugia support populations of species with low dispersal capacity (as well as strong dispersers) because they persist for long periods of time (millennia) being decoupled from local climate (and climate change) and supported by deep/ancient groundwater. In contrast, some habitats rely on surface water coming from local rainfall (e.g. claypans), only persist for relatively short time periods (weeks–months) and support only the most mobile species. Lastly, some refuges may also form refugia, such as large riverine waterholes that receive both surface and groundwater, persist for long time periods (centuries–millennia) and support species with a wide range of dispersal capacity.
Source: from Davis et al. 2013, with permission John Wiley and Sons.

Topic Box 4.1 *Continued*

In catchments with intermittent rivers and ephemeral streams (IRES), the most obvious and widely studied drought refuges are perennially flowing reaches and perennial pools [waterholes; Box Figure 4.1b(a) and (b)]. These support species absent from intermittent reaches, providing refuges even during supraseasonal drought (Chester et al. 2015). Perennial reaches are the only drought refuge type that almost all stream biota can use, and they sustain diversity in many catchments. Perennial pools also support a high proportion of the local species assemblage, except taxa that require flow (e.g. some net-spinning caddisflies). Such refuges are important because they sustain species with widely dispersing adults, that can travel out across the landscape and repopulate other streams after the drought

has broken (Chester et al. 2015). Other refuge habitats include tributary junctions and groundwater springs or seeps. Tributary junctions may create thermal refuges in a mainstem river (Ebersole et al. 2015) and groundwater flow sustains surface pools or perennial waterholes or merely moistens the river bed.

Particular species traits may promote survival in other types of drought refuge such as the hyporheic zone, accumulations of woody debris and inside animal burrows (e.g. those of some crayfish). The hyporheic zone comprises permeable sediment and groundwater beneath the stream bed that may be accessed by surface-dwelling species capable of moving downwards into the substratum. The most common species that may make effective use of the hyporheic zone

Box Figure 4.1b Examples of drought refuge pools in headwater streams in south-western Australia. Panels (a) and (b) show groundwater- (spring-) fed perennial refuge pools that supported populations of invertebrates when streambeds were otherwise dry. Panel (c) shows a spring-fed weir pool that is perennial in most years and provides an ecological refuge for a wide range of species with differing dispersal capacity. Pools in (a) and (b) supported exceptionally large nymphs of the mayfly *Bibulmena kadjina* (panel d): the left-hand (grey) image shows the maximum size recorded historically; images from summer (December) 2016 and autumn (April) 2017 show much larger nymphs present in these refuge pools, some of which were not yet in their final instar (red arrows show the developing wing buds on the April specimen compared to the mature black wing buds on the December specimens).
Source: from Carey et al. 2021, with permission John Wiley and Sons.

Topic Box 4.1 *Continued*

---

in this way are amphipods, mayflies, oligochaetes and cope-pods (see e.g. Vander Vorste et al. 2016). The assemblages of the hyporheic zone (the 'hyporheos') and of animal burrows may overlap (both may contain oligochaetes, cope-pods, ostracods, mayflies, chironomids), although larger crustaceans such as isopods and non-burrowing crayfish may also use burrows as refuges.

Dormant stages of a broad range of river species occur in or upon sediments or in-stream wood. Freshwater algae may form part of a dry biofilm on stones or wood that is protected from heat and ultraviolet light by the dark pigments that form as streams dry (Robson et al. 2008). Animals may take refuge in sediments as dormant propagules (eggs, cysts), larvae or adults. For example, some adult frogs, isopods and crayfish construct an aestivation chamber beneath large stones in the stream bed (Bogan et al. 2017) while bivalves may close their shells or burrow into sediments and may thus resist short-term drying. Larvae of some taxa, including dragonflies (e.g.Telephlebiidae), may simply retreat beneath stones or woody debris and sit dormant until flow returns (Bogan et al. 2017). Some final-instar damselfly larvae are capable of emerging as adults in the absence of surface water for up to a month (Chester et al. 2013). A few taxa (leeches, some gastropods) are capable of anhydrobiosis, whereby body tissues slowly lose water during dormancy. By far the most common dormant stage in stream invertebrates (particularly crustaceans and insects) consists of desiccation-resistant propagules that together form an 'egg-bank'.

Many catchments have a mixture of perennial and intermittent streams and ecologists have wondered whether intermittent streams harbour only a subset of the fauna of perennial streams that is tolerant to drying. Alternatively, are there specialist species present only in the intermittent streams? Rehydration studies examining stream egg-banks provide the answer. Hay et al. (2018) compared perennial pool and sediment refuges in temperate and semi-arid zone intermittent rivers and found taxon richness in rehydrated sediments was almost twice as high in the semi-arid areas, despite lower total richness there. Similarly, Bogan et al. (2013) showed that the fauna of intermittent streams in an arid zone was not just a subset of that in perennial streams but also contained specialist species adapted to drying. Datry et al. (2017b) found that the egg-bank contributed more taxa in streams in regions where stream drying was more prevalent. However, in their review, Stubbington & Datry (2013) found that the proportion of taxa contributed by the egg-bank to river assemblages was lower in warmer, drier climates because sediment moisture was lower there. It appears likely that the resolution of

this contradiction lies with the evolutionary and environmental history of the stream fauna and the historical frequency and duration of riverbed drying (Bogan et al. 2013; Datry et al. 2017b). Rivers with a more 'arid zone' fauna (often abundant benthic microcrustacea and Diptera) have more species that contribute to the egg-bank than do rivers dominated by more familiar stream insects (especially Ephemeroptera, Plecoptera and Trichoptera). Despite the importance of the egg-bank and sediment drought refuges, most studies have shown that perennially flowing reaches and perennial pools promote the survival of species that lack desiccation-resistant stages and therefore these refuges become increasingly important as the climate dries and flow recedes.

Global warming is causing more frequent and prolonged drying of rivers and streams, some of which have no history of intermittency (Crabot et al. 2020; Piano et al. 2020; Carey et al. 2021). Under those circumstances, perennially flowing reaches and perennial pools assume an even greater importance for conserving biodiversity because dispersal from these refuges is the main way that streams recover from novel intermittency (Gauthier et al. 2020). This requires avenues for connectivity to be maintained and researchers have called for connectivity to be a key focus of conservation (Crabot et al. 2020). Similarly, many researchers have called for the protection of perennial drought refuges (e.g. Chester et al. 2015; Vander Vorste et al. 2020). Limits on extraction of surface or groundwaters can prevent drying of perennial refuges and sustain hydrological connectivity along rivers, but what can be done where there is no water extraction, yet rivers are still drying out under climate change? One novel potential solution is the use of freshwater ecosystems created by humans that could be managed as drought refuges to sustain biodiversity in drying landscapes (Chester & Robson 2013). Potential locations include perennial weir pools and farm ponds or dams [Box Figure 4.1b(c)]. Some may already lie on stream channels and may therefore be hydrologically connected to stream networks for at least part of the year. Others (e.g. farm ponds) are isolated and would therefore only provide refuge for fauna, such as insects and frogs, capable of crossing dry land. Connectivity could be enhanced through vegetated corridors to facilitate movement. Globally, research into these anthropogenic refuges is in its infancy, but it is now a focus in many countries facing prolonged and more frequent dry periods in rivers and streams.

Dr Belinda Robson is Associate Professor at the Centre for Sustainable Aquatic Ecosystems, Harry Butler Institute, Murdoch University, Australia

Figure 4.2 (a) Megalopteran *Neohermes filicornis* (Megaloptera: Corydalidae) prepupa in a pupal chamber in the sediment below an intermittent stream in California; (b) pupa and larval exuvium (arrowed).
Source: from Cover et al. 2015, with permission from Matthew Cover, Bringham Young University and Western North American Naturalist.

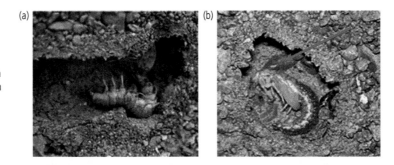

from drought disturbances and recolonise formerly dry habitats. This results in rapid population growth, although, in terms of individual fitness, such adaptations presumably allow access to a substantially 'empty' habitat. Key aspects may include a life stage with enhanced dispersal ability—a strongly flying adult or one easily carried passively on the breeze, or a propensity to drift in the current or swim. A short life span, asexual reproduction (as in Naididae worms) and eggs laid on land (as in Hydraenidae beetles) can promote faster recolonisation once flow resumes (see Chapter 5, Section 5.3.4). Some of the respiratory adaptations to low oxygen discussed earlier enhance survival in remnant low-oxygen pools during low flows, an example of 'resistance' to adverse conditons.

In general, fish are less well adapted to drought and their diversity in intermittent rivers and streams is low (Kerezsy et al. 2017). Those species characteristic of periodically dry channels have usually recolonised from elsewhere, sometimes migrating long distances. Others can, however, persist through the drought conditions. For instance, the African annual killifish, *Nothobranchius* (Figure 4.3a), has an embryonic diapause within the drought-resistant egg, while the adults are small and grow rapidly to maturity. They may spawn daily, with the fertilised eggs deposited

into the substratum throughout the rainy season. Features such as early maturity, high fecundity and multiple (*iteroparous*) spawning are characteristic of fish adapted to intermittent rivers and streams. A few species, often found on seasonal floodplains, show the most remarkable ability to breathe air (Figure 4.3b). The climbing perch (e.g. *Anabas testudineus*) can travel short distances overland and, like other members of the family Anabantidae, can extract oxygen from the air via a specialised 'labyrinthine' organ in the head. Various members of the speciose Gobiidae family and certain catfishes such as the Clariidae (Kerezsy et al. 2017) can also breathe air. The best-known air-breathers are the lungfish (Protopteridae), which can aestivate in cocoons in the completely dried substratum during the dry season, until the river refills floodplain channels with water. Aestivation is triggered by receding water, declining water quality and rising temperature.

4.2 Body form, size and other features

In spite of phylogenetic constraints on the body plan of organisms, many external environmental factors are highly influential, flow being an obvious factor for lotic organisms. Size and shape interact with the environment, substantially explaining the lifestyle

Figure 4.3 (a) The southern rainbow killifish *Nothobranchius pienaari*, and (b) the air-breathing African sharp-tooth catfish *Clarias gariepinus*.
Source: from Kerezsy et al. 2017, with permission from Elsevier.

and microhabitat choice. For a small invertebrate, for instance, it is important whether it lives on the substratum surface or interstitially, is exposed to fast currents or resides in slack waters, or if it swims. In animals directly exposed to water flow, the near-bed hydrodynamic forces (drag, lift and shear) affect the ability to move over the substratum and, not least, their susceptibility to dislodgement. Passive filter-feeders depend directly on the current to provide their food, and there may be a trade-off between maximising food delivery and the physical hazards of the flow. Whether grazing or filter-feeding, taxa in exposed microhabitats have to cope with the fluctuating forces of flow.

4.2.1 Shape

Body shape plays an important role in the biology of running-water organisms. Even in diatoms, which are generally very small, morphology is variable, reflecting the trade-off between environmental stresses and accessing resources. These morphological characteristics can be used to distinguish several different 'guilds' (e.g. as seen in three tributaries of the Kentucky River, USA; Molloy 1992), which show clear longitudinal patterns, from headwaters to high-order rivers, with respect to resource availability (e.g. light and nutrients) and exposure to sources of mortality (e.g. from scouring flows and grazing) (Figure 4.4). Hence algal growth form and shape is recognised as a major ecological and evolutionary response to the steep environmental gradients in lotic systems (Passy 2007). Generally, there is a switch from low- to high-profile species with a decrease in flow—or grazer-induced losses or a decrease in the availability of nutrients and light, and an increase in motile species in naturally or anthropogenically perturbed conditions (such as eutrophication) (Passy 2007; Berthon et al. 2011; Tsoi et al. 2017) (Figure 4.4).

Although commonly dispersed via plant fragments carried in the flow, rooted macrophytes cannot easily escape adverse conditions. Rather, individual plants are remarkably variable in response to the trophic and physical characteristics of flowing water (Ali et al.

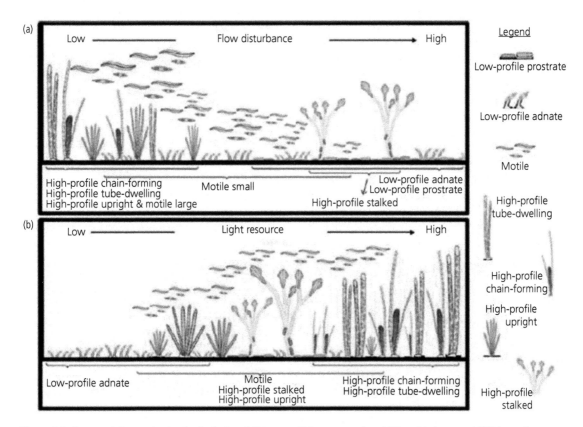

Figure 4.4 Conceptual diagram showing the distribution of diatom growth-form groups along (a) flow disturbance and (b) light gradient. *Source:* from Tsoi et al. 2017, with permission of CSIRO.

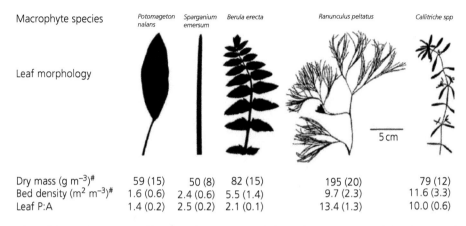

Macrophyte species	Potomageton natans	Sparganium emersum	Berula erecta	Ranunculus peltatus	Callitriche spp
Dry mass (g m^{-3})$^{#}$	59 (15)	50 (8)	82 (15)	195 (20)	79 (12)
Bed density (m^2 m^{-3})$^{#}$	1.6 (0.6)	2.4 (0.6)	5.5 (1.4)	9.7 (2.3)	11.6 (3.3)
Leaf P:A	1.4 (0.2)	2.5 (0.2)	2.1 (0.1)	13.4 (1.3)	10.0 (0.6)

Figure 4.5 Leaf morphology in the top 20 cm of five macrophyte species. Data below each species are means (SE) and represent the dry biomass and bed density (volume per m^2) in typical plant beds in the study streams. Leaf P:A refers to the leaf perimeter to area ratio and reflects the complexity of the leaves.
Source: reproduced from Levi et al. 2015, with permission from Springer Nature.

1999). One of the key adaptations relates to the size and shape of the leaves.

Generally, fresh waters have a low and fluctuating supply of carbon dioxide and light, often limiting plant growth in streams. Therefore, plants that can reach the water surface and gain access to atmospheric carbon dioxide may have a strong competitive advantage in terms of growth (Battrup-Pederson et al. 2015). *Heterophylly* (possession of more than one type of leaf) provides ecological flexibility in response to declines in water level (e.g. in summer and/or the dry season), when species populations often develop floating or aerial leaves. This is likely to provide a clear advantage for species growing in the land–water transition zone or in intermittent or ephemeral systems. Heterophyllous species not only have access to atmospheric carbon dioxide, they also have submerged leaves with lower surface-area-to-volume ratio (as in *Callitriche palustris*, *Alisma plantago-aquatica* and some *Ranunculus* species), which may contribute to maximising the carbon dioxide uptake both below and above the water surface.

In slack water in deeper channels, species able to concentrate their photosynthetic active biomass at or near the surface, either by growing from apical meristems or by producing floating leaves, can form dense, continuous surface canopies and thus maximise light capture (e.g. *Potamogeton natans*). On the other hand, *Ranunculus peltatus* and *Callitriche* spp. have more complex, finely dissected leaves (Figure 4.5), and spread via rhizomes to form thick beds within the stream channel (see Figure 3.4, p. 69). In contrast, *Sparganium emersum*

has simple leaves originating from the base of the plant (Figure 4.5) and it grows in less dense beds.

Turning to lotic animals, Statzner (2008) pointed out that stream ecologists are still struggling to understand fully how (or even whether) invertebrates are adapted to the somewhat chaotic near-bed flow conditions (illustrated in Chapter 2), which create such a diversity of constraints and opportunities. Matching all of them is physically impossible. Indeed, the paradox is that most body shapes do not seem to be particularly well adapted to hydraulic stress at all (Waringer et al. 2020) and, as such, morphology should not be interpreted as adapted solely to flow but as an ecological trade-off; for example, between physical environmental constraints, resource acquisition and avoidance of predation (all within limits imposed by phylogeny). However, there are some global generalities in morphology amongst stream and river invertebrates inhabiting similar microhabitats, even when they are phylogenetically unrelated.

Many lotic invertebrates, for example, have a strikingly flattened body that is held against the substratum, as exemplified by heptageniid mayfly larvae (Figure 4.6). Originally it was believed that such flat-bodied animals could avoid the impact of the current by crouching inside the boundary layer, or even that the current helped to press them down against the substratum. Using laser-doppler anemometry, Statzner & Holm (1982, 1989) changed this view by showing that not only does flow 'separate' above a flattened animal, but it is also much more complex than was first thought. Flow separation reduces lift, but at a cost

Figure 4.6 A heptageniid mayfly nymph, illustrating the broad, flattened head and femora.
Source: image © Jan Hamrsky at lifeinfreshwater.net.

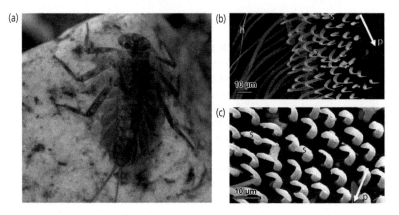

Figure 4.7 Attachment devices of the heptageniid mayfly *Epeorus assimilis*: (a) the larva; (b) and (c) micrographs of setae of the pads on the ventral side of the gill lamellae. The setae on the lateral part of the lamellae (b) are a different shape and bordered by long setae compared to other areas (c).
Source: photo from Le Monde des insectes, www.insecte.org; micrographs from Ditsche-Kuru & Koop, 2009, reproduced with permission of Taylor & Francis Ltd.

of increased drag, although it is more likely that the individual will remain attached. For the heptageniid larvae, the flattened body and large head shield may in fact lead to negative hydrodynamic lift in flowing water. This is accomplished by lowering its head shield and by using its femora as spoilers to direct flow, thus pressing the body against the substratum (Weissenberger et al. 1991).

The most flattened animals in streams are, somewhat unsurprisingly, 'flatworms' (Tricladida, Figure 3.6, p. 72) and 'water pennies' (larval beetles of the family Psephenidae). Flatworms, however, are probably flat because of evolutionary constraints on body shape and, while some are found in torrential streams, most avoid fast-flowing microhabitats (Hansen et al. 1991). The water pennies seem to adhere to the substratum through suction, and active pumping of water through lateral slots of the carapace and from under their body reduces turbulence around the body and decreases drag in high flows (McShaffrey & McCafferty 1987).

What was previously assumed to be a sucker-like structure formed by the overlapping gills of some grazing dorsoventrally flattened heptageniids again turns out to be rather more complex. There are additional structures such as setose (i.e. covered with setae) pads on the ventral margins of gill lamellae, thus increasing friction, and areas with spiky cuticular projections (acanthae) on the abdominal sterna (plates of cuticle on the underside of the abdomen), plus strong, hook-like claws on the first leg (Figure 4.7; Ditsche-Kuru et al. 2010).

True 'suckers' are in fact rather rare in lotic invertebrates. Leeches do have suckers, but they are not particularly adapted for life in flowing water, and they appear to be unable to move against the current. Six ventral suckers are found in blepharocerid (net-winged midge) fly larvae which enable them not only to withstand very rapid currents in torrential streams, but also to move against currents of up to 2.4 m s^{-1}. Each sucker consists of a suction disc fringed with fine hairs and

(a) 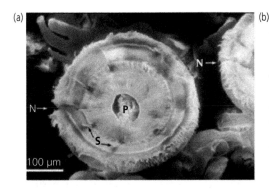   (b)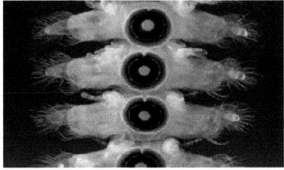

Figure 4.8 (a) Suction disc of *Hapalothrix lugubris* (Diptera: Blephariceridae), N = V-shaped notch; S = sensilla; P = piston. (b) Suction discs on the ventral side of body segments.
Source: (a) from Frutiger 2002, with permission of John Wiley & Sons; (b) photo Fabrice Parais, from https://www.nikonsmallworld.com/galleries/2013-photomicrography-competition/abdominal-segments-of-diptera-blephariceridae-larvae.

a cup-shaped cavity with a piston (Figure 4.8). The suction discs attach to the substratum by negative pressure produced by upward movement of the piston. The adhesive forces, as well as the size of the suckers, varies with body size (increasing through the instars) and between species, and is related to the hydraulic stress to which they are exposed in their usual habitats (Frutiger 2002).

In contrast, baetid mayfly larvae (see Figure 3.11a, p. 80) have a streamlined shape and the body is held clear of the substratum. As streamlining reduces drag and lift, this shape enables these mayflies to live in fast flows. Some cased caddis larvae are also streamlined (e.g. *Drusus bigutattus*), particularly as they align the longitudinal axis of their tapering cases with the direction of flow. Despite the fact that these larvae are large enough to project well above the viscous sublayer and into the turbulent layer above the substratum, they can withstand instantaneous flow velocities of up to 0.7 m s^{-1}. In addition to the streamlining, case mass plays a small role in overcoming the forces of drag (reducing it by just over 5%), while strong muscles and claws plus the possible use of silk to fasten the case to the substratum make up the deficit (Waringer et al. 2020). Lateral ballast in the cases of larger caddis larvae such as *Silo nigricornis* can contribute up to 40% towards overcoming drag.

Larval blackflies also have a distinctive streamlined body shape (see Figure 3.15a, p. 85) and are attached via a large number of small hooklets that are anchored onto a pad of silk produced by the larva on the substratum (see Section 4.2.4.1). The abdomen is widest at about one-fifth of the distance from the hind end, which may help reduce drag, but it probably also determines where vortices will form from which the larvae feed

(Chance & Craig 1986). In contrast, the middle part of the blackfly larva is narrow, which facilitates bending and rotating. The current deflects the body in a downstream direction; the deflection is almost zero in a larva feeding in very slow currents, whereas in very fast currents, the body is held almost parallel to the substratum. Blackfly larvae seem to be able to control their feeding posture, balancing the conflicting demands of drag and feeding (Hart et al. 1991). By bending the body, larvae can hold their head (labral) fans into the current, where particle flux is higher.

Streamlining in fish is well known and, with the aid of swimming, enables some species to hold station in midwater or near the bed, as clearly seen in the salmonids. In contrast, the ventral sucker disc of armoured catfish (Loricariidae) offers a structural adaptation to benthic attachment. In the smaller (7–8 cm long and 1–2 cm deep) North American stream-dwelling 'darters' (e.g. *Percina pantherine*; Percidae, Figure 3.18b, p. 90), holding station is facilitated both by living within the region of reduced velocity near the complex substratum and, to some extent, by the negative lift forces generated by expanded pectoral fins. These produce a pocket of reduced flow downstream of the fin and weak counter-rotating vortices associated with a small but downward-directed force (Carlson & Lauder 2011).

4.2.2 Size

The body size of lotic organisms is extremely important, with respect not only to biological features such as life history, respiration and metabolism, but also to ecological factors such as predation risk, position in the food web, population size and contribution to ecosystem processes such as nutrient cycling and decomposition (as we will see in Chapters 5–9). Body size also relates to hydrodynamics and the ability to move within the flowing water or over the stream bed. For instance, lotic cladocerans, ostracods and chironomids rarely exceed a few millimetres in length, while fish or crayfish are normally 10–100 times larger. We discussed in Chapter 2 (subsection 2.3.1) how the relative importance of viscous and inertial forces of the flow around organisms is described by the Reynolds number (Re) and that Re increases with increasing body size. Thus, small animals live in a world of low Reynolds numbers, experiencing an environment where viscosity predominates—'living in treacle'. Movement is impeded and gas exchange slowed down, although they benefit from a relatively still water layer surrounding the body that provides a protective shield from the buffeting of the current. In contrast, large animals, such as fish, experience higher Reynolds numbers, which means that inertial forces are much more important than viscous forces and so large animals have relatively few problems related to locomotion and gas exchange but must expend energy to retain position in flowing water. The effects of flow will, therefore, change not only with the size and type of organism, but also with the body size of a single individual as it grows. Life for a small insect larva with a Re of 1–10 is dramatically different from that of a fully grown larva with a Re of around 1,000. Such ontogenetic changes have implications for microhabitat selection by different stages of a particular species, and in fact may help to explain some of the distributional patterns of species observed in the field.

In general, the larger inhabitants of a stream dwell either in the water column or on the stream bed while the smaller ones, to a greater extent, live within the substratum. A diverse meiofauna (Chapter 3) resides mainly within the stream bed and are all small organisms including nematodes, mites, rotifers and microcrustaceans (see Figure 3.5, p. 70). This produces a variable vertical gradient in body size. We also expect a difference in the body size of organisms living in the spaces between substratum particles of differing coarseness—larger body size in coarser substrata. Of course, many relatively large organisms, such as larval lampreys, can burrow into a finer substratum, while the plankton of the water column of larger rivers consists of very small organisms.

Finally, body size is also related to life history, as we will see later in this chapter. Large animals develop more slowly; for instance large stonefly or odonate larvae may take several years to complete their larval growth, whereas small dipterans can reach maturity in a few weeks or less. Overall, body size is probably the single most important feature of lotic organisms at almost all levels of organisation in biology and ecology (see Hildrew et al. 2007), and patterns and processes related to body size are prominent throughout this book.

4.2.3 Hooks, bristles and hairs

Animals living in streams and rivers have a variety of morphological features which may be of particular selective value in running water, such as aiding attachment to the substratum, and gathering food (often by passive filter feeding). Such characteristics also sometimes help us distinguish species taxonomically.

The tarsi of aquatic insects with legs all have terminal claws. In riffle beetles, the claws are large and stout, enabling them to retain their position on the surface of substratum particles even at high current velocities. The posterior prolegs of some caddis larvae, such as the free-living *Rhyacophila* (Figure 4.9) and net-spinning hydropsychids and polycentropodids, the megalopteran *Corydalus*, and lotic chironomids, in

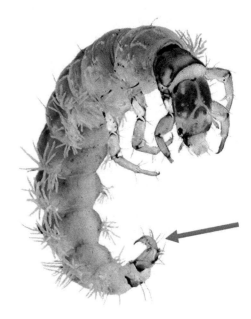

Figure 4.9 Free-living caddis larva (*Rhyacophila obliterata*) with anal proleg and terminal claw (arrowed, one of the two prolegs is clearly visible in this shot). These are important in the larva maintaining position in fast flows.
Source: from Rinne & Wiberg-Larsen (2017), photo by Aki Rinne, with kind permission.

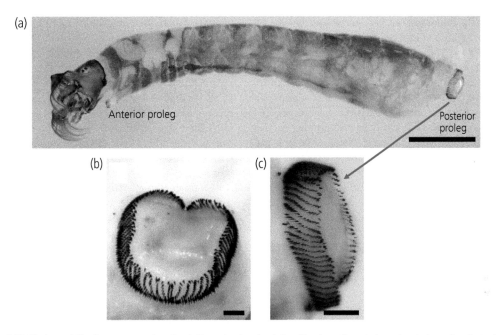

Figure 4.10 The larva of *Simulium coreanum* from South Korea showing the circles of hooks on the anterior and posterior prolegs that grip onto the silken pad spun on the substratum. Scale bar = 1 mm. (b), (c) Higher-magnification of hooks of the posterior proleg of *Simulium coreanum*. Scale bars = 0.1 mm.
Source: from Kim (2015) under Creative Commons Attribution Non-Commercial Licence.

addition to larval blackflies (see Figure 4.10), have hooks or hooklets that similarly help to stop larvae from being swept away when they move over exposed surfaces. The smaller hooks on the last abdominal segment of cased caddis larvae secure them within their case.

Most insects also have bristles and hairs of various kinds, often of uncertain significance, that are scattered in intricate patterns across their legs, head, thorax and abdomen. These presumably assist the larvae to monitor its immediate surroundings. For instance, a cased caddis larva can assess whether the case is in place; hairs and smaller setae with associated nerves also detect flow. Burrowing species often have a dense covering of bristles and hairs helping to keep the sediment particles away from the body surface and, in the stonefly *Capnopsis schiller*, even the eyes are covered. In the limnephilid cased caddis *Allogamus auricollis*, hairs on their legs are used to filter-feed on FPOM, and the front legs of burrowing mayflies also filter particles, whilst limnocentropodid (Trichoptera) larvae are filtering predators using spiny legs to capture drifting prey (Morse et al. 2019).

Cerci (the 'tails' extending from the tip of the abdomen of many insects such as stoneflies and mayflies; Figures 3.11 and 3.12, p. . .) help provide sensory information and probably assist in swimming and other movements. In ephemerellid mayflies, the end of the abdomen and cerci are often raised over the head and pointed forwards following disturbance. This is known as the 'scorpion posture' and may deter predators (Peckarsky & Penton 1988). In *Baetis bicaudatus* a similar tail-curl posture is adopted following detection of the tiny hydraulic disturbances caused by predatory stoneflies, after which the mayfly actively swims, drifts or crawls away (Peckarsky 1987). Some net-spinning caddis larvae have long brushes of quite coarse hairs on the tip of their abdomen. These provide warning of a predator or competitor crawling into the rear of the larval shelter (the net is built at the front end), enabling the resident to turn its armoured head and thorax towards the intruder, either to defend itself or to allow backwards retreat and escape into the drift.

4.2.4 Insect silk: an important evolutionary development

Silk is a generic term used for all fine, chiefly fibrous, protein threads extruded by all sorts of arthropods, both terrestrial and aquatic. In lotic animals, silk is particularly important to the larvae and pupae of blackflies, chironomids, caddisflies and Lepidoptera (moths). In caddisflies at least, silk is composed of long, unbranched polypeptide chains of fibroin and it has the unusual property of being liquid when stored in the animal but sticky in water. There is a substantial cost to silk production. For example, a large fraction of the production, particularly in net-spinning caddis larvae, goes into silk so it must therefore be of great survival value. The silk is used for attachment; constructing tubes, cases and feeding nets; and sometimes serves as a lifeline. Silk strands, although pliable, are very strong. Recent studies have also confirmed that silk of net-spinning caddis larvae (e.g. *Arctopsyche* and *Ceratopsyche* species of the family Hydropsychidae) can increase the critical shear force required to initiate movement of gravel substratum particles by 20% or so on average, a case of small-scale 'ecosystem engineering' (Albertson et al. 2019; see also Chapter 10). In view of its importance to lotic insects, we delve a bit deeper into the ecological and biological role of silk in three major insect groups.

4.2.4.1 Blackflies

Blackfly (Simuliidae) larvae, even when very small, produce large amounts of silk. This is important in terms of anchorage and helps explain the extraordinary capacity of blackfly larvae to remain attached at high flows. The larvae have numerous hooks, arranged in characteristic circular rows, on the tips of both the anterior (thoracic) and posterior (abdominal) prolegs (Figure 4.10, p. 113), which help larvae to anchor firmly onto silk pads attached to the substratum. In species dwelling in particularly fast currents, the number of hooks on the abdominal proleg may exceed 8,000, compared to only 500 in species living at slow velocities (Crosskey 1990). By sticking a new silken pad to the substratum in front of the body, to which it can carefully transfer its grip, the blackfly larva can move forward in a 'looping' movement, holding its body nearly parallel to the substratum. Just before looping starts, silk is secreted onto the thoracic proleg, and the larva then swings forward and sticks the new material firmly to the substratum. When anchored at two points, the larva enlarges the new silk pad, releases its hold on the previous anchor point, and swings the abdomen forward, placing its abdominal proleg onto the new silk pad. It can then begin to filter-feed by releasing the grip of its thoracic (anterior) proleg, or repeat the forward movement. They are exquisitely sensitive to small-scale changes in hydraulic pressure

near the substratum, and filter-feed in places where it is most profitable.

Blackfly larvae can also drift away by quickly releasing the abdominal hooks' grip from the silken pad. A larva about to drift almost invariably attaches a strand of silk—a lifeline—to the substratum. This is only a few micrometres in diameter, but enables the larva to minimise the risk of drifting too far or for too long. When drift is caused by a brief disturbance, such as contact with a predator, the larva can return along its lifeline to its original position. Silk is also an important prerequisite for the construction of pupal cocoons (Figure 3.15a, p. 85). Interestingly, the chemical composition of silk in blackfly silk glands appears to change when approaching the last larval instar (Barr 1984). This can be seen in a shift of silk colour, apparently associated with the changed functions of the silk through the life cycle.

4.2.4.2 Chironomidae

Many species of midges build larval and pupal tubes using silk which can vary in shape and in the materials that are glued together by silk (plant fragments or mineral particles). *Chironomus ramosus* for example constructs silken tubes around 1.6 cm in length, and individuals appear to be able to recognise their own 'home' (Thorat & Nath 2018). The larvae of some other chironomids build nets for filter feeding that can either be placed inside the living tube (in the tribe Chironomini) or on arms extending from the tube (in Tanytarsini; Figure 3.15c p. 85). One such species, *Rheotanytarsus muscicola*, uses three different kinds of silk (Kullberg 1988). A silk of fine texture is used for constructing and lining its tube. The second and third types are used in the catch net; one is thicker and is used for making the main framework, while the other is somewhat finer and used to spin the net. Thicker strands are produced when silk is ejected through the mouth, while finer strands are produced when silk is forced through grooves on the anterior edge of what is effectively the 'lower lip' of the larva—the ventromental plates. The catch net consists of extremely sticky irregular strands which capture colloids and particles in the size range 0.01–10 μm. Collected particles are either eaten or incorporated into the tube. Detailed studies on the silk of *Chironomus* species indicate that the salivary gland secretions contain proteins that differ in molecular size and are encoded by several genes (Thorat & Nath 2018).

4.2.4.3 Trichoptera

The silk produced by caddis larvae plays an important role in net construction, case-building and attachment.

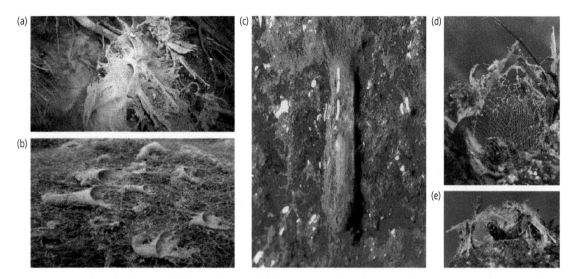

Figure 4.11 Nets of web-spinning caddis larvae: (a) *Plectrocnemia conspersa*; (b) *Neureclipsis* sp.; (c) a philopotamid larva visible within its elongate net; (d) and (e) *Hydropsyche* sp. (Hydropsychidae), (e) with resident.
Source: (a) and (c) photos Alan Hildrew; (b) Morse et al. 2019; © W. Graf under a Creative Commons Attribution Licence; (d) and (e) © Jan Hamrsky at lifeinfreshwater.net.

It has many similarities with that produced by lepidopteran larvae, and strands are extruded through an opening at the tip of the labrum (the 'upper lip').

Nets

Nets are built predominantly in three families of caddis; the Hydropsychidae, Polycentropodidae and Philopotamidae. They collect living or dead food items (depending on species), mostly as passive filter-feeders of particles entrained in the flow, but sometimes the resident is a more active predator, attacking prey trapped among or simply disturbing the silken strands. The nets include a tubular shelter into which the larva can retreat and where it spends most of its time.

Most polycentropodid larvae construct relatively large nets with irregular and coarse meshes. *Plectrocnemia conspersa* is probably the best studied and is a predator (Townsend & Hildrew 1979). Typically, larvae sit in a silken tube, from both ends of which funnel-shaped catch nets widen and are often orientated towards the current. They only exploit microhabitats (i.e. those immediately around the larva) with currents below 0.2 ms^{-1} and, in very slow flow (< 0.05 ms^{-1}), the net type constructed depends on the depth. At depths of 5 cm or less, silken threads are attached to the surface film (and where they can detect and attack small prey items falling into the water surface), while in deeper water, the larva spreads out

strands radially from the openings of the retreat tube onto the stream bed, resulting in an area of meshwork that traps small animals moving across the substratum, rather akin to the way spiders trap prey (Figure 4.11a). The adult female lays a mass of eggs arranged in a flat layer covered in a thin layer of 'jelly'. This is usually on the underside of a large, partially submerged rock or piece of wood. Upon first hatching from the mass, the siblings spin and occupy a 'colonial' net for a few days, before dispersing away. Such nets may be large (10 cm^2) and support more than 300 larvae (Hildrew & Wagner 1992). The net probably helps the larvae to obtain their first meal—prey somewhat larger than themselves can be subdued—and affords protection from the flow. Similar aggregations in communal nets have also been observed in blackfly larvae.

Larvae of other Polycentropodidae build large nets of somewhat irregular, 'bag-like' shape, like those of *Polycentropus flavomaculatus* which are often positioned on the lower sides of rocks. Very large (up to 20 cm long), beautifully crafted funnel-shaped nets (Figure 4.11b) are found in *Neureclipsis bimaculata*, a common European species of lake outlet streams where they feed on the relatively dense supply of particles drifting from the lake. The size of the aperture of the funnel appears to depend on both current velocity and food density. Thus, larvae inhabiting outlets from eutrophic lakes will construct much smaller nets, especially when occupying fast-flowing microhabitats, than larvae in oligotrophic (or slow-flowing) outlets.

The elongated nets of the Philopotamidae are shaped like a 'wind-sock' and attached to a stone surface at their upper end (Figure 4.11c). Typically, they are hung in vertical water films trickling through large stones in steep, headwater streams, though some species can be found beneath stones in less-steep and larger channels. The nets are not exposed to rapid currents and have extremely fine meshes, often only about 1 μm, and a very small front aperture. The water oozes through these fine meshes and ultra-fine particles are brushed from the inner surface of the net using the modified, hairy labrum. The larva lives within its net and there is no separate retreat.

Hydropsychids are probably globally the most widespread net-spinning caddis larvae. They typically inhabit faster-flowing parts of streams and rivers. Their filter nets are relatively small (Figure 4.11d) and built at the anterior entrance of a silken retreat tube attached to stable stones or rocks or built between the stems of tightly attached mosses on large rocks. They can be positioned beneath a stone in spaces big enough for water to flow freely, at the sides or on top of the stone (particularly where the net can be built among moss, and the living tube secreted within the moss mat). The filter net of *Hydropsyche siltalai* has an area of about 40 mm^2 and can filter around 500 L of water daily (Morse et al. 2019). When in high densities (up to 10,000 individual larvae m^{-2}) in moderately eutrophic/enriched conditions and/or in lake outlets, the filtering hydropsychid larvae can intercept substantial amounts of organic matter from suspension. These nets are often supported with small plant fragments and gravel, and they have a characteristic bilaterally symmetrical configuration and are tended continuously (Figure 4.11e, p. 115).

Meshes of hydropsychid catch nets are rectangular and typically measure in the order of 300 × 200 μm, although mesh size varies substantially between species. This variation has received a great deal of attention, since it has interesting implications for larval distributions in relation to the size range of suspended particles and current velocities, and thereby habitat partitioning (Loudon & Alstad 1990; Statzner & Dolédec 2011; see Chapter 7). Species with coarse net meshes typically live in fast flow and are predominantly carnivorous, feeding on small drifting animals. Species spinning smaller meshes are primarily detritivores and live in somewhat lower–velocity flow (see e.g. Wallace et al. 1992). Mesh size, diet and distribution also shift similarly between the instars of the same species—that is, *ontogenetically* as the larva grows.

Cases and galleries

Silk is also used for the construction of the cases and retreats of larvae in many caddis species, and in some instances is the sole building material. More commonly, however, it is used for lining cases and binding/sticking pieces of plants or mineral particles together. Caddis cases are usually made into species-specific, elongated shapes and frequently taper towards the posterior end (Figure 4.12a, p. 117). The fixed cases of larvae of some species of Hydroptilidae such as *Leucotrichia* are generally attached to relatively stable substrata such as larger stones, woody debris, aquatic plants or exposed roots of riparian plants (Morse et al. 2019). In species with portable cases, the case is usually tubular and constructed with different materials and in different arrangements that tend to be distinctive for the various taxa (Figure 4.12), although in some families, such as Glossosomatidae, the cases are 'armoured' enabling life on top of stones (Figure 4.12b). Cased caddis live in these constructions throughout their larval existence, usually adding material as they grow to the gradually enlarging anterior end (as evident in Figure 4.12a) and discarding case material from the narrower posterior end. The posterior end of the case commonly has a silken membranous 'sieve' that protects the larva from intruders but allows the flow of water through the case, facilitating respiration (Morse et al. 2019).

There are a number of possible benefits that caddis larvae may derive from their cases. They make respiration more efficient, since larvae are able to create an active current through the case. Cases may protect larvae from predators or camouflage them. Caddis cases sometimes have larger pieces of stones or sticks, making the larvae difficult to ingest by predatory fish and reducing the number of successful attacks by large invertebrate predators such as larval dragonflies. Sometimes, notably in Psychomyiidae, larvae construct attached tubes ('galleries') several times as long as their bodies. The larvae lead an active life inside the galleries and rarely emerge from the case completely—though the head and thoracic segments do regularly protrude from the front as they graze from the surrounding rock surface or gallery surface or carry out repairs and extensions. 'Gardening' of algae

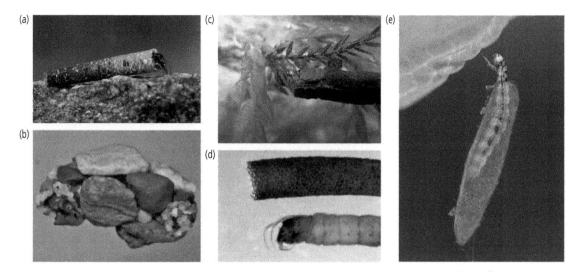

Figure 4.12 Case-bearing caddis larvae: (a) *Oecismus monedula* (Sericostomatidae) in its gravel case—note the increasing diameter and change in case particle content to the front end as the larva grows; (b) *Synagapetus* sp. (Glossosomatidae) with a dome-shaped (armoured) case; (c) *Oligostomis reticulata* (Phryganeidae) in its case built with pieces of leaf; (d) *Beraeodes minutus* (Beraeidae) has a case of fine sand; (e) *Agraylea multipunctate* (Hydroptilidae) in its silken case.
Source: (b), (c) and (d) from Morse et al. 2019, © W. Graf under a Creative Commons Attribution Licence; (a) and (e) with permission of Jan Hamrsky, lifeinfreshwater.net.

has also been discovered inside and on the galleries of psychomyiids (e.g. *Tinodes waeneri*; Ings et al. 2010). Some Hydroptilidae also incorporate algae within their silken, seed-like cases (Figure 4.12 e).

4.2.5 Colour

Colour in animals is generally involved in signalling and communication, camouflage and thermal regulation. In running waters, however, almost all invertebrates are rather drab. These shades may of course help animals escape detection by visual predators as camouflage against a background of dark rocks, shadows and detritus on the stream bed. Taxa exposed to fish predators within stream vegetation, such as several chironomid and simuliid species, are frequently greenish in colour, which may make them less visible. Amongst the vertebrates, however, colour becomes much more important, and in fish, amphibians and reptiles it is associated with multicomponent signalling through pigment-based and structural elements associated with the dermal chromatophore (Price et al. 2009). Colour is most evident in fish, particularly related to mating communication, but amphibians and some reptiles also often adopt mating colours. Some fish seem able to alter their general shading to match the substratum. For example, in the same stream one can find both light and dark trout depending on where the

individuals were caught. Presumably this too is an example of crypsis, reducing predation risk.

4.3 Behaviour and life in running waters

Morphological adaptations are of selective value only when linked with appropriate behaviour. Broadly speaking, behaviour is any activity that alters the relationship between an organism and its environment (both abiotic and biotic). In plants, behaviour as such is evidently restricted and innate; for example, motile algae can travel towards or away from light, while the growth form and structure of macrophytes can respond to the forces of flow. Animals have more complex behavioural patterns that tend to be innate in 'lower' invertebrates, but involve more learning in 'higher' forms, especially the vertebrates. Most of our understanding of behavioural adaptations in river and stream animals comes from studies on aquatic insects, larger crustaceans and fish. Here we concentrate on a few themes of importance in the biology of running-water animals.

It is true to say that current is the most significant factor for life in running waters. The very fact that we do not find masses of animals collecting at the downstream end of river systems attests to the ability of organisms to cope in some way or another with the forces exerted by flowing water. Stream animals must (normally) maintain their location and feed in a unidirectionally flowing environment. In addition to the broader general habit and microhabitat selection of animals, at a finer scale many taxa change their body posture and activity in response to changes in flow. When water velocity increases, for example, crayfish alter body posture to enhance streamlining and thus counteract the effect of drag (Maude & Williams 1983). As mentioned earlier, *Ecdyonurus* mayflies lower their large head shield, and *Simuliim* larvae bend their body to lie parallel to the substratum as flow increases. Many individuals temporarily shift their position (actively or passively) during spates into patches of reduced flow and shear stress. Such areas offer refuge from the flow and may be of great significance in the population and community dynamics of lotic organisms—of which more later.

A lot of the active responses to the environment are controlled by *taxes* (innate directional movements in response to environmental stimuli) and are presumably adaptive in a changing environment. Response to light (*phototaxis*) is widespread. Most invertebrates (especially clingers and burrowers) are negatively phototactic, avoiding bright light by moving into the substratum or under stones (and indirectly out of the current). Examples include crayfish, mayfly larvae and flatworms. At night, many insects emerge onto the surface of the substratum to graze, a diel rhythm controlled by light—although this does expose them to potential dislodgement by the current, as we discuss in the context of drift in Chapter 5, sections 5.3.5 and 5.3.6. However, under respiratory stress, these insects move towards the light to exposed surfaces allowing access to faster currents and more oxygen (Wiley & Kohler 1980).

Two behavioural taxes that reduce the risk of inadvertent dislodgement are *rheotaxis* (orientating towards currents) and *thigmotaxis* (active response to touch). Most amphipod and isopod crustaceans, for example, are negatively phototactic and positively rheotactic and thigmotactic such that they avoid bright light, hide in crevices or under stones and crawl or swim mainly upstream. This reduces drag and the risk of being transported downstream, possibly to less favourable circumstances or exposure to drift-feeding predators (Covich & Thorp 2010). In flatworms, thigmotaxis plays a role in a range of behaviours, such as choosing substratum, hunting prey and avoiding predation. Positive rheotaxis in fish involves mainly vision (positioning in relation to surrounding habitat features) and the lateral line system (e.g. Suli et al. 2012) and orientates them facing into the current where their streamlined

shape reduces the effort required to maintain station and, in drift feeders, increases encounter rate with floating food particles. Less-streamlined (often benthic) species avoid currents through negative rheotaxis.

Whilst a range of behavioural mechanisms may help under 'normal', albeit challenging, conditions in streams and rivers, additional mechanisms are needed in the face of environmental extremes. We return to the concept of 'disturbance'—the ecological effects of physicochemical events—and the response of populations and communities, in subsequent chapters (particularly Chapters 5 and 6). Here we deal mainly with responses and adaptations of individuals.

From laboratory studies (see Statzner et al. 1988) it seems that many lotic species can actually withstand quite high water velocities (> 1.5–2.0 m s^{-1}) without being dislodged and washed away, even though such velocities near the stream bed (i.e. where most animals live) are fairly rare in natural channels. So how can they do it? Body shape, size and a range of other adaptations such as hooks and grapples, friction pads, ballast and silk are important in this context. But these cannot provide the complete answer. First, simply resisting the flow appears to be energetically 'expensive'. For the caddisfly *Micropterna*, for example, a large part of the entire energy budget is expended simply by moving against the current (Bournaud 1975). Second, flow forces fluctuate. In turbulent conditions, there are frequent brief (seconds) accelerations of flow at small spatial scales (cm). There are then larger-scale and longer-term increases in velocity and shear stress during spates (many of them more or less seasonal), which would exceed the ability of most organisms simply to resist the drag and lift forces imposed. Escaping or avoiding the highest velocities is an alternative.

For a long time, it was generally assumed that the boundary layer (Section 2.3.3, p. 37) enabled the smaller stream organisms to escape the harshest flow forces. For instance, unicellular algae, growing in crustose or felt-like 'turfs' a few tens of μm thick, live mainly in the boundary layer. Crevices and roughness of the substratum surface also offer shelter to tiny microorganisms, plants and animals. Many of the morphological characteristics of benthic invertebrates described earlier also appear to allow them to exploit the reduced flow forces close to the bed. However, most lotic biota do not seem particularly well adapted to hydraulic stress (Waringer et al. 2020). Further, it is now clear that the viscous sublayer is probably thinner than previously thought and that most benthic macroinvertebrates of 'hard' substrata, even including streamlined and dorsoventrally flattened taxa, probably experience rather complex turbulent flows, and consequently must endure the forces of flow rather than escape them.

Use of, or movement into, different microhabitats that offer some refuge against the current is a common behavioural response to changes in flow conditions. Many macrophytes, which themselves do not live in areas of very high water velocity, provide protection and shelter from the current for a variety of animals. *Myriophyllum*, with finely divided leaves and large surface area (similar to *Ranunculus peltatus*; Figure 4.5), *Potamogeton crispus* with its crenulated leaves, and clumps of mosses, all have high invertebrate densities. On the other hand, plants with smooth, linear leaves like *Vallisneria* and *Sparganium* (Figure 4.5) are generally the least populated. Similarly, use of leaf packs and marginal areas can offer protection from the current during normal flow conditions. As discussed earlier, flow is weaker downstream of individual large substratum particles (e.g. cobbles, boulders) or other obstructions (such as woody debris and riparian tree roots), and small recirculating eddies are established. Larger animals, including fish, use such areas behind larger structures whereas smaller animals can reside in the lee of smaller particles, or within crevices and holes. As we discuss in some detail in Chapter 5 (section 5.3.4), the use of flow refugia at a variety of scales is commonplace, including so-called hydraulic dead zones, accessing the hyporheic zone and at a larger scale lateral extensions of channels onto floodplains.

We have seen that stream animals can respond to high flows, but there are also periods of very low flow or when the channel may dry partially or completely—conditions which may be hazardous events for the fauna. The concept of refuges/ia is again important here, as discussed by Belinda Robson in Topic Box 4.1 above. Moving into the substratum (for small animals via spaces between substratum particles or for larger animals by actually burrowing) to take advantage of subsurface water offers one mode of escaping very low flows and high summer temperature or surface drying. Some species have tolerant life stages (eggs or terrestrial adults) which survive the conditions if produced 'at the right time'. Alternatively, there are physiological adaptations involving dormancy or diapause in the larval or egg stage, as seen in many invertebrates (see Section 4.4.3, p. 123). Hynes's classic book (1970b) reviews much of this literature. We will come back to the use of refugia to cope with stream drying again in Chapter 5 (section 5.3.4).

Where adverse conditions are more or less predictable (e.g. when related to seasonal changes), or can be 'forecast' in some way, individual evolutionary behavioural or other adaptations can be of great selective advantage. For instance, the large flightless belostomatid bug *Abedus herberti* lives in the desert streams of the American south-west, which are susceptible to seasonal drying and violent flash floods. Remarkably, before the flood arrives, the bug crawls out of the stream into riparian areas, presumably in early response to a minor increase in flow caused by the onset of rainfall (Lytle et al. 2008; Figure 4.13). Mortality from flash floods in *Abedus* populations is around 15%, compared to the 95% displacement of mortality of most other taxa. Life-cycle events, which we turn to in

Figure 4.13 The large flightless belostomatid bug *Abedus herberti* can crawl out of its desert stream home onto the banks before flash floods, and walk from stream to stream. This photo is of a male bearing eggs on his back.
Source: photo with kind permission of Ivan Phillipsen.

the next section, can be 'keyed in' to predictable events. For example, the emergence of adults of the sycamore caddis (*Phylloicus mexicanus*) occurs ahead of the flash-flooding monsoon season in the Chihuahuan Desert, Arizona, USA (see Lytle 2008), avoiding catastrophic larval mortality.

4.4 Reproduction, life cycles and life histories

The terms 'life cycle' and 'life history' have been used in distinct ways by stream ecologists (e.g. Butler 1984; Resh & Rosenberg 2010). A *life cycle* is the general sequence of morphological stages and physiological processes through which an individual of a species passes during its life, effectively linking one generation to the next. The qualitative and quantitative details of events associated with the life cycle make up the *life history*, and these include at least nine elements: recruitment, mortality (or survival), growth, development, dormancy, reproduction, dispersal, voltinism (the number of generations per year) and phenology (the timing of a natural event). The life history of a species is therefore the combination of these processes into a complex adaptation which, in turn, may be said to form a *life-history strategy* (Wilco et al. 2008). It is wise to be a little wary of the word 'strategy' in ecology, since it may bring with it concepts of 'purpose' or 'conscious intention'. Rather, a life-history strategy is simply a combination of biological features that may together be of fitness value to an individual, may vary among individuals and may be subject to natural selection.

The study of life cycles and comparisons of life-history patterns have a very long history in freshwater ecology, particularly on the larger benthic invertebrates, due partly to their high diversity and ecological importance and partly, it has to be acknowledged, because they are 'relatively' easy to work with! In temperate streams, for example, many species have a burst of recruitment as eggs hatch in spring or summer and the juveniles grow quite quickly, producing a clear sequential appearance of progressively larger individuals over the season as they move through their life cycle. There is, of course, a lot of variation in this process, as different species progress at different rates and may be more or less prominent in samples taken at different times. Even in the largely aseasonal tropics, life-cycle patterns are often evident as organisms grow and mature.

4.4.1 Life-history patterns

The general features of the life cycle are essentially fixed, such that all aquatic Diptera, for example, have egg, larval, pupal and adult stages, whereas the details of life history vary within and between species. Thus, the duration of stages, the number of larval instars, the activity of the pupa (where present) and the emergence and flight period of the adult can all vary, linked to the nature of the prevailing environment. The number of larval instars varies among the insect orders, from three in Neuroptera, to four to five in Hemiptera, four to seven in Diptera, five to eight in Trichoptera, 12 to 22 in Plecoptera, and 15 to 50 in Ephemeroptera. The duration of each instar also varies but, generally, the later instars last longer. In other arthropod taxa, such as Crustacea and Arachnida, and in many Mollusca, life cycles are more complex (see Chapter 3).

Two aspects of life history are especially important: *voltinism* and *phenology*. To determine life histories, biologists try to follow the development and progression of individuals derived from one reproductive period (a *cohort*) through the various life-cycle stages or size classes over time (usually based on linear measurements of body parts such as head capsule width, body length, etc). There is a vast body of literature based on such approaches, although the life history of some groups is much better known than that of others. We will highlight some key aspects here but readers should consult Merritt et al. (2019) and Thorp & Rogers (2014), who provide a wealth of information of wide applicability.

Variation in life-history patterns can be attributed to both intrinsic and extrinsic factors. Intrinsic factors, including physiology, morphology and behaviour, tend to restrict life-history characteristics within certain inherited ranges. Thus, freshwater crustaceans, rotifers, fish and molluscs spend their entire life cycle in water, whereas most aquatic insects and some amphibia are only aquatic during larval stages. As discussed in Chapter 3 (section 3.4.7) insects show either complete or incomplete metamorphosis from larva to adult. In the former, holometabolous, life cycle (Megaloptera, Neuroptera, Diptera, Coleoptera, Lepidoptera and Trichoptera), individuals pass through egg, larval and pupal stages before metamorphosing to the adult. The larva does not resemble the adult. In the latter, hemimetabolous, life cycle (Ephemeroptera, Plecoptera, Odonata and Hemiptera), individuals pass from egg through larval stages (instars) that increasingly resemble the adult, while the reproductive organs and wings develop gradually (the latter externally to

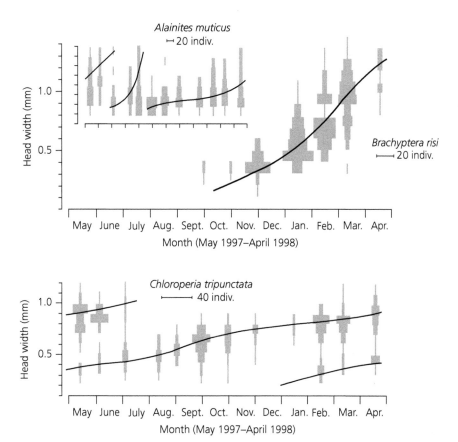

Figure 4.14 Life history of a mayfly (*Alainites* (formerly *Baetis*) *muticus*) and two stonefly species (*Brachyptera risi* and *Chloroperla tripunctata*) in the River Araglin catchment in Ireland, shown as changes in head width distribution of cohorts over time.
Source: Giller 2020, with permission from UCD Press.

the body). Mayflies actually enter a 'sub-imago' stage, before the adult, where they are able to fly, but only become mature after a further moult. The larval stages of lotic insects are usually the most prolonged element of the life cycle and the instars are progressively larger (Figure 4.14), each shedding the old exoskeleton and the larva subsequently growing into the new (larger) one such that individual growth occurs in a series of step-like increases in size.

Extrinsic environmental factors such as temperature, photoperiod, nutrition, degree of habitat permanence and presence of other taxa can influence most life-history parameters. Temperature influences the rates of growth, development and metabolism, as well as indirectly affecting food availability. Diverse life-history patterns within the insect community determine the exploitation of seasonally available food (e.g. peaks in primary production, or allochthonous leaf fall), and the emergence of adults is timed in relation to the

terrestrial environmental conditions and to possible competitors and predators (Wallace & Anderson 1996). Even the tropics, the environment is seasonal in terms of precipitation (see e.g. Wolda 1987) and life cycles often relate to the rainy season or the filling of the floodplain (Butler 1984).

4.4.2 Voltinism and longevity

Very small organisms normally have a short life cycle and can complete many generations relatively rapidly, whereas the development of large organisms naturally takes longer. However, even within a species, prevailing conditions, and especially temperature, can determine how many generations can be completed within the year, and there is often significant life-history plasticity. Thus, the short summers at high latitudes allow fewer generations than at lower latitudes. For instance, in some dragonflies voltinism ranges along

a latitudinal gradient in Europe between two generations per year (bivoltine) to one every two years
(semivoltine) (see e.g. Flenner et al. 2010).

Univoltine species (one generation per year) tend
to be seasonal and have non-overlapping generations.
Examples can be seen among sponges and flatworms
(Frost 1991; Kolasa 2000). Pulmonate gastropod snails,
such as *Lymnaea* and *Physa*, reproduce in spring with a
complete separation of generations. This single lifetime
reproduction is known as a *semelparous* life cycle. Most
amphipods and isopod crustaceans also tend to be
annual and sometimes semelparous. Among the northern temperate insects, including many mayflies, stoneflies and true flies, an annual cycle is again the most
common, although the life cycle itself can vary from a
few weeks to several months. An extreme example is of
the baetid mayfly *Fallceon quilleri* which develops from
egg to adult in 9–11 days (Gray 1981).

Some other groups are clearly univoltine and annual
but have partially overlapping generations. That is,
new recruits in one generation are present at the same
time as older individuals of the previous generation
Thus, many Trichoptera species are present as larvae
in benthic samples for most of the year, a situation
also found among amphipods, freshwater limpets and
some leeches.

Multivoltine species have several generations per
year. Very small-bodied taxa like protozoans, rotifers
and microcrustaceans generally have rapid development, while hermaphrodite flatworms are usually multivoltine (Kolasa 2000). Most pulmonate snails tend to
reproduce once (semelparity) and live no more than
a year, whereas caenogastropods (a large and diverse
group including freshwater pulmonates) commonly
reproduce several times (iteroparity) and have a life
cycle of up to 4–5 years (Thorp & Rogers 2011). Tropical
and subtropical gastropods also often have two or three
bouts of reproduction per year, with various degrees
of overlap in generations (Brown 1991). In insects,
multivoltinism is very common amongst the Diptera,
especially chironomids and small mayfly species, but
in other groups it is less common and largely depends
on the geographic location and favourableness of the
climate—it increases towards lower latitudes. A number of tropical mountain rainforest stream insects, however, show aseasonal life cycles with continuous hatching and larval growth (Marchant & Yule 1996); in these
cases, the larger the insect, the longer the life cycle.

Some taxa have life cycles exceeding one year,
either reproducing several times during their life (an
iteroparous cycle) or having a relatively long pre-
reproductive period before a single reproductive event.

Some tubificid worms mature in their first year, reproduce, resorb their gonads, mature and reproduce again
in the second year, and then die (Brinckhurst &
Gelder 1991). Many gastropod molluscs are perennial,
especially prosobranchs (e.g. Hydrobiidae), and the
unionacean bivalves can live for over a century and
do not mature for over 20 years. Decapod crustaceans,
like crayfish, live for several years, moulting up to 11
times and tripling in length in the first year. Among the
insects, it is usually the relatively large, often predatory
or wood-feeding taxa, that live for more than one year.
Perlid stoneflies and some caddis species may live for
two to four years, large dragonflies even longer. The
length of the life cycle can vary with latitude, as seen in
the perlid stonefly *Dinocras cephalotes*, which takes just
two years to complete its life cycle in Spain, three in
central Europe but four to six in Scandinavia (Bonada
& Dolédec 2018).

At a smaller spatial scale, subtle water temperature differences can advance emergence of adults by
increasing the growth of larvae, as in the earlier onset
and peak emergence of the stonefly *Amphinemura
nigritta* in the warmer (~ 1.5 °C) headwater streams
in the north-eastern USA (Cheney et al. 2019). Annual
differences in temperature can also shift emergence,
as seen in the earlier peak emergence of a number of
stonefly species in warmer years in the Rio Conejos
of southern Colorado (DeWalt & Stewart 1995). These
clear temperature-related changes in life history of
populations raise the spectre of similar shifts associated with human alterations of river systems (Cheney
et al. 2019) through, for example, logging of riparian
trees, stream regulation, thermal pollution from effluents and, of course, progressive global warming (see
Chapter 10). Indeed, in the famous Breitenbach stream
in Germany, an increase in the mean duration of insect
emergence by over 15 days and an earlier peak in emergence by 13.4 days—associated with a increase in mean
temperature of >1.8 °C over 42 years—is perhaps a
foretaste of what might be to come (Baranov et al. 2020).

Overall, stream invertebrate life cycles are very
variable, from multivoltine to long-lived, sometimes
with pauses in development in the egg or larval
stages, usually semelparous reproduction, but sometimes iteroparous and sometimes synchronous other
times not. Despite this variability, Hynes (1970b)
attempted a simple classification of life cycles of invertebrates from temperate streams, distinguishing three
main types:

1. *Slow seasonal cycles* show a distinct change in size
   distributions with time. The eggs may hatch soon

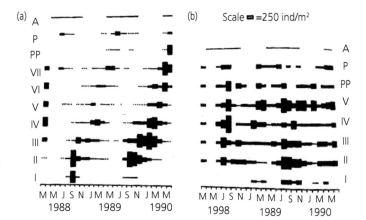

Figure 4.15 Life-history patterns of two caddis species over a three-year period in an Irish Stream. The diagram shows the number of individuals m⁻² from monthly benthic samples. (a) *Agapetus fuscipes* shows a clear univoltine seasonal cycle, (b) *Rhyacophila dorsalis* has a complex bivoltine life history with overlapping generations.
Source: N. Sangpradub and P. Giller, unpublished.

after laying, the larvae growing towards maturity nearly a year later (e.g. Figure 4.15a), or hatching may extend over a relatively long period resulting in continuous recruitment over time (e.g. univoltine crustaceans, and many stoneflies, mayflies and caddisflies).

2. *Fast seasonal cycles* show rapid growth following a long egg or larval diapause (a genetically programmed dormancy initiated in a particular life-cycle stage) or after one or more intermediate generations. Typical fast-cycle insect species include representatives of the mayfly genus *Baetis* and many blackfly genera, while some caddis show rapid growth after a long egg diapause.

3. *Non-seasonal cycles* occur where individuals of several stages or size classes are present in all seasons (e.g. the caddis *Rhyacophila*; Figure 4.15b). This may result from lifespans exceeding one year (e.g. large stoneflies) or taxa which have multiple overlapping generations (e.g. many molluscs). This type of life cycle is the predominant one in the southern hemisphere and tropics (Wallace & Anderson, 1996).

Among vertebrates, small amphibians and fish are mainly annual; larger fish and reptiles are perennial and often long-lived. Most vertebrates have distinct breeding seasons, and both iteroparity and semelparity exist. Generations frequently overlap, and even individuals within the same species may differ, as we discuss in section 4.4.4.

4.4.3 Phenology and life-history responses to environmental conditions

If the prevailing environment plays such a major role in the timing of various life-cycle processes, we might

expect a particularly clear pattern in species inhabiting streams and rivers that flow intermittently. As discussed earlier, small size, rapid development and an egg and/or larval diapause are commonly cited as adaptations to conditions in such habitats. A couple of examples will illustrate the point. Eggs of mayflies in temporary streams in western Oregon begin to hatch in late autumn, with the onset of flow, but the hatching period is long, and larval development extends over 5–7 months with some larvae delaying hatching until the spring (Dietrich & Anderson 1995). These survive summer drought in the few remaining permanent pools. Emergence of adults also tends to be extended, although peak emergence precedes the summer drought. In New Mexico, the eggs of stoneflies are deposited when the stream is flowing and remain in the substratum for several months when it dries, resuming development when flow returns (Jacobi & Cary 1996). However, mortality is high during this period. Dormancy is thus an important feature of the life cycle under such conditions, as we saw for example in Megaloptera in the pupal stage (Figure 4.2, p. 107).

Even in more benign conditions, there are seasonal variations in temperature, oxygen and food availability, and periods when conditions are more likely to be favourable for dispersal and reproduction. Two taxa may have similar life cycles and voltinism, yet their phenology can differ substantially. There are often long periods of the year when many taxa are absent from samples (usually winter or summer) or are present only as small individuals. This often involves a diapausing egg or pupal stage, or cessation of growth during either summer or winter. Phenological differences between species may also be related to interspecific interactions. For example, Elliott (1995) suggested that differences in egg biology (in terms of development rate, timing of egg hatching etc.) among 12 species of carnivorous European Plecoptera might be important in avoiding competition. Such differences also occur among species of mayflies and caddisflies such that, at any one time, immature larvae or nymphs of similar species are at different stages of growth and hence one species always has larger individuals than another. Whether this temporal partitioning really does 'avoid competition', however, remains uncertain and is difficult to test. The timing of life history with that of the major prey organism, as shown by *Rhyacophila*, might also explain phenological differences between predator species.

4.4.4 Life-history plasticity

In addition to the differences in life cycles and life history between species, there is evidently tremendous intraspecific variation and flexibility in life history, both within and between populations, particularly in lotic insects. This may result from intrinsic differences among individuals within a population or in response to variation in environmental factors. The duration of egg development in non-diapausing invertebrates, for instance, is inversely related to temperature, and the development of arthropods is faster at higher temperatures. Nutrition can also influence voltinism (Butler 1984). Voltinism is thus rather labile, responding directly to the environment through local adaptation or phenotypic plasticity, within broad phylogenetic constraints. In their review of the life cycles of over 300 aquatic insects in Europe, Bonada & Dolédec (2018) found a greater prevalence of multivoltine life cycles in the Mediterranean Basin but more uni- or semivoltine life cycles in Scandinavia. This plasticity is most apparent in widely distributed species such as the mayfly *Baetis rhodani*, where life cycles vary from relatively synchronous and univoltine in Swedish Lapland to multivoltine cycles with overlapping cohorts from two or more generations a year in Austria, western Norway, southern Spain and Belgium (Sand & Brittain 2009). As a word of warning on phenotypic plasticity, we need to be aware of the possibility of cryptic species (genetically distinct taxa identified through molecular analysis but without obvious morphological differences) in very widely distributed species. Cryptic species are present for example in *Baetis rhodani* and other species of *Baetis* (Williams et al. 2006; Leys et al. 2016). Similar variation in latitudinal and altitudinal voltinism has been found in the related mayflies *Alainites muticus* and *Baetis alpinus* (Lopez-Rodriguez et al. 2008). While such

variation does occur in most mayfly species, this is not universal. Genetic constraints may limit voltinism, as in the mayfly *Leptophlebia*, which remains univoltine over a wide climatic and geographical range (Brittain 1982). Stoneflies, on the other hand, tend to have stable life-history patterns, although in *Nemoura trispinosa* the life cycle varies from a univoltine, slow seasonal type, to a univoltine fast seasonal type with extended egg development; the differences depend on maximum annual water temperature (Williams et al. 1995). This species exhibits 'eurythermal egg development'—there being major differences in the number of degree-days needed for egg development among local populations (Lillehammer et al. 1989)—and it switches between one- and two-year generation times depending on the local temperature regime and food supply.

This kind of ecological generalisation or plasticity may be of fitness advantage for life in waters with unpredictable flow. The basic components of life-cycle plasticity are prolonged hatching and emergence periods and a wide range of larval stages being present at any one time. This 'spreads the risk' of local extinction; some individuals are found in the different stages at any one time, making it more likely that the species will be relatively invulnerable to any one physical event (i.e. an aerial adult may be able to persist through drought; Dietrich & Anderson, 1995). Drought during the summer is predictable and the life-cycle adaptations described earlier, such as drought-resistant eggs, enable rapid recolonisation following the drought. Unpredictable winter or spring droughts do not allow such adaptations, however, and hence life-cycle plasticity is then a distinct advantage. Plasticity is also evident in temperate systems that are subject to unpredictable spates. For instance, coexisting caddis in the Glenfinnish River in Ireland have a wide degree of life-cycle flexibility (Sangpradub et al. 1999). *Agapetus fuscipes* eggs hatch fairly synchronously, but larval development is variable (Figure 4.15a). *Silo pallipes*, *Plectrocnemia conspersa*, *Philopotamus montanus* and *Sericostoma personatum* show great variation in development rate between individuals, some extending to a two-year cycle and leading to split cohorts. *Glossosoma conformis*, *Drusus annulatus*, *Potamophylax cingulatus* and *Halesus radiatus* have variably delayed egg hatching and first instars appear over several months. *Rhyacophila dorsalis* (Figure 4.15b) and *Odontocerum albicorne* both have two cohorts a year and overlapping generations. Flight periods also vary among the caddis species from one month to nine months, with most extending over five months. These variations in rate of development, the wide range of size classes present at any one time, the ability to overwinter in different larval stages and the asynchronous, extended flight periods, mean species can variously tolerate erratic year-to-year differences in conditions.

Similar plasticity is evident in other lower animal groups and often involves a combination of sexual and asexual reproduction, as in protozoa and sponges. Among the vertebrates, variable growth rates within populations of salmonids are well known, leading to some individuals being able to smoltify and go to sea after just one year, while others take two years. Atlantic Salmon (*Salmo salar*) are highly variable and can remain at sea for between one and, in extreme cases, five years, thereby spawning at vastly different sizes. Individual salmonids usually die following reproduction but others, even within the same species (e.g. steelhead, *Oncorhynchus mykiss*, and Atlantic salmon), may return to sea a second time and then back to fresh water to reproduce a second time (Bayley & Li 1996). Both sedentary and anadromous phenotypes occur among charr (*Salvelinus alpinus*) and brown trout (*Salmo trutta*) (e.g. brown trout and 'sea trout' are the same species, but only the latter individuals go to sea). Such flexibility is generally much less evident in longer-lived, higher vertebrates, apart from the ability to vary the number of offspring and social controls on the individuals that can reproduce (e.g. in territorial mammals like the otter, or birds like the dipper).

4.5 Foraging and trophic adaptations

Food for consumers in streams and rivers has been widely categorised into four groups: (1) detritus, dead organic matter of 'several' size classes—coarse, fine and dissolved; (2) various living green plants and photosynthetic microbes; (3) heterotrophic microbes, prominent in biofilms and as a part of the detrital/microbial complex; and (4) other animals, including living invertebrates and fish. We explore these in more detail in Chapter 8 in relation to organic matter dynamics, secondary production and the support of stream food webs. As far as concerns the foraging and feeding adaptations we consider here, we can concentrate on how animals use coarse and fine organic particles in the sediment and water, periphyton (algae and microbes incorporated in biofilms) and animal prey (Figure 4.16). With regard to the exploitation of these resources, stream ecologists have grouped animals (particularly invertebrates) into several *functional feeding groups*.

4.5.1 Functional feeding groups

A large proportion of stream-living invertebrates are polyphagous (able to feed on a variety of food), although different animals access the various kinds of food present in running waters in different ways. On this basis, Cummins (1973) categorised animals into a number of functional feeding groups (FFGs). The

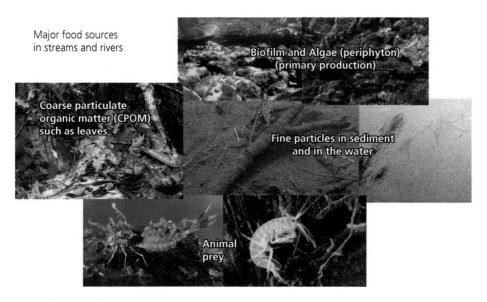

Figure 4.16 Major food sources for stream and river organisms.

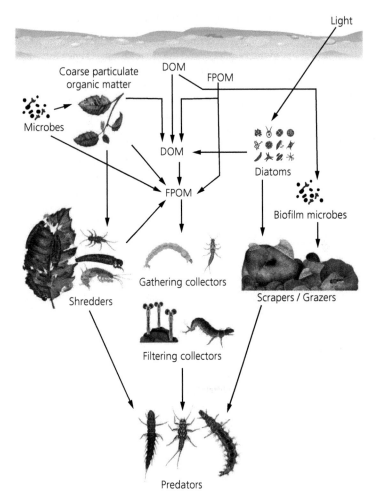

Light

Coarse particulate
organic matter

DOM

FPOM

Microbes

DOM

Diatoms

FPOM

Biofilm microbes

Shredders

Gathering collectors

Scrapers / Grazers

Filtering collectors

Predators

Figure 4.17 A conventional conceptual model of food sources available to macroinvertebrates, organised into functional feeding groups. Scrapers/grazers ingest algae (e.g. diatoms) and biofilm. CPOM from the riparian zone and instream macrophytes is consumed by shredders once it is conditioned by microbes (hyphomycete fungi and bacteria). Shredders convert this coarse plant tissue into FPOM, which is transferred to collectors (filter feeders or gathering collectors). Predators feed on all FFGs. *Source:* modified and redrawn from Feeley et al. 2020, adapted from Cummins 1974, 2016; with permission from UCD Press.

concept was initially developed for insects but has since been extended to include other invertebrates and focuses on the morphology and behaviour involved in feeding. That is, it focuses on 'how' the animals feed rather than on exactly 'what' they feed on. This is schematically illustrated Figure 4.17 and described in more detail in Table 4.1.

Despite the appeal of this general scheme, and its widespread use across the literature, it is not without problems in its application. Although some taxonomic groups tend to be specialised with respect to feeding modes and diet, such as blackfly larvae and unionid bivalves as filterer-collectors, many other taxonomic groups contain genera, or even species within genera, that exploit widely disparate food resources. As Covich (1988) suggested, neotropical streams are dominated by generalist consumers, clearly seen in the study by Tomonova et al. (2006) on a number of rivers

in the foothills of the Bolivian Andes. Indeed, dietary generalism (regardless of functional feeding group) seems common among many stream and river animals (Chapters 7 and 8). In addition, some species seem to switch from one functional feeding group to another during their development, some of the limnephilid cased caddisflies being a prime example. Indeed, it is likely that most aquatic insects, including predators, are facultative gathering-collectors as early instars (Merritt et al. 2017). Lastly, food availability and hence a species's diet can also change as a function of habitat, season or even sex. This further complicates any accurate classification of species into functional feeding group and poses the question how does one categorise such clearly omnivorous species?

Whilst the value of this concept has thus been challenged, especially on the basis that often individual species thought of as belonging to a particular FFG

Table 4.1 The feeding mechanism and likely dominant food of functional feeding groups of aquatic macroinvertebrates (larval stages and adults), based on Cummins's (1973) classification with some examples of invertebrate taxa.

Functional feeding group (general category based on feeding acquisition adaptations)	Likely food	Feeding mechanism	Examples of invertebrate Taxa
Shredder herbivores	Living vascular hydrophyte plant tissue	Herbivores—chewers and miners of live macrophytes	Trichoptera: Phryganeidae, Leptoceridae
Shredder detritivores	Decomposing coarse particulate organic matter: vascular plant tissue (allochthonous and autochthonous) and wood	Detritivores—chewers, wood borers and gougers	Plecoptera: Nemouridae, Peltoperlidae. Diptera: Tipulidae. Trichoptera: Limnephilidae, Lepidostomatidae. Crustacea: Amphipoda, Decapoda.
Filtering collectors	Decomposing fine detritus, bacteria, algae and animal fragments suspended in the water column	Detritivores—filterers or suspension feeders	Trichoptera: Hydropsychidae. Diptera: Simuliidae. Unionid molluscs
Gathering collectors	Decomposing fine particulate organic matter and algae deposited on the substratum	Detritivores—gatherers or deposit (sediment) feeders (includes surface film feeders)	Ephemeroptera: Ephemeridae. Diptera: Chironomidae. Plecoptera: Leuctridae. Oligochaetes.
Scrapers	Biofilm including periphytic algae and cyanobacteria, heteroptorhic microbes, detrital particles, very small animals, incorporated in a polysaccharide matrix	Herbivores—grazing scrapers of mineral and organic surfaces	Trichoptera: Glossosomatidae. Coleoptera: Psephenidae. Ephemeroptera: Heptageniidae. Gastropoda: Planorbidae.
Piercer herbivores		Herbivores that suck the contents of algal cells	Trichoptera: Hydroptilidae.
Predators	Living animal tissue	Carnivores—attack prey, pierce tissues and cells, and suck fluids	Hemiptera: Belostomatidae, Naucoridae, Notonectidae. Platyhelminthes: Tricladida.
	Living animal tissue	Carnivores—ingest whole or parts of animals	Odonata. Plecoptera: Perlidae. Megaloptera: Corydalidae, Sialidae. Trichoptera: Rhyacophilidae, Polycentropodiae.

(Modified after Merritt, Cummins and Berg 2017.)

are often found to take food other than what they are 'assumed to' based on their classification, Cummins (2016) has strongly defended the original concept. The example he uses highlights the key role of the local environment in determining the actual food resources that are collected in the same way. The caddisfly *Glossosoma nigrior* feeds on the exposed surfaces of stones and uses its mandibles to scrape the attached epilithic biofilm (or 'periphyton') from rock surfaces. In an open-canopy stream, green algae dominated both the periphyton and the gut contents of all five larval instars, consistent with its classification as a scraper.

In a closed-canopy stream, the biofilm contained some diatoms and a higher component of fine detrital particles. This was reflected in the gut contents. Therefore, based on gut contents alone, larvae from the shaded stream might be classified as gathering collectors. Thus larvae in two different streams with the same feeding mode and in the same general habitat type, and with the same morphological adaptations, had different gut contents. It is important to recall that it is the mode of foraging that is being classified—the actual food consumed by the different functional groups is not defined.

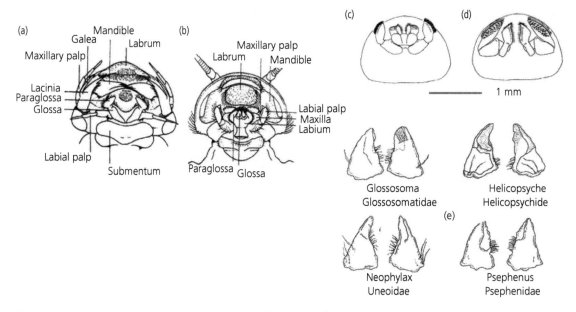

Figure 4.18 Detailed structure of the mouthparts of a larval stonefly (a) and mayfly (b) illustrating the major anatomical elements; ventral view of main mouthpart structures of *Rhithrogena pellucida* showing (c) the labium and brush-like labial palps and (d) maxillae and maxillary palps with special pectinate (comb-like) setae. (e) The mandibular structure of four different scraper taxa representing three families of Trichoptera (Glossosomatidae, Helicopsychidae and Uenoidae) and a coleopteran family (Psephenidae). All the mandibles have a flat, sharp edge used to scrape attached periphyton from hard substrata, while the concave inner surface has setal brushes that move food into the mouth.
Source: (a) and (b) from Giller & Malmqvist 1998; (c) and (d) modified from McShaffrey & McCafferty 1988, with permission from University of Chicago Press; (e) from Merritt et al. 2017, with permission of Elsevier.

Despite the challenges, the functional group concept has been popular for a number of reasons. Foremost, it offers to provide a way of linking the structure of aquatic communities (in terms of relative abundance of different functional feeding groups) with the general availability of certain types of resources and ecosystem processes (but see Lauridsen et al. 2014, discussed in Chapter 8, section 8.6.3). This has led to the view that functional group structure of macroinvertebrate communities has the potential to be used to assess stream ecosystem attributes (e.g. Cummins 2016; Merritt et al. 2017). It also illustrates the coupling between morphological adaptations of the animals and their use of various food resources. It is accepted that within each FFG there are obligate and facultative members which can relate to different species or different stages in the life cycle of a given species. It remains to be seen the degree to which the functional feeding group concept can provide a reliable and useful measure of real ecosystem processes.

4.5.2 Morphological aspects of feeding

Since morphology is intimately related to ecological relationships of a species (Ricklefs & Miles 1994), a similar morphology between species may reflect common ecological relationships and capabilities, although it may also result from a shared ancestry (e.g. Douglas & Matthews 1992). In relation to the mode of feeding and diet in stream organisms, mouthpart morphology is of particular importance, and has been the basis for comparison among taxa.

The mouthparts of lotic insects, for example, vary greatly, but are 'built' on the basic insect morphology shown in Figure 4.18a and b. Different groups have structures suited to piercing, biting, sucking, scraping, brushing, browsing, filtering, holding and grinding their food. As mentioned above, mouthpart organisation does reflect phylogenetic relationships, but significant variation also occurs within orders. For example, a comparison of the labral fan morphology of blackfly larvae used for filtering FPOM (Figure 4.19) showed that fan area, ray length and thickness and other measures, varied systematically among species inhabiting a range of habitats from slow streams to large rivers (Zhang & Malmqvist 1996). Detailed anatomical studies have also shown a variety of feeding behaviour and associated mouthpart adaptations within the mayflies of the Macae River basin in Brazil (Baptista et al. 2006). In the Leptophlebiidae

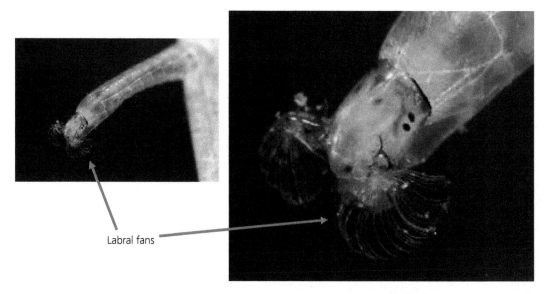

Labral fans

Figure 4.19 A *Simulium* larva and close-up of its head capsule illustrating the primary labral fans used as a filtering organ. *Source:* photo with permission from Jan Hamrsky, lifeinfreshwater.net.

for example, the distal part of the maxilla ends in a tuft of brush-shaped setae which collect food particles from the substratum. The maxillary palps remove food particles from the brushes, taking them towards the mandibles and hypopharynx, while the labrum and labium assist in retaining food. The labial palps are important in producing a water current towards the prebuccal cavity. There is an exception in *Hylister plaumanni*, which has reduced glossa and paraglossa and long fringes of setae which, together with the labial palps and maxilla, filter suspended organic material in the stream. The Baetidae have long labial palps and articulated maxillary palps used to gather and manipulate detritus and algae. The paraglossae and glossae have a few short setae and a crown of chitinous teeth on the tip of the maxilla, and the toothed tips of both mandible and galea-lacinia act as a 'reaping' rather than a 'scraping' device—that is, they remove taller or less closely attached material. In the Leptohyphidae, the distal part of the mandibles bears two chitinous wedge-shaped teeth and the molar part is covered by robust short 'spicules' (tough, spine-like structures) appropriate for scraping more closely attached periphyton.

Even within a single species, different elements of the mouthparts can be used in distinct ways. In *Rhithrogena pellucida*, a dorsoventrally flattened heptageniid mayfly, McShaffrey & McCafferty (1988) describe two feeding cycles. In a 'labial brushing cycle' larvae used the labial palps (Figure 4.18c) to brush loosely accreted

material from the substratum, while in a 'maxillary scraping cycle' special comb-like setae on the maxillary palps (Figure 4.18d) removed more tightly attached biofilm. In both cycles, the legs also removed material from the substratum.

As might be expected, there are cases of convergence of mouthpart morphology among unrelated taxa that exploit similar food resources. For instance, the larvae of several families of grazing Coleoptera and Trichoptera have almost identical mandibular structures used in scraping periphyton (Figure 4.18e). Similarly, in freshwater snails, periphyton is rasped from rocks and macrophytes with a file-like *radula*. This consists of many minute, hard teeth arranged in regular patterns, specific to particular species.

Predatory insects, including many of the large stoneflies and free-living caddis larvae like rhyacophilids, seize and cut up prey in simple jaws (Martin & Mackay 1982) whilst others, such as polycentropodid (web-spinning) caddis larvae, often engulf their prey in one piece. In stonefly larvae the laciniae (shown in Figure 4.18a) are used to capture prey (Tierno de Figuero & Lopez-Rodriguez 2019). Alderfly larvae, including *Sialis*, also have biting mandibles with sharp teeth and usually engulf prey items.

Piercing mouthparts are found among a variety of predatory insects. The mandibles of the dytiscid beetles have a tubular channel through which enzymes can be injected into the prey and the dissolved tissues sucked out. Similarly, prey of Hemiptera like notonectids

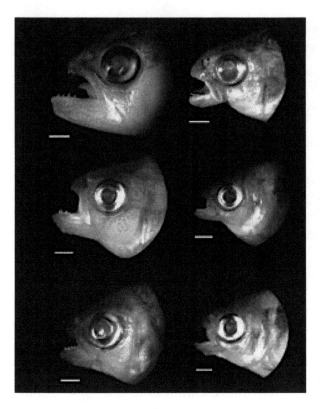

Figure 4.20 Variation in the morphology of feeding apparatus in coexisting species pairs of Characidae fish in neotropical rivers: top row—Perequê-acú River, *Hollandichthys multifasciatus* and *Bryconamericus microcephalus*; middle row—Ubatiba and Caranguejo Rivers, *Astyanax janeiroensis* and *Astyanax hastatus*; bottom row—Guapiaçú River, *Astyanax taeniatus* and *Astyanax* sp. Scale bars: 5 mm.
Source: modified from Portella et al. 2017, with permission from John Wiley & Sons.

and gerrids are typically 'harpooned' by mandibular stylets. The long, flexible and barbed maxillary stylets reach deep into the prey, lacerating the tissues. Predatory water bugs have toxic saliva that paralyses the prey, and the larval ibis fly (*Atherix ibis*) also immobilises its prey with a toxic injection before sucking out the contents. Not all piercing/sucking lotic insects are predators, though. For instance, larvae of the hydroptilid caddis *Hydroptila consililis* can breach individual living cells within the filaments of the alga *Cladophora* and suck out the contents and its relative, *Ochrotrichia spinosa*, has robust and cusped mandibles that can both pierce *Cladophora* and scrape algae from the substratum (Keiper & Foote 2000).

Filter-feeding invertebrates show extensive morphological specialisation, with some of the most striking examples found in blackfly larvae. The filtering fans, which are modified from labral mouthparts, are situated anteriorly on the head (Figure 4.19) and 'passively' remove fine organic particles from the water. These organs are quite complex, with each labral fan comprised of secondary, median and 'scale' fans in addition to the large primary fan (Zhang & Malmqvist 1996). The primary fan, when fully extended, has the shape of a hemisphere, concave against the direction of flow, and has 30–70 fan rays (depending on instar and species). The fans are periodically folded in and cleaned by the mandibles and labrum. Most other lotic filter-feeders trap food particles with special organs or devices (such as silken nets in net-spinning caddis; Figure 4.11, p. 115).

Among the vertebrates, the lotic fish of neotropical streams illustrate the close relationships between morphology, habitat and food type. The Characidae are a diverse group of neotropical fish, with a variety of body forms and behaviours related to the exploitation of different microhabitats and environments. Morphological features related to feeding behaviour and diet include head length, mouth width and height (fish with larger mouths can normally capture larger prey), eye diameter (smaller prey can be detected with bigger eyes) and position on the head (influences the ability to detect items on the water surface), and intestinal length (longer intestines indicate a plant diet) (Portella et al. 2017). In each of four different South American coastal rainforest streams, coexisting pairs of characin species are distinguished mainly by body size (Figure 4.20). In each case, there is one slightly larger species (with a bigger head and mouth) taking larger items of allochthonous origin (arthropods or plant litter) and a second smaller species (with a smaller head but relatively large eyes and greater visual sensitivity and acuity) taking smaller autochthonous items (again arthropods or algae). This may be significant in niche partitioning and coexistence.

Dentition, gill raker architecture and stomach and intestinal organisation have also been identified as adaptations to food selection and feeding in Amazonian fish. Ramirez et al. (2015) illustrate the significant variation in morphology amongst the fish and their diet. They showed that herbivores and piscivores have long and short intestines, respectively. Carnivorous fish have fang-like teeth, omnivores have terminal mouths, while aquatic, invertivorous fish, such as characins, have multicuspid teeth. Of the 19 species analysed, several could be characterised as opportunists with respect to diet and morphology, species with similar morphology also had similar diets, whilst others had unique morphology and were food specialists not 'sharing' their diet with others.

As a final example, catfish (Siluriformes, Osteichthyes) are one of the most abundant groups in tributaries of the Paraná River in Brazil and can be separated into three trophic guilds (detritivores, insectivores and omnivores) and two ecomorphological groups; detritivores (Lorocariidae) and insectivores/omnivores (Heptapteridae) (Pagotta et al. 2011). Detritivores have long caudal peduncles (the tapered region behind the dorsal fins) and large caudal and pectoral fins that increase the area of contact with the substratum that, together with dorsoventral flattening, allow maintenance of position and balance while the fish feed on the encrusted organic material on stones in the turbulent rapids. The insectivores/omnivores, in contrast, have dorsoventrally compressed bodies and large anal fins, enhancing manoeuvrability, and are able to swim continuously among the rocks and backwater zones where they feed.

4.5.3 Generalists and specialists

Based on simple visual examination of gut contents, many non-predatory stream invertebrates would seem to be rather generalist and opportunistic feeders, as indicated earlier. It is often very difficult to identify exactly what the food is, and certainly what they assimilate. Algae are normally recognisable but much of the rest is often rather amorphous detritus. We revisit the question of exactly what food resources support secondary production and food webs in Chapter 8.

However, experimental studies suggest that (predominantly) detritivorous taxa, such as *Gammarus*, *Asellus* and limnephilid caddis larvae, can discriminate between autumn-shed tree leaves of different species, between leaves colonised by different fungi, and even between patches of different fungi on a single leaf (see Topic Box 3.1, p. 62).

Predatory species also often overlap in diet and are apparently opportunistic in the sense that they appear to eat whatever they can catch of a suitable size. Streams and their living communities are dynamic, with a continuous redistribution and change in abundance of animals, seasonal life cycles and variation in activity. Thus, the prey available is constantly changing, militating against dietary specialism, as is often reported for lotic predators (with the generalist brown trout being a good example). The physical environment and flow can intervene in predator–prey interactions. For instance, blackfly larvae (Simuliidae) are usually abundant at high current velocities, which may preclude predators unable to hunt in very fast flow, such as perlodid stonefly larvae. Predatory rhyacophilid (free-living Trichoptera) larvae, however, can cope well in the flow and capture blackfly larvae over a wide range of velocities (Malmqvist & Sackmann 1996). In fact *Rhyacophila obliterate* larvae are specialists and their phenology is 'tuned' to the abundance of blackflies in Finnish lake outlets near the Arctic Circle (Muotka & Penttinen 1994). The most specialised stream-living species are probably found among parasitoids, such as hymenopterans of the genus *Agriotypus* that are almost exclusively found on caddis hosts in the family Goeridae.

4.5.4 Ontogenetic changes

Ontogenetic dietary shifts—the change in diet of a consumer with size and age—are widespread across the animal kingdom, and certainly in streams and rivers (Sanchez-Hernandez et al. 2019). The maximum size of food items that can be ingested usually depends on 'gape'—essentially the size of the consumer's mouth—suctorial predators with piercing mouthparts being the main exception. Not surprisingly, therefore, maximum food particle size tends to increase with the size of these consumers—filterers of fine particles are of course an exception. Most engulfing animals also avoid food that is much smaller than they can readily collect, handle or ingest. The upper limit is, however, better defined than the lower size limit, and sometimes the range of food sizes simply expands with increasing consumer size (Figure 4.21). The relationship between body size

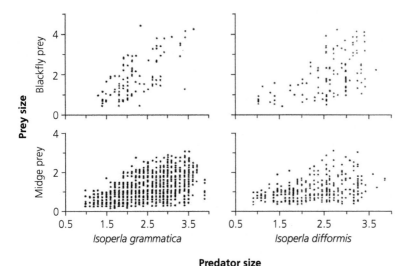

Figure 4.21 Changes in prey size with predator size (head width, mm) in nymphs of two isoperlid stoneflies feeding on chironomid (midge) and blackfly larvae. *Source:* from Malmqvist et al. 1991, with permission from Springer Nature.

and prey size is strong in fish, particularly in piscivores and other predators, largely because they do not have any appendages to manipulate prey. Their ability to handle prey thus generally scales with the gape which, in turn, scales allometrically with body size. Typically, dietary switches involve distinct shifts in prey sizes from millimetre to centimetre and finally to decimetre orders of magnitude (Sanchez-Hernandez et al. 2019).

Many lotic animals do not eat the same food throughout their development. Predatory invertebrate species often start out as herbivores or detritivores, as in predatory chironomids (Tanypodinae; Hildrew et al. 1985). In the stonefly genus *Isoperla* there are species that change from complete herbivory to carnivory as they grow. Similarly, a study of two sympatric *Isoperla* species in southern Sweden indicated that the diet not only differed between the species, but also according to season, locality and sex (Malmqvist et al. 1991). There are also clear ontogenetic dietary shifts in many Trichoptera, especially among the cased caddis, where a change from mainly detritivory to predominantly predatory behaviour has been recorded in some species (Giller & Sangpradub 1993).

In the case of fish, early in the life cycle, many species prey on phytoplankton, zooplankton or small macroinvertebrates, but may switch to larger macroinvertebrates, fish, plants or detritus later in development. Some species, such as many neotropical characins, undergo ontogenetic dietary shifts from terrestrial insects to fruits and leaves. A detailed study on nine sympatric piscivores in rivers in the lowland tropical grassland plains of Venezuela showed distinct shifts

from invertivory by small juveniles to piscivory by subadults and adults (Winemiller 1989; Figure 4.22). The pirhana *Pygocentrus* showed two distinct ontogenetic dietary shifts from primarily microcrustaceans to aquatic insects (at body length 20–40 mm), and then from aquatic insects to fish (at body length 40–60 mm). Similarly, between 50 and 60 mm, the tetra *Charax* shifted from primarily aquatic insects to feeding on small characins.

In their review, Sanchez-Hernandez et al. (2019) suggested that competition and predation risk are the major drivers of dietary shifts in fish. These drivers may not in themselves directly affect trophic ontogeny, but may have an indirect effect through ontogenetic shifts in habitat use and prey availability.

Many animals change their habitat at certain points in their life history. For example, following metamorphosis, amphibians and aquatic insects move between water and land. In diadromous or amphidromous fish (e.g. many salmonids, lampreys and galaxiids), developmental shifts accompany or anticiptate migration between freshwater and marine environments. Clearly, this phase involves a complete change in most aspects of their biology, including feeding and diet (prey size and species composition). A few examples will illustrate this.

Most insects shift from one food resource to another when entering the adult stage. Many species of blackflies shift from a larval diet of fine organic particles, including colloids and bacteria, to the ingestion of blood by flying adult females, required for egg production. Adult blackflies, particularly the males, also ingest energy-rich nectar. Many adult chironomids

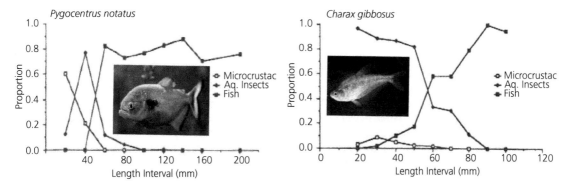

Figure 4.22 Examples of ontogentic dietary shift with body length in the tetra *Charax gibbosus* (glass headstander) and *Pygocentrus notatus/cariba* (black spot piranha) in a swamp-creek of the Rio Apure-Orinoco drainage in the western llanos of Venezuela. Graphs show relative proportion of diet of microcrustaceans, aquatic insects and fish.
Source: modified from Winemiller 1989, with permission from Springer Nature; photos of *Charax gibbosus* from www.fishbase.se by M. Casacuberta, and of *Pygocntrus* from https://www.georgiaaquarium.org/animal/black-spot-piranha/.

Figure 4.23 (a) Ammocoete larva of the lamprey *Lethenteron reissneri*; (b) adult western brook lamprey, *Lampetra richardsoni*.
Source: (a) by Loki Austanfell—own work, CC BY-SA 3.0, https://commons.wikimedia.org/w/index.php?curid=18814616; (b) Wikipedia Creative Commons, author USFWS Pacific Region.

appear to feed on nectar, but fresh fly droppings, pollen and aphid honeydew are also included in their diet (Armitage 1995). The adults of most stoneflies move after emergence to the riparian area where they stay from a few days up to several weeks. Species which have short adult lives mate and oviposit without feeding, whereas the more long-lived species feed on algae, leaves, twigs and dead insects (Lillehammer 1989). Adult caddisflies have sucking mouthparts and some species feed on nectar. In contrast, adult mayflies do not feed at all, whereas larval and adult dragonflies are both predatory. Semi-aquatic bugs and dytiscids are also predatory in both larval and adult stages.

A final example of drastically changing diets during life is found in lampreys (Malmqvist 1993; Kelly & King 2001). Larval lampreys (Figure 4.23a) live buried in the substratum of running waters. Currents of water created by the pumping movements of the pharynx provide oxygen and also bring small food particles

which are trapped in a network of mucous strands in the pharynx. After several years larval lampreys metamorphose, a process lasting several months, during which they cease to feed. After transformation, adult lampreys (Figure 4.23b) migrate downstream, either to the sea or to lakes or rivers, depending on species. Most species assume an ectoparasitic (blood-feeding) life on suitable species of fish. Non-parasitic lampreys do not feed in the adult stage but spawn in spring the year following the onset of metamorphosis, after about 9 months without food.

4.6 From Adaptations to species traits

4.6.1 Adaptations, traits and selection

In this chapter, we stress the biology and ecology of individual species—an approach sometimes called 'autecology'—and how they match the (often harsh) physical templet of rivers and streams. Natural selection is widely accepted as the mechanism that gives rise to adaptations. It is important to remember that natural selection is a retrospective and 'blind' process. It 'chooses' the fittest from the biological variation available and in relation to the environmental templet of rivers and streams. It modifies over time those characteristics that are heritable and within developmental limits set by the fundamental phylogeny of the group. In considering adaptation we normally compare the features of related species—classically through 'adaptive radiation' from a common ancestor. Perhaps not surprisingly, however, natural selection can also sometimes produce similar adaptive 'solutions' in species that are more or less unrelated—a process known as *convergent evolution*. The long, thin and flexible body shape of some interstitial crustaceans and insect larvae has converged with that of true 'worms', for instance.

The composite biological features of organisms have been called *species traits* and may together be adaptive in particular situations. Species traits are of great interest to ecologists because they point to the basic evolutionary relationship between living things and particular characteristics of their natural environment. Further, if we can match the occurrence of species traits with particular environmental conditions, that match should be consistent across large spatial scales, even if the actual species composition of the community differs biogeographically. That is, stony streams everywhere should have a similar representation of traits in the biota, even if in different biogeographic realms and thus with differences in species identity. If the environment changes due to human activities, moreover, then we might be able to match the representation of particular species traits with the incidence of pollution, for instance. Thus, species traits can be important in explaining community composition and potentially for biological monitoring—topics which we develop further in Chapter 6.

4.6.2 Traits, templets and the use of words

Here we consider further the habitat templet we introduced in Chapter 2 and the species traits of lotic organisms that relate to that templet. The habitat templet concept in general ecology was developed by Southwood (1977, 1988) and, in relation to the 'life strategies' of terrestrial plants, by Grime (1977, 2006). In freshwater ecology, it was adapted for the phytoplankton by Reynolds (1984), while Townsend & Hildrew (1994) defined a templet specifically for streams and rivers,

which has been further developed, modified and redefined by, among others, Statzner et al. (2001), Poff et al. (2006), Verberk et al. (2008, 2013), Statzner & Bêche (2010) and Schmera et al. (2015). Southwood's (1977, 1988) habitat templet concept suggested that, through their effects on the fitness of individual organisms in ecological time, certain combinations of adaptations maximising survival and reproduction are selected along gradients of habitat 'favourableness' and disturbance. He also plotted the likely impact of 'biotic agents' (such as predation and competition) on the templet. The representation of effective life-history strategies then lay along these gradients (Figure 4.24a), and included the well-known 'r' and 'K' strategies (see below) and a strategy appropriate at high adversity but low disturbance—in plants these latter are known as *ruderals*.

We use the term 'disturbance' at many points in this book, and a precise ecological definition in all contexts is elusive (see Chapter 6), but most simply we can think of it as a discrete environmental event that removes organisms and opens up space or other resources that are then available to others. Sharp peaks and troughs in flow are obvious examples in streams and rivers. Disturbance is seen as an axis in almost all theoretical concepts of a habitat templet. A second axis in Southwood's (1977, 1988) templet defines a gradient of 'adversity' or 'stress'. This essentially relates to the opportunity for rapid growth and production; in terrestrial or aquatic plants, for example, this might relate to the availability of mineral nutrients in the soil or water respectively. Three life-history strategies are then feasible at different points on the templet: r, when the environment favours rapid population growth but is frequently disturbed; K when disturbance is minimal and populations can increase to the point where resources become limiting; A ('adversity') when disturbance is rare but conditions for growth are poor. No strategy is feasible in highly disturbed *and* very unproductive environments (Figure 4.24a).

In devising their templet specifically for streams and rivers, Townsend & Hildrew (1994) figured axes of disturbance (they termed it 'temporal heterogeneity') and 'spatial heterogeneity', the latter inferring the availability of refugia from disturbance (Figure 4.24b). In this model, a group of life-history traits underpinning reproduction govern the return time of populations after disturbance. In frequently disturbed environments, species with short lifespans and rapid population growth would be favoured. In more constant habitats, longer individual lifespans, larger body size, iteroparity (repeated episodes of reproduction

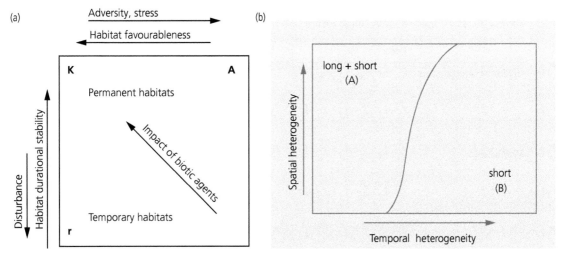

Figure 4.24 Two conceptual models relating specific strategies and traits to various environmental gradients: (a) Southwood's (1988) habitat templet concept plotting life-history strategies (r, K and A (adversity) selection) along gradients of habitat favourableness and disturbance. Biological interactions are strongest in potentially productive and least-disturbed circumstances—where there is selection (K) for competitiveness—rather than in highly disturbed environments—where there is (r) selection for fast population growth; (b) Townsend & Hildrew's (1994) templet for streams with predictions for the trait lifespan of stream organisms in habitats varying in the frequency of disturbance (temporal heterogeneity) and the availability of refugia (spatial heterogeneity). Relatively short lifespans are favoured in very disturbed habitats with few refugia; longer life spans are feasible in relatively constant or highly heterogenous habitats. Both long and short lifespans were seen as possible in stable and/or highly heterogenous habitats, as long as short-lived species were not outcompeted there.
Source: (a) reproduced with permission from John Wiley & Sons; (b) also with permission from John Wiley & Sons.

throughout life) and parental care would be feasible. Rapid reproduction can be attained by asexual or vegetative means, which would be expected in variable habitats. A further group of traits, imparting resistance (by surviving *in situ*) or resilience (features which favour rapid recovery) to populations, would be particularly favoured in habitats with fluctuating flow, and might include firm attachment mechanisms, a flexible body form, streamlining or flattening, behavioural 'escape' strategies or timing of critical and/or susceptible life-cycle phases.

The terms *species traits* or *biological traits* have been adopted to describe some of the biological features of living things we have been discussing—including lifespan/-history, body size and shape, mode of locomotion, how organisms gather resources and deal with biological and physical threats and stresses, mode of respiration and reproduction etc.—that determine individual fitness in any particular environment. However, the field is rather confused because terms have unfortunately been used in inconsistent ways (see Schmera et al. 2015). Commonly, a trait describes, for instance, the way in which an organism gathers food (the trait is 'feeding habit'; e.g. Usseglio-Polatera et al. 2000). Then, any one species may feed in one, or sometimes more than one, way (e.g. deposit-collector,

filter-feeder, predator, parasite). Thus, the trait 'feeding habit' for a species can fall into one or more *modalities* (e.g. filter-feeder). Synonyms for 'modality' in common use are 'category' and 'trait state'.

The proposition is then that the representation of modalities for species traits in a particular habitat can be tested against theoretical predictions, such as those made for species in the River Rhône by Townsend & Hildrew (1994). These traits can actually be measured at the individual organism level without reference to the external environment (Verberk et al. 2013). In this way, we could hope to test our theory of what biological features of organisms ('trait modalities') should be adaptive in different environments. Schmera et al. (2015) argued, however, that for reasons of the mathematical analyses of this sort of data, we need to redefine terms. They propose the term 'grouping feature' for what we have described as a species trait (e.g. feeding habit), a 'species trait' is then its mode of feeding (e.g. filter-feeder), and a species can then be given a quantitative score of its fidelity to that trait. That is, if the species feeds exclusively as a filter-feeder, it would be scored 100%, and this is its 'membership state' (on a scale of 0–100%). This disparity is unfortunate. As elsewhere in science, it is perhaps a forlorn hope to expect ecologists and others to use

Table 4.2 The macrophyte traits used in the study of 772 European lowland streams by Battrup-Pederson et al. (2015) examining functional trait differences with stream size and trophic status.

	Traits	Modality
Life form characterisitics	Life form	Free-floating, surface Free-floating, submerged Anchored, floating leaves Anchored, submerged leaves Anchored, emergent leaves Anchored, heterophylly
Morphology	Leaf area	small (< 1 cm^3), medium (1–20 cm^3), large (20–100 cm^3) and very large (> 100 cm^3) areas
	Meristem growth point	Single apical growth point Single basal growth point Multiple apical growth point
	Morphology index = (height + lateral extension of the canopy)/2	Classified into categories (2, 3–5, 6–7, 8–9 and 10) with values ranging from 1 to 5
Dispersal/ reproduction	Reproduction mode	By seed By fragmentation By rhizomes
	Number of reproductive organs per year and individual	Number (total individual flowering heads, fruiting structures, seeds, turions, tubers or other relevant structures)
Survival	Overwintering organs	Low, high and very high
Ecological preference/ indicator value	Ellenberg light	
	Ellenberg nutrient	

words in consistent ways. The older terminology (trait and modality) arose largely for linguistic reasons—it is just the way ecologists have described and thought about traits and it is intuitive. The newer terminology is probably more correct analytically (grouping feature/trait/membership state) but is linguistically more cumbersome and less intuitive. We raise the issue here mostly as a warning to the reader to beware of such differences in the literature. That said, in the following subsections we specify some of the traits that have been distinguished for three different taxonomic groups—macrophytes, macroinvertebrates and fish.

4.6.3 Macrophyte traits

The key environmental factors affecting lotic plants include discharge, water velocity, channel depth, the nature and extent of riparian cover (that determines shading), water chemistry (especially nutrient status) and substratum type. Naturally, these change with stream size and distance from the source as well as with catchment land use. Key features of macrophytes relating to these factors fall into three broad categories;

morphological (including life-form, leaf morphology, dispersal mode and growth form), *physiological* (including growth rates, photosynthetic pathways and leaf chemistry) and *phenological* (including life history and timing of flowering and reproduction).

As an example, Battrup-Pederson et al. (2015) used 18 traits to explore plant trait characteristics across 772 European small and medium-sized lowland streams with varying intensity of agricultural land-use catchments (Table 4.2). The traits related to morphology, dispersal, life form and survival. The authors also included ecological preference/indicator values of nutrients (Ellenberg nutrient) and light (Ellenberg light) which give a general indication of preference for nutrient and light availability.

The macrophyte communities of small streams were characterised by a higher abundance of light-demanding species growing from single apical meristems (e.g. species that proliferate on the water surface, like *Callitriche*), reproducing by seeds and rooted to the bottom with floating and/or heterophyllous (more than one type of) leaves (e.g. *Alisma plantago aquatica*). Medium-sized streams on the other hand had a higher abundance of productive macrophyte

Table 4.3 Summary of the major traits used under three principle grouping features in various studies of lotic macroinvertebrates (based on Townsend & Hildrew 1994; Poff et al. 2006; Verberk et al. 2008; Statzner & Bêche 2010; and Schmera et al. 2015).

Morphology/physiology	
Trophic habit	Body flexibility
Body shape/streamlining	Respiration technique
Maximal size at maturity	Degree of body armouring
Resistance form against unfavourable conditions (e.g. dessication)	Flow, drag or silt adaptations
Degree of attachment to benthos	Dimorphism
Diapause and quiescence	Thermal preference

Mobility	
Adult dispersal distance (water/air)	Ability to exit aquatic environment
Dispersal medium	Habit/locomotion and substrate relations
Dispersal mode	Occurrence in drift
Active flight	Passive transport

Reproduction, life cycle and development	
Number of aquatic life stages	Fecundity (descendants per reproductive cycle, number of eggs)
Development speed/pattern	Length of egg phase
Egg diapause	Longevity of adults
Number of reproductive cycles per individual	Voltinism
Emergence synchrony	Parental care
Reproductive method	Emergence season (beginning–end)
Emergence behaviour/location	Number of reproductive cycles/year (iteroparity/semelparity)
Diapause stage/resistance form	Egg type and per capita investment

species growing from multi-apical and basal growth meristems forming large canopies. With an increasing proportion of agriculture in catchments, the abundance of species growing from single-basal-growth meristems decreased in both small and medium-sized streams, and in small streams ecological preference for light decreased (tolerance for low light conditions increased) and dispersal by fragmentation increased, whereas seed dispersal decreased. The life-form characteristics also changed with agricultural intensity as the abundance of species with submerged leaves increased in small streams, whereas the abundance of species with floating and emergent leaves decreased. The mode of dispersal can also affect macrophyte population distribution within floodplains dependent on the level of connectivity to the main river, as we discuss in Chapter 5 (see Table 5.1).

4.6.4 Macroinvertebrate traits

Most advances in the use of functional traits have been made with macroinvertebrates, often linked to

their possible use in water quality monitoring tools and in exploring responses of invertebrates to natural and man-induced stressors and other environmental drivers (see Chapters 6 and 10 and Topic Box 6.1 by Dolédec & Bonada). Two major trait databases currently exist, one centred on North American systems (with 20 traits and 54 modalities; Poff et al. 2006) and one on Europe (with 20 traits and 108 modalities; Usseglio-Polatera et al. 2000). Statzner & Bêche (2010) provided a summary of the biological traits compiled in large databases up to the time of their review.

In Table 4.3 we have listed some of the major traits under the grouping features morphology and physiology; mobility; and reproduction, life history and development, as a way of illustrating the kinds of characteristics that have been considered in the literature. Each of the individual traits would include a number of trait states or modalities (such as semi-, uni-, bi- or multivoltine life history under the trait voltinism; or burrow, climb, sprawl, cling, swim, skate or attach, under the trait habit).

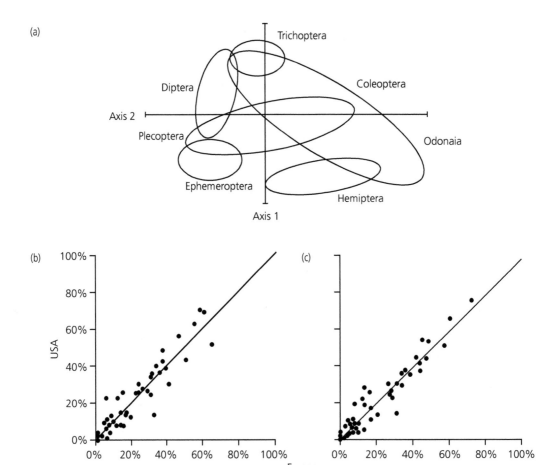

Figure 4.25 (a) Principal coordinates analysis ordination plot of 311 North American macroinvertebrate genera and 20 biological traits, showing 95% confidence interval ellipses for genera within the main lotic orders. (b) and (c) Plots comparing the overall mean category values of 61 trait categories in the USA versus Europe for (b) common invertebrate genera and (c) invertebrate communities (dashed line is y = x; R^2 of the regression slopes (solid lines) were c.0.9 with a slope of close to 1.0).
Source: (a) from Poff et al. 2006, with permission from University of Chicago Press; (b) and (c) from Statzner & Bêche 2010, with permission from John Wiley & Sons.

As an example of the use of species traits in the study of macroinvertebrates, Bonada et al. (2007) explored data from 609 sites in 527 non-impacted streams in the Mediterranean basin and temperate streams elsewhere in Europe, from Scandinavia to Morocco and Turkey. Compared to the temperate communities, the Mediterranean communities contained more macroinvertebrates with reproduction through terrestrial egg clutches, summer diapause and with specialised respiration techniques, as well as smaller body size, more frequent reproduction and dispersal via active flight (favouring rapid and widespread dispersal). These traits are considered to provide better resistance against and resilience after drought, reflecting the nature of the Mediterranean climate.

On a larger scale, a multivariate analysis of trait distribution amongst 311 insect genera in North America (Poff et al. 2006) produced a strong separation of groups of genera along two PCA (principal components analysis) axes (Figure 4.25a). The first axis represents genera of Hemiptera and Odonata, and some Coleoptera and Plectoptera, highlighting trait modalities of high crawling rate, long adult lifespan, semi-voltinism, strong adult flight and predatory feeding. Some genera also showed lack of attachment to the benthos and large adult body size. The second trait gradient primarily represents Ephemeroptera and some Plecoptera genera, showing fast seasonal life cycles, no body armouring, very short adult lifespan and small adult size. These genera are also common or

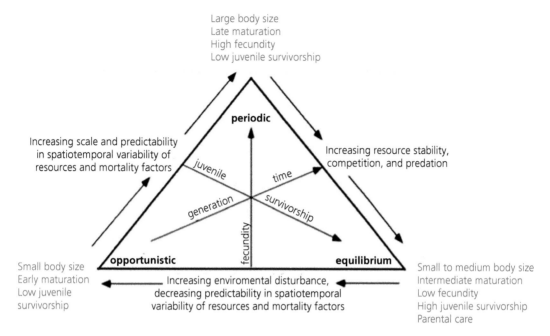

Figure 4.26 Triangular life-history model illustrating environmental gradients selecting for endpoint strategies (periodic, opportunistic and equilibrium) in fish species. The inside arrows summarise fundamental trade-offs between the traits of juvenile survival, generation time and fecundity that define the three end-point strategies. The outside arrows summarise how selection pressures may favour certain strategies in relation to biotic and abiotic factors (modified key trait modalities are shown for each of the three life-history strategies).
Source: inside arrows from Winemiller 2005, with permission from Canadian Science Publishing; outside arrows after Heino et al. 2013.

abundant in the drift. A third trait gradient in the analysis mainly describes Trichoptera and some Diptera and Coleoptera, genera largely characterised by no swimming ability, some attachment, body armouring, and a short adult lifespan.

We return to the analysis of traits in the context of community assembly and habitat filters in Chapter 6 (section 6.3), but it is worth emphasising here that trait patterns of lotic macroinvertebrates do indeed seem to be quite robust and stable across biogeographic regions (as seen in the Stazner & Bêche (2010) comparisons of 61 trait categories between the USA and Europe (Figure 4.25b, c).

4.6.5 Fish

The trait approach to studying fish ecology has been growing in importance, particularly in relation to fisheries management, and is often focused on life history (e.g. Winemiller 2005; Mims et al. 2010). A different kind of conceptual trait model approach to that used for macroinvertebrates was proposed by Winemiller (2005), one based around trade-offs between the life-history traits of juvenile survival,

generation time and fecundity, that define three primary life-history strategies; periodic, opportunistic and equilibrial in relation to environmental variation (Figure 4.26). Recent work on 603 fish species across 350 North American catchments showed significant support for this model (Mims et al. 2010). Opportunists such as killifish (Cyprinodontidae) and topminnows (Fundulidae) can cope with frequent and intense disturbances, with small body size and early maturation but low juvenile survival. Equilibrial species including catfishes (Ictaluridae) and sunfishes (Centrarchidae) are favoured in stable habitats with low environmental variation. They can be small to medium sized, with moderate age at maturity (relative to other species), small clutch sizes and high juvenile survival, largely associated with parental care. Periodic strategists, such as salmon and trout (Salmonidae) and suckers (Catostomidae), are usually associated with seasonal environments, achieve a large body size, mature late and are highly fecund, although juvenile survival is poor. Not all families of fish sit exclusively within any one of these strategies and several, including the most diverse family (the Cyprinidae), include representatives of all three.

Table 4.4 Trait groups, traits and modalities for Australian freshwater fish (based on Sternberg & Kennard 2014). Note that in this study actual quantitative values rather than categories were used for some modalities (e.g. longevity and length at maturation). Categories such as short, medium, long or < 1 yr, 1–3yrs, > 3 yrs etc. could also be considered.

Types of traits	Traits	Modalities
Life history	Longevity	Maximal potential life span (yr)
	Length at maturation	Mean total length at maturation (cm)
	Age at maturation	Mean age at maturation (yr)
	Movement classification	No spawning movement, potadromous, amphidromous, anadromous, catadromous
	Spawning substratum	Mineral (e.g. gravel), organic (e.g. plants), various (mineral and organic), pelagic, other (e.g. mouth rearing)
	Spawning frequency	Single spawning per season, batch/repeat/protracted spawner per season, single spawner per lifetime
	Reproductive guild	Nonguarders (open substratum spawners), nonguarders (brood hiders), guarders (substratum choosers), guarders (nest spawners), bearers (internal), bearers (external)
	Total fecundity	Number of eggs or offspring per breeding season
	Parental care	Metric representing total energetic contribution of parents to offspring
Morphology	Maximum body length	Maximun total body length (cm)
	Shape factor	Ratio of total body length to maximum body depth
	Swim factor	Ratio of minimum depth of caudal peduncle to maximum body depth
	Maxilla (jaw) size	Ratio of maxilla length to total body length
	Eye size	Ratio of eye diameter to total body length
Ecology	Vertical position in the water	Benthic, non-benthic
	Trophic guild	Herbivore-detritivore (> 25% plant matter), omnivore (5–25% plant matter), invertivore, invertivore-piscivore (> 10% fish)

Other types of traits, in addition to life history, have also been explored in freshwater fish, including morphological and ecological ones. A good example includes 17 traits examined across 194 species from 35 families of Australian freshwater fish (Sternberg & Kennard 2013; Table 4.4).

The variation in traits of Australian freshwater fish species can be explained by two major ordination gradients. On the primary gradient, the Australian fish fauna separated into (a) benthic invertivores with a relatively high degree of parental care, and egg-guarding reproductive strategies, amphidromous spawning migrations (between fresh water and the sea) and fusiform body shapes, and (b) non-benthic omnivores, with little parental investment in brood survival. The second gradient represented a 'periodic'—'opportunistic'—life history gradient (in Winemiller's sense; Figure 4.26), and separated (a) 'periodic' highly fecund, late-maturing, long-lived, large pelagic-spawning species from (b) non-migratory, batch-spawning, herbivore-detritivore, 'opportunistic' species (which actually made up the majority of Australian fish species examined). The nature of the Australian fish fauna, with its marked endemism and long history of isolation from other areas, and a climate that has been increasingly arid over the last 500,000 years, provides a plausible explanation for the broad taxonomic differences in

species functional traits here (Sternberg & Kennard, 2014).

While the finer-scale ecological processes and traits contribute to interspecific differences within the larger taxonomic units of the Australian fish fauna, traits related to habitat and feeding mode help to explain patterns of species along large river systems (Heino et al. 2013). The upstream fish communities were found to have the lowest diversity of trait combinations and contain mostly benthic, rheophilic invertivore species, with an opportunistic life history. Downstream, the diversity of trait combinations increases for any of the traits but the relative importance of traits may differ between habitats (e.g. oxbows, main channel) and hydrological regimes. The upstream trait sets may form a nested subset of downstream ones. We provide some more examples of these kinds of longitudinal patterns within rivers in the community context in Chapter 6.

4.7 Conclusions

This chapter has focused on how lotic animals and plants are adapted to the special environmental constraints and challenges posed by stream and river environments. We have highlighted some of the unique physiological, morphological and behavioural adaptations that help the biota to overcome the

Table 4.5 Modes of existence of lotic animals (from Giller & Malmqvist 1998, modified from Cummins & Merritt 1996).

Category	Description	Examples
Skaters	Adapted for life on the water surface of low-order streams or margins of high-order rivers, where they feed on organisms trapped in the surface film or coming to the surface	Water striders/pond skaters (gerrid bugs), adults and juveniles
Planktonic	Inhabiting open water, slow-flowing or still, in high-order rivers	Planktonic crustaceans (Cladocera, Copepoda)
Divers	Insects adapted for swimming in slow-flowing pools, by 'rowing' with hind legs, coming to surface to replenish air bubbles, and often clinging to macrophytes or submerged objects.	Water boatmen (Corixidae and Notonectidae) and diving beetles (Dytiscidae), (adults and juveniles)
	Semi-aquatic vertebrates including those that forage underwater but spend most time on the water surface (birds) or land (mammals).	Diving ducks and other waterfowl, dippers, otters, mink, platypus, desman
Swimmers	Insects adapted for 'fish-like' swimming, clinging to the substratum between short bursts of swimming.	Streamlined mayfly nymphs (Baetidae, Leptophlebiidae).
	Fully aquatic vertebrates that maintain position by swimming or using flow refugia.	Fish, lotic amphibians and reptiles, freshwater dolphins.
Clingers	Possess physical (e.g. silk nets, pads or fixed retreats) or morphological (claws, dorsoventral flattening, suckers) adaptations for attachment to substratum surfaces, or that are sedentary, truly sessile and/or colonial	Net-spinning caddis larvae, simuliid larvae, heptageniid mayflies, gastropod snails (Ancylini), leeches, Bryozoa, sponges
Climbers	Living and moving on vascular plants or detrital debris (e.g. overhanging branches, roots and vegetation)	Damselfly (Coenagrionidae), mayfly (Ephemerellidae) and midge (Chironomidae) larvae
Burrowers	Inhabiting fine sediment (and hyporheic zone), small enough to move between sediment grains, or burrowing (some constructing discrete burrows), or ingesting their way through the sediments; either very small-bodied or filiform (long and thin, e.g. cylindrical shape)	Microcrustacea (copepods, ostracods), rotifers, larval stoneflies (Leuctridae), burrowing mayflies (Ephemeridae) and Chironominae midges, oligochaetes, bivalve molluscs (Sphaeridae, Unionidae), lampreys (Cyclostomata)

physical challenges of life in running waters as well as to gather the necessary resources for survival. This led on to a consideration of more complex traits (suites of species adaptations) in exploring further species–environment relationships at the individual level of organisation.

Traits and adaptations should not be considered in isolation because they interact, often synergistically, to influence the success of the individual as well as the species population within its habitat. As we will see in Chapter 6, they also influence the assembly of communities. The importance of trait interactions is exemplified by reference to the study of Poff et al. (2006), who identified over 1 million possible ways in which the 20 traits and 59 trait categories or modalities could be combined amongst the 311 lotic insect genera studied. However, only 233 of these combinations (or 'functional trait niches' as they were described) were actually found, reflecting the evolutionary linkages amongst traits through trait interactions.

Finally, ecologists have distinguished a number of *modes of existence*, a generalised scheme that incorporates habit, locomotion, attachment and concealment, while behaviour is evidently an intrinsic part of all of these. The scheme was developed initially by Cummins (see Cummins & Merritt 1996), specifically for insects. Giller & Malmqvist (1998) added other taxa and life forms (Table 4.5). Seven modes of existence can be readily indentified, which describe where and how the organisms live in the lotic habitat and illustrate the key features that contribute to their success. Various taxonomic groups can be assigned to each of the modes, although, as with all such classification schemes, there will be exceptions to the 'rule'.

The study of species adaptations and traits in relation to the environment, underway for well over 100 years, has made good progress, although uncertainties remain (Schmera et al. 2015). The nature of the habitat templet of rivers and the representation of species traits on that templet are essential to our understanding of the dynamics of populations and the composition of communities—subjects to which we now turn in the following two chapters.

CHAPTER 5

Population ecology

The population is a fundamental 'level of organisation' in biology and ecology, and populations are the subject of much modern ecological research, with applications in fields such as conservation, bioassessment of pollution, fisheries policy and many others. In this chapter we focus on the particular features of populations in rivers and streams—largely in relation to the river habitat templet discussed in Chapter 2, though lotic populations are like any other in terms of the basic processes.

There are a great many definitions of populations, the one by Turchin (2003) being among the most complete: '*a group of individuals of the same species that live together in an area of sufficient size to permit normal dispersal and migration behaviour, and in which population changes are largely determined by birth and death processes*'. The essential feature here, common to all definitions of the population, is the focus on a single species and its distribution and abundance within a specified area—ideally the range of the population. Only four processes affect population size: *fecundity* (birth rate), *mortality* (death rate), *immigration* and *emigration*. Note here that immigration and emigration refer to the arrival and departure of individuals, and not of species. The problem of determining the spatial limits of populations and of whether population dynamics (i.e. changes in numbers over time) are determined mainly by births and deaths '*in situ*' (as in Turchin's definition) or by movements in or out of the study area (dispersal) is challenging. It is particularly difficult in the context of spatially complex and divided habitats such as river systems.

Populations have characteristics not evident from the study of individuals, such as density and changes in numbers over time ('dynamics'), age distribution, spatial patterns (distribution in space) and patterns in gene frequencies. All populations are capable of exponential growth (i.e. increasing with some constant doubling time) and potentially of increasing without limit. In practice, however, real populations have fluctuations in density that are normally 'bounded' (i.e. within lower and upper limits) and often vary in relation to resource availability and other biotic and abiotic factors. The classic issue in population ecology is whether or how the numbers of organisms are *regulated*—the tendency for populations to persist and to fluctuate around this (albeit often 'fuzzy') central band of values. The main question is whether such population persistence requires that *density-dependent factors* operate; that is, restraints on numbers increase in intensity as population size (density) increases. Such mechanisms include primarily intraspecific competition for a limiting resource and, potentially, parasites, pathogens, grazers and predators as well as interspecific competitors. Alternatively, population size may depend on largely abiotic environmental factors that simply enable numbers to increase during favourable times or limit numbers at others. These latter, *density-independent* factors operate without reference to population size itself, and they can determine the spatial range of a species as well as its numbers within that range. This is essentially because habitats vary (from place to place) in how much of the time the conditions favour population growth.

Here, we need to distinguish what are more or less single, 'closed' populations from those that are patchy and divided (and may be formed essentially of a *metapopulation* of local populations, potentially linked by occasional dispersal). The persistence of closed populations depends only on births and deaths. Theoretically, a closed population will eventually wander to extinction, by the chance occurrence of a long period of unfavourable conditions for instance, unless regulated by density-dependent factors. Populations that occupy patchy, divided habitats that can exchange individuals may have local dynamics that depend on dispersal and mobility more than on mortality/fecundity within each patch. Such a system of local (or sub-) populations

The Biology and Ecology of Streams and Rivers. Alan Hildrew and Paul Giller, Oxford University Press. © Alan Hildrew and Paul Giller (2023).
DOI: 10.1093/oso/9780198516101.003.0005

may then persist longer than does any of its components. Rockwood (2015) presents a useful introduction to the general features of biological populations and their dynamics.

5.1 Introduction—what is special about populations in rivers and streams?

Two features of rivers and streams make them special in terms of the population ecology of their biota. The first is that river habitats are *highly dynamic* (i.e. temporally variable). Flow is the architect of river channels and continuously creates and destroys channel features, moving sediment, particulate organic matter, and solutes as it does so. It can be a source of disturbance to river organisms apparently so pervasive that some river ecologists have doubted whether the size of lotic populations can ever be determined by density-dependent biotic interactions. Surely, this dynamism must test the limits of natural selection to match species traits to such harsh conditions? Second, river systems are *physically complex and patchy* habitats. Thus, their features vary along the channel, both over short distances between habitat features (such as riffles and pools) and over longer distances from source to mouth. Further, alluvial sediments vary in coarseness and depth beneath the riverbed and offer a vertical 'extension' of the channel habitat to some organisms. In addition, lateral to the mainstem, river dynamics determine the extent and nature of the *riparian* system (the biota and physical nature of the river banks, frequently interacting with the main channel) and can also create complex systems of side channels, floodplain pools and swamps with extremes of conditions for the river biota. Combining the dynamics of flow, sometimes predictable and seasonal but often much more erratic, with the patchiness of the physical habitat creates a shifting mosaic of conditions that can scarcely be matched by any other natural habitat. It is against this habitat templet that the dynamics of lotic populations (and those of organisms that live within the physical influence of the river) are worked out. This chapter deals with both the spatial distribution and abundance of populations of river organisms in relation to these habitat features.

5.2 Distribution

5.2.1 Scale, dimensions and the habitat templet

There are very large-scale spatial and long-term patterns in distribution (e.g. continental) which are in the

realm of biogeography. In this section we deal with the distribution of populations in what we might refer to as 'ecological' space and time—that is, at the scale(s) at which distribution is determined by the basic population processes, of births, deaths and dispersal. No species occurs everywhere, but all have a pattern of distribution (sometimes referred to as 'habitat preferences') that we can try to relate to the environment (i.e. are the conditions right?), and/or to the ability to disperse (i.e. can the organism reach all suitable habitats?). The pattern of spatial distribution observed also depends very much on the scale at which observations are made (for a simple example, see Figure 5.1). There are two related but distinct meanings of scale that are widely used in ecology: one refers to the overall spatial or temporal 'extent' of the study (e.g. are we studying a whole river system or a single pebble, and will our study last for decades or just a few days?). The other meaning of scale refers to the resolution or 'grain' of the measurements made. This can be thought of rather like the gradations on a ruler or tape measure—are we measuring to the nearest μm or the nearest metre—and are we taking measurements annually or every few minutes? Extent and environmental grain are related because studies of wide spatial extent, or undertaken over the long term, rarely also include very fine-scale or frequent measurements; that is, there is normally a trade-off between extent and resolution. The main lesson here is that the scale of the study determines the pattern detected.

Further, we have referred already to the three-dimensional spatial nature of whole-river systems, with patterns in distribution recognisable longitudinally (downstream sequences of species), laterally (from the centre of the channel towards the margins and into the riparian habitat or floodplain) or vertically from the water column into the stream-bed and down into the contiguous groundwater. If we add the fourth dimension of time (Ward 1989), we can also see that many species use different parts of the habitat at different stages in their life cycle—as in the conspicuous longitudinal and/or lateral migrations of many fish and some invertebrates, or the emergence of aquatic insects from the channel when they become flying adults. The movements of water and organisms along these physical dimensions also constitute linkages, in terms of organic matter and nutrients, among different parts of the whole-river ecosystem (see Chapters 8 and 9).

In Chapter 2 we described the physical and chemical habitat templet, and in Chapter 4 the adaptations of species living in running waters and the overall species

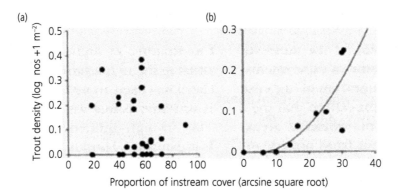

Figure 5.1 The relationship between the density of young (in the first year of life) rainbow trout (*Oncorhynchus mykiss*) in a Californian stream and cover (i.e. any habitat feature offering refuge for a young fish—e.g. large rocks, coarse woody debris etc.). Data were collected at the scale of individual riffles and pools and plotted both at that scale (a) and when they were combined into estimates of density over whole reaches (a few hundred metres long, and each containing several riffles and pools) (b). The relationship is weak at the scale of individual riffles and pools, but much stronger at the overall reach scale.
Source: from Cooper et al. 1998, with permission from John Wiley & Sons.

traits that match various parts of the habitat templet. We now try to relate these adaptations and traits to the spatial distribution of populations (a) at various scales and (b) along the different habitat dimensions. The most prominent aspects of the templet in this respect are flow, substratum, water chemistry, oxygen supply and temperature, though this is not an exhaustive list. While we deal with these in sequence, it should be evident that the various factors interact strongly (e.g. flow and substratum), and also that spatial distribution changes in time (i.e. it is dynamic).

5.2.2 Flow and distribution

We begin with flow because it is *the* key characteristic feature of rivers and streams and intervenes in many ways in the ecology of stream organisms—which may show specific adaptations to particular hydraulic regimes (velocity, shear stress etc; Chapter 4). Not surprisingly, many of the most intimate (small-scale) relations between flow and distribution are to be found among suspension- (i.e. filter-) feeding invertebrates. The main difficulty here has been to match the scale of measurement of flow to that at which the organism experiences hydraulic forces, recalling that these forces themselves vary with body size and shape. Among the most abundant filter-feeders in running waters are the larvae of blackflies (in the family Simuliidae; see Figure 4.19, p. 129). Hart et al. (1996) measured velocity profiles, over stones bearing attached *Simulium* larvae, at heights above the stone surface of 1–10

mm. These profiles were variable and velocity 2 mm above the stone surface (the approximate height at which the filtering apparatus of a late instar larva operates) ranged among stones between 7 and 59 cm s^{-1}. The relationship between larval density on five stones and velocity at 2 mm above each stone was also variable but consistently positive, whereas there was no relationship between velocity at 10 mm height above the stone surface and larval density and, indeed, no consistent relationship between velocity at 2- and 10-mm height above the same stone. Such fine-scale measurements of flow are technically challenging but important if we wish to match the scale of observation to patterns in the distribution of small organisms.

Of course, correlations between flow and distribution do not prove that flow is the cause of the pattern observed. However, in a simple but elegant early field experiment, Edington (1968) used a baffle to manipulate water velocity over moss-covered bedrock in an English stream where the filter-feeding caddis *Hydropsyche siltalai* was abundant. He first counted the canopy-shaped nets of the larvae that protruded from the moss stems—the larvae themselves live in silken tubes within the moss cover—and measured water velocity at the same time (Figure 5.2). He then removed the nets, without disturbing the resident larvae themselves, and added the baffle, thus reducing flow over part of the rock surface. He inspected the area 43 hours later and found that nets had been rebuilt in areas where the velocity remained high, but not in areas where the flow had been reduced; indeed, the larvae themselves had left this latter area completely. In an

(a)   (b)

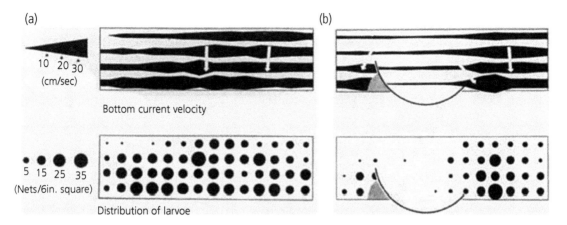

Figure 5.2 Edington (1968) examined the distribution of nets of *Hydropsyche siltalai*, assessing water velocity (a, upper) and the number of larvae (a, lower) in a series of small quadrats on bedrock. Having added a flow baffle and destroyed the feeding nets (but leaving the larvae in place), he then remeasured velocity (b, upper) and the number of nets (b, lower) that had been rebuilt after 43 hours—finding that the relationship between net-building and velocity had been maintained.
Source: Edington (1968) with permission from John Wiley & Sons.

extension of this experiment, a baffle was again used to reduce velocity over an area of bedrock, but this time water was also redirected via a pipe over an adjacent mossy area of the bed that had previously been left 'high-and-dry' by a natural reduction in stream discharge. After five days the larvae had: (i) left the area of flow reduction, (ii) colonised the formerly dry stream bed over which water now flowed and (iii) increased in density in areas where rapid flow had been maintained or augmented. This example shows not only that flow can be decisive in the small-scale distribution of filter-feeding larvae, but also that they are sensitive to and can respond rapidly to changes in their environment. Indeed, *Simulium* larvae may adjust their position almost continuously in relation to very short-term flow variability (in the order of minutes).

It is not only filter-feeders that respond to flow, however, and it has been argued that the forces of flow are implicated in setting the distribution, at a variety of spatial scales, of most lotic organisms. Thus, Mérigoux & Dolédec (2004) showed that the distribution of almost 70% of benthic taxa in a Mediterranean river in two seasons could be related to hydraulic forces on the bed (Figure 5.3). Of course, such relationships do not mean that flow is the *only* factor determining the distribution of stream species, since flow is inextricably linked to other environmental features, such as the nature of the substratum (coarse or fine), oxygen availability, food supply and others. It is this whole suite of factors that seem to determine habitat suitability and thus distribution.

5.2.3 The substratum and lotic organisms

There are a number of features of the substratum that are important for the biota of streams and rivers. For instance, Ledger & Hildrew (1998) found that the amount of biofilm and epilithic algae increased with substratum particle size in a southern English stream. They ascribed this to the higher profile of large particles, keeping their surfaces free of deposited particles, that are also less likely to be shaded by neighbouring stones and are perhaps more stable. Experimental studies by Hart (1978) showed significantly more species, and individuals, of macroinvertebrates on large than on small rocks. However, the density was actually lower on the larger rocks, smaller objects having a greater surface area relative to volume. However, when the larger rocks were combined into larger groups, they supported higher densities than groups of small rocks. This may be due to the larger interstitial spaces between bigger rocks.

The roughness of individual particles also seems to influence the diversity of colonising animals and algae; the more complex the surface, the greater the number of species (e.g. Hart 1978). For instance, crevices on the surfaces of rocks provide refugia for algae during flow disturbances (e.g. Bergey 2005; see section 5.3.4). Single stones also present several surfaces to organisms (front, top, sides, back and underneath) and each surface is under a different small-scale flow regime. The shapes and mixture of particle sizes and the degree of embeddedness will influence what is called the *porosity*—the

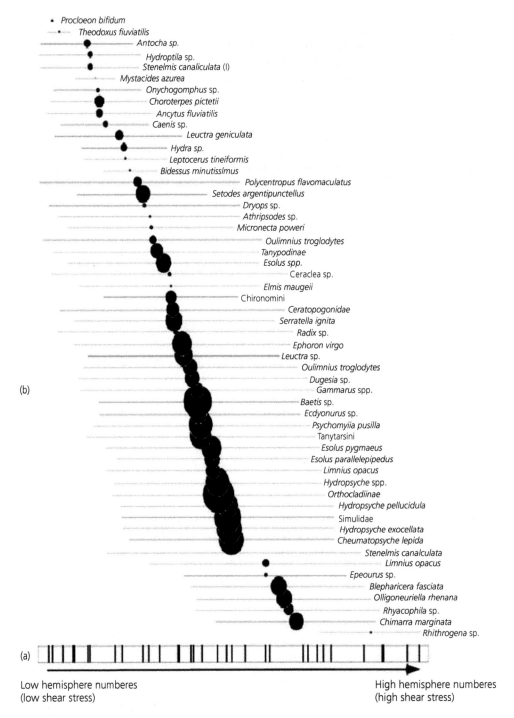

Figure 5.3 The position of stream benthic taxa (b, the size of circles is related to abundance, horizontal lines show standard deviations) along an axis of hydraulic forces (a, 'shear stress', assessed using different sized hemispheres) on the bed of the Ardèche River, France in autumn. This shows the occurrence of both strongly rheophilic ('current loving', at high shear stress) and more lentic species in the same stretch of river.
Source: Mérigoux & Dolédec 2004, with permission from John Wiley & Sons.

extent and calibre of subsurface pores, tunnels and spaces within the substratum. Larger and more extensive pores can allow greater subsurface water flow and hence supply of oxygen. A very large proportion of lotic animals can live in these microhabitats.

While most benthic invertebrates can be found on a variety of substrata (i.e. they are generalists in this respect), many occur mainly on broad categories of bed materials. Ward (1992a) based this largely on factors including: (i) whether the substratum is living (algae, moss, macrophyte) or dead organic matter (leaves, woody debris), or mineral particles; (ii) the amount of silt on, and organic content of, the substratum (including moss and algal covering of cobbles, etc.); (iii) the size and heterogeneity of substratum particles and the texture and porosity of the substratum; and (iv) the stability of the substratum, related to its tendency to move downstream. Some taxa, particularly some algae, are overrepresented on silt/clay substrata, sand, stones and/or vegetation (e.g. Bere & Tundisi 2011). Sediment conditions are also very important for rooted aquatic and riparian plants (Wohl et al. 2015).

Particle size and texture of the overall substratum also influence the number of species found and their individual density (Downes et al. 1995). Aquatic organisms may also be sensitive to the mobility of bed materials, such that the phenology of the life history may relate to the typical timing of bed disturbances, such as seasonal floods and spates (e.g. Lytle et al. 2008). As a general rule, stable substrata (boulders and cobbles) support higher densities of stream populations, and a greater number of species, than less-stable ones (gravel and sand) (e.g. Epele et al. 2012). Sandy substrata are usually the least rich in species and population densities, while stony riffles normally have a greater range of invertebrates than pools. These differences may not be related mainly to substratum stability itself, however. The deposition of silt reduces habitability of the substratum for animals by clogging interstitial spaces, reducing interstitial water movement and the supply of oxygen. In general, substratum heterogeneity (the variety of substrata available locally) influences the abundance and diversity of benthic organisms (both animals and plants). Mixed substrata provide a greater variety of surfaces available and of near-bed and interstitial flow patterns. Note also that the surface area and space available to organisms within the substratum increases as body size declines, for similar reasons that the apparent length of a coastline increases as we measure it more finely (i.e. with greater resolution). This is one possible explanation for the positive relationship between body size diversity and abundance in 'sedimentary environments' such as both stream beds and soils (Giller 1996).

It should be evident that patterns of flow and the nature of the substratum are extremely closely interrelated and can produce larger-scale gradients in the habitat templet that lead to marked distributional discontinuities in stream organisms along the various habitat dimensions. We deal here with two cases: the distribution of rooted aquatic and semi-aquatic plants along the lateral habitat dimension, and the distribution of the interstitial fauna (mainly) along the vertical dimension.

Among lotic organisms showing the clearest patterns *laterally* to the main channel are rooted aquatic macrophytes (mainly angiosperms). Their distribution is governed by the physical forces of flow and the consequent deposition and erosion of sediments of differing grain size (with a consequent 'cascade' of other environmental factors). At the same time they are 'ecosystem engineers' of both the river substratum and the larger-scale morphology of river channels and floodplains (for more on 'ecosystem engineers' see Chapter 10). Thus, they determine to a large extent the nature of the channel and neighbouring floodplain by altering the resistance of the river to the current (e.g. Gurnell 2014). They can also be considered creators of the habitat for a myriad of other creatures. Particular features of the different species of plants (Chapter 4) determine their position along the lateral habitat gradient, from the centre of the channel itself out to the remotest positions at the margins of the floodplain. For instance, Leyer (2006) studied floodplain vegetation in relation to the frequency of its connection ('connectivity') to the main River Elbe (a large river in central Germany). Depending on their mode of dispersal, different species of seedlings were observed emerging in experimental plots placed in four different categories of floodplain habitat (Figure 5.4; Table 5.1). Species occurring in the better-connected habitats were dispersed mainly by water, whereas those more isolated from the channel mainly emerged from the sediment seed bank. The lateral spatial dimension of rivers—the extent of inundation of banks and the floodplain (or indeed in drying)—evidently is temporally highly dynamic. We can think of this as space and time interacting. The habitat at any one point can change radically from time to time and, as we shall see, this is associated with reproductive behaviour and life cycles (as discussed in Chapter 4), particularly amongst populations of a range of fish species in rivers with large floodplains.

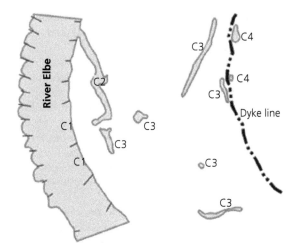

Figure 5.4 A schematic of a section of the River Elbe and four categories of associated water bodies with differing connectivity to the main channel: C1 river margin, C2 permanently connected side arm, C3 isolated water body on the recent floodplain, C4 isolated water body behind flood defences (Leyer 2006; see Table 5.1).
Source: Leyer 2006, with permission from John Wiley & Sons.

Table 5.1 The occurrence (X) of macrophyte seedlings across the floodplain of the River Elbe (Brandenburg, Germany). C1, C2 etc. are the connectivity classes (see Figure 5.6) of habitats in relation to the main channel; (W) and (S) denote cases where seedlings emerged mostly from water-dispersed (W) seeds or from the sediment-based seed bank (S) (after Leyer 2006).

Species	Connectivity class			
	C1	C2	C3	C4
Eragrostis albensis	X (W)			
Juncus bufonius	X			
Spergularia echinosperma	X			
Cyperus fuscus		X		
Rorippa palustris		X (W)		
Rumex crispus		X (W)		
Chenopodium glaucum	X	X		
Limosella aquatica	X	X (W)		
Pulicaria vulgaris	X	X (W)		
Chenopodium rubrum	X	X	X	
Plantago major	X (W)	X (W)	X	
Gnaphalium uliginosum	X (S)	X	X	X
Persicaria lapathifolia	X	X	X (S)	X (S)
Alopercurus geniculatus			X (S)	X (S)
Chenopodium polyspermum			X	X
Glyceria fluitans			X (S)	X (S)
Oenanthe aquatica			X (S)	X (S)
Atriplex prostrata			X	
Rorippa amphibia			X	
Bidens tripartita				X (W)
Hottonia palustris				X (S)
Ranunculus aquatilis				X
Rumex maritimus				X

Our second example of a marked gradient in the habitat templet is that along the vertical dimension between the surface of the substratum down into the stream bed, largely determined again by the interaction between the nature of the substratum itself and the forces of flow. In this vertical habitat dimension (i.e. into the stream bed and contiguous alluvial sediments) we are dealing with a rather different biota—the interstitial fauna (mostly in the meiofaunal size range). In pioneering work on the French River Rhône, for instance, Dole-Olivier et al. (1994) synthesised characteristic distribution patterns found for this interstitial fauna (so-called hypogean species; see Chapter 1). There is a consistent vertical distribution of species along the depth gradient in the alluvial sediments of the Rhône. Remarkably, this sequence of hypogean species is apparently also repeated along short longitudinal gradients (within a stream reach) from downwelling to upwelling zones, and laterally from the channel out to the edges of the floodplain (see Figure 5.5). Dole-Olivier et al. (1994) related this vertical sequence to habitat stability provided by the bed sediments, envisaging that the deep groundwater is the most stable (in terms of factors like temperature and flow) and the sediment surface the least stable habitat. Similarly, they saw habitat stability declining longitudinally from upwelling to downwelling patches (the latter reflecting the dynamic nature of the river channel), and laterally from the floodplain margin to the channel itself.

5.2.4 Water chemistry and distribution

Because of the general lack of small-scale spatial variation in water chemistry in the water column, this factor normally does not explain distribution patterns of larger animals and plants at small spatial scales (although we do need to note that there can be sharp, small-scale gradients in water chemistry and microorganisms within sediments; see Chapter 9). Thus, whereas differences in fish guild composition and richness among sub-basins in regulated and unregulated rivers of India's Western Ghats mountains could be explained by differences in water chemistry, those in smaller segments within rivers could not (Atkore et al. 2020). In general, there are quite strong relationships between the biota at larger spatial scales (e.g. within and between catchments and regions) and water chemistry. Acidity and nutrients

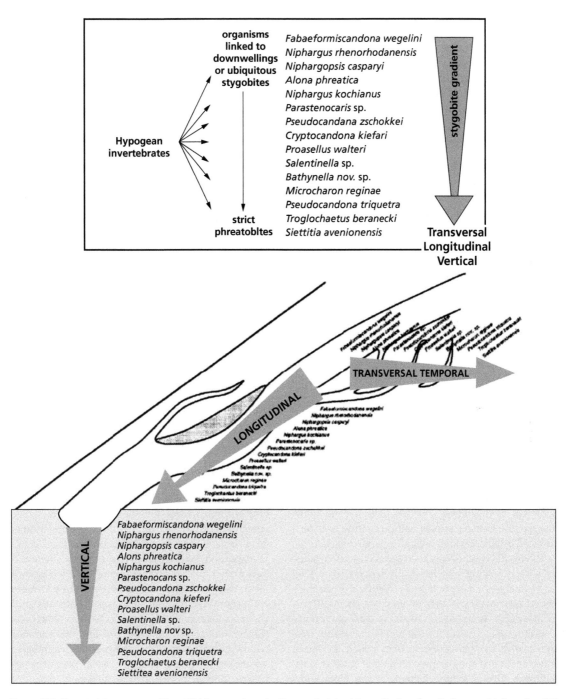

Figure 5.5 Characteristic sequences of interstitial (hypogean) species (the so-called stygobite gradient) are found in three spatial dimensions (NB the sequences are the same along all three gradients) (indicated by the large arrows) and along gradients of habitat stability—see text for further details. Only two of these species indicated are not crustaceans—the annelid worm *Troglochaetus beranecki* and the beetle *Siettitia avenionensis*. *Source*: Dole-Olivier et al. 1994, Ward et al. 1998, with permission from John Wiley & Sons.

seem particularly important. The strength of these relationships has enabled the development of many biotic indices of water quality throughout the world (e.g. Eriksen et al. 2021; Chapters 9 and 10). Note that most such relationships are developed for various multi-species assemblages rather than for single-species populations (the focus of this chapter). Nevertheless, we deal with several of them here rather than fragment the material and, of course, patterns in assemblages are underpinned to a large extent by those of individual species.

One major approach to the study of patterns in the animals and plants of streams and rivers has essentially been to correlate the distribution of species with one or more environmental factors that vary among the streams. Samples from a series of sites can be classified or projected onto axes of variation (ordination) in relation to species composition and abundance using various multivariate techniques, and the resulting patterns can in turn be compared with gradients of environmental factors including water chemistry. There is a huge literature presenting data on this approach as applied to macroinvertebrate communities and examples can be found from all continents (except Antarctica). As a flavour of the kinds of studies and locations we can point to the central plateau in Mexico (Rico-Sanchez et al. 2022), south-east Brazil (Buss et al. 2002), Puerto Rico (Uriarte et al. 2011), Zimbabwe (Dalu et al. 2017), north-west North America (Corkum 1989) and Great Britain (Moss et al. 1987), to mention just a few. Caution is needed, however, with interpretation of these kinds of studies. Many of the underlying environmental factors are intercorrelated and are probably surrogates for both regional (including history and chance) and local effects (including flow, productivity and detritus inputs). Detritus supply and processing is in turn affected by climate, land use and riparian vegetation.

Where the benthos has been sampled from a range of sites extending into the acidic range (pH < 4–6), such analyses characteristically distinguish samples based on stream pH and/or its close chemical correlates such as hardness, alkalinity and aluminium concentration. The results are highly consistent, irrespective of the spatial extent over which the study sites are located and compared (Hildrew & Giller 1994; Hildrew 2018). Various elements of the biota show remarkably similar patterns when studied over similar suites of sites. These include benthic microarthropods (microcrustacea and mites; Rundle & Hildrew 1990); macroalgae, bryophytes and higher plants (Ormerod et al. 1987); and epilithic algae (Planas 1996; Layer et al. 2013). Mean species richness and the total species pools

increase with pH (Hildrew & Townsend 1987). Several invertebrate families are absent from low-pH sites (e.g. particularly mayflies and molluscs) while others are usually better represented (e.g. stoneflies, blackflies). Similarly, many algae, mosses and higher plants seem to be confined to either hard or soft waters. Fish are often absent from streams of low pH and high concentrations of aluminium, both of which have clear physiological effects (e.g. Ormerod et al. 1987; Bowman & Bracken 1993). In addition to physiological effects, ecological factors are influenced by pH. For example, there are effects of acidity on the quantity and quality of food available to non-predatory invertebrates, while the absence of fish may release acid-tolerant species from their predators (Hildrew 2018; see more in Chapter 6).

In contrast to acidic waters, the invertebrates of streams which are broadly circumneutral to slightly alkaline (pH > 7) seem less clearly influenced by chemical factors—at least directly. Surveys of macroinvertebrates conducted over wide areas naturally show biogeographic groupings of sites, for instance by biome for all macroinvertebrates (e.g. Corkum 1991) or drainage basin for Simuliidae (Corkum & Currie 1987). Sandin (2004) found that both locally measured variables (e.g. water chemistry, substratum composition) and regional factors (e.g. latitude, longitude and an ecoregional delineation) were important for explaining the variation in assemblage structure and taxon richness for stream benthic macroinvertebrates across over 600 randomly selected streams across Sweden. On an equally large scale, Min et al. (2020) examined distribution patterns of macroinvertebrates across over 2,600 stream sites in the Republic of Korea and identified gradient, substratum, flow velocity, biological oxygen demand (BOD) and altitude as the major environmental variables influencing distributional patterns. Across England and Wales, analysis of national monitoring data from a total of 48,000 samples from the Environment Agency's biology database (Vaughan & Ormerod 2012) confirmed that the north-west–south-east gradient in the invertebrate fauna reflected a combination of climate, topography and land use, and that variations within individual river systems reflected variations in stream size, channel character, water chemistry and water quality and catchment urbanisation.

Freshwater plants are clearly influenced by water chemistry. For example, diatom community structure and species richness across an urban-to-rural gradient in southern Finland were related to local-scale variables such as water temperature, aluminium concentration and conductivity, which were in turn influenced by patterns in catchment land use and land cover (Teittinen et al. 2015). In this case, the change in communities along the land-use intensity gradient was accompanied by a distinct decline in species richness. Conductivity was also found to be important for diatom species richness and community composition in lotic sites in Ethiopia, along with dissolved oxygen, pH, soluble reactive phosphorus and turbidity (Shibabaw et al. 2021). Other evidence suggests that the frequency and magnitude of flood events or grazing might be more influential for benthic algae than inorganic N and P (e.g. Biggs 2000).

Conductivity is important for species distribution patterns of Characeae (green macroalgae—stoneworts) in Europe (Rey-Boissezon & Auderset Joye 2015) and in the north-eastern US (Sleith et al. 2018). Higher plants (macrophytes) are also influenced by water chemistry (e.g. Sleith et al. 2018 and references therein). The species distributions in streams and rivers do usually respond strongly to nutrient enrichment, though other factors are also influential. Thus, in a large-scale study across over 160 sites in north-east Scotland, south and east England, Demars & Edwards (2009) found that the best predictors of the variability in the species data in circumneutral sites were BOD_5, conductivity, a group of physical variables (temperature, altitude, various substratum elements and the occurrence of rapids), NO_3 and alkalinity. In alkaline sites, conductivity, pH, $EpCO_2$ (measuring the excess partial pressure of dissolved CO_2 relative to atmospheric partial pressure) and geographical isolation were most influential (although the lack of physical variables in this list was possibly an artifact). Their overall conclusion was that the role played by nutrients (nitrogen and phosphorus) was either mostly indistinguishable from other site attributes (e.g. conductivity and nitrates possibly being surrogates of each other) or subordinate to them (e.g. soluble reactive phosphorus, ammonium).

Where anthropogenic impacts on water quality are evident, water chemistry is naturally important. For instance, Buss et al. (2002) identified a gradient of environmental and water quality conditions related to chloride, alkalinity, conductivity, BOD and hardness that, in turn, were associated with a loss of macroinvertebrate taxa across a river basin in the Atlantic forest region of south-east Brazil. Similarly, Rico-Sanchez et al. (2022) showed that high concentrations of heavy metals, nutrients and salinity (related to mining activities) limit the presence of several families of seemingly sensitive macroinvertebrates in the central plateau of Mexico. The long-term (two-decadal) data analysed by Vaughan & Ormerod (2012) showed that temporal changes in macroinvertebrate assemblages across England and Wales correlated most strongly with (improving) water quality and variations in discharge.

Overall, distribution patterns in streams and rivers are clearly increasingly affected by the impacts of human activity, rather than being solely a consequence of natural environmental variations. The occurrence of species of stream organisms in relation to environmental factors that are altered by organic pollution, toxins, pharmaceutical products, metals, microplastics and nanoparticles has, quite appropriately, received an enormous amount of attention. Much of this has been concerned with the detection of pollution and the use of organisms as indicators of what is often known as 'stream health'. River pollution and eutrophication are dealt with in more detail in later chapters (mainly 9 and 10).

Finally, note that there are (indirect) effects of pollution, via species interactions, 'multiple stressors' (interactive effects among several stressors) and through food webs (e.g. Gessner et al. 2004; Gessner & Tlili 2016). An example makes this point clearly. Lotic ecologists have usually been concerned with species that spend all or most of their lives in the water although, as we have seen, there are a great many semi-aquatic or terrestrial species that live in what we may loosely term 'the river corridor' or floodplain and depend on the river for their food and habitat at some time of the year or other. Many such species are birds, such as the European dipper (Cinclus cinclus: Figure 3.22 and Topic Box 3.3), one of a small but widely distributed family of mainly upland species. It is territorial and lives its entire life along stony streams and rivers, feeding almost exclusively on benthic invertebrates and small fish captured while walking or briefly swimming under water. It is thus entirely dependent on aquatic production. In a survey of Welsh catchments, Ormerod et al. (1986) found that the dipper had a strangely patchy distribution, being abundant in some streams but absent from others, even though all seemed physically suitable habitats (stony beds with abundant riffles). Research showed that this was associated with water chemistry and land use, the birds being scarce or absent from streams that drained plantations of exotic coniferous trees and had low pH with a high concentration of aluminium (Figure 5.6). The low stream pH in this area of Wales is largely a legacy of pollution by acidic atmospheric deposition ('acid rain'), whose effect is magnified where there are conifer plantations. Ormerod et al. (1991) subsequently found that dippers laid eggs later and had smaller broods on acidic than on circumneutral streams, and that nestling

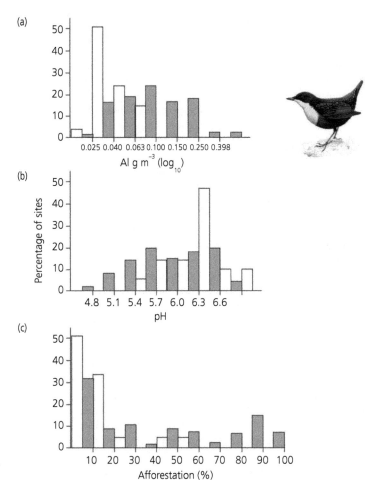

Figure 5.6 The frequency distribution (as a percentage of sites) with (open columns) or without (shaded columns) breeding dippers (*Cinclus cinclus*) on 74 Welsh streams in 1984 in relation to (a) mean filterable aluminium concentration, (b) stream pH and (c) percentage of the catchment with coniferous afforestation. *Source*: after Ormerod et al. 1986, with permission from John Wiley & Sons.

growth was slower. Adult mass before the breeding season was also lower on acidic streams. This and other evidence suggests that dippers resident on acidic streams struggle to provide enough prey, and particularly calcium-rich prey (such as larger crustaceans and small fish), for the growth of their chicks to successful fledging. Low body mass at fledging reduces the chicks' chances of survival over the first winter of life (Tyler & Ormerod 1992). Non-breeding dippers on acidified streams appear to spend significantly more time foraging, with reduced prey capture success reflecting the lower prey densities in these streams (Logie 1995). A similar process explains the distribution of another species in the USA; the Louisiana waterthrush (*Seiurus motacilla*). Like the dipper, it occupies linear territories along watercourses and feeds primarily on aquatic macroinvertebrates. Mulvihill et al. (2008) showed that the density of breeding pairs was

lower on acidic than on circumneutral streams, egg laying was delayed, clutches were smaller, and the young were smaller—all apparently a result of a lack of preferred prey (acid-sensitive mayflies) in the former. The distribution of these birds is thus an indirect effect of anthropogenic acidification, in this case mediated via food supply.

5.2.5 Temperature, oxygen and distribution

Water temperature varies in streams and rivers over a variety of scales in time and space (see Chapter 2, section 2.6.1, p. 53) and this imposes a number of biological patterns on the biota that we bring together here under the umbrella of 'distribution', though temperature effects are fundamental to many biological processes and are discussed throughout the book. As oxygen concentration and temperature are so closely linked in freshwater systems (see Chapter 2, section 2.6, p. 53), it is difficult to untangle their individual effects on the biology and ecology of stream and river organisms. Thus, respiration rates can increase by 10% or more per 1°C temperature rise and, according to Van't Hoff's rule, biological activity of ectotherms doubles for every 10°C increase (Caissie 2006—the so-called Q_{10} *coefficient*) but can vary quite widely among species (Seebacker et al. 2015). Thus, increasing temperature reduces oxygen availability (warm water holds less oxygen than cold) but it also raises oxygen demand and, therefore, physiological stress. Oxygen limitation is therefore an important physiological constraint for ectotherms, especially at high temperatures. Based on both field and laboratory studies, Verberk et al. (2016) concluded that oxygen limitation (associated with e.g. organic pollution) both impairs survival at high thermal extremes and restricts species abundance at temperatures well below upper lethal limits.

Temperature sets ultimate limits to where species can live. Species that tolerate a wide range of temperature are called *eurytherms* (or eurythermic), whereas those of narrow tolerance are *stenotherms*. Such differences may be reflected in their distribution. For example, temperature is an important variable for stream diatoms, influencing them directly through metabolic processes and indirectly through concomitant changes in the physicochemical properties of water (Teittinen et al. 2015). Low temperature at high altitude restricts growth and reduces the light saturation coefficient of benthic algae, while special adaptations enable them to survive the freeze–thaw cycle (Rott et al. 2006). The restriction of many species to cold water may reflect a requirement for a high concentration of oxygen rather than being an effect of temperature *per se*. Thus, most stoneflies are cool stenotherms and require abundant oxygen, hardly any occurring where temperature exceeds 25°C (Hynes 1970a). Cold-water stenothermal fish like salmonids also have a high metabolic rate and oxygen demand and an upper incipient lethal temperature of around 24–28°C depending on the species (Elliott 1994). In Canadian rivers, trout populations are sparse where weekly maxima exceed 22°C (Mackay 1995).

Most components in the life history of insects, and in fish, such as egg development, growth rate, emergence time, adult size and fecundity, are affected significantly by temperature (as described in Chapter 4). Influences on growth rate and life cycle are often related to a combination of temperature and the time period over which it is applied—termed the accumulated *degree days*. Degree days over any period may be calculated by summing the daily mean temperature above some threshold, often 0°C. There are clear biogeographical patterns in annual degree days, which decrease with increasing altitude and increase with decreasing latitude towards the tropics. In Quebec, eastern Canada, the annual degree days increase from 1,702 to 2,219 from first- to ninth-order streams (stream order being a surrogate for altitude; Naiman et al. 1987), while in the eastern United States, annual degree days increase from 3,000 in northern New York to 7,000 further south in Georgia (Webster et al. 1995). These differences lead to spatial variation in the number of generations per year (*voltinism*) for individual species of insects (see Chapter 4, section 4.4.2, p. 121).

Temperature makes its impact on the physiology of individuals, and thus on the population ecology of single species. Not surprisingly, therefore, spatial patterns in temperature also have profound effects on assemblages of species. Thus, the diversity of insects usually declines with altitude and latitude (since temperature declines with increasing altitude and towards the poles). For instance, Jacobsen et al. (1997) found that small lowland streams in the tropics have, on average, a two- to fourfold higher species richness than temperate lowland streams. While temperature plays its part in such patterns, however, other biogeographic factors, such as geological history and other aspects of climate, are also influential (assemblage-scale patterns are discussed further in Chapter 6, section 6.3).

Examples of clear distribution patterns of species along rivers are to be found in practically all groups of lotic animals and plants and have frequently been ascribed to temperature. Among the most obvious are those where longitudinal gradients in conditions are particularly marked. For instance, streams originating at the 'snout' of glaciers present very harsh conditions in the form of low temperature, frequent physical disturbance as ice and snow melts, limited food availability and a large load of inorganic sediment (e.g. Milner & Petts 1994), but fairly quickly become rather more 'benign' downstream. In many parts of the northern hemisphere, larvae of the midge *Diamesa* and its relatives (subfamily Diamesinae, Chironomidae) are the first to colonise such habitats closest to the glacier and in many instances are lost once temperature exceeds 2°C; other taxa then appear in a downstream sequence (see Figure 5.7 for results for European streams, though similar sequences are found elsewhere). In effect the

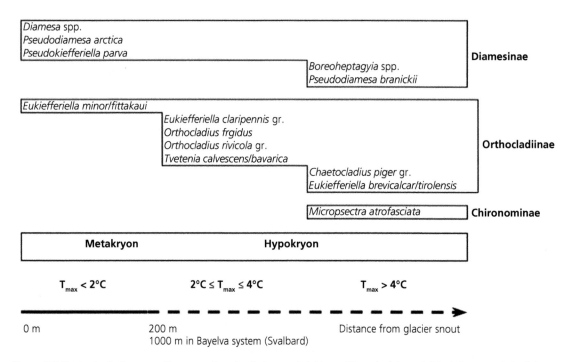

Figure 5.7 The longitudinal sequence (downstream from the glacier snout) of chironomid taxa (mainly species) in six European glacier-fed streams (from the Pyrenees, the Alps [two streams], western Norway, Iceland and the Norwegian Arctic island of Svalbard).
Source: Lods-Crozet et al. 2001, with permission from John Wiley & Sons.

reverse is seen in streams that originate from hot springs, with highly adapted species found near the source that are replaced by other species as the water cools downstream.

Under rather less extreme conditions, downstream sequences of species within guilds of related taxa occur and are widespread. There are particularly well-known patterns, for instance, among the globally distributed family of web-spinning caddis, the Hydropsychidae. In the relatively species-poor island of Britain, the sequence begins with *Diplectrona felix* in extreme headwaters (temperature did not exceed 15°C), while it is replaced by a variety of species of *Hydropsyche* further downstream (on the River Usk Figure 5.8). Similar longitudinal sequences of hydropsychids have been described elsewhere in Europe, North America and Asia (e.g. Hildrew et al. 2017; Ficsór & Csabai 2021), many of them applying to much larger rivers than the Usk. Species in the downstream sequence seem increasingly tolerant of higher temperature, finer sediment and more eutrophic conditions, as well as less hydraulically 'rough' flow in larger rivers. At least in some cases, this is accompanied by an adaption in the metabolic rate–temperature relationship such

that species have the highest net growth efficiency in the river reach in which they are found (Hildrew & Edington 1979).

Global air temperatures are rising under ongoing climate change and, even though water temperature fluctuates less than that of the air (see Figure 2.24 a), the temperature of rivers and streams will inevitably follow. The complex effects of climate change on streams and rivers are the subject of intense research (see more in Chapter 10). In terms of longitudinal sequences of species, changes in the distribution might follow the warming trend patterns, with cool-water species shifting upstream (e.g. see Jacobsen 2020). Changes observed so far are rather inconsistent in this respect, however. Based on previous work in the very high Andes of Equador, Jacobsen (2020) described the potential changes in the altitudinal distribution of insects that would be required in order for particular species to remain within their preferred temperature and oxygen regimes. Essentially, insects 'seeking' cooler water at higher altitudes as temperature rises may actually be prevented from moving upwards. Although the solubility of oxygen increases at lower temperature, its concentration in water at very high

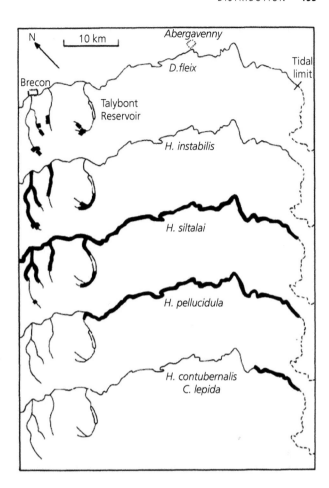

Figure 5.8 The longitudinal sequence of species of hydropsychid larvae in the River Usk (south Wales, UK). *Diplectrona felix* occurs only in extreme headwaters (usually less than 2 km from the source), and is replaced by *Hydropsyche siltalai*, *H. instabilis*, *H. siltalai*, *H. pellucidula* and *Cheumatopsyche lepida* further downstream.
Source: Hildrew & Edington 1979, with permission from John Wiley & Sons.

altitudes (> 4,000 m) may fall, because of the reduction in air pressure and the partial pressure of oxygen—that is, oxygen saturation declines, and this counteracts the effect of decreasing temperature alone. For example, the stonefly *Claudioperla* sp. has a high Q_{10} of 4 (meaning a 4°C increase would result in a 74% increase in oxygen demand). Simply to keep within its current temperature range, the species would be required to shift upwards in altitude from 3,450 m asl to 4,750 m asl under a 4°C temperature increase in a warmer world, but based on the balance of oxygen demand and the oxygen supply at this atlitude, the species would be required to shift further up to 5,600 m asl, to a temperature regime 3°C cooler than at present. This species could therefore be exposed to a 'summit trap' where available habitats are rare at this altitude. Jacobsen (2020) contrasts this with another species, the beetle *Anchytarsus* sp., which has a much lower Q_{10} of 1.6 and technically would need to move to a

lower altitude to balance oxygen demand and supply, which might take it far outside its upper temperature tolerance.

There are other reasons why lotic species may be unable to shift upstream as temperature increases. Thus, Hildrew et al. (2017) resurveyed hydropsychid distributions in both the River Usk (Wales) and the River Loire (France), but found no consistent evidence of upstream movement among the Hydropsychidae in either river, despite variable increases in summer maximum temperature over the study (in the case of the Usk, distribution was compared between the late 1960s and 2010). Hildrew et al. (2017) concluded that either the temperature increase had not yet been sufficient or that other features of the habitat, not related to temperature, precluded a simple shift upstream. For net-spinning filter-feeders, such as hydropsychids, this could include the hydraulic habitat—to which they are very sensitive. In effect, this points out again that distribution is influenced by a suite of interacting habitat features (reflected in the whole habitat templet) rather than by single factors acting alone.

5.3 Abundance: population dynamics in streams and rivers

In terms of population distribution, a frequent approach in stream ecology has been simply to correlate physicochemical factors with the distribution of a particular species or, mainly at small scales, to undertake experimental manipulations in the field. However, population ecologists are often more concerned with what determines the number of individuals in a population, rather than simply its spatial occurrence. Abundance and distribution could simply be 'two sides of the same coin'—they could relate to the same factors that determine both the location of suitable habitat and the temporal persistence of suitable conditions. It is also possible, however, that population density is affected by a partially separate set of factors, at least some of which feed back on density itself (i.e. these latter factors are density dependent). In the case of the dippers (Figure 5.6), for instance, we might suspect that stream acidity determines the potential food supply and thus the distribution of the birds, whereas it is intraspecific competition for territories that regulates their abundance. The effect of acidity on density would thus be indirect and mediated by food supply and territorial behaviour. In this example, density-independent factors (acidity, food supply) and density-dependent factors (intraspecific competition for territory) interact.

5.3.1 Long-term patterns and the decline of the 'megafauna'

Our perception of changes in population size depends on timescale. Probably the best 'long-term' information on lotic populations relates to fish, and particularly to salmonids. Salmonids are extremely valuable economically, both for commercial fishermen and as highly prized sport fish. Perhaps not surprisingly, therefore, we have some substantial records of salmonid populations based both upon catch data and from scientific monitoring. Such records often suggest that climatic variability has widespread effects, although contemporary data usually span too short a time to define the nature and frequency of the patterns that emerge.

One of the longest ongoing monitoring schemes is for the Burrishoole catchment in Co. Mayo, Ireland, an 'index site' for diadromous fish in the North Atlantic, where every salmon migratory movement (at the individual fish level) to and from the river catchment has been recorded since 1958 using traps. Climate affects salmonid survival in this system (de Eyto et al. 2016). The North Atlantic Oscillation (NAO) index is the major indicator of climatic conditions in winter in north-west Europe, and is measured as the difference in air pressure between the Azores (high) and Iceland (low). A positive NAO relates to a bigger pressure difference and brings warm, wet winters in western Europe that coincided in the Burrishoole with poor survival in brown trout and Atlantic salmon, while cold springs saw greater survival of trout.

Techniques that allow longer-term reconstruction of populations (or more likely surrogates of population size) can be useful. For instance, Finney et al. (2002) depict a 2,200-year record of sockeye salmon (*Oncorhyncus nerka*) abundance from a salmon nursery lake on Kodiak Island in Alaska, reconstructed on the basis of the concentration of the stable isotope of nitrogen (^{15}N) and a diatom marker in the lake sediment. The idea here is that nutrients transported upstream from the ocean by the salmon during their spawning runs ('salmon-derived nutrients', see Chapter 9, p. 345) are significant in otherwise oligotrophic freshwater systems. Therefore, periods of high sedimentary ^{15}N, and of diatoms, indicating enriched conditions, mark times when the salmon run was particularly abundant (Figure 5.9).

In the oldest sediments in this lake, about 200–100 BC (i.e. 2,200–2,100 years before the present), marine-derived nutrients were abundant, as was the diatom that indicated enrichment. Conditions were similar to the period around 1880 AD, when commercial fishing began and an estimated 3 million sockeye salmon entered the lake annually. Beginning around 100 BC there was a marked decline in ^{15}N, indicating a low salmon run continuing for almost 200 years, when a gradual increase began that was sustained to around 1200 AD. These two marked shifts in salmon numbers coincided with shifts in ocean circulation in the northeastern Pacific, changes which appear to have altered the relative marine productivity of Alaskan waters and those further south (Finney et al. 2002). Note, first, that this infers that the size of the salmon run is determined by marine productivity and, second, that shifts to very low abundance can occur even in the absence of fisheries or other human impacts. A very long historical perspective can thus be valuable for interpreting more

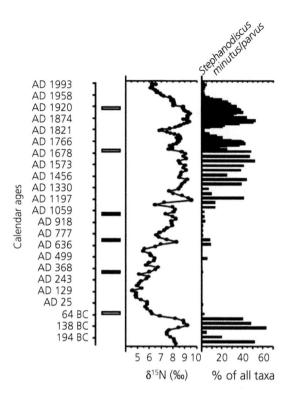

Figure 5.9 Palaeolimonological evidence of changes in the sockeye salmon run upstream into Karluk Lake, Alaska over the past 2,200 years; $\delta^{15}N$ (‰) indicates the fraction of the stable isotope relative to terrestrial sources (i.e the size of the marine nutrient subsidy transported by the salmon) and % of all taxa is the percentage of all diatom remains made up of two *Stephanodiscus* spp. (indicators of mesotrophic/eutrophic conditions).
Source: from Finney et al. 2002, with permission from Springer Nature.

modern changes in 'ecological' time. However, note also that the apparent decline in ^{15}N since 1880 coincides with an increased fishing take and a reduction in the size of the salmon run. Similar significant variation in sockeye salmon production has also been identified in the Kenai River catchment in Alaska, some 400 km or so north-east of Kodiac Island (McCarthy et al. 2018), suggesting that regionally synchronous multicentennial production shifts in salmon may result from shifts in conditions in the north-eastern Pacific.

More recent historical records of salmon populations often suggest a widespread and profound diminution in numbers under various kinds of human impacts. Among the most famous examples of decline in formerly enormous runs of salmon is that of the Columbia River basin of north-west North America. One large tributary of the Columbia is the Snake River of Idaho,

on which four dams were constructed between 1961 and 1975. These restricted access by migrating Chinook salmon (*Oncorhyncus tshawytscha*)—whose migration had already been hindered by dams (built in the early 20th century) further down the Columbia system. All indices of salmon stocks on the Snake River show a decline, from highs in the 1950s of around 130,000 salmon and steelhead (*Onchorhynchus mykiss*), with spawning populations in 2000 averaging only 10% of their values in the 1950s (Figure 5.10a: numbers were down to 10,000 fish in 2017). Removal of the dams would undoubtedly have improved survival, although it was calculated that other measures would be necessary to reverse population decline (Kareiva et al. 2000).

Other anthropogenic effects can also be important, as shown by very long-term records of fisheries yield in Norwegian rivers. In the two southernmost counties of Norway, the total Atlantic salmon (*Salmo salar*) yield from nine rivers in 1885 was 81,000 kg, falling to 11,000kg in 1925 and just 1,650 kg in 1968 (Jensen & Snekvik 1972). In some of these rivers there had been occasional, unexplained mass mortalities of adult salmon over many years; in the Kvina river in 1911 and 1948 and in the Mandel in 1914 and 1921. In November and December 1948, there was particularly mild weather and a snowmelt in the headwaters of the Kvina and Frafjord rivers was followed by the appearance of dead adult salmon: pH in the Frafjord river at the time was between about 3.9 and 4.2 (Jensen & Snekvik 1972). Of course, there could also have been mortality of eggs (incubated in nests in the riverbed) and alevins (larvae) that might have escaped notice. Subsequently, Leivestad et al. (1976) compared salmon catches in seven of these southern Norwegian rivers with those from 69 rivers from elsewhere in the country (Figure 5.10b). Catches in the southern rivers had declined from around 1910 onwards and were practically zero by the mid-1960s. Measurements of pH made in 1975 showed that these seven rivers were much more acidic (overall mean pH 5.1, with mean minima of 4.5 during acid episodes, normally after snow melt) than rivers with thriving salmon fisheries (pH values 6.6 and 5.1, respectively). Evidently, the early decline of salmon in southern Norway, and in many other places, had been another consequence of 'acid rain'.

More widely, reviews have indicated that the total abundance of Atlantic Salmon (*Salmo salar*) has declined significantly over the last 30 years or so, attributable to a range of anthropogenic factors (Forseth et al. 2017). In Norway, the number of farmed

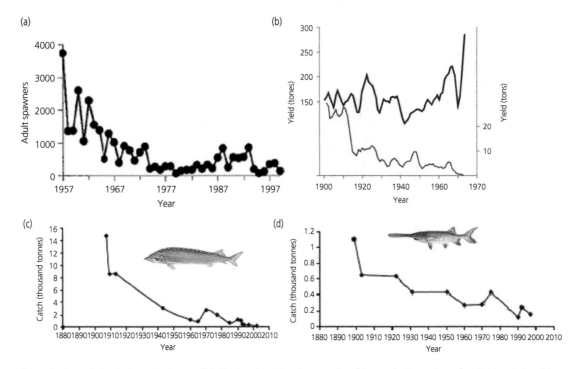

Figure 5.10 Population declines, or indicators of decline, in salmonids and two species of the true freshwater 'megafauna': (a) An index of the adult (four and five years old) spawning Chinook salmon (*Oncorhynchus tshawytscha*) in the Snake River, USA 1957–1999, based on nest (redd) counts; (b) annual mean yield for Atlantic salmon (*Salmo salar*) fisheries in seven acidified rivers in southern Norway (lower curve, right-hand scale for yield) compared with that from 69 rivers in the rest of the country (upper curve, left-hand scale); (c) catch of Beluga sturgeon (*Huso huso*), the largest of all freshwater fish, from the Caspian Sea; and (d) commercial catch of paddlefish (*Polyodon spathula*) from the Missisipi basin.
Source: (a) Kareiva et al. 2000, with permission from the American Association of the Advancement of Science; (b) from Leivestad et al. 1976, free public access; (c) and (d) from Pikitch et al. 2005; with permission from John Wiley & Sons.

Atlantic salmon in the 600 farms along the coast is over 730 times larger than the abundance of returning wild salmon. Unless farming practices change radically, escaped farmed salmon and salmon lice (a parasitic copepod, *Lepeophtheirus salmonis*) from Atlantic salmon fish farms are likely to cause further reductions in wild fish. Further ongoing threats come from the introduced parasite *Gyrodactylus salaris*, expanding hydropower regulation, residual acidification and other habitat alterations.

More generally, the modern fate in the world's great rivers of populations of the so-called freshwater megafauna—loosely defined as large (adult wet body mass exceeding 30 kg and thus even bigger than most salmonids), often migratory fish and the fully aquatic or semi-aquatic larger mammals—is bleak indeed (Carrizo et al. 2017). This megafauna includes the largest of all freshwater fish, the beluga sturgeon (source of the most valuable caviar, trading at up to $5,000 per kg in 2017), which is classified as critically endangered by the IUCN (International Union for the Conservation

of Nature). Like all sturgeons, the beluga is anadromous, migrating up into rivers to spawn. The species is long-lived and late to reach sexual maturity (c.16 years in females). Estimates of the number of spawning females entering the Volga river from the Caspian Sea (by far the most important extant population) suggest a decline from about 26,000 annually between 1961 and 1965 to about 2,500 in 2002 (Pikitch et al. 2005), while the (known, legal) catch of beluga fell effectively to zero in the early 2000s (Figure 5.10c). Another very large migratory species is the American paddlefish (*Polyodon spathula*), which once inhabited 26 US states, while its range included the Great Lakes and Canada. It is now found in 22 US states and its range is restricted to the Mississippi basin and some smaller coastal rivers flowing into the Gulf of Mexico. Paddlefish numbers have certainly declined in much of its present range, while fishing effort is now legally constrained (Figure 5.10d).

Of 132 known species of freshwater megafauna in the IUCN 'red list' (Carrizo et al. 2017), no fewer than 27 are critically endangered and well over half are

Table 5.2 Total number and percentage of the 132 megafaunal species classified in each IUCN Red List category (from Carrizo et al. 2017, under Creative Commons CC BT licence).

	Red List category								
	EX	EW	CR	EN	VU	NT	LC	DD	Not evaluated
Number of species	0[a]	1	27	18	17	6	32	6	25
Percentage of assessed species	0	0.9	25.2	16.8	5.6	29.9	5.6	5.6	Not applicable

Categories of threat: EX, extinct (note [a] Baiji and Chinese paddlefish are critically endangered, probably extinct); EW, extinct in the wild; CR, critically endangered; EN, endangered; VU, vulnerable; NT, near threatened; LC, least concern; DD, data deficient; NE, not evaluated/assessed (from Carrizo et al. 2017).

at least 'vulnerable' or worse (Table 5.2); the Yangtse River dolphin (*Lipotes vexillifer*, last confirmed sighting 2001) and Chinese paddlefish (*Psephuris gladius*) are both extinct (Zhang et al. 2020). Threats to megafaunal fish include overexploitation, pollution and, probably overwhelmingly, the construction of dams that prevent their migration and cut off spawning grounds (see Topic Box 2.1 by Christianne Zarfl). For instance, dam construction on the Volga river has denied access for sturgeons to between 30 and 90% of their former breeding sites (depending on species). The decline of larger species and individuals of animals in rivers has been referred to as a 'great shrinking', and we return to this and other aspects of biodiversity loss in Chapter 10).

5.3.2 Evidence for population regulation in stream animals

Populations in general decline, like those of the megafauna described above, are evidently not effectively regulated, and extra (usually anthropogenic) sources of mortality, such as fishing or habitat destruction, exceed the ability of population processes to compensate for the population loss. Next, we deal with the more common cases in which population size is not inevitably subsiding towards local or global extinction. We begin with examples where we have some expectation that population density is set mainly by births and deaths within a population, rather than by exchange (immigration and emigration) with other (sub-) populations nearby, and the population is in this sense 'closed'.

5.3.2.1 Fish populations

Much evidence suggests that fish populations, about which we have most data, are often regulated by density-dependent factors. Among the best data are those from J.M. Elliott's long-term study of a population of migratory brown trout (*Salmo trutta*) in a small stream in the English Lake District (Elliott 1994).

The life cycle of the brown trout is variable; it can be a stream-resident species, sometimes undertaking small-scale migrations within fresh water, or fully migratory, in which all (as in this example) or part of the population undertakes migrations to the sea for one or more years. Spawning fish in Black Brows Beck are mainly males in their third year of life, having spent one summer at sea, and larger males and females in their fourth year, having spent two summers at sea. The eggs (500–1,800 per female, in two to five batches depending on maternal body size) are laid in gravel nests (*redds*) in the late autumn and early winter and hatch in February to early March (after 444 degree days) into yolk-sac fry (*alevins*). These remain in the gravel for an additional 408 degree days after which they swim up into the water column as young fish and are called *parr* (usually in late April to early May). They start to feed, normally remaining in the stream for about 2 years, after which they migrate downstream to the estuary as *smolts* and begin the marine part of their life cycle.

Basic census data were collected in Back Brows Beck between 1966 and 1990 and consisted of electro-fishing catches in each year in late May/early June and again at the end of August/early September. These data revealed a remarkably clear 'stock-recruitment curve' for the early part of the life cycle. Thus, the number of young (0+) fish (parr) caught in the early summer, just after they have dispersed from the redds, rises in line with the total number of eggs laid by females in the previous autumn/winter, to reach a maximum at just over 2,000 eggs m^{-2} (the area of stream bed censussed) (Figure 5.11). This is a 'hump-backed' relationship, the number of recruits to the parr population falling with further increases in the number of eggs (i.e. beyond 2,000 eggs m^{-2}), and indicates a density-dependent factor acting on the young fish. It was possible to partition the 'loss rate' of fish (which is probably mainly mortality but could also include emigration) through the life cycle into a number of stages (e.g. eggs hatching successfully to 'swim up' as parr, parr surviving through

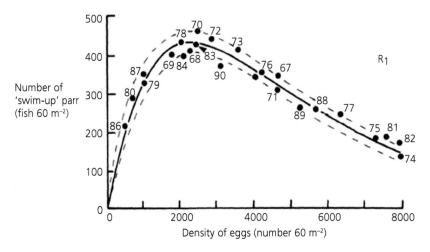

Figure 5.11 A 'Ricker' stock-recruitment curve for a population of migratory brown trout (*Salmo trutta*) in Black Brows Beck (English Lake District) from 1967–1990, showing the density of swim-up parr in the first early summer of life in relation to the density of eggs that were laid in the previous autumn/winter.
Source: from Elliott 1994, with permission.

their first summer, etc). A loss rate is given as the natural logarithm of the numbers at the beginning of a life stage minus those surviving to a later stage. One advantage of calculating such loss rates on a logarithmic scale is that the values for the various life stages are additive and their total equals the loss rate over the whole life cycle (Elliott 1994).

Plotting these loss rates at different stages of life against density at the beginning of the stage confirmed that survival from egg to young parr (k_1) was intensely density dependent—nearly all the variation in loss rate at this stage was accounted for by egg density (it was independently shown that there was almost no mortality between the eggs and successful hatching of the alevins). Survival as parr over the first summer was also density dependent, though with more unexplained variation. Loss rate in later stages was clearly density independent (Figure 5.12). Years with markedly high loss rates (1969, 1976, 1983, 1984 and 1989) had severe summer droughts in which the area of stream bed habitable was reduced, growth was poor and the smolts migrating to the sea were on average small (in turn reducing their survival). The source of the density-dependent loss in young parr was shown to be due to the failure of many of the fish to begin feeding. Up to 80% of parr are lost, and moribund fish drift downstream, mainly at night, and die (Elliott 1994).

This rather beautiful example of the power of long-term research shows that there is strong density dependence in this population of migratory brown trout,

but that there are also density-independent sources of mortality associated with climatic variations, in this case severe droughts. Comparisons with other local streams suggested that Black Brows Beck was an excellent habitat for trout, while some others held much lower densities of non-migratory fish where density dependence was apparently absent. The long-term persistence of fish in these latter more marginal habitats could depend on occasional dispersal from local populations in more favourable habitats elsewhere. This leads us to the possibility that such local populations (often in habitats of differing size and/or quality) may be 'linked' by dispersal and that there could thus be networks of local populations. We return to this spatial aspect of populations in rivers later in this chapter (section 5.4).

Other migratory salmonids have been shown to exhibit density-dependent survival of young fish and a significant 'stock-recruitment curve' of one shape or another. For instance, Jonsson et al. (1998) showed that the production of Atlantic salmon (*Salmo salar*) smolts in the Norwegian River Imsa reached an asymptote with an increasing number of eggs laid (Figure 5.13a). This pattern differs somewhat from the dome-shaped curve for Elliott's (1994) migratory brown trout, for reasons that are not entirely clear but, as in Elliott's trout population, density dependence operates in the young, freshwater stages of the salmon. In contrast, survival at sea in River Imsa salmon was density independent, the number of returning adults simply

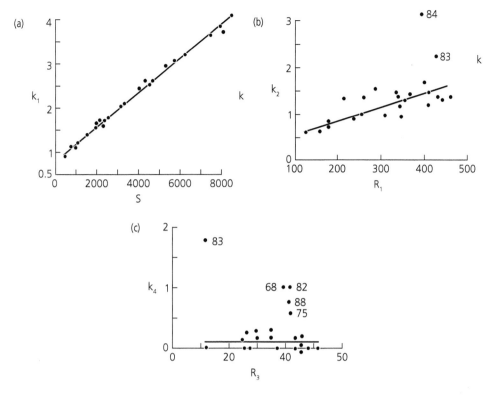

Figure 5.12 Relationship between loss rates (mortality plus possible emigration) from a population of migratory brown trout (*Salmo trutta*) in Black Brows Beck, English Lake District, and the population at the previous stage: (a) k_1 is the loss between eggs laid (S) and resultant swim-up parr in May/early June (e.g. Ln Egg density/number of young parr), (b) k_2 is the rate of loss of parr from early (initial number R_1) to late summer in their first year of life, (c) k_4 is the rate of loss between parr in their early, second summer of life (R_3) (i.e. age one year) and those in their second winter plus those returning after migration to sea (note the anomalous drought years picked out, where loss rate was high).
Source: from Elliott 1994, with permission.

increasing with the number of smolts that had left the river (Figure 5.13b). The population of adult salmon at sea was far below any marine carrying capacity, unlike the situation among young fish in the river, and is probably due to low marine survival for reasons other than food supply.

As discussed previously, declines in the return of salmon (both Atlantic and Pacific) over the past few decades are widespread, with reports of decreased marine growth rate and survival. Jonsson & Jonsson (2004) related this, for River Imsa Atlantic salmon between 1981 and 2000, to a declining sea temperature off northern Norway, and showed a reduction in the marine growth rate, and an increase in mortality. The North Atlantic Oscillation (NAO) winter index (see section 5.3.1) has mainly been positive over recent decades, bringing largely warm, wet winters in western Europe. Somewhat anomalously, this apparently leads to *lower* sea surface temperatures off northern

Norway, because of an associated weakening of the warm Gulf Stream (the North Atlantic drift). This apparently influences salmon growth and survival at sea, and the proportion of fish that reach a size threshold in their first sea winter enabling them to return to the river to spawn (Figure 5.13c).

In contrast to migratory species and populations, several studies of resident salmonids indicate density dependence among adults rather than juveniles. Elliott & Hurley (1998) showed this for non-migratory adult brown trout in a small English stream, as did Lobon-Cervia (2012) in a long-term study of brown trout in north-western Spain, and Grossman et al. (2017) for a population in Bruce Creek Pennsylvania from 1985–2011, where the brown trout is a non-native species. In all these cases the mechanism at work seemed to be intraspecific competition for nest sites among adults. This was sufficiently strong to regulate the population in the face of disturbances and

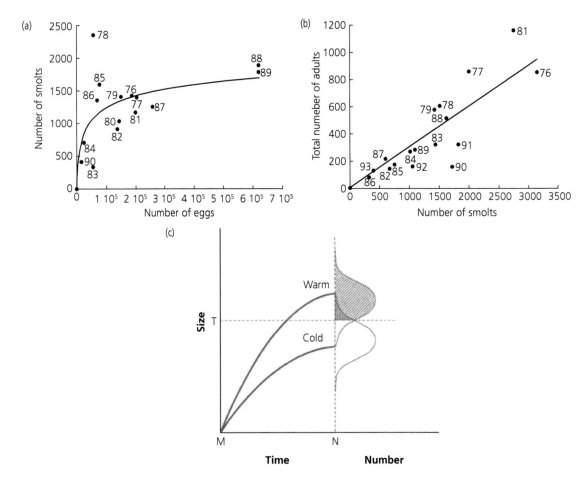

Figure 5.13 For the River Imsa, Norway, (a) the number of Atlantic salmon (*Salmo salar*) smolts leaving the river in relation to the number of eggs laid in 1976–1990, fitted to a 'Cushing' type stock-recruitment relationship, (b) the number of returning adults increased linearly with the number of smolts that had earlier left the river; (c) the influence of sea temperature (warm or cold) in the first summer (M to N, May to November) on the number (relative frequency) of fish reaching the size threshold (T) necessary to become a two-sea-winter fish.
Source: (a) and (b) after Jonsson et al. 1998, with permission from John Wiley & Sons; (c) after Jonsson & Jonsson 2004, with permission from the *Canadian Journal of Fisheries & Aquatic Sciences*.

environmental fluctuations that acted independent of density.

As can be seen, we have a wealth of information about regulation in stream-dwelling salmonids, but there is also evidence for strong density dependence in populations of other groups of fish. The Cottidae is a family of benthic fish, commonly called sculpins or bullheads, which lack swim bladders and live among stones on the bed of fast-flowing streams. They are pugnacious little predators of benthic invertebrates and are fiercely territorial, the breeding males defending plaques of eggs laid by the female beneath large stones. Perhaps unsurprisingly, populations of cottids are often very stable, probably due to strong

intraspecific competition for space. Grossman et al. (2006) found this in their 12-year study of a population of the mottled sculpin (*Cottus bairdi*), the most abundant fish in streams draining the renowned Coweeta experimental catchments in North Carolina (see Topic Box 1.1). They also carried out a removal experiment, finding that juveniles rapidly moved into sites from which adult fish had been taken.

The second most abundant fish at Coweeta was a small cyprinid, the rosyside dace (*Clinostomus funduloides*), a drift-feeding insectivore. In another long-term study, Grossman et al. (2016) showed that both the instantaneous per capita rate of population change (r) and individual growth were density dependent

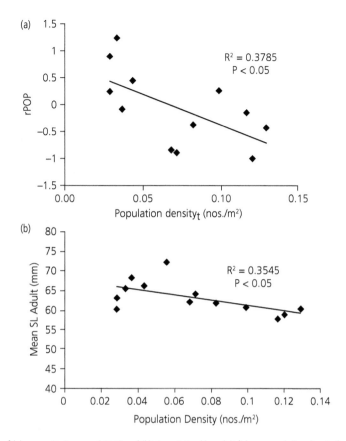

Figure 5.14 Regressions of (a) per capita increase (rPOP) and (b) size attained by adult fish on population density for rosyside dace (*Clinostomus funduloides*) in a stream at Coweeta.
Source: Grossman et al. 2016, with permission from John Wiley & Sons.

(Figure 5.14), sufficient to regulate the population, though once more flow variations were associated with fluctuations in population size. The most likely source of density dependence was again intraspecific competition, this time for the best feeding sites. We can conclude overall, therefore, that where it has been looked for rigorously, populations of stream fish are often regulated by density-dependent factors. Populations in more marginal habitats may be more affected by harsh environmental conditions, and their persistence is more likely to rely on recolonisation by dispersal from elsewhere.

5.3.2.2 Invertebrates

While there are few data on the population ecology of lotic fish of more limited economic importance than salmonids, even less is known about the dynamics of the diverse benthic invertebrates of streams and rivers. This is largely because, with a few exceptions,

life tables (based on counts of recruitment and mortality over successive generations) of stream invertebrates have not been developed, while they have for many, mostly terrestrial, species of agricultural or medical significance, for instance. Nevertheless, there are case studies available for some of the better-known, larger (macro-) invertebrates. The principles of population ecology, of course, remain the same whatever the species.

There have been a few instances where strong local density dependence has been demonstrated. Thus, Elliott (2013) studied the common European mayfly *Baetis rhodani* in a small English stream, sampling larvae and drifting adults (both newly emerged and 'spent' females) plus eggs over 39 months. There were two generations per year and a variable number of instars (up to 26 for males and 27 for females). As for the migratory brown trout (see Figure 5.12) he calculated 'loss rates' for differing stages in the mayfly life cycle and showed a density-dependent relationship between the density of first instar larvae ('stage 1') and that of instars 2–5 ('stage 2'); that is, the loss rate of early instar larvae increased with the density of the first instars. However, loss rates for all other stages were relatively constant and independent of initial density. Overall, the 'optimal' number of eggs—that is, the number yielding the most emerging adults in the next generation—was 2,700–2,750 m^{-2}, depending on the generation. Intriguingly, therefore, both fewer and more eggs laid yielded fewer emerging adults, although it is not clear what the source of this density dependence is.

Strong density dependence was also demonstrated for a population of a predatory insect, *Sialis fuliginosa*, in Broadstone Stream, southern England (Hildrew et al. 2004). This is an alderfly (Sialidae) and has a complete metamorphosis (i.e. it is holometabolous, with a pupal stage) and a two-year life cycle. When fully grown, the larva crawls from the stream at night and pupates in soft, moist soil on the stream bank. Following emergence from the soil, the adults mate and the female lays her eggs in a thin 'plaque' on the underside of tree leaves overhanging the stream. Upon hatching, the first instar larvae fall into the stream to begin their aquatic lives. Broadstone Stream was the subject of long-term research (25+ years at the time of this experiment), and the background density of *Sialis* larvae over that period did not differ consistently among three 150-m stretches in which the number of eggs was manipulated. Leaves with eggs were clipped from a 'depletion reach' (where recruitment was strongly reduced) and transported to an 'addition reach', where they were taped to tree branches above the stream, thus approximately doubling the supply of eggs and first instar larvae. The third reach was left unmanipulated as a control. This was done over three summer egg-laying seasons. Despite reducing the number of hatchlings in the depletion reach and increasing it in the addition reach, the effect was short-lived and, within a few months of recruitment of young larvae (i.e. the time of peak population density), the numbers were similar in all three reaches. Density-dependent mortality seemed to have returned numbers to the control value. That is, mortality was very high in the addition reach, somewhat lower in the control reach, and very low indeed for the few survivors of egg removal in the depletion reach. In these two examples, of *Baetis rhodani* and *Sialis fuliginosa*, there is fairly clear evidence of density-dependent mortality during parts of the larval stage that could regulate the local population in each stream, but how widespread are such phenomena?

Peckarsky and colleagues have carried out extensive research on recruitment and dynamics of mayflies in Colorado (USA) streams. In a groundbreaking study, Peckarsky et al. (2000) measured emergence and oviposition in a common mayfly *Baetis bicaudatus* at one stream site over three years and at a number of other sites more qualitatively. This and other studies of *Baetis* mayflies take advantage of the unusual (for mayflies) behaviour of the females in laying a single plaque of eggs on the underside of large emergent stones; particularly suitable sites sometimes accumulate aggregations of many egg masses. In the Colorado Rockies, potentially suitable stones for oviposition are submerged during snow melt in spring, but then become available as flow recedes into summer. The times such stones appeared was also measured. At least three important conclusions can be drawn from this initial study. First, that the local (i.e. at each site) emergence of adults was independent of the number of eggs laid there initially. This infers the operation of density-dependent processes acting on the larval stage, as is conventionally thought. Second, however, the number of females ovipositing at a site depended on the 'regional' supply of females, not only on those that had emerged locally. That is, females laying eggs at a particular place had not necessarily emerged close by but often had flown in from elsewhere. Third, early emerging females at sites where suitable stones were still under water must have dispersed to sites where oviposition sites were already available as flows receded. In particular, flow normally recedes earlier in smaller streams high up in the catchment, and last in larger channels further downstream. This may explain the predominantly upstream flight of females emerging from downstream sites. This would be further enhanced if adults emerged earlier from warmer streams at low altitude. Peckarsky et al.'s (2000) study was an important step in showing not only that local, post-recruitment density dependence does occur (egg supply does not usually limit local density) but also that dispersal can link together (sub-) populations throughout a complex catchment, certainly in terms of gene flow at that scale. Such a process of dispersal could of course also 'rescue' the population at a local site if some perturbation had eliminated it.

Subsequent surveys of recruitment (egg masses laid) and larval densities of *Baetis bicaudatus* in the same part of the Rockies added a level of complexity, in that streams both with and without fish were considered (representing high- and low-'risk' environments) (Encalada & Peckarsky 2011). Although dynamics differed somewhat between the two, the most general conclusion was that post-recruitment density dependence (i.e. among the larvae), in the spring and early summer leading up to emergence, eventually obscured differences in the number of recruits (i.e. the number of egg masses laid) at any one site. Encalada & Peckarsky (2012) then also manipulated recruitment in three reaches ('addition', 'depletion' and controls) in each of four *B. caudatus* streams, more or less as was done for *Sialis* as explained earlier. They added oviposition sites by reorientating rocks so that more protruded from the water and were thus available for egg-laying, and depleted oviposition sites by fully submerging rocks that had previously protruded through the water surface (this had to be done repeatedly as flow receded). This manipulation had a radical effect on recruitment, as egg density in the addition site was increased by about four-fold and reduced in the depletion sites almost to zero. In contrast to the field survey (Encalada & Peckarsky 2011), and to the egg-manipulation study of the alderfly *Sialis* (Hildrew et al. 2004), the results of the *B. caudatus* manipulation did persist for at least a year. Encalada & Peckarsky (2012) suggested that really radical differences in egg supply could overwhelm post-recruitment processes, but that both (i.e. the supply of eggs and subsequent density-dependent survival among the larvae) could occur and were likely to be important. More generally, it is of course acknowledged that density dependence can be overridden by perturbations of sufficient strength (such as the complete failure of recruitment), in which case long-term persistence of species can be maintained only by dispersal from elsewhere.

Recruitment of *Baetis rhodani* was also studied by Lancaster et al. (2010) in four Scottish streams. They showed that the number of egg masses laid in each stream again increased with the local density of emergent rocks (those potentially suitable for oviposition), but also scaled that up to show an overall positive relationship between egg mass density (at the scale of 1 km of channel) and the availability of emergent rocks across the four streams (Figure 5.15). Egg masses became more crowded onto fewer rocks when the latter were in short supply, although this did not entirely obscure the relationship between recruitment of eggs and overall space available. This is consistent with the view that intraspecific competition for space, this time among egg-laying females, could regulate the population.

Further recent and convincing examples of density dependence in stream insects come from (a) a study of two 'armoured' grazing caddis species from south-west Victoria in Australia (Marchant 2021) and

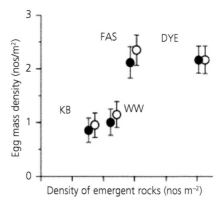

Figure 5.15 The positive relationship between the density of emergent rocks and of egg masses of the mayfly *Baetis rhodani* in four Scottish streams (distinguished as FAS, DYE, KB and WW). Open and closed symbols simply show estimates based on regressions constrained to pass though the origin (closed) or not (open).
Source: from Lancaster et al. 2010; with permission from Springer Nature.

(b) a manipulative study of a large, shredding caddis in New Zealand (McIntosh et al. 2022). Marchant (2021) showed that regulation of one species (*Tasiminia palpata*) clearly occurred at the reproductive stages (mortality in the pupal and eggs stages). The number of young larvae recruited was clearly related to the number of eggs laid, via a 'hump-backed' recruitment curve, and reached a maximum at an egg density of just over 1,000 m^{-2}. There was no density dependence during the subsequent larval life. In contrast, density dependence occurred only in the larval stages of a second species (*Agapetus kimminsi*). Larval growth was density dependent and probably related to competition for algal food and/or space, possibly including interspecific competition with the larger *Tasiminia palpata*. As we shall see, stream ecologists are often quick to ascribe sharp fluctuations in population density to flow disturbance causing density-independent mortality. In this case, however, two substantial (over-bank) high-flow events during Marchant's (2021) five-year study did not cause obvious declines in population density, larvae apparently being able to shelter beneath large stones. Rather, population fluctuations were ascribed to variable predation by large stonefly larvae (*Cosmioperla kuna*), possibly related to complex dynamics in this long-lived species (which has a three-year life cycle).

McIntosh et al. (2022) changed the density of late instar larve in nine populations of *Zelandopsyche ingens*, removing individuals from 200-m stretches of three streams (halving subsequent adult emergence), adding them to similar stretches of three other streams (doubling emergence), while leaving three other streams as unmanipulated controls. This replication of streams was a strong aspect of the design. The results suggested that density dependence occurred, late in larval life via intraspecific competition for detrital food. Larvae tracked food resources via dispersal plus possible mortality related to food shortage. In addition, large-scale stochastic processes also caused large variations in abundance.

What are the likely biological sources of density dependence in other cases where it has been observed? In the *Sialis* manipulation (Hildrew et al. 2004), it was shown that young larvae in the addition stretch had the largest proportion of empty guts, so that starvation (and hence intraspecific competition for food) could have been important. Further, predation of young *Sialis* larvae was itself density dependent, such that the percentage of larvae eaten by other predators rose with *Sialis* density. Intraspecific competition for space and territories can also be important, as in the case of several of the examples of fish populations referred to earlier. Animals that build at least temporarily fixed shelters need space or sites to do so. Experimental supplementation of such sites, if they are in short supply, should lead to increased local density. Thus, Lancaster et al. (1988) added structures that were readily used as 'web-spinning sites' by predatory larvae of the caddis *Plectrocnemia conspersa* and found that the local density was indeed increased. Of course, although this finding is consistent with the view that intraspecific competition for resources could regulate the population, such evidence falls a long way short of showing that it actually does so. This is because this experiment was short-term and small-scale and could not demonstrate persistent (intergenerational) density dependence at a spatial scale approximating that of the whole population.

Overall, evidence is accumulating for an important role for density dependence in such populations of stream insects, where the different stages usually occupy different habitats. Stochastic processes clearly also have a role and we turn our attention to these in the next section.

5.3.3 Density independence and the population-level consequences of disturbance

To many ecologists, the dynamics of flow in rivers have seemed so overwhelming that physical environmental factors in stream ecology have been thought to dominate all else, including density-dependent (biotic)

factors. This has created something of a 'tension' in the field (between proponents of equilibrial and those of non-equilibrial processes), reflecting an enduring debate in general ecology. Flow is the architect of river habitats and very rare, though profound, events such as tectonic movements, headwater capture (where the flow of a tributary stream switches from one river system to another) and climatic changes causing advance and recession of glaciers, and the most extreme floods and droughts, undoubtedly leave an indelible fingerprint on biotic diversity and lotic ecosystems. More frequent events of lower magnitude, however, act as shorter-lived disturbances to which the biota may be 'adapted' in some way. Disturbance is a very prominent concept within stream ecology, and we return to it throughout this book in the context of the habitat templet (Chapter 2), the adaptations of individuals and species traits (Chapter 4), its role in population dynamics (this chapter), assemblages and communities (Chapter 6) and ecosystem processes (Chapters 8 and 9). A single definition of disturbance is elusive in itself (see Chapter 6, section 6.5.2.1), though in the present context of populations we can think of it simply as a discrete event in time that removes organisms and opens up space that can be colonised by individuals of the same or different species (Townsend 1989). In their influential review of disturbance in stream ecology, Resh et al. (1988) concluded that 'to some of us, disturbance is not only the most important feature of streams to be studied, it is the dominant organising factor in stream ecology'.

In terms of lotic populations, disturbances are generally thought to act as density-independent factors, contributing to fluctuations in abundance but not affecting 'equilibrial density'. As an example, see the long-term data collected for the numbers of breeding herons (*Ardea cinerea*) in river catchments in the UK between 1928 and 2012 (Figure 5.16), where sharp population declines coincided with particularly cold winters (1947 and 1963) when mortality was high. Nevertheless, in this example of a semi-aquatic bird, and in a much wider range of stream research, the countervailing role of density dependence in stream organisms is persuasive, and a more nuanced (i.e. balanced) view of disturbance and biotic processes in streams and rivers is necessary (see review by Stanley et al. 2010).

The most obvious natural sources of disturbance to populations in streams and rivers are high- and low-flow events. Episodic high flows either rise over the banks—a 'flood'—or remain within the banks—as in lesser events more properly termed 'spates'; droughts cause a loss of flow at the surface, either

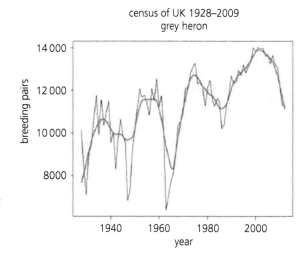

census of UK 1928–2009
grey heron

Figure 5.16 Fluctuations in the UK population (breeding pairs) of the heron, *Ardea cinerea*, between 1928 and 2012.
Source: after Woodward et al. 2016, based on data from the British Trust for Ornithology, published under Creative Commons Licence.

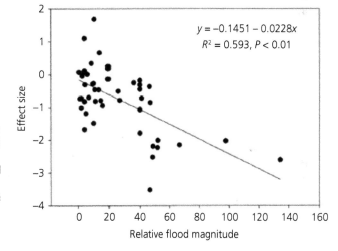

$$y = -0.1451 - 0.0228x$$
$$R^2 = 0.593, P < 0.01$$

Figure 5.17 Overall effect size of floods on total aquatic invertebrate density on hard substrata (gravel and cobbles) in riffles and runs of rivers and streams worldwide (41 studies on six continents) in relation to the size of the flood event.
Source: McMullen & Lytle 2012, with permission from John Wiley & Sons.
See text for explanations of axes but note that more negative values on the *y*-axis indicate stronger effects.

partial or complete. Both can have at least temporary effects on population density (and/or biomass). As we have seen (Chapter 2), flow regimes differ widely throughout the world in terms of the total range of discharge, the temporal dynamics and intermittency of flow, and the predictability of the occurrence of floods and droughts, factors that largely determine the nature of the biota (i.e. which species traits/life cycles are of advantage in different flow regimes) and thus the response of species to disturbance.

Tracking population density of lotic organisms through the course of a flood or spate is difficult for practical reasons, although it has often been attempted. In a meta-analysis of such studies, McMullen & Lytle (2012) found that the effect size of any event, perhaps not surprisingly, depends on its relative magnitude.

'Effect size' here refers to the natural logarithm (ln) of the ratio of total invertebrate density (all taxa) after the flood/before the flood (somewhat counter-intuitively the lower this value of effect size measured in this way the greater the effect), while relative magnitude is measured as the peak discharge/mean discharge or baseflow (Figure 5.17). This is an 'assemblage-scale' analysis (it covers many species), not a population study *per se*, but the pattern is clear that on average invertebrate numbers are reduced by at least a half immediately after flood events (within 10 days), with a greater effect in pools than in runs and rapids. Further, species-specific changes are well known (e.g. Giller et al. 1991; Woodward et al. 2015). We should not assume that density differences before and after floods are all due to mortality, however, since movements out of the study area (i.e. dispersal) are also possible and, as we shall see, numbers can subsequently recover quite quickly via a number of pathways.

Stream drying is also a powerful stressor and source of disturbance, and is now attracting much more research attention. This is in the context of climate change (with alterations in rainfall patterns), hugely increasing demand for water (much of it for irrigation agriculture) and an increased recognition of the global extent of channels with intermittent surface flow (see Datry et al. 2016b). This theme of drought and adaptations to drought (Chapters 2 and 4), its community and ecosystem effects (Chapters 6 and 8) and issues of water shortage and water security (Chapter 10) are represented throughout this book.

A common observation in studies of apparently quite substantial disturbance events in streams and rivers is that recovery to pre-disturbance values of population density (a measure of *ecological resilience*) is relatively rapid, often occurring well within the generation time of the species concerned (e.g. Death 2007). Exceptions to this can be found in events of an extreme magnitude, and/or where several events occur over a short period (e.g. Woodward et al. 2015). Where recovery is relatively rapid, however, this source of resilience cannot always be due to reproduction by survivors and a good deal of attention has been given to the shorter-term sources of this resilience, to which we now turn.

5.3.4 Refugia, environmental heterogeneity and species traits—keys to ecological resilience in lotic populations?

How is it that population density of stream organisms often 'bounces back' so quickly after disturbance events? Most attention has been focused on the idea that there are various kinds of 'refugia' that reduce the scale of density-independent losses to populations, and thus make it more likely that density-dependent processes will come into play at certain times (see Topic Box 4.1, p. 104). In this area we must take particular care not to lapse into 'teleological' language by arguing, for instance, that 'stream organisms exploit refugia (*in order to . . .*) to maintain population densities at high levels'. It is essentially the fitness of *individuals* and the frequency of their genes that is maximised by natural selection, so the use of refugia (whether passive or active) must be of adaptive value to the individual and its progeny.

Refugia rely on some aspect of the spatial and/or temporal heterogeneity of streams and rivers (i.e.

differences in the habitat from place to place or from time to time). An organism's 'eye view' of such heterogeneity depends on its biological traits (see Chapter 4) and, in particular, the generation time and the typical area occupied over a lifetime. Both these are associated with body size, larger organisms typically living longer and having wider ranges than small ones. The first, and undoubtedly most important, aspect of surviving physical or chemical 'events' is the long-term evolutionary match between the organism and its environmental conditions and fluctuations. As we have seen, many organisms have morphological and/or physiological features that have fitness advantages in the physically demanding world of rivers and streams. We have introduced such adaptations in Chapter 4. Speciation over evolutionary time has led to the accumulation of diversity (and of 'adaptive traits' or suites of traits) in the various taxonomic groups (and extinctions in others) so that 'pools' of species arise in different regions. These are available to colonise new environments as and when they arise during the history of the earth and as river drainage basins form, coalesce or divide. Such generally long-term changes in species are the result of many generations of repeated population processes that are the mechanism for natural selection and evolution.

Ecological explanations for persistence in the face of disturbance refer to environmental heterogeneity that is smaller in scale than the major historical events in the earth's history, such as glaciations, mountain-building etc., that are involved in evolution and biogeography. Various refugia that operate in 'ecological time' have been distinguished more formally (e.g. Lancaster & Belyea 1997; see Figure 5.18). First, envisage situations in which individuals do not survive the disturbance but are killed, in a flood or drought for instance. If the species concerned has a complex life cycle, and a fraction of the population is in some other habitat or resistant life stage when the disturbance in the stream occurs, that fraction of the population that survives can then recolonise the stream when conditions return to normal. For instance, most benthic insects of streams and rivers have adults that live for some time on land. This requires that the life cycle of all individuals is not synchronous (i.e. not all individuals are in exactly the same life stage at the same time). Lancaster & Belyea (1997) called this first type 'refugia through complex life cycles'. It falls in the right half of the graph space in Figure 5.18 and operates at a timescale exceeding the generation time of any individual. A second mechanism depends on large-scale spatial heterogeneity, such that at least some individuals survive the

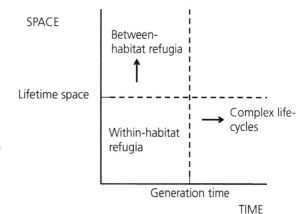

Figure 5.18 A schematic showing how persistence of populations despite disturbances may depend on various refugia that operate at different spatio-temporal scales, related to the normal range and generation time of the species concerned (see text for further explanation and examples, and see Lancaster & Belyea 1997).
Source: Winterbottom et al. 1997a, with permission from John Wiley & Sons.

event in at least one part of the overall range and are available gradually to repopulate the whole range over subsequent generations after the disturbance. Populations distributed throughout different tributaries of a river system, for instance, may persist, providing that not all the tributaries are affected by the event at the same time, and that recolonisation can occur via dispersal. This 'between-habitat' refugium hypothesis operates at spatial scales exceeding the normal life-time range of an individual and is placed in the upper half of Figure 5.18.

Finally, we can distinguish refugia that depend on at least a fraction of the population surviving disturbances via smaller-scale spatial variations in the habitat. This can involve, for instance, active movements into and out of more favourable habitat patches, or even passive movements that have the same population consequences—that is, a fraction of the population is saved and can re-disperse and or recruit at a later time. These 'within-habitat' mechanisms occupy the lower-left quadrant of the graph space in Figure 5.18.

A good deal of empirical work has tried to identify these within-habitat refugia in streams and rivers. For the smallest organisms, very small-scale heterogeneity may suffice. For example, tiny crevices and surface roughness can shelter attached algal (and cyanobacterial) cells on stones during scouring disturbance (as might be caused by high flows, e.g. Dudley & D'Antonio 1991). Bergey (2005) demonstrated this by growing algae on hard substrata ranging from smooth to rough in experimental channels and showed that scrubbing (effectively mimicking the abrasion caused by high flows) removed 95% of algal cells from the smoothest (glass) but only 20% on the roughest surfaces (pumice), thus providing a remnant assemblage of populations that could recover rapidly

under normal conditions. Surveys including smooth and rough substrata in natural streams were consistent with these experimental findings.

At a somewhat larger spatial scale, 'microform bed clusters' (see Chapter 2, section 2.4.1, p. 40) are patches of stable substratum in the bed of gravel rivers formed around an 'anchoring stone' and have been proposed as refugia for both algae and invertebrates. Thus, Francoeur et al. (1998) studied algae and cyanobacteria on substrata of three kinds (microform bed clusters, bedrock and more mobile cobbles/gravel) in a New Zealand stream in relation to high-flow events. Floods produced the least loss of biomass on the microform bed clusters, followed by bedrock and cobbles/gravel—inferring a refugium effect. Note that microform bed clusters provide more shelter than bedrock, which of course is itself very stable, since there are interstitial surfaces within the cluster that may harbour epilithic organisms. In terms of populations, floods reduced the relative density of the filamentous *Diatoma hiemale* while that of the small diatom *Gomphonema minutum f. syriacum* and the tightly attached blue-green bacterium *Amphithrix* increased. Filamentous diatoms are particularly susceptible to the high shearing forces that occur during floods and spates, and, indeed, filamentous forms are often associated with more stable substrata. Remaining cells on or within microform bed clusters presumably are available to establish more widely distributed algal populations between floods. In the case of microform bed clusters, the refugium effect would thus seem to operate in the 'between-habitat' mode of Lancaster & Belyea (1997).

For benthic invertebrates, several categories of potential refugia from both high and low flows have been investigated. These include, hydraulic 'dead zones' (patches of stream bed maintaining low shear forces even at high discharge; see Chapter 2), stable substratum particles (including microform bed clusters), stream margins and the riparian zone, the sediment below the stream bed (the hyporheic zone) and, in the case of drought, remnant damp patches in or on the stream bed (the special case of invertebrates with life-history stages resistant to drying is dealt with in Chapter 4) (see Topic Box 4.1, p. 104 by Belinda Robson for more on drought refugia).

Hydraulic dead zones were identified by Lancaster & Hildrew (1993a, b) as possible refugia during high flows. They identified patches (at the scale of about 0.05 m²) on the bed of a small southern-English stream that retained low shear stress even at high discharge—these were flow 'dead zones'. Shear stress immediately rose with increasing discharge in other patches, while yet others were intermediate ('slow', 'fast' and 'variable' in Figure 5.19a; Lancaster & Hildrew 1993b). Some, though not all, species responded by moving into the dead zones at high discharge in the spring. An abundant stonefly *Nemurella pictetii* did so, along with the larger-bodied (older) cohort of another (semivoltine) species, *Leuctra nigra* (Figure 5.19b). In both cases this presumably sheltered a fraction of the population from the disturbance. Winterbottom et al. (1997a) then created experimental dead zones in the same southern-English stream by putting out fine and coarse mesh

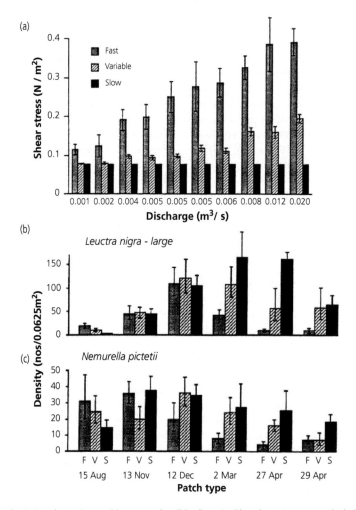

Figure 5.19 Refugial patches in Broadstone Stream: (a) some patches ('slow') retained low shear stress even at high discharge (s = 'slow', v = 'variable', f = 'fast') and were the candidate refugia. Some organisms moved into these refugia at high spring discharge (increases on six sampling occasions, from left to right). (b) The large-bodied cohort of the hemivoltine stonefly *Leuctra nigra* and (c) the stonefly *Nemurella pictetii*.
Source: from Lancaster & Hildrew 1993b, with permission from the University of Chicago Press.

cages over a range of discharges (the fine-meshed cages reducing flow forces inside the cage relative to the coarse-meshed cages, thus constituting 'dead zones'). Both species of stoneflies colonised the refugium cages preferentially when there was a high-flow event during a week-long trial (Figure 5.20). Along with other evidence of rapid movement into refugia during high flows, this field experiment supports the role of dead zones in the population resistance of at least some benthic invertebrates (e.g. Borchardt 1993).

Bed movement—when at least some substratum particles (stones and/or organic matter) become unstable and roll over or are transported downstream—can be a severe ecological disturbance, particularly during really high-flow events. Its prevalence depends upon the interaction between the intrinsic stability of the bed and the intensity of precipitation and catchment

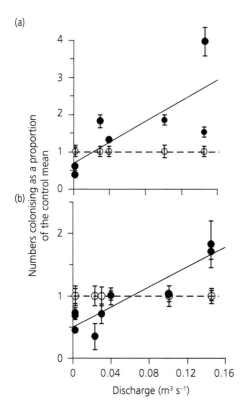

(a)

(b)

Figure 5.20 Experiments on flow refugia in Broadstone Stream: Winterbottom et al. (1997a) exposed cages of substratum over weeks with differing prevailing discharge, some of coarse mesh which did not reduce flow substantially ('control cages', open symbols) and some of fine mesh which did ('refugium cages', filled symbols). Two species of stoneflies—(a) *Leuctra nigra*, (b) *Nemurella pictetii*—colonised the refugium cages differentially at high discharge.

storage (i.e. how much precipitation infiltrates into the soil and how much runs off rapidly; see Chapter 2). Particularly stable particles, either singly or in groups (such as microform bed clusters), can provide refugia for invertebrates (as they do for algae; Francoeur et al. 1998). Thus, Matthaei et al. (2000) sampled well-embedded (stable) and loose stones before and after a spate in a New Zealand stream. The densities of several taxa were similar on both stone types six days before the spate but had declined on unstable stones when assessed approximately three days afterwards; these differences had subsided 19 days after the spate. This again implies the operation of refugia that facilitate the survival, via short-term redistribution, of a substantial fraction of the benthos during spates.

On a larger scale, lateral extensions of channels onto floodplains can offer refugia during flood disturbances. These are known to be very important for fish (Schlosser 1991) and can be supplemented by longitudinal refugia in large river systems where delta habitats also offer refuge for fish during flood disturbances (Carlson et al. 2016). However, there is relatively little information on the extent of their use by invertebrates. There is, of course, always the risk of stranding when the flood subsides and, in any case, in headwaters constrained by the geology, extensive lateral habitats are often lacking and valleys are relatively narrow and steep. In large gravel-bed rivers in northern latitudes, however, a broad diversity of benthic invertebrates have been documented as moving from deep water to the shallows of the inundated shore during annual flooding, which can run for up to 4 months (Rempel et al. 1999).

The idea that the sediments beneath the stream bed—the hyporheic zone—may act as a refugium for stream organisms at both high and low flows is evidently more than 60 years old (see reviews by Dole-Olivier 2011 and Stubbington 2012). Current speed declines markedly within the substratum, there are lower-amplitude water temperature cycles (daily and seasonally) and there is greater substratum stability (Krause et al. 2011); thus, organisms living beneath rocks or in the hyporheic zone will be less exposed than those on the surface.

Depending on the coarseness of the substratum and other factors, the subsurface sediments of rivers frequently harbour large numbers of what are normally regarded as animals of the surface benthos (so-called epigean animals), as well as species normally more or less confined to subsurface habitats (hypogean animals: see Chapter 1). However, this hyporheic refugium hypothesis proposes that some benthic organisms occur within the bed sediments only temporarily, during disturbances at the surface. Of course, the substratum surface offers resources for epigean lotic invertebrates (e.g. periphyton for the grazers, coarse detritus for shredders and access to suspended particles for filter-feeders), and hence the hyporheic zone cannot be a permanent home for such species. There is a large body of evidence showing diel patterns of activity on the substratum surface due to vertical migrations in lotic invertebrates (see below and section 4.3; Text Box 4.1).

Stubbington (2012) concluded that the hyporheic zone is one of a number of potential refugia from flow disturbance. In her review, however, Dole-Olivier (2011) wrote that 'support for the hyporheic refuge hypothesis is equivocal, often inconsistent and spatially variable'. The occurrence of true hyporheic refugia probably depends on flow paths through the bed sediments, the greatest refugium capacity being found in areas of substantial upwelling (during droughts) or downwelling (in spates). For instance, in a field study on a section of the River Rhône, Dole-Olivier et al. (1997) had shown that benthic forms, including species of *Gammarus*, moved into a downwelling zone of a large gravel bar. This suggests that the refugium affect was 'patchy'—variable in space—and thus that the physical heterogeneity of rivers was crucial in population persistence in such a disturbed habitat. In a well-known, and somewhat contrary, example, Palmer et al. (1992) tested the hyporheic refuge hypothesis for the meiofauna in Goose Creek, Virginia. They did not report strictly population-level data, but dealt with more coarsely identified taxa, including rotifers, oligochaetes, copepods and chironomids, but found little evidence that downwards movements during either natural or experimental (in a flume) floods could prevent substantial losses of the meiofauna in this sandy stream.

Stream drying is a characteristic of many streams globally and represents a form of disturbance that is likely to increase with climate change and water withdrawals for agriculture and domestic water supplies (Ledger & Milner 2015). With respect to drying and the epilithon, Robson et al. (2008) surveyed three types of candidate refugia for biofilm-dwelling algae and cyanobacteria in seasonally flowing streams in southeastern Australia: remnant perennial pools, dry leaf packs and dry biofilm. All species were found in at least one of the refugium types, the species complement in perennial pools being closest to that in the streams after flow resumed, although most species were also found in the other refugium types. No species persisted in the assemblage that was not found in at least one refugium type. Clearly, resilience to drought of these photosynthetic microorganisms is facilitated by their tolerance of drying—an evolved species trait.

Since it is so difficult to observe directly the behaviour of benthic, and particularly hyporheic, animals during droughts, there have been some attempts at manipulative experiments. In the laboratory, Vadher et al. (2017) used transparent (acrylic) sediment columns and six sediment treatments (mixtures of different sized particles) with water standing initially 5 cm above the surface. The level was subsequently reduced to 20 cm below the sediment surface to mimic drying. The responses of the five experimental species differed, while the distance that animals moved downwards in the sediment column often increased with pore volume (and sediment 'calibre', i.e. particle size). A stonefly (*Nemoura cambrica*) often found in temporary streams moved furthest into the substratum and this movement was little affected by the sediment type. Two other and larger surface-dwelling insects (a filter-feeding caddis larva, *Hydropsyche siltalai*, and the grazing mayfly, *Heptagenia sulphurea*) were less able to migrate, were more affected by the sediment and more frequently became stranded at the dry surface. Two benthic crustaceans, *Gammarus pulex* and *Asellus aquaticus*, also displayed some ability to move into coarser sediment at least, suggesting they can use hyporheic refugia from drought. These and other observations suggest fairly widespread, though by no means universal, potential survival of stream animals in hyporheic sediments.

Further, in perhaps more realistic experiments, artificial stream-side channels were connected to an English river and benthic animals allowed to colonise. Lancaster & Ledger (2015) then imposed a series of 10 drying events, each of 5–6 days, over the course of a year. They focused on 12 abundant 'core taxa'. Some species were resistant to drying, thus evidently having appropriate resistance traits. These included, for instance, two species of aquatic snails (the prosobranch snail *Potamopyrgus antipodarum* and the pulmonate *Radix balthica*), whose densities did not differ before and after drying. The prosobranchs have an operculum, which can be closed off to reduce water loss, while the pulmonates can breathe air (see Chapter 4, section 4.1.1). Larvae of three arthropod species, plus the midge subfamily Tanypodinae, showed density-independent losses varying between 57 and 100% of pre-event numbers (Figure 5.21). Intriguingly, losses of two further subfamilies of midges, the Tanytarsini

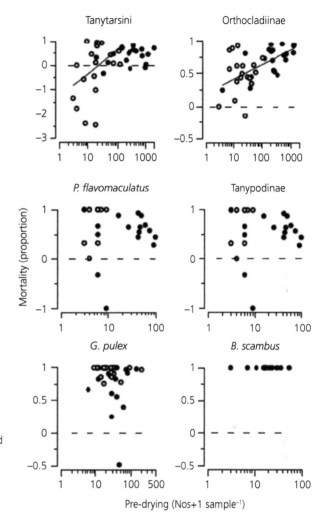

Figure 5.21 Proportionate mortality in six taxa of stream invertebrates in relation to pre-drying density when exposed to drying events in experimental stream channels: three species and one coarser taxon showed density-independent losses, the web-spinning caddis *Polycentropus flavomaculatus*; a group of midge larvae (the largely predatory Tanypodinae); the amphipod 'shrimp' *Gammarus pulex*; and the mayfly *Baetis scambus*), while in two further taxa of midge larvae (Tanytarsini; Orthocladiinae) losses were density dependent.
Source: Lancaster & Ledger 2015, with permission from John Wiley & Sons.

and Orthocladiinae, appeared to be density-dependent (percentage mortality went up with increasing density before the event), which at first sight seems to infringe the hypothesis that disturbance events impose density-independent (not density-dependent) mortality (Figure 5.21). It is likely that this is due to the survival of a fraction of the population in remnant wet patches in the bed that provide refugium from drying. If the numbers of organisms that can 'fit' into such refugia is limited, then the resultant losses would of course be density-dependent. Recall, however, that in this case it would be competition among individuals for limited space that is the density-dependent factor, not the disturbance itself—which occurs without reference to population density.

Books about the ecology of standing waters are often dominated by studies of the plankton—small plants and animals that live in the water column and have fairly limited mobility. Intuitively, the relentless one-way flow of rivers and streams should preclude the development of self-sustaining populations of truly planktonic organisms in rivers, particularly during floods and spates when the transport time of entrained particles through the system is much less than that required to complete even brief planktonic life cycles. This must be an example in which physical conditions—including the periodic disturbance of high-flow events—ensure that there is no niche for truly planktonic organisms. Nevertheless, planktonic algae (and indeed some small zooplankters) are often abundant in larger rivers with long, meandering lowland reaches. It has often been assumed that their occurrence is explained by recruitment from floodplain ponds and lakes which may be hydrologically connected to the main channel, and by occasional entrainment from the river bed. Such mechanisms can be important, although it is now clear that a true *potamoplankton* (river plankton) can be maintained by population processes occurring purely within the main channel itself, and that this is explained by the occurrence of significant spatial heterogeneity in the flow. Thus, refugia for planktonic organisms are found in retention zones where the water flows sufficiently slowly (or in eddies where flow is actually 'backwards') for population growth to occur, providing 'inocula' of cells by fluid exchange with the more freely flowing parts of the channel. The species concerned have traits associated with rapid population growth, such as small cell size and frequent cell division, and their persistence requires that the volume of retention zones relative to the discharge of the river is sufficiently large, and that nutrient and light supply are sufficient to sustain rapid production. This last criterion is usually satisfied in temperate rivers of the agricultural lowlands, but may not be in some more pristine tropical areas, particularly in parts of South America, where a river plankton may be very limited (Lewis et al. 1995).

The retention properties of river channels has been studied by releasing a dissolved tracer into the flow and measuring its concentration as it passes two downstream sensors at a known distance apart. The rise and fall in concentration changes shape between the two points, the peak being highest and sharpest at the upstream station, and becomes lower and more 'spread out' downstream, this being more marked in spatially heterogeneous channels (see an example in Figure 5.22a). This is because the marker, and in nature solutes or small particles, are by chance entrained in water flowing at different rates, some flowing rapidly and being washed early from the reach, some going more slowly and being caught in eddies near obstructions or close to the banks. A model can be applied to the data that conceives of the total volume of a reach simply as an actively flowing fraction and a non-flowing volume of 'dead water' (this being referred to as an *aggregated dead zone*: ADZ, also known as the 'dispersive fraction'), the two parts exchanging water via normal diffusion rather than by active flow or advection. This model described the passage of the marker through a reach of the British River Severn very well (see 'expected output, downstream site' in Figure 5.22a). Reynolds (2000) proposed that it is in such retentive dead zones that, somewhat paradoxically given the name, true planktonic life can persist. Thus, the concentration of chlorophyll and some photosynthetic planktonic organisms across a transect of a lowland reach of the River Severn (UK) reached a peak in a sheltered bay at the margin of the main channel, where even the blue-green bacterium *Oscillatoria* was found (blue-greens normally require a relatively stable water column; Figure 5.22b).

5.3.5 Drift and population persistence

Downstream transport in the current entails great risks for almost all river organisms, and can be part of density-independent mortality driven by disturbance. The risks come from predation, via filter-feeders of many kinds and drift-feeding fish, from physical damage, and from transport to downstream areas where the environment may be less favourable (for instance warmer water or where the substratum is not suitable). In our discussion of the effects of high-flow disturbance, we presented refugia mainly as parts of the habitat in which benthic organisms can avoid being entrained in the flow. Yet large numbers of organisms are indeed swept downstream, particularly during peak flows and when parts of the stream bed are mobilised. This includes benthic algae scoured from the stream bed, very young fish, planktonic organisms of various kinds and invertebrates that are normally benthic. The downstream transport of benthic invertebrates actually occurs all the time, and particularly at night (see the next section), not just during flow disturbances, and is referred to as *drift*. This phenomenon has been widely studied over many years, yet its significance is still not entirely clear.

In earlier studies of the stream drift of benthic invertebrates, it was clear (perhaps not surprisingly) that the distance of transport of entrained particles (including organisms) in any single event was positively related to water velocity (e.g. Allan & Feifarek 1989). Thus, animals accidentally dislodged from the stream bed might be swept a relatively long way at the high flows prevailing during a spate or flood, increasing the risk of mortality. Further, early studies of drift distance found that organisms released at any one point returned to the stream bed at a rate that could be described by a negative exponential model (e.g. McLay 1970; Elliott 1971). There are theoretical reasons to question the general applicability of the exponential model to downstream transport, however, because there is not only a simple single downstream transport vector of particles (including organisms) in natural, turbulent flows, but also complex vertical and lateral vectors. Further, water velocity in any channel declines as we approach bounding surfaces (the bed and bank), so there is a distribution of velocities through a cross-section of the

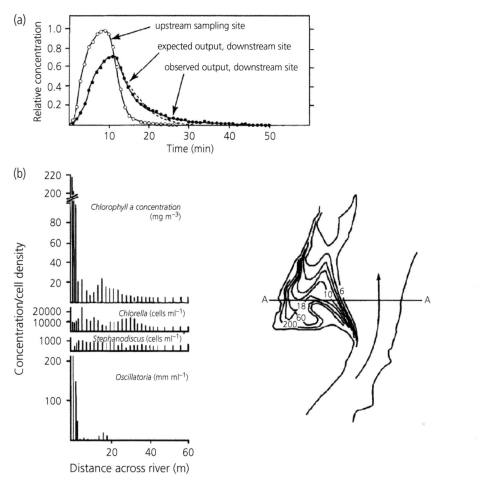

Figure 5.22 (a) The transport of a dissolved marker, injected directly into the River Severn (UK), past two sampling sites, and the output of an aggregated dead-zone model assuming there are two fractions of water, one flowing actively and one 'dead'. (b) Left panel, spot surface concentrations of chlorophyll and some planktonic 'algae' across (right panel) a transect (A–A) on the River Severn.
Source: after Reynolds 2000, with permission from John Wiley & Sons.

channel (see Chapter 2, Figure 2.14). The particular range of velocity experienced by a drifting organism depends, for instance, on the height above the bed at which the organism is released and other stochastic factors (e.g. McNair & Newbold 2012).

In the context of downstream losses of benthic organisms accidentally entrained in the flow during flow disturbances, Lancaster et al. (1996) studied the instantaneous return rate (the exponential model again being an acceptable fit in practice) to the stream bed of invertebrates that were experimentally dislodged in four different natural channels at a variety of discharges. Two were swiftly flowing, upland streams in south-west Scotland and two were more sluggish lowland streams in southern England, and in one of

the latter almost 50% of the reach volume consisted of 'dead water' (this dispersive fraction was around 30% in the other three) (Figure 5.23a). The highly retentive (lowland) Broadstone Stream had very high standing crops of woody debris and a high density of debris dams, creating many 'step pools' with intervening short riffles. In a series of separate experiments, the number of animals remaining in the drift declined approximately exponentially (measured as an instantaneous rate of return to the stream bed, β) with distance downstream from the point of disturbance (Figure 5.23b shows an example of a stonefly in one of the upland streams). The return rate to the stream bed of benthic animals varied: (i) among the four channels and (ii) with reach mean velocity (Figure 5.23c).

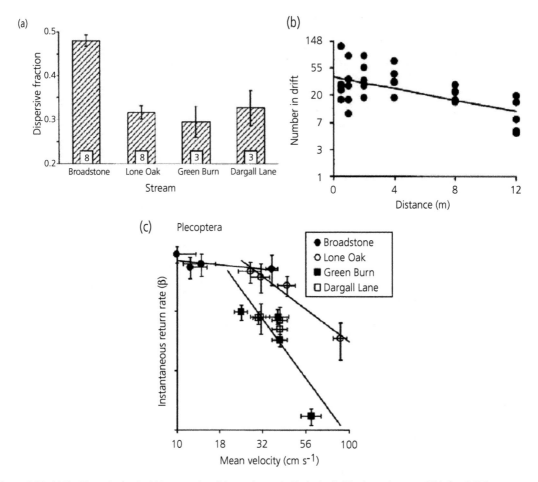

Figure 5.23 (a) The 'dispersive fraction' (the proportion of the reach occupied by hydraulic 'dead zones', mean ±SE) in four British streams (numbers at the base of bars are the number of field measurements in each stream). (b) Drift density (logarithmic scale) of the stonefly *Amphinemura sulcicollis* downstream from an experimental disturbance in Dargall Lane (a stream in south-west Scotland). (c) The instantaneous rate of return (β) to the bed of stoneflies entrained in the flow in four British streams in relation to mean stream velocity at the time of the experiment (a common line is fitted to the two Scottish streams, Green Burn and Dargall Lane). Both axes are logarithmic.
Source: Lancaster et al. 1996, with permission from the *Canadian Journal of Fisheries & Aquatic Sciences*.

The return rate of Plecoptera was most rapid in the most retentive channel (Broadstone Stream), in which nearly 40% of animals had regained the bed within 1 m, this rate being almost invariant with mean reach velocity prevailing during the trial (suggesting active and directed movement of the animals). Return rate was lowest overall in the two upland channels, and declined steeply with increasing velocity (i.e. animals were swept a long way when dislodged during spates). Return rate was intermediate in the lowland channel with a lower dispersive fraction, and also declined with mean reach velocity. These results are partly consistent with the view that the dispersive fraction is a measure of how effectively benthic animals can be retained in natural, complex stream channels, although other factors must come into play that could be related to hydraulic features, such as channel depth (animals returning to the bed relatively quickly in shallow reaches), or perhaps due to the traits (in morphology and or behaviour) of the exact species concerned. These examples show how stream drift has been studied in relation to disturbance, and this leads us next to consider its significance in terms of population persistence in lotic organisms, and then much more generally as a pervasive form of mobility and dispersal among patchy river habitats.

An initial and apparently obvious question about drift is that if there is an overall tendency for individuals to be swept downstream, how can stream populations persist and 'maintain station' in the river network? It is sometimes postulated that there must be more or less exactly compensatory upstream movements *'in order for populations to maintain themselves'*. This notion was the basis of the 'colonisation cycle' of stream insects, which postulated that immature stages drifted downstream while the adults flew upstream again (Müller 1954: Figure 5.24a). The colonisation cycle attracted a good deal of interest. Note that casting the hypothesis in terms of upstream flight 'in order to compensate for downstream drift' is apparently based on 'group selection' (i.e. populations in which it occurs persist whereas those in which it does not are lost). A better question, based on 'natural selection', is to ask whether there is a fitness benefit to individuals or genotypes that fly upstream. A number of conceptual, theoretical and empirical examinations of the hypothesis have been undertaken and attempts made to model various scenarios.

First, and on observational evidence, Waters (1966) and others argued that there might be no 'need' for upstream flight, because drift consists only of surplus individuals that are crowded out by superior competitors for space or food (the latter remaining upstream). In other words, drift was seen as strongly density dependent (Figure 5.24b). Anholt (1995) also argued for density dependence but found in his model that at least some upstream migration was also necessary, although an exact balance to drift was not required, as it was in Müller's (1954) original concept. Dispersal of the adults could essentially be random but at least some individuals must move upstream, essentially to rescue upstream populations lost stochastically (Figure 5.24c). Speirs & Gurney (2001) used a different model structure and showed that density-dependent drift was not necessary in advective environments in general, as long as there was subsequent random diffusive dispersal and that some individuals could return upstream (Figure 5.24d). The persistence of headwater populations in their model required that the distance drifted was limited, as perhaps satisfied by the operation of refugia, or if only brief periods of drift occurred and mean downstream water velocity was not too high. Humphries & Ruxton (2002) sought to include arguments for species that had no flying stage (non-insects) yet also maintain upstream distributions. Results from their model suggested that a combination

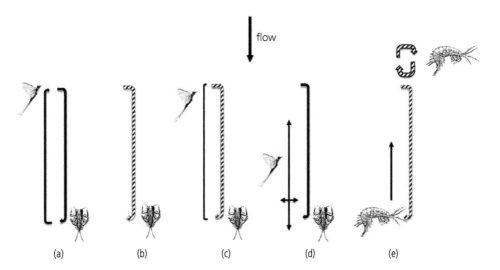

Figure 5.24 Conditions for population persistence of stream organisms under different conceptual or modelling scenarios: (a) the original colonisation cycle hypothesis of Müller (1954)—drift is balanced by upstream flight; (b) Waters (1966) thought downstream drift was strongly density dependent (denoted as a cross-hatched arrow) and consisted of 'surplus' individuals—therefore upstream compensatory flight was 'unnecessary'; (c) Anholt (1995) considered that density-dependent drift was necessary but also required some return upstream, though not an exact demographic balance; (d) in the Speirs & Gurney (2001) model density-dependent drift is not required, but there needs to be subsequent random dispersal so that at least a few (but not all) adults move upstream; (e) Humphries & Ruxton (2002) widened the argument to include non-flying species and concluded that even modest upstream movements (e.g. crawling) by a few individuals would be sufficient, along with density dependence at some stage in the life cycle (possibly but not necessarily density-dependent drift).

of density dependence at some stage in the life cycle (not necessarily in drift itself) and even very modest (cm-scale) upstream movement—achievable by walking, crawling or swimming—by at least some individuals would be sufficient to prevent populations being lost from upstream areas (Figure 5.24e). Most recently, Lutscher et al. (2010) suggested that the only strict requirement for persistence in an advective environment is that the population can invade upstream—that is, that some individuals move upstream at some stage in the life cycle—a condition identified persistently in these theoretical studies. Fecundity is also often high in many stream insects.

Empirically, many studies have asked whether there is indeed an upstream bias in adult flight, but the results are equivocal. In some circumstances, and for some species, predominantly upstream flight has been demonstrated but quite commonly flight seems to be unbiased in direction or in a few cases even predominantly downstream (e.g. Müller 1982; Petersen et al. 1999). In most of the examples supporting upstream movement assembled by Müller (1982), he noted that the traps were set at the downstream end of riffles, which are the oviposition sites of most stream-dwelling insects. This would seem to bias the results towards upstream movement, since the females have to seek suitable places to lay their eggs (normally in riffles). In some cases, it is obviously beneficial to fly upstream to oviposit for reasons of fitness rather than population persistence, for instance at the confluence of rivers with tidal reaches (Speirs & Gurney 2001) or, for the *Baetis* females studied by Peckarsky et al. (2000), because large emergent rocks first became available for oviposition in upstream reaches. Thus, while the evidence does not point to universally upstream and compensating flight, as required by the colonisation cycle hypothesis, at least some individuals do disperse upstream to some extent, apparently satisfying the main theoretical criterion for population persistence. Further, in a neat and unusual experimental study, Elliott (2003) showed that instream dispersal by larval stages was predominantly upstream in nine of ten species of benthic invertebrates (nine of them insects), although to variable extents. The animals were free to drift, crawl or swim in these experiments, suggesting that active upstream dispersal within the benthos can certainly provide some potential to reinvade upstream, as required for persistence in models. Many stream organisms are 'positively rheotactic—that is, they tend to move into the prevailing current (see Chapter 4).

Persistence is somewhat easier to understand where transport downstream in the flow ('advective transport') is limited and where drift is density dependent—though neither seems to be a formal theoretical requirement. A great deal of attention has been paid to the distance which stream invertebrates actually drift in nature, both in single events and, much more rarely, over their lifetime. In their review, Naman et al. (2016) concluded that the distance drifted per event generally was quite short (about 2–10 m on average—as in the above example by Lancaster et al. 1996) and, except in cases of extreme bed movement, was more or less restricted to a subset of taxa particularly prone to drift. Instances of much longer drift events, of 100 m or more, have been measured however (Brittain & Eikeland 1988). In terms of lifetime drift, there are very few estimates but Humphries & Ruxton (2003) calculated *c*.1.5 km for *Gammarus pulex*, an estimate subject to a good many assumptions about the time spent in the drift, generation time and number of drift events per lifetime. It does not take into account possible active movements back upstream, though of course *Gammarus* has no flying adult. Using isotopic tracers, Hershey et al. (1993) estimated a net distance drifted downstream of 2.1 km for up to one half of the population of *Baetis* in an Arctic river (i.e. including upstream crawling/swimming as well as drift downstream). They also estimated that a similar proportion of the population flew upstream by 1.6–1.9 km. If these estimates are anything like realistic, displacement downstream in drift-prone taxa like these could be substantial, particularly during high flows, although not sufficient to compromise the persistence of their populations in upstream areas.

There seems to be little consistent evidence on whether dispersal of the benthos by drift or crawling is density dependent or not (Naman et al. 2016), different conclusions being drawn in the various studies completed. It clearly can occur in some taxa and under some circumstances, but by no means all. Thus, we cannot conclude that density dependence alone is the answer to the 'drift paradox', though it can contribute to population persistence in some circumstances. Overall, we can conclude that there is no real 'paradox' in the persistence of invertebrate populations in advective environments, and neither is there a colonisation cycle in the original sense of the term (drift exactly balanced by upstream flight). Limited upstream dispersal clearly occurs in stream invertebrates and this seems sufficient to explain population persistence.

Downstream 'drift' has also been implicated in the population dynamics of fish in rivers and streams. Recall that young brown trout emerging from the redds in Black Brows Beck either found territories and began to feed or, if they could not find territories, were carried in the current downstream where most eventually died due to starvation, predation or other causes (Elliott 1994). In this instance, density-dependent intraspecific competition for space determined the population density of 0+ fish and the carrying capacity of the stream. Of course, the salmonids are well adapted to life in swiftly flowing rivers and streams, and their use of gravel nests beneath the bed which harbour the larvae until they are well developed young fish, plus the provisioning of the eggs with yolk, could be regarded as adaptations to the lotic life. Many other fish (non-salmonids) in larger streams and rivers produce numerous small larvae which seem much more susceptible to passive drift, and potentially to density-independent mortality. In fact, rather little seems to be known of the passive or accidental downstream drift of larval fish as an erratic source of mortality (Lechner et al. 2016). On the contrary, although large numbers of larval fish, embryos and eggs of fish are taken in drift nets in rivers, most attention has been focused on drift as a form of dispersal, and as a normal part of the life cycle. In fact, drift in plant propagules, invertebrates and the young stages of fish is clearly an agent of dispersal in the population dynamics of very many species in running waters, and we now turn to drift and other forms of mobility in that light.

5.3.6 Migrations and mobility

In this chapter we have touched on the thorny issues of scale, in the context of the spatial extent and resolution of measurements (section 5.3.1), and of disturbance and refugia (section 5.3.5). We emphasised in Chapter 2 that river systems are 'patchy', physically complex and potentially divided habitats, and that this habitat heterogeneity must be viewed at a variety of spatial and temporal scales in relation to the longevity of different fractions of the biota and to the space over which individuals move during their life cycles. We begin this section on migrations and mobility with the habitat used within the usual lifetime of a single individual.

Where individuals in a population regularly make a journey to different parts of their overall habitat (or between areas of suitable habitat), often more or less synchronously, then such mobility may more properly be called *migration*. A large number of river species are migratory at one scale or another, and such migrations normally involve a return journey later in life. Some of these migrations are at large spatial scales and integrate whole-river systems. The spawning (return) migrations of adult salmon into northern rivers from

their marine feeding grounds are legendary in this regard. The migrations of larval Atlantic eels from the Sargasso Sea to the east coast rivers of North America and the west coast of Europe, as well as the return of adults from fresh water to the sea for breeding, at first seemed scarcely credible. These are known as *anadromous* and *catadromous* migrations, respectively; that is, upstream to breed in fresh water (salmon and others) or downstream to breed in the sea (eels). Other salmonids may migrate, either entirely within fresh water or involving visits to the estuary or further afield, and several species of eels migrate to the ocean to breed. Many other river species also migrate, but are less well known. These include some of the largest species of all freshwater fish (sturgeons, paddlefish etc) in the larger rivers of Central and Eastern Europe, Asia and the Americas, and some (such as characins, see next paragraph) that undertake spectacular longitudinal and lateral migrations to exploit nursery areas on the great tropical floodplains. We have also seen how such migrations often make them vulnerable to habitat deterioration and overexploitation (see section 5.3.1; Table 5.1), including pollution and river damming.

Seasonality in the tropics is characterised more by the occurrence of regular and predictable wet and dry seasons than by fluctuations in day length and temperature, and fish in tropical rivers often have two distinct centres of concentration in their wet- and dry-season habitats, sometimes with a long 'commute' between them. Among the most famous are the migrations of species of the family Prochilodontidae (order Characiformes, commonly known as 'characins', a large South American group of river fish). These have been described as typically moving upstream, sometimes hundreds of kilometres, at the end of the dry season to spawn at the head of the river during the onset of the rains. As the flood swells, the adults move back down to their wet-season habitat. Eggs and larvae are swept downstream in the drift (here a form of free downstream transport) and out onto the expanding floodplains, which are productive and relatively safe from larger predators. There the young fish grow, eventually moving or drifting back to the main river at the onset of the dry season. Upstream migrations by *Prochilodus platensis* of almost 9 km d^{-1} have been reported in the Paraná of Argentina and those of *Salminus maxillosus* (another characin) up to a spectacular 21 km d^{-1} (see Welcomme 1979 for details of these and many other migrations in floodplain rivers).

Less well known are very long-distance migrations by some river invertebrates. *Amphidromy* is a distinctive form of migration in some river species in which both freshwater and marine habitats are used within the lifespan of an individual. It is considered to be another form of diadromy, although it is distinct in various ways from both the better-known catadromy (as in eels) and anadromy (as in salmon). In amphidromy, the adult grows, mates and spawns in fresh water, the planktonic early larva then drifts downstream to brackish estuarine or marine coastal waters where development continues until it reaches some 'postlarval' stage, which migrates back upstream where growth and development are concluded.

Amphidromy occurs in at least nine different families of fish, particularly gobies (McDowall 2010), and in many species of crustaceans. In the crustacea, it is found mainly in caridean shrimps in the families Atyidae and Palaemonidae and is particularly common in streams on tropical and subtropical islands (see Topic Box 1.2). Most freshwater crustacea (such as the common gammarids) have 'abbreviated larval development'—that is, larvae are retained in the female and are released only at an advanced larval or even post-larval stage of development. Clutch size is then relatively small and the few eggs are well provisioned with yolk. In amphidromous species on the other hand, many first-stage larvae are released by non-migratory females in upstream areas and drift towards the sea. After development in the sea, post-larvae or juveniles migrate back upstream to the adult habitat. An obvious possible feature of this kind of life history is that it could facilitate dispersal over wide geographical distances, perhaps explaining the overrepresentation of amphidromous lifestyles in isolated fresh waters on often remote islands. The New Zealand freshwater shrimp *Paratya corvisrostris* is a good example.

The number of larval amphidromous shrimps drifting downstream in streams on Puerto Rico increased with total length of stream above the sampling site. There was also a nocturnal peak in numbers drifting through reaches containing potentially predatory fish, but the number drifting was aperiodic in fishless streams, consistent with the view that nocturnal drift confers some protection from predation (March et al. 1998—see Chapter 6). In some very long rivers draining larger continents it can take a long time for poorly provisioned larvae to drift to the sea (they need saline water to begin development). For instance, newly hatched ('stage-1') larvae of *Macrobrachium ohione* (Palaemonidae) living in a branch of the Mississippi River in Ohio would not moult to the first feeding 'stage-2' unless exposed to saline water (Rome et al. 2009). Larvae that did not moult died within 11 days. In such large rivers it is likely that the gravid females

migrate at least some of the way downstream before releasing their young, thus reducing the time needed to drift without feeding. It is nevertheless incredible how the vulnerable juveniles then manage to migrate the sometimes hundreds of km back upstream to the adult habitat in the main river.

We have so far concentrated on regular bouts of mobility, some of it involving drift in river currents, often over quite large distances and 'taking advantage of' habitat heterogeneity—using part of the habitat as feeding areas, part as a 'nursery' and/or sometimes as a physical or biotic refuge—yet still taking place within the normal lifetime of an individual. We call this migration, which is common in terrestrial, freshwater and marine habitats. However, many benthic invertebrates are also in a state of more or less 'continuous redistribution' on the beds of rivers and streams. This involves much more frequent, smaller-scale movements, including bouts of drift, resulting in short-term shifts in the patchy local density of populations of invertebrates (see Section 5.3.4). In brown trout and other fish more directed longitudinal movements of up to a couple of kilometres (Bridcut & Giller 1993), or on a smaller scale between pools and riffles, occur often on a daily basis and are related to feeding and predator avoidance (e.g. Greenberg & Giller 2001). These movements are not all simply accidents resulting from disturbance, nor are they significant merely in the context of density-independent population losses.

As we suggested before, drift is not simply a passive phenomenon, as inferred by a number of lines of evidence (reviewed in Brittain & Eikeland 1988). First, the likelihood of drifting ('propensity to drift') varies greatly among taxa (and sometimes among different size classes of the same species), which of course could be explained by a variety of traits, and is often a consequence of some behavioural control. There are also well-known daily rhythms in drift, most invertebrates drifting at night (as in the shrimp larvae above) when visually feeding fish are least effective. These rhythms are not related to flow but could be due to more active foraging at the surface by benthic animals during darkness, when relatively safe from fish, despite many predatory invertebrates also being most active then. High nocturnal drift rates by prey can therefore either be a result of accidental dislodgment (while moving about on the substratum surface at night), and thus also to some degree 'passive', or be due to active escape from invertebrate predators, or be part of locating high-quality resource patches. Experimental comparisons of drift transport have also been made between live and dead individuals of the same taxon, the two thus having the same morphology and density but differing in behaviour. In many of the more common taxa found in the drift, particularly insects and amphidods, live animals apparently are able to modify the time spent in the drift (and therefore the distance drifted) compared with the dead ones—suggesting active behavioural control (e.g. Oldmeadow et al. 2010).

Exactly how mobile are benthic organisms? This evidently differs enormously among taxa, and some are very mobile indeed, at least at the small scale. Elliott (2003) described dispersal in ten species of stream invertebrates, finding that upstream dispersal by crawling was common, and the predominant type of movement for nine of the ten species. The most mobile were three carnivorous insects (two stoneflies and one free-living caddis), that travelled up to 9.5–13.5 m d^{-1}, while the cased caddis *Potamophylax* was least dispersive—few larvae moving much at all and the maximum distance travelled being 3.5 m d^{-1}. The most detailed study so far remains that of Freilich (1991) who marked 1,000 individual larvae of the large, omnivorous American stonefly *Pteronarcys californica* in a stream in Wyoming, recapturing individuals over a period of 3 months. Most individuals moved short distances (remaining within a few metres up- or downstream; mean 1.8 m downstream), but with a few moving rapidly up- or downstream (eight individuals moving a remarkable 6–22 m d^{-1}).

In early field experiments, Townsend & Hildrew (1976) measured the colonisation of trays filled with gravel substratum in the sluggish, fishless Broadstone Stream. Some trays were raised off the bed (reachable only by drift) and others were resting on it. Considering all taxa, they estimated that about 3.6% of individuals moved their position per day. Two species were particularly highly mobile, the predatory net-spinning larva of the caddis *Plectrocnemia conspersa* and the largely detritivorous stonefly *Nemurella pictetii*, 20% and 43% if which moved their positions per day, respectively. Overall, 80% of colonisation of trays was attributable to drift, even though mean water velocity was very low (< ~ 0.05 ms^{-1}). Interestingly, most other estimates of the proportionate contribution to colonisation by drift are rather lower than that (~ 50% or somewhat less). It was concluded that species with patchily distributed resources on the stream bed (aggregations of prey and leaf packs for these two species, respectively) were particularly mobile and in a 'continuous state of redistribution'. In some longer-term, follow-up experiments in Broadstone Stream, Winterbottom et al. (1997b) measured colonisation of boxes of cleaned, natural substratum in 26, week-long exposures—under

varying conditions. Two species of stoneflies, *Nemurella pictetiii* and *Leuctra nigra*, showed very different patterns of mobility, the first driven by very active movements, more or less independently of flow, the second much less mobile and largely related to flow.

Clearly, benthic invertebrates in rivers and streams range between highly mobile species, that crawl actively and/or drift frequently during their brief life cycles, and fairly immobile species that rarely drift and then only when moved by high flows or an unstable substratum. A few members of this immobile group, such as some freshwater pearl mussels (Margaritiferidae), are very long-lived (up to 100 years or more), but even they have ectoparasitic larvae that attach for a time to highly mobile fish. High mobility at small scales in highly disturbed systems encouraged views that the population dynamics of benthic species are overwhelmingly controlled by disturbance, recolonisation from refugia, and by 'weedy' species traits (small body size, short life cycles, highly fecundity). Put simply, at small spatial scales (typically that of the sampling device that is used, often a fraction of a square metre), instantaneous population density may be determined by the number of individuals arriving and departing—that is, by 'mobility'. These numbers will also be affected by the disturbance history of the particular small patch of habitat. This reasoning thus emphasises density-independent, non-equilibrial processes and disturbance as the most important determinants of the small-scale population dynamics of stream invertebrates. However, it is widely accepted that, as the spatial and temporal scales of a study are increased (for instance to encompass larger areas of stream bed and/or more than one generation), the processes of mortality and recruitment become more dominant and it is more likely that density dependence will be apparent (e.g. Anderson et al. 2005). We return to the concept of 'mobility control' in the context of communities and assemblages in Chapter 6.

5.4 Abundance: 'open' populations

5.4.1 Open populations and the river hierarchy

So far, we have restricted our discussion of mobility and dispersal to movements that occur within the normal lifespan and lifetime range of an individual (including regular migrations and the more frequent within-habitat movements). We now move to mobility that may take organisms or their propagules outside their 'normal' (individual) range and persist beyond their lifetime (i.e. from which those individuals may never return). What is the significance of movements on that scale for populations and their dynamics? In terms of regulation and persistence, we have so far considered mainly populations whose numbers are determined by the balance of births and deaths and are in that sense 'closed' (see section 5.1 above). We now turn to the contrary case where populations are subdivided spatially, and where the dynamics of the (sub-)populations are to some extent independent of each other, but where dispersal (arrivals and departures) between them can occur and can affect population dynamics.

Here we must consider further the heterogeneity of the physical world—and, increasingly, the progressive fragmentation of habitats via human activity (including that of streams and rivers). This physical structure can in turn impose spatial structures on biological populations. A number of different scenarios in the field of 'spatial ecology' have been considered (Rockwood 2015). Perhaps the best known is the classical 'island–mainland' arrangement of populations, in which the species on a small island have come from a larger mainland source; species richness on the island is then determined by the balance between the arrival of new species (via colonisation) and local extinction of the often small populations on the island (related to island size). 'Islands' here can refer to conventional patches of land in a body of water, patches of water (or rivers) in the landscape or habitat patches set within a matrix of unsuitable territory. A more classical 'metapopulation' consists of a series of habitable patches which are equally likely to contribute colonists to each other and in which the rate of extinction is finite and equal. A species can persist in such a metapopulation as long as uninhabited patches can be 'rescued' by recolonisation events and the overall rate of recolonisation events exceeds the number of extinction events among the patches. If the rate of colonisation is high enough, then effectively the patches act as part of a single population (a metapopulation), regulated by births and deaths across the entire set of habitat patches. Finally, patches can be differentially productive, some consistently producing potential colonists and with a very low rate of extinction while others have a higher rate of extinction and rarely produce new colonists for elsewhere. In this case, the patches are referred to as sources and sinks, respectively. Do any of these scenarios fit the physical arrangement of habitats in river systems, or does river architecture impose different kinds of population structures on inhabitants?

Rivers are examples of a rather unusual kind of habitat arrangement in the landscape called a hierarchical dendritic network (see Chapter 2); 'hierarchical', because there is a hierarchy of branches in which the smallest tributaries coalesce to form larger streams and eventually major rivers; and 'dendritic', because the whole network resembles the root network of a tree below the ground. Other examples of such habitat networks in nature could include certain kinds of mountain ranges—where there are long ridges dividing and/or coalescing along the ridge—and cave systems—which are themselves mainly produced by subterranean water flow (Altermatt 2013). Human activity has added further dendritic networks to the landscape, as in the branching of roads (and roadside verges) and hedgerows. In addition, rivers are 'advective' (flowing) environments, on the face of it encouraging unidirectional dispersal, with a variety of patterns of longitudinal distribution and relatedness in biological populations (e.g. Campbell et al. 2015). Acknowledgement of this network structure and its potential to affect populations and assemblages, as well as ecosystem processes, has now largely superseded concepts of river systems as linear systems, such as the River Continuum Concept (see Chapter 6).

It is difficult to demonstrate unequivocally that lotic populations do actually function as open, metapopulations in their broadest sense. Nevertheless, evidence of dispersal between sub-populations, a prerequisite for this process, has first been gathered by what we might call 'conventional' methods, using field experiments, isotopic markers, and various kinds of trapping and sampling (Peckarsky et al. 2000, see section 5.3.2). As an example, the riparian zones of rivers are amongst the most disturbed of all habitats, the vegetation of soft alluvial sediments frequently being scoured back to the bare substratum by floods. The vegetation that redevelops is then largely a product of recolonisation, often by propagules from upstream.

Soons et al. (2017) suggested that, in larger rivers and riverine wetlands, plants that grow submersed in the main channel have heavy, non-buoyant seeds that may be carried downstream as bedload (Figure 5.25). One might expect that their population persistence in such sites (in the face of frequent disturbances) must depend on resistant plant parts, such as rhizomes in the bed, plus a more or less continuous supply of propagules from populations upstream—such that local populations can be 'rescued' by water-based dispersal ('hydrochory') from upstream. They found that plants growing at the margins of rivers and in associated wetlands (sites that are particularly prone to disturbance) have very buoyant seeds that tend to be trapped

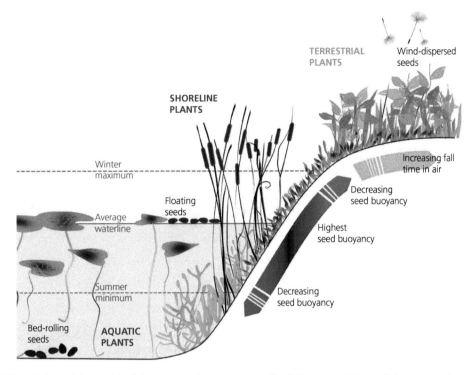

Figure 5.25 Seed dispersal characteristic of plants growing in, or close to, river/floodplain margins. Submerged plants produce heavy, sinking seeds dispersed as bed-load by river flow. More terrestrial plants growing towards the landward margin of the floodplain produce light, wind-dispersed seeds. Emergent plants in the riparian zone produce floating seeds dispersed by flow to shoreline sites.
Source: Soons et al. 2017, with permission from John Wiley & Sons.

in the strandline, so it appears probable they are dispersed predominantly by water, and that their seed traits maximise the likelihood of being transported to suitable sites. Again, their (sub-) populations must occasionally be founded afresh from refugia upstream. Finally, plants growing out on the floodplain under increasingly terrestrial conditions tend to have very light seeds that can be dispersed on the wind, giving them some chance of transport to suitable habitats (Figure 5.25). Such plant species thus can achieve a degree of what might be called 'directed dispersal'—in the sense of increasing the probability of finding a suitable site rather than it being a matter of mere chance.

The dispersal of winged adult insects with an aquatic immature stage has mainly been studied using traps and nets set at various distances from the stream. Commonly, the numbers caught decline sharply with distance, implying that most do not venture far from the channel itself. For instance, Petersen et al. (2004) placed traps ('Malaise traps') up to 75 m away from seven stream channels in upland Wales, catching more

than 29,000 adult aquatic insects. Half of all stoneflies were taken in the traps set up to 18 m from the channel banks, while 90% had travelled less than 60 m. Caddisflies and mayflies travelled even less far on average (Figure 5.26). This and other evidence suggests that the overwhelming majority of such insects 'stay at home', dispersing mainly along, rather than away from, the channel before laying eggs in their natal stream. Of course, there remains the possibility that a few individuals do go further, perhaps dispersing over the watershed from their 'home' catchment, and even recolonising any neighbouring channels without a resident (sub-)population. There is a lot of anecdotal evidence for such long-distance dispersal, particularly for strongly flying adults of some dragonflies and damselflies (Odonata), water beetles (Coleoptera) and true bugs (Hemiptera) but also for very small insects dispersed by the wind and even for small invertebrates transported by larger animals.

Habitat-restoration measures of various kinds provide an opportunity to study recolonisation of rivers and streams by species that had formerly been lost

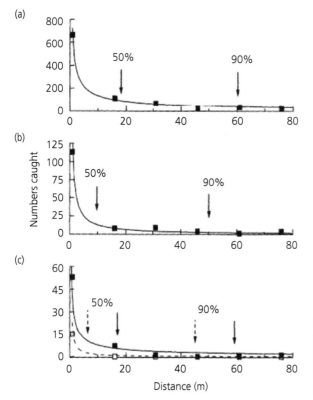

(a)

(b)

(c)

Distance (m)

Figure 5.26 Dispersal of adult aquatic insects away from the channel of a small, upland Welsh stream: (a) stoneflies, (b) caddis flies, (c) mayfly males (solid line, closed symbols) and females (dashed line, open symbols). Arrows depict the distance reached by 50% and 90% of the total animals caught.
Source: Petersen et al. 2004, with permission from John Wiley & Sons.

from those systems. For instance, acidification of streams and rivers in catchments of low buffering capacity resulted in the loss of many species of acid-intolerant aquatic insects, particularly in Europe and north-east North America, at a time when industrial atmospheric emissions (and subsequent deposition back to the landscape) of sulphur dioxide were high (see Hildrew 2018). As stream water chemistry recovers following measures taken to reduce pollution, could a lack of adult colonists (i.e. flying adults) prevent reestablishment of such species? Masters et al. (2007) tested this possibility by trapping adult insects alongside acidified streams in Wales. Eight acid-sensitive species in three insect orders (Ephemeroptera, Plecoptera and Trichoptera) were found close to acidic streams in which their larvae had not been detected in 21 years of benthic sampling. This strongly suggests that the adults can move quite widely across watersheds and, therefore, that the absence of their larvae from the stream itself, should it recover chemically, could not be ascribed to a lack of adult dispersal.

More direct evidence of cross-watershed flight has been obtained through the use of isotopic labels. Thus, Briers et al. (2004) labelled larvae of the stonefly *Leuctra*

inermis with the stable nitrogen isotope ^{15}N, before they emerged from a Welsh upland stream. Subsequently, a few of the istopically 'tagged' emerging adults were found at distances of 0.8–1.0 km from the labelled source, including a stream in a different river system unconnected by any purely aquatic route, thus demonstrating unequivocally dispersal across the intervening terrestrial landscape. Macneale et al. (2005) demonstrated something very similar for a different species of *Leuctra* from a headwater stream at the Hubbard Brook Experimental Forest in New Hampshire. Most *Leuctra* adults indeed remain close to their natal stream, but these studies do show that a few individuals disperse much further—and could thus feasibly establish, or re-establish, populations in suitable habitats elsewhere (as required by the metapopulation model).

Colonisation by aquatic insects of completely new (previously uninhabited) and unconnected systems provides further evidence of dispersal to new habitats by flight. For instance, Ladle et al. (1985) set up two very large, open-air recirculating channels in south-west England and found that they were initially colonised by a previously undescribed midge (Chironomidae; *Orthocladius calvus*) that had extremely rapid larval growth and a short generation time (< 16 days in the laboratory but probably *c.*10 days in the channels). Colonisation by flying adults was the only feasible mechanism and larvae were abundant by day 16 of the experiment but had almost gone, replaced by more common species, by day 33. This rare insect, and others similarly adapted, must persist in the landscape by exploiting short-lived 'gaps', presumably normally created by disturbance, reproducing rapidly and then 'moving on'. It is thus an excellent coloniser but a poor competitor. Much larger-scale cases of colonisation of newly created habitats are to be found where glaciers are retreating—as is the case in the Glacier Bay National Park of Alaska, where a new stream, Wolf Point Creek, began to emerge from the ice in the mid-1940s, and whose community development was observed from 1977 (Milner et al. 2008; Brown & Milner 2012). Dispersal constraints were important in development of the community, non-insects requiring 20 years or more to colonise, though other biotic processes, such as development of riparian vegetation and the arrival of spawning Pacific salmon, were also occurring while temperature increased.

What of amphibians, which of course cannot fly, but are found in streams in various parts of the world? Lungless salamanders (Plethodontidae) are the most diverse of all the salamanders, centred in the New World, and are common in streams in the south-eastern states of the USA and in central America. Some even have fully terrestrial development, although in the dominant genus *Desmognathus* there is an aquatic larval stage followed by terrestrial juvenile and adult stages. In the Shenandoah National Park of Virginia (USA), Campbell-Grant et al. (2010) used mark-and-recapture techniques to infer the probability of dispersal between streams by larvae, juveniles and adults, either directly overland ('as the crow flies') or along channels (of which there were pairs, the two coalescing downstream). Juveniles were the most dispersive stage overall, showing biased upstream movement within the channel but also substantial movements across the terrestrial catchment (Figure 5.27). They then modelled the persistence of a theoretical metapopulation consisting of 15 patches, assuming various probabilities of extinction of the individual sub-populations and using the measured rates of dispersal among patches overland and within the aquatic network. Dispersal was sufficient to permit long-term survival of the metapopulation, even with a probability of extinction per patch of 10% per generation. This study provides an explicit example of a persistent metapopulation spread across a stream network.

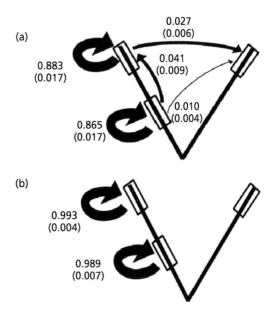

Figure 5.27 The mean (SEs shown) probability of dispersal by (a) juveniles and (b) adult salamanders (*Desmognathus* spp.) within and between two branches of a stream network in Virginia (USA). Juveniles are the more dispersive, most moving within a reach or upstream within the channel, but with also evidence of overland movements between channels.
Source: from Campbell-Grant et al. 2010, with kind permission of Evan H. Grant.)

5.4.2 Open populations: lessons from genetics

The potential importance of the riverine network for the dispersal and spatial structuring of riverine populations has been emphasised by landscape/population geneticists, who have the means to measure genetic relatedness (and to infer gene flow) among populations in relation to the straight-line (cross-country) distance between sites and to the longer distance along fully aquatic pathways within networks. Genetic approaches offer significant opportunities for progress in this area and should be in the 'toolbox' of any aspiring lotic ecologist of the future.

Four different 'zoogeographic' models of connectivity, inferring different modes of dispersal, have been distinguished for river networks based on gene flow and exchange of individuals (Figure 5.28; Meffe & Vrijenhoek 1988; Finn et al. 2007; Hughes et al. 2009). For organisms that must disperse along watercourses, that is those that cannot fly or do so only weakly, we expect differences between neighbouring tributaries without an intervening aquatic route but closer relatedness among neighbouring tributaries within the same river

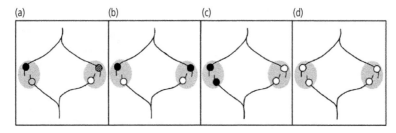

Figure 5.28 The four 'zoogeographic models' of Finn et al. (2007), Hughes et al (2009) and Meffe & Vrijenhoek (1988), conceptualising gene flow and exchange of individuals among populations distributed in the headwaters of two river systems (one flowing 'north' and one 'south') with separate outflows to the sea: (a) the 'Death Valley' model, (b) the 'stream hierarchy' model, (c) the 'headwater' model, (d) the 'widespread gene flow model'. See text for details of each of the models.
Source: Hughes et al. (2009), with permission from Oxford University Press.

system—the 'stream hierarchy model' (Figure 5.28b). The so-called Death Valley model differs only in that the species concerned have extreme habitat specificity and/or very limited powers of dispersal—such as those limited to extreme headwaters or small springs— and hence all populations are strongly differentiated genetically (Figure 5.28a). Where the powers of dispersal are also fairly limited, but where travel overland (e.g. via weak flight) is possible, we expect a common distribution and high relatedness among closely neighbouring channels, regardless of whether a fully aquatic pathway is available, as in the 'headwater model' (Figure 5.28c) (Finn et al. 2007). Finally, where species have fairly broad habitat requirements and good powers of dispersal both within the stream network and across the landscape, we expect widespread distributions and weak genetic structuring of populations— termed the 'widespread gene flow/dispersal model' (Figure 5.28d). As well as population-level patterns, these models have been extended to predict patterns in (multispecies) assemblages of organisms (Tonkin et al. 2018).

Examples of these four models of distribution and relatedness of populations living in dendritic stream networks were discussed by Hughes et al. (2009), each related to the mode and powers of dispersal of the species concerned. For instance, two species of snails (*Caldicochlea harrisi* and *C. globosa*) live in isolated springs in an Australian semi-desert, the springs only being connected by surface water after intense rainfall. As predicted by the 'Death Valley model' there was no evidence of contemporary gene flow among springs for either species, nor any effect of catchment boundaries or any accrual of genetic isolation with distance. In contrast, the relatedness among subpopulations of a fully aquatic salamander (the 'mudpuppy'—*Necturus maculosus*) in three sub-basins of the Ohio River across

eastern and central Kentucky (USA) was closer to expectations of the 'stream hierarchy model' (Murphy et al. 2018; Figure 5.28b). This salamander is paedo-morphic (i.e. attains sexual maturity while retaining its juvenile characters and aquatic habit) and is restricted to dispersal within the stream channel. The population was 'structured' (i.e. the greatest genetic distance between sites) between the Kentucky River basin and two others (the Licking and Kinniconick basins), with a secondary level of structuring within both the Kentucky and Licking River basins, and with an isolated and particularly distinct subpopulation ('Sturgeon', Figure 5.29) within the Kentucky basin. Overall, significant genetic isolation was associated with the presence of dams, which act as barriers to dispersal, and with river (fully aquatic) distance between sites. Murphy et al. (2018) concluded that the genetic structure of this (meta)population was at least partially constrained by river architecture, and by the exchange of individuals within, but not between, river basins.

Several headwater specialists with a limited ability to cross the terrestrial catchment have been shown to conform to the 'headwater model'. These include several taxa of flying insects (e.g. a weakly flying black-fly, *Prosimulium neomacropyga* (Finn et al. 2006) and an Andean mayfly, *Andesiops peruviana* (Finn et al. 2016)), and even some taxa that can only crawl between neighbouring headwater streams. Examples of the latter include some cold-adapted crayfish on Australian mountains (*Eustacus* spp.; Ponniah & Hughes 2004, 2006) and a flightless giant water bug on mountains of the US south-west (*Abedus herberti*, Finn et al. 2007, see Topic Box 5.1 by Debra Finn below). In such cases there is the potential for groups of similar, neighbouring headwaters to exchange individuals and together to act as a classical metapopulation. Alternatively,

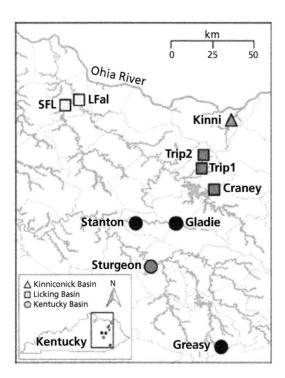

Figure 5.29 Sampling sites of a fully aquatic salamander (the mudpuppy, *Necturus maculosus*) across three independent tributaries of the Ohio River in Kentucky. Symbol shape denotes sites in different basins (triangles, Kinniconick; squares, Licking; circles, Kentucky). Colours denote assignment to genetic clusters.
Source: from Murphy et al. 2018, with permission from John Wiley & Sons.

if one local stream is particularly productive, it could act as a source population and maintain others nearby in which the species would not otherwise persist (i.e. 'sink' populations).

Some aquatic insects seem to disperse sufficiently strongly across the landscape that gene flow is widespread and genetic isolation is confined to sites at a considerable distance from each other—conforming to the 'widespread gene flow' model (Figure 5.28d). For instance, the widely distributed European web-spinning caddisfly *Plectrocnemia conspersa* inhabits mainly small scattered seeps, springs and headwaters. Genetic structuring between sites in the UK was found only at intervening scales greater than about 50 km or more and was most apparent where there were 'gaps' in the landscapes, consisting of areas with a low frequency of freshwater habitats suitable for larvae, such as highly permeable geology with limited permanent surface water (e.g. chalk), large urban areas or the sea (Wilcock et al. 2003). Thus, genetic differentiation between samples of larvae at pairs of sites at

progressively greater intervening distances than 50 km increased within Britain but was much greater when sites on continental Europe were included, thus introducing a sea gap (Figure 5.30a). The inference here was not that individual females make single very long flights between larval habitats but that suitable larval habitats can act as (intergenerational) 'stepping-stones' for gene flow through the landscape, a lack of stepping-stones acting as a barrier to dispersal. Schultheis & Hughes (2005) found a similar lack of genetic structuring between sites separated by less than about 20 km for the Australian caddisfly *Tasimia palpata*, suggesting a lack of such stepping-stones at distances > 20 km.

A graphical model of the relationship between genetic and geographical ('Euclidian') distance between populations by Phillipsen et al. (2015) is an excellent summary of the processes involved (Figure 5.31 upper panel). In species with very poor powers of dispersal, genetic drift in local populations is more influential than gene flow between them. Thus, two neighbouring populations can accumulate genetic differences just as much as those further apart and the variance in genetic distance is great, even for proximate populations. At the other extreme, in powerful dispersers gene flow outweighs drift, genetic differences between populations are small even at large geographic differences and hence variance is low. For intermediate species, there is an equilibrium between drift and gene flow, and genetic differences increase with geographic difference and thus so does variance. Three species of stream-dwelling insects from the American south-west with different dispersal abilities fit these theoretical expectations extremely well (Figure 5.31 lower panel). One of them is our friend the giant flightless waterbug, which fits the headwater model.

A corollary of this is that for many lotic organisms exploiting 'insular' habitats, such as headwaters scattered through the landscape, a minimum density of patches is required for widespread distribution of species and the 'penetration' of genes across geographic distance. In landscapes where this minimum density is not available, the species will not be able to disperse or even colonise initially. Suitable patches for habitat specialists will, by definition, be scarce but for generalists they will be abundant. This difference could explain the well-known pattern that common species (with a high rate of occupancy of habitats patches) also tend to be widely distributed (Figure 5.30b). Evidently, the extent of gaps that can be bridged also depends on the powers of dispersal of the species concerned. A possible example of this kind of

Topic Box 5.1 Lessons from genetics—flightless giant water bugs and the Headwater Model

Debra S. Finn

Abedus herberti is a giant water bug (Hemiptera: Belostomatidae: see Figure 4.14) whose geographic range occupies portions of the arid south-western USA and northern Mexico. The species has persisted through the hotter and drier cycles of interglacial periods in these landscapes for millions of years, despite being obligately aquatic at all life stages and lacking the ability to fly. Populations of *A. herberti* principally occur in headwater streams on 'sky islands'— isolated mountains that rise to more than 2,000 m above the surrounding desert. Permanent water is most common in headwaters at intermediate elevations on sky islands and declines downstream approaching the surrounding 'sea' of desert. Most of this bug's geographic range experiences a summer monsoon, during which storms of small spatial extent but high intensity regularly bring flash floods. *Abedus herberti* and several other invertebrates exhibit what is known as 'rainfall response behaviour' (RRB) in which individuals crawl from the stream to higher ground in response to heavy rainfall. This behaviour dramatically increases survival during flash floods and is most prevalent in *A. herberti* populations in which heavy rainfall is indeed a reliable cue for impending flash flood (Lytle et al. 2008). Most individuals move temporarily up to 20 m from the source stream, then return to the same stream after the storm has passed.

However, this response to rainfall during monsoon storms is also likely to be the driver of occasional dispersal by the flightless *A. herberti* across catchment divides (i.e. watersheds) between neighbouring headwater streams. Population-genetic analysis of *A. herberti* has revealed that landscape-scale population structure is strongest at the among-mountain scale (Finn et al. 2007), an observation that led to the first formal description of the headwater model (Figure 5.28c). Further studies with additional molecular markers showed that the most likely pathways for terrestrial dispersal are 'concavities' (low points) in the landscape (Phillipsen & Lytle 2012)—which primarily include hydrologically connected valleys but also probably low passes across watersheds. Given the large size and long lifespan of the adult bug, its obligate aquatic nature, and large eggs that must be firmly attached to 'Dad's back' for embryonic development (Smith 1976), it is unlikely that individuals are transported by birds or other large and strongly dispersing organisms. With rainfall response behaviour as the probable sole driver of overland dispersal, it is important to note that headwater populations of *A. herberti* are quite isolated from one another, with significant population-genetic structure even among populations occupying nearby headwaters

on the same sky island. This isolation allows adaptive evolution to take different directions in local populations (Lytle et al. 2008), but it also means that populations face a risk of long-term local extinction should aquatic habitat in headwater segments change from permanent flow to seasonally intermittent.

Climate change and an increasing rate of groundwater pumping in recent years are therefore clear concerns for *A. herberti*—both contribute to the conversion from permanent to intermittent surface waters. Some local *A. herberti* populations have indeed been lost (Bogan & Lytle 2011), and it is unknown how long recovery might take if and when perennial surface-water habitat returns. According to the Headwater Model, successful recolonists would probably originate from persistent source populations in nearby headwaters on a shared sky island. In 2007, we mapped the areal extent of aquatic habitat during the dry season in several hydrologically independent headwater segments where populations of *A. herberti* occurred and found the quantity of permanent habitat to vary extensively (Finn et al. 2009). Some *A. herberti* populations persisted in streams in which surface water contracted down to one or two small and stagnant bedrock pools ('tinajas') while others occupied headwaters with permanently flowing water along entire stream segments throughout the dry season. With the same molecular markers used to infer the headwater model, we ran additional analyses to make inferences about long-term population demographic stability and genetic diversity of each local population. We found a strong signal of population bottlenecks in some marginal localities, an indication of either a past local extinction followed by recent founder event or a population that dwindled to just an individual or a few that recently recovered in census numbers (counts of individuals). The most stable and diverse populations tended to occupy habitats with a greater areal extent of dry-season surface water. We inferred that these populations would be the most likely to persist through future change and therefore to serve as source populations that might provide dispersers with the potential to recolonise nearby localities following periods of extensive drought. Due to their potential as source populations through continuing climatic uncertainty, these robust populations should be given special conservation priority with the overarching goal of making the species less vulnerable to climate change.

The same general approach can be used to support conservation planning for other stream-dwelling taxa in locations affected by rapid environmental change. For example, alpine streams are experiencing hydrological change thanks to dwindling ice and snow in high mountains. The readily

Topic Box 5.1 *Continued*

observable decline in surface glaciers and long-term snow-pack has made these systems more visible to policymakers and the general public and, in the USA, two 'meltwater' stoneflies occupying the upper limits of alpine headwaters in the Rocky Mountains are now federally listed in association with the Endangered Species Act. An ongoing search for cold, stable 'refugia' habitat in isolated alpine headwaters

(e.g. Brighenti et al. 2021) follows a similar strategy as that for identifying and placing conservation emphasis on the most robust populations of *A. herberti* in its arid mountain streams.

Debra Finn is an Associate Professor in the Department of Biology at Missouri State University, Springfield, USA.

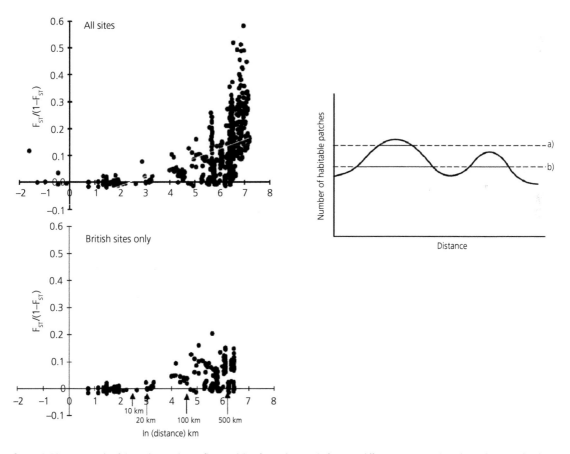

Figure 5.30 An example of the widespread gene flow model. Left panel: accrual of genetic differentiation—$F_{ST}/(1-F_{ST})$—with geographical distance between pairs of populations of *P. conspersa* in (upper) Britain, Germany and France and (lower) British populations only (Wilcock et al. 2003; with permission of John Wiley & Sons). Right panel: conceptual relationship between the varying number of suitable habitable patches over the landscape (distance) for two species of differing dispersal ability. A weak disperser ('species a') requires a greater density of patches to spread and persist, and a strong disperser ('species b') requires a smaller density and has a wider range. Species ranges are denoted by the unbroken horizontal lines (Hildrew 2009).

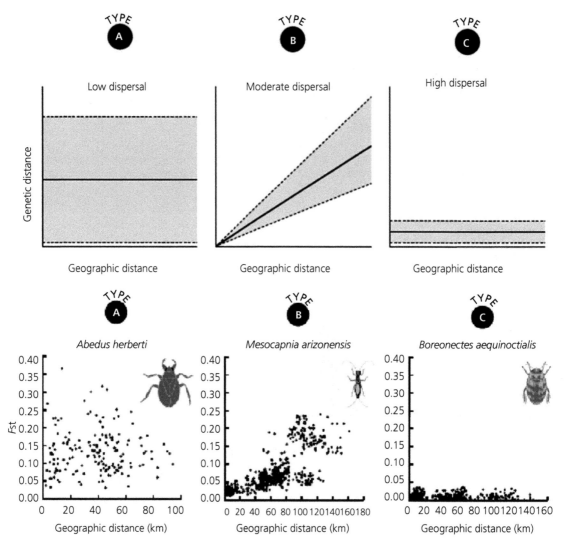

Figure 5.31 Upper panels: predicted relationship between genetic and geographic difference between pairs of populations for species of differing dispersal ability (A, low; B, moderate; C, high). Lines are regressions and grey areas are variance. Lower panels: measured relationships for three real species conforming to expectations (A, *Abedus herberti*, giant flightless waterbug; B, *Mesocapnia arizonensis*, stonefly; C, *Boreonectes aequinoctialis*, water beetle).

Source: Phillipsen et al. 2015, with permission from John Wiley & Sons.

process is found in two regions of central and western Australia, separated by about 1,500 km of predominantly dune desert (Razeng et al. 2017). Here, weakly dispersing mayflies showed no sign of recent gene flow between the two regions, whereas strongly flying dragonflies evidently could pass this considerable barrier. Further, it was found that mayfly lineages had diverged, and new species arose, at the time of a drying climate some 13 million years ago (Razeng et al. 2017).

5.5 Conclusions—beyond population biology

This chapter is about populations, focusing therefore on the distribution and abundance of single species, rather than initially on interactions with others. It is thus a null hypothesis that distribution and abundance depend on the fundamental habitat, resource requirements and powers of dispersal of the individual species. Attention in this context is given to the physicochemical environmental factors, such as temperature, the forces of flow, oxygen supply, the coarseness of the substratum and the rest (Chapter 2), plus density-dependent biotic factors involving intraspecific competition. We have seen that high mobility is a feature of many lotic organisms and this interacts with the complex, physical network that is a river, bringing about the structure of lotic populations that is observed. Some local populations are strongly regulated by density dependence and their persistence rarely, if ever, relies on recolonisation from elsewhere. Others seem to require frequent 'rescue' by recolonisation. Thus, both density dependence and 'spatial factors' seem to play important roles in lotic populations.

The apparently overwhelming physical forces of flow (i.e. as a potential source of disturbance and density-independent mortality) dominating rivers and streams may be ameliorated by habitat heterogeneity.

This creates a range of habitat conditions in different rivers, differing in physical harshness, against which populations are selected from the species available in the regional species pool. Species can therefore be thought of as passing through a 'filter' based on their suites of traits, thereby resulting in different multispecies communities. A 'minimalist' view is that communities are simply collections of species with independent dynamics, which can both access (by dispersal) a particular habitat and can tolerate the conditions there. Alternatively, the realised distribution and resource use of species is frequently restricted (or facilitated) via interactions with others, and such species interactions may play a role in (a) the patterns of distribution and abundance that we observe and, ultimately, (b) the assembly and diversity of lotic communities. We move on to such multispecies assemblages in the next chapter.

CHAPTER 6

Living communities in rivers and streams

6.1 Introduction

In the previous chapter we dealt with the population biology and ecology of lotic organisms, which by definition concerns single species and what determines their abundance and distribution. In natural habitats, however, single species almost always occur embedded within groups of other species, which we can call communities or assemblages. These two terms are used interchangeably in practice, although they do have somewhat different connotations to some. *Assemblage* is a more neutral term, inferring nothing about interactions among the species but merely that they occur together at some place or time. It is more likely to be used for a group of coexisting and related species—as in a 'fish assemblage', for example. A *community* on the other hand more correctly refers to a group of coexisting species of all taxa and trophic positions (from decomposers to large predators, for instance), as in the 'river community', with a full suite of species interactions potentially represented. Such distinctions are not universally adhered to, however. In this chapter, therefore, we begin to address phenomena and patterns that occur at this 'multispecies' level of organisation. We are interested in diversity (most simply, but not exclusively, how many species are there?), in multispecies patterns in space and time (how do communities vary from place to place and from time to time?) and in what processes bring about these patterns (e.g. whether they are largely driven by the environment, and/or by the intrinsic biology of species, or by interactions between species).

Rich assemblages of species live in all parts of river systems and throughout what we have called the physical habitat templet. Despite the limited spatial extent of lotic systems, the biota of rivers is clearly conspicuously diverse. A difficulty, however, is knowing where to set physical limits on what we might consider part of the river community. There are certainly no neat habitat boundaries within which river communities live, despite what one might think at first glance. As we show in all other aspects of the biology and ecology of rivers and streams, the lateral, longitudinal and vertical linkages with surrounding habitats blur these boundaries considerably. River communities are thus open, dynamic systems. A further, more unfortunate, characteristic related to the 'openness' of river communities is that they are so profoundly threatened by humans and their activities. As humans we 'ask' an enormous amount of rivers systems—usually without realising or acknowledging it—so it is fortunate that river communities are also highly resilient. Nevertheless, there are definite limits to this resilience and we are presently engaged in a kind of unintended 'test to destruction' of rivers, despite their biological diversity being an important part of the 'life-support system' of our planet (a topic we return to in Chapter 10).

In Chapter 1, we introduced different assemblages that inhabit different parts of river systems (e.g. the benthos living on or near the substratum surface, and the hyporheos living interstitial lives below the surface), have different 'lifestyles' (e.g. free floating or attached) or cover different parts of the overall size range of living things in rivers (e.g. meio- and macrofauna). Putting names on the different parts of the overall biological community of rivers gives us a handy vocabulary (once we have learned the language!) and reminds us of the overall range of organisms and habitats that are represented. However, it is ultimately a descriptive part of science, useful as a first step and perhaps in prompting some further questions. However, a persistent emphasis on 'typologies', the classification and naming of different (sub)communities, lacks explanatory power as to the processes at work (the goal of community ecology as in any other science). A proliferation of more and yet more names in which to pigeonhole the variety of natural communities, whether from river systems or anywhere else, has

The Biology and Ecology of Streams and Rivers. Alan Hildrew and Paul Giller, Oxford University Press. © Alan Hildrew and Paul Giller (2023).
DOI: 10.1093/oso/9780198516101.003.0006

dogged (and clogged!) riverine community ecology and its application. Therefore, in this chapter we concentrate on the processes that may underlie the assembly, composition and persistence of communities.

6.2 Assembly, filters, traits and 'strategies'

Here we conceptualise the processes by which local communities of species may be assembled, adding layers of potential constraints on local diversity. In Chapter 4 we said 'no species occurs everywhere'—which seems obvious. From the viewpoint of ecological communities, however, it is also clear that the set of species that can be found at any particular site (think of this for now as some conveniently small study site, such as a riffle in a stream) does not include every species that occurs in the surrounding locale (such as a longer reach of the river). This simple observation has led many authors to conclude that species must pass through some sort of 'filter' before being part of a local community—and we can think of such filters as being the match between the features and adaptations of the species concerned and the conditions in our local site. Thus, some species that occur within a longer river reach might not be able to survive the particular conditions of a riffle (such as the hard, stony substratum and turbulent flow with high shear stress). There could also be biotic filters as well as abiotic ones. Thus, some species that could tolerate the abiotic environment of our riffle might be excluded by the presence of particular predators or competitors there. In general, we can most simply conceptualise a local community as comprising species which are represented in some wider, 'regional' species pool, and have passed through habitat 'filters' (Figure 6.1a).

To refine this concept of assembly from a wider regional species pool to a local community we should recognise that it is 'scale dependent'. To do this, we return to the idea of streams as hierarchical networks of physical habitats (see Chapter 2; Frissell et al. 1986). That is, the characteristics of the river habitat at any level in the catchment hierarchy are determined by the characteristics of the level above. For instance, the characteristics of a stream reach are determined by those of the overall catchment. Hence, if a catchment has low relief and a uniformly base-rich geology, it cannot contain a stream reach with a steep profile, highly turbulent flow and low alkalinity ('soft' water). All these local characteristics of the reach—turbulence and flow forces, plus water chemistry—are likely to

select for a particular set of species well-adapted to such conditions. As an example, Poff (1997) combined the physical framework with the notion of 'filters' that select from the wider species pool, as shown in Figure 6.1b. Evidently the regional species pool, which encompasses one or more river catchments, is made up of successively smaller local communities (i.e. species occurring in the catchment, the stream reach, the channel unit or the microhabitat, etc) as we pass through the environmental filters relevant to each descending scale.

So, what determines whether a particular species will pass through a filter at any particular level in the hierarchy? As we saw in Chapter 4, species have biological characteristics, or 'traits', that determine their ability to persist in a particular environment—at least theoretically. These traits include features such as final body size, generation time, the number and size of propagules typically produced, aspects of body shape and many others. Any one species usually displays only one or a few 'modalities' of each trait, e.g. the maximum body size of a species may be small, medium or large. Repeated attempts to relate the trait modalities of species to the nature of their environment (this last typically characterised by habitat type, disturbance regime, physical complexity and others) have been made, with variable success, and have generated a good deal of discussion, much of it focused on difficulties with the data and in analysis (e.g. Resh et al. 1994; Statzner & Bêche 2010; Verberk et al. 2013). For instance, species with short generation times and asynchronous life cycles are expected to be overrepresented in disturbed habitats (see Chapter 5, section 5.4.1, p. 181). 'Short generation time' is a modality of the species trait 'generation time', which of course varies amongst species. The development of an ability to predict community composition and structure on the basis of an organism's traits is of more than academic interest, because it is the basis of promising methods of bioassessment, and the detection of particular stressors (e.g. chemical pollutants, fine sediments, morphological degradation) (see Topic Box 6.1 by Sylvain Dolédec and Núria Bonada for more details, and Chapter 10).

In order to improve our ability to predict community composition, we need to appreciate that individual species traits, of which there are many, are not independent of one another, but occur in combinations that may offer alternative 'solutions' to the challenges of a particular environment (Chapter 4). 'Trade-offs' between individual traits may occur and often have a strong phylogenetic signal, since related species may necessarily (because of developmental constraints) possess similar suites of traits and modalities.

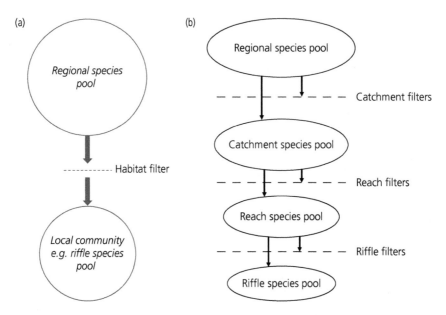

Figure 6.1 (a) Local communities can be most simply envisaged as a group of species recruited from a wider regional species pool by passing through a 'habitat filter' of abiotic and/or biotic conditions (see Keddy & Weiher 1999). Habitat filters are scale dependent and may be 'nested'. Thus, in a river system (b), species that occur in a stony riffle must first find suitable habitat represented somewhere in the catchment, then in a particular reach, then in the riffle (see Poff 1997).

A 'trade-off' occurs when the possession of one trait alters the adaptive value of another. Thus, various alternative combinations of traits may provide successful 'solutions' for a particular set of environmental conditions. The following examples demonstrate how very different traits address the challenges of living in temporary streams. Some stoneflies have a diapause stage in their life cycle, allowing them to survive dry periods and live in intermittent streams. For instance, larval (and possibly egg) diapause enables *Mesocapnia arizonensis* to survive during long, dry periods in temporarily flowing desert streams in the south-west of the USA (Bogan 2017). Its development is fairly rapid when the stream is flowing during wet periods, but completion of the life cycle may take over a year. In contrast, the life cycle of the midge *Orthocladius calvus* includes no drought-resistant stage, at least in its pre-adult life stages, and its generation time is extremely short (< 20 days). Having resistance traits (such as diapause in the stonefly), and an unremarkable rate of growth, is therefore one 'good solution' to the challenge of living in a temporary stream, while rapid growth, a very short generation time and no drought-resistant stage (as in the midge) is a second. Verberk et al. (2013) pointed to the notion that community assembly occurs through the ability of *species* with various *combinations* of traits (not separate species traits selected

independently)—and which, by implication, represent a more limited number of overall life-history 'strategies' (Figure 6.2). This idea is rooted in previous work, including that by Southwood (1977), Hildrew (1986), Resh et al. (1994) and others.

So far, we have viewed community assembly as involving what is often known as 'species sorting'—in which species do or do not occur depending on their fitness in the abiotic environment and on their ability to persist alongside other species (via species interactions within the local assemblage). The ability of any species to *disperse* to a local site that is environmentally suitable for it also places further potential constraints on the composition of a local assemblage. Sometimes called 'spatial effects', such dispersal constraints are a further explanatory variable, beyond species sorting, for community composition (see Chapter 5, section 5.3.4, p. 168).

To summarise, species sorting can occur in two main ways: (1) habitat constraints—do the combined traits of a species enable it to pass though the abiotic habitat filter?; and (2) biotic constraints—does the species pass biotic filters based on an ability to coexist with others (be they competitors or natural enemies)? Those familiar with ecological jargon will recognise that these two criteria equate to the *fundamental niche* and the *realised niche* of a species,

---

Topic Box 6.1 The use of biological traits in biomonitoring

Sylvain Dolédec and Núria Bonada

The effective assessment and management of aquatic ecosystems facing many natural and anthropogenic stressors requires reliable biomonitoring tools that (i) discriminate accurately among degrees and types of stressors and (ii) are easy to implement. Over the past 40 years, biomonitoring tools have focused on the taxonomic richness and environmental tolerance of whole biological groups (e.g. macroinvertebrates, diatoms, macrophytes or fish) and several methodologies based on single biotic indices, multimetric or multivariate approaches have been developed (Bonada et al. 2006). Ideally, such biomonitoring tools should (i) be ecologically sound with predictable responses to known stressors, (ii) yield similar results in sites which are only slightly impacted, with reliable and repeatable responses to overall or human impacts and (iii) have low costs in terms of taxonomic identification, sampling and sorting (Bonada et al. 2006). Traditional methods based on taxonomic richness and composition do not fulfil all these criteria because taxa vary across regions and thus reference conditions are region-specific (Menezes et al. 2010). The increasing availability of biological trait information in public databases for aquatic macroinvertebrates (e.g. Schmidt-Kloiber & Hering 2015 in Europe) has promoted diagnostic tools based on the use of multiple traits related to life history (e.g. life-cycle duration, number of reproductive cycles per year), physiology (mode of respiration), morphology, feeding behaviour (e.g. scraping or shredding), dispersal ability and reproduction. Based on the habitat templet concept (developed for streams and rivers by Townsend & Hildrew 1994), trait-based approaches are deeply rooted in the ecological theory that the representation in the assemblage of various species traits depends on characteristics of the habitat. Anthropogenic stressors modify the habitat and thus may change the proportion of individuals with given traits in assemblages (Box Figure 6.1a), which in turn could affect ecosystem processes. A few real examples of the effect of stressors on the trait distribution in assemblages are shown in Box Figure 6.1b. Assessment based on multiple traits is thus a promising area (Bonada et al. 2006; Culp et al. 2011; but see also Verberk et al. 2013) because biological traits represent general characteristics that are linked to ecosystem processes and are widely applicable (Statzner & Bêche 2010). In addition, most traits are affected predictably by various natural and anthropogenic stressors, permitting a comparison between a response predicted a priori and the observed response to human impairment (Statzner & Bêche 2010). For example, overall disturbance should favour resilience traits, enabling rapid population

growth and thus rapid recovery (e.g. many descendants per reproductive cycle, short life cycles). To compensate for the action of flow forces (high discharge, water velocity and shear stress), we would expect a high proportion of individuals with firm attachment to the substratum, streamlined shape and/or having a small size. In contrast, low flows should favour less-streamlined, larger organisms with good swimming ability. Siltation should exclude taxa susceptible to smothering by sediment (e.g. without any means of egg protection) while favouring traits that enable penetration of fine substrata (e.g. burrowing) and particular feeding habits (e.g. fine detritus eaters). Organic pollution and oxygen depletion should increase the proportion of individuals with aerial respiration. Heavy metal contamination should select against a high body surface-volume ratio (e.g. against small size, gill respiration). Such mechanistic linkages between biotic responses and environmental conditions represent one critical advantage over purely taxonomically based methods.

Previous studies of unaltered situations have shown the stability of the biological trait composition in assemblages, suggesting the potential use of mean European trait profiles as reference endpoints in the development of trait-based biomonitoring tools (Statzner et al. 2001). Studies using disturbed situations to examine the reliability of assessments have shown the ability of biological traits to: (i) discriminate significantly upstream and downstream of a waste-water effluent (Soria et al. 2020); (ii) disentangle the effects of natural and anthropic disturbances in large rivers (Usseglio-Polatera & Beisel 2002) or temporary streams (Soria et al. 2020); (iii) demonstrate the significant effects of sediment contamination associated with feeding habits, resistant forms, dispersal, respiration and mode of reproduction of taxa (Colas et al. 2013); and (iv) demonstrate how multiple stressors, in combination with water scarcity, involved the selection of taxa with specific characteristics (e.g. egg protection, indicating a potentially higher risk for egg mortality; Kuzmanovic et al. 2017). Biomonitoring methods based on multiple traits thus offer good opportunities for management and conservation and should be incorporated in large-scale routine programmes to assess reliably the ecological status of freshwater ecosystems and to link taxonomic patterns to ecosystem processes.

Sylvain Dolédec is a Professor at the Université de Lyon, Université Claude Bernard Lyon 1, Villeurbanne, France.
Nuria Bonada is a Serra Húnter Professor at the Departament de Biologia Evolutiva, Ecologia i Ciències Ambientals, Facultat de Biologia, Institut de Recerca de la Biodiversitat (IRBio), Universitat de Barcelona (UB), Catalonia, Spain.

Topic Box 6.1 *Continued*

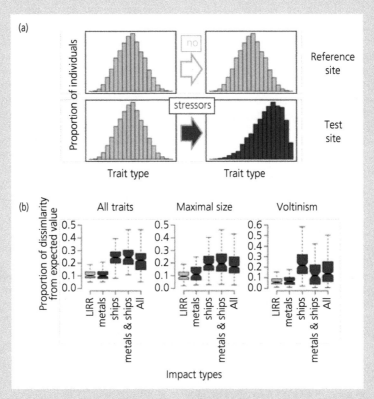

Box Figure 6.1 (a) A conceptual example showing the proportion of individuals expected at reference and test sites. After exposure to a stressor (date shown in red rather than blue for no stress), the selection of individuals involves a selection of those traits at the test site conferring the ability to survive. This results in a shift of the distribution of traits in the assemblage in comparison to the reference site. (b) Example of three trait variables showing the deviations (Bray–Curtis dissimilarity) from the expected trait value (mean of all categories of a trait in all Least Impacted River Reaches (LIRRs)), reaches with heavy metal pollution (metals), with cargo-ship traffic (ships), with both cargo-ship traffic and heavy metal pollution (metals & ships) and with a mixture of impacts (All). *Source*: (a) redrawn from Culp et al. 2011; (b) redrawn from Dolédec & Statzner 2008.

concepts long recognised in general ecology (see e.g. Begon et al. 2005). The balance between 'dispersal constraints' (spatial effects) and the latter 'niche-based' processes (species sorting) is attracting most contemporary attention. This is also related to the density-dependence/density-independence controversy in population ecology (Chapter 5) and thus to the relative importance of equilibrial and non-equilibrial processes in population and community ecology. We develop these ideas in the following sections, but first we consider patterns in river assemblages in space and time, including interactions between the two.

6.3 Community patterns in space

6.3.1 'Gently down the stream'—longitudinal changes in river assemblages

The study of the distribution of species and assemblages along habitat and environmental gradients is a long-established approach in ecology—often known

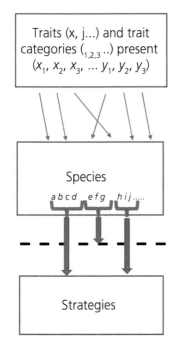

Regional species pool

Trait categories are combined (non-randomly) in real species (a, b, c...) with trade-offs and constraints

Regional species pool

Each species is allocated to a smaller number of 'strategies'

Environmental filter (biotic and abiotic conditions) through which a limited number of strategies can pass

Local community

Figure 6.2 Here we depict trade-offs and constraints in the combination of various categories ('modalities') of species traits (e.g. large body size is not usually combined with a very short lifespan) that may allow species to pass habitat filters. Thus, there is a wide range of species traits and categories of those traits in a region, that are variously combined together in real species. The characteristics of these species can then be grouped into a smaller number of 'strategies' which may or may not fit them for a local community, which they can potentially join. *Source:* based on Verberk et al. 2013.

as 'gradient analysis'. In the study of rivers, ecologists once spent a long time debating whether biotic assemblages change abruptly along the length of rivers, or do so only gradually. These opposing views have been described as the zonation and continuum concepts, respectively (Illies & Botosaneanu 1963; Vannote et al. 1980).

Biologists love to classify and pigeonhole things—creating order out of what might seem to be chaos. This instinct to classify lives on, and in freshwater ecology is still reflected in the workings, for instance, of the European Union's 'Water Framework Directive' (and its derivatives applying elsewhere) and the IUCN Global Ecosystem Typology 2.0, which classify 'types' of fresh waters (see Chapter 1). If we could discern distinct multispecies communities in rivers that are genuinely discontinuous—for example whole suites of species reaching their longitudinal limits at the same place—it would be of great management interest, since it would reduce the complexity of managing almost infinitely variable living systems.

Early in the history of community ecology, two contrasting views of communities had already emerged.

The first suggested that communities are highly interactive natural groupings of species that normally occur together within repeatable associations, behaving almost like a single 'superorganism' (Clements 1916). The second (individualistic) view countered that communities are just variable mixtures of species that happen to occur together in similar circumstances, but are just as likely to be found elsewhere and 'in the company of' other species (Gleason 1926).

The Illies & Botosaneanu (1963) school of river ecologists thought that there are river zones corresponding with real and genuine biocoenoses[1]—clearly adopting a 'Clementsian view' of river communities. They attempted to devise a comprehensive classification of river communities based around their concept of

[1] 'Biocoenosis' is a term not widely used today, but still appears occasionally in the literature. It postulates the existence of a discrete association of species (a 'biocoenosis') found in a particular habitat type—then called a 'biotope'. These terms are thus similar, though not identical, to the more familiar 'community' (or assemblage) and 'habitat', except that they are more loaded with meaning. Biocoenoses are purported to be distinct assemblages living in a particular type of biotope.

biocoenoses. Thus, they recognised assemblages corresponding to the 'crenon', 'rhithron' and 'potamon' (essentially springs, streams and rivers, see Chapter 1). However, the overlaps in species composition that were apparent required these zones to be subdivided into, for instance, the 'epi-rhithron', the 'meta-rhithron' and the 'hypo-rhithron'. In attempting to produce a universal classificatory system, they allowed for differences in altitude and latitude in deriving particular boundaries between these types of zones, but all were supposedly represented everywhere. Habitat features, including the nature of the riverbed, were also included in their classification, further dividing their river zones and subzones into more and smaller biocoenoses, each with a separate name based on substratum. As can be imagined, this led to a blizzard of terms and complexity, although the authors asserted that their scheme was 'sufficiently flexible to allow the assimilation of new information as it became available'. Thus, if exceptions arose, they could be accommodated by erecting even more subdivisions of biocoenoses and biotopes. This process is ultimately non-scientific, since a testable (falsifiable) hypothesis never arose. It was a controversial idea, even in its day, with trenchant opposition to it expressed by a number of very prominent contemporary river ecologists (e.g. Macan 1961; Hynes 1970b) who took a 'looser' view of river communities. The approach of Illies & Botosaneanu (1963) was never widely adopted, particularly by English-speaking ecologists, but its ghost lives on in contemporary quantitative (statistical) attempts to classify and group river assemblages. Such groupings are certainly useful for management, but it would be a mistake to believe that they are 'real and genuine individual biocoenoses'.

The debate over zonation is a part of the history of stream ecology, conducted at a time before the rather complacent dominance of the English language. Neither did ecologists have the computing power available today. Its arguments may seem remote to the modern student, but it did reflect a rather fundamental stage in the search for the 'true' nature of lotic communities (and of communities more generally).

A starkly different view of longitudinal patterns in river communities, particularly of benthic invertebrates, arose in the United States and is known as the *river continuum concept*. The original paper (Vannote et al. 1980) became the most highly cited in all of river ecology but is also among the most difficult to unravel in all its aspects. Like the zonation concept it is based on an equilibrial view of communities, on knowledge of the environmental changes occurring along

rivers (to which whole communities were in some way 'adapted') and on the recognition of functional feeding groups of invertebrates (described in Chapter 4). The river continuum concept not only addresses communities, but also ecosystem processes, which are discussed elsewhere in this book (see Chapter 8, section 8.3). Nevertheless, its predictions about community structure are among the easier aspects to understand and to test. It also has the advantage of invoking processes that might bring about the communities described, rather than merely being descriptive.

In terms of community structure, the river continuum concept proposes that shifts in resource inputs and availability along the course of a river determine gradual changes in the relative abundance of different functional feeding groups of benthic invertebrates (inferring a gradual change in species composition along the 'continuum'). Thus, headwater streams were seen as depending primarily on inputs of terrestrial leaf litter, which was consumed by 'shredding' detritivores, the litter having first been softened and penetrated by aquatic fungi (see Chapter 8). The feeding activities of shredders produced particles (partly faecal pellets) which in turn would be microbially conditioned and consumed by collectors, either deposit or suspension feeders. Both shredders and collectors could support a guild of predators, but grazers of epilithic biofilms were considered to be a minor component of headwater stream communities since shading of the narrow channel would limit algal growth. However, as the stream channel widened downstream, and shading by riparian vegetation declined, one would expect algal production to increase along with the relative frequency of grazers. At the same time, the biomass of shredders would be expected to fall as inputs of tree litter declined and were more poorly retained on the beds of larger streams or have largely been broken down to FPOM by shredders upstream (Figure 6.3; see also Chapter 8 for more details of the concept and criticisms of it). In the lowland, turbid reaches of larger rivers benthic grazers were predicted to decline again, to be replaced mainly by collectors (deposit-feeders and filter-feeders) on organic material on the bed and in suspension.

Other conceptual approaches to serial (longitudinal) patterns in river communities have also been made. Most prominent among them was Ward & Stanford's (1983) *serial discontinuity concept*, which gave recognition to the role of large tributaries and impoundments in 'resetting the continuum' or, more simply, by bringing about abrupt changes in environmental conditions such as temperature, coarseness of the substratum

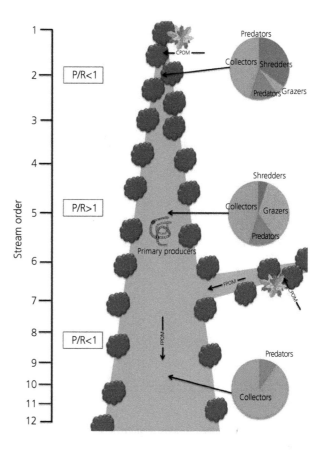

Figure 6.3 Schematic of the river continuum concept of Vannote et al. (1980) emphasising shifts in the functional feeding groups of invertebrates as food resources switch from leaf litter in the shaded headwaters to primary producers further downstream. P/R refers to gross primary production and total ecosystem respiration which are postulated to shift in relative magnitude from up- to downstream.
Source: from Doretto et al. 2020, under a Creative Commons licence and with kind permission of the authors.

and hydraulic forces and, consequently, in community structure (see Jones & Schmidt 2018 for a recent example). Statzner & Higler (1986) also stressed the role of discontinuities in the physical environment by collecting together impressive evidence of abrupt changes in the near-bed flow forces along a wide variety of rivers. These abrupt changes coincided with changes in the benthos, leading them to propose that flow forces were the primary factors determining longitudinal biotic patterns in stream animals (Figure 6.4).

Although there is no question that hydraulic forces are extremely important, stream animals are also sensitive to a great variety of other factors that affect their distribution. The important conclusion for the modern freshwater ecologist is to recognise that where there are distinct points of change in the environment (such as changes in slope) there may also be abrupt longitudinal changes in community composition, a non-controversial proposition in the analysis of communities along gradients. This conclusion also meshes well with the view that communities consist of species with similar environmental requirements that are distributed along gradients in an individualistic way (i.e. close to the Gleasonian view of communities). This does not mean that species do not interact along environmental gradients (see Chapter 7), or that sharp boundaries in the distributions of similar species may be produced by biological interactions, but it does indicate that, overall, lotic communities are flexible, highly variable assemblages of species that share similar general environmental requirements.

6.3.2 High diversity in small tributaries

The river continuum concept suggests that, when rivers are considered as a simple linear thread, species richness should peak in 'mid-order' (around orders 3–5) channels. The basis for this hypothesis is rather speculative, and predictions should strictly be applied only to rivers of the kind on which the concept was based (i.e. relatively undisturbed systems with forested headwaters as in the north-west of the United States); nevertheless, an analysis of 11 studies collated by Clarke et al. (2008) lends it some support in other places, and only

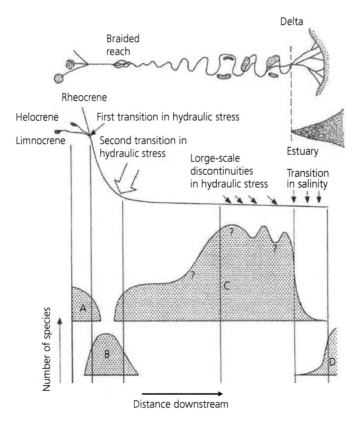

Figure 6.4 Statzner & Higler (1986) depicted a 'typical' river in a natural state. At the top we have an aerial view of the river planform showing three different kinds of sources followed by an abrupt increase in slope; then, as the slope declines again when the river enters its lowland reaches, there is a braided (multi-channel) section, followed by meanders of increasing magnitude as the river cuts its way through an alluvial floodplain (typically with some 'abandoned' floodplain water bodies) before entering its estuary (a delta) and, finally, the sea. Below that we have a profile showing typical slope changes. Here the sources are shown as a (surface-fed) 'rheocrene', a groundwater spring (a 'helocrine') and a 'limnocrene' (the source is a pond). Discontinuities in hydraulic stress are associated with these slope changes and, in the lowland reach, with meanders and the presence of side-channels and oxbows. A transition in salinity occurs as the river enters its estuary. Statzner & Higler (1986) envisaged sharp changes in the benthos (and thus 'zones') at points where the hydraulic stress and salinity change. Species A are those in the extreme upstream reaches, at or very near the source, giving way to species B in the turbulent steep stream, and to species of group C in the lowland reach.
Source: modified after Statzner & Higler 1986, with permission from John Wiley & Sons.

one of the studies found the highest species richness in a headwater site.

All these earlier concepts treated rivers as linear systems, extending from source to mouth, and did not recognise them as branching networks. In reality, almost all river systems comprise a large number of small tributaries, several larger ones and a single main stem (see Chapters 1 and 2). Estimates suggest that 'headwaters', (loosely) defined as first- and second-order channels (as well as 'zero-'order, intermittent channels, upstream of first-order ones), make up around three quarters of the total length of river networks (e.g. 73.4% in England and Wales and about 77% on the island of Ireland; Riley et al.

2018). Bishop et al. (2008) expressed their extent in a slightly different way and reported that around 90% of the stream network in Sweden drained catchments of less than 15 km². This figure represents the lower size limit at which water courses are monitored, and indicates that an astonishing proportion of Swedish streams, and most streams elsewhere, are essentially 'unknown'. When viewed from a network perspective, it has often been claimed that headwaters hold the majority of the total diversity in river systems, certainly of benthic macroinvertebrates (Meyer et al. 2007; Bishop et al. 2008; Clarke et al. 2008; Finn et al. 2011; Cilleros et al. 2017; Richardson 2019). The reason for this is that, despite the relatively low species richness in an individual tributary (its α diversity), tributaries differ from each other in their species composition. Consequently, β diversity (the ratio between regional and local species richness) can be high and greater than that in larger mainstem channels.

What contributes to this pattern? Three main factors may be involved. First, and probably most important in our view, is that the smaller sub-catchments of low-order headwater streams are likely to be environmentally more diverse (when viewed over the whole river system in question) than are the progressively smaller number of higher-order channels. For example, small-scale geological variation within a large catchment might produce a number of small headwater sub-catchments differing in water chemistry (e.g. basic versus acidic) and therefore different biota (see examples in Chapter 2, section 2.5.7). Environmental conditions further downstream are inevitably 'averaged out' as catchment size increases, and tributaries join together and the biota then becomes less distinct than in individual headwater streams. In environmentally more uniform catchments this is less true, as the biota of small tributaries is also likely to be more uniform in composition and there is less variation among them—producing a phenomenon known as 'nestedness' (where the fauna of tributaries is merely a subset 'nested' within the larger community downstream; Figure 6.5). An increase in nestedness can also be caused by homogenisation of conditions and biota resulting from anthropogenic stress, as found by Larsen & Ormerod (2010) in a comparison of the benthos of streams in Welsh catchments converted to agriculture (with its associates stressors) with that in semi-natural catchments.

A possible second factor accounting for the high β diversity of small headwater streams is isolation by distance (via a fully aquatic route), presumably making it difficult for species to disperse among streams, and thereby leading to potential differences in species complement (e.g. Clarke et al. 2008). We are somewhat less convinced here, since small headwaters are relatively common in the river network, and can even act as 'stepping stones' over the generations (actually easing dispersal, at least for species that can disperse over land, particularly flying insects). Large channels are rarer, even though they are well connected via aquatic routes from upstream. Therefore, species restricted to the most downstream reaches may face long 'commutes' to find similar conditions elsewhere. We await further evidence for both these putative mechanisms (diversity of conditions and dispersal constraints).

A third possible reason for the high diversity of headwaters lies in the fact that the tributaries may constitute a greater surface area of habitable channel, and certainly have a greater overall combined length, than higher-order channels. Hence the concept of the species–area relationship potentially contributes to the high diversity of the lower-order channels overall. This touches on the biogeography of streams and rivers, which we discuss in the next section.

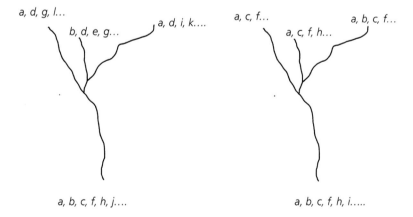

Figure 6.5 A simple river network with one downstream community (letters represent species: a, b, c etc) and three tributaries. In both networks, α diversity (here simply species richness) is greater in the larger stream. The network on the left shows low 'nestedness'—the headwater assemblages are not simply subsets of species also occurring downstream. β diversity (turnover) among headwater sites is then relatively high. The network on the right shows greater nestedness—headwater assemblages are subsets of the species occurring downstream and β diversity is low.

6.3.3 Larger-scale patterns in space

Patterns in species richness and composition are found at much larger spatial scales than we have discussed up to now, and are the province of biogeography (e.g. see Chapter 3, Figure 3.8 for freshwater bivalves, and Figure 3.17 for freshwater fish). They involve global and continental differences in the representation of taxa (including higher taxa) and are attributable to long-term and large-scale physical and evolutionary processes. Nevertheless, these patterns are important in contemporary community ecology, mainly because of the relationship between regional and local species pools. Thus, if local community richness simply increases in proportion to regional species richness, we can infer that those local communities are not 'saturated' with species (i.e. local dynamics do not preclude further species from entering the community). On the other hand, if local species richness reaches an asymptote despite further increases in the size of the regional species pool, it is likely that species interactions are preventing further colonisation by new species.

One of the most pervasive patterns in biogeography is an increase in the diversity of many groups towards lower latitudes (i.e. nearer the equator) (e.g. Rohde 1992 and many others). There is no single explanation for this pattern and there is some question as to whether it holds for lotic organisms (Boyero et al. 2009). Pearson & Boyero (2009) compiled data on the diversity of seven freshwater animal taxa and plotted these against latitude (having allowed for differences among regions in the area sampled). For several taxa, and probably for aquatic species overall, diversity was indeed greatest at low latitude, as is usual for most terrestrial and marine groups. For dragonflies, fish and anurans (frogs and toads) there was a strong tendency towards higher diversity nearer the equator, but caddisflies and salamanders showed no latitudinal trend and diversity was actually higher at higher latitudes in mayflies (Ephemeroptera) and stoneflies (Plecoptera) (Figure 6.6). Although the data are somewhat variable in quantity (we know far more about temperate than tropical diversity in many cases), it seems that

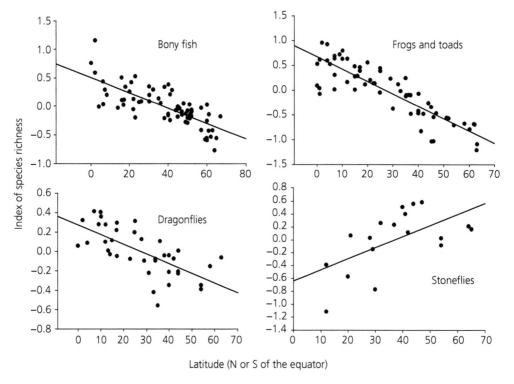

Figure 6.6 Diversity–latitudinal gradients for four groups of freshwater animals (bony fish, Osteichthyes; frogs and toads, Anura; dragonflies, Odonata; stoneflies, Plecoptera). Diversity is represented by species richness adjusted for sampling area.
Source: from Pearson & Boyero 2009, with permission from the University of Chicago Press.

not all groups become more diverse towards the equator. Boulton et al. (2008) and Pearson & Boyero (2009) pointed out that semi-aquatic groups with a consistently long aquatic phase in their life cycle and brief terrestrial lives (Plecoptera and Ephemeroptera) were more diverse in temperate than tropical areas, whereas several species-rich groups of invertebrates which were more diverse in the tropics (Odonata, in particular) tended to have a fairly long terrestrial phase. We can only speculate why this might be the case.

Perhaps not surprisingly, far less is known of latitudinal or other large-scale biogeographical patterns in the diversity of microorganisms, but the adoption of new molecular techniques for evaluating biodiversity is beginning to rectify this. For instance, the diversity of soil fungi peaks in the tropics (Tedersoo et al. 2014). There is no similarly detailed latitudinal assessment for aquatic fungi, although Seena et al. (2019) assessed fungal richness on discs cut from alder (*Alnus glutinosa*) leaves placed in 19 streams at latitudes ranging from 69° north (Norway) to 44° south (New Zealand). These data suggested that richness peaks at mid-latitudes around 40–50° north and south and appears to be lower in the tropics, although the distribution of sites is very uneven which might bias the results (Figure 6.7). Distinguishable communities were found in different water temperature bands (the most diverse at intermediate temperature (8.9–19.8°C). Whereas soil fungal diversity peaks in the moist tropics, and water stress

is a key factor, this is less likely to be true of fungi in permanently flowing streams.

Regional patterns in diversity often reflect environmental history, particularly the extent of past glaciations and, on a longer time frame, continental drift. Molecular tools also make it possible to assess subtle aspects of genetic diversity in addition to simple species richness as well as phylogenetic relatedness and the timing of the divergence of species. For instance, Pinkert et al. (2018) examined the distribution of dragonflies across Europe using both conventional species identifications and molecular assessments of phylogenetic diversity of the 122 species found there. In Europe (latitudes ~35–70° N), species richness peaked at central latitudes, was very low in the far north (particularly above the present 0°C isotherm— where annual mean temperature is below freezing), and also declined in the south. However, various measures of phylogenetic diversity (essentially the variety of different lineages present in one area) declined monotonically from the warm south and south-east to the cold north and north-east, diversity being particularly low in areas north of the position of the 0°C isotherm at the time of the last glacial maximum (about 21,000 years ago). Further, the ratio of lentic to lotic species (with larvae inhabiting mainly still and running water, respectively) changed sharply along the latitudinal gradient, with a preponderance of lentic species in the north and relatively more lotic species in the south.

It is believed that the dragonflies originated in the tropics, where they are most diverse, and that those in the northern hemisphere were confined to relatively warm southern refugia during the time of the glacial maximum. A few lineages of lentic species that were apparently able to tolerate lower temperatures moved north as the climate ameliorated after the last glaciation and underwent a temperate radiation. Just two speciose groups (Coenagrionidae and Libellulidae, the largest and youngest families of dragonflies) account for the peak in *species* diversity at mid-latitudes in Europe (and in North America). Nevertheless, overall *phylogenetic* diversity remains lower in mid-latitude Europe than it is further south. Environmental history and dispersal ability appear to have driven the present-day patterns of diversity in dragonflies. A few lineages had the ability to adapt to low temperatures and this trait has been phylogenetically conserved (remains restricted to those lineages). It also seems that lentic species are more effective colonisers than lotic species, as might be expected of taxa occupying truly patchy, divided and often temporary habitats. An increase in the representation of lentic species with

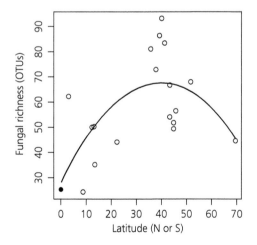

Figure 6.7 Latitudinal–diversity relationship for fungal assemblages developing on standard discs of alder leaves in 18 streams between 69° north and 44° south. One site on the equator (solid symbol) is in Equador.

Source: from Seena et al. 2019, with permission from Elsevier.

increasing latitude was also found by Hof et al. (2008), who assessed information for 25 biogeographic areas of Europe, over 14,000 species and more than ten major taxa (including disparate groups such as fish, water mites, insects and rotifers, amongst others) (Figure 6.8), a situation they also ascribed to the greater dispersal ability of lentic species.

A consequence of some of these large-scale patterns of diversity in lotic animals is that they point to possible geographic differences in the representation of functional feeding groups and therefore in stream ecosystem processes (Chapter 8). Thus, many of the cool-adapted lotic insects found in temperate streams, including some Plecoptera, which are important shredders of coarse organic matter, are absent from tropical streams. Does this signal that the shredder guild, and thus the breakdown of leaf litter by animals, is less important in the tropics than in the temperate zone (e.g. Irons et al. 1994; Dobson et al. 2002)? Boyero et al. (2012) addressed this issue by assembling data for stream-dwelling shredders at 156 local sites from 17 regions (mainly large parts of individual countries plus some states of larger countries) at latitudes ranging from 67° north to 41° south. They found that the species diversity of shredders at individual sites (i.e. α diversity) did indeed increase with latitude in both the northern and southern hemispheres, although overall regional (γ) diversity did not, implying greater species turnover among sites in the tropics (greater β diversity). Furthermore, the relationship between mean α diversity and regional diversity differed between temperate and

tropical regions. The number of shredder species per site did not increase in the tropics when more than 15 were present in the regional species pool, but this did not seem to be the case in temperate regions, where the number of shredder species per site continued to increase. These contrasting scenarios might indicate that species interactions in the tropics, but not in the temperate regions, limited local (per site) diversity.

Although Boyero et al. (2012) found that the numbers of shredder species did not differ between the tropics and temperate regions overall, differences may occur in the representation of particular taxonomic groups. In a tropical study, Yule et al. (2009) sampled leaf packs from 12 stream sites in peninsular Malaysia at altitudes ranging from 55 to 1,560m above sea level. Packs from the cooler highland sites were dominated by a shredder fauna (nine to 15 species per site) rather similar to that in the temperate zone (cased caddis of the genus *Lepidostoma*, stoneflies and tipulid fly larvae), plus some semi-aquatic cockroaches. In contrast, packs from the warmer lowland streams had fewer shredder species per site (three to eight) and they were predominantly different species—crabs, snails, cockroaches, tipulid larvae and calamoceratid caddis larvae. Some of the lowland shredders were large-bodied species, including snails of the genus *Brotia*, which are voracious shredders up to 7 cm long. Overall, the studies of Boyero et al. (2012) and Yule et al. (2009) suggest that although the shredder fauna of tropical streams may be less diverse than in temperate ones, it differs in taxonomic composition, is more variable from stream to stream and includes a larger fraction of larger-bodied species that lack a flying stage. It therefore seems unlikely that shredding is more limited in tropical-stream food webs, although it may be particularly vulnerable to local loss of species. We examine the further consequences of these kinds of differences in shredder diversity and taxonomic composition on organic matter dynamics in Chapter 8 (section 8.5.5).

6.4 River communities in time

We now turn to the temporal aspects of community structure in rivers and streams. How persistent or variable are lotic communities over various timescales? This question is of rather more than just academic interest as community structure (particularly of benthic animals and plants) is widely used as a key feature of the assessment of ecological status of rivers and streams (see Topic Box 6.1). In doing so, we tacitly assume that

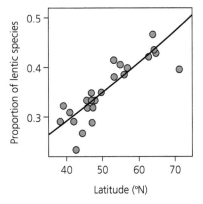

Figure 6.8 Lentic (standing water) species make up a greater proportion of the overall freshwater fauna as we move north in Europe. The data span 25 regions and over 14,000 species of freshwater animals from taxa ranging from rotifers to fish. *Source:* from Hof et al. 2008, with permission from John Wiley & Sons.

the spatial environment/community structure signal is robust and reasonably stable. Further, assessments of the persistence of natural communities are necessary if we are to be able to distinguish fluctuations which might be termed 'normal' or 'natural' from directional or persistent changes. Substantial and longer-term shift in species composition and/or relative (or absolute) abundance signal the formation of a new community that will not necessarily return to the former long-term average condition (e.g. Matthews & Marsh-Matthews 2016). The latter might well occur as a result of large-scale anthropogenic activities, such as those that have brought about long-term increases in fixed nitrogen and other nutrients, increased fine sediment loads, acidification and, of particular contemporary interest, climate change (see Chapter 10).

6.4.1 Persistence and change

A key initial question is: how persistent are lotic communities in relation to a 'natural' disturbance regime, and in catchments where human perturbations are modest? This mirrors the debate over the role of density dependence in regulating populations, within limits, around some steady density. Are natural communities essentially persistent and stable (remaining around some consistent average condition, in terms of species composition and abundance) or do they fluctuate stochastically (with frequent and apparently random changes in species complement and rank abundance)? There are still relatively few long-term studies of lotic communities that rely on consistent methodology and have produced data of sufficient quality to resolve this issue reliably, but there are some.

There have been a number of assessments of fish assemblages in rivers, some of which have come down unequivocally on the side of the stochastic view. Grossman et al. (1982) examined the dynamics of a fish assemblage in Otter Creek, a stream in Indiana (USA), over a 12-year period, and found a 'total lack of persistence for the ranks of species abundances and of . . . trophic groups for all seasons'. That is, common species (or trophic groups) on one sampling occasion were often rare or absent at other times, and *vice versa*. They suggested that such apparent stochasticity in this community was likely to be characteristic of other assemblages in streams and rivers where there were unpredictable environmental conditions, such as floods and droughts. These data for Otter Creek (1962–74), were subsequently compared with information on two other well-characterised stream fish assemblages—Sagehen Creek, California (1951–61)

and Coweeta Creek, North Carolina (1984–95) (Grossman & Sabo 2010). The assemblage from Otter Creek emerged as the least persistent of the three, while that at Sagehen Creek was the most persistent. These community data varied consistently with measures of predictability in the timing of catastrophic (high and low) flow events (most predictable in Sagehen Creek) and with other characteristics of flow magnitude. Thus, there may be an inverse relationship between the importance of deterministic factors in fish communities and environmental 'harshness'.

In a persuasive study, Matthews & Marsh-Matthews (2016) showed that fish communities in upland streams (the Piney Creek catchment of the Ozark Mountains) in Arkansas (USA) were in what has been called 'loose equilibrium'. This does not infer a fixed single state, but suggests community changes are limited to a somewhat broader, bounded 'space' (as in an ordination, for instance) within which they move about and to which they return after perturbation. Their study covered 42 years in all, so was genuinely 'long term', and included 12 individual stream reaches, each surveyed 11 times (giving 'decadal' scale data), and a subset of five of these sites was surveyed 16 times. The study period encompassed two extreme flood events and one 'exceptional' drought, and the catchment was only slightly affected by human activities and perturbations. At both scales of this study (i.e. more sites censused less frequently vs fewer sites censused more frequently), communities remained broadly similar in terms of species composition and relative abundance (i.e. common species remained common throughout), and a persistent longitudinal distribution of species was found along the stream continuum (with occasional 'strays'). Overall, these and other case studies suggest that lotic fish assemblages are in a loose equilibrium, at least when viewed at a sufficient spatial scale. In local sites subjected to a regime of frequent and severe natural disturbances (such as the Otter Creek example, above), species representation must depend on dispersal and recolonisation from refugia elsewhere in the catchment, and at the smaller local scale assemblages may therefore be more stochastic.

Other than fish, the persistence of communities of benthic invertebrates was assessed in one fairly early study by Townsend et al. (1987), who sampled 27 stream riffle communities in 1976 and again in 1984—that is, two 'snap-shot' samples eight years apart. The sites were on first- to fourth-order streams, varying in pH and size, in one small area of south-east England. Persistence varied among streams, whether measured as species composition, rank abundance of species, or by an ordination of sites on the two occasions, but was consistently greatest in small (low discharge) headwaters that were cool and acidic. In particular, samples from sites of similar pH but from different years were close together in ordination space, which is reassuring in terms of using invertebrate assemblages to indicate environmental conditions. More data are necessary to assess persistence under various conditions and are becoming available at the time of writing, particularly from various monitoring schemes.

One example compared a long-term (> 30 years) study of benthic invertebrates at a fairly small number of sites (10 streams in the Welsh uplands, assessed annually) with regional assessments (three sampling occasions between 1984 and 2012) of diversity at 58 streams spread over a larger area of Wales (Larsen et al. 2018). There was no change in taxonomic diversity (or 'functional diversity', assessed from the occurrence of particular species traits) at either spatial scale (neither α nor β diversity) in the study, suggesting a remarkable degree of persistence. Nevertheless, changes in species composition appeared to indicate some decline in 'mean community specialisation'. This latter was essentially measured first as the average, over all taxa, of the number of modalities of each trait represented in that taxon. The traits included features of body size, length of life cycle, modes of locomotion and feeding, with a specialised taxon scoring highly for its affinity to single categories (e.g. large body size) of the various traits. A second measure assessed the range of environmental conditions (nutrients, pH etc.) under which taxa were found—a wide range for generalists. Larsen et al (2018) suggested that the decline in specialisation reflected a relative increase in the incidence (abundance) of generalists over specialists, although this was masked overall by a lack of change in taxonomic and functional richness.

In another study based on monitoring data, Floury et al. (2018) analysed an impressive data set from 305 sites on rivers throughout France, each of which was sampled 15 or 16 times within the period 1992–2013. They used trait data to define two 'functional indices' that assessed vulnerability to climate change and degree of feeding specialisation, arguing that climate and productivity are the two most important environmental drivers of species diversity. In this case, trends in both indices increased, indicating an overall increase in functional diversity. Geographical trends were also found in the data, with climatic vulnerability decreasing at lower latitudes—that is, the proportion of species vulnerable to climate change declined in the south but increased in the north, and

feeding specialisation increased most in headwaters. The apparent decline in the proportion of specialists in the study of Larsen et al. (2018), but an apparent increase in functional diversity in that of Floury et al. (2018), appear to be contradictory, and are difficult to reconcile, though many factors could play some role. For a start, the environmental stressors at work in the two areas were somewhat different (acidification and a decline in acidification being particularly prominent in the Welsh streams), the geographic scale was much greater in the study of Floury et al. (2018) and the analytical approaches differed as we have seen.

Overall, invertebrate communities in rivers seem relatively persistent where there is no marked and ongoing environmental change. In the face of natural short-term environmental disturbances these communities seem resilient, as we discuss below (section 6.4.2.2). Where an environmental change is prolonged and directional, however, river communities are highly sensitive and respond. Undoubtedly, the most apparent causes of change in river communities, ultimately caused by a burgeoning human population, are organic loading by sewage, animal wastes and plant nutrients. Organic pollution of waterways has been a more or less global phenomenon, but has occurred at different times in different areas, depending on their history of population growth, urbanisation, industrial development and intensive agriculture. In more developed countries, crude forms of domestic and industrial pollution have been addressed by effluent diversion and treatment at vast expense, largely on the grounds of safeguarding human health. As an example of a country which has among the earliest history of industrial development and rapid urbanisation of its population, the UK has some of the longest-term records of the chemical and biological state of rivers, in some cases dating back to the mid- to late-19th century (e.g. Woodiwiss 1964; Langford et al. 2009). A partial recovery of benthic macroinvertebrates from the grossly polluted state of the industrial revolution has been little short of spectacular, as seen in the River Tame (Figure 6.9). Although industrial and urban/suburban rivers of densely populated areas have not achieved 'high' ecological status, and there are some more recent indications of reversal in condition (due to a lack of investment in infrastructure, despite increased demand), over the longer timescale UK rivers at least have shown substantial improvement from conventional sources of urban pollution. A caveat to this comes from a large-scale survey of data from

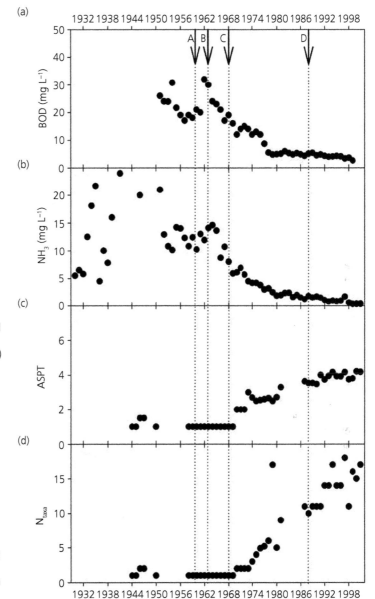

Figure 6.9 A record of biological and chemical parameters related to industrial and urban pollution in an English Midlands River (the Tame) from around 1930 to 2000. It had been grossly polluted during the industrial revolution in the 19th century and remained so between 1930 and 1950, when ammonia concentrations sometimes exceeded 20 mg L^{-1}. Chemical determinands are biochemical oxygen demand (5-day BOD) and ammonia concentration. Indicators of biological quality are average score per taxon (ASPT, a measure of the sensitivity to organic pollution of different invertebrate taxa within the site) and the number of taxa (N$_{taxa}$). Measures that reduced pollution in the river are marked by vertical arrows: A, cessation of coal gas production; B, implementation of the Rivers Act; C and D, start and completion of settlement lagoons for pollution mitigation.
Source: from Langford et al 2009, with permission from Elsevier.

British rivers by Whelan et al. (2022). They suggest that, indeed, conventional markers of organic pollution, including faecal organisms, biochemical oxygen demand, ammonia, plus heavy metals and catchment acidification have shown improvement from industrial revolution maxima. However, some other pollutants have shown deterioration. These include 'personal care products', pharmaceuticals, nitrogen and phosphorus, and plant protection products. These in some cases reflect a great increase in use and disposal through

sewage works, plus diffuse pollution via agricultural intensification. We return to pollution issues in Chapter 10.

Gradual improvements in water quality in the older industrialised countries of Western Europe and in North America (at least from conventional urban point-sources), have been ongoing and have been extensively monitored by various environmental agencies. For instance, Vaughan & Ormerod (2014) analysed a remarkable data set on benthic macroinvertebrates (identified to family level) from > 2,300 rivers across England and Wales between 1991 and 2011. The most obvious 'signal' in these data has been an increase in the prevalence of most taxa in response to improving water quality. The authors had expected ongoing climate change to counteract improvements in water quality, but, as of 2011, changes in water quality had masked any deleterious effects brought about by a warming climate at that time. In an equivalent study from France, Van Looy et al. (2016) analysed data from 91 sites over 25 years (1987–2012), a 35-year data set from two rivers each with seven sites, and 51 annually sampled 'reference sites' largely free from local human influences and changes in water quality, but not from possible climate change. They too found a substantial increase in taxonomic richness (24% on average), which could be largely attributed to improvements in water quality. Unlike Vaughan & Ormerod (2014), they also found evidence of a more abrupt shift in species diversity starting in about 1990, which was associated with climate warming and had induced an increase in water temperature and productivity. The climate-induced effects on the benthic communities could be distinguished because water quality did not change at the reference sites whereas water temperature did. This latter effect actually reinforced the positive effects of water quality on richness rather than opposing them, at least given the relatively modest increase in temperature at the time of the study. To demonstrate even further the value of harnessing sustained, professional-standard monitoring over long periods, a further recent French study revealed decadal trends in water quality and biodiversity of river biota at over 200 river sites throughout France (Tison-Rosebery et al. 2022). Nutrient loadings have declined on average, and the richness of diatoms declined (particularly that of planktonic rather than benthic diatoms) while richness of macroinvertebrates increased. This supports the view of relative community persistence in the face of discrete disturbances but shifts and changes in the community in the face of sustained environmental change (as in water quality in this case).

The gradually warming climate also seems to have had effects on the invertebrates and fish of the River Sâone in France. Daufresne et al. (2003) found that the upper Sâone had warmed by about 1.5°C over a 20-year period to 1999, an increase that was at least partly attributable to a warming atmosphere. They also found evidence of replacement of northern, cold-water fish species, such as the dace, *Leuciscus leuciscus*, in the upper reaches, by more warm-adapted species including the chub, *Leuciscus cephalus*, that

had only been present downstream. The invertebrate fauna also showed equivalent changes with greater numbers of species tolerant of warmer water and incidentally of more eutrophic conditions. These faunal changes seem to be a response to a sustained change in conditions induced by climate warming, although perhaps enhanced by the effects of nuclear power plants (that release warm water) being built upstream and modifications of flow due to the construction of dams. Regardless of the primary source of river warming, the influence of temperature on the fauna is clear.

While it is evident that river assemblages shift in response to directional and sustained environmental change, biotic responses are not always immediate and can often be delayed. For example, Langford et al. (2009) found that recovery from pollution was most delayed at sites remote from a source of 'clean-water' colonists and, like a number of other authors, invoked restrictions in dispersal of potential colonists as a likely explanation. The onset of air pollution from the industrial combustion of fossil fuels, particularly coal and oil, which produce oxides of sulphur and nitrogen, also resulted in widespread changes to the water chemistry and biota of rivers and streams in north-west Europe and north-east North America. Salmonid fisheries were lost in southern Norway (and elsewhere), where soils and rivers were very low in base cations that could neutralise the acidity (see review in Hildrew 2018; and Figure 5.10b), and there were extensive losses of acid-sensitive species in groups ranging from microbes to streamside birds. International agreements to reduce such emissions have reduced acidic deposition to the landscape, and in turn acidity in rivers and streams has ameliorated. It was expected that restoration of 'clean' water would be followed by a reversal of ecological damage, but this has again been substantially delayed. In one, well-studied acidified stream in south-east England, there was no simple return of acid-sensitive species, such as herbivorous mayflies and other species feeding low in the food web. Rather, there was a series of irruptions of successively larger-bodied invertebrate predators over several decades, as the stream deacidified (see Figure 6.10). This eventually culminated in a return of brown trout (*Salmo trutta*) as the top predator. Such changes at the top of the stream food web, rather than lower down, had not been predicted and are an example of an 'ecological surprise' in the course of community change. Its cause is still uncertain.

Some kinds of environmental change are not continuous but involve climatic cycles that can drive similarly

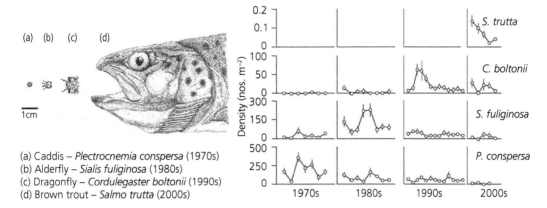

(a) Caddis – *Plectrocnemia conspersa* (1970s)
(b) Alderfly – *Sialis fuliginosa* (1980s)
(c) Dragonfly – *Cordulegaster boltonii* (1990s)
(d) Brown trout – *Salmo trutta* (2000s)

Figure 6.10 Patterns of predator abundance in a southern English stream (Broadstone) that from the 1970s as acid deposition declined. The left-hand panel shows the head widths (to scale) of four successively larger predatory species (three invertebrates, one fish) that were dominant in four successive decades. The right-hand panel shows their average seasonal population densities in the same four decades.
Source: after Layer et al. 2011, with permission from Elsevier.

Figure 6.11 The relationship between the North Atlantic Oscillation (winter index, circles) and mean invertebrate community similarity (Jaccard's Index, triangles, estimated between pairs of neighbouring years) in eight Welsh streams. A high value of the NAO indicates wet, stormy winters, while a Jaccard's Index of 1 is obtained for identical communities. Data for 1991 in the plot of Jaccard's Index were interpolated (indicated by the dashed line).
Source: from Bradley & Ormerod 2001, with permission from John Wiley & Sons.

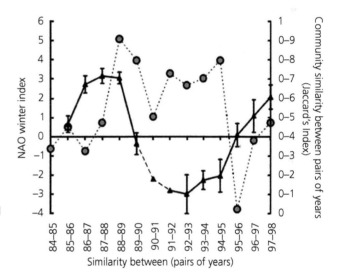

cyclical changes in river assemblages. One prominent climatic cycle is the North Atlantic Oscillation (NAO), a fluctuation in the air pressure difference between Iceland (north-east Atlantic) and the Azores (towards the mid-Atlantic, south-west of southern Portugal). The NAO has an approximately decadal cycle and drives weather patterns in western Europe, being associated with cold, dry winters when pressure is relatively low in the Azores (the index is negative) and stormy, relatively mild, wet winters (when the index is positive). Bradley & Ormerod (2001) compared the NAO index from 1985 to 1998 with between-year community persistence in the benthos of eight streams in south-west Wales (Figure 6.11). They found an intriguing,

emerging pattern of apparently lower year-on-year community persistence (in species composition and rank abundance) when the index was positive (mild, wet and stormy winters), and perhaps therefore associated with higher winter flows. The comparisons are short term, however, and the exact mechanism producing this pattern was not clear.

6.4.2 Disturbance in a community context

The search for long-term patterns in assemblages, dealt with in the previous section, looked at what might be termed shifts in 'average' conditions and largely gradual changes. Environments change not only in that way, however, but also in the frequency and severity of more extreme shorter-term events, to which communities respond. We saw in Chapter 5 how short-term events can be a source of disturbance and density-independent mortality at the level of the population. Not surprisingly, then, disturbance can also have a pervasive influence on multispecies assemblages, in terms of their species and trait composition. A key question is whether their community composition and trajectory are mainly products of such episodic environmental disturbances, or whether ecological processes (such as species interactions) in more benign intervening periods are more influential. Almost certainly, the influence of the two will vary depending on the assemblage concerned and the particular environmental regime—as in the fish communities of the streams compared by Grossman & Sabo (2010) and discussed earlier (section 6.4.1).

6.4.2.1 Defining disturbance

We have mentioned disturbance quite extensively already, particularly in the context of the lotic habitat templet (Chapter 2), adaptations and traits (Chapter 4) and density-independent factors in population dynamics (Chapter 5). However, it is at the community level that the development of ideas about disturbance has been of greatest interest and focus, and at this point we need to be more precise about its meaning. This is more easily said than done, and there have been a great many influential papers written by stream ecologists as they struggled to agree on some common meaning for the term as it applies to running waters, and to relate the concept to that adopted in ecology more generally. Even now, different views prevail. In their review of the development of ideas about disturbance, Stanley et al. (2010) pointed out that the prevailing view of communities before the mid-1980s was an overwhelmingly 'equilibrial' one, in which biotic interactions were seen as controlling community composition. An increasing appreciation of the role of disturbance in general ecology (e.g. Sousa 1984, Pickett & White 1985) then ushered in an era in which disturbances, particularly those associated with floods and droughts, were seen as dominant in the ecology of rivers and streams. Pickett & White (1985) defined disturbance as 'any relatively discrete event in time that disrupts the ecosystem, community, or population structure and changes resources, substrate availability, or the physical environment'. This definition applies to populations, communities and ecosystems, and indeed we discuss it in all three contexts in this book. Notice also that they defined disturbance as incorporating a

'biological response'—that is, if there is no biological effect of some physical event, there is no disturbance—and limited it to a 'discrete' excursion from average conditions. Thus, a disturbance in the eyes of Pickett & White is more or less episodic and is interspersed with longer periods without disturbance. Of course, the term 'discrete' is itself open to interpretation and is a matter of timescale.

This basic definition has been adopted, broadened and even partially contradicted in various ways. In their discussion paper, Resh et al. (1988) tried to reach a consensus on the criteria defining a disturbance as applied to rivers. Their agreed definition is similar to that of Pickett & White (1985) except that it specifies the magnitude of the event and requires that it should be 'unpredictable' to be considered a disturbance. Their definition of an 'unpredictable event', for example an increase in stream discharge, was that it should lie outside a more or less arbitrary two standard deviations of the median value. Poff (1992) criticised this addition of a statistical criterion about the size of the event on various grounds, and stressed that disturbance is by definition an ecological event. Resh et al. (1988) argued for the 'unpredictability' criterion because they asserted that organisms could adapt to predictable events, via natural selection, given sufficient time. Of course, a more rapid response to disturbance in ecological time would be that different species could be recruited from the regional to the local species pool if they have ecological characteristics (combinations of traits) that fit them for that particular environment (including its disturbance regime). Further, we have pointed out previously that the *same* environment is likely to be 'perceived' differently by various fractions of the biota, partly determined by lifespan, body size and resistance traits (Hildrew & Giller 1994). At its simplest, very small, very short-lived organisms experience only a very minor part of the flow regime of a river during their individual lifetimes. For them, no particular level of discharge might be regarded as 'predictable'. In contrast, very large and/or very long-lived species will experience the full range of flows expressed in that system. Thus, different fractions of the whole community might be structured in different ways as the effective disturbance regime differs amongst these fractions. For this reason, attempts to define any particular event as a disturbance statistically (as in the argument of Resh et al. 1988) may be ineffective.

Lake (2000) proposed that the term 'disturbance' be used only for the (usually physical) event, and considered the ecological response to be its effect.

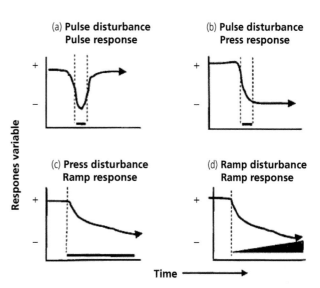

(a) Pulse disturbance
Pulse response

(b) Pulse disturbance
Press response

(c) Press disturbance
Ramp response

(d) Ramp disturbance
Ramp response

Respones variable

Time

Figure 6.12 The time course of disturbance and ecological response as proposed by Lake (2000). We generally restrict the term disturbance to describe a 'discrete' event (panel (a) here distinguished as a 'pulse'. Lake (2000) separated the physical event (the disturbance) and the ecological response, the two together being a 'perturbation'. The physical event in each panel is shown by the solid, bold horizontal bar on the *x*-axis. This is short (a pulse) in panels (a) and (b), prolonged in panel (c) (a 'press') and increasing in panel (d) (a 'ramp'). In Lake's usage the (ecological) response variable on the *y*-axis may also be a pulse (a), prolonged (press, (b)) or increasing (ramp, (c) and (d)).
Source: with permission from University of Chicago Press.

He proposed that the two together should then be called a 'perturbation'. This places undue limits on the useful general term perturbation. Various other definitions have also described disturbance as a 'discrete' event—an episode of limited duration with a definite beginning and an end. Lake (2000) elaborated upon this view by specifying the temporal course of both the environmental event and the ecological response, and describing disturbances as 'pulse', 'press' and 'ramp' events (Figure 6.12). A pulse describes a brief event with a temporary ecological response, as might be caused by a single flood pulse, while a 'press' is a more prolonged change with a gradual but persistent response such as the result of eutrophication or acidification. Finally, a 'ramp' disturbance gradually becomes more intense and prolonged—as in an extended period of drought—and the response also becomes gradually stronger. This does stretch the usual meaning of disturbance, perhaps rather too far.

In practice, ecologists are generally rarely so scrupulous in the use of such terms, particularly where words such as perturbation and disturbance have common English meanings; we need to be aware of this in reading the literature. Nevertheless, it is useful to understand that real ecological issues underpin the development of ideas in this field. Stanley et al. (2010) have traced the further use and development of the term disturbance and note that 'Many of us now use disturbance to refer to virtually any human activity that has a measurable effect on some facet of a stream'. This is true, but a pity, because by broadening the (ecological) concept of disturbance too far the word loses both precision and explanatory power. It clearly makes

sense to include human effects under the 'umbrella' of disturbance, where they share the characteristics of a natural disturbance—that is, where they can be characterised as 'discrete events', such as erratic pulses of pollutants cause by spillages. Where the effects are chronic (prolonged) and directional, however, as in the case of extended ramp and press changes, we feel an alternative term should be used, such as simply 'perturbation'.

6.4.2.2 Disturbance at the multispecies level

We should not let this wrangling over words distract us from investigating the real effect of episodic environmental fluctuations on ecological communities (and on populations and ecosystem processes—see Chapters 5 and 8). There are now some reasonably long-term records of lotic communities that have been subject to extreme events of various magnitudes, including high and low flows and episodes of high temperature. For instance, Woodward et al. (2015) analysed 13 years of macroinvertebrate data for the Glenfinish River in Ireland, during which there was one 'catastrophic' summer flood in 1986 (a one in 50-year event; Giller et al. 1991), followed two to three years later by two contrasting events of very low and high flow.

Despite these large disturbances in this Irish river, the invertebrate community was relatively persistent, with a core of 15 taxa being present throughout the record (35–45 taxa were there most of the time) with changes being mainly in relative abundance. Regular seasonal fluctuations in presence and abundance were the most prominent patterns. The single catastrophic summer flood reduced overall abundance of benthic invertebrates by about 95%, but most taxa returned to pre-flood densities in less than three years (Figure 6.13a). Taxon richness also declined to about 70% of pre-flood values immediately after the event. As we might expect, the larger-bodied, slower-growing species were most affected, while small-bodied species with short life cycles (such as midge larvae: Chironomidae) recovered very quickly. This scenario is largely in accord with known effects of population-level disturbances discussed in the previous chapter. However, it was also striking that some community-level metrics behaved rather differently. Thus, measures of taxonomic similarity between samples from adjacent months showed marked differences when the large flow event intervened between samples, although differences were even greater around the time of two combined events (one low-flow event and one flood) that occurred in 1989 (Figure 6.13b), but were less marked when 'functional' similarity (looking at the representation of different species traits in the community) was considered. These findings suggest, first, that there may be stronger effects of disturbance when contrasting events follow closely upon one another. Second, the relative stability of 'functional' attributes might suggest that there is some

'redundancy' in the representation of traits among species in the community: that is, species with similar traits can replace one another over time such that the overall representation of such traits changes relatively little.

A highly unusual opportunity to study the assembly of river communities in newly created streams was provided in Glacier Bay, Alaska (see Milner et al. 2000), as glaciers receded due to climate warming and formed new drainage channels fed by meltwater. A long-term study of Wolf Point Creek, a stream which first appeared in the mid-1940s, began in 1977 when the present catchment was still 70% glaciated. A succession of macroinvertebrate species colonised the stream over the subsequent decade after the initial dominance by chironomids. The riparian area of the catchment was colonised successively by mats of mountain avens (a ground-covering arctic-alpine flowering plant; *Dryas* sp.) and larger and longer-lived riparian trees, beginning with willows (*Salix* spp.) and alders (*Alnus* spp.). Migratory salmonids, Dolly Varden (an arctic char, *Salvelinus malma*), were first seen in 1987 and were followed in 1989 by pink salmon (*Oncorhynchus gorbuscha*) and coho salmon (*O. kisutch*). The development of the macrobenthos was followed throughout this period,

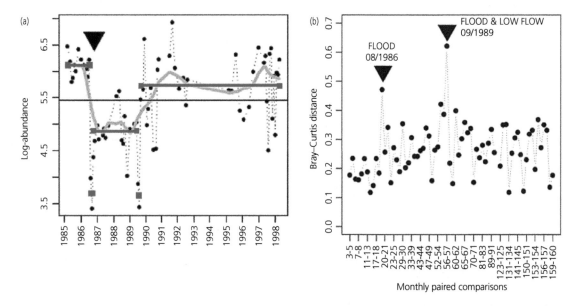

Figure 6.13 (a) A 13-year time series of macroinvertebrate density (log$_{10}$ scale) in the Glenfinnish River (Republic of Ireland). Significant break points are identified by horizontal red lines, the single black horizontal line shows the overall mean density over the entire period, while the grey line is a Lowess 'smoothing' function. The 1986 flood is identified by a black arrow-head. (b) Community similarity (Bray–Curtis distance, in which high values indicate greater taxonomic dissimilarity in community composition) between monthly samples over the 13 years. The flood of 1986 and the flood and low-flow events of 1989 are indicated by black arrow heads.

Source: Woodward et al 2015, with permission from John Wiley & Sons.

and of meiofaunal microcustacea from 1994, and the effects of floods on the invertebrate community monitored. An extreme winter flood with a recurrence interval of < 1 in 100 years in 2005 set back the development of the macroinvertebrate community by about 15 years, and it had not recovered 9 years later when it was set back even further by summer floods (Milner et al. 2018). The small-bodied and largely interstitial meiofauna was affected far less—another example of species traits apparently affecting vulnerability to disturbance. The timing of disturbance events was also important, subsequent recruitment to sea by pink salmon being much more affected by the winter flood, when eggs had been laid in redds, than by the summer floods, when eggs had yet to be laid. We can only speculate whether such a 'young' system (still developing both physicochemically and biologically after glacial recession) would be more or less susceptible to disturbance than longer-established ecosystems.

Recall that Daufresne et al. (2003) attributed biological changes in the French River Sâone to an increase in temperature over a fairly long period (a 20-year record to 1999). Subsequent surveillance of that river included the year (2003) of an extreme heatwave (at that time the hottest summer in Europe at least since 1500; Mouthon & Daufresne 2015), which constituted a true, pulsed disturbance to a system already stressed by more gradual warming, and was followed by a return of the river temperature to more normal values in subsequent years. These authors focused on the mollusc communities of the river and they had expected that their recovery would take a few years—judging by other examples of similar disturbances. However, recovery was not complete after eight years and it may be that, along with invasions by exotic species with high temperature tolerance, the community had entered into a different 'stable state'. This example shows that the resilience (the ability to recover) of river communities, while certainly a strong feature of their ecology, can be exceeded by particularly extreme events and if accompanied by species invasions.

Drought is another important form of disturbance, as indicated above, that has attracted increasing attention, particularly in parts of the world most prone to it, and in the context of projected climate change and human population growth. Drought is defined as a prolonged period (seasons to years) of low precipitation compared to the long-term average (e.g. Boulton & Lake 2007). Thus, a meteorological event leads to a hydrological drought in rivers and streams, where normal flow is not sustained. Hydrological droughts in running waters can also be caused by

abstraction (sometimes called water withdrawals), the construction and operation of dams or by land-use change, for instance where normal infiltration of rainwater into the soil and groundwater is reduced by impaction of the soil surface (the latter leading to brief but intense spates and prolonged low or no flows). Groundwater droughts arise when rainfall fails to recharge aquifers to normal levels, in turn leading to reduced flows in groundwater-fed springs and streams.

Droughts can stretch our view of disturbance as a 'discrete' event, although this of course depends on the timescale involved. Lake (2003) distinguishes two kinds of drought in fresh waters. The first are the predictable and regular seasonal events that occur in climatic zones where there are dry seasons of varying severity (e.g. in intermittent rivers of Mediterranean climates and the seasonal tropics). The second are much more gradual accumulations of rainfall deficits that may be progressive over several years— and may be termed supra-seasonal droughts. In terms of the time-course of the reduced discharge, seasonal droughts can be thought of as 'press' (sustained for a period) events whereas supra-seasonal droughts are 'ramped' (increasingly severe) perturbations. The regular occurrence of seasonal droughts places a 'filter' on the organisms that can persist at the affected local site, and well-adapted species, such as those with resistant phases in their life cycle, may not perceive the event as a disturbance at all (see Chapter 4, section 4.1.3). On the other hand, temporary colonists appearing during wet periods may be lost in the next dry season.

The effect of droughts on lotic communities, particularly invertebrates, has been well described (see e.g. Lake 2003; Boulton & Lake 2007; Lake 2011). A meteorological drought causes water to be lost progressively from a river channel in at least three ways as described by Lake (2003). Downstream drying (Figure 6.14a) occurs when perennial headwater springs (often from a 'perched' aquifer) keep flowing but water is lost further downstream, either by evaporation or percolation through a permeable stream bed. The springs can provide refugia for biota as long as they are sustained by ground water. In other channels, upstream flow is lost as the overall height of the water table declines and the 'spring-line' moves down the slope. Springs, sometimes locally called 'winterbournes', then may reappear further up the slope when rainfall resumes and the ground water is replenished (Figure 6.14b). In such situations, refugia for aquatic invertebrates may be in downstream perennial reaches, or perhaps deep in the stream bed. Yet other stream systems

Longitudinal drying patterns

(a) Downstream drying

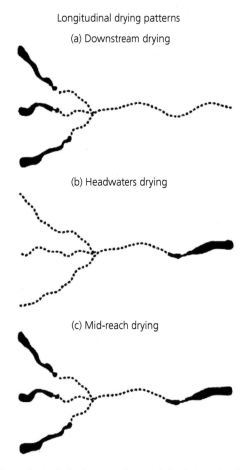

(b) Headwaters drying

(c) Mid-reach drying

Figure 6.14 Idealised patterns of stream drying during a drought: (a) downstream drying (upper stretches sustained by perennial springs), (b) headwater drying (surface flow retreats downslope as water table declines), (c) mid-reach drying (e.g. a permeable 'losing reach' in mid-channel occurs between permanent upstream springs originating from a perched aquifer, and the downstream water table). *Source:* after Lake 2003, with permission from John Wiley & Sons.

may have both perennial upstream springs and permanently flowing reaches much further downstream, while reaches between the two run dry during the drought but maintain flow at other times (Figure 6.14c; see Arscott et al. 2010 for the example of the Selwyn River, New Zealand). Variations on these three basic patterns may also occur in temporary river systems (see section 6.6.1)

Hydrological droughts have both direct and indirect effects on stream communities. Direct effects result from loss of habitat, as water recedes, and from associated physicochemical environmental changes, including reductions in water velocity, increased temperature, sedimentation as entrained particles settle out,

and reductions in oxygen supply. Indirect effects occur when there are resultant shifts in interspecific interactions, particularly predation and competition, or in food supply. These may occur when predators and prey become crowded together in a reduced area of habitat or, as expressed by Boulton & Lake (2007), forming a 'predator soup'. This wide range of direct and indirect effects manifests in losses of populations of susceptible species, as well as shifts in whole communities (in terms of species richness and body size, for instance), as species better able to live in warm, lentic conditions are favoured over those requiring cool, fast-flowing water.

Communities may pass through 'thresholds', as conditions become progressively more severe during droughts with consequent dewatering of channels. In a recent example, Aspin et al. (2018) simulated prolonged drought over 18 months in stream mesocosms representing lowland headwater streams. A novel aspect of the study was that they analysed the representation of invertebrate 'functional traits', rather than simply the species complement *per se*, as the mesocosms were subjected to progressive water loss. They asked whether drought 'pushed' ecological communities beyond critical thresholds at which a small change in conditions brought about a disproportionally large biotic response, and indeed found abrupt responses in the representation of many (12; 70%) of the 16 individual species traits analysed and in 'trait profile groups' (combinations of individual traits that commonly occurred together) (Figure 6.2, section 6.3). The responses of trait groups ranged from collapses in non-aerial dispersers, as water in the mesocosms fragmented to isolated pools, to eventual irruptions of small eurythermal (temperature-tolerant) dietary generalists as channel dewatering neared completion (Figure 6.15a–c) and a more terrestrial community began to invade. Note, however, that the response of biological communities to climate-driven changes in flow may not always be so neatly related to species traits. In a study of Australian streams in a mediterranean climate, some of which had switched from perennially to intermittently flowing, predictions of the fate of several benthic invertebrates based on life-history traits were not realised because timing and growth proved much more flexible than originally understood (Carey et al. 2021).

As rainfall resumes, recovery is normally quicker in seasonal than in supra-seasonal droughts (Lake 2003). Thus, recolonisation occurs either from spatial refugia or through organisms having complex life cycles, where resilience is provided by drought-invulnerable

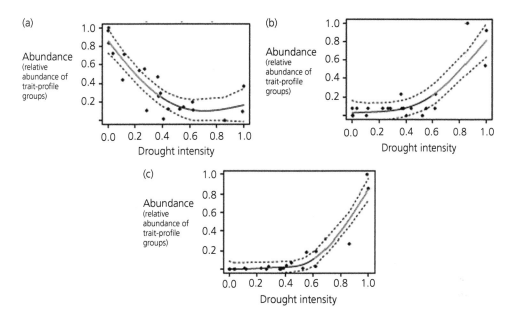

Figure 6.15 The effect of experimentally imposed drought in outdoor mesocosms (channels) on invertebrate community structure (trait-based): (a) tegument ('skin')-breathing aquatic dispersers, (b) spiracle breathers, (c), small-bodied, eurythermal generalists. A 'Drought index' (x-axis) measures the intensity of water withdrawal, ranging from sufficient flow to maintain the most wetted habitat to little or no surface flow. *Source:* from Aspin et al. 2018, published under a Creative Commons CC BY licence.

stages of the life cycle, such as terrestrial adults, seeds, spores or cysts (e.g. Stubbington et al. 2016 and Topic Box 4.1). Chapter 5 (Section 5.3.4, p. 168) gives a fuller discussion of the operation of refugia during and after disturbance. The rate of recovery after supra-seasonal droughts seems to vary widely among species and may also be rapid in some cases (for examples, see Lake 2003). However, where particularly vulnerable species are locally lost it may take some time for colonists to 'find' the newly restored habitat. In some cases, the community goes through a process of succession, in which (for instance) there may be interactions between rapidly colonising pioneers and slowly colonising but more competitive species.

We know most about the effects of drought on lotic animals, but in an unusual laboratory study on the effects of drying on the microbial component of stream communities, Gionchetta et al. (2019) subjected columns of sediment from a stream in Catalonia to three treatments. These were: continually wet, a prolonged drought (5 months) and a prolonged drought broken twice by temporary 'storms'. The dry treatments were subsequently rewetted and estimates were made of microbial assemblage structure on sediment at the surface and subsurface, and on plant litter at the surface. Rates of processing of litter by heterotrophic

microbes were also examined by measuring enzyme activities and decomposition. Microbial communities in the subsurface sediments survived best during prolonged drought, as they retained a little more moisture than sediment at the surface or leaf litter, suggesting that in nature the hyporheic zone could similarly be a refuge for microbes and promote overall resilience to drought. Perhaps not surprisingly, treatments in which the drought was broken by 'storms' recovered more quickly than those that were drier for longer, at least in terms of processing rates and bacterial viability.

A common pattern in the recovery of invertebrate communities after disturbances, or indeed after the creation of new habitats, is the 'short-term high abundance of otherwise rare species which briefly flourish and then disappear', which Lake (2011) saw as possible examples of disturbance-dependent fugitive species. We have already mentioned the midge *Orthocladius calvus* that rapidly but briefly colonised a newly exposed stream bed (see Chapter 5.) while some other midges (mainly of the genus *Zavrelimyia*) rapidly proliferated in the bed of a southern English stream upon rewetting after a seasonal drought before rapidly declining again as slower colonists increased in numbers (Ledger & Hildrew 2001).

Succession—which we can think of in this context as a more or less orderly and predictable sequence of species becoming established (even if only temporarily) in a community after disturbance—has been a controversial issue throughout the history of (mainly plant) ecology, though little of the evidence comes from freshwater communities, or from rivers and streams in particular. One early and influential example of community recovery and succession in a stream following a disturbance comes from the desert in the south-west of the USA (Sycamore Creek, Arizona; Fisher et al. 1982). Here, occasional flash floods erode organisms from the coarse, sandy sediment of the stream bed and their recovery was monitored following a summer flood that had virtually eliminated algae and reduced invertebrate biomass by almost 98%. A clear succession of algal assemblages was found as recovery proceeded. Thus, bare sand was replaced first by a flora of diatoms, followed by the filamentous alga *Cladophora* and its epiphytes, then by blue-green bacteria, and finally by a mat of *Cladophora* with associated epiphytic diatoms and blue-greens (Figure 6.16). These changes did not occur simultaneously throughout the 500 m-long reach studied, but in a series of patches which could switch from one state to another between sampling occasions (up to 63 days after the disturbance) and before winter flooding disturbed the stream bed once again. This appears to be a clear case of succession, though it is brief and truncated compared with the longer-term and more complete successional sequences in many

terrestrial systems or in the gradual establishment of a riverine community in the newly emerging pristine rivers in the Arctic we described earlier, this latter example also being accompanied by fairly rapid climate change (Milner et al. 2000).

6.5 Scale, space and time and river communities

So far, and for simplicity, we have tried to deal separately with spatial and temporal patterns in community ecology. However, rivers are not only inherently patchy, physically complex systems but also highly dynamic, such that the spatial arrangement of habitats is mobile according to fluctuations in river flow—that is, the spatial and temporal habitat axes interact. This inevitably is reflected in river communities, which consist of organisms that match this dynamism in various ways. This dynamism is an important feature of the 'habitat templet', a concept that recurs throughout this book. We also have to consider the importance of scale, both in space and time, as it affects the relationship between the habitat and organisms large and small (but in rivers predominantly small) and organisms with lives brief and long (but in rivers predominantly brief) (see Giller et al. 1994; Hildrew et al. 2007). Exceptions to this generalisation of small size and brief lives of river organisms, such as the freshwater pearl mussels (some with lifetimes exceeding 100 years) and the 'megafauna', naturally gain much attention. Neither should we give the impression that a dynamic habitat, perhaps inferring frequent disturbance, inevitably means that interactions between species are negligible in rivers. We have already seen that river populations can rebound quickly from disturbance, such that they can soon begin to face resource shortages, competitors and natural enemies. Here we begin by returning to perhaps the most obvious sign of spatio-temporal dynamism in streams and rivers, the fact that there are very large numbers of streams and rivers in which flow is not maintained through the year.

6.5.1 Intermittent river and stream communities

Probably 50% or more of the length of rivers and stream channels worldwide have intermittent flow, a figure which varies with the map resolution adopted (i.e. it increases as smaller and smaller channels are included) (e.g. Datry et al. 2016b, 2017a) and in different geographic areas. Intermittent flow will also become more common as humans withdraw ever

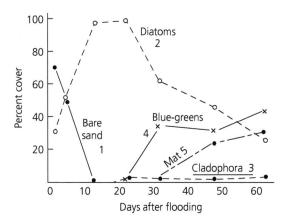

Figure 6.16 Percentage cover of different patch types of algae and cyanobacteria (blue-greens) following flooding in an Arizona desert stream). 'Mat' refers to a mixture mainly of the filamentous green alga *Cladophora* and blue-green bacteria.
Source: from Fisher et al. 1982, with permission from John Wiley & Sons.

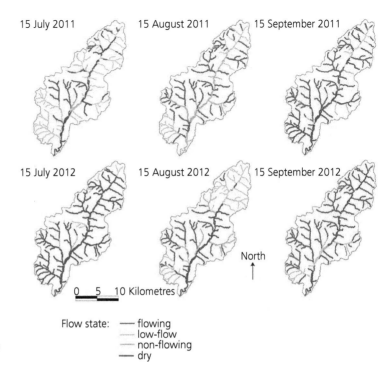

15 July 2011   15 August 2011   15 September 2011

15 July 2012   15 August 2012   15 September 2012

North ↑

0   5   10 Kilometres

Figure 6.17 An example of changing flow conditions in a partially intermittent river (the Tude, Poitou-Charentes, France).
Source: from Datry et al. 2015, with permission from John Wiley & Sons.

Flow state: —— flowing
                        ---- low-flow
                        ···· non-flowing
                        —— dry

more water, while climate change is likely to contribute via an increase in the frequency of extreme events probably resulting in more intermittent rivers. These habitats, which are temporarily 'disconnected' parts of river networks, are scattered in the landscape but also highly variable in time as the aquatic (flowing) phase comes and goes, ultimately following rainfall (Figure 6.17). As conditions change in stream reaches, from flowing to non-flowing to dry and back again, there are associated changes in the biota from purely lotic species, to lentic and terrestrial organisms (Figure 6.18). The community-assembly processes at work, primarily environmental 'filtering' and dispersal across the landscape, are then likely also to vary with conditions. For instance, species interactions (a component of filtering) may be particularly intense as lotic populations become crowded into drying pools, and as predatory species adapted to lentic conditions, such as some water beetles, colonise the channel and feast on the prey that are trapped.

A major question in the community ecology of rivers and streams is whether there are overall differences in diversity between systems with permanent flow and those with intermittent flow. Considering only aquatic biodiversity (there are also relatively poorly known assemblages of terrestrial organisms that are characteristic of dry and drying channels), Soria et al. (2017) found in a metanalysis of 44 published studies that the diversity of invertebrates, fish, macrophytes and diatoms was higher in perennial streams than in intermittent streams. These studies all had replicated data, although coverage was biased towards North America, Europe and Australia, with very few data from Africa or Asia and none from South America. There were also some exceptions to the overall pattern. Nevertheless, flow intermittence appears to be a significant feature that reduces overall freshwater biodiversity, which is likely to be further compromised as intermittent flow conditions become more widespread. However, the biota of intermittent systems also contains some specialists with adaptations that enable them to survive through dry periods or to recolonise rapidly (see Chapter 4, section 4.1.3, p. 103). The communities of intermittent streams exist in a patchwork of occasionally flowing reaches that may be connected (to each other or to more permanent reaches) by overland dispersal and during wet seasons or other hydrological events. From this particular situation, we now consider this spatio-temporal patchiness in river habitats and communities more generally, and at several scales.

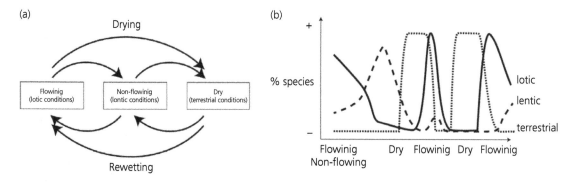

Figure 6.18 Conceptual figure illustrating: (a) alternating cycles of flowing, non-flowing and dry conditions in intermittent rivers; and (b) associated changes in relative abundance of lotic, lentic and terrestrial species.
Source: from Datry et al. 2015, with permission from John Wiley & Sons.

6.5.2 'Patchiness', scale and river communities

The terms 'patch' and 'patch dynamics' should carry a public health warning for ecologists and need careful use. At its least restrictive, patchiness simply and conveniently refers to the obvious spatial heterogeneity of habitats and living communities of any kind, and at any scale (whether viewed 'through a microscope' or 'from a satellite'). Thus, environmental patchiness is evident at many scales relative to that at which an organism interacts with, or 'perceives', its environment (Downes 1990; Hildrew & Giller 1994). In rivers, spatial patchiness is particularly obvious at all scales, and is also variable in time, driven largely by the dynamics of flow. However, its significance has been much debated in river ecology (e.g. Tokeshi 1994; Winemiller et al. 2010)—as it has in the general field of ecology.

6.5.2.1 Small-scale heterogeneity

We have already mentioned (Chapter 5) the extreme mobility of some benthic invertebrates, at least at small scales. As Townsend (1989) remarked, 'We have all been struck by the speed at which colonisation (of experimentally exposed substrata) occurs and by the recognition of an ever-present source of colonists'. This has led to the idea that what determines the species composition of small areas of stream bed at any instant is predominantly the largely stochastic processes of disturbance and dispersal, rather than biotic interactions. The 'continuous redistribution of the stream benthos' (Townsend & Hildrew 1976) has consequence for species interactions, including predation, and its detection at various scales (see Chapter 7). For the short-lived (typically one to a few generations per year) invertebrates concerned, an individual

of a mobile species will range quite widely over the stream bed during its aquatic lifetime, with uncertain consequences for overall community composition and persistence.

An example of this mobility is seen in an experiment by Connolly & Pearson (2018) who exposed streamside channels containing packs of tree leaves to colonisation by drift from an Australian stream over a period of 3–38 days. The leaf packs were colonised rapidly, the mean number of invertebrates (per replicate channel) reaching an asymptote of about 550 in only around 12 days, while taxon richness did so after 24 days when 21 taxa (the mean number per channel) were present (Figure 6.19 a, b). However, there was variation in taxon richness among replicate channels and a persistent gradual increase in the *total* taxon richness (i.e. the cumulative number of species observed, over all channels), reaching 50 at day 38, (Figure 6.19 c). Thus, while the total instantaneous number of taxa over the four replicate channels reached an asymptote of 32 taxa at day 12 (versus only 21 as a mean per channel), the total cumulative number of taxa recorded among all channels continued to increase with a mean of one new taxon added per three days for the final 26 days of the trial.

The total number of individuals and taxa thus apparently fitted a classical equilibrium model, which disguised a turnover of species with different colonisation rates. Numbers were broadly dominated by some highly mobile chironomid midge larvae and copepod species which arrived early and then declined as less mobile species, predominantly caddis larvae and mites, colonised leaf packs. The colonisation rate and time for mean richness to stabilise was typical of those found in these kinds of short-term experiments

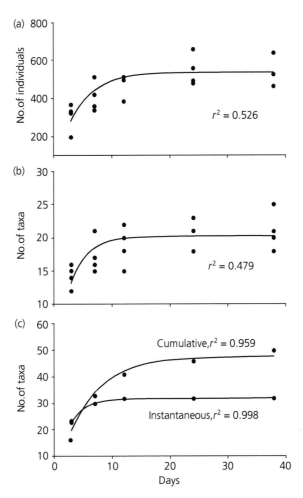

Figure 6.19 Colonisation by drifting invertebrates of experimental stream channels over 38 days: (a) the mean total number of invertebrates per channel, (b) mean taxon richness per channel, (c) total number of taxa over all four channels—'instantaneous' is the total number of taxa on each of the five sampling dates; 'cumulative' is the total number recorded over all channels over the entire period elapsed.
Source: from Connolly & Pearson 2018, with permission from John Wiley & Sons.

(Connolly & Pearson 2018). This again suggests that, at these small scales (both spatial and temporal), the stream zoobenthos is a mix of species consisting of early colonisers of newly available vacant patches (produced by disturbances of various kinds), often followed by rapid departure, and species turnover as less mobile (but speculatively more competitive?) species arrive.

At only slightly longer timeframes, some submerged macrophytes produce an annual fresh 'crop' of many, similar stems that can be colonised by small epiphytic algae and their invertebrate (midge larvae) grazers. These 'patches' (the apices of the plants) can be thought of as rather similar, ephemeral, divided habitats, that last about as long as the aquatic lifetime of the grazers

and on each of which individual larvae may live out their lives before emergence. Tokeshi (1994) described the situation of nine chironomid species coexisting on the apices of a water plant, the spiked water milfoil (*Miriophylum spicatum*), in a small English river. He took monthly samples of apices (the sampling replicate) of the plant and counted the number of larvae of each species per stem. His initial hypothesis was that if interspecific competition was strong, then the species, which are all grazers of epiphytic diatoms, should be spatially segregated among stems. That is, pairs of species should be found on the same stem less frequently than expected from a random model. Most species pairs did not segregate significantly on different plant stems when compared with a null ('neutral') model in which observed numbers of larvae (i.e. the number of individuals of each species found per stem) were simply reallocated stochastically among stems in the model (see Tokeshi 1994). The observed data (the real numbers per stem), however, showed that each species was mainly, and independently, aggregated among stems in nature, probably due to the initial laying of single egg masses by females, which by chance endowed pairs of species with a low level of spatial overlap. Added to this, the intervention of periodic disturbances, such as flow, resulted in density-independent mortality over the lifetime, leading Tokeshi to argue that interspecific competition for space was weak in such assemblages. A similar argument in the case of temporal segregation—that is, that interspecific competition had led to avoidance of overlap in the timing of maximum resource use among species pairs—was similarly dismissed. Thus, a combination of stochastic dispersal and mortality, together with some degree of independent patch formation, reduces interspecific competition in such systems and promotes coexistence. A similar situation of strong intraspecific (intrinsic) aggregation but much less common interspecific aggregation was found amongst macroinvertebrates colonising experimental leaf packs across a range of spatial scales (from several metres to kilometres) and different patch sizes along a 2 km section of a third-order stream (Murphy et al. 1998). We discuss this further in Chapter 7 (section 7.6.4, p. 249). Again, we have to conclude that assemblage structure and diversity is by no means universally related to interspecific competition in river communities. This may be particularly true of systems where habitat patches are similar, ephemeral and highly divided—as in these examples in macrophyte beds and leaf packs.

6.5.2.2 Larger-scale patch dynamics

The term patch dynamics was initially used in the context of sessile or at least sedentary organisms that occupy locations which were occasionally cleared by disturbance and were gradually replaced by a succession of species until something like the original biota was restored. The juxtaposition of such patches at different stages in succession leads to higher overall diversity than would otherwise be the case in a uniform environment (Pickett & White 1985). Key processes in patch dynamics of this sort act mainly over the generations, and there is time for a true succession of species to occur. Propagules arrive, populations grow and may then decline and be replaced by others before the next disturbance. A situation of this kind

applies to the attached algae and blue-green bacteria of Sycamore Creek, as discussed earlier in this chapter (Figure 6.16; Fisher et al. 1982). On a much grander physical scale (and for much bigger and longer-lived organisms), disturbances in the upper floodplains of the Amazon River produced by fluvial dynamics clear spaces by the scouring of rooted plants, and create a heterogeneous series of patches of woody vegetation at different stages of succession giving the floodplain forest enormous overall diversity (e.g. Salo et al. 1986).

Where river morphology is unaltered by humans, and there is a supply of sediment and large woody debris, the interactions between fluvial dynamics and woody plants can create an active floodplain of bewildering biological diversity, as well as physical complexity. This is the case in what is probably the last remaining, largely unaltered, Alpine river catchment, the Tagliamento River, where several species of willow (Salicaceae) act as 'ecosystem engineers'. The Tagliamento, which arises in the Alps of north-east Italy and flows 172 km to the Adriatic, includes stretches of wide (up to 1 km), active floodplain with many vegetated islands in a braided planform of multiple gravel channels, gravel bars and scour holes. Edwards et al. (1999) described the floodplain after an extreme flood as being littered with 'thousands of trees and other pieces of large woody debris'. These trees were the sites where succession occurred, leading to the formation of new vegetated islands. They accreted material from the flow, including sediment, plant propagules and further woody debris, sometimes growing to be substantial wooded islands in the floodplain, but also eroding at the margins to form deep, swiftly flowing scoured channels. Islands were thus habitat 'patches', which underwent succession via a process of 'facilitation' by pioneer plants. The main pioneers were willows which, through their growth characteristics and ability to resist flow, created the conditions in which other species, both aquatic and terrestrial animals and plants, could thrive. On 22 islands they found 162 plant species, (a mean of 26 species per island; mean island size 44 m^2). The interaction between fluvial dynamics and the willows that created the overall floodplain habitat enables the system to be reset periodically, by further scouring floods.

Similar floodplain processes were probably common elsewhere where mountains give way to flatter areas, often near the coast. However, most of those active floodplains have probably been lost due to management activities that have aimed to restrain the lateral flooding of rivers onto valuable agricultural land, often by 'snag removal'. This has transformed thousands of kilometres of braided rivers worldwide into single, constrained channels. A striking example provided by Sedell et al. (1982) is that 80,000 snags were removed from 1,600 km of the Mississippi River (USA) over a 50-year period from 1870, and many similar river works were carried out in forested areas following the advent of European colonisation in the United States and elsewhere. Comparable river transformations were carried out much earlier in Europe and much of Asia, where settled agriculture and high human population density have been present for much longer, and unsurprisingly, we have many fewer detailed records of the environmental degradation of the great rivers of the 'Old' World (e.g. Tockner et al. 2021).

6.6 The river hierarchy and 'metacommunities'

Many of the earlier studies in community ecology assumed that local communities were essentially closed systems that were isolated from each other. Within such communities, species were seen to interact, thereby affecting each other's rates of birth and death (Leibold et al. 2004). Population density was believed to be determined almost solely by births and deaths, rather than by immigration and emigration. In the previous chapter we saw, in contrast, how the physical structure of river networks divides up populations of lotic organisms into what can be regarded as 'metapopulations'—that is, made up of local populations, each having a finite probability of extinction, but also capable of being 'rescued' by dispersal from local populations elsewhere. This idea has been expanded to the community level, in which separate habitat patches may each hold a variable number, and a variable complement, of species. Each species has a finite probability of extinction from a local community but can again be 'rescued' by dispersal from other variously linked communities. Thus, even if the probability of local coexistence between any species pair is low (for instance if they compete strongly within the local community or if predation between them is severe), they may be able to coexist within a complex of local communities, which can be described as a 'metacommunity' (Leibold et al. 2004). The fact that local communities are embedded in a larger metacommunity affects their diversity and species composition, and in turn these local phenomena (dispersal, local extinction and recolonisation) may affect the much wider regional biota.

Leibold et al. (2004) discussed from a general theoretical perspective four 'modes' of metacommunity. Whether one or more applies depends on the nature of local sites (e.g. whether identical or different) and on species traits, for example how closely the species 'match' the conditions of the sites, whether or not they are equally competitive and whether they have similar or different dispersal abilities. Together, these factors will determine local and regional diversity and species composition and, at different sites, different processes may dominate. Whilst not specific to lotic communities, it is instructive to think about how each of these modes might reflect the situation in different riverine environments.

The first 'mode' or view is a *patch dynamics* perspective (though here taking a particular view of 'patches') (Leibold et al. 2004). Here the patches are local sites that are identical and sufficiently large to house multispecies assemblages of the organisms concerned. Patches may be occupied or unoccupied by any particular species and local extinctions can occur stochastically (e.g. from disturbances) or by deterministic interactions (e.g. competitive exclusion at the patch scale). In such systems coexistence of species can occur but requires a trade-off between competitiveness and dispersal rates (poor competitors are good dispersers), and local diversity is limited by dispersal.

The second mode is the familiar notion of *species sorting*, which we introduced previously in the context of overall community assembly. This can operate where there are qualitative differences among local sites (e.g. along a length of river or between separate tributaries within a catchment), and species have different competitive or survival abilities depending on the nature of the patch. Patch quality can then affect local species composition and each patch has its complement of well-adapted species. Dispersal is important in that it 'allows' appropriate species to reach sites that are suitable—in contrast, if there is insufficient dispersal, community composition may vary among patches that are of similar quality but differ in their remoteness from well-adapted colonists. A similar argument was used earlier in the context of 'filters' determining community assembly, such dispersal constraints often being termed 'spatial effects'.

The third mode is '*mass effects*'. Here, an uncompetitive or poorly adapted species can persist within a local community as long as immigration (from local sites where it does well) is sufficiently frequent to overcome its 'handicap'. This is sometimes called 'source–sink dynamics', inferring that some local communities will be consistent sources of immigrants of particular species, whereas others cannot sustain emigration and are therefore 'sinks' for the same species (e.g. Pulliam

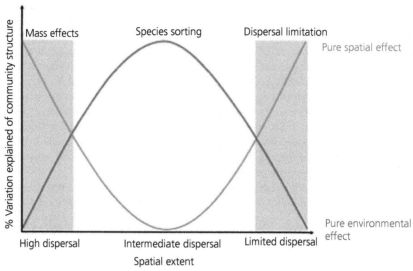

Figure 6.20 Schematic of basic metacommunity processes in river networks, suggesting how variation in community structure between sites might be explained with increasing intervening distances between them (x-axis, 'spatial extent'). At the largest, biogeographic scales 'spatial effects' predominate because there are severe dispersal constraints between sites. At extremely short intervening distances, high rates of mobility may homogenise communities (i.e. 'mass effects' predominate). At intermediate distances, species sorting (environmental effects) predominates because dispersal is sufficient for well-adapted species to reach suitable sites but not for poorly adapted species to persist simply because of rapid immigration. The exact scales at which these processes may operate will vary within taxa depending in species traits such as body size, lifespan and mobility.
Source: from Heino et al. 2015, with permission from John Wiley & Sons.

1988). In lotic ecology, stable headwater spring systems might be considered as sources for spring-adapted species whereas sites downstream might act as sinks.

Finally, Leibold et al. (2004) included '*neutral processes*' (Hubbell 2001) as a possible fourth mode, which sets out the consequences of a system of identical habitat patches and a set of species of equal competitive and dispersal abilities. This is a kind of null model, against which real communities could be assessed, and (according to Leibold et al. 2004) would eventually lead to a random walk to only a single species persisting in the regional community, unless balanced by speciation (or presumably by invasions from outside the region).

A great deal of attention is currently being paid to the assemblages of river systems as metacommunities and, in the light of the theoretical possibilities outlined above, much of the rapidly growing literature in this field of metacommunity ecology is concerned with using statistical methods to discern the relative influences of 'spatial' and 'environmental' processes on the composition of river communities for a range of taxa and at a range of scales. In particular, what are the possible effects of the network structure of

rivers on biodiversity (e.g. Brown et al. 2011; Altermatt 2013; Heino et al. 2015)? The basic proposition here is that the 'dendritic, hierarchical nature of river systems' imposes a divided structure on river communities and could mediate the relative influence of species sorting (i.e. 'niche-control' of ecological communities) and dispersal on biodiversity. Thus, if the dispersal of species is constrained (hindered) by the river network—for instance if dispersal can only take place within water courses—then there should be a strong 'spatial' influence on species composition and diversity (i.e. distant patches will be more dissimilar in species composition than neighbouring patches). Where patches more readily exchange colonists, then species sorting will prevail until mass effects begin to have an influence if the rate of immigration is sufficiently high. Expectations as to the relative effects of the spatial and environmental control of communities in relation to the spatial extent of a study are shown in Figure 6.20. Essentially, it is proposed that limits to dispersal across increasing distances (relative to the mobility of the species concerned) will eventually result in 'spatial' effects, whereas mass effects and source–sink dynamics will predominate over very small spatial scales; species sorting should prevail at intermediate spatial scales.

The metacommunity concept has been applied successfully by ecologists to situations where there are clearly separable and discrete habitat patches in the landscape, such as ponds, islands, forest fragments and rotting wood, with 'non-habitat' in between. River networks are rather different for a number of reasons (Brown et al. 2011). First, they are linear, hierarchically branching structures without the discrete patches of these other systems. Thus, the branches are connected by confluences (sometimes called 'junctions' or 'nodes') which are themselves potential habitats. Second, the most available and obvious form of dispersal in river systems is linear and one-way, that is, predominantly downstream in the water flow via drift for passively dispersing organisms, and for actively dispersing organisms restricted to the stream channel. Third, the mode and strength of dispersal of most stream organisms differs strongly over the course of their complex life cycles; for instance, passive drift in insect larvae and active flight in the adult. In addition, a specialised dispersal phase is found in some river-dwelling animals such as anadromous salmon and catadromous eels.

To sum up, spatial (or 'regional') processes purport to reflect mainly the influence of dispersal, whereas 'environmental' (or 'local') processes relate most obviously to species sorting, such that only well-adapted species survive in a patch. Unequivocal conclusions have yet to be drawn from this burgeoning field of research, but Heino et al. (2015) concluded that 'most stream studies have found that environmental control . . . prevails over spatial constraints'. An example comes from a study by Lopéz-Delgado et al. (2019) in which species sorting was identified as the principal process structuring assemblages of river fish in a near-pristine tributary of the Orinoco River in Columbia. An important corollary of this tentative conclusion from an applied perspective is the need for the maintenance of habitat heterogeneity (i.e. a variety of habitat patches) in order to maximise the opportunity to conserve biodiversity, rather than placing emphasis solely on 'connectivity' (i.e. determining the ease of dispersal). The prevalence of species sorting may also vary amongst taxa and be context dependent (e.g. Heino et al. 2015). For instance, Thompson & Townsend (2006) found that models including both distance (dispersal constraints) and environment best explained the species composition of invertebrate assemblages at ten sites differing in environmental conditions and intervening distance

in New Zealand. Their results differed somewhat for 'freely dispersing' species, for which there was some evidence of mass effects, and those with more modest powers of dispersal, but the overall conclusion was that species must (self-evidently) be able to reach suitable sites to be present there and persist.

It has been proposed that the main effect of a dendritic network on river communities is that its physical structure *per se* increases diversity (e.g. Altermatt 2013; Tonkin et al. 2018a), regardless of habitat heterogeneity. This is encapsulated in the idea of the 'blue network' (Figure 6.21). This concept suggests that diversity (at least of organisms dispersing along the channel) is greatest in large, downstream channels, simply because they receive so many propagules and immigrants from upstream, whereas headwaters being spatially more isolated have lower α diversity (although there may be high species turnover between headwaters, and thus high overall β diversity). This argument is also the basis of the so-called *network position hypothesis*, which proposes that upstream sites are structured mainly by environmental factors (species sorting), while central mainstem rivers are structured both spatially (by mass effects, a high rate of immigration from upstream) and by environmental factors (Schmera et al. 2018).

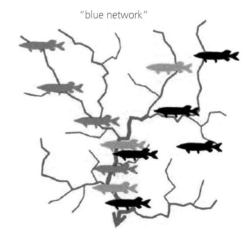

"blue network"

Figure 6.21 Theoretical expectations of diversity in a so-called blue network of a river system. Tributaries are relatively isolated from each other, at least for species restricted to dispersing along the river channel, by large intervening distances, whereas aquatic dispersal downstream to network confluences and mainstems is relatively easy. Thus, we would anticipate higher α diversity in mainstems than in headwater tributaries.
Source: Altermatt 2013), with permission from Springer Nature.

Two empirical tests undertaken to date have found only equivocal support for this hypothesis, however, one dealing with several taxa (Schmera et al. 2018) and one based on a large data set for French river fish (Henriques-Silva et al. 2018). This limited support is perhaps not surprising because, even in principle, this hypothesis can only apply to species whose dispersal is restricted to the channel (and the 'wet' distances between headwater sites are long), rather than more directly, overland. As yet, we know too little about the dispersal of most species, although it is apparent that many, especially flying insects, can disperse overland, even if only fairly rarely (see Chapter 5, section 5.4.1). In addition, as mentioned before, headwaters are not in all respects more 'isolated' in the landscape than mainstem channels given they are common and frequently close together, whereas large rivers are rarer and further apart. Furthermore, as well as headwater specialists, there may also be specialists adapted only to the most downstream freshwater stretches of a mainstem, which can be characterised by low shear stress, fine-grained substrata and relatively high temperature. The population size of such downstream specialists may be very low because suitable habitat is limited, and they may therefore be vulnerable to extinction, or threatened by climate change as sea level rises. Rescue (by dispersal) of such populations is unlikely because the nearest sources of colonists are very distant and there is unlikely to be any fully aquatic (freshwater) route between them.

The role of a branching network in the maintenance of riverine diversity has been examined, theoretically, by Holt & Chesson (2018) who found that branching itself had little effect on community dynamics in the absence of environmental heterogeneity (i.e. variation in habitat conditions among the branches). For this reason, they also suggested that the maintenance of environmental heterogeneity, rather than branching and connectivity *per se*, should be the primary emphasis of conservation. In the 'cartoon' of a blue (entirely aquatic) network (Figure 6.21), the higher α diversity of the mainstem may not be simply because of 'mass effects'—as in the network position hypothesis—but because of habitat features, such as size (e.g. width of channel) and physical heterogeneity. We suspect that in real rivers, both habitat heterogeneity (through species sorting) in a catchment, and the branching structure (affecting dispersal) of the water course play important roles, although the scientific evidence is as yet inconclusive.

6.7 Conclusions—from multispecies patterns to interactions, food webs and processes

This chapter has been about patterns in multispecies assemblages, their diversity and variation in space and time, as well as the processes that might account for them. So far, we have referred mainly to those interactions among species that have played important roles in the development of ideas about biological communities and the conservation of biodiversity. However, species interactions also include processes by which living things influence biological productivity, encapsulated in the flow of energy through food webs, and contribute to further ecosystem processes such as the transformation and transport of carbon and nutrients. In the next chapter we deal in greater detail with the biology and ecology of species interactions in and around river systems, build these into food webs, and set the scene for subsequent chapters on ecosystem processes in running waters.

CHAPTER 7

Species interactions and food webs

7.1 Introduction

In the two previous chapters on populations and communities we referred to the potential importance of species interactions. Partly in an attempt to 'simplify' the subject, early theoreticians and practical ecologists concentrated on the idea of a 'closed' ecological system, generally resource limited and in a stable equilibrium, and in which predation (in its broadest sense, including grazing etc), interspecific competition and resource partitioning are particularly important. However, streams and rivers are very much 'open' systems. As we have seen in earlier chapters, they are spatially complex, along their length, laterally from the main channel to bordering floodplains and riparian systems, and with strong vertical gradients from the water column to the groundwater. Exchanges of organisms and materials along these dimensions are extremely important. Most rivers and streams are also subject to particularly strong abiotic 'forcing factors'—including unpredictable disturbances—and are in general highly dynamic. Can biotic factors possibly be important in determining abundance and diversity in such systems? We might expect the effect of biotic interactions to be most evident in the least disturbed systems, such as those with relatively stable flow regimes, like groundwater-/spring-fed streams or in lake outlets. Also, the presence of some strongly interactive species—we can call them *keystone species*—may determine the overall species composition of the community.

It has been challenging to identify and demonstrate rigorously the role of species interactions in running waters—just as it has been to demonstrate the regulation of lotic populations (Chapter 5). The interactions can be pair-wise, but more often than not involve a number of species. They are embedded within the network of trophic interactions that form the food web, which we discuss later (section 7.8). These interactions often involve species feeding at different 'heights' in the food web and may be 'indirect' (e.g. species A affects species C via an effect on species B). Thus, to make the problem simpler, we often break the food webs down into subcompartments (*guilds* or more recently sometimes called 'modules' or 'motifs') of potentially interacting species. This can be done on the basis of organism type or size (e.g. concentrating on macroinvertebrates or the meiofauna), location (e.g. benthic or hyporheic habitats or subhabitats such as pools and riffles), or focusing on the food web compartments with the strongest potential linkages within the ecosystem (Estes et al. 2013; Beauchesne et al. 2021).

Direct observations of interactions in lotic systems are frequently difficult to make—the organisms are often small and/or cryptic, and the strength of the interaction is often difficult to measure. Inferring interactions is also problematic—just because two species share a resource, this does not mean that they compete, and the strength of a predator–prey interaction cannot be determined solely from the frequency of occurrence of a prey species in the diet of the predator. Further, scale plays a part in our perceptions of the importance of species interactions. For example, small-scale experiments on fish may suggest the importance of competition, whereas large-scale studies often emphasise abiotic controls on communities (Jackson et al. 2001). As Estes et al. (2013) point out, an experimental approach using controlled perturbations or following a 'natural experiment', where the perturbation occurs naturally, can be useful. We can therefore study the responses to removal, addition, or manipulation of the density of the various actors across a range of spatial and temporal scales (from small-scale, highly controlled laboratory or field experiments to large-scale natural experiments caused, for example, by a catastrophic flood or drought, or by a disease).

Interactions between species are often thought of as negative, at least for one of the interactors, as is the case for predation, herbivory, parasitism, disease and competition. However, there is a growing realisation that (at least partially) positive species interactions, such as facilitation, mutualism, commensalism and symbiosis,

The Biology and Ecology of Streams and Rivers. Alan Hildrew and Paul Giller, Oxford University Press. © Alan Hildrew and Paul Giller (2023).
DOI: 10.1093/oso/9780198516101.003.0007

can also play a significant role in streams and rivers. Holomuzki et al. (2010) provide a nice summary of the history of research on, and our understanding of, biotic interactions in freshwater benthic habitats including streams and rivers. As we will show in this chapter, biological interactions are important processes in lotic communities, just as they are in other types of ecosystem, although these biotic processes are often modified or even triggered by abiotic factors, and in streams this factor is frequently flow.

7.2 Predation

A significant fraction of the species in streams and rivers are predators (exclusively or partially). Whilst the taxonomic composition of the predator assemblage may change biogeographically, there are consistent differences in the nature of the predators between channels of different size. In small, fishless streams, insects or other larger invertebrates are often the dominant predators, but further downstream larger-bodied fish and aquatic and semi-aquatic vertebrates become important. In bigger rivers, large fish and large reptilian predators like crocodilians occur, and even cetaceans (including river dolphins) are found in tropical and subtropical habitats.

Some predators attack prey at the water surface (as adult insects emerge into the air or return to lay eggs, or as terrestrial invertebrates fall in). Many species of 'drift-feeding' fish take prey from the surface—most classically the salmonids of swiftly flowing streams—while some predatory insects 'patrol' the surface and

will attack prey of an appropriate size. These latter are uncommon in turbulent streams, being confined to sheltered river margins and pools and floodplain water bodies, and include the well-known 'water striders' (Gerridae, Hemiptera; Figure 3.14 b, p. 84) and 'whirligig' beetles (Gyrinidae; Figure 3.16 c, p. 88). Fish are generally the main predators catching prey in the water column, along with some filter-feeders. Many fish also search for prey on the stream bed, along with the majority of predatory invertebrates, like the larvae of Perlidae and Perlodidae (stoneflies; Figure 3.12 a), Odonata (dragonflies and damselflies; Figure 3.13 c, d), rhyacophilid caddisflies (Figure 3.14 b), alderflies (Sialidae, Figure 3.16 d) and dobsonflies (Corydalidae, Figure 3.16 e, p. 88), plus some dipterans such as the tanypod Chironomidae (Figure 3.15 b, p. 85).

Lotic predators use a variety of different cues to detect their prey. Most fish are visual predators, although species living in turbid or coloured water (or that forage at night) use touch (via their lateral-line system) and olfaction, following chemical or hydrodynamic cues left by prey. On the other hand, Amazonian electric eels such as *Electrophorus electricus* (Figure 7.1) produce low-voltage electric pulses, which operate much like radar. Electric eels can also generate exceptionally strong discharges of up to 860 volts, to stun or kill their prey at distances of 30 cm (de Santana et al. 2019). The predominantly nocturnal (or 'crepuscular'—dawn and dusk) activity of many benthic invertebrate prey may be an innate response that avoids drift-feeding fish (see Chapter 4). Invertebrate predators more rarely use vision. Predatory

Figure 7.1 The electric eel *Electrophorus electricus* (Pisces, Gymnotidae).
Source: National Institute of Ecology, KOGL Type 1, http://www.kogl.or.kr/open/info/license_info/by.do, via Wikimedia Commons.

stonefly larvae, for example, detect prey primarily via their antennae and leg hairs, despite having rather well-developed eyes (Sjöström 1985). Hydrodynamic cues may also be involved (Peckarsky & Wilcox 1989). Surface-feeding insects either use vision (gyrinid beetles) or, as in Gerromorpha (Veliidae and Gerridae), perceive ripples created by the prey on the water surface through sensitive hairs on their legs. Some net-spinning caddis, including several polycentropodids such as *Plectrocnemia*, detect movements in their net (see Figure 4.11 a) and rush to attack the prey, rather like spiders (Edington & Hildrew 1995). Most predatory birds and mammals forage visually, but the platypus (*Ornithorhynchus anatinus*, Figure 3.22 d, p. 98) finds its invertebrate prey through electro-location by sensing weak electric currents produced by muscular activity of its victims. The Amazon river dolphin (*Inia geoffrensis*) uses sonar in its turbid river habitats.

Most predators show some degree of specialisation with respect to diel (24-hour) activity, in how they detect, attack and capture prey, and in exactly where they feed. Specialisation towards particular prey taxa appears to be less common, however (Thompson et al. 2012; and see below, section 7.8). For example, recent studies on the Pyrenean desman (*Galemys pyrenaicus*; Figure 3.22 b, p. 98), using next-generation sequencing molecular analysis to detect traces of prey in the faeces (Biffi et al. 2017; Hawlitschek et al. 2018), have identified between 150 and 220 invertebrate genera or operational taxonomic units (OTUs) respectively, including terrestrial prey taken directly on the river bank or from the exploitation of terrestrial insects entering the river. It is also important to note that during the lifetime of the predator, prey consumed will usually change in relation to shifts in predator size, prey availability and perhaps habitat.

7.2.1 Mechanisms of predation

The two basic predatory mechanisms are active foraging or sit-and-wait ('ambush')—the second sometimes involving the use of various kinds of traps. Actively foraging predators often aggregate in patches with high prey density (the *aggregative response*). However, when prey are also highly mobile there may be no positive correlation between the abundances of predators and prey. Widely foraging predatory invertebrates include various stoneflies, alderflies and dytiscid beetle larvae. Others, including predominatly drift-feeding fish like the brown trout, are more passive, often selecting a feeding station from which they wait for drifting prey to come close enough to be captured. These

stations are often in slower-flowing water close to a faster current, thus maximising the supply of drifting prey while minimising energy expended in maintaining position against the flow. There may be competition for the best feeding stations that offer benefits in terms of growth (e.g. Greenberg & Giller 2001). Dominant individual fish can remain at or close to such locations for a considerable time (Bridcut & Giller 1993). Classic sit-and-wait predators are the larvae of dragonflies and damselflies (Figure 3.13 c, d, p. 82) with their specialised labial mask used for prey capture. Some forage from 'perches' on the stems of macrophytes while others wait, hidden among sediments and detritus, on the stream bed. The net-spinning caddis *Plectrocnemia conspersa* tends to abandon sites where prey have not been captured within a threshold time since the last meal but remain at sites with a high prey capture rate, hence demonstrating an aggregative response (Hildrew & Townsend 1980). Larvae also contest ownership of nets (in the laboratory at least) with the outcome determined largely by body size.

Predators generally catch more prey per unit time with increasing prey density—the rate of increase is termed a *functional response*. In the case of filter-feeding predators, this increase can be linear up to some threshold prey density delineated by the minimum time taken to deal with individual prey. This is known as a *Type 1* functional response. In other predators, the response gradually levels off (a curvilinear *Type 2* response), as the predator becomes satiated and as handling time (the time to capture, kill and consume a prey item before resuming the search) takes up an increasing amount of the foraging time available. Under some circumstances the functional response curve may follow an 'S-' or sigmoidal shape (*Type 3*), where the predation rate is low at low prey density, either because it is difficult to locate and capture prey at low density, or because the predator switches to another more abundant species of prey or reduces its searching activity. The rate then increases rapidly with increasing density until the curve starts to level off again for the same reasons as in the Type 2 response. Both Type 2 and 3 curves have been found among stream-living predators (Malmqvist 1991; Giller & Sangpradub 1993), and increasing habitat complexity can lead to a switch in the functional response curve from Type 2 to Type 3 (Hildrew & Townsend 1977). There is increasingly strong evidence that functional response parameters like handling time and search coefficient, which determine the response shape and prey capture rate, may depend on body mass of the predator relative to prey, temperature and habitat complexity, as well as prey attributes (morphology) and defensive behaviour (Dodd et al. 2014; Kreuzinger-Janik et al. 2019).

When predator density is high, interactions between individual predators can reduce capture rates, as seen from experiments involving predatory stonefly larvae and baetid mayfly larvae prey (Sjöström 1985; Malmqvist 1991; Peckarsky & Cowan 1991). In experiments involving both a vertebrate and an invertebrate predator (a sculpin, *Cottus bairdi*, and a perlid stonefly, *Agnetina capitata*), complex responses were observed, depending on which mayfly prey taxon was available (Soluk 1993). Interference between the fish and the perlid was observed when the more surface-active and drifting baetid prey were used (with the fish reducing the activity of the perlid and its ability to capture *Baetis*), whereas facilitation occurred when ephemerellids (a less active, crawling mayfly) were the prey, as they became more conspicuous to fish as they escaped the predatory perlid. Interestingly, further studies (Soluk & Richardson 1997) showed that young trout that shared artificial streams with predatory stoneflies gained an average of 2.4% of original body mass over a month whereas fish living without stoneflies lost 2.6% of body mass.

In an unusual example, cooperation among predators of the same species was detected at the inflow of the Iriri River into a small lake in the Amazon basin. Bastos et al. (2021) reported coordinated hunting (or *social predation*, normally associated with mammals and some birds) by the electric eel *Electrophorus voltai*. This is normally a nocturnal solo hunter on vertebrates or large invertebrates. In the low-water season, however, up to 100 individuals could aggregate, herding shoals of small fish into a 'prey ball'. Groups of two to ten eels launched a series of high-voltage strikes around dawn and dusk. Prey hit by the shocks leapt from the water, only to fall back stunned and be quickly consumed by the eels (videos of this can be seen in the supplementary information to Bastos et al. 2021). Other fish species seemed to take advantage of the stunned shoal of prey—again a form of facilitation.

7.2.2 Anti-predator adaptations

The substantial risks to prey from predators are borne out by the wide variety of anti-predator adaptations and traits, both morphological and behavioural, as well as the use of spatial and body-size refugia. Particular microhabitats are relatively safe from the larger predators; e.g. under stones and cobbles or within sediment crevices, within thick macrophyte beds or in deep water or fast flow (Holomuzki et al. 2010).

Heavy sclerotisation ('armouring'), cryptic and disruptive colour, spines and bristles, and strong, transportable cases are frequent morphological adaptations that may reduce predation (see also Chapter 4). Behavioural adaptations are also many and varied, including nocturnal or reduced activity, drift escape, noxious exudates, 'playing dead' (*thanatosis*, e.g. the surface-dwelling bug *Velia*), defensive threats (e.g. 'scorpion posturing' by ephemerellid mayflies), and retaliation. Specific, *trait-mediated defences* occur when prey change behaviour, morphology or a chemical trait when exposed to predators (Holomuzki et al. 2010). There is often a fitness cost to such defences in terms of reduction in growth or reproduction (e.g. through redirected energy allocation or loss of foraging opportunity). For example, nymphs of the mayflies *Baetis bicaudatus* (Peckarsky et al. 2002) and *Drunella coloradensis* (Dahl & Peckarsky 2002) mature at a smaller size and have lower fecundity when exposed to trout or even trout-tainted water. *Drunella* has longer caudal filaments per unit body size and relatively heavier exoskeletons in streams with trout, which experiments suggested could be induced by water-borne fish chemical cues. These additional structural defences led to reduced mortality when exposed to predation by trout compared to mayflies originating from fishless streams.

7.2.3 Direct effects of predation

It has often been difficult to demonstrate a substantial effect of predation on the prey population density. This effect is influenced by a wide range of factors, including the size of the predator population, their mode of attack, the relative body size of predator and prey, prey defences, the wider community (presence of other predators, alternative prey) and the environmental context (habitat heterogeneity and the availability of prey refuges, physical disturbance) and many others. Predation can no doubt limit the abundance of prey and influence the age/size structure of their population. On a small scale, changes in habitat use that reduce predation risk can lead to different assemblages in particular pools or riffles, depending on the presence or absence of predators (for example, amongst fish assemblages; Jackson et al. 2001). At a larger scale, indirect effects can propagate widely or 'cascade' through food webs (see Section 7.8.4.2).

Many experiments have been carried out on the dynamical effects of predation, often using predator enclosure and exclosures. For example, tadpoles of salamanders eat macroinvertebrates in small fishless streams across many North American headwater systems. Keitzer & Goforth (2013) manipulated the presence of two salamanders, *Desmognathus quadramaculatus* and *Eurycea wilderae*, in enclosures (Figure 7.2). Both prey abundance and taxon richness were reduced (by 87% and 57%, respectively) when both predators were present, particularly of the Chironomidae. The effect of either predator alone was negligible, which Keitzer & Goforth (2013) attributed to niche complementarity or facilitation. Other studies, however, also based on predator-removal experiments, have shown that single salamander species can affect invertebrate populations (e.g. Huang & Sih 1991)

In another example, predation experiments in a small stream in western Canada were carried out in 'flow-through' channels containing a set of small enclosures, each with a 2.5g alder leaf pack and together creating a gradient in density of predatory chloroperlid stoneflies (an assemblage of several genera including

Sweltsa and *Sumallia*). The biomass of a range of meiofauna associated with leaf packs, including rotifers, nematodes and chironomid larvae, was reduced (Majdi et al. 2015). The colonisation of leaf litter by the prey was also inhibited, perhaps associated with predator-induced bioturbation of soft sediment deposited on leaf surfaces.

The overall results of such experimental studies have been variable, however, ranging from very strong effects to having none whatsoever. What can contribute to this variability? Many experiments have involved mesh enclosures or exclosures where the mesh size is small enough to stop predators leaving or entering but large enough for prey to move in and out. Differences in prey density between treatments with and without predators (or with different densities of predators) are then ascribed to predation. A lack of statistically significant effects could indicate a real lack of predator impact or point to various confounding factors,

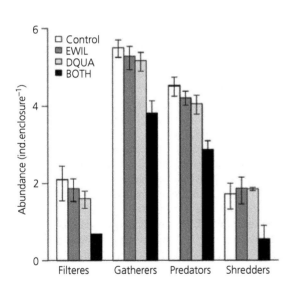

Figure 7.2 The bar chart on the left shows the effects of salamander predation on the mean abundance (+/-SE) of different macroinvertebrate functional feeding groups in an enclosure experiment (top right) in a headwater in the southern Appalachian Mountains (USA). Prey abundance values are log(x + 1) transformed to accommodate the large differences in density among functional groups. Two species of salamander were studied separately: DQUA—*Desmognathus quadramaculatus* (photo bottom right); EWIL—*Eurycea wilderae*; and together (BOTH).
Source: from Keitzer & Goforth 2013, with permission from John Wiley & Sons. Photo of *Desmognathus quadramaculatus* by Tom Ward, from https://appalachian.org/salamanders/.

including the notoriously patchy nature of streams and rivers. That is, background variation in prey density may be so great that it is extremely difficult to replicate enclosures sufficiently to detect an effect of the predators. Further, Cooper et al. (1990) suggested that the range of outcomes of such studies could be explained by the mesh size of the enclosure or exclosure cages. Small meshes might inhibit the movements of prey in and out of the cages, resulting in a stronger apparent impact of predators than when larger meshes are used that allow prey to move more freely. Similarly, Lancaster et al. (1991) tested the effect of mesh size directly, focusing on the predatory net-spinning larvae of the caddis *Plectrocnemia conspersa*. In predator enclosures in Broadstone Stream (southern UK) they found that the rates of both prey arriving and prey leaving the enclosures could exceed direct predator consumption, thus masking its apparent influence. Reduction in the 'exchange rate' of prey through the cage walls (both in and out), achieved by reducing mesh size, did increase the apparent effect of the predator on prey density.

It is worth noting that the doubtful effects on prey density of manipulating predator density in relatively small cages do not mean that predation is insignificant in the wider system. As an analogy, imagine trying to lower the water level of a small patch of the surface of a large swimming pool using a ladle. Water would immediately rush in from the surrounding pool area to replace the water scooped out—the effect on overall water level would be undetectable. But, if that effort was repeated over a large proportion of the pool's area, the water level of the whole pool would drop. This is thus a scale effect; if sufficient predators are 'at work' over a large area of stream benthos, prey numbers will be reduced, regardless of small-scale prey mobiity (Englund & Cooper 2003).

In this context, flow is important for the predation process in lotic habitats. Lancaster (1996) showed that small patches, which act as flow refuges to both predators and prey (see Chapter 5 section 5.3.4), also acted as centres of intense predation by *Plectrocnemia conspersa*, probably through increased prey encounters. In contrast, the more actively foraging predator *Sialis fuliginosa* did not aggregate in these patches, and their impact was therefore lower during periods of high flow. Predatory stonefly larvae like *Isoperla grammatica* and *Diura nanseni* have a high attack rate in slowly flowing water but are severely inhibited even at moderately high velocities. *Rhyacophila* larvae, in contrast, show no reduction in predation success at least up to a near-bed velocity exceeding 0.5 m s^{-1} (Malmqvist & Sackman 1996; Figure 7.3).

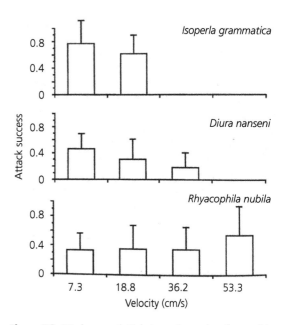

Figure 7.3 Attack success (ratio between the number of successful attacks and the total number of attacks + 1 SD) for three predatory insects on feeding on *Simulium* prey at four different flow velocities. *Source*: after Malmqvist & Sackman 1996, with permission of Springer Nature.

While small-scale experiments have proved instructive, we are interested in the extent to which predators influence their prey populations overall, and clearly this is best approached by studying the effect at a larger scale. Many studies, for example, have shown strong effects of predators on fish populations (Jackson et al. 2001). Similarly, both the European dipper (*Cinclus cinclus*; Ormerod & Tyler 1991) and the American *Cinclus mexicanus* (Harvey & Marti 1993) can reduce the density of their benthic prey, particularly caddis populations. In this latter example (Figure 7.4), excluding dippers from areas of about 40 m^2 revealed a significant effect on populations of the large limnephilid caddis *Dicosmoecus gilvipes* and on heptageniid mayflies, but none on large baetid nymphs found on exposed surfaces. The effect on small baetids was greater in deeper than shallower stream reaches. In Welsh streams, dipper pairs consumed 0.59–1.11g m^{-2} y^{-1} dry mass of Trichoptera, and 10.5–11 kg dry mass of total fish and invertebrates were ingested per year, equivalent to an annual exploitation of secondary production of 0.93–2.35 g dry mass m^{-2} (Ormerod & Tyler 1991; see also Topic Box 3.3 in Chapter 3). In the shorter-term experiment of Harvey & Marti (1993) an estimated 5,520 individual prey were consumed over an area of 1,831 m^2 during a 16-day experiment, which amounted

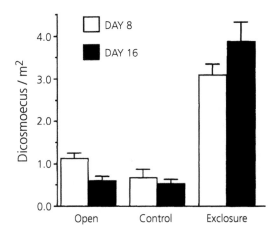

Figure 7.4 Large cased caddis (*Dicosmoecus gilvipes*) density (mean + 1 SE) in open stream sections, exclosure controls (with roofs but no sides), and American dipper (*Cinclus mexicanus*) exclosures (with roof and side netting covering 116.3 m² of stream) after 8 and 16 days, in Wheeler Creek, Utah, USA.
Source: from Harvey & Marti 1993, with permission from John Wiley & Sons.

to a similar intensity of predation to the Ormerod & Tyler (1991) study. It is not surprising, therefore, that a number of studies have suggested density-dependent relationships between dippers and certain groups of benthic insects (see references in Harvey & Marti 1993).

In another study, Forrester (1994) observed the effects of brook char (*Salvelinus fontinalis*) on mayfly larvae in stream sections 35m long and 2.5 to 3m wide. In this case, much of the predator's effect was caused by increased drift out of the area by *Baetis* and *Paraleptophlebia* mayfly larvae. Other mayflies, such as *Ephemerella*, *Eurylophella* and *Stenonema*, were unaffected. Both increased and decreased activity is possible in invertebrates in the presence of predators. If prey actually avoid activity in the presence of predators such as fish, this may result in them being concentrated where predators are present, thereby erroneously suggesting that particular predators have weak impacts. This has been termed the 'paradox of danger' (Sih & Wooster 1994) and points to another of the pitfalls in the interpretation of experimental studies on the strength of predator influences in lotic environments.

Nature occasionally offers us situations in which predators are 'manipulated' in various ways. These 'natural experiments' can be informative. The invasion of a top predatory dragonfly species, *Cordulegaster boltonii*, into Broadstone Stream in the southern UK offered such an opportunity (Woodward & Hildrew 2001). As the food web for this stream was already well worked (see section 7.8.2, Figure 7.22, p. 255) changes in the overall food web structure could be identified

and included increases in mean food chain length and the number of trophic links within the food web but no species loss. Similarly, dispersal barriers often provide instances where streams naturally differ in predation risk. One such example was discussed by Peckarsky et al. (2008) in examining loss rates of two co-occurring mayfly species in high-altitide streams that differed in the presence or absence of brook trout. Some quite complicated relationships arose that depended on both the prey species and time of year. *Baetis bicaudatus* developed during snowmelt, and neither the overall population loss rate nor abundance was related to the presence of predatory fish, although predatory stoneflies could explain a higher proportion of the natural loss rate than non-predatory losses. In contrast, another *Baetis* species (*Baetis* B) developed during baseflow and natural loss rates could be explained through predation in streams with fish, while a significant proportion of natural losses in fishless streams could be explained by stonefly predation.

In contrast, Allan (1982) reduced trout (*Salvelinus fontinalis*) density in quite long stream sections, finding that fish apparently had insignificant or only modest effects on the density of benthic animals. In this study, patchy prey distributions, as well as high prey mobility and recruitment, possibly obscured the impact of predation (Cooper et al. 1990). Further, only about 75% of the fish were actually removed, possibly allowing the remaining 25% to increase their feeding rate. In a study in southern Ireland, invertebrate densities were decreased dramatically (by 95%) by a natural flood disturbance, yet prey intake rate and condition of trout remained unaffected (Twomey & Giller 1991), though there were short-term decreases in trout numbers following the flood. In such a situation, the top predator population was clearly not food-limited (though possibly habitat-limited). Where prey are not a limiting factor, it is unlikely that a predator can influence prey communities unless it is strongly selective—and most aquatic predators (including brown trout) are polyphagous (see later). As Peckarsky et al. (2008) concluded, predation alone may not explain variation in prey population dynamics in streams and rivers. This is in part due to the strong effects of flow, disturbance and habitat complexity and to the openness of the systems. Thus, the relative importance of predation is clearly species- and environment-specific.

7.3 Herbivory

A range of living plant material is available to grazers in streams and rivers. Sources include macrophytes, algae (benthic, epiphytic and, in larger rivers, planktonic) and mosses. Grazing of both bacteria and algae within biofilms also occurs. The exact mode of grazing varies amongst herbivores (see Chapter 3 for some details for the various groups and also section 4.5.2). Mechanisms include rasping (as in snails), gathering and brushing (as in heptageniid and leptophlebiid mayfly larvae), cutting (as seen in the caddis family Conoesucidae), browsing (e.g. mobile browsers such as cichlid fish, or macrophyte canopy-living retreat-dwelling chironomids) and scraping (found amongst glossosomatid cased caddis and benthic feeding fish like the cyprinid stoneroller (*Campostoma anomalum*) and loricariid (e.g. genus *Corydoras*) armoured catfish). The raspers and scrapers generally have a greater impact on periphyton biomass, structure and composition than other types of grazers; some snails, for example, can graze biofilms almost down to the bare rock (Holomuzki et al. 2010). Grazing within biofilms is carried out by many different micrograzers including heterotrophic flagellates, ciliates, nematodes and rotifers, and there is growing evidence of significant grazer-related flux of matter (Weitere et al. 2018).

There is still uncertainty over the extent of direct grazing on macrophytes in lotic systems. Many plants possess secondary compounds such as alkaloids, flavonoids and phenolics like tannins or lignins that deter and reduce herbivory (Lodge 1991; Holomuzki et al. 2010) and induced chemical defences have been

documented, for example in *Cabomba caroliniana*, an initially palatable subtropical–temperate submerged macrophyte of eastern North and South American fresh waters (Figure 7.5; Morrison & Hay 2011). It is generally thought that, overall, less than 10% of macrophyte production is directly consumed, mainly by invertebrate herbivores such as gastropods and crayfish and, in larger rivers, by vertebrates such as carp, some birds, hippos and manatees. Nevertheless, grazing still has the potential to reduce plant biomass and many macrophytes in lowland streams and rivers (such as *Potamogeton natans*, *Callitriche obtusangula* and *Sparganium emersum*) show signs of grazing and herbivore damage that can be tracked, using stable isotope signals, through to herbivores. In one example, a Belgian lowland stream, consumption of macrophyte tissue was demonstrated in a number of herbivorous and omnivorous macroinvertebrates (including baetid mayfly and orthoclad chironomid larvae and crayfish) and fish taxa (e.g. the gudgeon, a cyprinid), and was found to be the most important food source for the lepidopteran *Nymphula nitidulata*. The calculated consumption by all herbivores of the standing macrophyte biomass overall was relatively low, however (Wolters et al. 2018). Mosses like *Fontinalis* are also chemically defended (through C_{18} acetylenic acid) against large generalist herbivores, including crayfish and Canada geese (*Branta canadensis*), although it can be consumed

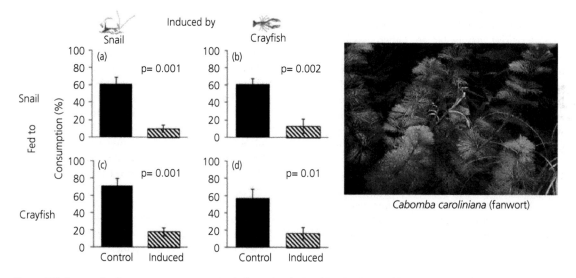

Figure 7.5 Consumption (mean percentage eaten + 1 SE) of control and induced (previously grazed by either crayfish or snails) plants eaten in paired assays using live *Cabomba caroliniana* (a subtropical freshwater submerged macrophyte). In all treatments induction showed strong (71–83%) depression of consumption of the macrophyte by the two generalist herbivores.

Source: from Morrison & Hay 2011, with permission from Springer Nature. Photo of *Cabomba caroliniana* (fanwort) from https://concordma.gov/768/Fanwort-Cabomba-Caroliniana.

by other macroinvertebrates, such as the amphipod *Crangonyx gracilis* and the isopod *Asellus aquaticus* (Parker et al. 2007).

In contrast to the macrophytes, grazing of periphyton is widespread in smaller, shallower streams and the effects on algal form, biomass and assemblage composition can be dramatic (Vadeboncoeur & Power 2017). In tropical upland streams, for example, grazing tadpoles can reduce algal biomass by 50%, as well as facilitating grazing insects by removing fine sediments thus exposing the periphyton (Ranvestel et al. 2004). Herbivory by caddisflies, mayflies and snails can similarly significantly reduce algal biomass by between 26 and 52% in microcosms offering algal food conditions representative of New Zealand streams (Holomuzki & Biggs 2006). Filamentous green algae can be eliminated more or less entirely by grazing minnows, snails, caddisflies and crayfish (e.g. Gelwick & Matthews 1992; Power et al. 1985; see section 7.8). However, some algae can 'defend themselves'. Algal morphology can mitigate against herbivores (e.g. through spines or a low-lying prostrate growth form). Macroalgae such as *Chara* and *Cladophora* produce secondary metabolites that deter some but not all herbivores (Holomuzki et al. 2010). Some cyanobacteria (often referred to as blue-green algae, although they are not actually algae) also produce allelopathic compounds including Anatoxin-a and microcystins that can poison grazers. A number of freshwater diatoms produce compounds such as polyunsaturated aldehydes and eicosapentanoic acid that can lead to decrease in fecundity or egg hatching success (or even death) in invertebrates (Leflaive & Ten-Hage 2007). Much of this kind of information is known from planktonic systems and the extent to which this applies specifically to lotic algae is less clear. The benthic diatom *Didymosphenia geminata* accumulates into blooms of mucilaginous 'river snot' in cold ultraoligotrophic rivers (Bothwell & Taylor 2017) and induction of chemical defences has been shown in a number of red algal (Rhodopyta) species (*Batrachospermum helminthosum, Kumanoa holtonii* and *Tuomeya americana*) common in North American rivers and streams. They reduced grazing on the algae by between 33 and 60% compared to controls (Goodman & Hay 2013). The advantage of these activated (induced) defences is that the fitness costs of a more permanent production of antiherbivore chemicals is avoided.

7.3.1 Effect of herbivory in lotic systems

Compared to predation, it has proved much easier to assess the impacts of herbivory on plants in streams and rivers, and small-scale experiments have readily demonstrated the direct effects. Using a combination of enclosures and exclosures and tiles colonised by the filamentous alga *Cladophora*, Hart (1992) showed that herbivorous crayfish virtually eliminated *Cladophora* from enclosures while algal cover was enhanced in exclosures (i.e. from which crayfish were excluded) (Figure 7.6). In the stream itself, *Cladophora* was absent from sites where water velocity was < 50 cm s^{-1} but present in mid-channel sites where velocity exceeded 50 cm s^{-1}, which was explained in part by impairment of crayfish feeding in fast-flowing water. In the absence of *Cladophora*, the hard substratum was covered by a 'microalgal lawn' (a result of the release from competition which we return to later in this section). Larger herbivorous crayfish tend to avoid shallow water and terrestrial predators, such as wading herons, kingfishers and racoons, and here *Cladophora* can dominate the substratum with > 75% cover (Creed 1994).

Grazing caddis larvae can have similar effects on the balance between macro- and micro-algae. In Big Sulphur Creek in northern coastal California, the algal community switches from upright *Cladophora* filaments in spring and early summer to epilithic diatoms and cyanobacteria in late summer and early autumn (Feminella & Resh 1991). Selective grazing on *Cladophora* by the sericostomatid cased caddis *Gumaga nigricula* accelerates this change, whereas grazing on microalgae by *Helicopsyche borealis* (Helicopsychidae), a caddis whose sand-grain cases resemble snail shells, dramatically reduced microalgal biomass. There is an indirect facilitation, with removal of *Cladophora* by *Gumaga* increasing microalgae thus benefiting *Helicopshyce*, although the two caddis possibly compete once *Cladophora* has been eliminated.

On a larger scale, natural experiments have again uncovered the nature and extent of grazing on the plant community. Small-scale experiments had demonstrated that herbivorous tadpoles can reduce the biomass of epilithic algae (e.g. Ranvestel et al. 2004), so it is not surprising that large-scale amphibian declines caused by the chytrid fungus *Batrachoytrium dendrobatidis* in Central American streams (see section 7.4) have led to dramatic changes to the appearance of the substratum. In upland sites in the neotropics over 75% of the amphibian populations have disappeared, particularly those that breed in the streams (Whiles et al. 2006, 2013). These include golden frogs (*Atelopus zeteki*) that graze periphyton in riffles, the brilliant forest frog (*Rana warszewitschii*), and treefrogs (*Hyla* spp.) that graze periphyton and remove FPOM from the

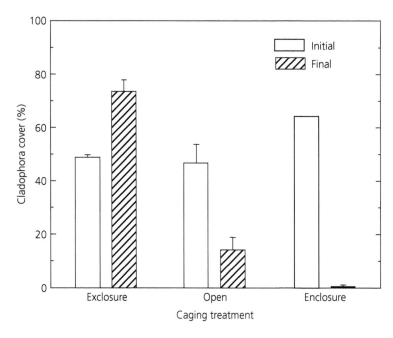

Figure 7.6 Effect of the crayfish *Orconectes propinquas* on *Cladophora glomerata* cover on colonisation tiles in a series of enclosure and exclosure experiments in unshaded reaches of Augusta Creek, Michigan (USA). Open cage treatment (with no sides) allowed natural movement of crayfish in an out of cages. *Cladophora* cover on tiles was assessed at the start and after 7 d of the experiment.
Source: Hart 1992, with permission from Springer Nature.

substratum in runs and pools. In the Rio Maria, a relatively undisturbed, wet-forest second-order headwater stream, Whiles et al. (2013) recorded the loss of 98% of tadpole biomass over a two-year period. This coincided with an almost doubling of the mass of algae and fine detritus on the stream bed. In 2006, prior to the amphibian decline, rocky substrata had little fine sediment and patches of prostrate microalgae (Figure 7.7a). Two years after the amphibian decline, the same rocks were covered with thick layers of living and senescent periphyton and organic sediments (Figure 7.7b). There was also an overall decreased uptake and cycling of nitrogen, largely due to a lower mineralisation rate (less ingestion of organic matter and excretion of mineral nitrogen by tadpoles) (Whiles et al. 2013).

Another way of examining the effects of grazing on plant populations in rivers is through close monitoring of the impact of new herbivores to a system. One example involves the Hudson River in eastern New York State in the USA. It was invaded by the exotic benthic zebra mussel (*Dreissena polymorpha*), as has occurred in many places around the world (see Chapter 10, section 10.3.2, p. 353 and Topic Box 3.2, p. 74). It was detected in May 1991 and rapidly became established in the lower 247 km (tidal) stretch of the Hudson and has been abundant throughout the freshwater portion of the estuary (Strayer et al. 2014). The mussel can reportedly reach densities exceeding 10,000 m^{-2} and made up more than half of

the heterotrophic biomass of the ecosystem by the end of 1992. Since 1993, growing season filtration rates by the zebra mussel population have typically been 10%–100% of the volume of the river per day, far exceeding the approximately 3% of river volume being filtered by all other suspension feeders prior to the arrival of the mussel (Strayer et al. 2011). Estimated grazing pressure on phytoplankton increased over ten-fold during the first few years after establishment, leading to an 85% decline in phytoplankton biomass in the tidal freshwater Hudson (Figure 7.8 a, b; Caraco et al. 1997). Although the zebra mussel population in the Hudson has changed since the 1990s (Figure 7.8c), with greater mortality, dominance of small, young individuals and a consequent fall by about 80% in filtration rates, the impact on phytoplankton populations is still substantial (Figure 7.8d, Strayer et al. 2014). The dominance of the zebra mussel on the river substratum and changes to phytoplankton abundance have had widespread effects, including a massive decline in zooplankton, and significant reductions in benthic invertebrates, particularly of other unionid and sphaeriid bivalves (Strayer & Malcolm 2018). There were also effects on fish, such as the striped bass (*Morone saxatilis*) whose condition, feeding success and early stage growth were negatively affected in upstream locations where the mussel now lives (Smircich et al. 2017). These hint at competitive interactions, to which we will return later (section 7.5).

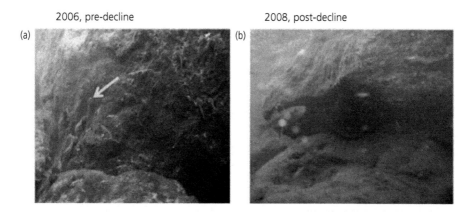

Figure 7.7 Photographs of an individual rock pile approximately 20 cm deep before (a, 2006) and after (b, 2008) the amphibian decline in the Rio Maria stream in Central America, illustrating the dramatic changes in periphyton and fine sediments over the short timescale. The yellow arrow indicates grazing tadpoles.
Source: from Whiles et al. 2013, with permission from Springer Nature.

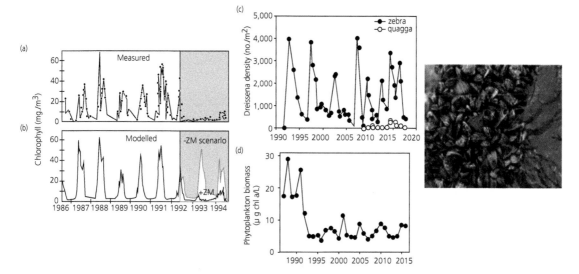

Figure 7.8 Temporal trends in phytoplankton biomass in the Hudson River over a nine-year period: (a) shows a clear annual cycle with one or two peaks in phytoplankton biomass before the invasion by zebra mussels (*Dreissena polymorpha*). The phytoplankton collapsed following establishment of the zebra mussel in 1992 during two years of invasion (stippled section of the graphs). (b) In the modelled output for the same period the heavy line (+ZM) represents modelled chlorophyll *a* using low estimates of zebra mussel density in the river and the light line (labelled (-ZM) represents what chlorophyll *a* would have been in 1992–1994 had the zebra mussel not actually invaded. (c) Since initial invasion, the mussel population has fluctuated significantly, and a new species, the quagga mussel (*Dreissena rostriformis*) has appeared, but (d) the phytoplankton biomass has remained significantly supressed.
Source: (a) and (b) Caraco et al. 1997, with permission from John Wiley & Sons; (c) and (d) Strayer et al. 2019, with permission from John Wiley & Sons Ltd. Photo of mussels from Missouri Department of Conservation website.

7.4 Parasitism and disease

Many taxa in running waters are hosts to parasites, the latter including fungi, microsporidians (single-celled, spore-forming parasites of many invertebrates and fish, probably related to fungi), nematodes, mites and insects. Whilst parasites can alter host behaviour and morphology, few studies have shown their population- or community-level effects (Holomuzki et al. 2010) although the outcomes can be quite spectacular.

Freshwater parasite life cycles are often complex, involving intermediate (invertebrate) and 'definitive' (final, usually vertebrate) hosts while the transfer between hosts is driven by feeding interactions (Figure 7.9). It is interesting that parasites with a wide range of definitive hosts often rely on a single intermediate host species. For instance, the trematodes *Stegodexamene anguilla* and *Telogaster opisthorchis* share a first intermediate host, the snail *Potamopyrgus antipodarum*. In the more complex life cycles, the adult parasite develops within the vertebrate definitive host and eggs are then released back into the water body.

Plant parasites in streams and rivers are less well known than animal parasites, but there are some interesting examples. Nematodes are well-known parasites of terrestrial plants, and evidence is emerging of

examples from river macrophytes to the extent that they may offer potential sources of biological control of aquatic pest weeds, such as the water hyacinth (*Eichhornia crassipes*, a serious problem species in large tropical rivers). *Justicia americana* is an emergent macrophyte found in riffles and gravel bars of streams in eastern North America and is parasitised by female gall-forming 'root knot' nematodes (*Meloidogyne* spp.) (Fritz et al. 2004). There was a negative relationship between stem density and *Meloidogyne* gall abundance in summer, but this was only a minor source of variation in the total biomass of *Justicia*. The infected plants lose vascular tissue, and water and nutrient uptake are reduced.

In large rivers, the dominant plants are often phytoplankton. Studies, largely from lakes, show that

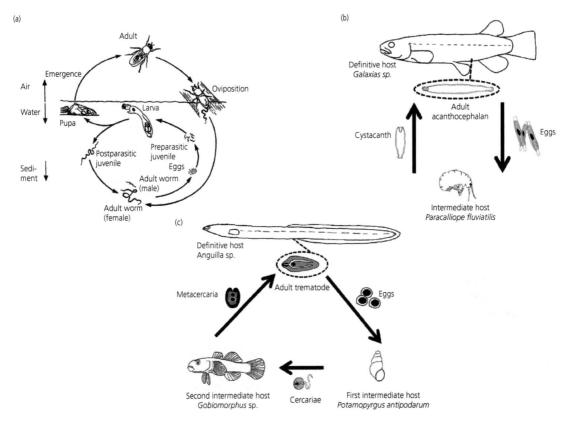

Figure 7.9 Life cycles of common trophically transmitted nematode, trematode and ancanthocephalan parasites: (a) the mermithid nematode parasite of blackflies; (b) the ancanthocephalan *Acanthocephalus galaxii* with an amphipod intermediate host that is eaten by, for example *Galaxia*, one of many potential definitive hosts; (c) trematode parasites *Stegodexamene anguilla* and *Telogaster opisthorchis* with a first intermediate host, the snail *Potamopyrgus antipodarum*, a second intermediate host (e.g. the common bully *Gobiomorphus cotidianus*) infected by free-living cercariae. The fish is eaten by the definitive host such as an eel (*Anguilla* spp.) or trout.
Source: (a) Giller & Malmqvist 1998, after Crosskey 1990; (b) and (c) from New Zealand streams—McIntosh et al. 2016, with permission from the New Zealand Freshwater Sciences Society and Dr Rachel Paterson.

chytrid fungal parasites infecting phytoplankton can transform functionally inedible colonial diatoms into nutritious fungal zoospores that are more readily consumed by cladocerans and copepods (a so-called mycoloop; Kagami et al. 2014). Free-swimming zoospores infect diatom cells and, following fungal development, create sporangia in which the next generation of zoospores develop. Much less is known from rivers and estuaries than lakes, although many species of large, pelagic diatoms in the Columbia River, north-western USA have been found to be infected by chytrids (Maier & Peterson 2017). Over a four-year study, the abundance of sporangia peaked in spring, where zoospores particularly infected the diatom *Asterionella formosa* and to a lesser extent *Aulacoseira* spp. Following a secondary peak of sporangia in summer, infection was predominantly of the diatom *Skeletonema potamos*. Prevelence of infection of *A. formosa* reached 30–45% during spring and it was estimated that an average of 10%, but up to 25%, of the total 'algal' carbon was actually made up of infected cells and the net growth of chytrid populations exceeded that of *A. formosa* under low-flow conditions. Summer infections were less severe and restricted to fewer diatom species.

There are also pathogens of benthic diatoms. In a study in the desert stream Sycamore Creek in Arizona, epilithic diatoms infected by pathogenic bacteria had less chlorophyll *a* and a greater percentage of damaged or reduced chloroplasts than cells in patches of the bed where there was no infection (Peterson et al. 1993). Patches of infected cells, found on all types of substratum, were evident as grey rings within the originally healthy algal patch which rapidly (1–2 weeks, depending on the month) spread from the point of initial infection to cover large parts of the substratum and led to algal sloughing. Peterson et al. (1993) point to other infections in Sycamore Creek and observations of similar patches of senescent algae elsewhere, so this may not be such a rare event.

Turning to animal parasites, parasitic hymenopterans (known as 'parasitoids') are well known in terrestrial systems but are rare in fresh waters. One fairly widespread genus, however, is *Agriotypus*, a species-specific ectoparasitoid. The life cycle and prey–host relationship between *A. armatus* and *Silo pallipes* (Trichoptera: Goeridae) have been well documented. This parasitic wasp oviposits underwater within the case of the caddisfly pupa. A parasitised pupa is easily recognised when the parasitoid has reached its final instar, because a long, springy, ribbon-like structure, probably functioning as a plastron (for respiration),

extends from the pupal case. The particular population of *Silo* studied by Elliott (1982, 1983) suffered a pupal mortality of about 10% from the parasite. Parasitic wasps in the genus *Trichogramma* are absolutely minute egg parasites—the adult parasitoid emerges from a single host egg! In streams and rivers they are known to infect alderflies (*Sialis*), which attach their eggs in groups on the leaves of trees and bushes overhanging streams (Elliott 1996). Rates of infection apparently reach 60 or 70% in some North American species of *Sialis*, but are probably normally much less than that. The dynamical effects on host populations are apparently unknown.

Blackfly (Simuliidae) larvae harbour various kinds of parasites (Crosskey 1990) such as mermithid nematodes, which often fill up a large part of the larval abdominal cavity (Figure 7.9a). Mermithids prevent host metamorphosis, cause sterility, reduce muscle tissue and reduce adult size and longevity (Molloy 1981). The non-feeding adult mermithids mate in the stream bed. Pre-parasitic juveniles then locate and infect blackfly larvae. The juvenile mermithids feed by absorbing blood from the host's haemocoel. Early instar blackfly larvae are the most susceptible to infestation, and the parasite completes development before its larval host, killing it as it escapes the body. When larger blackfly larvae are infested, the parasite continues to develop inside the blackfly pupa and adult, escaping during host oviposition. This carry-over of parasitism into the adult allows for local dispersal and upstream recolonisation of parasites (Molloy 1981). Simuliids are attacked by other parasites as well and a large study covering 115 stream sites in Southern Carolina, USA identified six different parasite taxa, common amongst 34 blackfly species throughout the world; mermithid nematodes (found in 23 species), the chytrid fungus *Coelomycidium simulii* (found in 17 host species) and four microsporidian species (found across 23 hosts) (McCreadie & Adler 1999). None of the parasites showed strong host specificity. There is little information on the effects of these and other parasites on populations of blackflies and, although mermithid parasitism inevitably results in death of the individual blackfly, prevalence is typically less than 15% as seen across a range of studies (McCreadie & Adler 1999).

Microsporidians can influence the population structure and life cycle of their hosts; trematodes can modify reproductive output, oviposition and activity of the intermediate hosts and cause fin and spinal abnormalities in the definitive host (fish); while acanthocephalens can effectively 'sterilise' their crustacean hosts and modify their colour and behaviour—making them more vulnerable to their predators and the definitive host of the parasite (Holomuzki et al. 2010; Grabner 2017; Preston et al. 2021).

The sampling and identification of many parasites is difficult, but the use of new molecular techniques has proved helpful and has revealed surprisingly high diversity, at least amongst the freshwater arthropods. For instance, Grabner (2017) identified 16 species of microsporidians and ten helminth parasites in 12 species of amphipods, mayflies, stoneflies, beetles and chironomids from a soft-water, upland stream in Germany. In contrast to the examples with blackflies and *Silo* described above, microsporidians and trematodes were found in practically all of the macroinvertebrates examined and in many cases at extremely high prevalence (Table 7.1). Most host species were infected by two or more microsporidian species and these parasites can be found in most groups of aquatic invertebrates. Apart from the cased caddis *Sericostoma*, all species tested were infected with trematodes and, overall, six different parasite species were distinguished and five identified. Parasitic nematodes were less abundant and diverse and only found in insects, whilst three species of acanthocephalans were only found in *Gammarus* but at low (5% or less) prevalence.

Not only can parasite diversity be high, but so too can biomass. In a fascinating study in three small, forested streams in Oregon, USA, Preston et al. (2021) found that a substantial part of the biomass of the dominant invertebrate, the snail *Juga plicifera* (mean density > 200 m^{-2}; mean biomass *c.*4 g m^{-2}) was actually comprised of parasitic trematodes. Six trematode

morphotypes were identified, with an overall prevalence (proportion of host individuals infected) in the three streams at 8.6%, 22% and 36.4%. The biomass of '*Juga*' exceeded the combined mean biomass of all other aquatic organisms including fish (total 3.1 g m^{-2}, with about 0.40 g m^{-2} of that consisting of trematodes) Remarkably, this biomass of parasites exceeded that of stoneflies, mayflies, caddisflies, dipterans, beetles, dragonflies and hemipterans combined. This raises questions about the role of parasitism in energy flow through lotic food webs.

If we expand our discussion from the impact of disease and parasites on individuals and populations, we can see impacts at the community and ecosystem level. Parasites that manipulate the behaviour of their prey (such as trematodes and acanthocephalans) can in turn influence energy flow by facilitating, shifting or causing new or enhanced predation interactions and hence energy-flow pathways (Sato et al. 2019; Figure 7.10).

In a classic example of the effect of a pathogen on a lotic food web, the population density of a dominant grazer (the caddis *Glossosoma nigrior*) in south-western Michigan streams collapsed following the infection by a microsporidian parasite *Cougourdella* sp. (Kohler & Wiley 1997). In this case, the pathogen caused strong indirect effects in the food web and demonstrated very strong competitive interactions among grazers, so we explore it in more detail in section 7.8. Clearly, however, *Cougourdella* could be termed a *keystone parasite*.

At a larger scale, a so-called crayfish plague has virtually eliminated some native crayfish populations

Table 7.1 The occurrence and prevalence (percentage of individual animals examined) of the three major groups of parasites amongst macroinvertebrates in a German stream. Acanthocephalans were found only in the amphipods.

	Host species	Microsporidia	Trematodes	Nematodes
Mayflies	*Ecdyonurus* sp.	100	80	0
	Ephemera danica	100	40	30
	Paraleptophlebia sp.	50	50	60
Caddisflies	*Agapetus* sp.	37.5	25	0
	Hydropsyche sp.	100	62.5	0
	Lepidostoma sp.	100	85.7	14
	Sericostoma sp.	100	0	0
Stonefly	*Nemoura flexuosa*	20	60	80
Beetle	*Platambus maculatus*	90	100	30
	Chironomidae	70	100	0
Amphipods	*Gammarus fossarum*	100	70	0
	Gammarus roeselii	90	70	0

Source: Grabner 2017, with permission of John Wiley & Sons.

Figure 7.10 Manipulative parasites can influence food chains in forested streams. Acanthocephalan parasitism of gammarid amphipods and trematode parasitism of fish can facilitate predation by fish and herons respectively. Nematomorph parasitism of crickets leads to their jumping into an adjacent stream where they are eaten by fish, with consequential reduced predation on aquatic prey. Light arrows depict energy flow in the absence of the parasites and dark arrows in their presence, whilst the width of the arrows is indicative of the strength of energy flow. The impacts on the functional response of predators can also cause changes to energy pathways, as in the case of the cross-ecosystem input of large numbers of terrestrial crickets leading to a decrease in predation on benthic insects.
Source: from Sato et al. 2019, with permission of Elsevier.

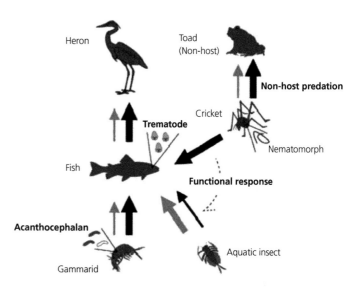

in Europe, including the white-clawed crayfish *Austropotamobius pallipes* (the only crayfish native to Britain and Ireland) and the European noble crayfish *A. astacus* (Holdich & Reeve 1991). Since these crayfish are of cultural importance and a source of food, their demise has been well documented. The plague was caused by an asexual fungus species (*Aphanomyces astaci*), itself native to various American crayfish, and was first recorded in Europe in Italy in 1860, from where it spread rapidly across the continent. Legal and illegal introductions of American crayfish further expanded the plague. The lack of a common evolutionary history between the plague fungus and the European crayfish species is a likely explanation for its devastating impact on their populations. Since crayfish are important omnivorous consumers, and often have a strong grazing role, the potential impact of their loss at the community level is clear. Somewhat similarly, the chytrid fungus *Batrachochytrium dendrobatidis* has infected amphibians in Australia, neotropical central America, the Caribbean, southern Europe and west America, creating waves of mass mortality and regional extinction (Whiles et al. 2006, 2013; Van Rooij et al. 2015). While some species and even a few individuals within a species are tolerant, they may function as carrier species/supershedders of the infection. Other species or individuals are extremely susceptible and develop severe and lethal disease related to vital respiratory and water-balance functions (Van Rooij et al. 2015). As important grazers in these tropical streams, the indirect effects on the stream community can be substantial.

Power et al (2013) issued a stark warning that the number of pathogens and parasites of freshwater vertebrates appears to be expanding in prevalence, virulence and even geographical range, probably associated with climate change, increasing eutrophication and global trade. For example, the introduced Eurasian ectoparasitic copepod *Lernaea cyrprinacea* ('Anchor worm') attacks native minnows, roach and tadpoles in the Eel River in California. In an unusually warm summer, the river-breeding yellow-legged frog, *Rana boylii*, was found to be suffering much greater prevalence and virulence with the onset of previously unseen limb deformities—this may be a prelude to climate-driven changes throughout the range of this frog. In another example of range expansion, until 2010, the chytrid fungus *B. dendrobatidis* coexisted in a kind of steady state in amphibian populations in northern Europe, with only rare evidence of mortality. However, from 2010, a serious decline was evident in fire salamanders (*Salamandra salamandra*) in the Netherlands and, by 2013, only 4% of the population remained. It turned out that this was caused by a new chydrid fungus, *Batrachochytrium salamandrivorans*, which has extended its range to populations in Belgium and Germany (Van Rooij et al. 2015). Globally, amphibians are probably the most threatened of all lotic taxa, probably due to spread of such chytrid infections (see Chapter 10). While the spread of this amphibian fungal disease appears to be independent of global warming, interactions between pathogens and parasites and their hosts are sensitive to temperature (Power et al. 2013), hence the fear that disease impacts are likely to increase, as 'emerging diseases'. For instance, the incidence of the hitherto mysterious 'proliferative kidney disease' (PKD) in salmonids, particularly in culture, increases with stress and high temperature. This is caused by a parasitic cnidarian (Myxozoa) with a (definitive) host in freshwater bryozoans (Okamura et al. 2011).

7.5 Competition

Competition can occur between individuals of the same species—*intraspecific competition*—a process dealt with in Chapter 5. Here we turn to competition between species—*interspecific competition*—a process involving a mutually negative influence on the fitness of both competitors (and can involve more than two species sharing a similar resource). It can limit abundance or distribution and/or alter resource or habitat use, though it is very often asymmetric in that one species is more affected than the other and the less competitive species can even be excluded. Demonstrating the occurrence, strength and effects of interspecific competition in natural communities has proved highly controversial, particularly in 'physically demanding' (frequently disturbed) and physically heterogeneous environments like streams and rivers. However, competition has a key place in the development of biology and ecology, even if its perceived role is now somewhat diminished (see example.g. Rockwood 2015). Here we concern ourselves with the particular features of competition and coexistence in running waters.

Intraspecific competition, inferring resource limitation, is theoretically a prerequisite for interspecific competition to occur. By inspecting patterns in the distribution, abundance and resource use of species, we can sometimes identify situations where competition could possibly be occurring. Great care is needed in making inferences, however—for example, the exploitation of a common resource does not mean that two species are competing, particularly if the resource is not limited. It is only through the investigation of responses to removal, addition or changes in the relative abundance of one or more potentially interacting species, or to a change in resource supply, that we can draw definitive conclusions. Experiments can be via direct manipulations, or sometimes via 'natural experiments' that fortuitously achieve the same thing. However, experimental studies on competition in lotic ecosystems have sometimes shown ambiguous results. The open, flowing and frequently disturbed nature of the habitat, and significant energy subsidies from the catchment, mean that the theoretical preconditions for competition of limited resources in a closed, equilibrial system are not readily met.

There are, however, clear examples of interspecific competition in lotic communities, although in general the effect on species abundance and distribution and on assemblage composition appears restricted to more local scales (or even to laboratory experiments). This may be because competitive interactions are themselves 'embedded' in a complex web of other biotic interactions (such as parasitism, herbivory, predation and various mutualisms) but it is more probably because of abiotic factors, such as disturbance. Further, rivers are highly divided, patchy habitats at a range of scales—making it highly likely that species distributed across such habitats will show a trade-off between competitiveness and colonisation ability, while complete competitive exclusion in this metacommunity will be less likely (see Chapter 6, section 6.6, p. 221)

The most convincing cases of competitive effects have involved sessile or at least sedentary organisms that primarily compete for space. Obvious space occupiers are the aquatic algae and cyanobacteria in the benthic periphyton. Benthic algae might compete for nutrients and light but are ultimately governed by space, and there is strong evidence for the competitive superiority of species with extended growth forms (mucilaginous stalks and filaments, such as *Cladophora*) or some level of mobility enabling individuals or colonies to access light and nutrients from the overlying water body (Holomuzki et al. 2010). On a smaller scale, inside biofilms, different species of prokaryotes and eukaryotes may compete for nutrients and photoautotrophic microorganisms might also compete for space to access light and nutrient-rich zones (Leflaive & Ten-Hage 2007). Within biofilms, the photoautotrophs (like cyanobacteria and algae) often produce allelopathic compounds involved in competition with each other. Allelopathic interactions are especially common in fully aquatic submersed macrophytes, as well as benthic algae and cyanobacteria. These compounds act through inhibition of photosynthesis (a very common mode) or growth, killing competitors or restricting them from the local vicinity of the 'donor' species. Gross (2003) provides a very comprehensive review but points out the problems of currents washing away such allelopathic compounds, largely restricting them to benthic algae and small-scale effects.

Vegetative growth and reproduction and the ability to regenerate from small fragments, thus facilitating colonisation, are important for macrophytes in shallow waters and along river margins, resulting in their ability to rapidly pre-empt newly denuded space and thereby making them more competitive (Barrat-Segretain 1996). For example, the sedge *Juncus* is often found along stream edges and in ditches. It has what can be referred to as a 'phalanx' vegetative growth strategy where modular units of the plant are spaced close to one another with a tightly packed advancing front of rosettes and radial spread. Many of the so-called pond weeds found in shallow streams and rivers, such as *Ranunculus fluitans*, rapidly produce extended shoots (up to 5–10 m in length; Figure 7.11), thus enhancing access to light. Wright et al. (1982) found that this rapid growth by *Ranunculus* in unshaded sites on the River Lambourn in southern England resulted in its overgrowing the 'less aggressive' *Berula* (water parsnip), whereas in shaded sites *Berula* grew well and dominated.

There are territorial benthic insects, including the larvae of caddis-like hydroptilids (*Leucotrichia*) and psychomyiids (*Psychomyia*), which have attached cases or galleries, or hydropsychids, which build a tubular silken retreat attached to their nets. Other temporarily attached animals include blackfly larvae (Simuliidae) (e.g. Hart, 1983; McAuliffe, 1984). For all these, space is potentially an important limiting resource, either directly or indirectly. Intraspecific competition is clear in these organisms, with evidence of aggression towards neighbours, and communication, as in some *Hydropsyche* that produce sound underwater via 'stridulation', thought to be used in territorial defence (e.g. Matczak & Mackay 1990). Similarly, interspecific competition occurs; for example, between different hydropsychid species (e.g. Gatley 1988), and between grazing caddis and moth larvae that compete for space on stream rock surfaces, on which they build their fixed retreats and graze the surrounding periphyton (McAuliffe 1984).

For sedentary filter-feeders, competition for attachment sites appears to be particularly important. Blackfly and hydropsychid larvae can reach very high densities, especially in stable and productive environments, such as lake outlets, or on rocky substrata (particularly on moss-covered rocks) in perennial streams where they can cover every bit of available substratum. Exploitative resource competition in such systems can be crucial if animals upstream also reduce the food available to those animals that live downstream. For example, hydropsychids can reduce zooplankton issuing from lakes, as was shown over a distance of less than 1 km in a lake-fed artificial stream in Sweden (Malmqvist et al. 1991). *Hydropsyche oslari* has been shown in field experiments to pre-empt space and to be directly aggressive to *Simulium virgatum* larvae in rocky habitats, leading to a depression of simuliid numbers and avoidance of hydopsychid nets and hence an altered microdistribution of the blackfly (Hemphill 1988). The competitive dominance of *Hydropsyche* was demonstrated via their experimental removal, which lead to rapid colonisation by *Simulium* of the vacated space. *Simulium virgatum* on the other hand caused significant decreases in time spent feeding (by 20%) and ingestion rates (by 60%) in the net-veined midge *Blepharicera micheneri* on rock surfaces in a fast-flowing Californian stream. The mechanism was direct interference competition (aggressive 'nipping') that led to decreased growth, increased time taken to pupate and increased mortality in the midge (Dudley et al. 1990). When *Simulium* was removed from natural substrata, blephaicerid density increased. This is a clear case of (asymmetric) competition for space, even though the two groups use different food—*Simulium* being a filter-feeder and *Blepharicera* a grazer.

Whilst the majority of these examples of competition are based on small-scale studies, competition can also be identified on a larger scale. Based on a long-term

Figure 7.11 The competitively dominant *Ranunculus*, showing the elongated trailing stems.
Source: from Ulla Niclaus, CC BY-SA 4.0 <https://creativecommons.org/licenses/by-sa/4.0>, via Wikimedia Commons.

(28-year) study, reductions in food availability in the water column through exploitative competition have been proposed as the reason for the dramatic decline of all, and for the possible local extinction of several, native unionid mussel species following the invasion and rapid spread of the zebra mussel (*Dreissena*) in the Hudson River and elsewhere (Strayer & Malcolm 2018).

7.5.1 Effects of competition

Similar to the challenges we described earlier in trying to identify the effects of predation, the mobility of animals influences the ease with which competitive effects can be identified, although evidence is accumulating based on correlational and some experimental data. For example, food limitation and interspecific competition may be significant among mobile grazers such as certain mayfly, chironomid and cased caddis larvae (e.g. Kohler 1992). Both laboratory and field experimental studies have shown that the grazers are quite capable of significantly suppressing algal resources, although some are more effective competitors than others, which can lead eventually to local extinction of the weakest competitor. In a classic study, it was clearly demonstrated that the caddis *Glossosoma*

is a strong competitor of other grazers. Working in streams in Michigan, Kohler & Wiley (1997) found that a number of populations of *Glossosoma nigrior* had been infected in the late 1980s by a specific microsporidian parasite, *Cougourdella* sp. This reduced densities of *Glossosoma* in a number of streams to a fraction (2.5–25%) of values before the epidemic, and densities remained lower due to recurrent pathogen outbreaks (Figure 7.12a). Fortunately, this was one of those rare situations where data both before and after the population collapse were available, providing a natural experiment to test what happened to other grazers when *Glossosoma* was essentially 'removed' from the community. The consequences included increased periphyton algal biomass (by several fold; Figure 7.12b), while the density of other grazers and filter-feeders, which had previously been rare before the spread of the parasite, increased (two- to five-fold). In particular, some other glossosomatids and the limoniid crane fly *Antocha* increased strongly (Figure 7.12 c, d), suggesting competitive release. Filter-feeders, including blackfly larvae and hydropsychid and *Brachycentrus* caddis larvae, also showed consistent increases (Figure 7.12 e, f), presumably following an increased drift of diatoms that become detached in greater numbers given their newly increased biomass.

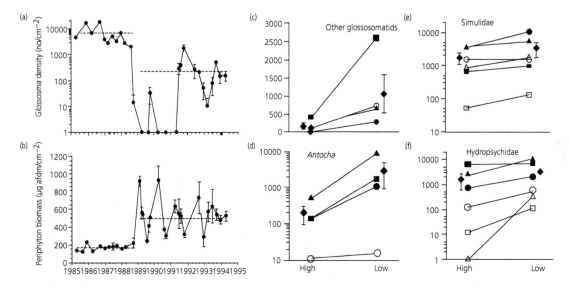

Figure 7.12 The impact of parasitism on a dominant grazing caddis species illustrates competitive release: (a) density (log scale) changes of *Glossosoma nigrior* in one the study streams (Spring Brook) over a 10-year period before and after the pathogen-induced population collapse; (b) periphyton biomass (ash-free dry mass) over the same period. Overall mean density or biomass before and after the *Glossosoma* collapse is shown by horizontal dashed lines. Examples of responses (densities, numbers m^{-2}) to the collapse of *Glossosoma* populations amongst other grazers (c, d) and filter-feeders (e, f). 'High' is before and 'Low' is during recurrent pathogen outbreaks in six streams from Michigan, USA.
Source: from Kohler & Wiley 1997, with permission from John Wiley & Sons.

One further circumstance makes this study remarkable. Kohler & Wiley (1997) had been conducting small-scale manipulations of *Glossosoma* density before *Cougourdella* arrived. This allowed them to compare the predictions from these small-scale experiments with the large-scale effects over several years following the *Glossosoma* population crash. Reassuringly, the small-scale experiments correctly predicted the direction of several of the larger-scale population changes, although they generally underestimated their magnitude. However, the small-scale manipulations could not have predicted some of the indirect effects, such as those on filter-feeding hydropsychids and *Brachycentrus*.

These kinds of significant, large-scale effects do not always occur following the removal of species, however. For instance, even though small-scale experiments show that tadpoles can significantly reduce algal biomass, Whiles et al. (2013) found no evidence of a significant response in other grazers following the decline of amphibian populations (through disease) across many river systems in central America. This suggests that either the invertebrates and amphibians do not compete or that other factors, such as predation or flow disturbances, maintain grazer populations below any limits imposed by food supply, and hence resources are not limited.

There is a significant literature, mostly based on field observations, that suggests competition is an important factor in the local organisation of fish communities (Jackson et al. 2001) and controlled experiments, on a small scale at least, demonstrate its potential. In small, lowland headwaters in much of eastern North America, the benthic sculpin *Cottus bairdi* and the fantail darter (*Etheostoma flabellare*) coexist. Manipulating densities in experimental streams showed that juveniles of the two species clearly compete but the species responded in different ways. The growth of *Cottus* was reduced in terms of both total length and body mass, but without any effect on relative condition (ratio of observed mass at a given length to expected mass based on the population length–mass regression) or survival, whilst relative condition declined in the darter, although growth and survival were unchanged (Resetarits 1997).

Non-native fish species that are more aggressive than native species tend to be better at acquiring resources, often leading to the weaker competitor shifting to suboptimal habitats and conditions. This was clearly demonstrated by Houde et al. (2016) through manipulations involving juvenile Atlantic salmon (*Salmo salar*) and non-native rainbow trout

(*Oncorhynchus mykiss*) in study streams near Lake Ontario, Canada. The presence of rainbow trout led to a shift to suboptimal habitats by salmon which resulted in reduced fitness in terms of reduced body length, mass and condition. From the other perspective, experimental removal of the non-native brown trout (*Salmo trutta*) from a study stream in Michigan, USA enabled native brook trout (*Salvelinus fontinalis*) to occupy more favourable daytime resting positions, suggesting the competitive dominance of the brown trout (Fausch & White 1981). Similarly, in the Horinai Stream in Hokkaido, Japan, Dolly Varden char (*Salvelinus malma*) is effectively excluded from drift feeding by the non-native rainbow trout and white-spotted char (*Salvelinus leucomaenis*) during summer and, although they can switch to benthic prey, they then have to compete with specialist benthic feeding cottids for this limited resource (fish consume almost 98% of aquatic invertebrate annual production in this system) (Marcarelli et al. 2020). As a result, Dolly Varden char production dropped almost to zero during the summer when living in sympatry while rainbow trout production was two-fold higher in summer than in other seasons, and production by white-spotted char was similar to that at other times of the year. It appears that, as a result, Dolly Varden is being displaced from many streams by rainbow trout across Hokkaido. In fact, many native trout populations have been threatened by invasions of introduced trout species, such as the brook trout (*Salvelinus fontinalis*) by brown trout (*Salmo trutta*) in Scandinavia (Ohlund et al. 2008) and in North America (Fausch & White 1981; McKenna et al 2013). Instances like these of the dominance of introduced or invading fish species have been taken to suggest that competition plays a significant role in the decline in populations and range, and even in the extinction of native freshwater fish globally (Buckwalter et al. 2018).

Evidence of competition is not just restricted to closely related species or similar kinds of organisms, and some intriguing examples can be found across phyla. One such involves Harlequin ducks (*Hisrionicus histrionicus*), which spend most of the year feeding in the intertidal zone of coastal marine habitats, before moving inland to clear, fast-flowing rivers during the breeding season in April and May. There they share aquatic insect prey with a number of fish species (LeBourdais et al. 2009). Densities of harlequins and fish were both positively related to insect abundance, but there was a strong negative correlation between the two predators across a number of rivers in the southern coastal mountains of British Columbia. Rather than being a result of direct competition, LeBourdais et al.

(2009) suggested that this is an indirect interaction. Anti-predator behaviour by insects in the presence of fish reduces insect availability to the ducks. The widespread introduction of fish to previously fish-less reaches of rivers in western North America may well have reduced the quality of streams and rivers for breeding harlequins. Another intriguing example involves asymmetric competition for terrestrial insect prey falling on the water surface between poecilid fish (tooth-carps like the well-known aquarium guppy, *Poecilia reticulata*), swordtail (*Xiphophorus helleri*) and platy (*Xiphophorus maculatus*) and gerrids (water striders). In both laboratory and field experiments, the presence of fish decreased the foraging success of water striders without any reciprocal effect (Englund et al. 1992).

7.6 Resource partitioning and coexistence

Classical early experiments in ecology led to the view that 'complete competitors cannot coexist'. This relates to the idea that each species has an 'ecological space' in a community—defined by its use of resources and its physical distribution—called an *ecological niche*. How similar can the niches of two species be, whilst allowing them to coexist indefinitely? Does this notion of 'limiting similarity' explain the diversity of communities? Such questions occupied ecology and ecologists through much of the history of the subject (e.g. Giller 1984; Begon et al. 2006; Rockwood 2015).

Rivers and streams have rich communities of coexisting species, so what mechanisms explain this? Part of the answer probably lies in the phenomenon of *resource partitioning*, which refers to the different ways in which species with similar ecological requirements share resources. The niches of most species do not overlap completely but differ in some way in the exact resources they exploit and/or in their environmental requirements and distribution. Sometimes, the niches of pairs of species differ when they are found together (in *sympatry*) whilst being more similar when apart (in *allopatry*)—frequently the 'niche space' of both is constrained in some way when they coexist, such that resources are shared. A classical instance from streams and rivers comes from Beauchamp & Ullyott (1932), who studied free-living flatworms in the Balkan Peninsula of south-eastern Europe. Two species, both predators, appeared to partition the stream habitat when occurring in the same system. *Crenobia (Planaria) montenegrina* was found from the cool spring head downstream to a point where the maximum temperature reached about 13°C, at which point *Planaria gonocephala*

was found. In streams where only one of the two was present, each was found over a wider range of temperature (a form of competitive release), *Planaria gonocephala* extending upstream to the spring head and *Crenobia montenegrina* downstream to a point where temperature reached 16–17°C. This example could constitute a kind of resource partitioning, the species sharing the overall stream habitat and both persisting in the regional species pool. At the smaller scale of the stream reach, of course, each does seem to exclude the other, the outcome depending on temperature.

The marked spatial heterogeneity of streams and rivers over both small and large scales (the individual stone to the whole stream), in terms of temperature (as in the example above), flow and substratum type, clearly creates a large number of microhabitats. Temporal changes through seasons can affect both the physical environment and the nature and availability of resources. There is also a great range of resource types, again at a variety of scales. There should, therefore, be plenty of 'room' and opportunities for the coexistence of species through resource partitioning. There are generally three ways in which resource partitioning can occur, dividing ecological space along three kinds of *niche axes* or *dimensions*: time (temporal separation), food and/or nutrition type (food partitioning) and space (habitat partitioning). There is a large, predominantly older, literature on resource partitioning, which we can only touch on here, and some of the key mechanisms involved are illustrated through a few examples.

7.6.1 Temporal segregation

Differences in life-history patterns and timing of maximum growth among closely related species can potentially enhance coexistence. This is because their greatest demands on resources, most obviously food but possibly others, are made at different times of the year. There are many examples of such temporal differences among benthic insects (see Hart 1983), particularly among the caddis (Oswood 1976; Hildrew & Edington 1979; Georgian & Wallace 1983). Among the stoneflies (Plecoptera), four predatory species, *Dinocras cephalotes*, *Perla bipunctata*, *Isoperla grammatica* and *Perlodes microcephalus* often coexist is small stony streams in the UK. Aggressive encounters between them occur and they show no clear overall differences in prey taken. Their life cycles and seasonal growth rates differ, however (Elliott 2000). Both *Dinocras* and *Perla* are large and slow growing with 3–4-year life cycles, while *Perla* hatches earlier in the year and the two species differ in body size over their early stages—potentially with size-selective predation occurring as a consequence. *Isoperla* and *Perlodes* are smaller and have much shorter (1-year) life cycles. Fish, especially tropical species, also have specialised life histories which may relate to resource partitioning (e.g. Watson & Balon 1984).

There are contrary examples, however. Thus Tokeshi (1994) rigorously examined temporal differences in the life cycles of an assemblage of nine species of midge (Chironomidae) living on stems of a water plant in a small river in eastern England. He compared observed numbers of larvae on the stems at different times with those expected from a random (neutral) model and found no significant differences in timing and thus no evidence of temporal resource partitioning. In terms of resource partitioning, any temporal division of limiting resources between potential competitors requires that their use by one species does not compromise the resource(s) available to the next. Otherwise, the 'early bird simply catches all the worms'! Thus, it has to be asked whether more minor differences in diel foraging reported among species do actually facilitate coexistence, though they may be significant for the fitness of the species concerned. The other two niche dimensions of food and habitat offer far greater potential for resource partitioning.

7.6.2 Food partitioning

Most consumers in lotic systems are polyphagous. Nevertheless, there are many examples of dietary differences among coexisting species within guilds of stream animals, such as scrapers of epilithic biofilms. Here, a *guild* is envisaged as a group of species feeding in the same way on overall similar resources. For instance, different filter-feeders often take suspended particles of different sizes, while predation is often size related (big predators taking big prey). Are such differences important for species coexistence? The size of particles consumed by coexisting hydropsychid caddis is influenced by mesh size of their nets, which increases between instars of the same species but also between species (e.g. Malas & Wallace 1977; Hildrew & Edington 1979; Edington et al. 1984). It is feasible that these differences facilitate coexistence of species in the same river, as long as their populations are food limited. Hildrew & Edington (1979) suggested that population density in some rivers at least is more probably limited by space for net-spinning, and favoured sites also differ between species, based on body size and flow preferences (see the next sub-section). In another example, Townsend & Hildrew (1979) studied the diet of two large-bodied insect predators in a small, southern-English stream (Broadstone). In midsummer to early autumn, invertebrate prey were abundant and the two predators overlapped almost entirely in the size of stonefly prey taken. Prey were scarce in winter, however, and then the caddis *Plectrocnemia* was able to take relatively more large stoneflies—by virtue of its net (Figure 7.13). The alderfly *Sialis fuliginosa* did not capture these larger stoneflies but had a winter 'food refuge' in the form of chironomid larvae which made up 57% of the prey taken, whereas the caddis took only 16% chironomids. Thus, differences in the food

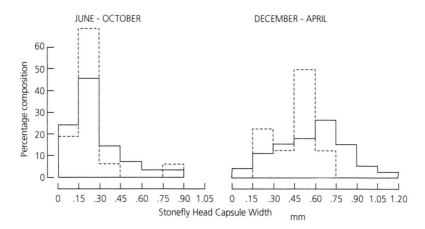

Figure 7.13 A comparison of the size of stonefly prey captured by the net-spinning caddis *Plectrocnemia conspersa* (solid line) and larval alderfly *Sialis fuliginosa* (dashed line) during the periods June to October when prey are abundant and December to April when prey are scarce. Note the shift in the size distribution in the latter period which suggests resource partitioning at times of shortage.
Source: from Townsend & Hildrew 1979, with permission from Springer Nature.

taken by these two voracious and abundant predators, in what was at the time a fishless stream, emerged only when food was in short supply.

Character displacement, when morphological differences are evident among coexisting species, is also often taken to infer resource partitioning between them. For instance, in the caddis family Hydrobiosidae from Australia, species-rich guilds of 10–20 species are frequently found, many differing in the size and shape of the prehensile forelegs which are used in prey capture (Figure 7.14). Among the seven species Lancaster (2020) studied, three showed the smallest niche breadth and greatest dietary overlap, with a predominance of chironomid prey. The other four species consumed a much broader range of prey, particularly including mayfly larvae, although two of them also consumed a significant number of caddis larvae (Figure 7.14). Overall, dietary overlap was smallest among hydrobiosid species which differed in the morphology of forelegs (such as *Ulmerochorema* spp. and

Apsilochorema obliquum or *Ethochorema turbidum*), and highest among species with similar forelegs (e.g. two congeners of *Ulmerochorema seona* and *U. rubiconum*). On an even finer scale, *A. obliquum* and *Koetingoa clivicola* have forelegs that resemble those of preying mantids and differ in the arrangement of spines along the femur as well as the curvature of the femur, which could result in differential ability to capture various prey types. As Lancaster (2020) points out, species with high dietary overlap may differ in other ways, such as foraging at different times of day or selecting different species within the main dietary food types (e.g. chironomids).

Among vertebrates, dietary differences may explain coexistence amongst tadpoles of widely distributed tropical lotic hylid frog species in south-eastern Bahia, Brazil, which are of similar size and morphological features (Santos et al. 2016). Topic Box 3.3 (Chapter 3, p. 95) also highlights an example among river birds from Himalayan streams. However, most attention has

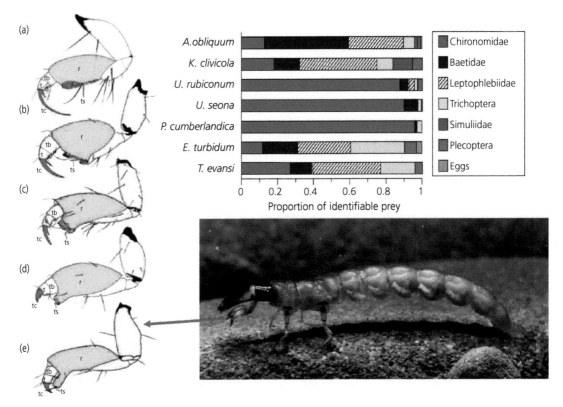

Figure 7.14 The prehensile forelegs of some Australian Hydrobiosidae, showing variations from the plesiomorphic (ancestral) condition of: (a) *Apsilochorema*, through several intermediate stages illustrated by (b) *Koetonga*, (c) *Ulmerochorema* and (d) *Psyllobetina*, to the chelate foreleg typical of the *Taschorema* complex (e) shown in the photograph. Abbreviations: f = femur, fs = femoral spine, tb = tibia, t = tarsus, tc = tarsal claw. Also shown are the proportions of major prey items in the diet of the seven species studied.
Source: from Lancaster 2020, with permission of John Wiley & Sons.

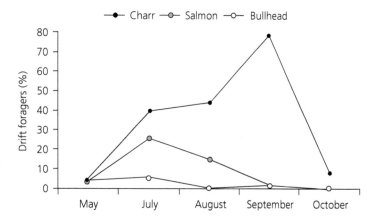

Figure 7.15 Seasonal shifts in the surface-drift foraging amongst three sympatric fish species (proportion of individuals with surface prey in their stomachs).
Source: from Sanchez-Hernandez et al. 2016, with permission of Springer Nature.

been focused on species coexistence and resource partitioning amongst fish, particularly in the salmonids. In Irish streams, Atlantic salmon (*Salmo salar*) and brown trout (*Salmo trutta*) frequently coexist. The larger gape size of brown trout and their tendency to hold station near the surface leads to encounters with both drifting and allochthonous terrestrial prey, whereas salmon tend to feed mainly on benthic invertebrates. Deciduous riparian cover may therefore facilitate vertical partitioning of feeding position within the water column between Atlantic salmon and brown trout (Dineen et al. 2007). In grassland streams the additional allochthonous source of terrestrial prey is much less abundant so dietary overlap is greater. Salmon parr are often excluded from pools through interspecific competition with the more aggressive trout and occupy shallower rapids (Kennedy & Strange 1982; Heggenes et al. 1999).

Seasonal shifts in the use of surface prey may reduce competition amongst sympatric fish species during summer in northern subarctic rivers. This form of resource partitioning is found in Arctic char (*Salvelinus alpinus*), which has a diet dominated by prey taken from the surface, whereas Atlantic salmon exploit surface prey rather less while the alpine bullhead (*Cottus poecilopus*) does so hardly at all (Sanchez-Hernandez et al. 2016; Figure 7.15). Seasonal changes in the availability of surface-drifting terrestrial prey can affect sympatric fish, as seen in mountain streams in northern Japan (Nakano et al. 1999). Two congeneric, and often sympatric, char species (Dolly Varden, *Salvelinus malma*, and white-spotted char, *S. leucomaenis*) are slightly different in head and body morphology. Char have two distinct foraging modes; either (i) ambushing drifting terrestrial invertebrates from relatively fixed positions, the larger fish holding the best positions,

or (ii) active searching for benthic prey over large areas of substratum. Both species switch from the former to the latter as drift density declines seasonally, although Dolly Varden shifts at a higher threshold of drifting prey (and thus the proportion of benthic prey is greater). This leads to reduced direct interference competition as the supply of terrestrial drift declines and reduces dietary overlap—while the two species still occupy the same stream reach at the same time (Nakano et al. 1999). This ability to adjust foraging mode in Dolly Varden (a form of flexible niche shift) might facilitate coexistence with the dominant, non-native rainbow trout (*Oncorhynchus mykiss*), although the growth rates and biomass of Dolly Varden were lower when coexisting with rainbow trout and the species is being displaced from many streams across Hokkaido in northern Japan (Marcarelli et al. 2020).

One further point is that greater prey diversity can allow resource partitioning by food type in instances where species do not appear to segregate by foraging time or habitat (Sanchez-Hernandez et al. 2017) and dietary overlap may be reduced and fish become even more specialised when resources become scarce, as in a Panamanian stream during the dry season (Zaret & Rand 1971).

7.6.3 Habitat segregation

It is clear that habitat seems to be by far the most important niche dimension as far as resource partitioning is concerned. Benthic organisms show clear preferences for particular microhabitat features, particularly among closely related species (e.g. Hildrew & Townsend 1987). For example, guilds that differ in growth form and flow preferences can be identified among diatom assemblages, with consequent differences in the substratum occupied. Similarly, separation of species by substratum preferences has been clearly demonstrated among macroinvertebrates (see Chapter 5, section 5.2.3, p. 145). On an even smaller spatial scale different species of the caddis *Hydropsyche* often use the distinct microhabitats offered by large and stable stones on the river bed, some occupying the upper (often mossy) surfaces and sides, some in smaller crevices between the stones and some in the smaller calibre sediments around them (Edington et al 1984; Harding, 1997; see Figure 7.16).

Among fish, there are many examples of habitat partitioning, such as the almost entirely separate distribution of congeneric darters (subfamily Etheostomatinae) living in pools within a single drainage basin (Page & Schemske 1978), tropical rainforest fish which live at different depths in the water column of streams (Watson & Balon 1984; Welcomme 1985), and different flow preferences leading to altitudinal niche partitioning between white-spotted char (*Salvelinus leucomaenis*) and masu salmon (*Oncorhynchus masou*) in Japanese streams (Figure 7.17; Morita et al. 2016).

In a fascinating field study, Fausch et al. (2021) continued their work on Dolly Varden and white-spotted char in northern Japanese streams, recording

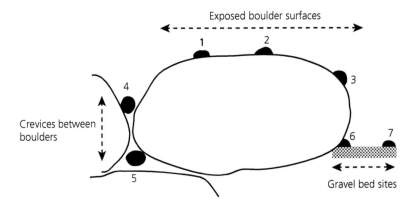

Figure 7.16 Microhabitat partitioning among Malaysian species of *Hydropsychidae* at the scale of a boulder. One species (*Hydropsyche annulata*) occupies the upper and flat lateral surfaces (1, 2 and 3), a second (*Hydropsyche*, new species) uses crevices between boulders (4 and 5) and a third (*Synaptopsyche klakahana*) the adjacent gravel bed sites or angles between the gravel and the boulder (6 and 7).
Source: from Giller & Malmqvist 1998, redrawn after Edington et al. 1984.

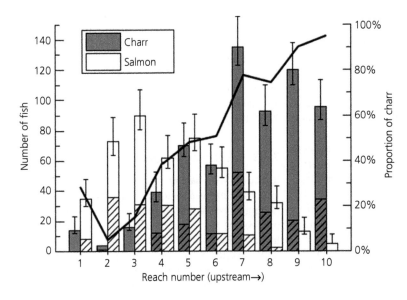

Figure 7.17 Habitat partitioning between white-spotted charr (*Salvelinus leucomaenis*) and masu salmon (*Oncorhynchus masou*) showing numbers (means ± 95% confidence intervals) and the proportion of charr (solid line) in each of ten 70-m study reaches from the river mouth to an impassable waterfall on the Ohkamaya River, Hokkaido, Japan. Hatched areas indicate young-of-the-year fish.
Source: from Morita et al. 2016, with permission of Springer Nature.

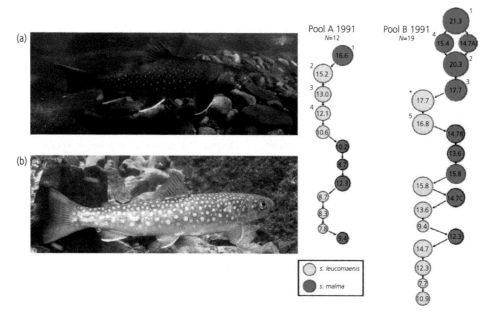

Figure 7.18 Mean interspecific dominance hierarchies between (a) Dolly Varden (*Salvelinus malma*) and (b) white-spotted char (*S. leucomaenis*) in two representative study pools (A and B) in the Poroshiri Stream in south-central Hokkaido, Japan. For each individual fish the circle is scaled for fork length and the numbers alongside the top-ranking fish are dominance ranks determined by knockout removal experiments. One fish in pool B did not interact and is marked by an asterisk.
Source: from Fausch et al. 2021, with permission of John Wiley & Sons. Photos (a) from US Fish and Wildlife Service and (b) from Morita et al. 2016, with permission of Springer Nature.

agonistic interactions and competitive hierarchies and performing 'knockout' removal experiments. These two salmonids are effectively equal competitors for foraging positions, taking drifting prey during the summer. Dominance hierarchies were actually based mainly on body length (larger fish being more likely to win contests) regardless of species (Figure 7.18).

7.6.4 Alternative explanations for coexistence of similar species

One clear pattern to arise in the study of communities in general is that where species overlap in resource use in one particular niche dimension, they generally differ in use of another; that is, they show *differential niche overlap* (Giller 1984). This kind of differential overlap can be seen very clearly among three coexisting *Hydropsyche* species in a western Malaysian stream (Edington et al. 1984). One species has a smaller mesh net than the other two, occupies fine gravel substrata and collects smaller-sized food particles. The other two species occupy similar substrata but differ in microhabitat, as illustrated in Figure 7.16. A large-scale review and analysis has highlighted the importance of niche

segregation in determining the structure of freshwater fish communities and that the interplay of habitat and environmental filters with niche specialisation mediates the strength of interspecific competition (Comte et al. 2016).

An important point to make here is that other species attributes, such as ability to cope with disturbances or colonisation ability, can contribute to successful coexistence of species that otherwise show little resource partitioning. In the face of relatively frequent disturbance, the unstable environment keeps populations well below their carrying capacity so resources are not limiting and competition is not an issue. One example is the *Leucotrichia* (caddis)–*Paragyractis* (moth) interaction we discussed earlier, where floods remove the competitively superior caddis species and allow colonisation of rock surfaces by the more resilient moth larvae (McAuliffe 1984). Colonisation ability could therefore be considered another niche dimension.

Finally, a number of ecologically similar species (with similar habitat and resource requirements) may theoretically be able to coexist if the resource is spatially patchy and renewable, as in the *aggregation model for coexistence* (see Atkinson & Shorrocks 1981; Ives

1988; Sevenster 1996). If species are independently dispersed across such patches in an aggregated pattern, then by chance they may avoid clumping together in the same patch. Thus, intraspecific competition across the system of patches is likely to be stronger than interspecific competition (a criterion for coexistence). Note that there is no 'need' for a qualitative difference in the nature of the patches, just that there is independent aggregation of each species—which thus occupy what have been termed *probability refugia*. Further, coexistence is possible even if interspecific competition is strong in any single patch. An example of this form of coexistence among ecologically similar species in streams may be found in the chironomids occupying the patchy and ephemeral resources of macrophyte stems described by Tokeshi & Townsend (1987). Further, stream macroinvertebrates colonising ephemeral patches of leaf packs were found by Murphy et al. (1998) to be distributed among the patches in a non-random manner across different spatial scales (from individual leaf packs 2 m apart to blocks of nine leaf packs distributed over 2 km of the stream). Almost all major colonising taxa exhibited strong intraspecific aggregation whilst interspecific aggregation among taxa was rare. This kind of intrinsic aggregation in space could be an important structuring force in the diverse assemblages of species associated with leaf packs.

7.7 Positive species interactions

Not all interactions between species are negative for one or both of the interacting species and nature is full of examples of either direct or indirect positive species interactions such as symbiosis and indirect facilitation. *Symbiosis* is effectively the umbrella interaction through which a prolonged, evolved interaction or close and intimate relationship occurs between species which are often phylogenetically widely separated. While it is sometimes described as including parasitism, most commonly symbiosis refers to positive interactions between species. It can be obligate, where the species have evolved to be interdependent, or facultative, where they can live independently although one or both benefit through the relationship. *Mutualism* is a form of symbiosis where both partners have beneficial fitness gains, for example through exchange of resources or provision of other fitness benefits in exchange for resources. *Commensalism* on the other hand refers to cases where one of the interacting species benefits, through access to resources, transport, protection or provision of habitat, whilst the other

is unaffected with no net cost or benefit. *Facilitation* occurs where the presence or activity of one species alters the environment in a positive way for one or more other species or reduces the extent of their interaction with natural enemies or competitors, thereby enhancing the fitness of these other species.

7.7.1 Mutualism and commensalism

Despite the ubiquity of positive species interactions in nature, Silknetter et al. (2020) conclude from their review that comparatively little is known in fresh waters, particularly of mutualism and commensalism. They do, however, describe some cases of commensalism from streams and rivers, including *phoresy*. For instance, an individual chironomid or gammarid may attach to a more mobile invertebrate, fish or aquatic bird, thus promoting dispersal of the commensal species while providing no apparent benefit to the host. A form of mutualism is ichthyochory where a fish gains food by eating the fleshy fruits of riparian trees that fall into the river and the trees benefit from dispersal of their seeds having passed through the digestive tract of fish.

A number of other intriguing cases can be found and it is worth looking at some of these in a little more detail. One of the best studied examples lies in the cleaning symbiosis relationship between crayfish and worms. North American, European and Asian river crayfish host multispecies assemblages of ectosymbiotic branchiobdellan annelid worms, while those in the southern hemisphere host temnocephalan flatworms. The ectosymbiont benefits from a safe breeding habitat and access to food, while they sometimes clean respiratory surfaces of the crayfish, removing biofilms and sediment. However, it appears that young crayfish may resist all worm species (by grooming themselves), because the symbiosis may not be beneficial at this age as the crayfish grows rapidly and moults frequently. There is also a potential cost as the worms may consume host tissue. However, adult crayfish moult infrequently and benefit more from the worm. They then frequently host both large and small species of the symbiont (Brown et al. 2012). This relationship can actually fluctuate between mutualism and commensalism, depending on the degree of environmental fouling. Benefits to crayfish are clear when under intense fouling while no effects are seen in cleaner environments (Lee et al. 2009). The relationship may become one of parasitism, however, when the worms feed on crayfish gill tissue when their densities are high and other food is short (Creed & Brown 2018).

Figure 7.19 The nest of the bluehead chub *N. leptocephalus* (top), and (below) an active nest with: (a) a host male bluehead chub initiating a spawning clasp with a female (not visible), and six nest associate species, including (b) mountain redbelly dace *Chrosomus oreas*, (c) rosyside dace *Clinostomus funduloides*, (d) central stoneroller *Campostoma anomalum*, (e) white shiner *Luxilus albeolus*, (f) crescent shiner *L. cerasinus* and (g) rosefin shiner *Lythrurus ardens*.

Source: Photos are from Toms Creek, Virginia, US by Emmanuel A. Frimpong. https://doi.org/10.1371/journal.pbio.2004261.g001 (Frimpong 2018, under Creative Commons Attribution Licence).

A surprising example of commensalism is nest association amongst stream fish. This appears common in a wide variety of taxa in Africa, East Asia and North America, where individuals of one or more species lay eggs in nests constructed by a host species (Peoples & Frimpong 2016). For example, the most common nest association in North America involves the cyprinid genus *Nocomis* (a chub) which facilitates the reproduction of over 30 other 'associate' species. Male *Nocomis* build large gravel mound nests and provide parental care in the form of egg guarding and nest cleaning to its own progeny and those of the associates (Figure 7.19). This commensal behaviour is nearly obligate for *Chrosomas oreas* (Mountain redbelly dace) and *Clinostomus funduloides* (Rosyside dace), two nest associates of *Nocomis leptocephalus* (the bluehead chub), throughout the New River basin in North Carolina, as their reproductive success depends on the nesting activities of the host (Peoples & Frimpong 2016). Interestingly, these nest associations may even be mutualistic as the bluehead chub appears to benefit from the additional eggs of an associate, the yellowfin shiner (*Notropis lutipinnis*), through a reduction (dilution effect) in predation risk to their own eggs (Kim et al. 2020).

A curious example is the obligate species-specific mutualism between chironomids (*Cricotopus* sp.) and the colonial cyanobacteria *Nostoc parmeliodes* and *Nostoc verrucosum* (reviewed in Holomuzki et al. 2010). The chironomids graze the cyanobacterium from the inside of a colony, gaining food and protection, while the cyanobacterium benefits through a reshaping of the colony to enhance light and nutrient uptake and resecuring detached colonies to the substratum.

Positive interactions can even stretch beyond the river channel, as seen in the relationship between riparian trees and riverine fruit-eating fish (Horn et al. 2011). Surprisingly there are at least 276 known frugivorous species, with examples found in six major biogeographical regions, with the most species (around 150) in the Amazon basin. The fish are usually large-bodied, long-lived, omnivorous neotropical Characiformes (e.g. piranhas and the related pacus) and Siluriformes (catfish) but also include Holarctic cypriniformes (carps and minnows). In the Orinoco, Amazon and Paraguay river basins the seeds and fruits of over 100 riparian tree, shrub and vine species are consumed by fish and transported long distances upstream and across floodplains during floods. The fish gain rich sources of carbohydrates, lipids and proteins and in turn disperse the seeds.

7.7.2 Indirect facilitation

It is clear that direct species interactions can lead to dynamic changes that can ramify more widely through the community. The phrase 'my enemy's enemy is my friend' captures this idea fairly simply. For instance, a predator (or pathogen, or grazer or competitor)—call it 'species A'—that reduces the impact of species B that itself reduces the fitness of species C, actually can benefit species C. Thus, species C can be said to be facilitated indirectly by species A. We have already referred to several examples of this type of indirect facilitation. Thus, the microsporidian parasite *Cougourdella*, by reducing the density of the competitively dominant grazer *Glossosoma nigrior* in southwestern Michigan streams, facilitated increases in other subdominant grazers (Kohler & Wiley 1997) (see section 7.5.1 and below). Such cases are widespread in nature—including in streams and rivers—and further instances are dealt with throughout this book.

A great variety of other ecological processes and phenomena (not all of them purely species interactions) involve some sort of indirect facilitation (see review by Silknetter 2020). Some species modify the physical environment, changing it for others—a process known as *ecological engineering* (see more in Chapter 10). For instance, reductions in deposited fine sediment on the substratum surface, as caused by the activity of fish, crayfish and others by 'bioturbation', may facilitate algal scrapers that feed on firmly attached epilithic algae that would otherwise by buried. Some filter-feeders increase the clarity of the water column, facilitating benthic algae by increasing light supply to the river bed. Beds of freshwater mussels in rivers can provide a hard substratum that might otherwise be absent.

It is evident that many forms of ecological interactions, between species or between species and the abiotic environment, can have 'knock-on' effects within ecosystems. How these are manifested depends not just on the actual identity of the actors involved, but, as we will see in the next section, also on how the interactions are interconnected within the complex 'webs' that we see within the ecosystem.

7.8 From species interactions to food webs

Hitherto, we have largely discussed 'pairwise' species interactions more or less independently. As encapsulated in concepts such as 'indirect facilitation' above, however, in nature such pairs of species almost always interact with others, potentially forming a 'matrix' of linkages which may be highly complex. Thus, pairwise feeding interactions are normally linked together 'vertically' to form a classic food chain (plant–herbivore–carnivore). Such food chains again usually have 'lateral' feeding links to other chains to form food webs, while non-trophic species interactions (such as competition and nutualisms) also have dynamic effects. While schematic food webs are depicted in almost every ecology text book, this is in reality an extremely challenging area to unravel, both practically and theoretically. Food webs essentially provide a 'map' of how the dynamics of individual species may have effects that 'spread' throughout the community and indicate the route taken by carbon and energy flowing through the ecosystem as well as the resultant transfer of nutrients and other materials (including pollutants). Once we have clear accounts of a sufficient sample of such webs, we can then seek to identify and explain recurrent patterns (both their structure and dynamics) between systems.

7.8.1 Food webs—some basic features

There are a number of ways in which we can attempt to characterise and depict feeding interactions in communities. *Connectivity webs* are simply maps of 'who eats what', recording all feeding interactions that have been observed (Figure 7.20a). While illustrating the potential complexity of food webs (as a form of network), connectivity webs effectively view all links as equally 'important' (a link that has been observed only once over a study period 'counts' just as much as a link that is recorded many times). Early compilations of food webs consisted of data gathered primarily for reasons other than describing the food web. In one of the most influential, Briand & Cohen (1987) examined 113 different data sets, from all kinds of habitats, of which nine were lotic. These were depictions of the simple presence of links and early estimates of *web connectance* (the fraction of all possible pairwise feeding links in a web that is realised) and other web statistics were made—despite the limitations of the data.

Alternatively, we can consider the relative strength of links by weighting them in terms of energy or biomass (or sometimes nutrients or even pollutants) passing along them, in so-called *flow webs* (Figure 7.20b). This is one way of identifying the most important pathways through the web in terms of energy, etc. However, even this second type of web alone does not necessarily capture the dynamical impacts of links (i.e. the effects of a predator on its prey) (Woodward & Hildrew 2002; Benke 2018).

A further approach, a *population dynamics web* (Figure 7.19c), portrays the impact of a consumer (e.g. a predator) on its food (e.g. its prey). To do this we must calculate the production of species or other taxa (i.e. the

(a) Connectance

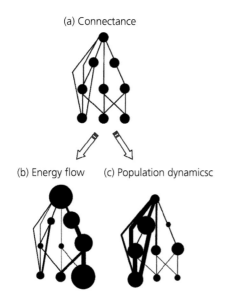

(b) Energy flow   (c) Population dynamicsc

Figure 7.20 Conceptual diagrams of the three major approaches to characterising food webs. In the connectence web (a), all taxa are of equal importance (i.e. the size of the 'nodes' are identical) and feeding links are also weighted equally (i.e. line widths are identical). In the energy-flow web (b) the structure is skewed among species and links, thus node size and link thickness vary according to the energy flow along that link. However, species that contribute most to the energy flow in the web do not necessarily have an appreciable effect upon the population dynamics of their prey. The population dynamics web (c) tries to reflect that effect. For instance, link thickness varies according to the ingestion of a prey species as a fraction of its production (I/P)—see text for further details.
Source: from Woodward & Hildrew 2002, with permission of John Wiley & Sons.

only on these primary producers, while secondary consumers (predators) feed exclusively on these herbivores, and so on—the basis of the classic trophic structure of ecosystems, with clear (whole-number or 'integer') trophic levels. Many food webs, including those in streams and rivers, however, are not really like that at all. Most species, other than primary producers, are omnivores, feeding at more than one trophic level. For instance, the carbon in detritus itself can originate at a variable number of links removed from the initial primary producer. As we have seen, many consumers have a mixed diet of animal prey and algae/detritus (sometimes more herbivorous or detritivorous when they are small, turning largely to carnivory as they grow). Therefore, the 'height' in the web at which species feed is variable and is often known as *trophic position*, based on estimates of the fractional amounts of each food type assimilated by each consumer and the trophic positions of all its food resources (Levine 1980; Benke 2018). Thus, clear whole-integer trophic levels in lotic food webs are unusual. The prominence of omnivory in real food webs, its significance and the circumstances in which occurs, have been much debated.

Characterising natural food webs in any of the above ways, in running waters as in any other system, is extremely demanding on data. The size of small invertebrates and consequently of their prey, for example, makes the identification of many prey species difficult without specialist microscopic techniques and taxonomic expertise, particularly in the case of the meiofauna. Consequently, the taxonomic detail is usually much coarser at the lower trophic levels (e.g. prey identified to family or order or even functional group) than at the higher ones (which can usually be resolved to species), and often many species are too rare to allow for suitably robust assessment of trophic relationships or production. At present we must ackowledge that no food web can accurately represent nature in full (Power et al. 2013) and there are probably none that can be considered as complete. Nevertheless, there are now some very well-described lotic food webs, and the problems encountered in their derivation are themselves informative. We begin with a few real examples, then confront the limitations and problems in this area, and finally use food web data to ask and answer some general questions about pattern and process in lotic food webs.

'nodes' in the web) and then assess the fraction of that prey production that is consumed by the predator—that is, the thickness of the links connecting nodes denotes ingestion/production (I/P). It thus identifies pressures placed by consumers on their 'prey'—'top-down' through the web. An alternative in the population dynamics webs is to express links as the fraction of production by a predator that is supported by a particular prey (i.e. P/I). This identifies possible limits placed on consumers and predators by their food supply—'bottom-up'. There are in theory even more sophisticated ways of expressing linkage strength, including testing or modelling the reciprocal dynamical impact of pairs of species on each other's population density (by removal experiments, for instance) but these are impossibly demanding on data in real, multispecies webs.

Primary producers, and other basal resources, by definition, lie at the base of the food web. In the simplest case, primary consumers (herbivores) then feed

7.8.2 Food webs in streams and rivers

Not surprisingly, ecologists working on connectivity food webs have often chosen what appear to be relatively simple systems in terms of limited species richness. In the Arctic National Wildlife Refuge on the North Slope of Alaska, Parker & Huryn (2006) compared the food webs of a mountain stream (subject to summer storm flows) and a spring (with relatively stable discharge). Both streams were species poor (the webs included 20 and 25 taxa, respectively, though there would certainly have been more). The biomass of bryophytes in the spring system was > 1,000 times greater than that in the mountain stream, yet the two food webs were relatively similar, apart from the addition of an extra top predator, the dipper (*Cinclus mexicanus*), in the spring. This feeds on young fry of the Dolly Varden char (*Salvelinus malma*), the other vertebrate predator in the system (Figure 7.21). The char is a generalist, consuming between 12 and 18 taxa across the two streams; the dipper less so, with four to five

taxa in the diet, while most of the invertebrates are omnivores.

One of the best-defined connectivity webs in the literature resulted from long-term studies of Broadstone Stream (Figure 7.22; Hildrew 2009). This is an acidic, forested stream in south-east England of relatively low diversity. At the beginning of the study in the early 1970s, there were no fish in upstream reaches, but some very abundant insect predators. The main basal food resources were leaf litter, FPOM and iron bacteria. Major feeding links among the insects were identified in these earlier years (Hildrew et al. 1985). The microcrustacea were then added and resolved (mainly) to species (Lancaster & Robertson 1995). A large-bodied predatory insect (the golden-ringed dragonfly, *Cordulegaster boltonii*; species 57 in Figure 7.22) was present at the outset of the study but was initially

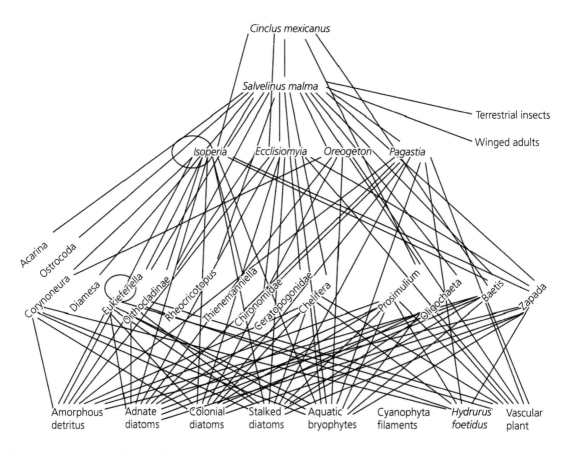

Figure 7.21 A connectance food web for the spring stream in the Ivishak River catchment in Alaska, sampled in August 2002. Algal taxa are summarised into functional groups to simplify the food web diagrams. Circles indicate cannibalism. The connectance food web of a comparative mountain stream was similar form the absence of the dipper (*Cinclus mexicanus*).
Source: from Parker & Huryn 2006, with permission from John Wiley & Sons.

extremely scarce. Its density irrupted fairly suddenly in the early 1990s, when it was the clear top predator (Woodward & Hildrew 2001). Much of the remaining meiofauna (including rotifers, oligochaetes and nematodes) was then added to the web (Schmid-Araya et al. 2002a). Later still, in 2006, Brown trout (species 129 in Figure 7.22) invaded the headwater system from areas downstream (as the acidity of the system declined—see below; Layer et al 2011). Even in this relatively low-diversity system, the food web as described becomes extremely complex because of the increased completeness and taxonomic resolution. There are also seasonal changes in basal resources, and species diversity. Many of the predatory insect species have a very high number of trophic links; for example, the predatory megalopteran *Sialis fuliginosa* (species 68 in Figure 7.22) and caddis *Plectrocnemia conspersa* (species 65 in Figure 7.22) fed on 48 and 56 different

prey species, respectively, and one of the predatory tanypod chironomids (*Trissopelopia longimana*, species 82 in Figure 7.22) included 73 species in its diet. Gut contents analysis of the brown trout showed they consumed 81% of the different invertebrate taxa in the food web (excluding the meiofauna), as well as terrestrial invertebrates, the adult stages of the aquatic insects and all the resident large invertebrate predators (Layer et al. 2011). Clear dietary changes with age were found amongst the insect predators, with ontogentic shifts from organic matter and meiofauna to dominance of meiofauna and then inclusion of macrofauna, contributing to seasonal changes in the food web through the year. This web probably approaches what is possible using conventional taxonomic techniques—more species and other prey taxa would undoubtedly be recognised and resolved using molecular methods.

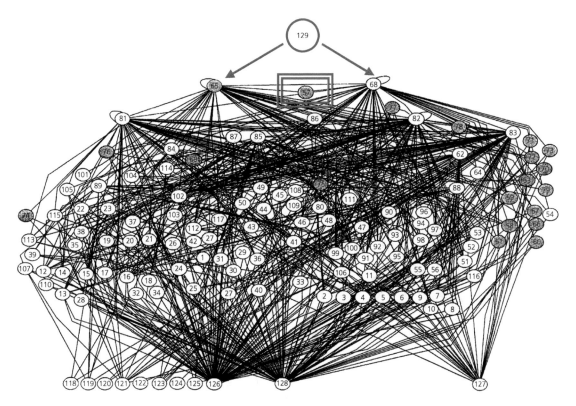

Figure 7.22 An integrated (summary) connectance food web in Broadstone Stream for all four seasons combined. The numbers are different food items/species. Open circles at the foot of the web (118–127) are the 'basal' species (118–121 algal species, 122 plant material, 123–125 various invertebrate eggs; 126 FPOM, 127 CPOM and 128 *Leptothrix ochracea* iron bacteria); other open circles are the meiofaunal species; and shaded circles the macrofaunal species.

Species identification can be found in the data appendix at Ecological Archives E083-020-A1. Invading top predators *Cordulegaster boltonii* (57) and Brown trout *Salmo trutta* (129) are highlighted.

Source: from Schmid-Araya et al. 2002a, with permission from John Wiley & Sons.

Dramatic changes to the resource base will naturally alter the shape of a food web, as seen following a litter-exclusion experiment in steep, deciduous and strongly heterotrophic forested catchments of the Coweeta Hydrologic Laboratory in North Carolina, USA (Hall et al. 2000; Wallace et al. 2015; see Chapter 8 and Topic Box 1.1). Litter exclusion had enormous effects on the connectivity web, but impacts were most apparent in the flow webs of reference and manipulated streams (Figure 7.23). In the litter-exclusion stream, the magnitudes of detrital flows were inevitably much lower, while a greater fraction of food web energy flow came from wood, largely as a result

of a switch in the diet of *Tipula* (a larval 'crane fly'). Some taxa were effectively lost and flows to predators were fewer and smaller in the litter-excluded stream compared to the reference stream (Figure 7.23 c, d), although these flows had higher per-biomass consumption coefficients, suggesting stronger interactions among the remaining common taxa.

The effect of changes in basal resources longitudinally within a single river are also reflected in organic matter flow food webs, as seen for instance along the Little Tennessee River, North Carolina, USA (Rosi-Marshall & Wallace 2002). The dominant resource changes from leafy detritus in upstream reaches

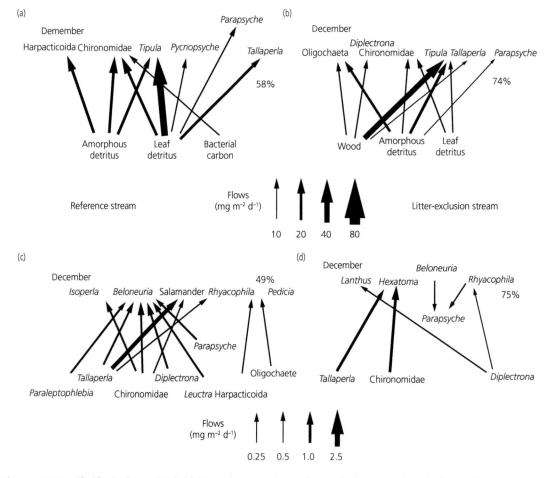

Figure 7.23 Simplified food webs associated with litter-exclusion experiments, showing the largest organic matter flows for December in the Reference (left side; a, c) and Litter-excluded streams (right side, b, d) at Coweeta Hydrological Laboratory. The top panel (a, b) illustrates flows from basal resources to invertebrate consumers and the bottom (c, d) from invertebrate prey to predators. The thickness of the arrows is proportional to flow magnitude, and the percentage values denote the total organic matter flow represented by the flow food web. Note the enormous change in scale in the upper and lower panels.
Source: from Hall et al. 2000, with permission from John Wiley & Sons.

(< stream order five) to algae in the mid-reaches (order six) and suspended FPOM in the downstream (seventh-order) site (Figure 7.24). In this study, these flow food webs were recognised as incomplete, with only eleven genera included, but these contributed between 50% and 66% of the total macroinvertebrate secondary production at each study site. The general point here is the structural stability in the web (the community at the sites is relatively similar), even though the webs at different sites were very different in terms of the food resources used, in the rates at which they were consumed and in the secondary production of the various taxa along the gradient. Of course, the changes in resources apparent along a river are much less abrupt than those imposed by the experimental interruption of litter inputs at Coweeta (Hall et al. 2000; Wallace et al. 2015), so the differences in outcome in the structure of the web in these two examples (i.e. the litter exclusion at Coweeta, and longitudinal shifts in the Little Tennessee River) are perhaps not surprising.

These and other studies of stream and river food webs affirm the lack of clear (whole) integer-based trophic levels in many systems, due to the extent of omnivory and the seasonal and spatial variation in the type and availability of resources. For example, a number of taxa from the grazer and collector-gatherer functional groups can vary from using mainly allochthonous energy sources to relying on entirely autochthonous sources depending on where they are found along the stream size gradient in both temperate (USA) and tropical (West Indies) streams (Collins et al. 2016). Instead, in flow and population dynamics food webs, *trophic position* depicts the mean number of steps that a species is removed from basal resources (a much more continuous variable). Thus, energy (or organic matter) flow pathways, in units of production and incorporating predator impacts on prey and their trophic position, provide a further advance in the exploration of stream and river food webs. This was achieved for much of the rather impoverished

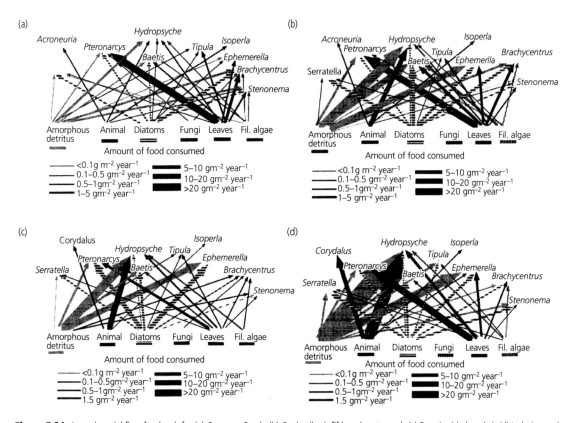

Figure 7.24 Annual partial flow food web for (a) Coweeta Creek, (b) Conley (both fifth-order streams), (c) Prentiss (sixth order), (d) Iotla (seventh order). The webs illustrate the rates of consumption (g m^{-2} y^{-1}) of each trophic interaction. The width of the arrows indicates the rate of flow (i.e. a thicker arrow is a higher flow rate), and the pattern of the arrow indicates the food resource (e.g. lightly shaded is amorphous detritus).
Source: from Rosi-Marshall & Wallace 2002, with permission from John Wiley & Sons.

Broadstone food web, excluding the meiofauna (Woodward et al. 2005). Benke (2018) also demonstrated this approach for the much more productive 'snag habitat' in the Ogeechee River, investigating the biota living on the surfaces of fallen trees (a relatively stable habitat in this meandering floodplain river). This latter study has provided one of the best examples to date of a flow food web from a stream habitat.

The complete flow food web for the Ogeechee snag habitat included bottom-up links with flows ranging over almost six orders of magnitude, ranging from 1 to 81,494 mg m^{-2} y^{-1} (Figure 7.25). The distribution of the 131 links between basal resources and 32 primary consumer/omnivore taxa was spread widely across flow magnitudes, with most ranging between 10 and 5,000 mg m^{-2} y^{-1} (82% of links). In contrast, most of the 331 predator–prey links consisted of small flows (260 flows < 50 mg m^{-2} y^{-1}) but the highest flows (including only 53 flows > 100 mg m^{-2} y^{-1}) represented 85% of total flow to predators. This was mainly through consumption of primary consumers (largely chironomids and mayflies) by omnivorous caddis larvae. The dragonfly *Neurocordulia molesta* was a major consumer of most taxa except the top predator *Corydalus cornutus* (another dragonfly). Trophic position (TP) analysis

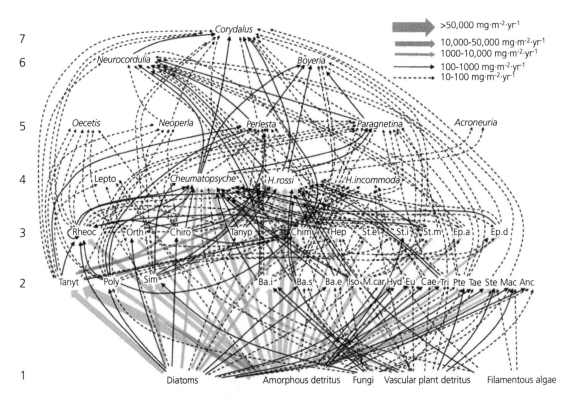

Figure 7.25 A 'simplified' flow food web diagram for the snag invertebrates in the Ogeechee River, including links with values > 10 mg m^{-2} y^{-1} (n = 272). The maximum integer trophic position is indicated by the numbers on the left. Line thickness indicates a roughly 10-fold range of values among flows. Brown arrows arise from amorphous detritus, green arrows from diatoms, blue arrows from vascular plant detritus, and red arrows from chironomids (thinner lines are not colour coded).

Abbreviations for complete names: *Corydalus cornutus*; *Neurocordulia molesta*; *Boyeria vinosa*; *Neoperla clymene*; *Perlesta placida*; *Paragnetina kansensis*; *Acroneuria* spp.; Lepto, Leptoceridae; *Cheumatopsyche* spp; Rheocr, *Rheocricotopus* spp.; Orth, miscellaneous Orthocladiinae; Chiro, miscellaneous Chironomini; Tanyp, Tanypodinae; Chim, *Chimarra moselyi*; Hep, *Heptagenia* sp.; St.e, *Stenomena exiguum*; St.i, *S. integrum*; St.m, *S. modestum*; Ep.a, *Ephemerella argo*; Ep.d, *E. dorothea*; Tanyt, Tanytarsini; Poly, *Polypedilum* spp.; Sim, *Simulium* spp.; Ba.i, *Baetis intercalaris*; Ba.s, miscellaneous *Baetis*; Ba.e, *B. ephippiatus*; Iso, *Isonychia* spp.; Ma.c, *Macrostemum carolina*; Hyd, *Hydroptila* sp.; Eu, *Eurylophella* sp.; Cae, *Caenis* spp.; Tri, *Tricorythode* ssp.; Pte, *Pteronarcys dorsata*; Tae, *Taeniopteryx lita*; Ste, *Stenelmis* spp.; Mac, *Macronythus glabratus*; Anc, *Ancyronyx variegata*. Zygoptera is not shown as all flows are < 10 mg m^{-2} y^{-1}.

Source: from Benke 2018, with permission from John Wiley & Sons.

yielded TPs from 1.0 to 3.69 over a maximum of approximately seven steps (based on following individual food chains within the web; Figure 7.25).

By constructing an *I/P* (ingestion/production) food web (which uses the same linkages as the flow food web but quantifies them in relation to the fraction of prey production ingested by the predator; Figure 7.26), Benke (2018) then showed that most (83%) of the prey sustained only weak or very weak impacts from *individual* predator species. However, 70% of the 40 prey taxa suffered *cumulative predator* impacts (i.e. summing the impacts of all predators) of more than half of their production, while 35% of them lost > 90% of their production to predators. Taking a 'top-down' view, the apex predator (*Corydalus cornutus*) had only weak impact on individual primary consumers, weak–moderate impacts on other predators and omnivores and highest impact on other dragonflies. Predatory

stoneflies consumed mainly primary consumers, as did omnivorous caddisflies, although the latter had a significantly greater impact (the caddisflies taking 23–44% of prey production as opposed to < 10% for predatory stoneflies).

In addition to the important allochthonous leaf-litter subsidies to many streams, subsidies of organic matter from terrestrial prey entering streams from adjacent riparian zones can also significantly influence the nature of riverine food webs. This is beautifully illustrated by the flow food web in the forested, spring-fed Horonai Stream in Hokkaido, Japan which has been the subject of significant work by the late Shigeru Nakano and colleagues. In a recent study (Marcarelli et al. 2020), the relative importance of terrestrial invertebrate subsidies to the five predatory fish in the stream is clearly demonstrated (Figure 7.27). What this cumulative picture hides, however, is the extent of seasonal

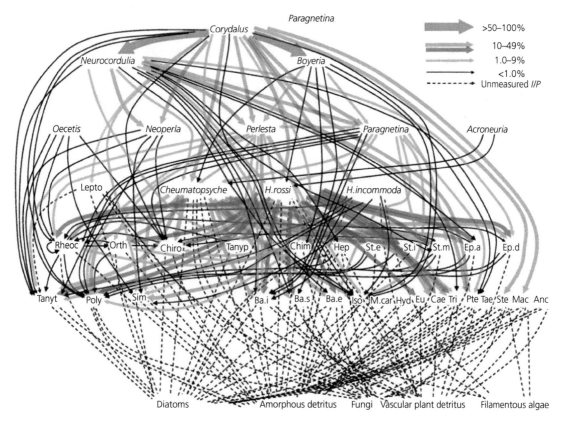

Figure 7.26 A simplified *I/P* (ingestion/production) food web for the snag invertebrates in the Ogeechee River, including links with values > 10 mg m^{-2} y^{-1} (n = 159). *I/P* values are shown only if original flows were > 10 mg m$^-$2 y^{-1} (n = 159). The thickness of arrows indicates the fraction of prey production ingested by predator (< 1 to 100%) and the taxa are as indicated in Figure 7.25 using the same placement. Dashed lines show connections to basal resources, but I/P values are undefined due to the absence of production for basal resources.
Source: from Benke 2018, with permission from John Wiley & Sons.

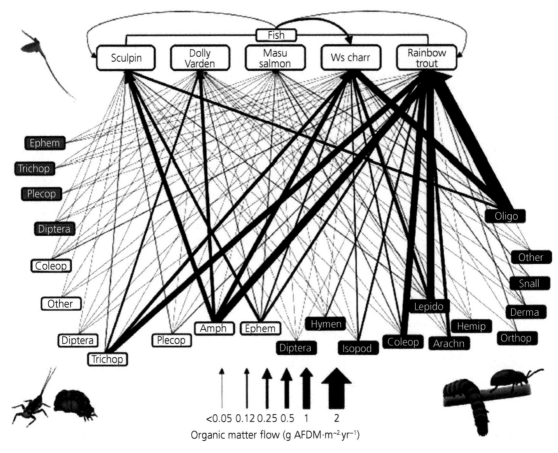

Figure 7.27 A simplified flow food web focusing on macroinvertebrate–fish trophic links for the Horonai Stream showing total annual organic matter flows to fish consumers from terrestrial prey (black boxes), aquatic benthic invertebrates (white boxes), adult aquatic invertebrates (dark grey boxes), and other fish (curved arrows on top row). Taxon abbreviations are: WS charr, white-spotted charr; Amph, Amphipoda; Arachn, Arachnida; Coleop, Coleoptera; Derma, Dermaptera; Ephem, Ephemeroptera; Hemip, Hemiptera; Hymen, Hymenoptera; Isopod, Isopoda; Lepido, Lepidoptera; Oligo, Oligochaeta; Orthop, Orthoptera; Plecop, Plecoptera; Trichop, Trichoptera.
Source: from Marcarelli et al. 2020 with permission from John Wiley & Sons.

changes in the food web and the switch in the main prey of the most productive predatory fish. All fish consumed aquatic macroinvertebrates during the winter and spring when terrestrial prey subsidies were not available and aquatic invertebrate biomass was high, but rainbow trout (*Oncorhynchus mykiss*) and white-spotted char (*Salvelinus malma*) switched to terrestrial prey (particularly insects and oligochaetes) during the summer. We will consider the importance of these kinds of terrestrial prey subsidies in relation to stream energetics, metabolism and carbon flow in Chapter 8 (section 8.7).

Interestingly, the energy subsidies are not just in one direction, and reciprocal subsidies between the river and surrounding terrestrial habitats are common and

important (see the review by Ballinger & Lake, 2006 and Chapter 8, section 8.7 in relation to energy transfers). This is particularly apparent in northern temperate regions, and can influence the food webs on both sides of the habitat interface. We have shown that the inputs of particulate organic matter from leaves along with the accidental inputs of terrestrial invertebrates are important in forested catchments, but conversely the emergence of aquatic insects from streams and rivers, albeit highly seasonal, can contribute significant energy transfers to riparian consumers such as birds, bats and spiders (Nakano & Murakami 2001). For the Horonai Stream, aquatic prey can contribute 25.6% of the total annual energy demand of the bird assemblage and terrestrial prey contributes 44% of the energy

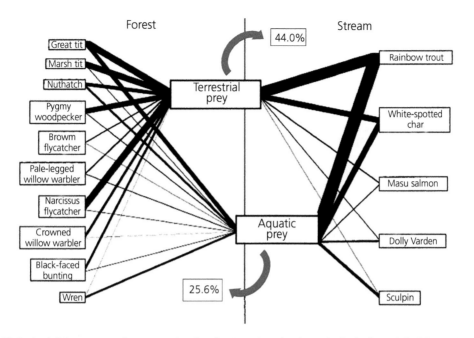

Figure 7.28 Food web linkages across a forest–stream interface showing reciprocal predator subsidies by fluxes of allochthonous invertebrate prey between the forest and the stream. Relative contributions of terrestrial and aquatic prey to the annual total resource budget of each species are represented by the line thickness. Annual total energy demand for the bird and fish assemblages accounted for by the aquatic and terrestrial prey respectively is shown by the % figures.
Source: modified from Nakano & Murakami, 2001, with permission of the (US) National Academy of Sciences.

budget for the stream fish assemblage through these reciprocal food web linkages (Figure 7.28).

An experimental study in Sycamore Creek, a Sonoran desert stream, used natural stable isotopes of nitrogen and carbon (^{15}N and ^{13}C) (see Chapter 8, section 8.6.3 for an explanation of their use) and an N^{15} tracer addition (that flows through autotrophs to aquatic insects) to track the exploitation of emerging adult aquatic insects by spiders of the surrounding riparian zone (Sanzone et al. 2003). Surprisingly, orb-web spiders derived almost 100% of their carbon and 39% of their nitrogen, and ground-dwelling spiders 68% of carbon and 25% of nitrogen, from emerging stream insects during the six-week study period. The longer-term impact of this subsidy was seen in the location of the highest spider diversity in the catchment being adjacent to the stream channel. Incidentally, semi-aquatic predatory bugs like the veliids (water cricket) and gerrids (pond skaters) that patrol the water surface in slow-flowing areas of streams and small rivers benefit from prey moving in both directions; they capture both emerging aquatic insects and terrestrial insects falling onto the stream surface (e.g. Brönmark et al. 1985).

Rivers can thus play a significant role in the surrounding landscape as we will discuss later (Chapter 8, section 8.7). As the review by Marcarelli et al. (2011) points out, it is clear that these subsidies of prey can have dramatic consequences for food webs in recipient habitats, not only supporting animal production but also influencing cascading trophic effects.

7.8.3 Food webs—practical challenges, conceptual difficulties

One practical challenge to constructing and comparing food webs in rivers and streams is that they are, firstly, incomplete—they almost always include only parts of the community, often mainly the larger organisms. Related to this is that they are often poorly resolved taxonomically; gross lumping of taxa is frequent, particularly among smaller organisms, basal species and 'difficult to identify' taxa (such as chironomid larvae). The effect of increasing taxonomic inclusivity and resolution on a stream web is exemplified by the study of Broadstone Stream (see Figure 7.22). Over the various iterations of web covering a period of some 20 years or more, the number of taxa included rose from 24 to 131, the known links from 90 to 841, and the links identified per species from 3.8 to 6.4 (e.g. Hildrew 2018). Despite these efforts, important taxonomic gaps remain even here.

Counting the links is also problematic, usually being achieved by the inspection of gut contents. For many predatory invertebrates and fish that engulf their prey, remains are identifiable and the approach can be useful (for an example see Figure 7.29). Often though, some components of the gut contents are unrecognisable, particularly for detritivorous species. Such amorphous material is common and could originate from a number of sources such as biofilm, digested periphyton, leafy detritus or FPOM. This requires other analytical methods which often do not allow specific identification. Some predators (such as bugs and many beetles) are 'suctorial' — and their guts contain the 'juices' of prey but no recognisable fragments. Even where prey fragments are visible, the identification of the minute prey of meiofaunal predators is extremely challenging, requiring specialist microscopic techniques and taxonomic expertise (Schmid-Araya et al. 2002a; Traunspurger & Majdi 2017). An underappreciated final difficulty is that many predator guts are often empty much of the time. Thus, around 300 guts of the dragonfly *Cordulegaster boltonii* were required for the number of its feeding links (the number of different prey taxa taken) in the Broadstone food web to reach an asymptote. The effort required for smaller predators was even greater (around 600 guts for the tanypod *Zavrelimyia barbatipes*) (Woodward & Hildrew 2001).

There are alternative methods for such problems. For instance, serological or electrophoretic analyses of gut contents have been used in the past for suctorial predators (e.g. Giller 1986). As in other fields, molecular identification of dietary items is clearly the way forward, though this has not yet been widely applied to stream ecosystems (Roslin et al. 2019). Radio-labelling of microbes can help identify microbial feeders, and use of stable isotope ratios and other dietary markers can separate and track pathways of various food types from autochthonous and allochthonous sources (see Chapter 8, section 8.6.3). However, the taxonomic detail is usually much coarser lower in the food web than for the larger engulfing predators (where prey can often be resolved to species) and many species are too rare to allow for suitably robust assessment of their trophic relationships.

Such methodological difficulties have frustrated the search for reliable quantitative patterns in connectance food webs. We also have to consider the design of food web studies. Studies sustained over long periods inevitably accumulate more species and links—and summary webs become more complex. This is partly a result simply of an increasing sample size—species richness increases with sample size in community ecology, up to some asymptotic value. It is also partly as a result of any sustained temporal changes in the community, over long periods and/or seasonally (Woodward et al. 2005). For instance, the Broadstone web is much simpler in winter than in summer, while the overall summary web is more complex than is ever realised at any one time (Schmid-Araya et al. 2002a).

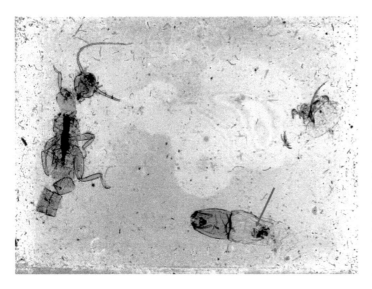

Figure 7.29 Gut contents of a caddisfly *Plectrocnemia conspersa*, itself found whole in the gut of a trout. A leuctrid stonefly is obvious (*Leuctra nigra*—far left) as is the head capsule of a tanypod larva (bottom). The tanypod also contains the tiny head capsule of a detritivorous midge (probably *Heterotrissocladius marcidus*). Thus, here we have three and four links (trout–*P. conspersa*–*Leuctra nigra*, and trout–*P.conspersa*–tanypod–detritivorous midge). The gut of *Leuctra* itself contains fine material, some of which is FPOM and iron bacteria. *Source*: photo by A. Hildrew.

Further, what is the real significance to community dynamics of very rare links between equally rare members of the web? Descriptions of webs quantifying energy flow along the various links, or the strength of species interactions (top-down or bottom-up) of those links, are surely more informative. While this seems undeniable, one potentially important feature of real communities and food webs is the degree of 'redundancy'—essentially communities with a number of species with broadly similar ecological requirements and trophic relationships. Rare species may lend robustness to the web, in terms of energy flow and overall productivity; if a dominant species is lost for some reason, a formerly 'subsidiary species' can assume its prominence in the web. This is an important point in concerns about the ongoing and widespread losses of species and their effect on ecosystem processes (see Chapter 10).

We see high food web redundancy in soft-water streams affected by acidifying atmospheric pollution, while streams affected by metals (which are often acidic too) seem to have similar, if more extreme, characteristics (Hogsden & Harding 2012). The soft-water systems have food webs with substantial detritus inputs, are very unproductive and have 'short' (limited trophic height, typically lacking fish) and 'wide' (high redundancy and omnivory) food webs. Omnivorous species may be favoured because the limited supply of available energy and nutrients hinders specialism. Simplified schematic interaction webs for

acidified and circumneutral 'trout' streams are shown in Figure 7.30.

Overall, there appears to be an abrupt change in the structure, diversity and composition of biotic assemblages and food web structure around the threshold of pH 5.5–6.0 (below which chemical alkalinity is exhausted and the main form of inorganic carbon is carbon dioxide; Hildrew 2018). In a series of papers, Layer et al. (2010, 2011, 2013) explored patterns in food webs associated with variation in pH across 20 streams in the UK. The acid streams were species poor and the macroinvertebrate assemblages dominated by stoneflies, chironomids and a few species of caddisflies. Fish were generally absent (especially at pH < 5.4), as were specialist grazers like snails and heptageniid mayflies (Ormerod et al. 1987; Layer et al. 2013). The grazer–algal link, while weakened, was not entirely lost, however, as it was sustained by some acid-tolerant generalists (such as leuctrid and nemourid stoneflies; Ledger & Hildrew 2005) that are usually assumed to be detritivorous shredders (see Chapter 8; Figure 8.21). The more acidic streams were more reliant on allochthonous detritus and, as mentioned above, had high levels of omnivory and generalism in the food webs. Further, it has consistently been found that the features of food webs from acidified/mine-drainage streams confer them with greater stability (Layer et al. 2010; Hildrew 2018; Pomeranz et al. 2020). It is thus striking that, as acid depositions from polluting industry have declined in Europe and north-east North America, the

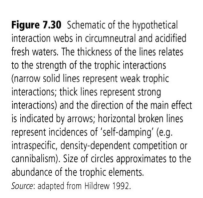

Figure 7.30 Schematic of the hypothetical interaction webs in circumneutral and acidified fresh waters. The thickness of the lines relates to the strength of the trophic interactions (narrow solid lines represent weak trophic interactions; thick lines represent strong interactions) and the direction of the main effect is indicated by arrows; horizontal broken lines represent incidences of 'self-damping' (e.g. intraspecific, density-dependent competition or cannibalism). Size of circles approximates to the abundance of the trophic elements.
Source: adapted from Hildrew 1992.

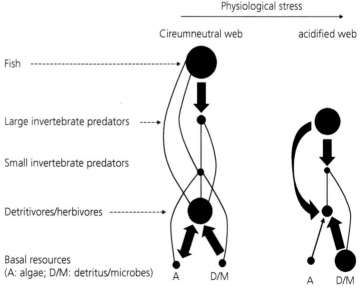

Physiological stress

Cireumneutral web   acidified web

Fish

Large invertebrate predators

Small invertebrate predators

Detritivores/herbivores

Basal resources
(A: algae; D/M: detritus/microbes)   A   D/M   A   D/M

chemistry of surface waters has improved, whereas biological recovery has so far been sluggish and weak, perhaps because of the extra stability of the communities in acidic systems (Hildrew 2018) (see section 6.4.1, Figure 6.10, p. 209).

7.8.4 Pattern and process in food webs

7.8.4.1 Size-structured food webs

Dietary generalism and omnivory are important factors underlying the complexity of many lotic food webs and in turn their customary lack of clear, 'integer' trophic levels. Body size of the interacting species is also implicated. In terms of simple biology, many stream animals grow radically in size over their life

cycle, their diet changing as they go, a phenomenon known as *life-history* (or *ontogenetic*) *omnivory*, and this is an important source of omnivory in general (see Figure 7.31 for an example).

There are many other patterns in stream food webs that may be size-related. In the Broadstone food web, dietary overlap between pairs of the six main invertebrate predators (three larger-bodied insects and three smaller tanypod chironomids) declined with the difference in their body size. This remained true even when comparing the diet of different sizes within as well as between species. That is, size seemed more important than taxonomy in determining diet in this 'guild' of predatory insects. There were many further corollaries of this effect of body size in the fishless Broadstone Stream. Thus, intraguild predation is common, as are

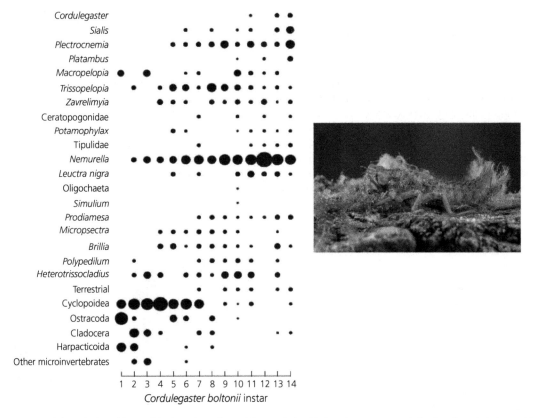

Figure 7.31 Shifts in the diet of larvae of the large dragonfly *Cordulegaster boltonii* in Broadstone stream as it develops through 14 instars (the area of each circle is the percentage by numbers of each prey type in the diet of each instar). Prey types are shown on the left. From the top, *Cordulegaster* down to *Zavrelimyia* are themselves predatory (in order of descending body size); *Nemurella* and *Leuctra* are detritivorous/herbivorous stoneflies; *Prodiamesa* to *Heterotrissocladius* and detrivorous/herbivorous chironomids Cyclopoidea to 'other microinvertebrates' are meiofauna (the smallest prey taken). This constitutes 'ontogenetic omnivory' as *Cordulegaster* is a consumer of both herbivore/detritivores (i.e it is a primary carnivore) and of other primary/secondary carnivores (such as *Sialis* and *Plectrocnemia*) and is also a cannibal, and is thus a secondary/tertiary carnivore (from instar 4 and beyond). The smallest prey are dropped from the diet of larger *Cordulegaster*. *Source*: from Woodward & Hildrew 2001, with permission from John Wiley & Sons.

cases where the usual pattern of large predators consuming small is reversed, in the sense that large individuals of the (on average) smaller predator can take small individuals of the (on average) larger predator. The strength of these 'reverse links' is weak, however. The result of this 'knot' of interactions is that not only is there high overlap in diet within this guild of predators (largely dependent on body size), but virtually every prey species shares at least two predator species with virtually every other prey species in the web, while many prey taxa are taken by all the dominant predators at some stage. This indicates high *trophic redundancy* (see previous section) and great potential for indirect interactions between prey species via their common predators. One such indirect interaction is known as *diffuse* or *apparent competition* (Holt 1977). This may occur where two coexisting prey species share a predator, one of the two being more susceptible to, or favoured by, the predator. The favoured prey in some way 'diverts' predation from the less favoured and may be diminished in abundance or even excluded—appearing as if the two prey were simply competing for limited resources (i.e. conventional competition). The occurrence and strength of these indirect prey–prey interactions is difficult to characterise but could be highly significant. Other indirect food web interactions are much more easily demonstrated, to which we return in the next section.

Within size-structured food webs there is normally a size disparity between the (larger) predator and the (smaller) prey. The extent of this disparity is important in estimating predator impact and interaction strengths. The average size disparity between predators and prey in stream webs (and other size-structured freshwater webs) depends on how it is measured. 'Species averaging' looks at all the feeding links in a web and simply compares the average body size of the predator with the average body size of the prey (e.g. Woodward & Warren 2007; Woodward et al. 2010). Estimates of the disparity (expressed as a ratio) between the two is thereby minimised and may be misleading—essentially because it assumes that every size class of a prey species is equally susceptible to every size class of the predator. As one might expect, this is usually far from the case. For each link, if the mean disparity in size between a predator and the prey actually found in their guts is calculated for the Broadstone web, this value is up to 10 times greater than the estimate based on species averaging (the predator is about 100, rather than 10, times bigger than its prey by mass) (Woodward & Warren 2007). This individual-based approach enhanced the ability to describe the structure of the

empirical Broadstone web, with a high proportion of observed links correctly predicted (see Woodward et al. 2010).

Clearly, predators are feeding on only a subset of the prey size spectrum, implying that there are 'size refugia' for prey. These are upper size refugia, when prey are too big to be vulnerable to the predator, and lower size refugia, when prey are too small (e.g. meiofaunal prey may be invulnerable to large invertebrate and fish predators). When the relative size distributions of prey and predators vary seasonally, then complex dynamics can arise. Such phenomena are poorly understood in lotic communities, although Woodward et al. (2005) found that the impact of the Broadstone predators on the prey species in the web, measured as the mass of each prey species eaten by all the predators in the web as a proportion of prey production, declines with increasing prey size—that is, predation fell disproportionately on smaller prey.

7.8.4.2 Indirect interactions and the occurrence of trophic cascades

What happens if a strong link within the food web is broken, for example through the loss of a key species or a basal resource, and what happens if a new set of linkages is introduced through the arrival of an invading species? We dealt earlier with the example of an epidemic disease affecting a dominant grazing caddis (*Glossosoma*) in streams in Michigan (USA), essentially almost deleting it from the community. This had effects that spread throughout the food web since it released other grazers from competition for algal resources (Kohler & Wiley 1997; see section 7.5.1). This was an excellent example of an indirect interaction (such as *indirect facilitation*, see section 7.7.2) in which one species, in this case a parasite, affected several others (those released from competition) more than one step away in the web. There are further very clear examples where such indirect effects change the overall food web profoundly. These can take the form of *trophic cascades* in which 'top-down' indirect effects of predators can cause changes lower down in the web, sometimes transforming the primary producers. The circumstances in which they occur have attracted a great deal of interest. Of course, the underlying (primary) productivity of the system (the availability of fixed carbon and nutrients) must ultimately limit, from the 'bottom up', secondary production higher in the web. Thus, primary productivity can be limited by light and nutrients (examples are given in Chapter 8) and the experimental reduction in the supply of leafy detritus to Coweeta streams changed the food web profoundly (see Chapter 8; Hall et al. 2000; Wallace et al. 2015). As we shall see, trophic cascades may also be less evident where top-down pressures are dissipated through the web via a complex network of indirect interactions.

Clear examples of trophic cascades are very well known from lakes, and although the situation is rather less clear in streams and rivers, there have been a number of studies exploring this topic. Peckarsky & Lamberti (2017) and Moerke et al. (2017) provide excellent overviews of the methods and approaches for studying the top-down effects of macroinvertebrates and larger macroconsumers in lotic systems, as well the main controlling factors on consumers (bottom-up).

A widely cited example of top-down regulation and a trophic cascade in streams is that of Mary Power, Bill Matthews and colleagues, using a combination of stream surveys and experimental manipulations in the Ozark mountains of the central, southern USA. They demonstrated that the piscivorous bass *Micropterus* indirectly affected the distribution of filamentous algae (dominated by *Melosira*) in stream pools by controlling the distribution of the algal-grazing minnow *Campostoma* (Power & Matthews 1983; Power et al. 1985, 1988). In the absence of the predator, but with a high density of *Campostoma*, stream cobbles were covered mainly by cyanobacteria (*Calothrix* and *Phormidium*

spp.), whereas in the presence of the piscivorous bass, and consequent absence of grazing fish, stone surfaces had profuse growths of filamentous diatoms (Figure 7.32). There was even evidence of downstream effects, as filamentous algae were more prone to dislodgement by the current and enhanced the food supply for suspension feeders (Power et al. 1988).

A study in the Eel River of California (Power 1990) found a similar, if more complex, cascade. When larger juvenile steelhead trout (*Oncorhynchus mykiss*) and Californian roach (*Hesperoleucus symmetricus*) are present, the former reduce smaller fish (sticklebacks and juvenile roach), and both suppress predatory damselfly larvae. This releases the dominant chironomids and other invertebrate grazers from predation, indirectly increases grazing of algae and leads to an assemblage of low, prostrate algal forms. The chironomids were invulnerable to the larger fish at the top of the food web but not to smaller fish and predatory invertebrates. In this system we can discern four reasonably clear trophic levels (large predators, small predators, grazers, primary producers; Figure 7.33). When the large fish were excluded experimentally, three trophic levels remained; small predators (damselflies and small fish) which suppress grazers and release from herbivory the dominant upright filamentous alga *Cladophora*

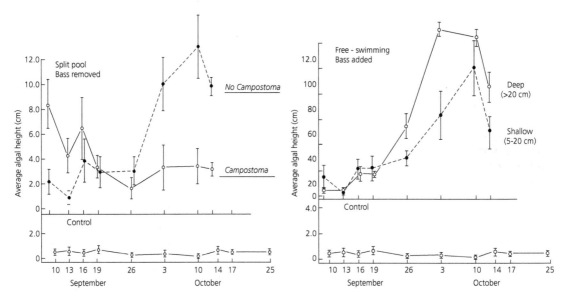

Figure 7.32 An example of the trophic cascade in Brier Creek, Oklahoma, USA. Left panel: average algal height in the two sides of an experimental pool, where bass were removed and *Campostoma* added to one side, and in an unmanipulated control pool, which contained a school of *Campostoma*. Right panel: algal growth in another experimental pool following the experimental addition of bass to this *Campostoma* pool. Both panels also show the situation in a control pool which naturally held bass. Vertical bars indicate means +2 SE.
Source: from Power et al. 1985, with permission from John Wiley & Sons.

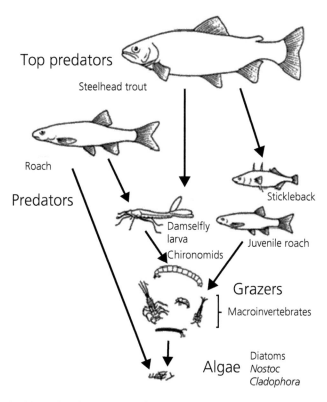

Top predators
Steelhead trout

Roach
Predators

Stickleback

Damselfly larva
Chironomids

Juvenile roach

Grazers
Macroinvertebrates

Algae Diatoms
Nostoc
Cladophora

Figure 7.33 The 'summer food web' in pools in the Eel River (California) showing top-down feeding interactions within the community. *Source*: redrawn after Power 1990, from Giller & Malmqvist 1998.

glomerata (along with its associated epiphytic diatoms and cyanobacterium *Nostoc*). The filaments can then reach several metres in length and cover most of the riverbed. Clearly this changes both the resource base and physical nature of the pool system. Winter flooding and summer drought in 'Mediterranean-type' rivers (Power 1992) facilitates spring growth of *Cladophora* (following the winter floods), preceding an increase in grazer densities in summer, which in turn crop the algae unless controlled themselves by predators. Biotic interactions are both direct and indirect, propagating through four trophic levels in the food web.

Species invasions have sometimes provided examples of trophic cascades. Thus, the introduction of the brown trout to New Zealand resulted in the substantial replacement of the native galaxiid fish species and a trophic cascade because trout are more effective predators of grazing insects than the native fish (thus releasing algae from herbivory: Townsend 1996; Huryn 1998; McIntosh & Townsend 1995).

It is now also clear that trophic cascades can occur in detritus-based systems, particularly where one detritivorous species dominates leaf breakdown but is

vulnerable to substantial predation pressure. Thus, Greig & McIntosh (2006) showed that predation by invasive trout reduced the population density of the large shredding caddis *Zelandopsyche ingens* resulting in a reduced rate of leaf breakdown. Further, Woodward et al. (2008) manipulated the density of the voracious benthic predator *Cottus gobio* in small-scale enclosures/exclosures in the productive southern English Bere Stream. The fish reduced the density of the dominant detritivore, the amphipod *Gammarus pulex*, and slowed the breakdown rate of oak leaves. Interestingly, *Cottus* also reduced the density of the grazing snail *Potamopyrgus antipodarum*, although there was no effect on algal production, which may therefore be constrained from the 'bottom up' (by light or nutrients). So, in this last case a cascade was apparent in the detritus food sub-web but not in the grazer food sub-web. This particular stream had a diverse community (142 species detected) and there were certainly ≫ 1,000 links in the web, and it had sub-webs of strong interactions embedded within it, so complexity alone evidently does not preclude the presence of cascades. Finally, the Bere Stream community has persisted, in terms of species composition, over at least 30 years. It is supported by a high rate of primary production (typical of chalk streams which rise from springs in landscapes with chalk bedrock (a very soft limestone)) but also has an important detritus food subweb (based on allochthonous leaf litter and decaying macrophytes).

These studies have shown that trophic cascades require there to be at least one species or guild per trophic level which could exert sufficiently strong topdown pressure on their food resources in the next lower trophic level (Power 1992). This is not always the case, particularly where there are no clear whole-integer trophic levels, as in many lotic systems (Cousins 1987; Jennings 2005). Thus, the irruption of the extremely polyphagous *Cordulegaster boltonii* in the Broadstone Stream food web did not bring about a clear trophic cascade and it consumed animal prey from all levels of the food web above the basal resources. Indeed, this was also the case with the subsequent invasion of brown trout, although there were changes and rearrangements in the relative abundance of benthic invertebrates (Woodward & Hildrew 2001; Layer et al. 2011). Where there is such trophic generalism and a high degree of omnivory, the top-down influence of predators is diffuse and trophic cascades weaker, if they occur at all. Further, the typical trophic cascade patterns are also only evident in stable habitats or during periods of relative physical/hydraulic stability. For example, in the Eel River, scouring floods during wet winters can kill or dislodge predator-resistant grazers, such as the large, cased caddis *Dicosmoecus gilvipes*, allowing blooms of *Cladophora* the following spring and summer, whereas winter droughts lead to dense populations of the caddis that graze back the *Cladophora* (Power et al. 2013). Thus, following drought winters, there is no evidence of top-down (indirect) influences of insectivorous fish on algae.

This suggests that physical factors can play a role in modifying food webs. Hart's (1992) study in Augusta Creek, Michigan (described earlier; section 7.3.1; Figure 7.6) is a case in point. In a similar way to fish in the Eel River, crayfish can eliminate the 'blanket weed' *Cladophora* from sites in Augusta Creek with velocities ≤ 50 cm s^{-1}, but higher flows provide a refugium for *Cladophora* by dislodging crayfish and restricting their foraging activity. If crayfish were the only limiting factor for *Cladophora*, we would expect sites with flows exceeding 50 cm s^{-1} to be monopolised by blanket weed, whereas cover rarely exceeds 50%. More sedentary grazers like the caddisflies *Leucotrichia* and *Psychomyia* can prevent such monopoly of space if they

have established prior residence. *Cladophora* in turn can inhibit the establishment of these grazers. The creation of bare space (such as by spates overturning stones, the emergence of sedentary insect grazers, or *Cladophora* mats sloughing off) and the nature of colonisation and propagule dispersal (related to life history patterns) are thus important in this system.

Similarly, Power et al. (1985) showed that biotic interactions became less important in their study streams when scouring floods or episodes of complete drying of the stream were frequent. Indeed, for invertebrates, the downstream drift and subsequent continual recolonisation of areas may override the strong topdown controls that lead to trophic cascades. Seasonal reductions in stream flow and drought can produce dramatic shifts in the main basal resources of riverine food webs, as seen in the South Fork Eel River when primary consumers like mayfly nymphs, larval chironomids and 'water penny' beetles shift from terrestrial detrital to algal carbon sources which are transferred up through the food web (Finlay 2001; Power et al. 2013).

Other physical factors can also have a strong impact on the nature of food webs. On a small, within-stream scale, water depth for example can be important, as shown by a neat study in the Rio Frijoles (Panama) (Power et al. 1989). This stream holds three species of armoured algal grazing loricariid catfish which can escape from gape-limited predatory fish but not from fish-eating birds (herons and kingfishers) which can feed in shallow (< 20 cm) water. The result is that the catfish tend to be restricted to the deeper waters that offer a refuge from bird predators, producing a pattern of algal-covered substrata along the shallow river margins but heavily grazed substrata in deeper water. At a larger scale, surveys of regulated and unregulated rivers in northern California showed some interesting changes in food webs attributed to the disruption of trophic cascades (Wooton et al. 1996). Regulated rivers had lower algal populations compared to unregulated rivers with natural flows. This appeared to be driven by weakening in the top-down control of algal grazers by the reduced populations of predators. Large-bodied, predator-resistant grazer populations (particularly large cased caddis), elsewhere susceptible to flood disturbances, became markedly abundant in the regulated systems. In their absence, predators can exert some control over smaller grazers, reflected in higher primary production. River regulation, which often involves a shift towards more lake-like systems and less terrestrial organic subsidies, will undoubtedly affect the nature of the food web, as exemplified for

Mediterranean rivers in the review by Power et al. (2013).

The effective fertilisation of streams through eutrophication can change the balance of basal resources and obviously affect the food web and associated energy flows. A similar change to basal resources will result from clearfelling of riparian vegetation, leading to reductions in allochthonous organic matter inputs (as in the experiments at Coweeta), as well as to an increase in incident light and autotrophic production. This again alters the balance of basal resources not only locally but also downstream. Increased sedimentation is a consequence of clearfelling of forests (Giller & O'Halloran 2004) and can lead to a shift from epibenthic to infaunal invertebrate communities, that are inaccessible to fish, as well as reductions in primary producers, again with significant consequences for food webs (e.g. Power et al. 2013).

7.8.4.3 The length of lotic food chains

The length of food chains is a parameter that has attracted a great deal of attention in the study of food webs and associated ecosystem processes, summarising a major dimension of the webs in terms of the vertical structure. One early speculation was that food chain length was limited by the inefficiency with which energy is transferred between trophic levels. Its calculation is fraught with difficulty, however, and estimates vary with taxonomic resolution and sampling effort. For instance, mean chain length in a number of descriptions of the summary Broadstone food web varied between $c.4.9$ and $c.5.4$, lengthening with the inclusion of the meiofauna and after invasion by brown trout (see Hildrew 2009). In the Broadstone web constructed for individual seasons, chain length in May was as low as 3.4, even when the meiofauna were included. Maximum chain length is the longest route through the food web and is obviously much longer than the mean chain length, and exceeded 10 even in the unproductive Broadstone. It should be evident that we need to be cautious about such web statistics.

Despite these potential flaws, it seems that food chain length in stream and river food webs tends to be relatively short, somewhat shorter than in lake and marine systems (Vander Zanden & Fetzer 2007). For example, in species-poor Arctic streams, even with vertebrate top predators, mean food chain length was 1.83 in a mountain stream and 3.04 in a more stable spring site (Parker & Huryn 2006). Data from 16 South Island New Zealand streams ranged between 2.6 and 4.2 (McHugh et al. 2010) and was highest in large, stable springs and lowest in small, fishless or disturbed

streams. In the ten third- or fourth-order tributaries of the Taieri River in New Zealand studied by Townsend et al. (1998), mean chain length varied between 1.8 and 4.4, but eight of the ten values lay between 2 and 3, and was lower still, between 1.57 amd 2.12, in a number of New Zealand pine forest catchments (Thompson & Townsend 2005). Even in the tropics, the average food chain length in a couple of neotropical streams was no more than 2.5, although a few individual chains in the food web reached five links (Ceneviva-Bastos et al. 2012).

What about larger rivers? Here we might expect longer food chains as they could involve both small phytoplankton and zooplankton and larger water bodies can support larger vertebrate predators. However, such systems are notoriously difficult to study and food web research often relies on indirect stable isotope analysis and modelling. At a number of sites in the regulated fifth-order Middle Rio Grande New Mexico (Turner & Edwards 2012), maximum food chain length averaged 3.99 (varying between 3.33 and 5.12). In a highly connected floodplain lake in the Middle Paraná River, South America (the second-largest river on that continent), food chain length only reached a value of 4 despite the high productivity (and omnivory was common) (Saigo et al. 2015). Zeug & Winemiller (2008) recorded up to five trophic levels in the Braxos River, in Texas (the ninth-largest river in the USA). Finally, across 66 sites in three river catchments in the wet/dry tropics of northern Australia, covering the main river, floodplain waterholes and tributaries, food chain length ranged from 3.2–6.1 and averaged 4.56 (Warfe et al. 2013), and tropical rivers seem equally variable with studies ranging from 2.6 to 4.35 and riverine studies in general ranging from 2.6 to 5.0 (Sabo et al. 2010; Saigo et al. 2015).

Productivity, disturbance and ecosystem size have all been proposed as controlling factors on food chain length (basically food chain length should increase with productivity and ecosystem size but decrease with disturbance), and some supporting evidence has been found, particularly but not always for productivity and ecosystem size (Vander Zanden & Fetzer 2007; Takimoto & Post 2012). In the seasonally connected rivers of the wet/dry tropics of northern Australia, for example, Warfe et al. (2013) have suggested that the movement of fish predators under the distinctive seasonal hydrology of this region links together isolated food webs, potentially creating a larger regional web that overrides the local effects of productivity, disturbance and ecosystem size. Sabo et al. (2010) reviewed data from over 30 major river systems and showed that intermittent rivers that dry during the summer had on average shorter food chain lengths (average 3.1) than perennial rivers systems (average around 3.9) of similar drainage area (ecosystem size). This is probably not unexpected as the top trophic species are normally large fish that would be most affected by the drying of the intermittent rivers. It is more likely that multiple factors interact to control food chain length. For example, Anderson & Cabana (2009), based on their studies in the St Lawrence River catchment, Canada, suggest that, when productivity is high and environmental stress low (in this case industrial pollution), lotic food chains can lengthen. We have seen how the addition or removal of the top consumer can change individual food chains, and that the degree of omnivory is also important. River regulation and engineering (see Brauns et al. 2022), climate-related changes to hydrology, pollution and changes to the riparian landscape will all have an effect on riverine food chains and especially on the top trophic species, a subject we will return to in Chapter 10.

7.9 Final remarks

In this chapter we have demonstrated the wide variety of species interactions found in streams and rivers. Interactions are often thought of as negative, at least for one of the interactors, as in predation, herbivory, competition, parasitism and disease. However, positive species interactions, such as symbiosis, mutualism, commensalism and facilitation, can clearly also be important. Streams and rivers are very much 'open' systems with high habitat complexity. The distinctive features of the lotic habitat, with its unidirectional flow, strong influence of the surrounding catchment and often pronounced disturbance regime, make it challenging to identify the interactions and modify or ameliorate their impact. However, creative, controlled *in situ* experiments and the ability to take advantage of *natural experiments* resulting from the natural or anthropogenic disturbances have provided a wealth of examples.

Body size is clearly very important in lotic food webs and scale plays a role in our perception of interactions. Technological advances have contributed enormously to our understanding of the nature and impact of the various types of interactions and will undoubtedly continue to do so into the future. Similarly, the ability to scale up from the substratum patch to the stream reach to the catchment has helped in the identification of key processes driving species interactions and their extent. This kind of information is becoming ever more important as we try to predict and moderate the effects of ongoing climate and general environmental change on stream and river biodiversity and ecosystem processes.

There are examples of strongly interactive (keystone) species that can modify the habitat and change the food web architecture through trophic cascades in streams and rivers, although these are not ubiquitous. The prevalence of omnivory may potentially increase the stability of food webs, particularly where most of the interactions are weak and high levels of redundancy are evident. At the small scale typical of field experiments, in-stream mobility (especially drift), patchiness in resource availability (e.g. leaf litter accumulations) and behavioural interactions between predators and prey can determine both predator and consumer impacts on the lower trophic level and hence the local food web structure. At the other extreme, processes operating at the landscape scale can also shape food web structure within individual streams and rivers (Woodward & Hildrew 2002). We saw in Chapter 6 (section 6.5) how river communities shift in response to directional environmental change, whether natural or anthropomorphic and directional, or as a result of disturbance events. Some lotic food webs have also changed profoundly in such circumstances, such as in the face of species invasions and chronic pollution.

The activities of species are key drivers of the ecosystem processes that we explore in the following two chapters. Food webs are the biological 'wiring' connecting species interactions and environmental forcing and through which energy and nutrients flow and are recycled in and between ecosystems. We now turn from the 'currency' of population dynamics (numbers) and biological diversity to that of the composite processes (energy and nutrient fluxes) which characterise ecosystems.

CHAPTER 8

Running waters as ecosystems

Metabolism, energy and carbon

8.1 Introduction—rivers as ecosystems

So far we have discussed the biology of river organisms mainly in terms of the species present, their adaptations, populations (density, dynamics and distribution) and diversity (the variety of river life), and have considered how these species interact with each other and their environment. This view presents only one side of the study of the ecology of rivers and streams, however. Like all living things, river organisms acquire energy and nutrients as they and their populations grow, while this energy is eventually dissipated (and nutrients recycled) during the processes of respiration, mortality and decomposition. This is the study of rivers as ecosystems, the second facet of all ecological systems, one which stresses processes (measured as the aggregated rates of various activities) rather than presence and abundance (measured as counts of individual organisms and species). There are of course intimate relationships between the two approaches. Organisms are integral to ecosystem processes, while process rates feed back on the biota. For instance, primary production ultimately determines the abundance and production of grazers and decomposers and their natural enemies.

The *ecosystem* is a well-worn concept in ecology, the term having taken a variety of meanings since it was first used in print in the 1930s, although the idea of a complex system involving interactions of the biota and the environment has been evident for much longer. The word has now 'leaked out' of ecology, moreover, and in daily life we can hear references to almost any complex system as an 'ecosystem', even extending to things like investment markets and the digital world. Here we use it to mean an ecological system characterised by the processes of energy flow and nutrient cycling, thus linking its biotic and abiotic components. To measure the rates of such processes it is not necessary to identify and count all the component organisms, and

indeed this is never achieved. An ecosystem is also best thought of as a 'level of organisation' in biology rather than a particular place. It is always almost impossible to 'draw a boundary' around an ecological system and to regard it as discrete (separate) from any other, because energy and materials invariably move across any such boundaries. Almost 90 years ago, Tansley (1935) wrote 'the systems we isolate mentally are not only included as part of larger ones, but they also overlap, interlock, and interact with each other'. This is true of riverine ecosystems perhaps more than any other, and we have already stressed their extremely 'open' nature and the close interactions between the lotic environment and the surrounding landscape. Rivers and streams are a dynamic part of the hydrological cycle while movements of water through river catchments are great vectors of organisms, detritus and sediments—transferring fixed energy (as organic carbon) and mineral nutrients to and from the river itself. These transfers of energy and materials between habitats are of great importance, and are often spoken of as *cross-system subsidies*, so some vestige of the concept of an 'ecosystem as a place' evidently persists. Here we deal with river ecosystems in two chapters. The first (this chapter) centres around energy flow through river ecosystems—what we can call river metabolism—and the second (Chapter 9) on how that energy flow drives the flux of nutrients and other activities (and also how nutrients in their turn can affect energy flow), although the separation of this material has sometimes proved a challenge!

8.2 River metabolism

We can think of river ecosystems as having a 'metabolism' since they fix energy and then dissipate ('spend') it, producing and consuming oxygen as they

The Biology and Ecology of Streams and Rivers. Alan Hildrew and Paul Giller, Oxford University Press. © Alan Hildrew and Paul Giller (2023).
DOI: 10.1093/oso/9780198516101.003.0008

do so (see Bernhardt et al. 2018). This is true of all ecosystems of course, although rivers are somewhat unusual in that the amount of energy fixed and used within them may be rather small compared with that contained in the organic carbon that is imported from their catchments (the surrounding landscape and from upstream) and in the material they export downstream or onto floodplains. *River metabolism* is most usually measured as changes in units of oxygen dissolved in the water column, generally increasing during peak photosynthesis in the daytime and declining at night when respiration predominates. This oxygen indicates the underlying energy flow, measured in units of joules, organic carbon or simply biomass, that is achieved as plants transform inorganic carbon (mainly carbon dioxide and bicarbonate) to energy-rich molecules, and *vice versa* as that carbon is respired by living things.

Almost all the energy powering ecosystems is ultimately created by organisms that can fix light energy to reduce inorganic carbon to energy-rich organic forms—during photosynthesis. The total energy fixed within an ecosystem is its *gross primary production* (GPP), while that dissipated (ultimately as heat) within the ecosystem measures its *ecosystem respiration* (ER). Organisms that fix light energy are called *autotrophs*: they also use some of that energy in maintaining themselves, but are normally net producers of oxygen (photosynthesis during the day exceeds their respiration over the 24 hours), at least during periods of growth. Ecosystem respiration then includes that due to the autotrophs themselves plus that of all the organisms that cannot fix their own light energy. These latter are called *heterotrophs* and include most bacteria, plus fungi, protists and multicellular animals of all kinds. Some aquatic microorganisms can act both as autotrophs (fixing light energy) and as heterotrophs (assimilating and respiring external sources of organic carbon) under different circumstances. These are known as *mixotrophs* and could be quite important, although little is known of them in an ecosystem context in rivers. The overall balance of GPP and ER is called *net ecosystem production* (NEP) and determines whether the system tends to accumulate (or export) organic carbon, or becomes depleted (or, if not, must import it). Finally, we can distinguish energy fixed within the ecosystem as *autochthonous*, while any that is imported is termed *allochthonous*. To reiterate the earlier point, the overall carbon budget of rivers may generally be dominated by allochthonous organic matter and by exports downstream (see Figure 8.9, below).

8.2.1 Primary production

The primary producers in rivers and streams were introduced in Chapter 3, section 3.3 (p. 65). Briefly, they consist of algae, cyanobacteria, mosses and liverworts (bryophytes), and higher plants (aquatic angiosperms). The bigger plants, commonly referred to as macrophytes, are the angiosperms, bryophytes and macroalgae (particularly those producing long filaments and/or colonies). The most important algae (at least in terms of their contribution to lotic food webs) grow attached to various surfaces as part of biofilms, sometimes forming what are variously called algal lawns, skins, felts and turfs (depending on thickness and composition; for a very useful review see Vadeboncoeur & Power 2017), but some are planktonic. Macrophytes mainly grow rooted into the substratum and may be fully submerged or emergent from the water surface. Apart from mosses and liverworts, macrophytes are most important in less erosive environments, such as the channel margins, in larger rivers with a soft substratum or in the floodplain.

In the majority of lakes, as in many terrestrial ecosystems too, primary production reaches a peak in spring as temperature rises, the days lengthen (at least at higher latitudes) and nutrients are available (having been regenerated via decomposition and returned to the euphotic zone). These factors together account for a 'spring bloom' in the phytoplankton. As Bernhardt et al. (2018) point out, rivers often do not follow this annual pattern. Increasing light supply and rising temperature may not coincide because the riparian canopy comes into leaf, shading small streams in the summer, or because of high sediment loads or coloured dissolved organic matter (in many larger rivers). These distinctive features have stimulated the study of the metabolism of rivers.

Which fraction of the plant community makes the greatest contribution to primary production in rivers? Macrophytes can account for much of reach-scale GPP under conditions in which they can persist (i.e. where the substratum and hydraulic habitat are suitable and light is sufficient). For instance, Alnoee et al. (2016) found that macrophyte habitats contributed around 50% of overall GPP in two Danish lowland streams (2.57 and 1.71 g O_2 m^{-2} d^{-1}), but only made up about 14% of total channel area. A substantial fraction (around 24%) of this production was due to their algal epiphytes. Indeed, McBride & Cohen (2020) found that the contribution of epiphytes to net primary production in the channel of two spring-fed rivers in Florida was closer to 75%.

Primary production by macrophytes themselves can be very high, possibly approaching the upper limit in lotic environments (in open, productive tropical rivers it probably ranges up to about 25 g O_2 m^{-2} d^{-1}). One system from the Pampas (grasslands) of Argentina was described as being 'as productive and slow as a stream can be' (Acuña et al. 2011). It drained an extremely flat catchment, with naturally nutrient-rich soils, and had an evenly high temperature, slow-flowing water, a relatively stable discharge and little riparian shading. Gross primary production was extremely high when discharge was low, up to around 22 g O_2 m^{-2} d^{-1} (equivalent to around 8.15 g C m^{-2} d^{-1}) with wild swings in dissolved oxygen concentration (0–25 g O_2 m^{-3}) between night and day. Unusually, the plants responsible for most production here were floating macroalgae, accounting for 30–90% of GPP. This is evidently only feasible in such a sluggish stream. Even these values may be exceeded where rooted emergent plants can establish, normally in fringing vegetation and floodplain water bodies. In units of biomass, for instance, Silva et al. (2009) estimated annual NET primary production in a lake in the east Amazon floodplain to range between 2.4–3.5 kg m^{-2} (equivalent to a daily mean of 6.6–9.7 g dry mass m^{-2} d^{-1}). In units of carbon this is about 3.3–9.35 g C m^{-2} d^{-1}, and in terms of oxygen to about 9–25.2 g O_2 m^{-2} d^{-1}). Over such a vast system as the Amazon floodplain this evidently sums up to an enormous amount of biomass and fixed carbon, sufficient to have a prominent role in the global carbon cycle. Macrophyte production over the whole Amazon floodplain could feasibly exceed 50 Tg C y^{-1} (1 Tg = 10^9 kg) (Silva et al. 2009). More generally, some of this material from highly productive riverine wetlands could be exported to floodplain rivers, stored indefinitely, transported further downstream or incorporated into food webs, and ultimately mineralised and emitted to the atmosphere as CO_2 or methane.

In the main channel of large, deep and turbid rivers, as well as in more erosive upper reaches and smaller streams, macrophytes usually play a much more minor role. We already noted, however, that at least the lower reaches of larger rivers can support a self-sustaining population of phytoplankton, where

hydraulic retention allows. This evidently also requires the ability to maintain a sufficient rate of net primary production in the turbulent and often turbid water column, in which planktonic cells spend some fraction of the time in relative darkness as they are swept deeper below the surface. This limitation can be severe. For instance, Ochs et al. (2013) estimated that net primary production by the phytoplankton at two sites in the lower Mississipi River was consistently negative at both, constrained by severe light limitation. The maximum time that cells were in the photic zone (where light is sufficient for positive NPP) was only 3 hours per day. The continued presence of phytoplankton in the main river was then attributed to transport from channel margins, side-arms and backwaters where cells could divide more rapidly. In the Murray River in Australia, on the other hand, the phytoplankton was responsible for much of the total primary productivity of the river [around 1 kg O_2 (or 370 g C) m^{-2} y^{-1}; Figure 8.1]. The dominance of autochthonous (phytoplanktonic) production of the Murray system may be due to its disconnection from the floodplain, its relatively constrained channel and regulated discharge. All are consequences of its management that may restrict the inputs of allochthonous carbon that are probably more characteristic of large rivers in more pristine catchments.

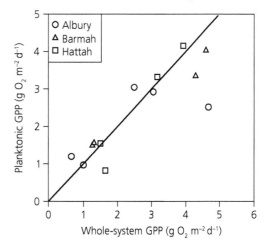

Figure 8.1 Planktonic and whole-system GPP at three sites (symbols denote sites, several estimates per site) distributed over 1,000 km of the Murray River (Australia)—the 1:1 line is shown (i.e. where planktonic production accounts for *all* of whole-system GPP).
Source: from Oliver & Merrick 2006, with permission from John Wiley & Sons.

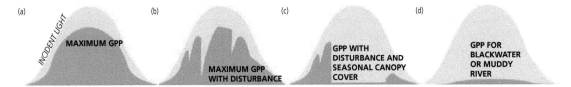

Figure 8.2 A conceptual model of four possible patterns of GPP (in green, vertical) over a year (horizontal dimension is time—no units are shown to enable visualization). This sees GPP as constrained firstly by (in yellow, vertical axis) incident light (a) and, secondly (b), by periodic disturbance (temporarily reducing photosynthetic biomass), then (c) by disturbance plus a further restriction of light by shading by riparian vegetation, and finally (d) in a turbid or highly coloured river (e.g. stained with organic matter) where light supply to the substratum is further reduced. This figure assumes rivers are at the same latitude and in the same terrestrial biome, such that background supply of incident light is the same.
Source: from Bernhardt et al. 2018, published under a Creative Commons CC-BY-NC Licence and with permission of Dr Emily Bernhardt.

In research terms, most attention has been focused on primary production by benthic microautotrophs (algae and cyanobacteria) in smaller streams, where a large biomass of macrophytes is lacking (apart from mosses of relatively short stature) and there is no true phytoplankton. In such circumstances, the attached (benthic) community is dominant. The main factor constraining primary productivity in these streams, by whatever fractions of the community, is undoubtedly light, determined primarily by the annual supply and seasonality of solar radiation reaching the earth's surface at any latitude and secondarily modified by riparian shading and water transparency. River flow is also influential in various ways. For instance, more or less frequent floods and spates, or episodes of drying, can reduce the biomass of primary producers, limiting their productive capacity, even though light may be sufficient. Figure 8.2 shows a conceptual model of the annual pattern of productivity of four rivers at similar latitudes and terrestrial biome. Apart from light and flow, nutrients and temperature can also act as limiting factors.

The fundamental effect of light on primary production by benthic microautotophs and mosses is evident from several excellent studies. Roberts et al. (2007) measured, daily, the ecosystem metabolism of Walker Branch in eastern Tennessee, located in the Oak Ridge National Environmental Research Park of the US Department of Energy. This is a first-order forested stream in the eastern deciduous forest biome of the Appalachians, with a stony bed and a 'mesic' or stable discharge. The daily supply of photosynthetically active radiation (PAR) fluctuates strongly with day length (season) and with the emergence of the forest canopy (Figure 8.3a), leading to a marked seasonal peak in GPP in March and April (Figure 8.3b). Gross primary production (GPP) in April, before the canopy is closed, is strongly correlated with light supply at that time. Storm flows also have a marked, if

transitory, effect on GPP, reducing it in spring because of scouring (and probably a reduction in the biomass of autotrophs) but increasing it in autumn (as fallen tree leaves on the stream bed are swept away, reducing shading).

Disturbance from scouring flows has widely been shown to affect GPP. For instance, Uehlinger (2006) found that bed-moving spates reduced daily GPP by 49%, assessed over a 15-year period in the Swiss River Thur, against a long-term annual mean of 5.0 g O_2 m^{-2} d^{-1}. Post-spate recovery was rapid in spring and autumn, but slow in winter. Bernhardt et al. (2017) compared GPP estimates over 60-day periods from four contrasting US Rivers (Figure 8.4). In all four records at least one major spate occurred, temporarily depressing GPP in each case. High flows can apparently have effects other than those associated with scour and bed movements, however. For instance, Hall et al. (2015) assessed GPP in the Colorado River at various locations as it flows through the Grand Canyon. There, benthic autotrophs dominate GPP, which is fairly low, varying between 0 and 3.0 g O_2 m^{-2} d^{-1} (overall annual mean 0.8). GPP was a function of turbidity along with other factors, mostly affecting light supply to the riverbed. Thus, winter conditions and cloud cover together greatly reduced GPP, while cloud cover alone had a lesser effect; the influence of temperature was modest. 'Hydropeaking'—sudden high flows caused by releases of water from dams upstream—also had an effect but only under turbid conditions, again suggesting light as a factor (because light to the bed had to pass through a greater depth of turbid water) rather than scour or bed movements.

Mulholland et al. (2001) present a cross-biome comparison of ecosystem metabolism, including GPP, in eight relatively pristine streams over a large geographic range of North America, with climatic regimes ranging from tropical to cool-temperate and from moist to arid. This confirmed the overriding effect of

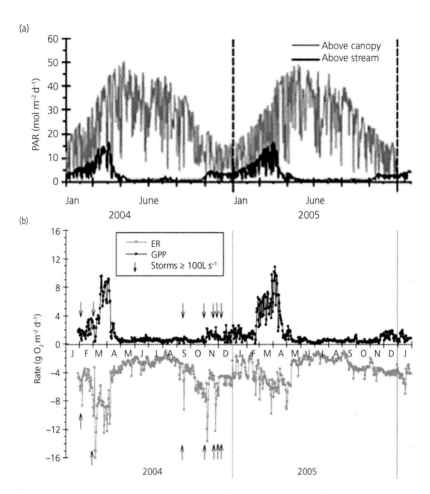

Figure 8.3 (a) Incident light above the forest canopy and at the water surface (above stream) at Walker Branch, 2004 and 2005, (b) daily gross primary production (GPP, positive values) and ecosystem respiration (ER, negative values), 2004 and 2005 (vertical arrows mark the occurrence of storm flows).

Source: from Roberts et al. 2007, with permission of Springer Nature.

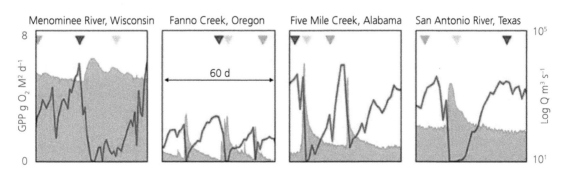

Figure 8.4 Estimates of GPP over 60-day periods (*x*-axis, green line) in four contrasting US rivers, plotted with the river hydrograph (discharge, Q, in grey) prevailing at the time: (a) Menominee River, northern Wisconsin—a large river with seasonality based on day length and temperature; (b) Fanno Creek, Oregon—a heavily shaded and frequently flooded small stream in the Pacific Northwest; (c) Five Mile Creek Alabama—southern river with limited seasonality; (d) San Antonio River Texas—a southern river with limited seasonality and some urban influence. The recovery rate differs among systems, while the shaded Fanno Creek has the lowest GPP.

Source: from Bernhardt et al. 2018, published under a Creative Commons CC-BY-NC licence.

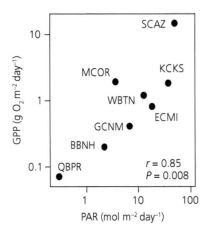

Figure 8.5 Relationship between mean daily rate of GPP and photosynthetically active radiation (PAR) in eight North American streams. (QBPR, Quebrada Bisley, Puerto Rico; BBNH, Bear Brook, New Hampshire; GCNM, Gallina Creek, New Mexico; MCOR, Mack Creek, Oregon; WBTN, Walker Branch, Tennessee; ECMI, Eagle Creek, Michigan; KCKS, South Kings Creek, Kansas; SCAZ; Sycamore Creek, Arizona).
Source: from Mulholland et al 2001, with permission of John Wiley & Sons.

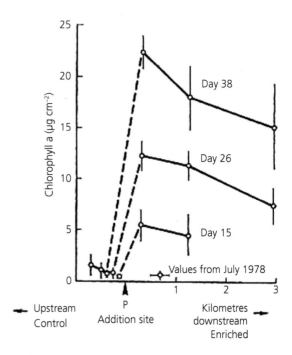

Figure 8.6 Peterson et al. (1985) dripped phosphoric acid into the Kuparuk River, northern Alaska, USA, raising phosphate-phosphorus concentration by 10 µg L-1, leading to a large increase in the biomass of epilithic algae (as µg chl *a* cm^{-1}).
Source: with permission from the American Association for the Advancement of Science.

light supply (Figure 8.5), with a further significant though lesser role for nutrients, in this case phosphorus. Light and phosphorus concentration between them explained 90% of variation in GPP. The relative effect of light and nutrients was also examined experimentally in artificial streams by Hill et al. (2009). Algal biovolume doubled over the range 5–300 µg L^{-1} of soluble reactive phosphorus (SRP), saturating at about 25 µg L^{-1}, a value very widely exceeded in nature in all but the most pristine streams. Light effects were much stronger, algal biovolume increasing 10-fold over the range 10–400 µmol photons m^{-2} s^{-1}, saturating at about 100 µmol photons m^{-2} s^{-1}. Overall, in landscapes dominated by humans, we can expect it to be difficult for management to reduce nutrient concentrations in river water sufficiently to limit algal growth, whereas limiting light supply by increasing riparian shading is much more feasible.

Despite this very strong evidence for the controlling role of light on primary production, Petersen et al. (1985) demonstrated the limiting role of nutrients in very oligotrophic systems. The Kuparuk is a meandering Arctic river draining the tundra landscape of northern Alaska. In this classic experiment, phosphorus (SRP) was dripped into the river during July and August 1983, sufficient to raise the concentration by 10 µg L^{-1}. The biomass of epilithic algae (as µg chl *a* cm^{-1}) increased by a factor of 10 or more

over the first two years (Figure 8.6), an effect extending far downstream. The biomass of epilithic algae at the control site immediately upstream of the dripper was 0.5 µg cm^{-2} compared with 6 µg cm^{-2} even 10 km further downstream. After eight years of P fertilisation mosses replaced epilithic diatoms as the dominant producers (Slavic et al. 2004). Free of restrictions due to light, it is clear that phosphorus alone limited primary production, and in this case its addition switched the metabolism of the whole system towards autotrophy, with very strong effects further up the food web as grazing insects increasesd, as did growth of young and adult grayling (*Thymallus arcticus*) (Petersen et al. 1985, 1993).

8.2.2 The dark side—ecosystem respiration in rivers

The collective dissipation of energy (from the mineralisation of organic matter) available in an ecosystem, both autochthonous and allochthonous, is termed *ecosystem respiration* (ER). Primary production produces oxygen whereas respiration consumes it, so the balance of the two can be used to measure river metabolism and to partition it into the two processes. Modern research into river metabolism relies on continuous measurements of oxygen concentration in the water column using the robust and accurate submerged sensors now available. Indeed, most of the measures of GPP mentioned above were made this way. The daily fluctuations in oxygen concentration in river water are modelled by the equation:

$$dDO/dt = (GPP + ER)/z + K(DO_{sat} - DO)$$

where dDO/dt is the rate of change in the concentration of dissolved oxygen, GPP is gross primary production (g O_2 m^{-2} d^{-1}), and ER is the rate of oxygen consumption (ecosystem respiration) by both autotrophs and heterotrophs (g O_2 m^{-2} d^{-1}, and a negative number, since it lowers oxygen concentration) (see Bernhardt et al. 2017). The term $K(DO_{sat} - DO)$, which is measured separately, refers to the net movement of oxygen between the water and the atmosphere and consists of the gas exchange rate (K, per unit time), with oxygen either tending to enter solution (if DO < DO_{sat}, where DO is the concentration of oxygen in the water and DO_{sat} is the saturation concentration at the same temperature and pressure), or to escape to the air (if DO > DO_{sat}). The term z is river depth (m) and converts from volumetric (m^{-3}) to areal (m^{-2}) rates. Various shapes of daily oxygen curves then reflect variation in the rates of photosynthesis, respiration and oxygen exchange (Figure 8.7). Gas exchange is relatively quick in fast-flowing, turbulent rivers and varies with temperature because oxygen is more soluble in cold water.

How does ecosystem respiration compare quantitatively with GPP in rivers? If it is less than GPP we describe the system as *net autotrophic*—fixing more carbon than it respires. If it respires more carbon than it fixes, it is *net heterotrophic*. It seems that most rivers are net heterotrophic for much of the time, particularly when subject to extra anthropogenic organic inputs, such as sewage or animal waste (e.g. Beaulieu et al. 2013), and so contribute to global emissions of CO_2. These emissions have been estimated at about 1.8 petagrams (=1.8×10^{12} kg) from streams and rivers annually (Raymond et al. 2013), though annual estimates have recently been increased to around 2.3 Pg C (Battin et al. 2023). This is substantially more than that from lakes and ponds, which occupy a greater overall surface area. Cumulative annual metabolism in four US rivers shows different patterns but all are heterotrophic

at least on an annual basis (Figure 8.8a). Even in the extremely highly productive stream 'La Choza', draining the fertile, flat grasslands of Argentina (Acuña et al. 2011), respiration exceeded GPP during periodic spates. Most respiration here was accounted for by hyporheic sediments (40–80%) and, even at baseflow when GPP was very high, there were 7–8 hours of anoxia d^{-1} during darkness and wild diel swings in oxygen concentration from 0–25 g O_2 m^{-3}. A recent review of measurements of ecosystem metabolism in tropical streams and rivers essentially confirmed the general pattern of an excess of respiration over primary production (Marzolf & Ardón 2021; Figure 8.8b). Gross primary production ranged between 0.01 and 11.7 g O_2 m^{-2} d^{-1} and ecosystem respiration between −0.2 and −42.1 g O_2 m^{-2} d^{-1}. Maximum ER was greater in the tropics. How rivers sustain the excess ecosystem respiration will be explained in the next section.

Ecosystem respiration is normally dominated by the microbiota—as are most ecosystem processes—and, like GPP, responds to various environmental factors. In fact, the supply of labile organic carbon from in-stream GPP is probably the closest single correlate with ecosystem respiration, and GPP and ER have a similar seasonal distribution (e.g. Roberts et al. 2007; Beaulieu et al. 2013) (Figure 8.3b). Driving environmental variables of ER include temperature, disturbance from spates and droughts, and nutrient concentration. Bed-moving spates and drying episodes have complex and not well-understood effects on ecosystem respiration. Spates can move, uncover, deposit or bury organic matter, depending on hydraulic patterns, bed morphology, woody debris and river vegetation, and this can change the quantity and quality of organic matter available to decomposers and, hence, ER. Droughts can allow the ingress of oxygen to formerly anoxic patches of sediment, briefly enhancing remineralisation on their rewetting.

Spates can depress ecosystem respiration in the short term and, in the Swiss River Thur, this reduction was by about 19% on average over a 15-year period from 1986 to 2000 (Uehlinger 2006). In the wooded Walker Branch, storms briefly depressed ER for a day or so, but then stimulated it, particularly in spring when it was increased three-fold over pre-disturbance values within three days of the event (Figure 8.3b). In the agriculturally impacted South Fork of the Iowa River, a spate that moved around 25% of the bed immediately stimulated ER (while GPP was reduced), profoundly increasing net heterotrophy from −5 to −40 g O_2 m^{-2} d^{-1}. Agricultural practices commonly increase the

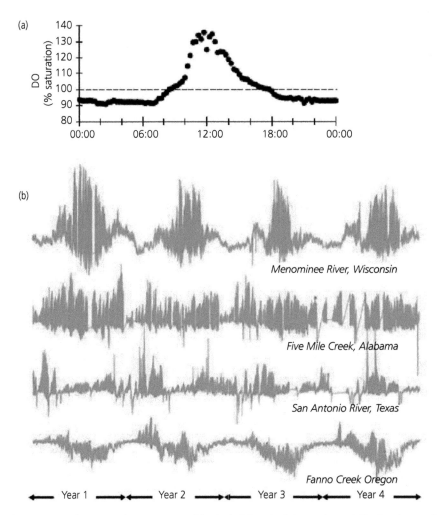

Figure 8.7 Daily oxygen curves for some north American rivers: (a) a single daily cycle of % saturation at Walker Branch, a small wooded stream in eastern Tennessee (10 March 2005) (Roberts et al. 2007); (b) four seasonal profiles of daily oxygen concentration (vertical axis) in four contrasting US rivers over four years (see Figure 8.4 for more site details: scales not shown to emphasise common patterns).
Source: from Bernhardt et al. 2018, published under a Creative Commons CC-BY-NC licence.

input of labile organic matter and nutrients to rivers. Therefore, sediments rich in organic matter that could be mobilised or exposed during a storm are readily available.

Eutrophication is probably the most widespread human perturbation of running waters worldwide, although we know less of its effects in streams and rivers than we do in lakes. Increased nutrients have been shown to increase ER in formerly pristine forested streams (e.g. Kominoski et al. 2018). In a two-year experiment in five, first-order, forested headwater streams at the Coweeta Hydrological Laboratory in North Carolina, these authors added both nitrogen and phosphorus to produce different molar (N:P) ratios ranging from 2–128, and concentrations from 96–472 µg N L^{-1} and 10–85 µg P L^{-1}. Such enrichment can reduce the standing stock of detritus in streams (through enhanced decomposition, see section 8.5.4), and here it was shown that ER was stimulated by such nutrient additions, in particular of N. Phosphorus had little effect and overall enrichment had little effect on GPP (probably because of light limitations). Enrichment in such detritus-based systems seems to increase heterotrophy by accelerating microbial respiration rates on various components of the detrital pool. Analysis of experiments at Coweeta (Benstead et al. 2021)

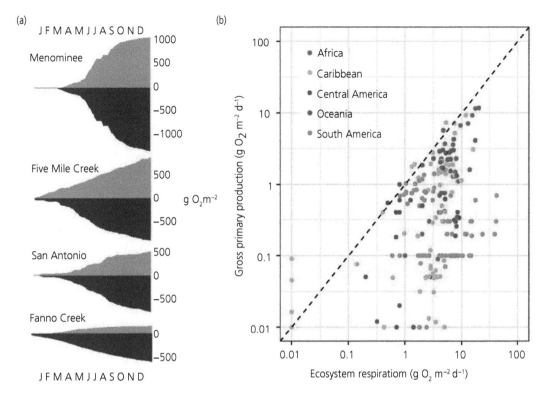

Figure 8.8 (a) Cumulative annual (months on *x*-axis) metabolism in four US rivers; details as in Figure 8.4), with GPP shown above the line in each case and ER below. All are heterotrophic (ER > GPP) annually and for most time steps. Metabolism is highly seasonal in the northern Menominee River, less so in the southern Five Mile Creek, 'stepped' in the frequently disturbed San Antonio River, and profoundly heterotrophic in the forested Fanno Creek. (b) Gross primary production plotted against total ecosystem respiration from streams in various tropical regions, which are colour-coded (dotted line plots primary production equals ecosystem respiration).
Source: (a) Bernhardt et al. 2018, published under a Creative Commons CC-BY-NC licence; (b) from Marzolf & Ardon 2021, with permission from John Wiley & Sons.

confirmed that the nutrient enrichment of one detritus-based forest stream resulted in severe losses of detrital carbon ($-576 \, \mathrm{g \, C \, m^{-2} \, y^{-1}}$, equivalent to 170% of annual inputs) over a two-year period, presumably taken from storage. By comparison, carbon flow through an unfertilised reference stream showed a nearly exact balance. The longer-term effects of such enrichment are less certain, although increasing the physical retention of detritus in stream channels might increase their resilience to enrichment. We return to the further effects of stream eutrophication in the next chapter.

Temperature is a fundamental driver of ecosystem processes. The temperature dependence of respiration in living things (i.e. on metabolic rate) can be expressed as an *activation energy*—the rate at which cellular metabolism increases with temperature (e.g. Brown et al. 2004; Demars et al. 2011). The response of ER to temperature has only fairly recently been measured in running-water systems but in some cases

seems to behave more or less as expected on theoretical grounds. Thus, Demars et al. (2011) took advantage of a remarkable set of small, closely neighbouring streams in Iceland that were variously affected by geothermal energy and ranged in mean annual temperature from 5–25°C but with little or no consistent differences in water chemistry and other confounding factors. As expected, ecosystem respiration increased with temperature across 13 sites. Hosen et al. (2019) also recently found an overall increase in ER with temperature at sites within the Connecticut River catchment of New England (north-eastern USA).

An increase in ER with temperature is to be expected, but temperature is also likely to affect the balance of GPP and ER. Yvon-Durocher et al. (2010) suggested that ecosystem respiration would increase with temperature more strongly than GPP, since its activation energy is higher. This would theoretically render ecosystems more heterotrophic with climate warming, such that they would emit more CO_2. This was the case in Demars et al.'s (2011) Icelandic streams. GPP increased with temperature, but rather more slowly than ER, and *net* ecosystem production became more negative as temperature increased across the sites. Song et al. (2018), in a study covering six biomes and ranging in latitude from 13° S to 68° N, also found that streams became more heterotrophic overall with rising temperature, with a decline in net ecosystem productivity and an increase in CO_2 emissions. Whether individual rivers react to climate change in the way predicted in theory is still uncertain, however (see Topic Box 10.2, and Chapter 10). Thus, while Hosen et al. (2019) found that ER was stimulated by higher temperature in the Connecticut River, GPP was stimulated most strongly (and more than ER) in larger channels, which thus became more autotrophic—contrary to the expectations based on cellular biochemistry. In this case, high temperature was associated with drought, and thus with factors other than temperature, and these counteracted the pure effect of temperature. These included lower turbidity during drought (the greater transparency increasing light supply to benthic autotrophs) and longer water residence time (reduced downstream loss of algal cells). The net effects of such complex environmental changes on the net heterotrophy of river systems in the real world, in the context of the global carbon cycle, thus remain unclear. This is of particular concern since we need to know if rivers are likely to become greater sources of CO_2 (and other greenhouse gases) in a changing world (see Battin et al. 2023).

8.3 Conceptual approaches to rivers as ecosystems

Ecologists have long pondered on what we would now call the metabolism of rivers, speculating on the sources of support of their biological productivity. The notion that most streams are essentially heterotrophic goes back a long way in the history of our subject, to the ideas of Thienemann, Margalef and others (see Hynes 1975). The apparent reliance of headwaters on imported organic matter was evident in the classic ecosystem studies of Likens and colleagues. Fisher & Likens (1973) used an 'ecosystem approach' to calculate an energy budget for a small, forested headwater stream in New Hampshire, Bear Brook (a simple compartment model is shown in Figure 8.9). More than 99% of total energy inputs to the system were allochthonous (imported), with autochthonous production (mainly

by mosses) accounting for less than 1%. Just over 40% of allochthonous inputs were in the form of leaf litter and throughfall (organic carbon dissolved in rain that has dripped through the leaf canopy) and just under 60% via stream flow (dissolved and particulate, also largely of terrestrial origin) and surface/subsurface runoff (mainly dissolved). The overwhelming quantitative dominance of terrestrial matter in the organic budgets in such forested headwaters is now well established. For instance, Webster et al. (1995), in reviewing organic matter budgets for stream of the eastern United States (predominately forested), found that particulate inputs can exceed 1 kg m^{-2} y^{-1}.

This raises the question of whether this dominance of terrestrial carbon in the organic budgets of forested river ecosystems holds true in most rivers and, within rivers, over the whole basin (headwaters and mainstem). An early and well-known conceptual view was articulated in the River Continuum Concept (RCC) (Vannote et al. 1980). We have come across this before in a community context (see Chapter 6, Figure 6.3, p. 199). The composition of the biological community, particularly of benthic invertebrates, was predicted to respond to the resources available along the length of river systems. Thus, for a typical river of the Pacific Northwest of North America, forested and shaded headwaters would contribute large amounts of tree leaves (particularly in autumn) and other allochthonous detritus, while primary production would be restricted by shading. As streams coalesced down the slope, and their channels became wider and more open, attached autotrophs (periphyton) would become more productive and contribute more to the food web via the increased representation of a 'guild' of grazers. As the river reached its lowland, alluvial reach, the deeper, turbid water would again restrict light to the bed and water column and authochthonous production would decline. In this theory, inputs of organic carbon to the river would now be dominated by fine organic particles generated in the headwaters, largely via the feeding of shredding invertebrates upstream. Aquatic macrophytes might play some role in the main channel, but they would mainly die ungrazed (as was thought at the time), their biomass entering the detritus food web, while phytoplankton would appear in the higher-order river system. Invertebrate assemblages of such downstream reaches would then be dominated by filter-feeders—both bottom-dwelling and planktonic—and by deposit feeders. At the ecosystem level, such changes should produce a strong pattern in the balance between total ecosystem respiration (ER) and gross primary production (GPP), the ratio

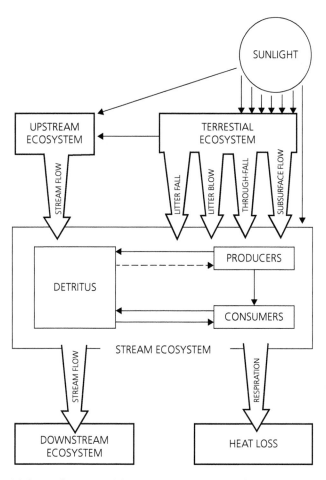

Figure 8.9 Simplified early model of energy flow in a small, forested stream, showing inputs from the terrestrial system and from upstream (allochthonous) and from instream (autochthonous) producers. In Bear Brook, the latter made up much less than 1% of the total ($\sim$ 10 kcal m^{-2} y^{-1}) compared with 6,000 kcal m^{-2} y^{-1} allochthonous. There was a large, steady standing crop of detritus retained within the system. About two-thirds of outputs were in stream flow, and one-third as heat from respiration.
Source: from Fisher & Likens 1973, with permission from John Wiley & Sons.

of GPP to ER (P/R ratio) being much less than 1 in the headwaters (i.e. strongly heterotrophic), shifting towards autotrophy further downstream (P/R $\sim$ 1), and then back to heterotrophy again (P/R < 1) in downstream reaches (see Figure 6.3; Chapter 6).

This influential view of the river ecosystem stimulated a great deal of research and debate about the nature of streams and rivers. It is now clear that the river continuum concept is very far from universally applicable and views have been modified in a number of ways. For instance, Ward & Stanford (1983) pointed out that the continuum of many if not most rivers was interrupted by natural or, increasingly, artificial dams and impoundments, which produce strong discontinuities in environmental conditions, such as

temperature and the supply of organic matter and sediment, and thus in community structure and ecosystem processes. Similarly, not all rivers originate in deciduous forested catchments (e.g. some arise above the tree-line or in grasslands) and in certain geographical areas are very short (e.g. on islands or draining small catchments at the edge of continents). More fundamentally, Junk et al. (1989) highlighted floodplain rivers and the enormous influence of the flood in all aspects of their ecology. This was termed the flood-pulse concept and asserted that the primary productivity of extensive floodplains is the major contributor of organic carbon to large rivers, either by direct grazing of floodplain plants or as detritus. These contributions could be made by 'hydrological vectors' (i.e. by water draining from the floodplain) or by 'biological vectors' (i.e. by animal migrations such as those involving fish). This is a major difference from the RCC which supposed that the mainstems of rivers depend on the supply of particles from headwaters and their terrestrial catchments.

None of these concepts challenged the view that rivers as ecosystems are essentially heterotrophic, depending mainly on imported terrestrial carbon. In contrast, this was questioned by the Riverine Productivity Concept (RPC) of Thorp & Delong (1994). This proposed that more of the carbon assimilated by lotic animals is actually of local autochthonous origin than had then been acknowledged. This falls short of saying that the overall ecosystem is autotrophic—and we know that normally it is not—but does suggest that the support of metazoan production is substantially more autochthonous (evidently with variation according to circumstances). It also suggested a role for local inputs of detritus from the riparian zones of many large rivers, rather than detritus transported from well upstream (as in the RCC). The RPC was originally seen as particularly applicable to large rivers that are constrained, increasingly artificially by engineers, to remain within their channels (rather than periodically spilling over a floodplain). We might add that streams draining non-forested catchments can also have very high primary production, with food webs based predominantly on grazing. A classic case here would be that of Sycamore Creek, in the dry, semi-desert Arizona (south-western USA) (Fisher et al. 1982).

8.4 Secondary production in rivers and streams

Primary production is conceptually straightforward (though biochemically complex), and is due to the reduction of inorganic carbon using light energy by green plants or photosynthetic bacteria. Secondary production is then the use of the resulting reduced carbon compounds (organic matter) to make new biomass of heterotrophic organisms—including that of microorganisms, protists and multicellular (metazoan) animals. Thus, the wonderful productivity of river floodplain fisheries, of terrestrial herbivores and predators feeding from rivers, the diversity and abundance of river invertebrates and the hotspots of microbial activity and growth in freshwater sediments, are all covered by the catch-all phrase 'secondary production'. Again, its units are those of quantity, area and time. Thus,

we could speak of production in units of energy (in joules) $m^{-2} y^{-1}$ or, much more usually, of carbon or total biomass (e.g. g dry mass) $m^{-2} y^{-1}$.

8.4.1 Microbial production

Bacteria are the most metabolically diverse group of organisms on earth and are found everywhere (Chapter 3, section 3.2.1, p. 60). Some are photosynthetic (Cyanobacteria) and thus contribute to primary production and can be important components of biofilms. The group also includes pathogenic and non-pathogenic forms, while many are more or less anaerobic. Others are obligate aerobes, which assimilate organic matter (particularly dissolved organic matter) of both allochthonous and autochthonous origin. Secondary production by heterotrophic bacteria, and its role in lotic ecosystems, is thus complex and a great deal remains to be discovered. Secondary production by aerobic bacteria that decompose organic matter is usually measured by the uptake of labelled amino acids by bacterial cells as they make proteins. Conversion factors can then be used to express this uptake in units of carbon per unit area per unit time, based on the mean ratio of protein to carbon in bacterial cells. Currently, leucine (mainly labelled with the weakly β-emitting tritium, ^3H, or with ^{14}C) is most commonly used, although older methods used thymidine which estimates DNA synthesis and therefore cell division (see Benbow et al. 2017).

Marxsen (1996) was able to estimate heterotrophic bacterial production in the famous German stream, the Breitenbach, by perfusing stream- or ground-water, enriched with leucine, upwards through cores of sediment (akin to upwelling of groundwater through the stream bed). As an early estimate he gives a figure of around 200 g C $m^{-2} y^{-1}$ for this unshaded, sandy stream, higher than the 26 g C $m^{-2} y^{-1}$ estimated by Bott & Kaplan (1985) for a woodland stream in the eastern USA. In a larger river, Fischer & Pusch (2001) separated bacterial production in the German River Spree into that by the bacterioplankton in the water column, on the leaf surfaces of rooted macrophytes (the epiphyton) and in the upper layer of the bed sediments. Sedimentary production was overwhelmingly the greatest fraction, exceeding that in the water column by 17–35 times. Epiphytic bacterial production was quite low on an areal basis, and less than that in the water column, though evidently it will vary with leaf area. Overall bacterial production exceeded total primary production, so that the river, as is usual, was net heterotrophic and subsidised by terrestrial organic matter, including dissolved organic matter (DOM) from soil, ground water and leaf litter.

Methane is a further source of highly reduced carbon (surprisingly abundant in stream- and ground-water) that can be oxidised by methanotrophic bacteria to make their own biomass plus carbon dioxide, their assimilation efficiency being about 50% (Trimmer et al. 2015; Bagnoud et al. 2020). These methanotrophic bacteria have only fairly recently been recognised as a possible resource for other consumers in stream food webs. Shelley et al. (2017) estimated secondary production by methanotrophic bacteria in 15 fertile, groundwater-fed English chalk streams (chalk being a soft form of sedimentary limestone) and compared this methane-derived carbon with that produced by photosynthesis. Photosynthesis is, of course, the overwhelming source of primary production, whereas methanotrophy is a heterotrophic process. However, methanotrophs do convert a gas (methane) into microbial biomass, which is potentially available to other consumers and thus can be considered as a source of 'particulate' carbon. Methanotrophic production across the 15 sites and sampling dates ranged from 16–650 nmol C cm^{-2} d^{-1}, compared with net photosynthetic production of 256–35,750 nmol C cm^{-2} d^{-1}. Methanotrophy was more important in shaded than in open reaches, making up to 13% of the total of these two sources of reduced carbon (photosynthate and microbial biomass). Methanotrophy thus may represent a small potential source of particulate carbon for stream food webs, of which we need to know more.

A wide range of fungi can be found in fresh water, by far the best known in streams and rivers being the aquatic hyphomycetes (see Chapter 3, section 3.2.2, p. 60), which are asexual forms of ascomycete and basidiomycete fungi. They are decomposers of coarse particulate detritus, particularly leaf litter, and have long been very prominent in the study of the biology of streams and rivers (e.g. Kaushik & Hynes 1968; Gessner & Chauvet 1994 and Topic Box 3.1, p. 61). Their biomass associated with leaf litter can be estimated by the quantitative analysis of ergosterol, a membrane sterol specific to fungi that decomposes upon the death of the fungus, so it measures their living tissue (see Gulis & Bärlocher 2017). Production can then be measured by the rate at which radiolabelled acetate is incorporated into ergosterol, and empirical factors used to convert that rate to units such as mg fungal biomass per mg leaf ash-free dry mass per day (Gessner & Newell 2002). This can then be converted to an areal rate of fungal production (mg m^{-2} d^{-1}) if the mass ('standing crop') of leaf litter is known.

Fungi usually account for greater than 95% of the total microbial biomass (bacteria plus fungi) associated with decaying plant litter in streams, and their production can even exceed that of invertebrate animals (Gulis & Bärlocher 2017). Suberkropp (1997) gives an annual estimate of 34 g AFDM m^{-2} in Walker Branch, Tennessee. Production was highly variable with time of year, mainly determined by variation in leaf standing crop, and the daily rate in terms of carbon varied between 0.003 and 0.25g C m^{-2}. Methvin & Suberkropp (2003) subsequently estimated annual fungal production at 31.7 and 26.6 g AFDM m^{-2} in two further streams in the south-eastern United States with a mean pH of 6.3 and 8.2, respectively. The first (soft-water) stream had a greater mean standing crop of leaf litter but a lower rate of production of fungi g^{-1} leaf litter than the second (hard-water) stream, consistent with what would normally be expected on the basis of water chemistry.

8.4.2 Metazoan production

To many river biologists the most obvious example of secondary production is probably that by metazoan animals—mainly fish and benthic invertebrates. Here, the two major approaches to ecology—population biology (including population dynamics, population genetics and community ecology) and ecosystem ecology (energy flow and nutrient cycling) evidently converge. Secondary production is often defined as the 'elaboration' (formation) of new biomass by heterotrophic organisms over time (e.g. Benke & Huryn 2017). In animals, for example, this would be all the new biomass produced in a year and would include the growth of individuals that survive over the year, plus the biomass from growth of animals that die during the period, plus the biomass of any propagules (offspring) that are produced. Single estimates of secondary production are of course just numbers, and in isolation are not that meaningful. However, together they are extremely powerful when comparing among species and systems. For instance, there are fundamental relationships between production, body size and longevity, and then between these species traits and population dynamics and the environment. Production also describes energy flow through ecosystems and the efficiency of energy transfer. Estimates of production enable the calculation of linkage strength in webs of species interactions and the overall quantification of food webs (Chapter 7).

Production in animals obviously requires food, some of which is assimilated after ingestion while a fraction is egested as faeces. Assimilation efficiency is thus assimilation (A) divided by ingestion (I) expressed as a percentage. In stream animals this varies enormously, depending on food quality and feeding mode and taxonomic identity, with values below 5% for some detritivores up to 90% or more for carnivores (Benke & Huryn 2017). The *assimilation efficiency* of carnivorous fish, for instance, varies between 70 and 98%, and in brown trout declines with increasing food intake (Elliott 1994). The energy in the food assimilated is then partitioned within the animal, some supporting individual growth and the production of propagules, some lost as heat as CO_2 is respired and a (usually) small amount in excretion (as nitrogenous waste, for instance). *Net growth efficiency* (NGE) is then defined as the energy in growth divided by the energy assimilated, while gross growth efficiency (GGG) is the energy of growth divided by the energy ingested (i.e. the food eaten)—both are expressed as percentages.

The 'mechanics' of secondary production by a stream population are most easily appreciated for the very simplest situation of organisms with clearly separable cohorts—that is, where a group (or 'cohort') of individuals are born (recruited) more or less at the same time. For, say, a benthic insect, the numbers alive at any time will decline through the aquatic life while the mean body size will increase as individuals grow

(Figure 8.10a). At the end of the aquatic life (most simply after one year), survivors emerge to mate in the terrestrial environment and females lay eggs back in the stream. Thus, production in the stream is made up of the biomass accumulated (as individual growth) by those who survive from egg to emergence plus the biomass accrued by those who die before the end of the life cycle. Individuals dying early in larval life will each have contributed little individual growth (though they are usually numerous), while those who die late in larval life will contribute more (though they are usually relatively scarce). This is because in most annual stream invertebrates, mortality rate is probably greatest early in life. Data on density and mean body mass in a cohort can then be presented as an 'Allen curve', and production over any period calculated as the area under the curve in a plot of population density against body mass (Figure 8.10b).

This is merely one (and the simplest) method of calculating secondary production in stream populations, and there are others applicable to cases where clearly separable cohorts are not evident—that is, where individuals of very different age are found simultaneously. The reader should refer to Benke & Huryn (2017) for an update on such methods, although the principles are the same as just described and essentially rely on the observation that abundance declines and body mass

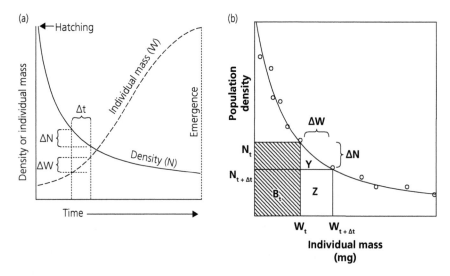

Figure 8.10 (a) Hypothetical cohort of a stream insect from egg hatching to emergence, in which mean individual body mass increases as density (N) declines but mean body mass (W) increases. Δ is the change in N or B over a short time interval (Δt). (b) The 'Allen curve' method for estimating (from successive samples) secondary production; between time t and $t + \Delta t$ density declines by ΔN while mean mass increases by ΔW, and production during that interval is equal to the area under part of the curve Y. Total biomass (B) at an instant t tends to increase as mean individual mass increases, but declines with mortality from t to t + 1.
Source: from Benke & Huryn 2017, with permission of Elsevier.

increases with age. It is also possible to approximate the production of whole assemblages of benthic animals, using 'short cuts' and empirical models. Benke & Huryn (2017) report that estimates of secondary production of whole assemblages of stream benthic invertebrates have ranged between ~ 2 and $\gg$ 100 g dry mass m^{-2} y^{-1}, with most < 20g m^{-2} y^{-1}. Individual populations (i.e. of the species that together make up a community) obviously range below these (whole-community) values, with most very much to the lower end. Extremely low production values are found in 'dystrophic' systems, often very humic or sometimes very acidic or metal rich. Outliers with extraordinarily high values of assemblage production, approaching 1,000 g m^{-2} y^{-1}, may be found in enriched systems or, in particular, lake outflow streams in which a few species of very productive filter feeders (such as *Simulium* and *Hydropsyche*) feed on particles washed from the epilimnion upstream—thus concentrating food resources derived from a wide area across a very narrow boundary between lake and stream (e.g. Wotton 1988).

Elliott's (1994) 24-year study of migratory brown trout in an English stream showed a mean production of little over 23 g m^{-2} y^{-1} (range 8.9–34 g m^{-2} y^{-1}), totalled over six age classes. Production was always greatest in age classes 0+ (first year of life) or 1+ (second year of life). Low annual production was associated with drought years. A review of studies of annual production in brown trout showed 38 values between 0.14 and 54.7 g m^{-2} y^{-1} (Mann & Penczak 1986), with almost all < 30 g m^{-2} y^{-1}. The relatively high production in Elliott's (1994) study is probably due to the migratory habitat of the trout in this system, since the older fish had exploited mainly food from the marine habitat and put on much growth there before returning to the river. Elliott (1994) suggested that the freshwater production in a variety of salmonid species probably ranges between about 9 and 34 g m^{-2} y^{-1}.

Estimates of production—which recall have units of biomass accrued per unit area per unit time—become more meaningful when we consider the standing biomass of the organism concerned. Put simply, some populations have a large biomass which produces relatively little over time, while others have low biomass at any instant but are very productive. This tells us a lot about the life cycle and general traits of the organism concerned and can be related to their environment. In terrestrial ecosystems, for instance, forests are often dominated by very large and long-lived trees, with a high biomass of metabolically rather inactive wood. Their total biomass thus 'turns over' very slowly. In aquatic systems, on the other hand, the plankton is dominated by small, short-lived algae, that are highly productive but of restricted biomass; the latter thus 'turns over' very rapidly. The relationship between mean population biomass (B) and population production (P) can be expressed as P/B and measures this rate of turnover. Since the units of production are mass per unit time per unit area, and those of biomass are mass per unit area, P/B takes units per time interval, often per year (y^{-1}), though it can be calculated over any period. Where we can discern clear cohorts and their lifespan, we can also calculate a 'cohort P/B'.

For many univoltine (a lifespan of one year) benthic invertebrates, the annual P/B is the same as the cohort P/B of about 5 with a range between 3 and 8. Organisms with two generations per year (biannual) will have an annual P/B of around 10 whereas a biennial (one generation per two years) species will average around 2.5. Some stream invertebrates have very short generation times (many generations per year) and can have annual P/B values up to 200 or more—that is, they are extremely productive relative to their population biomass. Examples include some small chironomids and mayflies, and small-bodied meiofauna (e.g. Huryn & Wallace 2000; Tod & Schmid-Araya 2009). Thus, the midge *Orthocladius calvus* had a P/B value of > 200 in southern England and a lifespan from egg to adult of only 16 days (Ladle et al. 1985), while the mayfly *Leptohyphes packeri* had a lifespan of only 12 days and an annual P/B of around 240 in a warm stream in Arizona (Huryn & Wallace 2000). These are almost certainly extreme values, although streams and rivers in general are characterised by mainly short-lived (< two years lifespan) and small-bodied invertebrates, such species traits reflecting the disturbed environment (see Chapter 4).

Stream and river fish also tend to be relatively short-lived, mostly less than five years or so, although the species occurring in high-order rivers can be much bigger and longer-lived (particularly migratory species) with much smaller P/B values down to 0.1 or less. In the population of migratory brown trout discussed earlier, Elliott (1994) reported P/B values varying between 2.3 and 4 over 24 years in Black Brows Beck (UK, Lake District). This is probably quite high for a stream-dwelling fish and, again, is because the freshwater year classes (0+ and 1+) of this migratory population are juveniles (and hence faster growing). There are also some long-lived and quite large lotic invertebrates, where conditions and their traits allow. These include crayfish and bivalve molluscs. The bivalve *Unio tumidus* had a lifespan of *c*.12 years in an English river and an annual P/B of *c*.0.1, while the white-clawed crayfish (*Austropotamobius pallipes*) can live for > 10 years and had a P/B of about 0.3 in an English stream (Huryn & Wallace 2000). Both crayfish (with a carapace) and mussels (a bivalve shell) have a substantial fraction of metabolically inactive tissues.

High P/B values for very short-lived and small-bodied animals suggests that they may account for a larger fraction of secondary production in river systems than at first appreciated. Relatively few studies have measured production by the meiofauna (metazoan animals passing through a 500 µm mesh: Chapter 3, section 3.4.1, p. 69) and compared it with production by the macrofauna in the same stream. Stead et al. (2005) did this for an acidic stream on the Ashdown Forest of south-east England, finding that the stream was unproductive overall (secondary production *c*.5g m^{-2} y^{-1}), but that the permanent meiofauna accounted for a substantial 15% of the total and the so-called temporary meiofauna (very small individuals of familiar benthic animals that grow out of the meiofaunal size range) for much more (around 50%). The permanent meiofauna was dominated by ostracods, copepods, rotifers and other taxa, often ignored in stream studies but diverse and numerous (Traunspurger & Majdi 2017). On the other hand, Tod & Schmid-Araya (2009) measured secondary production in a highly fertile English chalk stream, finding that the meiofauna (both permanent and temporary) made up less that 9% of the total production in both gravel and macrophyte habitats, though still substantial in absolute terms. It is still not clear what factors account for such variation (in the relative importance of very small metazoans) among different systems.

A related issue in stream ecology became known as the 'Allen paradox', in which the production by a population of brown trout seemed to exceed that of their presumed prey—macroinvertebrates living at or near the surface of the stream bed (Waters 1988; see Huryn 1996). In one study, it was only when sources of prey other than such macroinvertebrates were considered, such as terrestrial prey and small interstitial prey (almost certainly with high P/B), that production was sufficient. Even then, Huryn (1996) suggested that stream trout can harvest a high proportion of the prey available to them in some systems.

8.5 Organic matter dynamics

8.5.1 Allochthonous and authochthonous material

It is widely believed that allochthonous organic matter (i.e. imported from the terrestrial catchment rather than fixed within the stream) is by far the major 'fuel' for stream food webs (particularly in forest streams) and that this forms a clear example of an ecological subsidy of one system by another (Tank et al. 2010). On the other hand, the Riverine Productivity Model of Thorp & Delong (1994), mentioned in section 8.3, argued against this view and much recent evidence (see Section 8.6 later) has questioned sweeping generalisations about the overwhelmingly allochthonous (rather than autochthonous) support of animal production. How far we have to 'row back' from the view of the dominance of allochthony in lotic food webs (at least food webs including multicellular animals) is still uncertain. Nevertheless, ecological subsidies by terrestrial organic matter are certainly extremely important in many systems, and most lotic ecosystems seem to be net heterotrophic much of the time. So imported organic material is indeed a major feature of lotic habitats in anything like their pristine condition and dominates the overall carbon budgets of most stream and river ecosystems.

It is usual to partition organic matter in streams into size fractions (see Chapter 2, Table 2.4), each studied by a different suite of methods (Lamberti & Hauer 2017). These range upwards from the chemical miscellany that is 'dissolved organic matter' (DOM), to fine particles, coarse leafy particles, small pieces of wood and, finally, large fallen woody debris. The exact size fractions are largely a matter of convenience and have been variously defined.

Dissolved organic matter passes through a filter of 0.45 µm, but this category consists of a very wide variety of chemical compounds mainly made up of carbon, hydrogen, oxygen, nitrogen, phosphorus and sulphur in different proportions. It is largely responsible for water colour, though fine mineral sediments in suspension also play a role in overall turbidity (see section 2.5.4, p. 48). The term dissolved organic carbon (DOC) simply expresses the mass of carbon present in the DOM and makes up most of it. In smaller streams the standing stock of DOC ranges between about 0.5 and 25 g C m^{-2}, depending on soils and land use in the catchment (e.g. high in wetland and peaty streams). It is often the dominant component of detritus in transport in stream flow, usually exceeding all other forms of carbon (i.e. particulate) except at very high flows (Findlay & Parr 2017). DOM can be of allochthonous or authochthous origin, with the former generally more abundant and quantitatively stable, while the latter is usually labile, of lower molecular weight (such as amino acids and simple sugars) and much more variable in quantity.

Fine particulate organic matter (FPOM) conventionally includes material that is retained on an 0.45 µm filter but passes through one of 1,000 µm. Along with fine inorganic sediment, it is also known as 'seston' when in suspension. It is often of uncertain origin but FPOM includes leafy fragments colonised by fungi and bacteria, algal cells, protists, small metazoans, animal fragments and exuviae (cast 'skins' of moulting crustacea and insects), amorphous organic particles, faecal pellets of small animals and other fragments of detritus. It is actively transported in the current or deposited on the bed, largely depending on prevailing flow, so FPOM moves downstream in a series of steps of deposition and entrainment. It is important biologically as the food of filter-feeding animals or collectors that ingest organic deposits (see Hutchens et al. 2017). The amount of FPOM present on the bed at any one time can be extraordinarily variable; from river system to river system, between reaches within any one river system, in any reach over time, and even 'patch to patch' within a reach. For instance, in a large sample of streams and rivers in the eastern United States (the area east of the Mississippi River and coinciding with the eastern deciduous forest biome), Webster et al. (1995) report a mean benthic FPOM standing crop of about 400 g m^{-2}, though site means varied between about 50 and almost 2,000 g m^{-2}. FPOM made up an average of 57% of non-woody benthic organic matter, ranging widely from 8% to 91%. There was no significant trend of standing crop with stream order. Naiman et al.'s (1987) estimates from a subarctic Canadian river system ranged from 12.6–82.8 g m^{-2}, although the presence of beaver dams greatly increased these values.

Moving up the particle size scale, coarse particulate organic matter (CPOM) consists of particles greater than 1 mm. The 'non-woody' fraction of CPOM includes leaves, fruits, seeds and flowers from the catchment, plus any dead material of authochthonous origin such as larger algae, mosses and aquatic macrophytes. Wood includes anything from sticks and twigs up to branches and whole fallen trees. Organic particles tend to move downstream in the flow, unless they are retained in the channel by being trapped upstream of stable objects like cobbles, boulders and coarse woody material. These latter can cause 'debris

dams' of organic matter to form. Coarse organic particles can also be deposited in areas of slack flow at channel margins, side channels and other flow 'dead zones'.

Streams vary greatly in the capacity to retain coarse organic matter. For a range of stream sites in the USA, Webster & Meyer (1997) reported between 7 and 2,600 g C m^{-2} for non-woody CPOM and between 200 and 14,500 g C m^{-2} for woody material. The greatest standing crop is found in naturally heterogenous channels, of restricted width, and draining catchments with deciduous woodland, values generally decreasing downstream. In the past, and still today, much river management consisted of clearing logjams and organic debris from rivers and simplifying their channel form, on the grounds (often misguided) of flood prevention. It is only when CPOM is retained in the stream that its breakdown can occur. This occurs when particles lose mass, by physical abrasion, mineralisation mainly by microbes (i.e. decomposed, normally to CO_2, particularly by fungi) and consumption by detritivorous animals. Since partial decomposition by microbes renders that material much softer (a process known as *conditioning*; see Topic Box 3.1, p. 61), it is more liable to physical abrasion in the flow and is also more palatable to detritivores. Thus the fate of CPOM, as it is of all dead organic matter in streams, is: (i) to be transported downstream, ultimately to a floodplain, the estuary or the sea; (ii) to be mineralised through decomposition; or (iii) to be broken down to smaller particles (FPOM) or DOM (e.g. when the soluble fraction is leached from autumn-shed tree leaves) (see section 8.5.3).

The organic carbon entering fresh waters is of great significance at the global scale, and rivers and streams have been described as 'hotspots for exchange', referring to the evasion of CO_2 from the water to the atmosphere (Raymond et al. 2013; see also Battin et al. 2023). Measuring these vast fluxes in the global carbon budget is naturally very difficult, but Raymond et al. (2013) give an estimate of 1.8 Pg C y^{-1} emitted as CO_2 by rivers and streams, of which 0.6 Pg C y^{-1} comes from the Amazon and its tributaries alone (Pg is a petagram and equals 10^{15} g, or 10^9 tonnes—for context, a fully grown African savannah bull elephant only weighs about 7 tonnes!). This source of CO_2 is very large relative to the surface area of running waters, estimated at about 773,000 km^2 or 0.58% of the Earth's land surface (see Chapter 1, section 1.1). For comparison, the estimated 117 million lakes and reservoirs emit around 0.32 Pg C y^{-1} as CO_2 from an area of 5,000,000 km^2, about 3.7% of the global, non-glaciated land surface (Verpoorter et al. 2014). Rivers and streams are also supersaturated with methane, in part another end product of river metabolism and a powerful greenhouse gas once it is emitted to the atmosphere. Stanley et al. (2016) estimated global methane emissions at $c.27$ Tg CH_4 y^{-1} (Tg is a teragram and equals 10^{12} g) or about 20 Tg C y^{-1}, though Battin et al. (2023) quote a rather lower value. Such estimates lend a new importance to studies of river metabolism but are still approximate. However, it is now clear that outputs of greenhouse gases (which also include oxides of nitrogen) from rivers and streams are more important than appreciated formerly, and the role fresh waters play in the global carbon cycle was recognised for the first time in the IPCC report of 2014. Outputs of greenhouse gases are likely to increase due to anthropogenic landuse change and climate warming (see Chapter 10).

8.5.2 Ecosystem efficiency

The 'efficiency' with which streams and rivers process organic matter can be thought of as 'the extent to which energy inputs to a stream, both allochthonous and authochthonous, are used within the stream' (Webster & Meyer 1997). Here we mean all kinds of inputs of reduced organic matter, while to be 'used' that organic matter must be mineralised in the stream to produce carbon dioxide and heat. The most obvious measure of ecosystem efficiency is the fraction (or proportion or percentage) of the total inputs that is respired. The problem with this measure for running waters is that it is scale dependent—it depends whether the study unit is a short reach or a whole river and stream system (the

catchment). Thus, for a very short reach, most of the organic matter may arrive and leave by physical transport in the flow, while biological processing within the reach is relatively less important (because material is retained for a shorter time). That is, 'edge effects' of the small study area dominate. Deriving whole catchment budgets is more demanding in time and resources.

Alternative measures of ecosystem efficiency include 'turnover' length (and time). For carbon, *turnover length* is defined as the mean distance travelled by an atom of carbon between entering the stream in an organic compound until it is remineralised to CO_2. Biological processing of the carbon within the stream or river requires its retention and mineralisation. Turnover length (S) can then be calculated as:

$$F/R$$

where F is the downstream flux (export of carbon or energy per unit stream width per unit time; e.g. g m^{-1} s^{-1}), and R is ecosystem respiration rate (carbon or energy per unit area per unit time; g m^{-2} s^{-1}) (see Newbold et al. 1982; Webster & Meyer 1997). The shorter the turnover length, the greater is the 'ecosystem efficiency'. Webster & Meyer (1997) collected measures of carbon turnover length in 26 streams from a variety of settings (from tundra, through various forest types to arid lands, mostly in the USA) and found a strong relationship between this measure of ecosystem efficiency and discharge (Figure 8.11). Larger, deeper streams transport relatively more carbon than is used, while smaller channels are generally more retentive and 'efficient'. Anthropogenic simplification of channels,

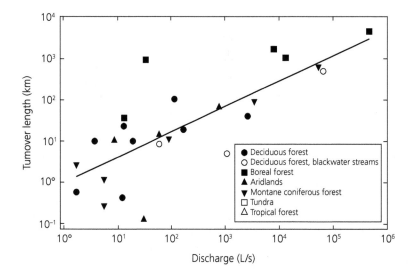

Figure 8.11 Carbon turnover length as a function of discharge for a collection of streams from different biomes.
Source: from Webster & Meyer 1997, with permission from the University of Chicago Press.

including the removal of large wood and other natural obstructions, will reduce retention and the efficiency with which carbon is processed in streams and rivers.

From studies specifically focused on allochthonous organic inputs to smaller streams, Webster et al. (1995) found that they can receive more than 1 kg of detritus m^{-2} y^{-1} in forested (deciduous) catchments. A very large amount of information on the fate of organic material 'falling into' streams at the Coweeta Hydrological Laboratory in the Blue Ridge Mountains of North Carolina (USA) has been collated (Webster et al. 1999). Coweeta is perhaps the most important 'outdoor stream laboratory' in the world, covering a 2,185 ha upland area drained by a network of streams, nearly all forested, and has been the subject of sustained and large-scale research (see Topic Box 1.1 by Bruce Wallis). Most of the streams are first or second order with a couple of larger channels (third–fifth order), and almost all the catchments have intact deciduous forest. The streams are high gradient (1–33%), very oligotrophic and heavily shaded. Webster et al. (1999) considered separately inputs of logs ('large wood'), sticks ('small wood'), leaves and fine particulate organic matter (FPOM).

Inputs of large wood—whole fallen trees and large branches—are notoriously difficult to measure because they are so erratic in space and time. At Coweeta, large logs appeared neither to be transported nor to break down over an eight-year period. In small channels, large wood is evidently moved only very infrequently, during extremely rare very high-flow events, and its breakdown may take decades. Perhaps unsurprisingly, therefore, the standing crop of large wood can exceed that of all other forms of organic matter, and is of greatest importance in terms of habitat structure rather than as an organic substrate *per se*. Sticks broke down at rates varying between 0.017 and 0.103% d^{-1} and were transported 0 to 0.1 m d^{-1} (depending on discharge). The mean rate of leaf breakdown (varied species over 40 studies) was 0.98% d^{-1} (range 0.16–3.16% d^{-1}), while transport varied widely depending on channel size and discharge. Fine organic particles broke down by 0.104% d^{-1}, while their transport in experiments increased with discharge but also with size and nature of the particles.

Using data from Coweeta, Webster et al. (1999) were able to calculate a turnover length of organic carbon. Their measure differed a little from the definition of Newbold (1992), in that it was calculated as particle velocity divided by breakdown rate (including decomposition, conversion of CPOM to dissolved organic matter or to FPOM). Based on a number of

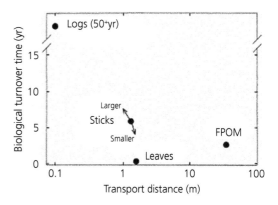

Figure 8.12 Transport versus breakdown for organic particles of different sizes at Coweeta streams.
Source: from Webster et al. 1999, with permission from John Wiley & Sons.

assumptions, there then appeared to be a broadly inverse relationship between this 'biological' turnover time (years) and transport distance (m) in a 'typical' stream at Coweeta (Figure 8.12). Note that biological turnover time includes any kind of mass loss of organic particles, including leaching, physical abrasion and the truly biological processes of animal consumption and microbial decomposition—though at Coweeta these biological processes probably predominate (Webster et al. 1999). A 'typical' stream at Coweeta was assumed to be of second order and about 1 km from the headwater, and to have a mean discharge of 20 L S^{-1} and an average depth of 10 cm. The relationship in Figure 8.12 is based on an average breakdown rate (k as % loss of weight d^{-1}) of 0.05 for sticks, 0.98 for leaves, and 0.10 for FPOM. No transport or decomposition of logs was observed but, to include such large woody particles in Figure 8.12, they assumed a transport distance of 0.1 m and a biological turnover time of 50 years. The particle turnover length (m) calculated for sticks, leaves and FPOM was 148, 108 and 42,400 (i.e. 42.4 km), respectively. Thus, while FPOM particles were readily transported in the flow, their breakdown rate is closer to that of wood (years) than leaves (months). Thus, the trend in Figure 8.12 shows a consistent decline in transport distance with an increase in particle size as expected (big particles move less), whereas an increase in biological turnover time with increasing particle size (leaves, to sticks to logs) is reversed for FPOM, which is often refractory.

Overall, Webster et al. (1999) suggest that small sticks and leaves that fall (or are blown) into natural streams generally break down close to the point of entry. That is, they are retained efficiently although their decomposition is highly inefficient (only a small fraction of this organic carbon is actually mineralised to CO_2). Some is converted to DOM but most is transported as refractory FPOM, much of it as the faecal particles of stream animals. As an example, in the Swedish Vindel River, from May through to August, faecal pellet loads from simuliid larvae peaked at 429 t dry mass d^{-1} and averaged between 47.5 and 93.7 t dry mass d^{-1} (1.9–3.7 t carbon) over a three-year period (Malmqvist et al. 2001). A proportion of these particles can settle on the bed, providing a food source for collectors. Aggregations of freshwater mussels can similarly transfer entrained FPOM to benthic habitats by production of faeces and 'pseudo-faeces' (Howard & Coffey 2006). This refractory FPOM material can also be transported long distances, to river floodplains, to lakes and reservoirs, or to the estuaries and ocean (thus serving to 'bury' organic matter for very long periods). Of course, FPOM can also come from erosion of soil organic matter (a product of terrestrial decomposition of vascular plants on land), but much of it is the product of inefficient processing of allochthonous organic materials entering forested streams around the world.

8.5.3 The process of detrital breakdown and the 'fate of organic matter' in streams

The breakdown of detrital carbon is a key biological process in many ecosystems and, along with downstream transport, largely determines the fate of such carbon in running waters. In this section and the next we deal with the biology of detrital breakdown, what factors affect it and why it varies, from place to place and from system to system. We know most about the breakdown of coarse particulate organic matter, which has been studied over a long period and in great detail (e.g. Webster & Benfield 1986; Chauvet et al. 2016; Benfield et al. 2017). Indeed, it is difficult to think of any process in stream ecology that has received so much attention and assumed such an important place in our view of the biology of running waters. The words 'breakdown' and 'decomposition' are often used rather loosely, but we really should use them more scrupulously. Breakdown of coarse particulate matter refers to any loss of mass over time, which is partly a result of physical processes. These latter include 'leaching' of any readily soluble components, such as simple sugars and amino acids, physical fragmentation in the flow and erosion caused by mineral particles (such as sand and gravel) entrained in the flow (almost like 'sandblasting'). This produces both dissolved organic matter

and smaller organic fragments (conventionally those < 1mm) which enter the pool of FPOM. However, these physical processes interact with the (mainly) biological process which is decomposition (or the synonymous term 'decay'). Decomposition thus refers to the mineralisation of organic to inorganic carbon (ultimately CO_2), as it is oxidised and its energy accessed by decomposing organisms (mainly microbes) and is a key part of breakdown.

The classic example of detrital breakdown in stream ecology is that of an autumn-shed leaf from a tree in the catchment that finds its way into the stream. This could be by directly falling in or by being blown in by the wind or washed in by overland flow. Autumn-shed leaves consist very largely of carbon, the trees having withdrawn a great deal of other materials before abscission. Leaves lose mass when submerged (initially by leaching) and, if a constant fraction of leaf mass is lost per unit time, then the rate of breakdown can be calculated from a negative exponential model and described by a single number (k), the instantaneous rate of mass loss with the units of 'per unit time' (usually d^{-1}), or (when multiplied by 100) as a percentage (% d^{-1}):

$$W_t = W_0 e^{-kt}$$

where W_t is the dry mass at some time t, W_0 is the initial dry mass (at time 0) and t is time (in days) (Figure 8.13a).

Breakdown is not a single process so it is not surprising that the negative exponential model is often not a perfect fit to real data. For instance, the softer portions of the leaf (between the leaf 'veins') break down faster than the veins themselves, which are often the last to go as the leaf is 'skeletonised'. However, the fit to this simple model is normally good enough to facilitate comparisons of the rate of breakdown among leaf types: for example aquatic versus terrestrial, woody versus non-woody, 'tough' versus 'soft' leaves etc, or to assess the effect on decay rate of environmental conditions such as temperature or acidity. The biggest source of variation is probably the intrinsic nature of the leaf itself, some species decomposing very rapidly, while others go very slowly. Webster & Benfield (1986) brought together data on processing rates of leaves in a variety of freshwater habitats, not just in streams (though streams exhibit faster breakdown than non-flowing waters in general). Leaves from woody plants break down more slowly on average than non-woody, although the non-woody plants are very variable. Of the non-woody plants, submersed and floating leaved aquatic plants break down rapidly, whereas emergent aquatic plants and

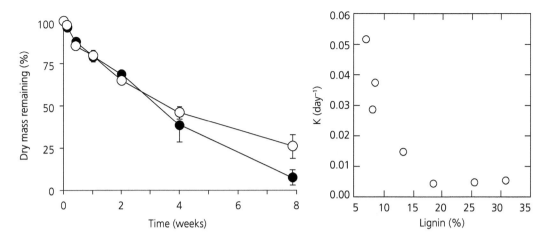

Figure 8.13 (a) Decomposition of alder (solid symbols) and willow (open symbols) leaves in a Black Forest stream (Germany), approximating an exponential decline in mass with exponents $k = -0.035$ and -0.027, respectively. (b) Breakdown rates of seven European tree leaf species with initial lignin contents ranging from 7–31% dry mass.
Source: (a) Hieber & Gessner 2002; (b) from Dodds & Whiles 2019, based on original data by Gessner & Chauvet 1994, with permission from Elsevier and from John Wiley & Sons.

terrestrial herbaceous plants break down more slowly. The latter group are more fibrous and include more supporting tissues than submersed or floating leaved plants. In general, breakdown rates relate positively to the nutrient content of the leaf (particularly of N and the C:N ratio; see section 8.5.4 and Chapter 9), and negatively to the content of fibre and lignin (Figure 8.13b) and to other inhibitory structures or secondary plant chemicals. These latter include thick, waxy cuticles of some leaves (including conifer needles), tannins (that form complexes with proteins) and polyphenolic compounds (Webster & Benfield 1986). Similar conclusions were reached by Hladyz et al. (2009) in their survey of leaf quality in native and exotic woody plants in Irish streams.

The various stages in the basic breakdown of CPOM in streams have been well described (Figure 8.14a). These include, in the first few days, leaching of soluble components that can account for around 5–25% of the initial dry mass, depending on species. The wetted leaves are colonised over the first 10 or more days by microorganisms, particularly by aquatic fungi (predominantly hyphomycetes) whose hyphae can penetrate the leaf. They are responsible for some mineralisation to CO_2 and also soften the leaf, making it more susceptible to physical abrasion and fragmentation in the flow. The fungi also take up nutrients from the water (see e.g. Gulis & Suberkropp 2003) and increase the relative nitrogen and protein content of the softened leaf, which is then more palatable to

'shredding' detritivores. Shredders then eat the microbially *conditioned* leaf, producing small particles as faecal pellets as well as non-ingested fragments from 'messy eating'. As the particle size is reduced, bacterial activity becomes more prominent as the relative surface area of the detritus is now larger. Further mineralisation to CO_2 is a product of fungal, bacterial and shredder metabolism. Of course, the prerequisite for these processes occurring *in situ* is that the leaf is retained in the reach rather than being washed downstream. The FPOM produced is more likely to be transported downstream than larger particles and, in the river continuum concept, these outputs from the headwaters were seen as the basis for the food web in larger rivers (Vannote et al. 1980). Benstead et al. (2021) made real estimates of these processes for an oligotrophic headwater stream at Coweeta (Figure 8.14b). This suggested a dominance of fungal metabolism in decomposition and, calculated by difference, of the production of fragments. Most of this latter appeared to be due to physical fragmentation to FPOM. Little CPOM was exported.

A rigorous experimental examination of the different processes involved in leaf breakdown was provided by Hieber & Gessner (2002). They used the usual method of exposing the leaves of two tree species, alder (*Alnus glutinosa*) and willow (*Salix fragilis*), in coarse-meshed bags in a stream (in the Black Forest of southern Germany), retrieving them at intervals to assess mass loss and the biomass

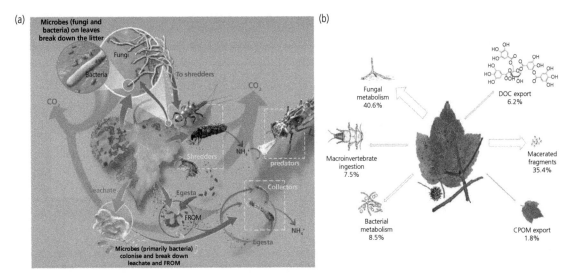

Figure 8.14 (a) Schematic of leaf breakdown, leading to the production of dissolved organic matter, fine particles and mineral nutrients Both microbes (bacteria and fungi) are involved at various stages; (b) Estimates of the fate of CPOM in an oligotrophic, forested stream at Coweeta (arrow thickness scaled to proportions). DOC is a tannic acid molecule.
Source: (a) from Marks 2019, by kind permission of Jane Marks, illustration by Victor Leshyk & Abigail Downard; (b) from Benstead et al. 2021, with permission from John Wiley & Sons.

of organisms that had accrued on the leaves. They also assessed the production of spores ('conidia') produced by aquatic hyphomycetes (fungi). The leaves of both tree species broke down exponentially (at rates of 0.035 d^{-1} and 0.027 d^{-1} for alder and willow, respectively; Figure 8.13a). They calculated that shredders accounted for 64 and 51% of the mass loss of alder and willow, respectively; fungi at least 15 and 18%; and bacteria around 7 and 9%. The mass loss unaccounted for was presumably due to leaching of DOM and physical fragmentation (other than via faecal production of shredders, which is included in the leaf material ingested).

A wider-scale study of litter breakdown in European streams showed mass loss rates (as k d^{-1}) varying from < 0.01 in northern Sweden to > 0.09 in Switzerland (Chauvet et al. 2016). Mass loss attributable to the 'microbial' component (i.e. in fine-mesh bags that excluded macroinvertebrates) ranged from < 0.005 to < 0.02, suggesting that macroinvertebrate activity (ingestion plus a contribution to physical fragmentation) predominated in all sites excepting northern Sweden and Portugal (Figure 8.15). The relative contributions of the different organisms ('microbes' and invertebrates) may vary in different systems but we can assume that both will be substantial.

A different but powerful approach to testing the role of shredding detritivores in leaf breakdown is through larger-scale field experiments. These are logistically difficult, require control of a suitable, secure field system and are most powerful in the context of long-term research, which is presumably why convincing examples are scarce. Cuffney et al. (1990) again used the outstanding facilities of the 'outdoor stream laboratory' at Coweeta. At the time of the experiments, they were permitted to apply an insecticide (methoxychlor, a synthetic organochloride insecticide since banned in both the USA and the European Union) to one of three small headwaters, using the two others as unmanipulated controls. They also used data collected before the manipulation. The insecticide was applied for a few hours on each occasion at approximately 3-monthly intervals between December 1985 and January 1988. As intended, the populations of all insect taxa were very greatly reduced, and shredders almost eliminated (there were no changes in bacterial density). Litter breakdown was substantially reduced (by 50–74% depending on tree species), and the concentration of FPOM in the water was reduced compared to control streams (Figure 8.16), while the export of FPOM was reduced to about a third of pre-treatment values in the manipulated stream over three years of study.

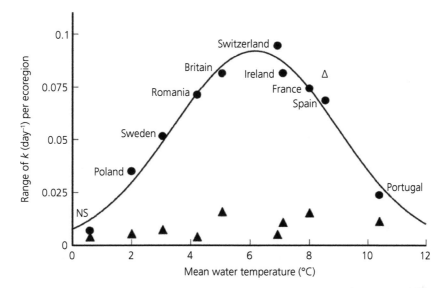

Figure 8.15 Among-stream variation in total (coarse-meshed litter bags, macroinvertebrate plus microbial, black circles), and microbially mediated breakdown rates (fine-meshed bags, black triangles) plotted against mean water temperature in reference streams across Europe. NS represents northern Sweden and the open triangle was considered an outlier.
Source: from Chauvet et al. 2016.

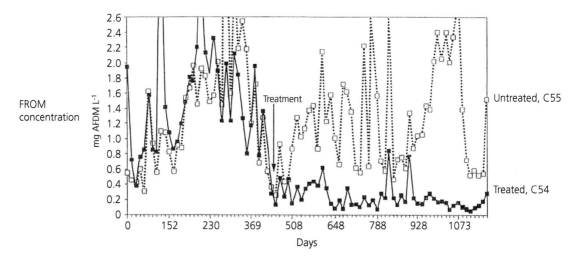

Figure 8.16 Reduction in FPOM output from a wooded catchment (C54) at Coweeta after insecticide treatment almost eliminated macroinvertebrate shredders, compared to that from an untreated control (C55).
Source: from Cuffney et al. 1990, with permission from John Wiley & Sons.

These and other experiments and surveys strongly infer a key role for shredders (both through direct ingestion and 'messy eating') in breaking down CPOM in forested headwaters and in greatly increasing 'leakage' of FPOM downstream—more or less as envisaged in the river continuum concept.

However, note that while shredders are often very important in terms of fragmentation into smaller particles ('comminution') and to mass loss of the leaves, it is the fungi and bacteria that account for most of the actual mineralisation of carbon to CO_2.

8.5.4 Patterns in the breakdown rate of plant litter

In summary, the breakdown rates (loss of mass) of organic matter are rather variable, largely depending on the intrinsic nature of the material (for instance, woody versus non-woody plant litter and the tree species from which the leaf litter arises), the composition of the decomposer community, and on features of the environment (e.g. temperature, nutrient supply and many others). In this section we try to deal with such variations more systematically and consider how they can bring about the spatial patterns in breakdown apparent at various physical scales. For instance, natural river floodplains (a feature of many unconstrained rivers) are a mosaic of aquatic, semi-aquatic and terrestrial habitats—such as the main channel, 'terrestrial' islands and floodplain ponds. These are not immutable features but are connected by high flows, their physical features being reworked in the natural dynamics of the river. Langhans et al. (2008) measured the rate of breakdown of black poplar (*Populus nigra*) leaves on the floodplain of the Italian Tagliamento River—breakdown was fastest in the main channel, slowest at terrestrial sites and intermediate in ponds. Rates in coarse and fine-meshed bags differed in main-channel sites, due to detritivore (mainly invertebrate) activity and/or extra abrasion in the former (only the coarse mesh allowing access to macroinvertebrate shredders). Leaves in terrestrial sites decomposed only very slowly, any mass loss being apparently due to leaching after rainfall. Such storage of organic debris in terrestrial sites would mean that a supply of particulate matter would become accessible, or could be mobilised, at the highest flows when such sites become reflooded. Such effects would not operate in rivers with regulated flow, since floodplains become more or less permanently isolated from the active channel.

A major recent trend in research on rivers and streams has been a new focus on channels with intermittent flow ('intermittent rivers and ephemeral streams', IRES). Recall that these may make up as much as 50% of the total length of the global network of running waters. Though uncertainties remain around such estimates, the extent of intermediate streams is certainly increasing with rising human demands for water. In a manner similar to temporarily terrestrial parts of natural floodplains, intermittent parts of the whole river network (often the extreme headwaters) can, when they are dry, accumulate terrestrial plant litter which then decomposes more quickly when flow resumes. Little was known about the contribution of such intermittent systems to the global carbon cycle, although Datry et al. (2018) recently carried out a remarkable wide-scale assessment in 212 intermittent river channels distributed over 22 countries. The total terrestrial plant litter accumulated in channels during the dry phase ranged from 0 to c.8.3 kg dry mass m^{-2} (mean 0.277), of which leaf litter made up from 0 to almost 1 kg m^{-2} (mean 0.088) and wood from 0 to 7.8 kg m^{-2} (mean 0.154). Not surprisingly, stored leaf litter increased with riparian tree cover and decreased with increasing channel width. The greatest quantity of plant material was found in low-order, temperate streams with forested catchments. Mineralisation was rapid upon rewetting. Datry et al. (2018) found that including IRES in global estimates of CO_2 emissions would substantially increase the apparent overall contribution of streams and rivers.

Most of what is known about decomposition in streams and rivers relates to temperate systems. More recently, however, comparisons have been made with streams and rivers in different climatic settings. There have been suggestions that in tropical streams (see Chapter 6) shredders are less prominent, that a higher proportion of terrestrial leaf litter is recalcitrant to decay and that, consequently, more is transported and ultimately stored as 'peat' in floodplains and deltas or exported to estuaries and the oceans (Wantzen et al. 2008). Boyero et al. (2015) found higher variability in litter breakdown rates in the tropics than in temperate regions, possibly as a consequence of substantial site-to-site variations in litter quality and in the complement of detritivorous animals in tropical systems.

There has also been a great deal of interest in how the complement of detritivores and microbial decomposers, and particularly the diversity of detritivores, might affect process rates. This is often set in the context of the very active wider debate about the general relationship between biodiversity and ecosystem processes. Some early experiments on litter breakdown were quite simple. For instance, Jonsson & Malmqvist (2000) used three species of shredding stoneflies (Plecoptera) common in northern Europe, feeding them on alder (*Alnus incana*) leaves. Experiments were run with one, two (three combinations) or all three species, all replicates with 12 individuals. Abundance was equal among species in replicates with species combinations; that is, six individuals of each in two-species combinations, four of each in three-species combinations. Process rate increased with the number of species and with the overall mass of animals. Such experiments are very difficult to design, however, particularly when more species are used, and there are many caveats to the results, and the effects of diversity (in the sense of species richness) itself have not proved universal (see

Gessner et al. 2010 for a review). For instance, Dangles & Malmqvist (2004) found that dominance (when one or a few species are much more abundant than others) strongly affected process rates, as did the identity of the dominant species. In these experiments there was also an effect of species richness, most clearly in assemblages with high evenness. On the other hand, Reiss et al. (2011) found no effect of diversity of detritivorous invertebrates, including various size classes of the crustaceans *Gammarus pulex* and *Asellus aquaticus*, and the caddis larva *Sericostoma personatum*, on leaf-processing rates in laboratory mesocosms. Mixtures performed simply additively, breakdown rates depending largely on the biomass of shredders, irrespective of diversity or species identity. More recently, Boyero et al. (2021) completed a global experiment on decomposition in 38 streams spread across six continents and involving 23 different countries. In this experiment, there was a positive effect of detritivore diversity, strongest in the tropics and weaker at high latitudes. In the latter, detritivore abundance and biomass had the strongest effects. They warned against species loss in the tropics, where detritivore diversity is already low and widely threatened.

Overall, there seems to be some effect of detritivore species richness itself on single ecosystem processes, such as litter decomposition. More subtle aspects of diversity, such as trait composition of the assemblage, identity and abundance of dominant species and species interactions in natural food webs, may be very important in maintaining what has been called in the jargon 'multifunctionality'—that is, the normal combination of processes characteristic of natural ecosystems (McKie et al. 2008; Reiss et al. 2009; Perkins et al. 2015).

In considering the possible effects of diversity on breakdown rate, we have also to consider the diversity of the leaf litter, as well as that of detritivores or decomposers (Gessner et al. 2010), and this aspect has also received a good detail of attention. For instance, Swan & Palmer (2004) estimated breakdown rates of six different leaf species, alone and in combination (of two, three, four and five species), in a stream in Maryland, USA. The total mass of leaves per litter bag was kept constant (3 g dry mass). In autumn, breakdown was simply additive, that is, mass loss in each bag was as expected from the single-species treatments. In summer, however, breakdown rates in most treatments containing American sycamore (*Platanus occidentalis*) were always less than predicted, that is, they were nonadditive. There was no effect of pure species richness itself, therefore, but there was an effect of species composition. It appears that the slower breakdown rates when the temperature was lower in autumn masked any effect of species composition, compared with the summer. American sycamore decomposes slowly when it is alone, but why it inhibits overall breakdown in mixed-species packs is not clear (Figure 8.17).

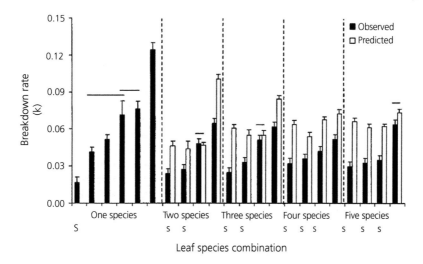

Figure 8.17 Estimates of breakdown rate (k + 1SE) for various leaf litter treatments (one, two, three, four or five riparian tree species) in summer 2000 in a stream in Maryland, USA. Treatments including American sycamore are marked 's'. Observed data, black bars; predictions (based on single-species data), open bars. For single species, horizontal lines connect treatments that are not significantly different between tree species or, for species combinations, between observed and predicted values.

Source: from Swan & Palmer 2004, with permission from University of Chicago Press.

The picture becomes more complicated on a biogeographical scale, where latitudinal differences in the effect of litter functional diversity on decomposition were found across 40 streams on six continents spanning 113 degrees of latitude (Boyero et al. 2021). Another study across five climatic zones found that the numer of functional types in litter mixtures actually had a negative effect on decomposition in subarctic and tropical streams but a positive effect in Mediterranean, temperate and boreal streams (Handa et al. 2014). In general, the question of the relationship between biodiversity and ecosystem process rates has to remain open for now, at least as far as litter decomposition in streams is concerned. Thus, Gessner et al. (2010) suggested that 'changes in species diversity within and across trophic levels can significantly alter decomposition', even though any such effect is apparently not ubiquitous.

A particular instance of the potential effect of litter diversity and species composition on litter decomposition relates to the possibility that exotic species may be different. The spread of exotic plants along river corridors can be spectacular, some species of which are intentionally planted, such as *Eucalyptus* and oil palm (*Elaeis* spp.), while others are naturally invasive, including salt cedar (*Tamarix*) in the south-western USA and *Rhododendron* in Ireland and other parts of north-west Europe. Does their leaf litter affect organic matter dynamics in stream ecosystems (e.g. Graça et al. 2002; Kennedy & Hobbie, 2004; Chellaiah et al. 2018)? If native detritvores are poorly adapted to exotic leaf litter, there might be reason to expect this litter to break down more slowly and be of lower food quality than native leaf species; however, Kennedy & El-Sabaawi (2017) found no such effect in a large-scale metanalysis of litter breakdown studies including information on exotic species. Overall, decay rates were similar between exotic and native species, although exotic species decayed faster at sites with higher temperature (largely tropical) and where litter bags with a coarse mesh were used. Exotic species at tropical sites had a lower mean C:N ratio (relatively more N, hence probably of higher food quality), which may explain their higher decay rate, even though any effect was evident only at higher temperature. Of course, where the exotic species produces particularly poor quality litter, as in the case of *Eucalyptus* and *Rhododendron*, the impact on the stream community can be substantial, often compounded by the intense shading of the stream channel (Hladyz et al. 2011).

Temperature is an obvious candidate as a driver of litter breakdown, and some quite powerful studies have exploited large-scale field experiments and/or surveys to test its effect. In an early example, Irons et al. (1994) exposed leaf litter of 10 species of varying 'quality' (judged by tannin content) in streams at three latitudes in the Americas (Costa Rica, Michigan and Alaska; 10 °, 46 ° and 65 °N, respectively) and compared their breakdown. As expected, breakdown was fastest in Costa Rica, although it was similar in Alaska and Michigan. When breakdown was expressed *per degree day*—thus removing the effect of temperature itself—leaves broke down faster in Alaska than in Michigan and Costa Rica. Comparison with extensive data from other North American studies confirmed that breakdown rate was actually faster per degree day at high latitudes, whereas there was no trend with latitude when expressed simply as mass loss per day. The litter bags in this study were of wide mesh, allowing access to macroinvertebrate shredders, and it was suggested that their consumption of litter increases with latitude, whereas microbial decomposition increases (with temperature) further south. Where both coarse and fine mesh bags containing the same leaf species were used across Europe, breakdown rates increased with temperature up to a maximum (when corrected for number of degree days) but then declined. At each end of the temperature range, there was little difference between microbial-only and total decomposition (Figure 8.15; Chauvet et al. 2016).

Boyero et al. (2016) conducted experiments at 24 stream sites at widely varying latitudes (48 °N in Germany to 43 °S in Tasmania), this time exposing litter mixtures of species local to each site, and thus of varying quality and phylogeny (PD, phylogenetic diversity), and also packs consisting of a single, normally highly palatable, species (alder, *Alnus glutinosa*). Litter bags were of both coarse and fine mesh in this experiment. For the single-species (alder) litter bags, microbial breakdown was fastest at the warmest (tropical) sites, with a greater role for detritivores (which could access the coarse-mesh bags only) at higher latitude. Litter quality and phylogenetic diversity were most influential in the bags with local litter mixtures, particularly in the warmer sites, confirming the strong influence on breakdown of intrinsic features of the leaves. A further strong influence was pH, with faster breakdown in basic (i.e. well buffered) streams, again particularly at warm sites. Thus, temperature modifies the effects of both leaf quality and stream acidity, presumably because temperature is the dominant driving variable of microbial activity. This also implies that variability in decomposition rate will increase as latitude declines (i.e. with increasing temperature, towards the equator), due to local factors such as litter quality, pH and nutrient supply having a greater effect.

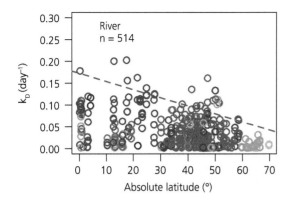

Figure 8.18 Decomposition rate (K_D, day^{-1}) of cotton strips at more than 500 river sites worldwide, spanning 140 degrees of latitude (70 degrees north or south), in all continents and including all the Earth's major biomes. Very slow decomposition can occur at all latitudes whereas rapid decomposition occurs mainly at low latitude, and not at all at high latitude. The dashed line shows the regression for the 95th quantile of decomposition rate. Colours refer to data from different biomes.
Source: (from Tiegs et al. 2019, published under a Creative Commons licence).

A further example of a global assessment of decay in streams is provided by Tiegs et al. (2019), who this time used standard cotton test cloth to assess cellulolytic decomposition at over 500 river sites (and at a similar number of neighbouring riparian sites) distributed among 11 major global biomes. These ranged across sites in the wet tropics, the tundra, temperate boreal forests, tropical savannah and others. This method of measuring decomposition works by measuring the decline in tensile strength of cloth strips as they are decomposed by microbes, and thus removes the effect of variable nature of the substrate (i.e. leaf quality). These biomes had different decomposition 'signatures', with the fastest decay in any biome largely related to its latitude (Figure 8.18). Thus, factors related to latitude, probably temperature, set an upper limit to decomposition, although slow rates can be observed at all latitudes. Since these data are based on the decay of a single standard substrate, variations here are probably due to local environmental factors, including nutrients and pH.

The influence of pH on decomposition, apparent in the studies by Boyero et al. (2016) and Tiegs et al. (2019), is not surprising since stream acidity has long been known to inhibit decay at low pH (< c.5.5). Slow decomposition and low litter quality could then be responsible for the low secondary production of most acidic, and anthropogenically acidified, systems (e.g.

Hildrew 2018). The acidity of many fresh waters has ameliorated since controls on sulphur emissions in Europe and North America were introduced in the later years of the last century. For instance, Hildrew et al. (1984) assessed cellulolytic decomposition in 1978, using the cellulose test cloth method, at 31 stream sites in an area of south-east England variably susceptible to acidification (the range in mean annual pH among all the sites was 4.8 to 6.9). This experiment was repeated using identical methods in 2010, when mean annual stream pH over all the sites had increased by 0.7 units (individually ranging from 6.0 to 6.7; Jenkins et al. 2013). In both studies, decomposition rate in winter, when pH reaches its seasonal minimum, was significantly related to stream acidity, while the mean rate among all the sites had increased over the intervening 32 years of gradual environmental change by about 18%.

At the other end of the spectrum from acidified streams lie those, much more numerous, that have been at least moderately enriched with mineral nutrients. An early discovery was that nitrogen in stream water was 'immobilised' (taken up) during the breakdown of autumn-shed leaves (Kaushik & Hynes 1971; Webster & Benfield 1986), a phenomenon mainly due to its incorporation into microbial protein. Phosphorus may also be immobilised, though this has been less frequently observed. This uptake of nutrients by decomposers leads to an overall improvement in the quality of the leaf litter as food for detritivores and is important in the overall process of breakdown (Marks 2019); in this way leaf litter can also have ecosystem-level effects on nutrient cycling and 'spiralling' in streams (Chapter 9, section 9.2.1). Experimental nutrient enrichment of yet another stream at the Coweeta facility resulted in an increase in overall ecosystem respiration, an increase (with N addition) in the respiration rate of leaves and wood, and a reduction in the standing stock of detrital carbon (Kominoski et al. 2018).

Severe eutrophication of streams and rivers is widespread and increasing at a global scale. Woodward et al. (2012) carried out a Europe-wide field experiment on leaf-litter breakdown in 100 streams covering a 1,000-fold gradient of nutrient concentration (both N and P), hypothesising that moderate nutrient enrichment would stimulate litter breakdown but that severe eutrophication would inhibit it as the effects of environmental degradation became more severe. These latter might include anoxia and physical smothering by excessive growths of various kinds, conditions that would exclude key shredding invertebrates. They used two sources of leaf litter (pedunculate oak, *Quercus robur*, a slow decomposer, and fast-decomposing alder, *Alnus glutinosa*) and both fine- and coarse-meshed litter bags. The rate of breakdown mediated by invertebrate feeding of both litter types was stimulated by increases in both dissolved inorganic nitrogen (DIN) and soluble reactive phosphorus (SRP) up to points of about $20\ \mu g\ L^{-1}$ (P) and $300\ \mu g\ L^{-1}$ (N) but was inhibited thereafter. In a subset of Irish streams covering the full nutrient gradient they showed that the peak breakdown rate (at least that mediated by invertebrates) coincided with the greatest abundance of large-bodied shredders, including amphipods and limnephilid caddis larvae, which declined in more severely polluted systems. This further underlines the major role of shredding invertebrates in leaf breakdown in many (but not all) systems.

Nutrients, particularly nitrogen, clearly stimulate microbial respiration on detritus in many running waters and can also enhance breakdown mediated by macroinvertebrates. Overall, therefore, the evidence suggests that the rate of litter breakdown increases with nutrient supply—but only up to a point. We return to nutrients and the effects of eutrophication in lotic ecosystems in the next chapter.

8.5.5 Dissolved and fine-particulate material

Leaf litter and other coarse organic material is a major source of fine particulate organic matter (FPOM) as the leaves become fragmented by various processes (including 'shredding'), and also of dissolved organic matter via leaching in the early stages of breakdown. There are numerous other sources of both these fractions, however (Allan & Castillo 2007). Dissolved organic matter is imported by 'hydrological vectors'—that is, in the various sources of surface runoff from the catchment (in upwelling groundwater, flow through the soil profile and litter layer and 'throughfall': see Chapter 2, section 2.5.3). This DOM itself derives from leachate from terrestrial plants—living or dead. Authochthonous DOM is released from aquatic plants (algae and macrophytes). Fine particles can come directly from the catchment—in runoff from the land surface and soils (particularly after heavy rain). In a reversal of the more usual reduction in particle size during breakdown, fine particles can also originate by (i) complexation of dissolved organic matter (whether allochthonous or authochthonous) with ions, such as calcium, iron and aluminium, to form particles; (ii) by microbial uptake of DOM (attached to suspended particles or on the substratum); and (iii)

by the action of filter-feeders (such as blackfly larvae and mussels, as mentioned earlier) that may be able to ingest colloids and exopolymers in the water and aggregate them in their faecal pellets (e.g. Wotton 1994, 1996).

The heterogeneous origins of FPOM account for its variable quality as food for animals, though this is often low. For instance, Callisto & Graça (2013) found a low nitrogen content of FPOM from streams in central Portugal, a high C:N ratio and a high lignin content, indicating a lower quality on average than decomposing leaves. They reared midge larvae (Chironomidae: *Chironomus riparius*) on FPOM produced by milling senesced but microbially unconditioned oak leaves, natural FPOM and sterilised natural FPOM, finding adult emergence rates of 80, 45 and 0%, respectively. This supports the view that FPOM is often refractory and likely to be exported rather than supporting substantially the metazoan food web. An exception may be FPOM issuing from lakes in stream flow, which is often rich in plankton and supports enormous populations of filter-feeding invertebrates (e.g. Wotton 1988).

8.5.5.1 Biofilm

Around fifty years ago, stream ecologists began to investigate the 'organic layers'—biofilms—that develop on practically all submerged (or even just wetted) surfaces, including stones, wood, dead leaves, living plants and sand grains (e.g. Madsen 1972; Geesey et al. 1978). The nature of the 'stone surface organic layer' was subsequently addressed by Rounick & Winterbourn (1983), who traced its development over two months in the dark and light in two New Zealand streams, one draining a forest and one a stony spring. The latter had low concentrations of DOC. Under natural light regimes photosynthetic organisms (diatoms and filamentous algae) dominated the layer at both sites. In experimentally darkened channels, a 'heterotrophic' layer developed at the forest site only (i.e. no organic layer developed at the spring site); this was made up of a 'slime' matrix containing fine particles, bacteria and fungi.

Further experiments showed that DOM was a prerequisite for layer formation in the dark, and that leaf leachate was actively taken up by biofilm microbes. Several stream invertebrates could graze the heterotrophic layers and assimilate it quite efficiently. These experiments showed a feasible additional element (in the form of grazing of epilithic biofilms) to the supposedly essential role of leaf shredders in stream food webs in forested headwater streams. The faecal pellets of stone surface 'browsers' could supply

FPOM to deposit and suspension feeders in streams where shredders were absent or scarce. Winterbourn et al. (1981) had previously pointed out that streams in New Zealand (as an isolated oceanic island) lacked many shredders. Further, native forests were dominated by southern beech (*Fuscospora* and *Lophozonia*), which have small, rather tough leaves that are inefficiently retained in the steep, frequently disturbed stream channels typical of the south island of New Zealand. More generally, such research pointed the way to a rather more pluralistic and flexible view of the energetic support of stream food webs and ecosystems, but this was controversial at the time.

The uptake of DOC by stream-bed sediments has now frequently been demonstrated. For instance, Fiebig & Lock (1991) demonstrated that DOC—particularly low-molecular-weight amino acids—in ground water upwelling through the stream bed was actively taken up by the biofilms on cores of stream-bed sediments in Welsh streams. Fiebig & Marxsen (1992) also carried out experiments on the absorption and mineralisation to CO_2 of amino acids in sediments from the first-order Breitenbach in central Germany. This research confirms an earlier speculation of Hynes (1983) that ground water could make a contribution to the supply of allochthonous organic carbon to steam food webs. Such carbon must originate largely from terrestrial vegetation and soil organic matter in the catchment that has infiltrated through the soil and into the ground water and may thus be quite 'old'.

Much progress has now been made in the study of stream biofilms, both of their structure and of the processes that occur. They are extremely complex 'ecosystems' in their own right and are major hotspots of metabolic processes in running waters, highly significant at a global scale. Lock et al. (1984) presented an early conceptual model of their structure—which they characterised as a polysaccharide layer (the 'slime') secreted by early colonising bacteria and phototrophs (algae and cyanobacteria) and in which their cells are embedded. This also contains exoenzymes produced by those cells (after cell lysis or by secretion in life), plus particles of detritus (Figure 8.19a). Biofilms can have a very wide variety of bacteria and larger algae, protists and small metazoans (see modern reviews by Battin et al. 2016, and Weitere et al. 2018 on the ecology, food web structure and biogeochemistry of stream biofilms; Figure 8.19 b, c). Clearly, biofilms are complex, and there is much still to be discovered about them. There are potentially many more kinds of species interactions within food webs, particularly mutualisms between microautotrophs and heterotrophic microorganisms (e.g. between fungi and algae), that have yet to be widely studied.

It has to be acknowledged more generally that the decay rate of DOM in streams—an enormously important part of organic matter pool in all aquatic systems—is poorly understood. Progress is being made, however, using methods specifically aimed at streams and rivers, in experimental systems that included a 'benthic zone' (see Kelso et al. 2020). Mixtures of labile (such as algal and plant leachates) and less labile sources of DOM (such as soil leachates) showed both positive and negative non-additive effects, and decay rates over the course of incubations were not constant, being more rapid initially. More progress in this field is certainly needed.

8.6 The support of stream and river food webs

As much of this chapter has stressed, for most of the history of stream ecology and biology the prevailing view has been that the typical woodland stony stream is heterotrophic and its secondary production is based substantially on terrestrial leaf litter. An accessible update on this paradigm is given by Marks (2019), who revisits the classic title of Kaushik & Hynes (1971)—'The fate of dead leaves that fall into streams'. Many early studies in stream ecology in temperate areas pointed to this conclusion, including Jones's (1950) study of the food of insects in a Welsh mountain stream, and Minshall's (1967) account of the role of allochthonous detritus in a small woodland stream in Kentucky, USA, and many others. Hynes (1970b) reviewed some of this early literature in his classic book *The Ecology of Running Waters*. The highly influential River Continuum Concept of Vannote et al. (1980) placed shredding invertebrates and a supply of terrestrial leaf litter to headwater streams at the very centre of the 'economy' of river ecosystems, and the long-term studies on the forested headwaters at Coweeta referred to above seem to bear this out. Questions and different views remain, however, not least on how far this model of a cool-temperate, forested stream represents the more general condition of larger rivers, and of streams and rivers in other biogeographic regions and in systems heavily modified by humans (now an almost universal condition).

In this section we deal with evidence on what sources of fixed carbon actually support secondary production in streams and rivers, particularly of metazoan animals. There have been a number of

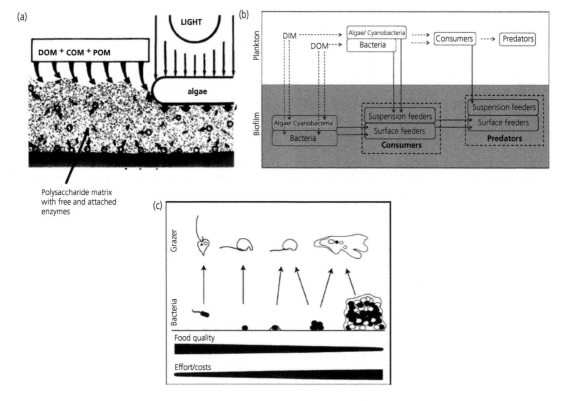

Figure 8.19 Developing views of biofilms in running water: (a) An early model of a layer consisting of a polysaccharide matrix, secreted by colonising microbial cells, in which those cells are embedded. Photosynthetic cells (cyanobacteria and algae) dominate near the surface of the biofilm where there is sufficient light. Exoenzymes are involved in the immobilisation and mineralisation of adsorbed DOM onto the matrix. (b) Processes involved in the biofilm food web, including interactions based on heterotrophic and photosynthetic production within the food web itself, plus grazing and predation on suspended (planktonic) cells by organisms in the biofilm (giving biofilm organisms access to overall river production). Diverse micrograzers (primary consumers) and predators are involved in this coupling, including both protozoa (particularly ciliates, flagellates and amoebae) and small metazoans (such as rotifers, turbellarians and nematodes). Biofilms are then very actively exploited by larger (free-living and sedentary) benthic grazers. (c) As biofilms mature over time, the predominant bacteria may switch from free-living single cells, to more firmly attached single cells, to attached complex colonies and consortia. These can be taken by different grazers—such as (from the left), *Spumella* (a chrysomonad, feeds on suspended bacteria), *Rhynchomonas* (heterotrophic flagellate, loosely attached single bacteria), *Planomonas* (heterotrophic flagellate, firmly attached bacteria) and *Acanthamoeba* (naked amoeba, microcolonies and mature biofilms). Food quality (e.g. nutrient to carbon ratio) declines with biofilm maturity (i.e. left to right) and the costs of grazing increase.
Source: (a) from Lock et al. 1984; (b) from Weitere et al. 2018; (c) from Weitere et al 2018; all panels with permission from John Wiley & Sons.

approaches. First, the 'trophic basis of production' allocates the secondary production of species, or groups of species, to particular kinds of resources based on gut contents analysis. Second, there have been a number of attempts experimentally to manipulate, by reduction or supplementation, the resources available to stream ecosystems. Third, some evidence suggests that the diet of stream animals often varies from expectations, while some novel 'food web markers' can be used to trace the origins of the food actually assimilated by consumers (see also Chapter 7, section 7.7.2). These latter

methods in particular, have given us pause for thought about the wider basis of river production.

8.6.1 The trophic basis of production

Benke & Wallace (1980) pioneered this approach when they measured secondary production in a guild of six species of web-spinning caddis larvae in a stony stream in the southern Appalachians of the eastern USA. This was combined with the microscopical analysis of gut contents—the latter involving inspection of about

270,000 particles! For half the species, the greatest percentage of the area of particles in the gut contents (and by inference of the food ingested) was made up by fine particles of detritus. In another species animal material and fine particles were about equal, while for the two largest species animal material predominated (in all six species, algal material and recognisable fragments of vascular plants were minor components making up the rest). Overall, the average for the six species was 36% animals, 52% detritus and 12% algae. These foodstuffs must be assimilated, and the 'assimilate' converted into somatic growth, to contribute to secondary production. Using values from the literature for net production efficiency (production/assimilation × 100%) and assimilation efficiency (assimilation/ingestion × 100%), they assessed the trophic basis of production as almost 80% due to animals (as an average over the six species), 13% detritus and 8% algae. All but the smallest species were mainly carnivorous. This was due to the much higher efficiency with which animals are assimilated (they assumed 70%) compared with detritus (10%), algae being intermediate (30%). This production was supported by consumption by the caddis of about 2.3 g m^{-2} of animals, 2.5 g m^{-2} of fine detritus and 0.5 g m^{-2} of algae. Of course, this method does not unequivocally allow us to discern whether the carbon supporting this production was mainly allochthonous or authochthonous in origin, since the non-predatory prey could have been exploiting either, and fine detritus could again have been terrestrial or aquatic in origin. Nevertheless, this approach was a major step forward in identifying major pathways in stream food webs in a more quantitative way.

Using similar logic, Benke & Jacobi (1994) assessed production and its trophic basis for a diverse guild (more than 15 species) of mayflies (Ephmemeroptera) in the Ogeechee, a 'blackwater' (stained with humic DOM) river on the coastal plain of Georgia. These systems are sluggish, sand-bottomed streams issuing from the floodplain swamp forests. Deposits of organic matter and large submerged wood are also common, the surfaces of the wood being centres of intense secondary production. Total annual production of mayflies on the wood was very high (20–42 g m^{-2}, equivalent to 7–12 g m^{-2} of riverbed overall). Fine detritus made up 87% of gut contents of wood-dwelling mayflies, and accounted for 70% of total mayfly production (about 18% being due to diatoms). One mayfly was a filter-feeder while the remainder were mainly collector-gatherers of deposited detritus. Benke & Jacobi (1994) concluded that the source of organic matter was overwhelmingly DOM, presumably immobilised in the

biofilm, and very fine particles from the floodplain, and in this case was almost certainly allochthonous in origin.

Knowledge of the energetic support of stream communities was advanced by the 'trophic basis of production' approach and certainly made the depiction of quantitative food webs a possibility (e.g. Benke & Huryn 2017; Chapter 7, Figure 7.26, p. 259). It also showed that, because of differing food quality (such as detritus versus animal material), gut contents analysis alone is unlikely to represent food assimilation. Evidence as to the actual source of the carbon (allochthonous or autochthonous) was still largely circumstantial, however, and based originally on visual examination of gut contents. Clearly, experimental evidence might help.

8.6.2 Experimental evidence

There have been several experiments that manipulated the detritus inputs to streams, including supplemenating CPOM to test for a response in invertebrate populations. Richardson (1991) set up experimental channels alongside a forested stream in British Columbia and diverted water through them. These were colonised by stream invertebrates. He added variable amounts of leaf litter over one summer season, finding that most shredding species were more abundant or grew faster where CPOM was added. Dobson & Hildrew (1992), and Dobson et al. (1995) increased the retention of coarse organic matter leaf litter over two years in two sets of streams (draining either moorland catchments or lowland forested catchments) in the UK, adding leaf litter to the moorland (Welsh) systems where there was no natural source. Where increased retention raised the supply of litter, the biomass of shredders increased. Detrital enhancement experiments at a larger spatial and temporal scale were more recently carried out in Hughes Creek, a sand-bed river in Victoria (Australia), over a five-year period (Lancaster & Downes 2021). Similarly to Dobson & Hildrew (1992), and Dobson et al. (1995), they increased retention of detritus in natural stream channels. They found persistent increases in the density of benthic invertebrates and species diversity. Responses of individual taxa differed, taken to infer that for some species detritus is an essential resource while for others it was a 'substitutable' resource—that is, it could be used by these latter species but as a 'subsitute' for alternative resources. The inference of experiments such as these, as well as others, is that allochthonous detritus is indeed a major, and often limiting, resource in headwater streams.

Most tellingly, the Coweeta Hydrological Laboratory was again the site of one of the most influential and long-term field experiments in stream ecology, referred to in a food web context in Chapter 7 (Figure 7.23; Hall et al. 2000). The overall results have been synthesised by Wallace et al. (2015), where experimental details and references to component studies can be found (see also Text Box 1.1, p. 6). These authors and colleagues carried out a remarkable 13-year manipulation of one first-order forested stream ('Catchment 55', C55). This was compared with two similar reference streams (C53 and C54), the total data encompassing 37 'stream years' (cumulative years of study in the three streams). Treatments in the experimental stream were:

(i) from 1993–2006, exclusion (using netting) of leaf litter by direct fall or by blowing in from the banks; (ii) in addition, in 1996 and 1998, small (< 10 cm diameter) and large (> 10 cm diameter) pieces of wood were removed, respectively; further (iii) physical structure (to replace wood as possible sites of retention), in the form of pieces of PVC and plastic, were added at an appropriate density in 2000–2001; then (iv) leaves of fast-decomposing species were added manually in 2001–2003, of slow decomposing species in 2003–2005, and of mixed leaves in 2005–2006.

The initial leaf exclusion reduced the standing crop of leaf litter (Figure 8.20a) which remained well below reference values until leaves were added

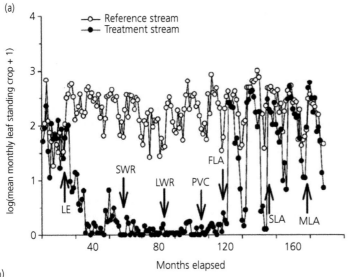

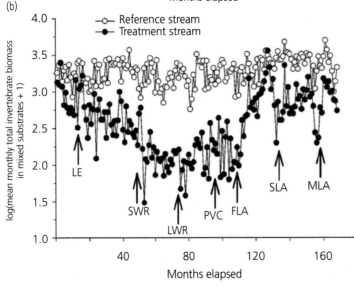

Figure 8.20 Monthly leaf litter standing crop in reference and treatment streams at Coweeta (on 'mixed substrates'). First 12 months untreated in both streams; thereafter (treatment stream only) leaf litter exclusion (LE) for 36 months, plus small wood removal for 24 months (SWR), plus large wood removal for 24 months (LWR), increase in physical complexity for 12 months (PVC), plus addition of fast-decomposing leaves for 24 months (FLA), then slow-decomposing leaves for 24 months (SLA), then, finally, mixed leaf addition for 12 months (MLA): (b) invertebrate biomass on mixed substrata in the reference and treatment streams from September 1992 to September 2006. Abbreviations as in (a). *Source:* both panels from Wallace et al. 2015, with permission from John Wiley & Sons.

experimentally after 9 years. Invertebrate biomass responded by declining steadily (Figure 8.20b), not increasing again consistently until leaves were added. This decline was observed on 'mixed substrates'—such as stones and cobbles—that included anything other than the surface of outcrops of solid bedrock. On the latter no effect was observed, probably because litter-shredding invertebrates were scarce there. Manipulation of physical structure alone (removing wood or adding plastic substrata) had no additional effect on invertebrate biomass (Figure 8.20b). The production of predatory invertebrates was also remarkably closely related to that of their prey in both experimental and reference streams. Even more remarkably, the effects of a reduction in the detrital subsidy ramified as far as the population size, growth rate and production of larval salamanders, *Eurycea wilderae*, in the same stream (Johnson & Wallace 2005). Further, these salamander larvae had fewer prey items in their guts than those in reference streams or in stretches downstream of the litter exclusion. Overall, there was very strong evidence that secondary production is tightly linked to subsidies of allochthonous detritus and thus that this detritus is indeed the main support of food webs in forested streams.

8.6.3 Diets, food webs and markers

One basis for the River Continuum Concept (RCC) was the 'functional feeding group' (FFG) concept of Cummins (1973, 1974). In the RCC predictable changes

in the representation of different FFGs in the benthic community are driven, for instance, by longitudinal shifts changes in food type. However, Cummins (1973) explicitly points out that 'most aquatic insects are best termed polyphagous or (dietary) generalists'. Rather, he proposed a classification of stream invertebrates into feeding groups based on the predominant way that animals acquire food (see Chapter 4, section 4.5.1, p. 125 for details of functional feeding groups). Nevertheless, the supposition that shredders feed mainly on leaf litter, scrapers on algae, and collectors (other than filterers) on deposited detritus often proves a tempting short cut in the analysis of energy flow in lotic systems, even though this was never the declared objective of the FFG scheme (see Mihuc 1997 for examples and a critique).

Acidified streams provide a further example of generalism and dietary switching among benthic invertebrates that we first introduced in Chapter 4, section 4.5.3, p. 131. For instance, Ledger & Hildrew (2005) showed that nemourid stoneflies, usually described as leaf-shredding detritivores, also included large but variable amounts of algae in their diets and are thus generalist herbivore-detritivores. A group of herbivore-detritivores (nemourid and leuctrid stoneflies) were present at stream sites across a gradient of acidity in 20 UK streams (Layer et al. 2013). They consumed an increasing percentage of biofilm in their diet as stream acidity declined—taken to infer a switch from allochthonous detritus towards algae (Figure 8.21a), as food quality of the biofilm increased

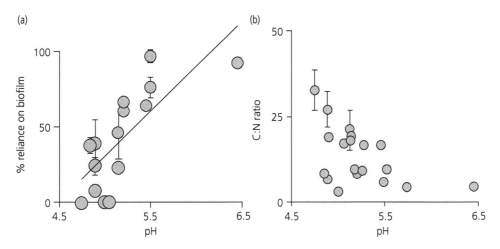

Figure 8.21 (a) The growing reliance of herbivore/detritivores (leuctrid and nemourid stoneflies) on biofilm along a gradient of increasing mean stream pH in UK streams, based on isotopic evidence; (b) C:N ratio of biofilm declines along the pH gradient
Source: Layer et al. 2013, published under a Creative Commons Attribution 2.0 International Licence.

(i.e. C:N ratio declines; Figure 8.21b). Further, Rosi-Marshall et al. (2016) reassessed the river continuum concept by resampling stations on the Salmon River of Idaho (USA). They compared gut contents of invertebrates in the these 'new' samples with archived specimens from the original RCC study (carried out in 1976). Diets were essentially unchanged over the 30-year intervening period. Thus, consumption of leafy allochthonous detritus declined downstream, although large amounts of algal material were also eaten (on average 35–75% of gut contents were authochthonous), even by a well-known shredder. Thus, the actual diet of invertebrates was not well predicted by functional feeding group—that is, diet was more generalist than expected. Lauridsen et al. (2014) showed that estimates of elemental imbalances (as C:N and C:P ratios) between the body tissues of stream invertebrates and their 'diet' were greater when the functional feeding group was used simply to infer the diet compared with actual gut-contents analysis. This again suggests feeding plasticity in benthic animals and means that the support of stream food webs may be rather more opportunistic and flexible than sometimes believed, such that the relative uptake of nitrogen and phosphorus is greater than might be expected from their functional feeding group.

Some techniques allow us to move beyond visual gut contents analysis, promising to reveal what animals actually assimilate from their diets, and in particular to distinguish allochthonous from autochthonous sources. These include the analysis of naturally occurring stable isotopes of several elements, which can act as food web markers (see e.g. Hershey et al. 2017). Most commonly used in food web research are isotopes of carbon and nitrogen. The fraction of the rather rare (compared with the common ^{12}C) stable isotope ^{13}C in organic matter changes only slightly when it is consumed by an animal—so that the stable carbon isotopic ratio (denoted as $\delta^{13}C$) of the consumer is similar to that of its food, there being slightly more ^{13}C in the consumer than in the food (it is said to be slightly 'enriched'). The ratio is calculated as:

$$\delta^{13}C = \left[\left(R_{sample} - R_{standard} \right) / R_{standard} \right] \times 1000$$

where R_{sample} is the ratio of ^{13}C to ^{12}C in the sample, and $R_{standard}$ is the ratio of ^{13}C to ^{12}C in a recognised standard material (in the case of carbon this is a particular kind of limestone). The units of $\delta^{13}C$ are 'per mille'—‰— and it is a negative number, because ecological samples are normally depleted in ^{13}C compared with the standard. The equivalent isotope of nitrogen is ^{15}N, the more common isotope being ^{14}N, while the standard for nitrogen is that in the atmosphere. In this case, the stable nitrogen isotopic ratio ($\delta^{15}N$‰) changes more substantially when food is consumed, consumers being markedly enriched in ^{15}N compared with the food—by about 3.4‰, though this is variable. This makes the isotopic ratio of N a useful marker of 'trophic position'—that is, how high in the food web is an animal feeding. The stable carbon isotopic ratios of allochthonous and authochthonous sources of carbon are often distinct, making it possible to tell whether a consumer is supported by one rather than the other, or by a mixture. If there is little difference between possible sources, or variability is great, then the simple stable carbon isotope method is of little use, though other stable isotopes are available.

Among the first to apply stable isotopes to stream ecosystems were M.J. Winterbourn and colleagues in New Zealand. Rounick et al. (1982) found a consistent difference in the $\delta^{13}C$ of leaf litter and other allochthonous sources (at about −27‰) and algae and macrophytes (about −35‰). Isotopic ratios of insects in a forest stream suggested that most depended on allochthonous carbon, while those from an open grassland stream used more autochthonous carbon. In catchments subjected to clear-felling of forests, those most recently felled (and containing much 'forest trash') retained the 'terrestrial signal' in invertebrates, whereas those in streams that had been felled five years earlier switched to a predominantly autochthonous signal. Winterbourn et al. (1984) then confirmed a terrestrial signal in forested sites and a more algal signal in open sites further downstream. The New Zealand mayfly *Deleatidium* appeared (from gut contents) to ingest few algae at forested sites, yet isotopic evidence suggested a substantially autochthonous diet, which could result from algal exudates in biofilms being taken up by heterotrophic microorganisms that were then grazed by the mayfly. This early isotopic evidence was more or less in line with expectations at the time; shaded, forested systems appeared mainly supported by allochthonous carbon, whereas algal carbon was important in open sites. However, there seemed to be few shredders of leaf litter. This difference was an important thread in questions about the river continuum concept itself, which saw shredding as perhaps the central process in lotic ecosystems that would generate fine particles (Winterbourn et al. 1981).

There have been many stable isotope studies of diets in the last 30 years or so, in a wide variety of environmental and geographic settings. Some have essentially confirmed the use of allochthonous carbon by metazoans, at least in forested headwaters, but sometimes

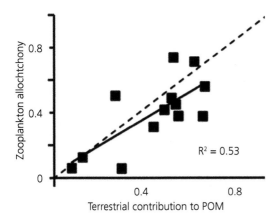

Figure 8.22 Allochthony of zooplankton in 13 rivers of northern Sweden increases with the proportion of suspended POM that is of terrestrial origin—the dashed line shows a 1:1 relationship, the solid line shows that actual regression.
Source: from Berggren et al. 2018, with permission from John Wiley & Sons.

more widely. For instance, Leberfinger et al. (2011) used carbon and nitrogen stable isotopes to assess the support of shredding detritivores at open and wooded sites on four Swedish streams. As expected, terrestrial carbon dominated at shaded sites, but it also dominated at more open sites, although in the latter more biofilm was also included in the diet, perhaps again indicating a degree of flexibility in feeding. Reliance on allochthonony can extend to the zooplankton. Thus Berggren et al. (2018) showed that the carbon intake of the crustacean zooplankton in northern Swedish rivers mirrored that of the allochthonous fraction of suspended carbon, with a relationship close to 1:1 (Figure 8.22). They attributed this to the difficulty of food selection (of more nutritious algal material) in the turbulent environment of the water column.

A substantial reliance on allochthonous carbon can extend to the wider community in larger rivers. Thus, Zeug & Winemiller (2008) assessed terrestrial and algal food sources in the main channel and oxbow lakes on the floodplain of the Brazos River, Texas. This is the 11th-longest river in the USA and has an active floodplain (frequently inundated) in the middle and lower reaches. Terrestrial C3 plants growing in the riparian zone were identified as the primary carbon source for almost all consumers in the main channel and most in the oxbow lakes. These included fish and invertebrates such as shrimps and crayfish, which seemed to access the terrestrial carbon via predation on smaller invertebrates (as judged by their trophic position using the $\delta^{15}N$). Smaller-bodied (< 100 mm) consumers in

the oxbows (but not in the main channel) did access more algal carbon. Zeug & Winemiller (2008) ascribed their results to the variable flow regime and frequent connection to the floodplain on the Brazos River, compared with more constrained rivers during prolonged periods of low flow.

Collins et al. (2016) widened the geographical setting, this time using the stable hydrogen isotope deuterium as a marker of terrestrial subsidies to stream food webs, in temperate and tropical areas. They included streams of a variety of differing sizes and canopy covers (factors that co-varied) in both the Adirondack Mountains of the temperate north-eastern USA and in tropical Trinidad and Tobago. In both systems, invertebrates and fish classed as shredders and predators had a mostly allochthonous signal independent of canopy cover, whereas grazers and collector-gatherers had an allochthonous signal in small, shaded streams but were strongly autochthonous in wider, unshaded streams. Overall, however, the tropical streams had a rather stronger autochthonous signal, while the allochthonous fraction of the diet increased more strongly with canopy cover in the tropics than in the Adirondacks (Figure 8.23). In a further study from the tropics, Neres-Lima et al. (2017) combined stable-isotope evidence with estimates of net primary production in five forest streams in Brazil, finding that the principal energy source for macroinvertebrates came from the riparian forest. They also found that this was unlikely to be a consequence of a low supply of autochthonous production, since this was more than enough to support secondary production—and annual ingestion of autochthonous material was less than 10% of annual NPP.

In a novel 'twist' to the more usual pattern, Atkinson et al. (2018) tested the carbon source supporting the food web in streams in the Ecuadorian Andes-Amazon (tropical) region over an enormous gradient of altitude, with some sites at over 4,000 m. Here, canopy cover increased down the slope, with more light and algal-based resources upstream (above the treeline) and more allochthonous detritus downstream. In this case there was evidence of an increase in the use of autochthonous (algal-based) resources with increasing altitude, as might be expected from the spatial pattern of resource availability. Exports of terrestrial detritus from forested rivers even support fish biomass at the interface between the rivers and downstream lakes, that is, in the freshwater delta (Tanentzap et al. 2014). In eight near-shore sites receiving inputs from separate rivers flowing into a lake in Ontario (Canada) they found that between 34 and

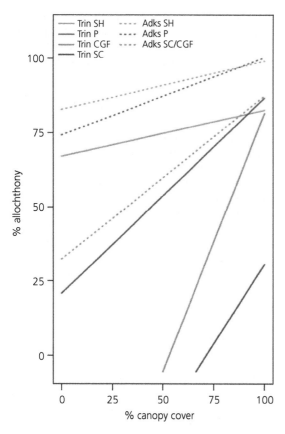

Figure 8.23 The regression lines of % allochthony versus canopy cover for streams in the temperate Adirondack Mountains, north-eastern USA (Adks) and on tropical Trinidad and Tobago (Trin) for several functional feeding groups: SH—shredders, P—predators, CGF—collector gatherer/filterers, SC—scrapers. The tropical streams generally showed less allochthony and switched to autochthonous support more rapidly as canopy cover declined.
Source: from Collins et al. 2016, with permission from John Wiley & Sons.

66% of fish biomass was supported by allochthonous carbon. Greater organic inputs from forested catchments increased bacterial growth, supporting heavier zooplankton and enhanced growth of planktivorous fish.

Despite the evidence shown above, some researchers have increasingly questioned the reliance of stream animals on allochthonous carbon, on the grounds of its relatively poor food quality compared with algal carbon (see Vadeboncoeur & Power 2017). In an update to their riverine productivity model (see section 8.3), Thorp & Delong (2002) wrote that: 'the primary, annual energy source supporting overall metazoan production . . . in mid- to higher trophic levels of most rivers (of order 4 or more) is autochthonous primary production entering food webs via algal-grazer and decomposer pathways'. This conceptual model therefore does not now confine itself to big rivers, or to those constrained within the main channel by flow regulation or river engineering. While conceding the overall heterotrophy of most river ecosystems, they see allochthonous carbon as recalcitrant, supporting only a 'microbial loop' and/or being exported. In their view, the metazoan food web is only weakly connected with this decomposer pathway (the microbial loop)—excepting only that detritus which is itself of autochthonous origin (i.e. dead algae etc) and of higher food quality (Figure 8.24).

Brett et al. (2017) martialled evidence that terrestrial carbon is mainly made up of lignocellulose, undigestible to most animals, and lacks biochemical compounds critical for their growth and reproduction—including essential fatty acids and particular amino acids. They again stressed that microalgal (autochthonous) production supports most animal production in fresh waters. The argument is based largely on knowledge about the nutrition of animals. Analysis of fatty acids in various food sources shows that microalgae are often rich in long-chain polyunsaturated fatty acids (LC PUFAs), including eicosapentaenoic acid, but also the shorter chain α-linoleic and linoleic acids. These PUFAs are in the tissues of invertebrates and seem to be strongly retained in stream food webs (e.g. Guo et al. 2016a; Guo et al. 2021). For the most part, it seems that these fatty acids must be obtained from the diet—though this requires further research. Algal fatty acids are virtually absent in allochthonous detritus—even in that conditioned by fungi and bacteria, leading to claims that algal carbon is dominant in the support of stream and river food webs. Somewhat similarly, Kolmakova et al. (2013) found that benthic invertebrates had a greater content of essential amino acids than was in the epilithic biofilm in a Siberian river, while the main fish predator (the Siberian grayling, *Thymallus arcticus*) concentrated them even more. This again suggests that limiting algal nutrients are concentrated and preferentially assimilated by consumers.

We are undoubtedly in a new phase of very active investigation of the old question of the main source of carbon supporting lotic food webs—which had seemed settled a long time ago. However, there is now a good deal of evidence that autochthonous carbon can be dominant in many stream and river food webs. Bunn et al. (2003) described the extremely interesting situation of the Cooper Creek drainage in arid central Australia. This covers a vast floodplain of anastomosing

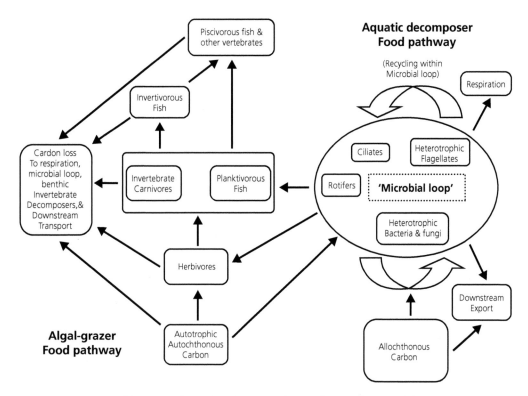

Figure 8.24 A conceptual view of the summary food web in a large river—the revised river productivity model. This sees two, modestly interconnected, pathways for autochthonous (the algal-grazer pathway, supported by planktonic and benthic microalgae) and detrital material (mainly allochthonous but including dead autochthonous carbon). Most metazoan biomass (invertebrates, fish and others) is derived from the algal-grazer pathway, whereas the overall ecosystem may be heterotrophic because of the decomposer pathway.
Source: from Thorp & Delong 2002, with permission from John Wiley & Sons.

channels that dry to remnant water holes in prolonged dry periods but are reconnected during periodic floods. The water holes are turbid and the expectation was that primary production would be low and the food web of the water holes based on large amounts of terrestrial carbon (wood and leaves) from the floodplain. On the contrary, stable isotope evidence suggested that algal carbon supported most consumers—including invertebrates (mainly crustacea) and fish. Researchers noticed that a green 'bathtub ring' of filamentous algae developed around the water holes, in which photosynthesis was intense, and it was this narrow band of primary production on which consumers apparently depended.

Lau et al. (2009) synthesised a number of studies of streams in the monsoonal tropics of Hong Kong that had collected stable isotope data, and found a consistent pattern of reliance of most consumers on algal and/or cyanobacterial primary production, rather than leaf litter, and indeed shredders were scarce on these

systems (see Chapter 5). Studies of temperate systems have sometimes revealed substantial reliance on algal carbon. For instance, Hayden et al. (2016) studied two river systems in eastern Canada, with sites arrayed from source to mouth in both cases. There were indeed some obligate shredders, for whom allochthonous leaf litter was the main resource, and this was maintained throughout both systems. However, all other invertebrate primary consumers, invertebrate predators and fish apparently relied mainly on autochthonous carbon throughout. Thus, secondary production in both systems was related mainly to autochthonous carbon, and not to terrestrial inputs—in contrast to predictions from the river continuum concept. As a final and somewhat unusual example of autochthony, Carroll et al. (2016) found that the macroinvertebrate community of three karst (limestone) springs in Missouri (USA) was supported primarily by mosses and water cress (*Nasturtium officinale*), rather than allochthonous leaf litter.

The use of stable isotope ratios to distinguish between carbon of allochthonous and autochthonous origin is not without its problems. This may be because bulk samples of organic matter—the easiest way to collect samples of possible food sources such as leaf litter, fine particulate organic matter and biofilm—may not represent what is actually ingested or assimilated by animals. Thus, the isotopic ratio of the real food source is not determined adequately, particularly where the signatures of allochthonous and autochthonous carbon are less distinct.

How are we to resolve these contrary and sometimes trenchantly expressed views? On one hand, there is the heritage of linking the 'stream and its valley' (Hynes 1975), the masses of evidence about the role of detritivores in leaf decomposition, the powerful influence of the river continuum concept, and experiments identifying leafy detritus as a limiting factor. On the other, there is accumulating evidence of the absolute requirement for algal fatty acids and other nutrients in the diet of animal consumers. Several authors have pointed to a middle way. Guo et al. (2016b) found that a biofilm of high-quality diatoms attached to leaf litter enhanced the growth of a stream shredder (the Australian caddis *Anisocentropus bicoloratus*), and suggested that the availability of essential algal fatty acids (absent even from microbially 'conditioned' leaves) enhanced the dietary use of riparian leaves. This allochthonous carbon supported much respiration but the animals retained essential fatty acids from the algae and this enhanced their somatic growth—that is, both components of the diet were used.

In an informative study, Twining et al. (2017) combined both isotopic and fatty-acid analysis to trace food web structure and food quality in an Adirondack stream (New York State, USA). In terms of fatty acids, they compared biofilm, conditioned detritus and fresh leaves, confirming that only biofilm contained the long-chain fatty acid eicosapentanoic acid (EPA), though fresh leaves did have its short-chained precursor alpha-linoleic acid (ALA). A few invertebrates may be able to convert terrestrial sources of ALA into long-chain highly unsaturated omega-3 fatty acids (HUFAs), particularly some midge larvae (Chironomidae; Goedkoop et al. 2007), but most seem unable to do so (Guo 2016a). However, Guo et al. (2021) more recently showed that brown trout, unusually among stream fish other than salmonids, seem able to synthesise one fatty acid (docosahexaenoic acid) found in their tissues but absent in dietary sources. Further, Pilecky et al. (2022) showed that *Daphnia* (a zooplankter mainly of standing waters) can compensate for a low dietary supply of

long-chain fatty acids by synthesising them from short-cahined precursors. We need to know more of how widespread this ability is.

In terms of the debate about the support for stream food webs (i.e. autochthony versus allochthonony), Twining et al. (2017) concluded that 'the answer probably lies somewhere in the middle', terrestrial sources supplying energy while aquatic resources (i.e. algae) provide essential nutrients, even in small quantities. Similarly, Crenier et al. (2017) confirmed experimentally that small quantities of long-chain PUFAs were essential for maintaining growth of the supposed shredder *Gammarus fossarum*, although terrestrial detritus alone permitted survival. Kühmayer et al. (2019) also found that *Gammarus fossarum* preferentially assimilated algae (when fed on diets that included leaf litter) and concluded that it gained both its carbon and long-chain fatty acids from algae rather than leaves or from the (heterotrophic) microbial biofilms on those leaves. In this it did not differ substantially from the grazing mayfly *Ecdyonurus* (see also Labed-Veydert et al. 2022). The inclusion of at least some algae in the diet of purportedly exclusively leaf-shredding detritivores recalls our debate about the fidelity of different functional feeding guilds to particular diets. Thus, small, detrivorous stoneflies included a higher fraction of biofilm (including algae) in their diet along the gradient from profoundly acidic towards more circumneutral streams (Layer et al. 2012: Figure 8.21a). This accompanies an increase in secondary production as acidity declines (Hildrew 2018), perhaps as long-chain fatty acids became more available. Further, Ledger and Hildrew (2000) showed that the diet of purportedly 'detritivorous' nemourid stoneflies included large amounts of algae when they were available in an acidic headwater stream seasonally shaded by deciduous leaves. For such species perhaps an annual 'pulse' of higher-quality (algal) food is sufficient.

In a few cases, sources of carbon other than either autochthonous photosynthesis *or* allochthonous detrital carbon have been implicated in stream and river food webs. Methane is a powerful greenhouse gas that is almost everywhere supersaturated in fresh waters, and can originate from the ground water (which we could view itself as allochthonous to the stream) or be generated by methanogenic bacteria in sediments rich in organic matter (common in streams affected by agriculture in the catchment) (Stanley et al. 2016). In some deep lakes, particularly those with anoxic hypolimnia, methane is generated in the sediments and then oxidised by methanotrophic bacteria (methane-oxidising bacteria; MOB), that can in turn be ingested by tube-dwelling midge larvae at the anoxic/oxic boundary (Jones et al. 2008). Since methane-derived carbon is very depleted in ^{13}C, the larvae themselves can become markedly depleted compared with other feasible food resources, such as deposited algae or detritus of terrestrial origin. We know less of this process in running waters, or even whether it exists, but a few stream and groundwater invertebrates have also been shown to be depleted in ^{13}C, and may also be taking methane-derived carbon (Trimmer et al. 2009; DelVecchia et al. 2016; Sampson et al. 2018). They assimilate this carbon either directly by eating the MOB, or perhaps indirectly by oxidation of the methane to CO_2—which is itself then depleted—its uptake by algae and subsequent grazing (as suggested by Bellamy et al. 2019 for a stream with a methane seep in Ohio, USA). Note that, as described earlier, Shelley et al. (2017) found that methane oxidation could be fairly substantial in UK chalk streams, accounting for up to 13% of the total particulate carbon production (via photosynthesis and mathanotrophy). The uptake of methane-derived carbon by stream animals would thus constitute a 'third way' in stream food webs.

Overall, stream ecology and ecologists need to move away from an almost ideological adherence to any one conceptual model of the support of lotic ecosystems. The supply of carbon to metazoans seems much more pluralistic than we thought previously, and variations between exclusive allochthony or authochthony seem to be the rule rather than the exception. Diets vary seasonally and spatially, with age and life-history stage, with disturbance and flow fluctuations—many species being able to respond opportunistically to shifting resource availability (i.e. they are ominivorous). For instance, recent evidence of the effect of forest practices on the relative importance of authochthonous and allochthonous carbon for some forest streams in eastern Canada suggests that allochthony becomes dominant as management intensity increases (road-building, increased sediment inputs, chemical changes) (Erdozain et al. 2019). Moreover, evidence is emerging of more intermingling of the so-called green and brown food webs (based on grazing plants and detritivory, respectively) (see e.g. Demars et al. 2020; Halvorson et al. 2018; Price et al. 2021).

There are also clearly reciprocal subsidies between autotrophs and heterotrophic bacteria and fungi in streams. These interactions may occur at the intimate scale of individual cells within biofilms (Battin et al. 2016). For instance, the decomposition of allochthonous carbon by stream heterotrophs within biofilms produces CO_2 (potentionally limiting within the boundary layer) which stimulates algal production in 'neighbouring' cells within that biofilm. Thus primary production in streams is stimulated by allochthonous carbon from the catchment—a new take on Hynes's (1975) classic perspective 'The stream and its valley' (see next section). Further, new research has used radioisotopes to trace fungal, bacterial and algal carbon and phosphorus assimilated by a detritivorous caddis larva (*Pycnopsyche*) fed on tulip tree leaves (*Liriodendron tulipifera*) (Price et al. 2021). They found that fungal carbon indeed supported most *Pycnopsyche* growth, even where algae were abundant growing on the leaves (in well-lit conditions). However, these and other results also suggest an intermingling of the 'green' and 'brown' food webs, because algae could provide labile carbon to the fungi that support animal growth.

Clearly, such research is taking us into a more sophisticated technical era, enabling us to answer 'old questions' in new ways. Nevertheless, streams and their catchments clearly *are* linked fundamentally, and we turn to some further examples of the exchange of carbon (and energy) across the aquatic–terrestrial interface in the next section.

8.7 Streams and rivers in the terrestrial ecosystem

In 'The stream and its valley', Noel Hynes (1975) brought together important emerging themes at the time about the way in which catchments 'rule' the streams and rivers that drain them, via the supply of dead organic materials, minerals and nutrients, and about the route by which precipitation may ultimately enter the surface flow in the channel. We have already said much about allochthonous (dead) organic matter from land and its fate in rivers, but this is just one kind of 'subsidy', of carbon and materials, and largely driven by physical (hydrological) 'vectors'. Hynes is silent about the possibility of possible reverse subsidies 'upslope', from water to the land, and says little about many of the other linkages (involving quite fascinating bits of natural history) between the two systems. An obvious and well-documented pathway for a reciprocal subsidy 'upslope' is found in the return from the ocean and subsequent death of spawning Pacific salmon and other migratory species, to what are otherwise fairly unproductive rivers. As well as carbon and energy, these migrations return marine-derived nutrients to rivers, riparian systems and further inland, and we deal with them in detail in the next chapter.

Writing thirty years after Hynes, Baxter et al. (2005) neatly summarised reciprocal interactions between water and land as 'tangled webs', which refers to the many other ways by which the 'stream and its valley' are linked and are really parts of the same overall (catchment) ecological system. Cross-system (land–water) subsidies driven by biological vectors can be addressed in the 'currency' of population ecology and species interactions (Chapters 5 and 7), in terms of the biotically driven nutrient flows across habitat boundaries (Chapter 9) and, as here, as transfers of biomass, energy and production. In this section we mention a few of the ways in which organisms transfer fixed carbon from one system to the other.

A somewhat different pathway by which organic matter, and nutrients, enters some rivers is exemplified by situations in which large populations of terrestrial and semi-aquatic vertebrates live in close proximity to the channel. For instance, Subalusky et al. (2018) worked on the Mara River in Kenya, where there are large populations of hippos (*Hippopotamus amphibius*). The river is also crossed each year by the biggest extant migration of grazing herbivores in the world, in which well over a million wildebeest (*Connochaetes taurinus*) in the Serengeti-Mara ecosystem of Tanzania and Kenya track the supply of grasses following the seasonal traversing of the rainy season. Many wildebeest drown or are killed in the attempt.

Hippos forage on terrestrial grasses during the night, sometimes far from the channel, but return to the river by day where they defaecate spectacularly and copiously, spreading their dung widely as they do so by whirling their tails. Subalusky et al. (2018) quote an estimate of > 3,000 Mg (1 Mg (megagram) = 10^3 kg) of faecal material into a stretch of the Mara of about 80 km, making up most of the allochthonous carbon entering the river, and sufficient locally to carpet the river bed at low flows. It decays over periods of around 80 days, similarly to leaf litter in wooded, temperate streams. More than 6,000 wildebeest carcasses enter the Mara River in Kenya every year, about 300 tonnes of dry mass (DM), consisting of 'high quality' organic matter (low carbon:nutrient ratios). Soft tissues decompose rapidly, in 15–75 days, the phosphorus-rich skeleton taking much longer. Such natural situations of enrichment may once have been much more common than present, since these kinds of accumulations of wildlife have largely been lost, though they are partly replicated by farmed livestock (see Topic Box 9.2 by Russell Death). Many freshwater species of aquatic animals tolerant of eutrophication must once have found suitable habitat in these naturally 'polluted' sites (Moss 2015).

In their review of reciprocal subsidies between streams and riparian areas, Baxter et al. (2005) highlighted invertebrate prey. Thus, terrestrial invertebrates that fall in can provide up to half the annual energy budget of drift-feeding fish (mainly salmonids; see Chapter 7 section 7.8.2 and Figure 7.28). Maximum inputs occur in summer to autumn and in streams with a closed canopy of deciduous trees. Such allochthonous inputs of prey are sometimes sufficient to change the expected allometric relationship between abundance and body mass. Within a closed community (with no energetic subsidies in or out) we expect a decline in abundance (N, density) with mean body mass (M) that fits a straight line (on a log–log plot) with a negative regression coefficient (Brown et al. 2004). This is essentially because the rate of metabolism per unit body mass declines with size, but it is also because the energy available declines with body mass through the food web (i.e. at approximately 10% per trophic level, energy transfer through the food web is relatively inefficient). An energy subsidy from outside can change the slope of the regression if that subsidy is size related. For instance, if the subsidy is available only to larger organisms, they are more abundant than expected and the M–N slope is shallower.

Perkins et al. (2018) assessed the M–N relationship for a sample of 31 UK streams, all with a population of brown trout (*Salmo trutta*), which is known to feed on terrestrial prey—a 'subsidy'. Trout density and dominance of the fish assemblage varied widely among streams. The trout exploits more terrestrial prey than any other fish in the community, most of which are omnivorous, taking mainly a variety of detritus and plant material or aquatic invertebrates. The overall pattern of abundance with body mass (arranged in six equal 'size bins) was shallower than predicted from theory (Figure 8.25a). When mean abundance per taxon was plotted against mean body size across all 31 streams, the brown trout was revealed as a relatively large, super-abundant generalist predator (Figure 8.25b). As trout dominance declined (i.e. when there were other fish and its fraction of the fish assemblage declined) its food niche width was reduced and its abundance was also less. This is thus a quantitative demonstration of the effect on a widespread pattern in food webs of a size-related subsidy (the subsidy was available to a single, large predator).

Somewhat similarly, Jardine et al. (2017) used isotopes to assess allochthony in relation to body size

(a)

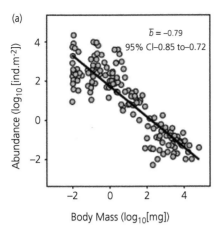

(b)

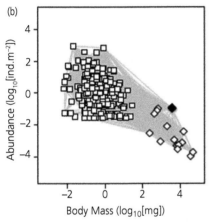

Figure 8.25 (a) The macroecological pattern from 31 UK streams of density (log N) plotted against body mass in six equal (logarithmic) 'size bins' on the x-axis and the mean abundance calculated for all species in each size bin (i.e. each point is for one size bin in one stream). In (b) mean abundance of each individual taxon is plotted against its mean body size across all 31 streams: squares are invertebrates, diamonds are fish and the black symbol is the brown trout—which is 'overabundant'.
Source: from Perkins et al. 2018, with permission from John Wiley & Sons.

in remnant water holes in the wet–dry tropics of Australia. Small organisms, invertebrates and small fish, were strongly supported by autochthonous (algal or macrophyte) carbon but the use of allochthonous resources increased by about 10% with every order of magnitude increase in body size. The largest animals in the web, estuarine crocodiles (*Crocodylus porosus*), derived 80% of their diet from allochthonous sources— from the marine system from where they migrate, and/or the floodplain. Such enormous animals could not persist in small water bodies without a large subsidy of suitable prey from elsewhere and there is presumably no negative relationship between abundance and body size for these webs!

As well as subsidies finding their way into the river, organic matter and prey moves in the reverse direction, from the river into the riparian zone and wider catchment (Baxter et al. 2005; see also Chapter 7, section 7.8.2). The 'decay' of this subsidy from the water out into the landscape is a matter of some debate, but Muehlbauer et al. (2014) speak of a 'biological stream width' as the distance from the channel at which the stream subsidy by aquatic resources reaches some specified fraction of its maximum value at the stream banks—say 50% or 10% remaining. Their metanalysis of the limited data suggests that the subsidy had declined to 50% as little as 1.5 m out from the bank, but that 10% remained at > 500 m away. This is similar, for example, to the rate at which catches of the adults of emergent aquatic insects often decline with distance (e.g. Petersen et al. 2004).

What are the origins of the carbon that leaves the stream? Kautza & Sullivan (2016) used stable isotopes to trace carbon fixed by photosynthesis in the stream compared with allochthonous sources (terrestrial sources of leafy detritus) in the tissues of emergent aquatic insects and in the rusty crayfish (*Orconectes rusticus*) along the Scioto River (Ohio, USA). They also analysed the tissues of five terrestrial predators: riparian tetragnathid spiders, rove beetles (Staphylinidae), damselfly adults (with terrestrial adults), swallows (aerial insectivores; Hirundinidae) and racoons (*Procyon lotor*). Spiders were most reliant on aquatic primary production (50%) (as also found in a study in Sycamore Creek described in Chapter 7, section 7.8.2.2, p. 253), followed by wider-ranging racoons (48%), damselflies (44%) and swallows (41%). Of the primary producers, phytoplankton contributed about 19% of the carbon to the riparian predators overall, periphyton 14% and macrophytes 11%. Phytoplankton was accessed via filter-feeding chironomids and hydropsychids (net-spinning caddis larvae), and macrophytes mainly via crayfish. It is evident that, as well as the well-known subsidy of streams by allochthonous organic matter, the terrestrial system is also supported by aquatic primary production.

In an experimental example from the tropics (Sao Paulo, south-east Brazil), Recalde et al. (2016), excluded aquatic insects from stretches of the riparian zone of streams and compared them with control stretches. In such exclosures, the biomass of terrestrial predators (largely spiders) decreased strongly, particularly those living on the vegetation rather than on the ground, while that of herbivorous terrestrial insects in the exclosures declined—possibly as a result of a trophic cascade as predators turned to terrestrial prey when faced with a shortage of aquatic insects, which may be of higher food quality.

One study even showed that those well-known ecosystem engineers of aquatic systems, beavers, can amplify the export of aquatic carbon to the riparian system. Working on stream sites in Montana (USA) with and without beavers (*Castor canadensis*), McCaffery & Eby (2016) used stable isotopes to track aquatic-derived carbon in terrestrial consumers (predatory wolf spiders, Lycosidae; and omnivorous deer mice, *Peromyscus maniculatus*). Many more aquatic insects emerged from beaver sites and tissues of the terrestrial consumers had much more aquatic carbon (Figure 8.26). It is likely that this effect was due to habitat changes wrought by the beavers, that increased populations of aquatic insects and their emergence, and that this increased the supply of aquatic prey to the terrestrial consumers.

It is now very apparent that, as well as ground-living predators, a wide variety of flying vertebrate predators—birds and bats—exploit aquatic production in the form of insects emerging from streams and rivers. Some 'trawling bats' are even adapted to pick up prey from the water surface, as well as taking those flying above or alongside the channel. These include,

in Europe, Daubentan's bat (*Myotis daubentonii*) and the pond bat (*Myotis dasycneme*), which both specialise in aquatic prey. In North America, the 'little brown bat' (*Myotis lucifugus*) is an important predator of chironomid adults (non-biting midges), mosquitoes, mayflies and caddisflies (Clare et al. 2014). Many species of songbirds (Passeriformes) migrate north in the northern summer, and feed on aquatic insects in riparian areas of stream and rivers. For instance, the neotropical species the Acadian flycatcher (*Empidonax virescens*), Louisian waterthrush (*Parkesia motacilla*) and wood thrush (*Hylocichla mustelina*) all nest (obligately or facultatively) near streams, have a breeding range in the Appalachian Mountains of the eastern USA, and are in decline (Trevelline et al. 2018). Aquatic resource subsidies are important for all three, but particularly for the Louisiana waterthrush. Most interestingly, it seems that the marked reliance of nesting insectivorous birds on aquatic insect prey relates to the greater supply of highly unsaturated omega-3 fatty acids (HUFAs) in aquatic rather than terrestrial insects. In the eastern USA, Twining et al. (2018) found that the chicks of tree swallows (*Tachycineta bicolor*) rapidly accumulate HUFAs after egg hatching up to the time of fledging, and that fledging success responded rapidly

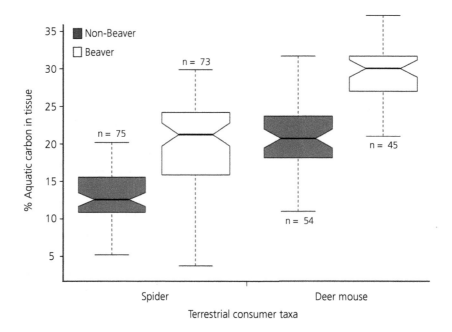

Figure 8.26 The percentage of aquatic carbon in the tissues of wolf spiders and deer mice within 45 m or 100 m of the water body, for spiders and deer mice respectively, at stream sites in Montana with or without beavers. The lower limit for each box is the minimum value observed, then the first quartile, the median (the notch in the box shows the 95% confidence interval around the median), the third quartile and the upper limit at each site.

Source: from McCaffery & Eby 2016, with permission from John Wiley & Sons.

to the biomass of aquatic prey available. Fledging success did not respond to variation in the supply of terrestrial insects.

8.8 Conclusions

In this chapter we have shown that river metabolism, the relative roles of autochthonous and allochthonous organic matter in secondary production, and the movements of carbon and the flow of energy within rivers and beyond their boundaries remain lively research topics. What we thought a few decades ago were clear patterns that fitted well into neat concepetual models are in fact somewhat blurred, with a marked mixing of the 'green' and 'brown' food webs. As experimental tools become more and more sophisticated, and as the benefits of precious long-term studies become ever more evident, we should be able to reconcile what are sometimes very different views. This will allow a better understanding of the dynamics of organic (reduced) and inorganic (oxidised) carbon in rivers—and between rivers and the rest of the biosphere—a topic of growing importance in the face of climate change. The movements of carbon through food webs and via hydrological processes also drive the cycling and transformation of key nutrients—the focus of the next chapter.

CHAPTER 9

Running waters as ecosystems

Nutrients

9.1 Introduction

The previous chapter was about the energy flow (involving transformations of carbon) that supports all ecological communities and drives the processes that characterise the living world. In rivers, as in other ecological systems, however, elements other than carbon are essential in the tissues and physiology of all living things. Energy can be said to 'flow' through ecosystems in the form of light energy being fixed as energy-rich organic carbon by photosynthesis and eventually being lost as heat during final decomposition and the mineralisation (oxidation) of that carbon. Energy is 'helped on its way' by many other elements, and in rivers these are principally nitrogen and phosphorus, essential components of the engine of life. These elements are present in a wide variety of biological molecules. Thus, amino acids contain large amounts of nitrogen, while lipids are richer in phosphorus. Nucleotides require both N and P, as well as carbon, which is used in all biological molecules. These elements are transformed in nature from one form to another but are conserved at the end of what we can conceive of as a 'nutrient cycle', driven largely by biological energy—and thus the cycling of carbon and of nutrients such as nitrogen are closely coupled. To repeat a theme of this book, rivers and streams themselves are not self-contained with regard to nutrients—nutrients and carbon are imported, mainly from their catchments (above and below ground), and exported, downstream to estuaries, the ocean and fringing floodplains or, for some elements, to the atmosphere (as in the nitrogen cycle). Water flow is a main vector of nutrients, in dissolved or particulate form, although sometimes nutrients can move against the flow, carried in the tissues of organisms migrating upstream or dispersing from the water into the terrestrial catchment.

Nitrogen and phosphorus are needed in relatively large quantities and can be referred to as *macronutrients*.

In river ecosystems, if there is nutrient limitation, it is normally N or P, or a combination of both, that is responsible. Other elements are also essential, but usually in somewhat smaller quantities. A scarcity of silicon can limit diatom populations, while a shortage of calcium reduces the presence of molluscs and crustaceans. Other essential elements, such as potassium, sulphur and magnesium, are not normally limiting. Elements required in only trace amounts, including iron, manganese, copper and others, can be referred to as *micronutrients*.

In this chapter we stress the basic biology and ecology of nutrients, essentially of nitrogen and phosphorus, in river systems. This includes a little about the fundamental background chemistry of nutrient cycling and its special features in running waters, outlining some of the (mainly microbial) processes involved. These are responsible for the retention of inorganic nutrients and their subsequent fate including transport downstream or loss to the atmosphere. We then consider how nutrients are taken up by the biota and transferred through food webs and how they affect food quality for animals. We touch on the widespread problem of an excess of nutrients in rivers as a major environmental perturbation as well as the crucial role played by the biota in transporting nutrients across habitat boundaries. Overall, streams and rivers are not simply pipes that receive nutrients from the land and transport them unchanged to the sea, but active ecosystems whose productivity is coupled to the transformations of nutrients.

9.2 Cycles and spirals

9.2.1 Nutrient cycles—essential features

Microbes are the 'heavy lifters' of the ecological world. As we have seen, they are responsible for the bulk

The Biology and Ecology of Streams and Rivers. Alan Hildrew and Paul Giller, Oxford University Press. © Alan Hildrew and Paul Giller (2023).
DOI: 10.1093/oso/9780198516101.003.0009

of primary production and almost all of the decomposition (i.e. actual mineralisation of the organic matter). Ecologists interested in larger organisms (particularly animals) have, until quite recently, been concerned mainly with identifying and counting their subjects and developing the ideas and concepts of population and community ecology. This was more difficult or impossible for earlier microbial ecologists who could, however, characterise and assess their subjects by measuring their activities, even though they probably knew less of their numbers, taxonomic identity and diversity. These microbial activities involve a vast variety of chemical transformation—so it is the contribution of microbes to ecosystem and biogeochemical processes that has been of most concern. If we are to understand rivers as ecosystems, therefore, we do need to pay attention to some of the more significant chemical processes and transformations that are microbially driven. New tools and methods are of course rapidly making it feasible for animal ecologists to measure processes, and for microbial ecologists to assess community diversity and abundance, with evident progress in whole-system understanding.

Nutrient cycles are essentially similar in all ecosystems, though the prevalence of particular processes can vary in space and time. In this short section we deal with the basics, concentrating on the nitrogen and phosphorus cycles before turning to the features of these cycles in flowing water. When CO_2 is converted to organic matter during photosynthesis it is said to have been 'reduced' (in this case, it loses its oxygen). Considerable energy is required. This comes from the light fixed by photosynthesisers, which they use to drive the reduction of CO_2, and which is the ultimate source of energy for nearly all ecosystems. Thus, the (reduced) organic matter produced is a store of this energy. In an oxic (oxygen-rich) environment, this energy can be released, just as it is when fire returns organic carbon to CO_2 once more (i.e. it is oxidised), with the release of (mainly) heat energy (only in biology it is done more gradually!).

Molecules can be placed on a gradient of relative oxidation and reduction, known as a *redox gradient*. All reactions along the redox gradient involve pairs of molecules, an oxidiser and a reducer (a redox pair), which can be thought of as exchanging electrons. The oxidising agent (in aerobic respiration, this is oxygen) 'accepts' electrons from the reducing agent. In this way, relatively reduced molecules can be oxidised to release energy (think of this as the 'heat from the fire') when

they find themselves in a relatively oxic environment. As free oxygen is used up, other molecules can in turn act as oxidising agents (i.e. accept electrons), though progressively less energy is available to heterotrophic organisms as we move down the redox gradient. In the final stages almost all the energy fixed by the photosynthesisers has been used up.

This is the brief story of energy flow and of mineral nutrients linked with the carbon cycle. Carbon is not usually thought of as a nutrient, but it is required by green plants in inorganic form (CO_2 or bicarbonate in solution, the supply possibly limiting photosynthesis in some circumstances) and by heterotrophs in a reduced (energy-rich), organic form. Nitrogen and phosphorus or other nutrients can similarly be taken up in various forms and incorporated into the tissues, an *assimilatory* process. The most common form of nitrogen is N_2 gas, which makes up almost 80% of the atmosphere. However, most organisms require nitrogen in a combined form, while only a few are able to access molecular nitrogen directly. These latter can 'fix' nitrogen, a process requiring a great deal of energy. In water, the forms of dissolved inorganic nitrogen (DIN) are nitrate (NO_3^-) or ammonium (NH_4^+), while nitrite (NO_2) is normally rare except in water low in oxygen. Organic nitrogen (in which nitrogen is combined with carbon) is present in dissolved or particulate form, some of it released from aquatic organisms by excretion as urea, as free DNA or as dissolved organic matter or, after death, by decomposition as amino acids and proteins.

Animals require particulate organic nitrogen in their diet, usually as protein which must be digested before assimilation across the gut wall. Protists can take up organic particles directly across their cell membranes. Some plants can also use organic nitrogen in dissolved form, but most primary producers require dissolved inorganic nitrogen (as ammonium or nitrate), which can also be taken up from the environment by heterotrophic microbes including fungi, bacteria and archaea. Ammonium is the form of inorganic nitrogen that can be used in biochemical pathways, so nitrate must first be reduced to ammonium enzymatically (requiring energy). Nitrogen fixation itself (atmospheric N converted to combined N) only occurs in anoxic environments, or where the enzymes concerned can by physically 'protected' from oxygen. Thus, many cyanobacteria fix nitrogen within specialised organelles called *heterocysts*, and nitrogen-fixing bacteria are sheltered as endosymbionts of other organisms, such as flood-tolerant alder trees. Nitrogen is also fixed by lightning in the atmosphere, so even clean rainwater contains a little nitrate

The nitrogen cycle is intimately related to the redox gradient in the environment. In lakes and other standing waters, there can be very clear vertical, physical gradients of redox conditions, from well-oxygenated surface waters to profoundly reducing, anoxic sediments in the bed (Dodds & Whiles 2019; Brönmark & Hansson 2017). Redox reactions are then clearly separable in space, sometimes over the millimetre scale in undisturbed sediments. Flowing and often more turbulent environments with (usually) coarser sediments are rarely so well ordered, perhaps excepting the more pond-like conditions out on the floodplain. The flowing water columns of all but the most polluted rivers and streams are rarely profoundly anoxic, with the possible exception of 'microsites' of anoxia within quite large suspended organic particles. To find most anoxic 'hot-spots' in rivers and streams we must look within accumulations of organic matter, such as leaf litter, in patches of soft sediment in pools and in the bed of large rivers and estuaries, or in parts of the hyporheic zone clogged by fine sediment. This makes for a rather complex spatial juxtaposition of patches of differing redox conditions and therefore of redox reactions.

Ammonium in well-oxygenated water has potential energy that can be released by its oxidation to nitrite and nitrate. Bacteria carry out this *nitrification*, using the energy to fix their own reduced carbon compounds. The nitrogen itself is not assimilated, however, so this is a *dissimilarity* process, distinguishing it from *assimilatory* processes in which the nitrogen is taken up. Nitrification does not occur in anoxic environments, because more potential energy is then produced by reducing nitrate rather than oxidising ammonium to nitrate. In anoxic environments, nitrate is reduced by *denitrifying* bacteria, eventually to nitrogen gas. Here, the nitrate is used as an electron acceptor to oxidise organic carbon to CO_2 and reduce the nitrate, nitrite or NO_x^- to nitrous oxide (N_2O) and, finally, nitrogen gas. The process can be stopped at nitrous oxide if the supply of reduced carbon is limiting. Denitrification does not occur in oxygenated environments, because oxygen can be used as the electron acceptor, a reaction yielding more energy than does the use of nitrate. Denitrification is important because it removes nitrate from river water, which contributes to reducing the eutrophication of estuaries and shallow seas, and nitrate from drinking water if the river water is used for domestic supply. We return to the eutrophication of rivers in rather more detail in section 9.4.

There is an alternative pathway from nitrate to nitrogen gas (i.e. other than denitrification; e.g. Trimmer et al. 2012). There is, first, a process of reduction of nitrate to ammonium—rather than 'all the way' to nitrogen—called *dissimilatory nitrate reduction to ammonium* (DNRA). It competes for nitrate with denitrification, and again nitrate is the electron acceptor in the oxidation of organic matter. In anoxic environments, ammonium can then be further reduced to nitrogen, in a process called '*anammox*' (carried out by 'anammox' bacteria) using nitrite as terminal electron acceptor. There remains uncertainty about the relative roles of nitrate reduction, DNRA and anammox in the removal of nitrate from rivers and the emission of nitrogen gas and of nitrous oxide to the atmosphere, so the nitrogen cycle is still an area of active research. This matters a great deal, not least because nitrous oxide is an extremely powerful greenhouse gas. All these processes put together give a conceptual view of the nitrogen cycle overall (Figure 9.1; Dodds & Whiles 2019).

Phosphorus occurs mainly as a single inorganic molecule, phosphate, and in unpolluted rivers is usually at very low concentrations—often below detection limits (1–10 µg L^{-1}). It is normally measured as 'soluble reactive phosphorus'. Its availability in natural waters often depends on its interactions with iron and calcium. Phosphate forms an insoluble precipitate with ferric iron (Fe^{3+}) in oxygenated water. In well-oxygenated rivers and streams this may limit the availability of phosphorus for primary producers and is important in iron-rich streams. Periods of anoxia in or near the bed can cause the complex to dissociate, releasing ferrous iron (Fe^{2+}) and phosphate into the water column. Phosphate also binds with calcium in high-alkalinity fresh waters (hard water, well buffered at pH 7 or above, with abundant carbonate). This can remove phosphorus from the water column in highly productive waters, though plant cells can often store phosphate when it is more abundant. Phosphate is regenerated by the mineralisation of dead organic matter by microbes, while plant cells can also take up organic phosphorus, with a phosphatase enzyme that releases inorganic phosphate from dissolved organic matter. Animals access organic phosphorus from their diet—in lipids and nucleic acids for instance. All heterotrophs excrete excess phosphorus in some form, as they also mineralise and excrete dietary nitrogen, so the release of nutrients from living things can complete nutrient cycles and sustain productivity.

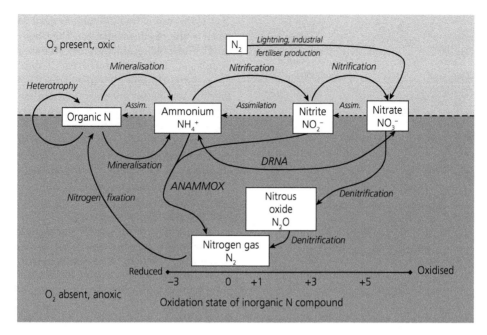

Figure 9.1 A schematic diagram of the nitrogen cycle showing the inorganic forms of nitrogen, with more oxidised forms to the right (arranged on the redox scale, E_h in volts). The top of the diagram in blue shows an oxidised environment—normally the water column—and the bottom of the diagram shows an anoxic environment—as may be found in some river sediments. The redox gradient shows the tendency for electrons to move from an electron donor (the reducing agent) to an electron acceptor (the oxidising agent) and is high in an oxidised environment. Thus, in the oxidised (blue) environment energy is generated by reactions going from left to right, while in the anoxic (brown) environment energy is generated by reactions going from right to left.
Source: from Dodds & Whiles (2019) with permission of Elsevier.

9.2.2 Nutrients 'spiral' in flowing water

What are the special features of 'nutrient cycles' in running waters? There are at least two. The first is that the catchment, including the river system that drains it, is an 'ecosystem unit' for which it is at least theoretically possible to construct nutrient (more strictly, elemental) budgets—what goes in and what goes out should be accounted for if all the vectors of the nutrients (hydrological, atmospheric, geological and biological) can be measured. There can be a net loss or even a net gain over a period (the latter if nutrients are retained), but the terms in the budget ± changes in storage in the catchment should balance. The second is that nutrients in flowing water, as well as cycling 'on the spot', are also carried downstream. This 'spreads out' the cycle into a *spiral* downstream—hence the term *nutrient spiralling* (Figure 9.2).

A wide variety of 'solutes' (dissolved inorganic chemicals) are present in stream and river water (see Baker & Webster 2017 and Chapter 2, section 2.5, p. 44). These are both anions (negatively charged ions) such as

sulphate, chloride, silicate etc, and positively charged cations (calcium, magnesium, sodium, potassium etc). It is helpful to divide them into *conservative* and *reactive* solutes. Conservative solutes essentially do not react chemically or biologically so they pass downstream unchanged, although they may be diluted with rainfall, from tributaries and from groundwater. Reactive solutes are changed by chemical or biological transformations and their concentration is affected both by flow (as for conservative solutes) and by their uptake and release. Nitrate and phosphate are reactive solutes. Chloride does react but behaves more or less as a conservative solute since its concentration in stream water is always far above biological needs.

Biological or chemical transformations of a reactive solute in the stream can be measured by comparing its transport with that of a conservative marker through the same reach, where the latter's concentration along the channel can be used to characterise flow and hydraulic properties alone. Truly conservative markers are lithium and bromide, although chloride is cheaper to use and easier to measure (Baker & Webster 2017

(a)

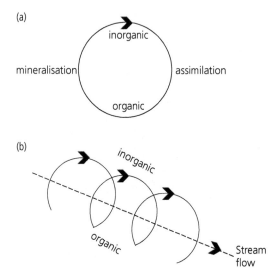

(b)

Figure 9.2 (a) A schematic of a simple nutrient cycle between inorganic (mainly in the water column) and organic forms (mainly in the sediments). (b) Adding longitudinal transport to the basic cycle creates a nutrient spiral.
Source: redrawn after Baker & Webster 2017, based on Newbold 1992.

give details of such methods). Conservative markers of flow can be added to a stream by prolonged injections of a solution or a single bulk (or 'gulp') application. The shape of the concentration peak of the marker as it passes one or more downstream sensors can then be informative. In the example in Figure 9.3 the pulse of the marker 'flattens out' as it passes through a stream reach from an upstream to a downstream sensor.

Following downstream transport of a conservative marker in the flow can help us understand what happens to nutrients. Physical transport is affected by two processes, *advection* and *dispersion*. The first is transport in the flow, while dispersion is the process of mixing, which in streams is mainly due to turbulence. Dispersion subjects solute ions to different rates of downstream transport—fastest in the main flow. The water column in a stream channel can then be visualised as consisting simply of a 'storage component'—with non- or slowly flowing water—and an actively flowing compartment. With this simplifying assumption, various models have been applied to the transport of conservative markers. The approach of Bencala & Walters (1983) can be used to estimate *transient storage*, the fraction of the flow that is slowly flowing, consisting of hyporheic flow paths, pools, backwaters and behind or in macrophyte beds or accumulated large debris jams (Baker & Webster 2017). Somewhat similarly, the *aggregated dead zone* model of Beer & Young (1983) and Wallis

et al. (1989) estimates the total flow dead zones in any given length of stream channel. The latter model was used by Reynolds et al. (1991) to explain the persistence of an algal plankton in rivers (see Chapter 6) and by Lancaster & Hildrew (1993a) as an estimate of (non-flowing) refugia for invertebrates from spates and to help to explain return times to the stream bed of drifting organisms (see Chapters 4 and 5). The modelling approach of Bencala & Waters (1983) has been widely used in North America in studies of solute transport of streams (as in this chapter). Both models can be used to compare the hydraulics of flow in different channels, to relate stream-bed complexity to transport, or to act as an index of habitat heterogeneity.

Estimating the transport of dissolved nutrients (*reactive* solutes) in stream channels requires, in addition to the above physical model of nonreactive solutes, an assessment of uptake and loss (largely associated with the stream bed). Biological and abiotic uptake is referred to as *immobilisation*. In streams this is dominated by photosynthetic and heterotrophic microbes. Physical adsorption is particularly important for phosphate. Measuring the uptake of a nutrient allows the calculation of an instantaneous dynamic uptake rate (see Baker & Webster 2017). Nutrients immobilised on the stream bed are eventually mineralised and returned to the water column (and in the case of dissolved gases thence to the atmosphere). Estimation of mineralisation requires measurement of the standing crop (mass of the immobilised nutrient per unit area of the benthos) and stream depth and the calculation of an instantaneous rate of mineralisation.

As nutrients cycle between the benthos (largely in organic matter) and water column (mainly mineral) they tend to move downstream to a varying extent, termed the *spiralling length*, consisting of two components: the *uptake length* (S_w) is the distance travelled in the inorganic form in the water column before immobilisation, and the *turnover length* (S_B) is the distance travelled in the organic form before mineralisation and release to the water column. Not surprisingly, uptake length dominates the spiral length, and varies with stream discharge and water velocity. Further calculations (see Baker & Webster 2017, and references therein) allow estimation of a theoretical *uptake velocity* as a nutrient atom moves towards the spot where it is immobilised, and *areal uptake*. The latter is the mass of solute immobilised by an area of stream bed per unit time.

The uptake length indicates nutrient retention because its calculation entails assessment of both physical transport and biogeochemical processes

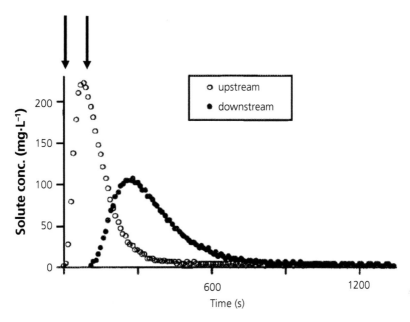

Figure 9.3 A 'pulse' addition of chloride (a marker) to a small stream passes two sensors successively downstream. The left-hand arrow marks the initial detection of the pulse (set at time zero) at the upstream sensor and the right-hand arrow shows the time (seconds elapsed) before it was first detected at the downstream sensor. The shape of the pulse changes as it goes.
Source: after Lancaster & Hildrew 1993a.

immobilising nutrients. Biologically active stream beds (e.g. with biofilm) require nutrients, which are taken up rapidly, tending to reduce uptake length. In particular, limiting nutrients (i.e. limiting primary production or heterotrophic activities) are expected to show shorter uptake lengths. Complex stream beds, with a relatively large benthic surface area, and streams beds with a high biomass of biofilm, are also likely to have shorter uptake lengths. Uptake can become saturated as nutrient concentration increases; that is, the fraction of nutrients retained declines at some rate with its concentration and uptake length then increases (Baker & Webster 2017). This is evident from comparisons of nutrient uptake lengths measured by supplementing the concentration above background (in so-called nutrient addition experiments) with those made by using a nutrient tracer. The latter involves the addition of just a small amount of the nutrient that does not perceptibly increase the overall concentration (see e.g. Mulholland et al. 2002; Figure 9.4) but in which the nutrient is 'labelled'. Labelled nutrients include radioactive forms of phosphorus or the stable isotope of nitrogen ^{15}N. These can be detected in stream water independently of overall changes in nutrient concentration – as applies in nutrient-additon experiments. In Mulholland et al.'s (2002) experiments,

the uptake length estimated from nutrient addition exceeded that measured with a nutrient tracer, and the disparity increased with the ratio of nutrient added over background concentration (i.e. the more that was added, the longer the nutrient remained in solution— because the larger amounts added were more likely to saturate uptake). Further, the two estimates of uptake length were more similar where ambient nutrient concentration strongly limited its uptake.

The fundamental concepts and methods briefly described above are a result largely of collaborative work by a group of mainly US scientists, beginning over 40 years ago (references in Baker & Webster 2017 and Webster et al. 2022; and see the Stream Solute Workshop 1990 and LINX collaborators 2014). The Lotic Intersite Nitrogen Experiments (LINX 1 & 2), led by (the late) Patrick Mulholland, and Jack Webster, Judy Meyer and Bruce Peterson, were pioneering and the first to characterise stream nutrient uptake and retention at the reach scale, and did so across biomes and a variety of land uses and with consistent methods involving a tracer (LINX collaborators 2014). This research has been critical to the realisation that streams and rivers are not merely 'pipes' that transport water and materials to the sea, but process and transform a great deal of organic matter and nutrients on that

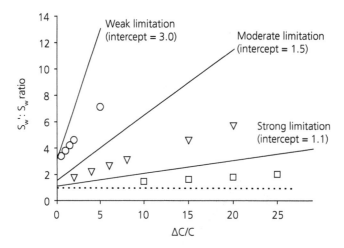

Figure 9.4 The expected ratio of nutrient uptake lengths measured by nutrient-addition experiments (S_W') and by tracer-only additions (S_W)—in relation to the relative change in nutrient concentration ($\Delta C/C$) caused in nutrient-addition experiments. For a weakly limiting nutrient, the disparity (S_W'/S_W) is large even with only small increases in the nutrient (i.e. at low values of $\Delta C/C$), whereas for a strongly limiting nutrient the disparity is less. The horizontal dotted line indicates a ratio of 1, that is, no difference in the two estimates of uptake length whatever the amount of nutrient added. Circles, triangles and squares indicate the results of field experiments under weak, moderate and strong nutrient limitation, respectively. *Source:* after Mulholland et al. 2002; with permission of the University of Chicago Press.

journey. This lends much greater significance to river systems in a global context, and processes such as denitrification can be seen as very important ecosystem services (see Chapter 10 and below).

9.3 Nitrogen and phosphorus in streams and rivers

9.3.1 Nitrogen retention and transformations

The first of the two large-scale collaborative LINX projects (LINX 1) included headwater streams draining catchments with mainly native vegetation. As such, the streams were fairly oligotrophic and the experiments were aimed at demonstrating the ability of such small headwaters to retain, assimilate and transform nitrogen. Labelled ammonium (^{15}N-NH_4^+) was dripped for six weeks into 0.2 km segments of 12 widely distributed headwater streams in a variety of North American biomes from Puerto Rico to Alaska (Peterson et al. 2001). Ammonium was rapidly removed from stream water, partly via assimilation by photosynthetic and heterotrophic organisms. The former were single-celled and filamentous algae and bryophytes, the latter bacteria and fungi. Uptake was also due to abiotic sorption to sediments, and via (dissimilatory) nitrification of ammonium to nitrate. On average, 70–80% of ammonium uptake was due to assimilation and sorption (but mostly assimilation) and 20–30% via

nitrification, though the relative amounts were highly variable among streams. Nitrification actually ranged from 3–60% of NH_4. That is, nitrate was released to the water column in addition to terrestrial inputs. However, nitrate was also retained by the opposing processes of biotic assimilation and denitrification, though it was less efficiently assimilated than ammonium. Thus, even though nitrate was on average around 10 times more concentrated than ammonium in stream water, the areal uptake of nitrate ($\mu g\ N\ m^{-2}\ s^{-1}$) to the stream bed was similar to ammonium uptake. Therefore, a molecule of nitrate on average travelled around ten times as far as a molecule of ammonium.

When addition of the ^{15}N ceased, it was possible to measure the rate at which the tracer was re-released as inorganic molecules (^{15}N-NH_4 and ^{15}N-NO_3) from the stream bed. Between zero and 63% of the tracer originally immobilised was released within one day in the different streams, showing that remineralisation of organic N can also contribute to inorganic N in stream water (Peterson et al. 2001). The experiments were carried out during the most productive part of the year, so it can be expected that retention would dominate at that time, whereas continued release at other times, in inorganic, gaseous or organic form, would prevent the long-term accumulation of N on the stream bed. Peterson et al. (2001) report that, typically, input and removal processes are balanced over the medium term such that profiles of the concentration of inorganic N along rivers are often relatively constant, which could be taken to suggest that rivers are indeed essentially unreactive 'pipes' of nutrients. This ^{15}N tracer technique, however, reveals that there is a dynamic equilibrium, including nitrification (of ammonium to nitrate), biological assimilation, sorption, denitrification and regeneration of inorganic N to the water column.

Plant nutrients are now normally plentiful in fresh waters draining modern 'working' landscapes, dominated as they are by agricultural, industrial, suburban and urban land uses. These nutrients come mainly from direct inputs of (hopefully) treated sewage, animal wastes and fertiliser runoff, while mean concentrations of nitrate in stream/river water range globally from < 0.1 µg L^{-1} to > 20,000 µg L^{-1}. Given widespread eutrophication of streams and rivers, much recent research on nutrients in rivers has moved into systems more affected by humans. This was recognised in the LINX programme, and the LINX 2 experiment covered an impressive 72 streams located over eight regions of the USA and Puerto Rico, including a variety of biomes and spanning a wide range of nitrate concentration (0.1–21,200 µg L^{-1}, median 100 µg L^{-1}). Land use was classified into three categories—reference, urban/suburban and agricultural—with nitrate concentration being least in 'reference' streams but higher in the other two categories (Figure 9.5a).

Areal uptake (mass per unit area per unit time) of nitrogen was greater in urban and agricultural streams, as uptake responded to the gross supply of nitrate (Figure 9.5b). However, uptake velocity of nitrate (cm s^{-1}) *declined* exponentially with increasing concentration (Figure 9.6a). Recall that uptake velocity

is the theoretical rate at which a molecule of nitrate moves downwards to be retained on the bed and is a measure of the efficiency of removal relative to supply. Thus, while the overall areal removal of nitrate increased with its concentration, the efficiency of this removal declined—indicating saturation of uptake pathways. This uptake includes assimilation of nitrate by autotrophs and heterotrophs, plus dissimilatory denitrification. The tracer method allowed separate assessment of denitrification, which also declined exponentially with increased nitrate, indicating a reduced efficiency of removal by this pathway (Figure 9.6b). Factors affecting overall nitrate uptake velocity included the gross rate of primary production, indicating the importance of photoautotrophs. Denitrification also responded to ecosystem respiration rate. Thus, high ecosystem respiration lowers oxygen concentration, at least in microsites in the stream bed, increasing demand for nitrate as a terminal electron receptor; in addition, ecosystem respiration reflects the availability of labile organic matter as a 'fuel' for denitrification.

Further modelling was able to account for almost 80% of variation in the nitrate uptake length in the LINX 2 experiment (Hall et al. 2009). Uptake length increased with specific discharge and increasing nitrate concentration. *Specific discharge* is a measure of discharge per unit stream width and goes up with increasing stream depth and mean velocity—both of which decrease contact between the water column and stream bed—while efficiency of retention goes down with increasing nitrate concentration. Gross primary production was influential in the model, uptake length

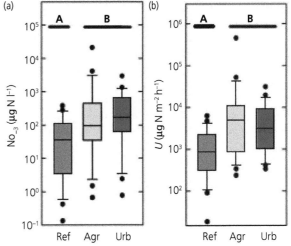

Figure 9.5 Box plots showing (a) nitrate concentration across the 72 North American streams, grouped into three land-use groups (reference—Ref, agricultural—Agr, urban—Urb), and (b) the total biological nitrate uptake per unit area of stream bed. Outlier data points outside the 10th and 90th percentiles are shown as single points. Capital letters and bold horizontal lines denote any significant differences between land-use categories—a common letter indicating no difference.
Source: after Mulholland et al. 2008, with permission of Springer Nature.

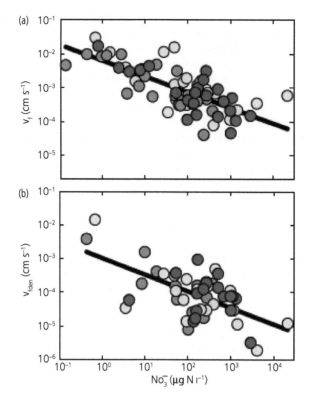

Figure 9.6 (a) Nitrate uptake velocity (the theoretical velocity with which a nutrient atom moves towards the spot where it is immobilised) declines with increasing nitrate concentration in 72 North American streams of differing land use. (b) Uptake velocity due to denitrification alone also declines with increasing nitrate concentration. Green—reference streams, yellow—agricultural streams, red—urban streams.
Source: after Mulholland et al. 2008, with permission of Springer Nature.

declining with increasing production. Land use *per se* actually had little effect on uptake length in the model, other than through its effect on GPP and nitrate concentration, both of which were greater in urban and agricultural streams and affected uptake length in opposite ways (low with high GPP, high at high nitrate concentration) thus cancelling each other out.

Considering nitrate dynamics at a whole-stream-network scale, Mulholland et al. (2008) suggested that there may be a progressive effect of the intensification of anthropogenic additions of nitrogen to the landscape and river systems. The only permanent removal process of fixed N from the ecosystem is via denitrification to molecular N, or to nitrous oxide (N_2O). This process is efficient in relatively unperturbed headwaters streams, as represented in LINX 1, when N loading is low. This can be overwhelmed by growing inputs from catchments, however, leading to increased export, first to larger streams, then to big rivers, lakes and estuaries, and to their eutrophication and/or *hypernutrification*

(situations where nutrients build up and do not limit biological production—such as turbid estuaries where light is limiting). At very high loading rates, the stream network exports virtually all catchment-derived nitrogen. Mulholland et al. (2008) therefore argued for the conservation and restoration of small streams as a focus of management of eutrophication of river systems. However, the concentration of nutrient inputs from municipal sewage-treatment works into the lower reaches of rivers circumvents the stream network and its capacity for denitrification.

More recent research on the uptake of inorganic fixed nitrogen in streams has added further details and answered some remaining uncertainties. Tank et al. (2018) widened the geographic scale of experiments, analysing the results of 17 ^{15}N-NH^4 tracer additions to streams across Arctic to tropical systems, the majority in the USA and Puerto Rico but including two in Central America, one in Spain, one in Iceland, one in Denmark and two in New Zealand.

Their aim was to partition assimilatory uptake, the process that usually accounts for most N removal from the water column over short periods, and to identify which organisms are responsible. Autotrophic uptake, by epilithic photosynthesisers, filamentous algae, bryophytes and macrophytes, was higher on average than by heterotrophic microbes (living on dead organic matter), although uptake was similar when heterotrophic assimilatory uptake was scaled to live microbial biomass rather than to bulk detritus. Assimilatory N uptake was also much greater in streams with an open (unshaded) canopy. Nitrogen uptake was closely related to gross primary production, linking metabolism and nutrient cycling, a topic to which we will return.

Dissimilatory transformation (i.e. not involving the assimilation of the nitrogen itself) accounts for some part of the uptake of ammonium and nitrate. In oxic environments this is mainly via nitrification of ammonium to nitrate. Day & Hall (2017) showed that 7–19% of ammonium added to three subalpine streams in the Rocky Mountains was denitrified immediately (within the range of values found by Hall et al. 2009), the efficiency of uptake declining with increased ammonium concentration in all three streams. The best-fitting model of this decline was not consistent among streams, however, some apparently having a greater capacity for denitrification than others.

Most of the research discussed so far has essentially been carried out in small streams—and constrained by the cost of adding labelled nitrogen to systems with high discharge (Trimmer et al. 2012). New methods are necessary, and Ritz et al. (2017a) used an 'open channel'

method for measuring the end product of denitrification, that is, nitrogen, directly in the water column of a moderately large European river, the Elbe. This is the third-longest river in central Europe (> 1,000 km), rising in the Czech highlands (as the Labe) and entering the North Sea at Hamburg as an eighth-order channel. It drains over half the Czech Republic, small parts of Austria and Poland and more than 25% of Germany (Pusch et al. 2021). In three long reaches of the Elbe, Ritz et al. (2017a) found overall denitrification rates of 18 and 13 mg N m^{-2} h^{-1} in summer 2011 and spring 2012, respectively (Figure 9.7). In context, these rates are very high compared to previous global mean estimate of 1.4 and 9.6 mg N m^{-2} h^{-1} quoted by Ritz et al. (2017a), although they point out there is extraordinary variability in space and time (high rates occurring in 'hot spots' and 'hot moments'), depending on the supply of nitrate and labile organic matter (particularly to the hyporheic zone, where much denitrification occurs). Estimates for the whole 582-km-long stretch of the Elbe downstream from the Czech/German border to the tidal weir about 30 km south-east of Hamburg suggest denitrification removes about 10% (10,000 t N y^{-1}) of the annual inputs of fixed nitrogen to this section of the river. Large, higher-order reaches can thus contribute substantially to the permanent removal of fixed nitrogen from river systems. A subsequent mass balance estimated for this stretch suggested about half of the total retention of nitrate was due to net algal assimilation, with the remainder attributable to seston deposition and denitrification (Ritz & Fischer 2019). Overall, the river retained almost 30% of TN (total nitrogen) inputs, much of it transformed to nitrogen

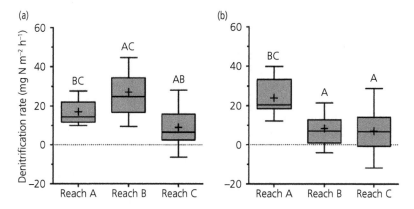

Figure 9.7 Denitrification rates in three reaches (A, B and C, up- to downstream) of the free-flowing German River Elbe in (a) summer 2011 and (b) spring 2012. Boxes are 5 and 95 percentiles, crosses are mean values.
Source: after Ritz et al. 2017a, with permission of Springer Nature.

gas. The Elbe is not impounded over this German stretch, with semi-natural flow dynamics and a permeable hyporheic zone, characteristics that maximise denitrification, are amenable to management, and can relieve eutrophication and nitrate pollution.

Recall that there is an alternative route to the permanent removal of fixed nitrogen, mainly known from estuarine and marine sediments, in which ammonium can be reduced to dinitrogen in a process called 'anammox', using nitrite as the terminal electron acceptor. Its role in fresh waters remains uncertain. However, Lansdown et al. (2016) found that anammox contributed up to 58% of N_2 production in permeable riverbeds in a southern English catchment, whereas denitrification dominated in less-permeable, clay-bed rivers. This is a little surprising since anammox is an anaerobic process (as is denitrification), although it may occur in tight association with dissimilatory reduction of nitrate to ammonium. The functional genes of the bacteria that carry out anammox were found to be present in all these rivers, though in variable amounts. This topic requires further research.

9.3.2 Light and the transfer of nitrogen through stream food webs

Use of the tracer ^{15}N has mainly highlighted microbial processes and the assimilation of nitrogen by primary producers or heterotrophic microbes—that together have been labelled 'primary uptake compartments' (PUCs). There is then the 'onward transfer' of assimilated N upward through stream food webs. Trophic transfer through food webs is normally assessed in terms of carbon and energy (see Chapter 8), although energy transfer itself can be limited by the relative availability of essential nutrients. The efficiency of transfer of essential nutrients can also be instructive. Norman et al. (2017) analysed data from 13 tracer experiments that used ^{15}N-NH_4 over different biomes (partly a subset of those analysed by Tank et al. 2018), mainly from North America but including one stream each from New Zealand, Denmark, Iceland, the Caribbean and Central America. As found for energy transfer, mean transfer efficiency of nitrogen across studies was lower from PUCs to primary consumers (e.g. grazers and detritivores; mean 11.5% range <1%–43%) than from primary consumers to their predators (mean 80%, range 5% – >100%). Apparent transfer efficiencies greater than 100% evidently require explanation but could be due to subsidies to predators from outside the system—such as drift-feeding fish taking terrestrial prey (e.g. Perkins et al. 2018).

A strong signal in these data was the relationship between canopy cover and total nitrogen flux through the food web. The rate of nitrogen uptake (Figure 9.8 a, b) was greater in open than in shaded streams, and light appeared to facilitate nitrogen transfer. This is probably because of greater assimilation by primary producers and the higher food quality provided by primary producers, rather than by detritus, to primary consumers. Transfer efficiency from primary consumers to their predators was in turn greater in open-canopy streams, suggesting a ramifying effect up through the web of higher basal food quality of algae relative to detritus (Figure 9.8 c, d).

These differences between shaded and open streams were also apparent in the overlapping data set of Tank et al. (2018); Figure 9.9. They showed that: (a) the ratio of gross primary production to ecosystem respiration (GPP:ER), a measure of overall heterotrophy, was lower in shaded streams, and (b) the biomass of nitrogen stored in 'green' and heterotroph primary uptake compartments also differed (much greater in heterotrophs in shaded streams, more similar between the two in open streams). Furthermore, (c) measures of assimilatory ammonium uptake rate and (d) percentage of the labelled N that was added that was stored in green and heterotrophic PUCs was progressively biased towards the autotrophic ('green') PUC. Thus, autotrophs played a disproportionate role in assimilatory N uptake and storage relative to their biomass. This evidence seems to bear upon the debate about the support of stream food webs, in favour of the view that autotrophic carbon is disproportionately important (Chapter 8).

9.3.3 Habitat, hydrology and nitrogen in streams

As indicated earlier, nitrogen transformations are highly patchy in space and time, with 'hot spots' and 'hot moments' for nitrogen transformations, such as denitrification (Ritz et al. 2017a). Much denitrification occurs in hyporheic or interstitial sediments, probably because that is where microbial communities are concentrated on the surfaces of particles (total surface area increases with a decline in particle size). Also, denitrification requires a supply of nitrate and labile organic matter, so at least some subsurface through-flow is required. The amount of water entering the hyporheic zone is determined by hydraulic conductivity—essentially a measure of the ease with which a fluid passes through a porous solid

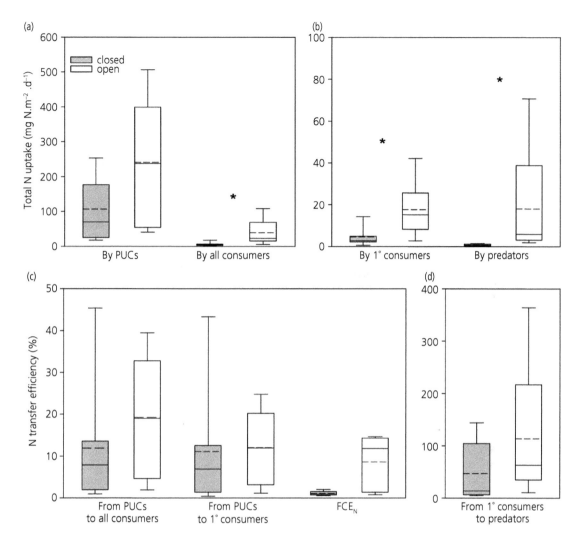

Figure 9.8 (a) and (b) Total nitrogen uptake measured in 13 ^{15}N-ammonium tracer addition experiments from various streams according to canopy type (open or shaded) for various elements of the stream biota: (c) and (d) N-transfer efficiency. PUCs are 'primary uptake compartments'—heterotrophic microbes and primary consumers; transfer efficiency between trophic levels (e.g. PUCs to primary consumers) is the total N uptake of the focal consumer level divided the N uptake of its food, as a percentage; FCE_N is the food-chain efficiency for nitrogen (calculated as total predator uptake of N divided by the PUC uptake, as a percentage). Asterisks indicate significant differences between open and close channels.

Source: from Norman et al. 2017, with permission of John Wiley & Sons.

medium. So, effective denitrification requires sufficient throughflow to provide nitrate and carbon, but not so great that there are no anaerobic spots and insufficient contact time between water and surface (see e.g. Mendoza-Lera et al. 2019). Thus, as Comer-Warner et al. (2020) showed in an English agricultural stream, nitrogen reduction was much greater in sandy sediments (water has a higher residence time) than in coarser gravels, while in the latter denitrification was

incomplete and nitrous oxide was abundant. Similarly, Pretty et al. (2006) showed that denitrification was restricted to organically rich, hypoxic sandy sediments (< 90 μM oxygen) in a chalk stream.

Nitrogen cycling is also variable in time and is closely related to flow fluctuations. Bernal et al. (2013) discuss the case of streams in Mediterranean-type climatic regimes (i.e. not restricted to the Mediterranean basin of southern Europe), which are often

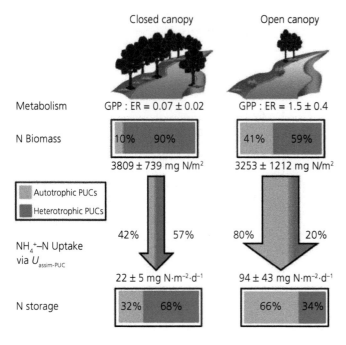

Closed canopy   Open canopy

Metabolism   GPP : ER = 0.07 ± 0.02   GPP : ER = 1.5 ± 0.4

N Biomass   10% | 90%   41% | 59%

3809 ± 739 mg N/m²   3253 ± 1212 mg N/m²

Autotrophic PUCs
Heterotrophic PUCs

NH_4^+–N Uptake via $U_{assim\text{-}PUC}$ 42% | 57%   80% | 20%

22 ± 5 mg N·m⁻²·d⁻¹   94 ± 43 mg N·m⁻²·d⁻¹

N storage   32% | 68%   66% | 34%

Figure 9.9 A summary of differences in the trophic transfer of nitrogen revealed in 17 tracer-addition experiments on streams (open versus closed canopies) worldwide. Autotrophs (green segments) played a disproportionate role in assimilatory N uptake and storage relative to their biomass. Thus, stream metabolism is overall strongly heterotrophic in closed (shaded) channels, and most nitrogen in biomass is stored in heterotrophic primary uptake compartments (PUCs—red compartments). However, uptake of ammonium tracer was more balanced between autotrophic and heterotrophic PUCs and storage of the tracer (as a percentage of N added) was greater for autotrophic PUCs. Open channels showed a consistent bias towards autotrophic PUC. Values are means ± SE. *Source:* from Tank et al. 2018, with permission of John Wiley & Sons.

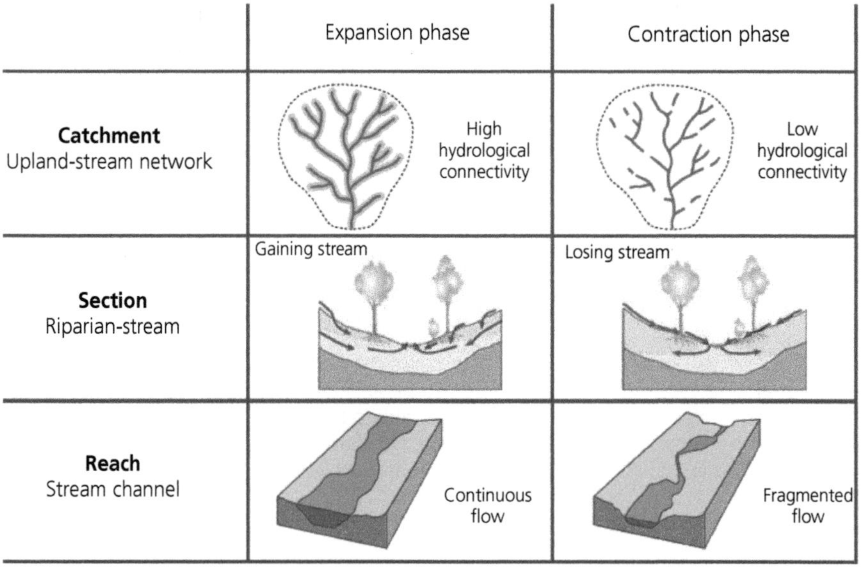

	Expansion phase	Contraction phase
Catchment Upland-stream network	High hydrological connectivity	Low hydrological connectivity
Section Riparian-stream	Gaining stream	Losing stream
Reach Stream channel	Continuous flow	Fragmented flow

Figure 9.10 Schematic of typical hydrological features of Mediterranean-climate streams during the seasonal hydraulic expansion and contracting phases and at different scales.
Source: from Bernal et al. 2013, with permission of Springer Nature.

intermittent and the stream network periodically fragmented (Figure 9.10). Mediterranean climates are characterised by strongly seasonal but erratic rainfall, usually with long, hot, dry summers. Precipitation tends to come in intense storms, which can occur at any time, but are concentrated in winter.

During dry periods, fragments of the stream network are disconnected from catchment sources of DIN, so differences in nutrient concentration develop. When reconnected during wet periods, nutrients are mobilised and inefficiently retained, leading to substantial export. Thus, annual DIN export is thought to

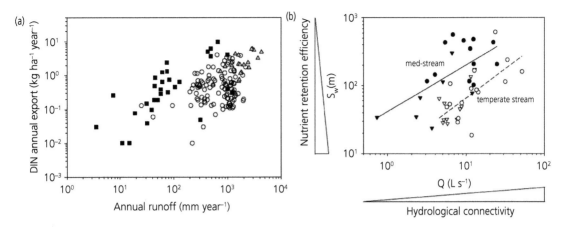

Figure 9.11 (a) Mediterranean-climate streams (black squares) export relatively more dissolved inorganic nitrogen per unit discharge than temperate (white circles) or tropical (grey triangles) streams. (b) Double logarithmic relationship between stream discharge (Q, related to hydrological connectivity of the network) and uptake length for ammonium (S_w, a measure of nutrient retention efficiency) for a Mediterranean and a temperate stream during the expansion (black and white circles) and contraction (black and white triangles) phases.
Source: from Bernal et al. 2013, with permission of Springer Nature.

be relatively high per unit annual runoff in Mediterranean streams, while the ammonium uptake length may be greater (i.e. nutrient retention is less efficient) than in temperate streams (reviewed by Bernal et al. 2013: Figure 9.11).

There are marked 'first flush' events, on the resumption of flow in intermittent streams, in which nitrate is exported intensely over a short period. Merbt et al. (2016) showed this for the Furiosos stream in north-east Spain, estimating that 50% of the total nitrate exported from the catchment during a six-day event was contributed by reflooded dry sediments (Figure 9.12). Around 5% of total annual export of nitrate came through these brief 'first flush' events. Dry sediments were shown to be hot spots for the production of nitrate by nitrification of ammonium; this nitrate is then mobilised in the first flush. Moreover, they showed that ammonium-oxidising Archaea were more abundant in dry sediments than bacterial ammonium oxidisers (Merbt et al. 2016). As intermittency in stream flow is expected to increase and extend to hitherto temperate areas, due to climate change and increased water withdrawals, such considerations about the reduction in nitrogen retention (which can be regarded as an ecosystem service) and processing in small streams may grow in significance.

Finally, patchiness in space and time is not always imposed by the environment, but biological processing can lead to a degree of 'self-organisation'. Thus, Dong et al. (2017) revisited spatial patterns of nutrients along 10 km of the well-studied Sycamore Creek

in Arizona. The catchment had historically been heavily grazed by cattle, which were excluded in 2000, the catchment then undergoing a succession from an open gravel/algal dominated system to a stream channel heavily grown over in patches by riparian vegetation. They carried out nutrient surveys at different times during this successional process, during which nitrogen became increasingly limiting. Early in succession, the longitudinal pattern of nitrate concentration was heavily dominated by the 'geomorphic template'— that is, strong upwelling zones introducing nitrate into the channel. Later in the succesion, there was still an effect of this 'external factor', but biological effects were just as strong in determining the pattern. This was mainly exerted via changes to channel morphology and upwelling/downwelling caused by macrophytes and the sediment they trapped, and by increased denitrification.

9.3.4 Nutrients, macrophytes and metazoans

So far, our discussion of biological nutrient (nitrogen) transformations in streams and rivers has overwhelmingly stressed that role of microbial communities. Most of the pioneering studies referred to above were done in relatively pristine systems, often forested, and where rooted higher plants are absent or scarce (LINX 1 and 2). We are increasingly aware of a contribution from larger plants and animals, however. Lowland streams with a high biomass of macrophytes are common in many agricultural areas (see Chapter 3,

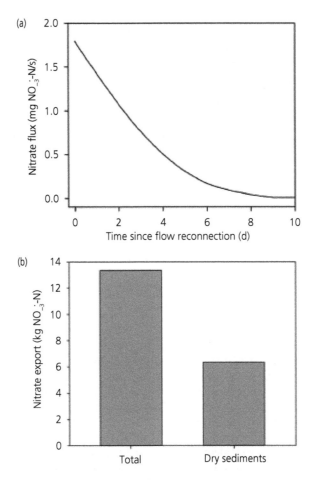

Figure 9.12 (a) Nitrate flux in a Mediterranean catchment (the Fuirosos stream) during the first days after flow resumption—a 'first flush'. (b) Total nitrate export (left) during a first flush event and the estimated contribution from nitrate stored in dry sediments (right).
Source: from Merbt et al. 2016, with permission of John Wiley & Sons.

Figure 3.4, p. 69), usually coinciding with high concentrations of fixed N from fertilisers, and progress is being made in our understanding of the role of such macrophytes in nutrient retention (see Topic Box 9.1 by Tenna Riis).

In some karstic areas of the world, strong springs fed by upwelling ground water feed in turn productive rivers with extensive and natural macrophyte beds. For instance, northern Florida (USA) has the highest density of large (discharge > 2.8 m^3 s^{-1}) springs in the world, now threatened by groundwater abstraction and pollution of aquifers with nitrate. McBride & Cohen (2020) assessed controls on production of submerged aquatic vegetation in two springs, one polluted by excess nitrate, the other with naturally low concentrations (in other respects they were similar). The emphasis here was not on how macrophytes affected retention and export

of fixed N, but on what environmental factors controlled the growth of the macrophytes. Growth of macrophytes was actually indistinguishable between the two systems, suggesting that water-column nitrate was not decisive. Further, while macrophytes made up most of the biomass of autotrophs, they contributed only about 25% of primary production, the rest being accounted for by epiphytic algae—short lived and small but highly productive compared to the relatively long-lived macrophytes. Multivariate techniques suggested that a combination of high light supply and oxidised sediment were associated with high macrophyte production while, somewhat counter-intuitively, high porewater phosphorus was associated with lower plant production. Free phosphorus in the pore-waters is likely to be associated with reduced conditions around the roots, which do not favour plant growth.

Topic Box 9.1 Stream macrophytes and nutrient retention

Tenna Riis

Submerged macrophytes often reach > 50% coverage of the channel in many low-gradient, open-canopy streams and rivers. They act as ecosystem engineers in streams, altering their physical structure and enhancing habitat complexity. Recent studies have also tested how macrophytes influence ecosystem processes, such as nutrient retention and stream metabolism. Given their high biomass, rapid growth and impact on the habitat, it is not surprising that macrophytes strongly influence nutrient dynamics (i.e. nutrient uptake, turnover and removal) (Box Figure 9.1). On the habitat scale, three main mechanisms have been shown to be responsible for enhanced nutrient retention in flowing waters dominated by macrophytes.

First, stream macrophytes, and the associated epiphytic biofilm growing on plant surfaces, directly assimilate inorganic nutrients from the water column. Studies show that inorganic nitrogen (N) retention, via this *assimilatory uptake*, range on average from c.150–600 mg N m^{-2} d^{-1} and is positively related to the biomass of the plants themselves and of the epiphytic biofilm (Levi et al. 2015). In a related field study, Riis et al. (2012) found that N uptake in macrophyte habitats is on average 481 mg N m^{-2} d^{-1}, compared to non-macrophyte habitats estimated at 27 mg N m^{-2} d^{-1}, while in a mesocosm study Olesen et al. (2018) found that uptake rates of phosphate (P) and N was three and five times higher, respectively, in habitats with plants compared to without plants. Not only the overall biomass of macrophytes but also leaf morphology is important for nutrient assimilation rates, and N uptake was on average 10-fold faster for species with high leaf complexity (i.e. perimeter:area ratio > 10) than for species with simple leaf morphology (i.e. perimeter:area ratio < 3; Levi et al. 2015). Overall, nutrient uptake kinetics are highly variable among species and seasons. It differs between groups of submerged and emergent life forms as well as among species within life forms, within and between seasons (Manolaki et al. 2020). These results imply that the presence of both submerged and emergent plants, and of more species within both life forms, would extend the period of nutrient uptake across the year, enhance nutrient uptake within any season and ultimately enhance annual nutrient uptake.

The *second* important mechanism at the habitat scale is that macrophytes provide *longer in-stream nutrient storage* than other autotrophs (e.g. epiphytic biofilm), with tissue nutrient turnover times often exceeding 30 days (Riis et al. 2012). Moreover, a high macrophyte biomass also increases the storage of nutrient-rich organic sediments within plant beds, due to hydraulic effects (Sand-Jensen 1998).

Third, anoxic conditions develop within organic sediment trapped beneath aquatic plants and enhance the removal of NO_3-N via denitrification (Petersen and Jensen 1997). Recently, we found mean denitrification rates across a range of stream habitats were 264 ± 867 μmol N m^{-2} h^{-1}, with a greater rate in vegetated habitats compared to bare sediments (Audet et al. 2020).

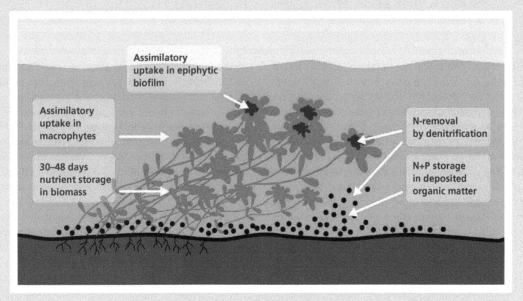

Box Figure 9.1 Role of macrophyte habitats in nutrient retention in streams. Darker green indicates epiphytic biofilm. N is nitrogen and P is phosphorus.
Source: modified from Riis et al. 2020.

Topic Box 9.1 *Continued*

The effects of macrophyte habitats on nutrient retention scales up to the whole stream reach, expressed as a habitat-weighted, reach-scale, nutrient-uptake rate. Studies show that macrophytes and their epiphytic biofilms can account for 71–98% of the habitat-weighted reach-scale uptake across streams (Levi et al. 2015). Through the seasons, macrophyte-driven nutrient uptake changes in a predictable manner, and is linked to macrophyte biomass via both biological activity (i.e. assimilatory and dissimilatory processes reflected in ecosystem metabolism) and physical processes (i.e. surface transient storage) (Riis et al. 2019). Autotrophic demand for ammonium and phosphate is highest in spring, when biomass accumulates, while nitrate demand is highest in summer indicating higher denitrification rates. In autumn, the demand for phosphate is high, which is probably linked to the enhanced heterotrophic activity associated with macrophyte senescence (Riis et al. 2019). Thus, stream nutrient uptake is not static, but highly seasonal, which has significant implication for modelling nutrient export from river networks.

Removal of macrophytes in streams and rivers (i.e. weed cutting) is a widely used management practice, aimed at enhancing drainage capacity and runoff from the catchment to prevent flooding of agricultural land, local housing and other human-made structures—though it can exacerbate flooding further downstream. It is well established that weed cutting has substantial short- and long-term effect on the physical habitats and on stream organisms including the composition of the macrophyte assemblage. In the short term, macrophyte biomass is evidently reduced by weed cutting, which also reduces the autotrophic metabolism and nutrient-uptake rates. In the longer term, regular weed cutting will change the macrophyte species composition, producing less diverse assemblages dominated by fast-growing species with basal meristem growth, rhizomes and high dispersal capacities (e.g. *Sparganium emersum*) while reducing the cover of slower-growing species with apical meristem and less dispersal capacities (Baattrup-Pedersen et al. 2002; 2016). Following Manolaki et al. (2020), a low diversity of macrophyte species and life forms should reduce reach-scale nutrient-uptake rate, although our knowledge of species complementarity in macrophyte communities is poor (Olesen et al 2018; Riis et al. 2019). Overall, results suggest that the current widespread removal of macrophyte biomass by direct cutting will reduce the positive effects of macrophytes on nutrient uptake.

In conclusion, macrophytes significantly enhance nutrient retention in streams via both assimilatory and dissimilatory nutrient processes and by physical processes within macrophyte habitats, scaling to integrated nutrient-uptake rates at the reach scale and across seasons. Placing the influence of macrophytes in the context of stream and river restoration highlights the potential of macrophytes for mitigating nutrient pollution. By restoring macrophyte habitats and supporting a diverse macrophytic vegetation in streams and rivers, we can enhance reach-scale nutrient retention and thus decrease nutrient transport to downstream lakes and coastal areas.

Professor Tenna Riis is in the Department of Biology at the University of Aarhus, Aarhus, Denmark.

The role of animals in the uptake of nitrate has not often been studied in an ecosystem context, although of course they do require dietary (organic) nitrogen. However, certain species traits of macroinvertebrates seemed to influence the uptake of nitrate from the water column of nine third-order European streams (Yao et al. 2017). Animals that were grazer/scrapers, inhabiting quite coarse substrata (from large boulders to pebbles) and/or living interstitially and feeding on detritus, were associated with high rates of nitrate removal. Yao et al. suggested that top-down effects on epilithic biofilms (that are responsible for much assimilatory uptake of fixed nitrogen), and/or bioturbation effects on sediment, were the underlying mechanisms responsible. Clearly some animals, similarly to the macrophytes mentioned above, are 'engineers' of the physical habitat in streams (see also Moore 2006), through activities ranging from nest-building by fish, stabilisation of sediments by species that spin silk, to bioturbation via the foraging of crayfish and larger predatory insects. All these may affect near-bed hydraulics, the flow path through stream-bed interstices and the hyporheic zone, and thus the exposure of water to reactive biofilms. We return to such ecosystem engineering in the context of 'ecosystem services' in the final chapter.

Finally, animals are sources of nutrients in streams, as in the substantial inputs of marine nutrients transported upstream by migrating salmon—which we deal with in detail later (section 9.6). All animals can contribute, however. Just as nutrients are eventually released at some stages of decomposition of organic matter by microbes, animals excrete mineralised nutrients and may also release nutrients in faeces. This is part of nutrient cycling in almost all ecosystems. Its importance in an ecosystem context varies and excretion rates may relate to organism size and the requirement to maintain a body content of different nutrients at some limited range of ratios—of carbon to nitrogen and/or to phosphorus, for example. This is the study of ecological 'stoichiometry', which we discuss in a bit more detail in section 9.5. Ecologists have available a database of nutrient excretion rates by 491 species of aquatic animals, in most aquatic phyla, and this includes information on body size, temperature, taxonomic affiliation and body composition (N:P ratio) (Vanni et al. 2017).

Nutrient excretion rates (E) scale 'allometrically' with body mass (M) (see e.g. Hall et al. 2007; Figure 9.13):

$$E = aM^b.$$

In words, as one would expect, total excretion increases with body mass (big animals excrete more than small). However, excretion goes up less than proportionately with body mass. Thus, the exponent (b) (slope) is less than one, such that large animals excrete *less nutrient per unit mass* than do small ones. Excretion rate thus behaves fundamentally as do many other biological features of organisms, like metabolic rate for instance, which also declines per unit body mass (as animals grow). Nutrient excretion rates also vary taxonomically, so that related taxa tend to be more similar

than unrelated ones. For instance, vertebrates such as fish with bony skeletons excrete less phosphorus per unit mass than do invertebrates. Diet affects excretion rates, and animals with a food source rich in nitrogen will excrete relatively more N than they do other, less abundant, nutrients.

In a lotic ecosystem context, the significance of nutrients released by animals will relate to the size of this input compared to, say, catchment sources. We can also consider nutrients released by animals in relation to the demand of primary producers (reviewed by Vanni 2002). Benthic insects and snails for example supplied 15–70% of the demand for N by algae in Sycamore Creek (Grimm 1988). In Walker Branch, a well-studied oligotrophic forest stream in Tennessee (USA), Hill & Griffiths (2017) estimated the balance between nitrogen assimilated and incorporated into growth in an abundant grazing snail (*Elimia clavaeformis*). Growth and assimilation add to the retention of N, whilst subsequent excretion and decomposition remobilise it. In this system, mobilisation dominated, the snails excreting a surprising 12 times as much N as they accumulated in their own biomass over the course of one year. Their assimilation efficiency, as N in growth/excretion, was only 8%. They accumulated most N in spring and autumn, when the trees were not in leaf, and lost it at times of summer food shortage (snail biomass was then five times greater than epilithic biomass). Overall, snails assimilated and recycled up to 50% of the dissolved inorganic nitrogen taken up by autotrophs and heterotrophic microbes. We know rather less about

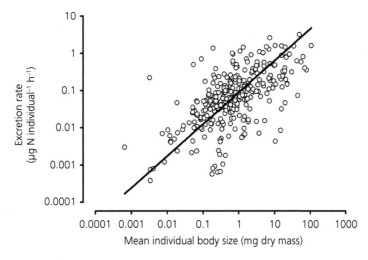

Figure 9.13 A regression of ammonium-N excretion rates per individual against body size for a variety of stream benthic invertebrates. Rates increase less than proportionally with body size (b < 1)—that is, larger invertebrates excrete less per unit mass than do small individuals. *Source:* from Hall et al. 2007, with permission of the author.

the wider significance of mineralisation of nutrients by animals in streams and rivers than we do for lakes, for instance. We might expect mineralisation by stream animals to be relatively less important in lotic systems, however, particularly in the many catchments with large human or livestock populations (which themselves contribute large quantities of excreted nutrients which find their way directly or indirectly into rivers and streams).

9.3.5 The particulars of phosphorus

We already noted that inorganic phosphorus in unpolluted inland waters is usually found at extremely low concentrations, and that organisms are extraordinarily efficient at taking it up. Total phosphorus is made up of soluble reactive phosphorus (usually assumed approximately equivalent to inorganic phosphorus, depending on the particular assay used), plus dissolved organic phosphorus and particulate phosphorus. It is tightly retained in clean streams and sometimes limits primary production (see Figure 9.14 for a representation of the phosphorus cycle in fresh waters). The concentration of soluble reactive phosphorus that limits algal growth in rivers is generally $< 15\ \mu g\ L^{-1}$ and frequently $< 5\ \mu g\ L^{-1}$ (Newbold 1992; see also Chapter 2, section 2.5.3). Algal production may not respond to experimental additions above those values, while mean global values of SRP

in unpolluted rivers are about $10\ \mu g\ L^{-1}$. Soluble reactive phosphorus is rapidly taken up by autotrophs and heterotrophic microbes, particularly those associated with the epilithic biofilm and particulate organic matter (including the seston). In classic experiments using radioactive $^{32}PO_4$ as a label in a secure research site (Walker Branch, Tennessee, USA), Newbold et al. (1983) measured a phosphorus spiral length of 190 m, with CPOM accounting for 60% of the uptake, FPOM for 35% and epilithic biofilm for 5%. Biotic uptake was dominant at Walker Branch. More recently, Ward et al. (2018) greatly increased the range of conditions to cover a large (seventh-order) oligotrophic river in Idaho (USA). This has a wide channel (mostly > 100 m) and is fast flowing, while discharge is up to three orders of magnitude greater than in other similar experiments. Adding nitrogen and phosphorus fertiliser, they found that P (and NH_4) was taken up rapidly, with an uptake length for total dissolved phosphorus of 5.7 km (which is short for such a large river), while uptake velocity was eight times greater than measured in smaller streams. Nutrient uptake was strongly associated with chlorophyll accrual and epilithic growth rate and was thus biotic. Abiotic immobilisation of phosphorus is often important, however, and is closely related to several cations, including iron, calcium and aluminium, in systems ranging from hard water (high calcium) to acidic and or iron-rich water.

Interactions between the iron cycle and phosphorus are among the best known biogeochemical reactions in

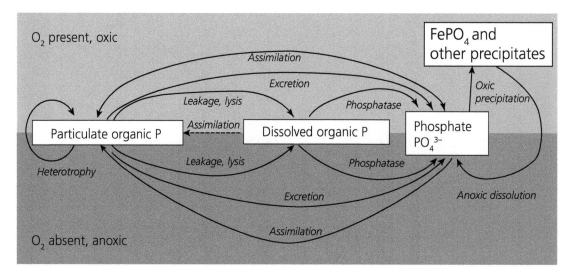

Figure 9.14 The phosphorus cycle in fresh water. The brown section represents the sediment (parts of which may be anoxic), while the blue is the water column (normally oxygenated).
Source: after Dodds & Whiles 2019 with permission of Elsevier.

lakes, but can also occur in running waters. Under oxic conditions, insoluble ferric iron (Fe^{3+}) is the main form and co-precipitates with phosphate (see Figure 9.14). Under anoxic conditions, the ferric iron is reduced to the soluble ferrous (Fe^{2+}) form and is released into solution along with phosphate. In rivers with large point-source inputs of phosphorus or high water velocity, such internal loading of P is unlikely to be very important. In lowland rivers, however, with sufficient iron (issuing from the ground water for instance) and hyporheic sediments rich in organic matter, internal release of phosphorus can be significant. For instance, Smolders et al. (2017) showed that the redox cycling of iron was largely responsible for a seasonal cycle of reactive phosphate concentration in the eutrophic lowland rivers of Flanders (Belgium). 'Legacy phosphorus' (phosphorus locked into sediments and remaining from former high inputs) was periodically released into the water column in rivers where there was sufficient iron in the sediments and at times (in summer) when flow and oxygen concentration were low (Figure 9.15).

In the wet tropics, Small et al. (2016) dripped phosphorus into a first-order stream in Costa Rica for eight years, finding that more than 99% of the excess phosphorus retained in the 200 m reach downstream was

stored in the sediment abiotically, this time bound to aluminium or iron, and accounted for 25% of the total phosphorus added. Less than 0.03% of retention was accounted for by biotic uptake. Sediment phosphorus had declined to baseline levels about 4 years after the end of the manipulation, so had been released eventually.

These streams in Cost Rica tend to be acidic with clay-rich sediments, conditions which favour abiotic retention of phosphorus with iron and aluminium. Calcareous streams are also often chronically low in soluble inorganic phosphorus. This is because deposition of calcium carbonate co-precipitates phosphorus. In paired streams in southern Arizona (south-west USA) Corman et al. (2016) shaded experimental stretches, only one of the two streams having active deposition of 'travertine' (mainly calcium carbonate; 'tufa' and 'calcite' are rather similar terms). Shading of the travertine stream reduced deposition of calcium carbonate by over 50%. This suggests that deposition in unshaded reaches is supported by photosynthetically induced reductions in the concentration of CO_2, tending to raise pH and increase deposition. Reduction in the deposition of travertine by experimental shading then increased phosphorus concentration in the water

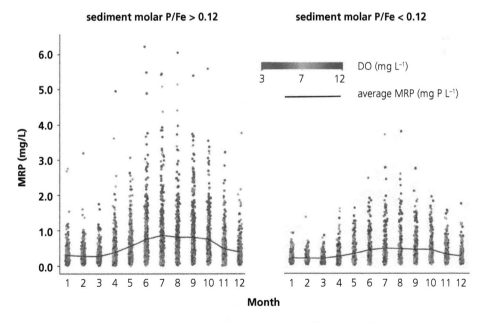

Figure 9.15 Dissolved phosphorus concentration (as molybdate reactive phosphorus, MRP) in eutrophic Flemish rivers over a seasonal cycle (mean values for a record covering 2003–2015) in relation to the P/Fe ratio in the sediment (higher in left panel, lower to the right) and water column dissolved oxygen (shown by colour of data points). As oxygen declines in the summer (redder data points) 'legacy' phosphorus is released, particularly where sediment P/Fe exceeds 0.12 (left-hand panel).
Source: from Smolders et al. 2017, with permission of the American Chemical Society.

column and the phosphorus content of the epilithic biomass. Again, therefore, abiotic uptake of phosphorus, this time with calcium carbonate, was dominant in the phosphorus cycle in this system and can restrict its biotic availability.

9.3.6 The 'coupling' of carbon and nitrogen

So far, we have considered the fate of organic carbon (Chapter 8) and nutrient spiralling largely separately, though the two are evidently closely linked—or coupled (e.g. Burgin & Hamilton 2007: and see Figure 9.16). Considering the spiralling of organic carbon and nitrate, 'respiratory denitrification' in Figure 9.16 refers to the use of nitrate as an electron acceptor to oxidise organic carbon. Biomass assimilation is the direct uptake of nitrate, along with organic carbon, by autotrophs and by heterotrophic microorganisms. 'Non-respiratory N removal' refers to denitrification by dissimilatory nitrate reduction to ammonium (DNRA). These are pathways in which the removal of nitrate and transformation of organic carbon are linked directly or indirectly.

Several research groups have addressed N–C coupling in various running waters. For instance, Heffernan & Cohen (2010) used high-frequency (hourly) *in situ* measurements of nitrate and dissolved oxygen, in a macrophyte-rich, spring-fed river in Florida (USA),

to estimate assimilatory demand for N and to assess the short-term dependence of heterotrophic assimilation of N and denitrification on gross primary production. They found that denitrification accounted for most nitrate removal (around 80% in both spring and autumn); these high rates were accounted for by the availability of labile organic carbon in the form of exudates from the macrophytes, suggesting that around 35% of denitrification was fuelled by the previous day's photosynthesis. This may be a rather unusual system, with very high autochthonous production, compared with small streams where allochthonous inputs are more likely to fuel heterotrophic respiration, but it is relevant to the many larger, open channels with abundant macrophytes.

Further, Wymore et al. (2016) analysed the LINX 2 data to identify drivers of nitrate uptake velocity. These data derive from 72 streams across eight regions of North America, with agricultural and urban streams in each region plus those with catchments of native vegetation—see Section 9.3.1 above. The ratio of DOC to nitrate, in particular, and photosynthetically active radiation, were the two most important predictor variables overall, suggesting a role for denitrification (using nitrate as a terminal electron acceptor to oxidise DOC), as well as autotrophic assimilation, in the uptake of nitrate. Nitrogen cycling was thus closely coupled to carbon transformation.

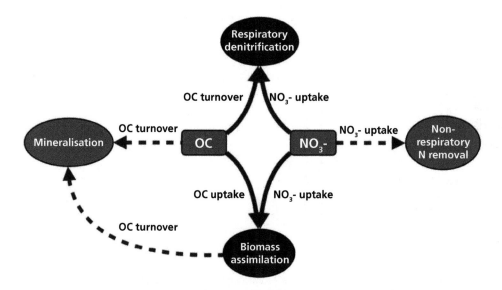

Figure 9.16 A schematic of the coupled organic carbon (OC) and nitrate (NO3⁻) cycles in streams. Boxes are compartments (solutes/materials), solid arrows are directly coupled (N and C) fluxes, broken arrows are indirectly coupled or uncoupled fluxes, ovals are fates.
Source: from Plont et al 2020, with permission from the University of Chicago Press.

Abril et al. (2019) carried out a simultaneous whole-reach addition of ^{15}N-ammonium and ^{13}C-acetate in a forested (and shaded) headwater stream in north-east Spain to examine the role of 'primary uptake compartments'—epilithic biofilms, heterotrophic microbes on leaves and small wood—on the uptake of N and C, and on their storage and transfer to consumers. Both N and C were efficiently retained, with heterotrophic microbes on allochthonous leaf litter making the biggest contribution, while transfer to consumers was rapid. Acetate is a highly labile source of DOC for uptake by microbes, which at the same time require nutrients such as ammonium, and not just carbon. The labelled N and C were also rapidly taken up into the tissues of consumers, suggesting their strong co-reliance on the labile DOC and nitrogen taken up by stream biofilms. Thus, unsurprisingly, combinations of nutrients and carbon are necessary for the growth and production of both primary producers and consumers. Most recently, Plont et al. (2020) again used the LINX 2 data to investigate the spiralling length and uptake velocities of organic carbon and nitrate. The spiralling length of organic carbon was shorter in the reference streams than in the agricultural and urban streams; that is, reference streams were more active in the uptake and mineralisation of organic carbon per unit channel length. Further, the rate of organic carbon mineralisation and nitrate uptake were positively correlated across all streams—although there was a good deal of variation between individual sites and land-use settings—and the transformations of organic carbon and nitrate were again coupled.

9.4 'Famine and feast'—nutrient limitation and the eutrophication of stream ecosystems

The increase in the supply of inorganic nutrients to rivers and streams is among the most pervasive of all current anthropogenic changes to fluvial ecosystems. We touched upon the effects of nutrients with a focus on energetics and river metabolism (including primary production, ecosystem respiration and detrital breakdown) in Chapter 8. In this section we revisit nutrient limitation (i.e. the idea that nutrients place limits on community structure or ecosystem processes) and then move on to the wider ecological consequences of adding nutrients greatly in excess of the requirements of an 'unperturbed' system, as is now so prevalent in running waters. In relation to river management,

this leads us to the tricky concepts of what a 'reference' system is—i.e. what might be an ecologically defensible and achievable target for nutrients in real rivers (we return to river restoration more generally in Chapter 10). These questions are ones of policy, and involve science mixed with politics and economics. Here, we begin by looking at field experiments at two scales.

9.4.1 Experimental additions of nutrients

We have seen evidence from the widespread use of nutrient-addition experiments at the reach scale earlier in this chapter. A deceptively simple and smaller-scale experimental approach to nutrient limitation of stream biofilms is to use 'nutrient-diffusing substrates', a technique introduced to freshwater ecology by Fairchild et al. (1985) but with earlier origins in marine systems. Essentially, containers of agar with various combinations of nutrients dissolved in it, or none in controls, are placed in the water and the nutrients gradually leach out through a porous surface of the container, on which a biofilm develops over three weeks or so. The biomass, species composition and activity of the biofilm community can then be assessed (see Tank et al. 2017 for experimental details). According to Beck et al. (2017), however, after 30 years of such research a rather complex picture arises. The response of the biofilm to nutrient supplementation can differ depending on (a) environmental factors (e.g. background in-stream nutrient concentration or temperature), (b) geography (e.g. variations in climate, ecoregion, land use) and (c) experimental details. They carried out a rigorous meta-analysis of published experimental results, confirming a significant 'effect size'—a measure of the response of the biofilm to nutrient addition—of NP treatments (i.e. substrates with both nitrogen and phosphorus) in running waters. The effect of N alone was higher than that of P alone, while the effect size of N and P together was greater than the sum of individual supplementation of either N or P—indicating an additive effect of the two nutrients. The explanatory power of the models (i.e. the percentage of experimental variation that can be accounted for) was always low, however, and many other factors apparently intervened, as mentioned. In fact, broader spatial factors like ecoregion, climate classification and stream order explained most variation, although even then the percentage was low ($< 20\%$).

Beck et al. (2017) argue for greater standardisation of experimental protocols, and the reporting of more environmental variables such as flow, light and grazing intensity. They also recommended deploying nutrient-diffusing substrates across defined environmental gradients. A good example of this approach is that of Myrstener et al. (2018), who placed nutrient-diffusing substrates with added N, P, NP or none in 12 streams in northern Sweden (around 200 km above the Arctic circle), four streams each with catchments of heath-tundra, birch forest or coniferous forest. Substrates offering either inorganic (a porous ceramic) or organic (cellulose sponge) tops were used as surfaces for colonisation. There was evidence of strong overall nitrogen limitation of gross primary production, community respiration and chlorophyll-*a* accumulation. The response to nitrogen addition in tundra streams alone was constrained by temperature, whereas light limited the response in birch and coniferous forests. In terms of the heterotrophic response (i.e. community respiration), the response to nitrogen addition was closely related to the background ratio of dissolved inorganic nitrogen to dissolved organic carbon (DOC:DIN), the response being strongest where background DIN was relatively low (Figure 9.17). This is consistent not only with the familiar idea that carbon availability can constrain uptake of N, but also suggests that the use of terrestrial dissolved organic matter by stream biofilms is facilitated by inorganic nitrogen.

Whole-stream addition of nutrients has also been used experimentally, most notably in the oligotrophic, forested headwaters at Coweeta (North Carolina, USA; see Topic Box 1.1). In one experiment, nutrients (N + P) were dripped into a single stream ('C54') for five years

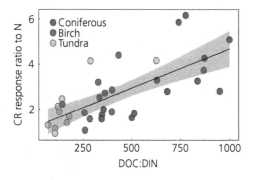

Figure 9.17 The strength of the response of biofilm community respiration to added N (in nutrient-diffusing substrates) in relation to the background (i.e. in stream water) ratio of dissolved organic carbon and dissolved inorganic nitrogen (DOC:DIN) in the Swedish Arctic. Thus, respiration goes up more when there is relatively more carbon than nitrogen in the water. Colours distinguish streams draining catchments of predominantly tundra, birch or conifer forest.
Source: after Myrstener et al. 2018, with permission of John Wiley & Sons.

from 2000–2005, with a similar, but unmanipulated, control (C53). Both streams were also observed over two years before the nutrient addition began (Davis et al. 2010a). After two years of fertilisation, there was an increase in the density, biomass and production of primary consumers (herbivorous/detritivorous invertebrates), an increase that was closely tracked by the production of their predators (Cross et al. 2006). The quality of detritus as food was increased by the added nutrients, even though litter biomass also declined. Over two subsequent years (years 4 and 5) of enrichment, however, while production of primary consumers was increasingly stimulated, this was decoupled from predator production—which unexpectedly declined (Davis et al. 2010a) (Figure 9.18). This was apparently due to an increase in body size of the consumer taxa, including an increase in the population of large, cased caddisfly *Pycnopsyche*, which were less vulnerable to predators than the formerly dominant small-bodied prey. An increase in nutrient availability does not necessarily increase 'food-web efficiency' (i.e. the transfer of carbon and nutrients through the web).

In further nutrient-addition experiments at Coweeta, Kominoski et al. (2018) enriched five first-order forested headwater streams for two years from 2011. They added both nitrogen and phosphorus to produce different molar (N:P) ratios ranging from 2–128, and concentrations from 96–472 μg N L^{-1} and 10–85 μg P L^{-1}. Here it was shown that ecosystem respiration (ER) was stimulated by nutrient additions, in particular of N, while the quantity of in-stream leaf litter declined with increasing N (due to more rapid decay). Phosphorus had little effect on ER and overall enrichment had little effect on GPP (presumably as the autotrophs were light limited). Bumpers et al. (2017) studied predatory salamanders in the same streams, with detailed dietary analyses before, during and after the two-year nutrient additions. Phosphorus supply resulted in a greater number, mean size and biomass of prey consumed by the larger of two salamander species (*Desmognathus quadramaculatus*). There was also a switch to more algivorous prey and away from detritivores, and thus towards 'green' (autotrophic) energy pathways to the predator rather than 'brown' detrital pathways. In the two nutrient-addition experiments at Coweeta, therefore, enrichment caused an uncoupling of predators and prey in one study (Davis et al. 2010a) but an increase in predator production in the other (Bumpers et al. 2017). It seems that ecological details such as species composition can be responsible for the variable dynamics observed, and there can be 'ecological surprises'. In an overall assessment

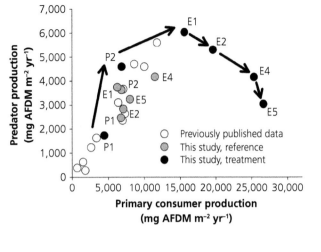

Figure 9.18 Relationship between the production of primary consumers and predators in the benthos at a stream in Coweeta in response to five years of nutrient enrichment (black circles). Data from an unmanipulated reference stream are shown (grey circles), as are 'previous' data (white circles; mainly from former years at the unmanipulated reference stream). P = pre-treatment data (two years, P1 and P2), E = experimental years (E 1–5).
Source: from Davis et al. 2010a, with permission from *Proceedings of the National Academy of Sciences*.

of these nutrient-addition experiment at Coweeta, encompassing 27 'stream years', Rosemond et al. (2015) found that the mean residence time of terrestrial organic matter in streams was reduced by about 50% and that terrestrial carbon was rapidly depleted where nutrients—particularly N—are supplemented, potentially offsetting carbon gains of fixed carbon via increased algal production.

9.4.2 Eutrophication of running waters

Most experimental additions of nutrients—whether by nutrient-diffusing substrates or by whole-stream additions—have been carried out in fairly small and oligotrophic systems. In much of the world, however, running waters (particularly in the lowlands) have long been subjected to very large additions of plant nutrients, arising from direct, point-source inputs of treated organic wastes such as sewage, and indirect, diffuse additions of fertilisers by runoff from agriculture. What are the ecological effects of these gross perturbations? Eutrophication of lakes has been extremely well studied, and there is a robust relationship between nutrients and the biomass of the phytoplankton of the open water in the growing season (see Dodds & Whiles 2019 and references therein). A more appropriate measure for the impact on streams is the growth of algae on the benthos, whose relationship with the concentration of nutrients in the water column is rather less 'tight'—that is, there is more scatter in the points of the relationship from individual streams (Figure 9.19), while both N and P are apparently influential. Of course, as we have seen, factors other than nutrients, including light, grazing and flow disturbances, frequently limit algal biomass

and productivity in streams and rivers, which presumably explains the greater variation for streams than for planktonic algae in lakes. Further, nutrients can stimulate both primary production and, particularly in running waters, the production of heterotrophic organisms and the processing of allochthonous carbon.

The control of nutrient loading has been a major objective in managing eutrophication in streams and rivers—where the excessive growth of algae, cyanobacteria and macrophytes can cause 'sags' in the concentration of dissolved oxygen, resulting in fish deaths and the unsuitability of river water for public supply and irrigation. Most data are available for North America and Europe. In some cases, reference conditions have been calculated for different ecoregions, taking into account expected variations in the natural background values (Figure 9.20a). In other cases, long-term records of nutrient concentrations are available, one example showing a particularly clear, almost linear, increase in nitrate concentration in the River Frome, an agricultural river in south-western England, from 1965 to 2003 (Figure 9.20b). In a recent survey of British rivers and streams, Jarvie et al. (2018) found, perhaps not surprisingly, that 'nutrient impairment' was greater in larger rivers than in headwaters—impairment being measured as exceeding concentrations needed to reach *good ecological status* under the European Union's Water Framework Directive,. Good ecological status is defined as a slight variation from undisturbed conditions, which would vary with the type of river system. Nutrient targets for good ecological status ranging from 0.5–3.5 mg L^{-1} TN and 11–105 μg L^{-1} TP have been established for various rivers (Nikolaidis et al. 2022). Nutrient impairment was related to the agricultural and urban influence

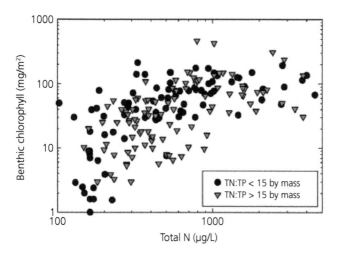

Figure 9.19 The relationship between benthic chlorophyll and total nitrogen concentration (log–log plots) from around 200 temperate streams in North America. Streams with a TN:TP of < 15 or > 15 are distinguished. *Source:* from Dodds & Whiles 2020, with permission of Elsevier.

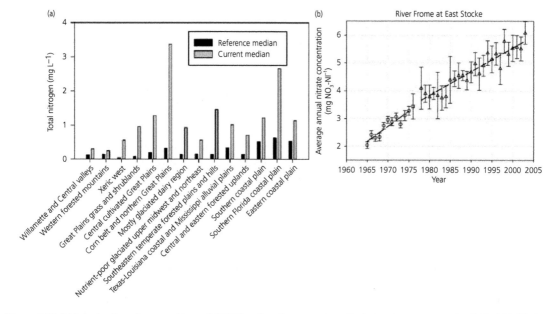

Figure 9.20 (a) Data showing reference and 'current' (1990s) median nitrogen concentration in streams across ecoregions in the USA; (b) shows the time course of mean annual nitrate concentration in the River Frome (south-west England) 1965–2003—separate regressions were fitted for 1965–1975 (circles) and 1978–2003 (triangles).
Source: (a) after Dodds & Smith 2016, with permission of Taylor & Francis; (b) from Howden & Burt 2009, with permission of John Wiley & Sons.

in the lowlands. On assessing nutrient limitation (i.e. the opposite of eutrophic impairment) Jarvie et al. (2018) found co-limitation by N, in particular, often with P, to be most common in the headwaters and recommended a focus on reducing N in headwaters and P in lowland reaches to maximise benefit per unit cost in river restoration. Overall, the nutrient reductions necessary were much greater for lowland reaches than for the uplands. In an impressive compilation of nutrient data from US streams, Manning et al. (2020) concluded that most streams and rivers had N and P

concentrations far higher than values required to 'protect ecological integrity' (roughly equivalent in Europe to 'good ecological status').

There is some better news on eutrophication in larger rivers as monitoring data accumulate, in continental Europe at least. In the French River Loire, for instance, records from 1980 show a continuous decline in phosphorus concentration from about 1991, and this seems to have led to a decline in the biomass of phytoplankton, although the invasive filter-feeding Asiatic clam *Corbicula fluminea* also played some role, as it has done elsewhere (Minaudo et al. 2015; Descy et al. 2021). Impressive reductions in nutrients are being achieved in several larger European rivers, that were previously hypernutrified (e.g. Figure 9.21), as a result of the tertiary treatment of waste-water from large urban centres (i.e. reductions in 'point-source' pollution). Diffuse pollution from over-fertilisation of agricultural soils is more difficult to counter.

Elsewhere, the trade-off between intensive agriculture and environment remains problematic. In New Zealand, for instance, a massive expansion of an intensive diary industry has led to the gross eutrophication of streams and rivers, fierce debates among scientists and policy-makers, and uneasy compromises in management (see Russell Death's Topic Box 9.2).

9.5 Elemental stoichiometry

Stoichiometry simply means 'measuring elements' and in ecology it refers to the elemental composition of living things and their resources. We have seen that the cycling of mineral nutrients is driven by energy from photosynthesis and the consequent transformations of carbon. The major nutrients, and a lesser supply of more minor ones, are essential components of living things and necessary for this flow of energy in ecosystems. We have spoken of nutrients limiting the biomass and activities of autotrophs and heterotrophs, and thus potentially determining the abundance of

living organisms and the diversity of communities. However, it may not be the absolute amount of a single element that is important but its supply relative to that of others. Thus, access to the energy stored in detrital carbon may be limited by the relative availability of fixed nitrogen—estimated by the ratio of C to N. Nitrogen or phosphorus alone can initially limit primary production, although once either is sufficient, further supplementation will have no further effect. Thus, a balance of nutrients is required, ultimately linked to the proportions of various chemical elements in the tissues of living things. In the oceans, the basic atomic ratio of carbon, nitrogen and phosphorus in the water and plankton is around 106C:16N:1P and is known as the *Redfield ratio* after the biological oceanographer who discovered it (see Sterner & Elser 2002). This is the basic stoichiometric ratio in (marine) ecology and reflects the overall requirements of marine phytoplankton for the main mineral nutrients. The actual body composition of living things, particularly in fresh waters, can vary very substantially around this basic Redfield ratio, while the ability of organisms to maintain a particular elemental ratio in varying environments (at least in the longer term) is known as *stoichiometric homeostasis*. It is the basic proposition of stoichiometry that differences (great or small) between the elemental composition of potential food resources and the elemental requirements of consumers affect processes such as consumer growth and the decomposition of detritus—substantial imbalances acting to constrain them.

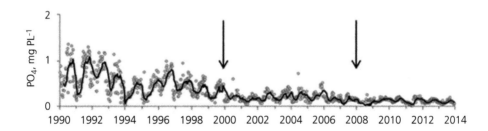

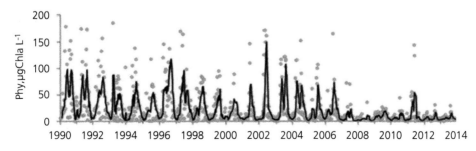

Figure 9.21 Interannual variation in phosphorus concentration and phytoplankton (Phy) biomass at the outlet of the basin of the River Seine (France). Points are observations, lines are 12-month moving averages. Vertical arrows indicate installation of new treatment (phosphate-removal) works.
Source: from Garnier et al. 2022, with permission of Elsevier.

Topic Box 9.2 Not so clean and excessively green - nitrogen in New Zealand streams

Russell Death

Eutrophication of streams and rivers is a global problem, and New Zealand is no exception (Larned et al. 2016). During the 1990s agriculture in New Zealand moved from a sheep and beef dominated industry to one focused on dairying (Foote et al. 2015). Land area in dairy farms increased by 46% from 1993 to 2012, dairy cattle numbers increased from 3.4 to 6.5 million between 1990 and 2012, and the average herd size increased by 147%. This massive shift in farming practice, and intensification, has led to widespread declines in water quality in New Zealand (Larned et al. 2016; Julian et al. 2017). Higher cattle stocking density has been supported by supplementary feed in the form of imported palm kernel, winter forage crops and (over-)fertilised pastures (Foote et al. 2015). Nitrogen, in highly concentrated urine patches, cannot be assimilated by the grass pasture and leaches into waterways. Measures taken to reduce this, including riparian planting and fencing, and redirection of dairy shed effluent to land away from streams, have resulted in declines in ammoniacal nitrogen and dissolved reactive phosphorus. However, nitrate-nitrogen concentration continues to increase in many streams (Larned et al. 2016); as a highly soluble chemical, it is much more mobile than phosphorus (which can bind to soil) and moves easily with subsurface flows, under riparian plantings and into adjoining waterways.

Despite the well-documented nationwide decline in water quality over several decades, and the obvious linkages with agricultural intensification, the dominance of agriculture as an export earner has delayed major policy intervention (Joy 2015). A few regions do have rules to limit nutrient inputs to iconic water bodies like Lake Taupo (Joy & Canning 2020). However, it was not until 2014, under pressure from the 'Land and Water Forum' (a sector-wide body set up to deal with increasing public unrest about the poor state of inland waters), that the government released the first set of limits for water quality in lakes, rivers and streams— the National Policy Statement for Freshwater Management (NPSFM) (New Zealand Ministry for the Environment 2019). Despite excellent guidelines for lakes, for rivers and streams there were only limits for dissolved oxygen associated with point discharges, ammoniacal nitrogen toxicity, periphyton and nitrate toxicity.

A change of government in 2017, elected in part on a platform to clean up New Zealand rivers, signalled a second review of the NPSFM. The NPSFM has numerical thresholds differentiating stream water quality in one of four states (from A to D, good to bad). The boundary between C and D is termed the 'national bottom line' and any water bodies currently in state D must be improved to exceed this threshold (New Zealand Ministry for the Environment 2019). A comprehensive assessment of the available data, using several lines of evidence (e.g. Box Figure 9.2), established the three nutrient concentrations differentiating rivers into the four states as 0.24, 0.50 and 1.0 mg L^{-1} for nitrate and 0.006, 0.010 and 0.018 mg L^{-1} for dissolved reactive phosphorus (Joy & Canning 2020). Not surprisingly, these nutrient thresholds were similar to those derived in other studies in New Zealand (Wagenhoff et al. 2017) and elsewhere around the globe (Poikane et al. 2019). A draft NPSFM was released with these management targets included; however, strong lobbying from the agriculture sector resulted in the nitrogen limits being removed from the final version, which became law in September 2020.

The New Zealand government has been praised for its use of science in dealing with the COVID-19 pandemic, and its hitherto highly effective response. Why then did it choose to ignore the substantial science on nitrogen enrichment of its waterways and the consequential impacts on ecological health (Joy & Canning 2020)? There were of course arguments that the cost to agriculture would be prohibitive, even though only 4% of pastoral catchments would be affected. There is also increasing evidence that farmers can actually increase profits by reducing stock density, milking frequency, supplementary feeding, fertiliser application and consequently nitrogen leaching (Joy 2015). Furthermore, the detrimental economic cost to tourism and export markets in the loss of New Zealand's '100% pure' image appears not to have been considered. As scientists, we are constantly challenged to make our research more relevant at the science–policy interface (Chasse et al. 2020). However, even an extensive body of rigorous science will not always sway government policy if other, often economic and political, pressures are too strong.

Russell Death is Professor of Freshwater Ecology in the School of Agriculture and Environment, Massey University, Palmerston North, New Zealand.

Topic Box 9.2 *Continued*

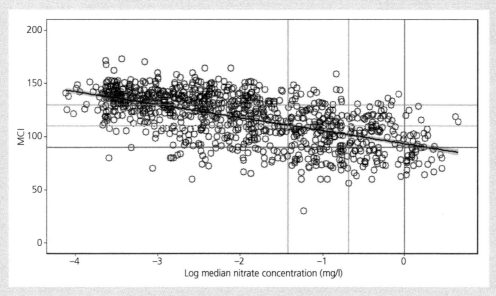

200 -

150 -

MCI

100 -

50 -

0 -

-4   -3   -2   -1   0

Log median nitrate concentration (mg/l)

Box Figure 9.2 A measure of ecological health, the Macroinvertebrate Community Index (MCI), from 963 stream sites in the North Island of New Zealand (Death et al. 2015) as a function of modelled median nitrate (Unwin & Larned 2013). Larger MCIs indicate greater ecological health. Regression line in black. Coloured lines are the A/B (green), B/C (yellow) and C/D (red) thresholds for the MCI (horizontal) in the NPSFM and for median nitrate (vertical) (New Zealand Ministry for the Environment 2019). Note that the MCI shows many sites in a bad condition even at nitrate concentrations far below the required threshold value.

The explicit study of stoichiometry in streams and rivers, as against marine systems and the pelagic zone of lakes, gained momentum only in the early 2000s. A first step is to describe the variability in elemental composition of both consumers and resources (Cross et al. 2005: Figure 9.22). It is evident that there is great variability in both C:P and C:N ratios in the resources available to consumers, particularly in biofilms. These latter contain variable amounts of algae, bacteria, fungi, 'slime' (technically, exopolymers produced by the microorganisms) and very small animals. It is also apparent that C:nutrient (both N and P) ratios of allochthonous resources (wood, leaf litter) are high relative to primary producers and bulk FPOM. The elemental composition of consumers is less variable than that of resources and they are relatively rich in nutrients (Figure 9.22). For instance, the mean N and P body concentration of invertebrate detritivores across a range of taxa is about 10 and 1% of body dry mass, respectively. Leaf litter ranges from

0.5–3% N and 0.01–0.2% P while the C:N, C:P and N:P ratios vary significantly across litter species (Hladyz et al. 2009). Of detritivores, crustaceans generally have the highest content of P and the lowest of N (see Frainer et al. 2015 and references therein). There are thus substantial elemental imbalances between consumers and what we presume they eat, particularly for detritivores.

We can now imagine a gradient in the relative supply of a nutrient that limits growth when it is scarce. As that nutrient becomes more available we reach a threshold at which there is a switch to another element which may then limit the rate of growth—this is known as the *threshold elemental ratio* (TER; Frost et al. 2006), as shown in Figure 9.23 for a gradient in the C:P ratio (Benstead et al. 2017). To the left of the graph (low C:P) phosphorus is relatively abundant and does not limit growth and may be excreted. To the right (high C:P) phosphorus is scarce and is retained. This ratio may vary among taxa and can have consequences for population

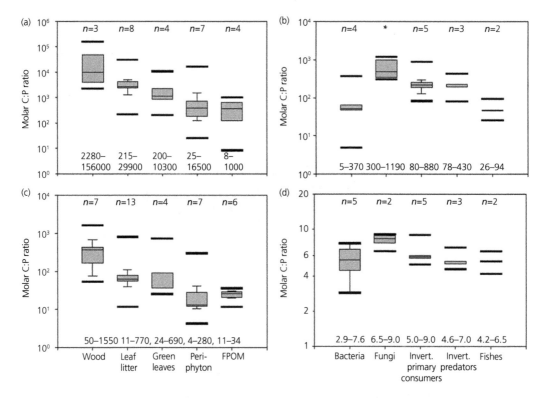

Figure 9.22 Molar C:P and C:N ratios (log scale) for a compilation of common freshwater benthic resources ((a) and (c)) and consumers ((b) and (d)). Grey boxes contain the median value and various percentiles of values, up to 90%. Black horizontal bars show the range of values (given in numbers below each box plot). Note differences in scale on vertical axes.
Source: from Cross et al. 2005, with permission of John Wiley & Sons.

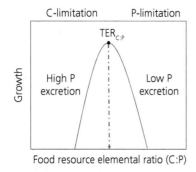

Figure 9.23 The postulated relationship between organismal growth and the C:P ratio of the food. Growth is limited by carbon and P is excreted actively at low values of C:P, while growth is limited by P and P-excretion minimised at high values of C:P. Growth is maximised at the threshold elemental ratio (TER$_{C:P}$).
Source: from Benstead et al. 2017, with permission of Elsevier.

dynamics and community structure. The species composition of a consumer community therefore can have consequences for the regeneration of nutrients (by animal excretion) available for primary consumers and decomposers.

As an example, Frainer et al. (2015) tested the effects of stoichiometry on resource consumption and growth rate of stream-dwelling detritivores. In experiments in flow-through containers ('microcosms') exposed in an upland French stream, they used three detritivorous consumers of differing body composition. For instance, N:P was lowest for the amphipod *Gammarus*, and higher for the caddis *Sericostoma* and the stonefly *Nemoura*. These were fed on tree leaves of four species differing in lignin content (leaves high in lignin are usually more resistant to decomposition—'recalcitrant'). Alder (*Alnus glutinosa*) and birch (*Betula pendula*) are low in lignin while oak (*Quercus robur*) and walnut (*Juglans regia*) are high. They also differed in N:P ratio, either high (alder and walnut) or low (oak and birch). These leaves were offered as four single-species treatments or two mixed-species treatments (walnut and oak; birch and alder). These six treatments were crossed with the three invertebrate treatments plus a 'no animal' control, giving 24 combinations, each well replicated. Litter consisting of alder alone or in a mixture had much the highest N:P ratio, with values ranging between about 80 and 130 compared with values of around 40 or less for other litter treatments (values were measured at the end of the experiment, after which leaves were variably conditioned by microbes).

Elemental imbalance (*EI*) can be estimated as:

$$EI_{X:Y}{}^{ij} = Ln\left(X:Y^i/X:Y^j\right)$$

Where $X:Y$ is the atomic ratio (N:P, C:N or C:P) of litter i and consumer j. $EI_{X:Y} > 1$ implies an imbalance between

resource and consumer of element X (i.e. relatively more of X in the resource than the consumer—normally relatively more carbon than N or P), while a ratio < 1 implies a greater imbalance of element Y (relatively more of Y in the resource). Growth of all three species of detritivore was related to N rather than to C or P (Figure 9.24), which Frainer et al. (2015) speculated was due to their production of N-rich chitin—which is lost when moulting—and, in the caddis *Sericostoma*, because of the production of silk (a protein) used in case-building and repair. In terms of resource consumption, all three species ate more of the preferred litter species (high N:P and low C:N), accelerating mass loss in litter mixtures containing the preferred species. Elemental stoichiometry is clearly important in driving the growth of detritivores—as they 'try' to

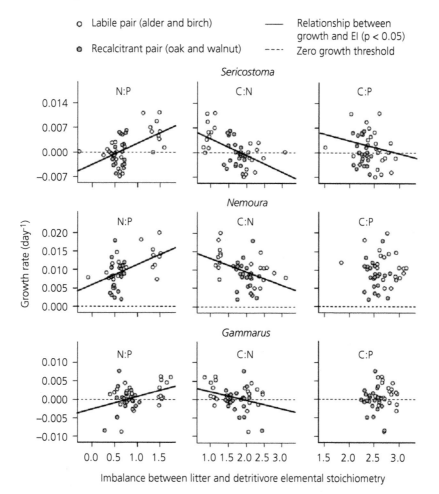

Figure 9.24 Growth rates of three detritivorous species (caddis *Sericostoma*, stonefly *Nemoura*, amphipod crustacean *Gammarus*) in relation to the elemental imbalances (N:P, C:N and C:P) between leaf litter and consumer. Recall that 'elemental imbalance' measures the difference in the stoichiometric ratio (e.g. N:P in the left column) of the litter and that of the detritivore. An elemental imbalance > 1 means relatively more of the first than the second element in the detritus than the detritivore. Growth is more constrained by N than by P or C.
Source: from Frainer et al. 2015, with permission of John Wiley & Sons.

maintain a particular body composition (i.e. homeostasis), although biological details of particular taxa are clearly also very influential.

The whole-stream nutrient-addition experiments at Coweeta referred to previously involved an experimental design explicitly aimed at revealing stoichiometric effects of N and P, with N:P ratios ranging from 2–128. Under these varying conditions, Manning et al. (2016) assessed experimentally the breakdown, both that mediated by microbes and that by detritivorous animals, of five types of detritus (leaf litter of four tree species and wood), with widely varying initial C:nutrient content. Under nutrient enrichment, detrital stoichiometry was reduced (i.e. N and/or P content was increased, presumably through uptake by *conditioning* of the detritus by microbes) while differences among detritus types were reduced. Detritus approached the nutrient requirements of detritivores (the threshold elemental ratio for nutrients was exceeded) and breakdown due to detritivores was increased relative to microbial breakdown.

In Chapter 8 we discussed the possibility that the diversity of detritivorous consumers could alter the rate of litter decomposition in some way—an example of a possible relationship between biodiversity and a key ecosystem process. Elemental stoichiometry offers a feasible mechanism by which this could occur. For instance, Ohta et al. (2016) proposed that differing C:nutrient ratios among an assemblage of detritivores (i.e. a high stoichiometric diversity) could increase the overall rate of litter processing compared with that achieved by a single species or by a group of species with more homogeneous C:nutrient ratios. Thus, detritivores with relatively nutrient-poor body composition could consume relatively nutrient-poor litter, whereas more nutrient-demanding detritivores require nutrient-rich litter—because their threshold nutrient ratios differ. They found some experimental support for this process using a series of detritivores from a Japanese stream. In developing this idea, Atkinson & Forshay (2022; Figure 9.25) suggest how the patchy distribution of freshwater mussels, and differences in species composition among local patches,

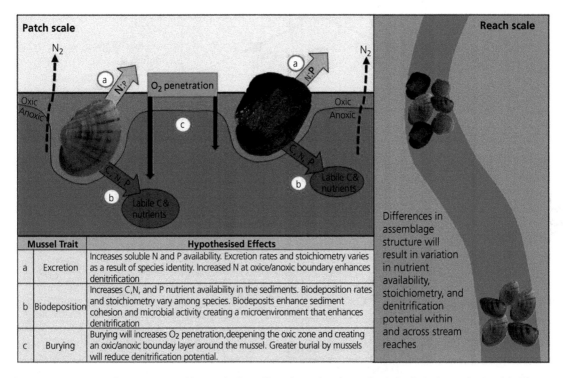

	Mussel Trait	Hypothesised Effects
a	Excretion	Increases soluble N and P availability. Excretion rates and stoichiometry varies as a result of species identity. Increased N at oxice/anoxic boundary enhances denitrification
b	Biodeposition	Increases C,N, and P nutrient availability in the sediments. Biodeposition rates and stoichiometry vary among species. Biodeposits enhance sediment cohesion and microbial activity creating a microenvironment that enhances denitrification
c	Burying	Burying will increases O₂ penetration,deepening the oxic zone and creating an oxic/anoxic boundary layer around the mussel. Greater burial by mussels will reduce denitrification potential.

Figure 9.25 A schematic illustrating potential direct and indirect effects of mussels on biogeochemical cycles in rivers—showing that different species can bring about different fluxes at the patch and reach scales. Effects include: (a) variations between species in the stoichiometry of excreted nutrients, (b) biodeposition and egestion of sediments at different rates, depths and elemental composition, and (c) burial by the mussels modifying biogeochemical interactions (e.g. denitrification) between the oxic and anoxic zones. Species depicted are (left) *Lampsilis ornata* and (right) *Amblema plicata*.
Source: from Atkinson & Forshay 2022; with permission of John Wiley & Sons.

could lead to different fluxes of nutrients and rates of denitrification in river systems at the patch and reach scale. Again, they obtained experimental results consistent with this view.

Overall, stoichiometry offers a unifying explanation linking food webs, production and nutrient cycling in ecology. However, biological details and differences among species can clearly account for a great deal of variation in real systems. This is an exciting and still quite novel area that stands at the crossroads between ecosystem processes and community ecology.

9.6 Nutrient subsidies across systems

A theme of this book is how 'open' river ecosystems are. In addition to the direct organic matter exchange between the river and its catchment, many species cross the land–water interface as a regular part of their lives. Further, many primarily terrestrial species feed on aquatic resources, and many aquatic species feed actively on terrestrial resources. Such exchanges can be studied as part of the 'tangled web' of dynamic species interactions (linking populations and food webs; Chapters 5 and 7) or be assessed as flows of carbon between systems (Chapter 8) or, as here, in terms of cross-system nutrient subsidies. Of course, most nutrients enter rivers along with the water, either dissolved or as suspended particles, or are blown or fall in as dust or dead organic matter. They leave rivers mainly hydrologically, by downstream flow, or are released into the atmosphere as dinitrogen gas or as nitrous oxide. Here we focus on nutrients entering or leaving by biological means.

As we have seen, the sheer array of biological interactions across habitat boundaries is immense—most of them involving the transport of organic matter and nutrients up- or downstream, or into the water from the land or from land to fresh water (see e.g. Baxter et al. 2005; Moss 2015). The migrations of fish, particularly salmonids, are the most thoroughly researched examples, though there are many others (see Chapter 8). For instance, emerging aquatic insects transport aquatic production into the riparian zone; they also transport nutrients. Small et al. (2013) measured nitrogen exported via insect emergence in seven Costa Rican streams, finding that it accounted for between about 0.4 and 1.25 mg N m^{-2} d^{-1}. In context, these values are within the range of other studies and accounted for between 2 and 16% of the rates of microbial denitrification in the same streams. Insect emergence appears to be a further pathway for the loss of fixed nitrogen from streams, though of lesser magnitude (in this case

at least) than microbial mechanisms. We referred previously to the import of terrestrial organic matter into an African river by hippos—which forage at night on land and spend the day in the water. Subalusky et al. (2015) estimated that hippos contribute annually almost 500 kg of N and 50 kg of P into the Mara River and major tributaries in the Masai Mara reserve in Kenya (the northern extent of the Serengeti ecosystem), amounting to 27% of the total loading of N and 29% of P from the upstream catchment. Most of the N was from hippo urine and about a third of the P, the rest coming from faeces. More generally, Moss (2015) argued that large herds of wild grazing mammals feeding on floodplains, before widespread decline due to hunting by humans and habitat loss, could have created locally 'eutrophic' conditions by concentrating nutrients in relatively small areas. These conditions would be the 'analogues' of the eutrophic state of many contemporary, highly modified, rivers.

It is the migrations of Pacific salmon that are best known in the context of the transport of marine-derived nutrients (see e.g. Gende et al. 2002; Rüegg et al. 2020). Five species spawn in the freshwater ecosystems of the Pacific seaboard of North America, all in the genus *Oncorhynchus*: chinook (*O. tshawytscha*), sockeye (*O. nerka*), pink (*O. gorbuscha*), chum (*O.keta*) and coho (*O. kisutch*). In all of them, adults return from the sea to freshwater spawning grounds, where they cease to feed, spawn and then die. Females create redds (effectively nests) in coarse gravel/stones in which they lay their eggs, the emerging young spending up to two years in fresh water before migrating out to sea. Adults may feed in the sea for 1–7 years, putting on 90% of their biomass before returning to fresh water. Runs of fish, while greatly diminished in most areas, can still be massive, with over 20 million fish being recorded entering rivers draining into Bristol Bay (south-west Alaska)—probably the world's largest extant salmon run, transporting upstream $> 2 \times 10^4$ kg of P and $\sim 2 \times 10^5$ kg of N of marine origin.

These marine-derived nutrients may be released directly into the river system by the excretion of the fish or, more importantly, after the fish die or their eggs are eaten or decompose. The nutrients may fertilise the stream itself and enter the stream food web, or they may leave the stream through predation or scavenging by terrestrial animals, including bears, wolves, eagles and others. The marine-derived nutrients in solution can also find their way into the riparian zone, via the hyporheic sediments, and be taken up from the soil by terrestrial plants (see below).

Dealing first with the effects of a marine nutrient subsidy on the stream food web, we find that, as well as a fertilising effect on biofilms, disturbance by large numbers of big salmon laying their eggs in gravel nests (redds) can actually depress productivity (at least while the salmon are there). In one study, Harding et al. (2014) sampled rock biofilms from 16 streams in British Columbia (western Canada) in the spring and summer, where spawning by Pacific salmon (mainly chum and pink) takes place in August to early November. Five of the streams had waterfalls that prevented the upstream migration of salmon. This provided an extra opportunity to infer the role of salmon; that is, the normal variation among streams in relation to the numbers of salmon, plus differences upstream (no salmon) and downstream (salmon present) of the waterfalls. Nitrogen and carbon of marine origin have a relatively 'heavy' isotopic signature (i.e. relatively more of ^{15}N and ^{13}C), compared with nutrients in fresh water and on land. This provides a 'label' of the prevalence of marine nutrients in streams receiving migrant salmon. Salmon density was the best predictor of biofilm $\delta^{15}N$ and was also an important predictor for $\delta^{13}C$, showing that marine-derived nutrients were assimilated (Figure 9.26). Notwithstanding this nutrient subsidy, biofilm chlorophyll a (a measure of algal biomass) was negatively related to salmon numbers in autumn, a time when redd building was underway, direct disturbance being a possible explanation. In the following spring, however, after the adult salmon had all died, chorophyll a was positively related to spawner density in the previous autumn.

The question of the relative roles of fertilisation and disturbance by salmon has now been investigated several times, with variable results. In a recent study, Rüegg et al. (2020) used relatively high-frequency sampling (intervals of eight days or less) to investigate biofilm attributes (including production, respiration, chlorophyll a, ash-free dry mass and stable isotope ratios) in an Alaskan stream from before the salmon run (early July) to its conclusion (end of September). Overall, the impact of the salmon on biofilm accrual was slightly negative (i.e. disturbance seemed to outweigh nutrient enrichment). However, the net effect of nutrients and disturbance seems to vary with 'environmental context'. That is, biological details and physicochemical differences between systems play a role. As examples, salmon density and competition for space among females can be important, substratum stability and particle size play a role, and early in the run males outnumber females (the males run first) so the enrichment effect may be most apparent at that time (since the males do not build redds). Background nutrient limitation will also determine whether the stream biofilm will respond strongly to enrichment by salmon. Of course, the disturbance effect is limited to the period of redd building, whereas the nutrient subsidy is much more prolonged and will be favoured in the longer term.

The web of possible interactions and ramifications of marine-derived nutrients is startling, involving some of the most spectacular sights in all of natural history, as large vertebrate predators gather around rivers as the salmon begin to run. The nutrients reach a remarkable range of organisms in riparian ecosystems. Helfield & Naiman (2006) conceptualised the various pathways for marine-derived nutrients in the riparian forests of salmon streams in Alaska (Figure 9.27), arguing that Pacific salmon and brown bears (*Ursus arctos*) together constitute a 'keystone interaction', both maximise the contribution of marine nutrients to forest productivity, and increasing the sutiability of streams for salmon spawning. Nutrients have been shown to influence such things as the fertilisation and diversity of riparian plants, terrestrial insects, the density of passerine songbirds, and even to affect tree-ring thickness in large forest trees. Hocking & Reynolds (2012) placed salmon carcasses alongside 11 streams in British Columbia, with variable natural densities of spawning salmon. The percentage N and the $\delta^{15}N$ (expressed as per mille or 'parts per thousand, ‰) of herbaceous plants and a moss living in the understorey alongside the carcasses had subsequently increased by 14–60% and by 0.5–3.3‰, respectively. Effects on $\delta^{15}N$ were

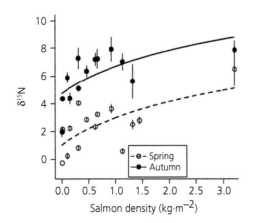

Figure 9.26 Bivariate plot of stable ^{15}N of stream biofilm versus salmon density in summer and autumn data from 16 catchments in British Columbia, Canada.

Source: from Harding et al. 2014, with permission of John Wiley & Sons.

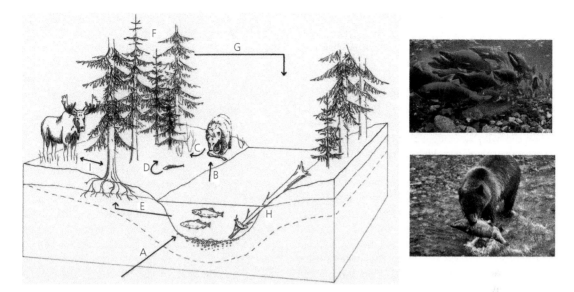

Figure 9.27 A conceptual schematic of cycling of marine-derived nitrogen (MDN) in rivers with Pacific salmon and brown bears. A, spawning salmon transport MDN upstream; B, bears and other consumers eat salmon; C, bears and others transport salmon-enriched wastes and partially eaten salmon carcasses into the riparian zone; D, insects and other invertebrates colonise salmon carcasses, aiding decomposition and disseminating MDN; E, dissolved MDN in stream water downwells into the hyporheic zone and is taken up by tree roots; F, MDN increases growth of riparian trees; G, riparian trees improve stream habitat and contribute leaves and wood to stream; H, large wood increases retention of salmon carcasses, and thus MDN; I, increased N content of tree leaves increases palatability of riparian plants, attracting herbivores. Insets (right) are migrating sockeye salmon in Alaska (a charr is in the foreground) and a brown bear with a captured salmon.
Source: after Helfield & Naiman 2006, with permission of Springer Nature; drawing by Lotta Ström. Photos by Jonny Armstrong and Alan Vernon, respectively, under Wikipedia Creative Commons Licence.

rather variable but greater for herbaceous species than for moss, with values varying from around 2‰ for understorey herbs at control (away from carcasses) sites to around 5‰ for experimental sites (plants growing next to carcasses). Hocking & Reynolds (2011) also compared riparian plant diversity in 50 catchments in a remote area of British Columbia, showing that the loading of nutrients from migrant salmon shifted the community towards more 'nutrient–rich' species while decreasing overall plant diversity (as nutrient-demanding plants could then suppress less competitive species).

Hocking & Reimchen (2002) sampled terrestrial invertebrates (soil and litter associated) using pitfall traps in riparian areas above and below barriers to salmon migration on two forest rivers in British Columbia, finding enriched nitrogen stable isotope ratios (implying the incorporation of marine nutrients) adjacent to stretches with salmon. They estimated that between about 20 and 70% of total nitrogen in terrestrial invertebrates was originally derived from salmon—although not by direct consumption of salmon carcasses but mainly after uptake of nutrients into soil/terrestrial food webs. On the same rivers, Christie et al. (2008) sampled the feathers and faeces of the ground-feeding winter wren (*Troglodytes troglodytes*) captured up- and downstream of waterfalls in autumns and summer. Feathers from birds captured below the falls had enriched $\delta^{15}N$ compared with values for birds feeding above them. Faecal samples became very much more enriched ($\delta^{15}N$ enriched by up to 14.3‰ in individual birds) between summer and autumn, probably due to the degree to which these birds had fed upon fly larvae from salmon carcasses during the salmon run. Christie & Reimchen (2008) further found that a number of songbirds (six species, including the winter wren) were most abundant close to salmon-holding reaches of these rivers. Wagner & Reynolds (2019) sampled songbird assemblages in 14 stream catchments in British Columbia, encompassing a wide range of salmon densities, finding that bird abundance and density increased with salmon density. These are just a few examples of many suggesting a role for marine nutrients in terrestrial ecosystems, derived from salmon migrating upstream.

These examples almost all concentrate on nitrogen, but Currier et al. (2020) widened the scope to include other elements. In their study of salmon streams in south-eastern Alaska they found evidence of an increase in stream-water concentrations of Ca, Fe, Mg and Na upon the arrival of salmon across seven study streams. In one intensively studied stream, Ca, Fe and Mg increased in stream biofilms near the end of the salmon run, with the evident possibility of the transfer of such nutrients more widely into stream food webs and riparian ecosystems.

Most of the attention has been given to Pacific salmon, probably because, at least historically, they run in such large numbers, are economically and culturally extremely important, and not least because they are semelparous—that is, they breed just once, and then die in or close to streams and rivers. Other examples of the migrations of semelparous species are those of the jawless, carnivorous lampreys that run up into streams and rivers from the coastal seas and estuaries (plus some very large lakes) of the northern hemisphere and whose carcasses are either transported, often by scavengers, into the riparian zone or decompose in the stream (see e.g. Dunkle et al. 2020). Most migratory fish are iteroparous, however—that is, they often survive spawning, returning to their adult habitat to feed and recuperate, and may come back to breed again, once or repeatedly. They thus provide less nutrients from decomposing adults than semelparous species. Their contribution via eggs and excretion can be large, however, as in the example of the long-nose sucker (*Catostomus catostomus*). The suckers are almost exclusively North American and can be very abundant. Childress et al. (2014) experimentally blocked fish access some way up an oligotrophic tributary of Lake Michigan, comparing reaches upstream of the barrier with reaches downstream containing thousands of breeding fish. Suckers raised the concentration of nitrogen and phosphorus by three to five times. Algal accrual in biofilms doubled and a cased caddis *Limnephilus* grew 12% larger in reaches with suckers, while an average of 18% of N in caddis tissues derived from these fish.

The single species of Atlantic salmon (*Salmo salar*) is iteroparous compared to the semelparous Pacific salmon, although mortality during the spawning run may be high. Their spawning is less synchronous and is more variable among and within populations than that of the Pacific salmon. Their numbers have greatly declined almost everywhere and it is suggested that this may have led to the 'oligotrophication' of headwater streams, reducing their productivity and potential as nursery areas for young fish if adults do spawn

there. It may be possible to restore productivity by the careful application of nutrients, usually by the addition of fish carcasses or artificial substitutes, such as pellets. In one recent experiment, McLennan et al. (2019) simulated the deposition of a small number of salmon carcasses at the end of the spawning period in five upland Scottish streams, comparing them with five unmanipulated reference sites. They compared the survival and growth of juvenile salmon following the addition of 3,000 salmon eggs to each reach—the progeny of the same 30 wild-origin clutches. They found that macroinvertebrate biomass and abundance were five times higher in the reaches with added carcasses ('high parental nutrient' treatment) than in those without ('low parental nutrient' treatment). This supported faster growth of juvenile salmon over the next two years and a greater genetic diversity in surviving salmon—without an overall increase in density. More fish reached the size necessary to migrate to sea after two years (rather than three) in high parental nutrient streams. Whether such addition of fish carcasses would be a beneficial management strategy for wide application remains to be seen.

9.7 Nutrients, rivers and new horizons

Our knowledge of mineral nutrients in flowing water has clearly been transformed in the last 50 years, partly due to fundamental research on the cycling of nitrogen (predominantly) in streams and partly by paying greater attention to larger and more nutrified rivers. These latter advances have been due partly to the accumulation of longer-term data and partly to the adoption of new, whole-channel methods.

Most research still regards nutrient pollution as something people 'do' to rivers—via domestic and industrial sources and from intensive agriculture. The question has been: how do rivers respond to excess nutrients and how do we assess it? It is apparent, however, that there are important feedbacks from rivers on people—that is, 'what do rivers "do" to or for us?'. Rivers can 'self-clean' a good deal of nutrient pollution, by producing nitrogen gas, but also produce disproportionate amounts of greenhouse gases, particularly nitrous oxide (N_2O) and methane. Such phenomena lend the study of the biology and ecology of rivers and streams a new urgency. As river ecologists or simply as interested citizens, we often think of rivers for their own sake—how fascinating they are, how rich in species and wildlife. It is not too much to talk of many people loving rivers with a passion. We love to walk their banks, we pay more to live with a river view (from a 'safe' distance of course), songs are written about them, they are gateways leading out to the wider world. We have concentrated so far on understanding how rivers 'work', from a biological point of view; this is what students need to be armed with into the future. Many will want more, however. How can the science help us to conserve and manage rivers sustainably in this rapidly changing world? What new issues lie ahead, what new knowledge and understanding is developing, and what new techniques may become available? We address some of these questions in the next (and final) chapter.

CHAPTER 10

New horizons

10.1 Introduction

Most of this book is devoted to the basic biological and ecological science of rivers—what biologists and ecologists need to know about the environment, the species composition and adaptations of the biota, distinctive aspects of the population biology and community ecology of rivers (including the consequence of the network structure of the habitat), how species interact and the nature of lotic food webs, and ecosystem processes involving energy flow and nutrient cycling. We have emphasised the biology and natural history of rivers, hopefully with a sense of curiosity about the diversity of river life and as part of the general human enterprise of understanding nature. As is abundantly clear, however, river ecosystems very rarely exist in what we could call a truly 'natural' state. Even the most remote and apparently least perturbed are subject to the outcomes of human activities, resulting from the long-range transport of pollutants and the changing climate. Many river systems with which we are familiar are almost entirely artificial, or very highly modified, and for them there is little or no prospect of returning to an unperturbed or 'reference' condition (even if that was understood or considered desirable).

Rivers and streams are a part of nature with which humans interact extremely intimately. We use and abuse them, either knowingly or unknowingly, frequently with conflicts of interest between different users (such as between the disposal and dispersal of pollutants and the public water supply, fisheries and recreation). Over the past several decades, science that focuses on rivers, including their ecology, has moved overwhelmingly towards understanding the way humans affect rivers and are affected by them. Science can play a role in the wise stewardship of river systems, in providing a deeper knowledge about rivers, in formulating environmental policy and in providing managers with tools for environment assessment, conservation and restoration—though we are a very long way from the nirvana of truly sustainable river ecosystems. By *sustainability* we mean river systems whose natural resources and their value to people are not degraded generation by generation, thus depriving our descendants of this 'natural capital'. In this final chapter we are not writing a whole textbook on the very many applications of science to river management. Rather, our object is to: (a) highlight key issues that are arising or gaining prominence in our view of river systems and the life within them, and (b) identify some of the newer methods that are becoming available to policy-makers and managers. Though we have entitled this chapter 'New horizons', many of these issues are not new, or very soon will not be new, though they are gaining a wider importance and appreciation. Moreover, most threats to rivers are multifaceted, cutting across the 'tram-lines' of disciplines and scientific traditions.

10.2 A freshwater biodiversity crisis?

The biodiversity of fresh waters is remarkable—with estimates that they contain around 9.5% of all species (including one third of vertebrates) in only 0.8% of the planet's surface area. There are almost as many species of freshwater fish as there are marine, despite the vast disparity in the extent of marine and freshwater ecosystems (Balian et al. 2008; Dudgeon 2020). Further, reviews have identified freshwater biodiversity as being more acutely threatened than marine or terrestrial systems—note this is for fresh waters overall, not just streams and rivers. Reid et al. (2019) revisited this topic 12 years after Dudgeon et al.'s (2006) original review, while Dudgeon's (2020) book gives an extended account. The 'Living Planet Index' expresses overall population trends of a 'basket' of vertebrates from marine, terrestrial and freshwater habitats from a baseline of 1970 to 2012 (Figure 10.1). The evident decline in abundance of species in general has been termed a 'great thinning' and even perhaps the commencement of a sixth global mass extinction.

The Biology and Ecology of Streams and Rivers. Alan Hildrew and Paul Giller, Oxford University Press. © Alan Hildrew and Paul Giller (2023).
DOI: 10.1093/oso/9780198516101.003.0010

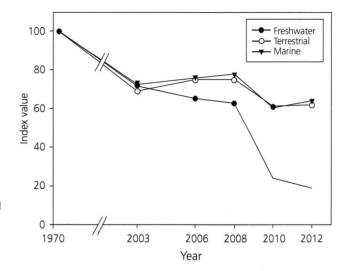

Figure 10.1 The 2016 World Wide Fund for Nature 'Living Planet Index' combines population trends for a collective group of vertebrates from terrestrial, marine and freshwater systems, relative to a common baseline (assigned an index value 100) of 1970.
Source: from Reid et al. 2019; based on WWF 2016, with permission of John Wiley & Sons.

For fresh waters, the Index contains data on 3,741 monitored populations, covering 944 species of mammals, birds, amphibians, reptiles and fish (WWF 2020). The mean population trend is a decline of 84% (ranging between 77 and 89% for individual species). This is a loss of approximately 4% per year since 1970, with the steepest declines in amphibians, reptiles and fish and, geographically, in Latin America and the Caribbean. It is also close to four times that of terrestrial systems. We have less detail on macroinvertebrates but there have been claims that at least four major freshwater taxa (Odonata, Plecoptera, Trichoptera and Ephemeroptera) have also suffered the loss of a considerable proportion of their species and some 35–45% are declining (Sanchez-Bayo & Wyckhuys 2019). On the evidence of the Living Planet Index, a global biodiversity crisis, for freshwater vertebrates at least, seems undeniable.

10.2.1 What are the drivers of biodiversity loss?—an initial census

Biodiversity losses may be attributed both to long-standing, well documented issues and to more novel ones (Reid et al. 2019). *Water pollution* has long been with us, the best-known examples being organic and toxic wastes derived from domestic, agricultural and industrial sources. Agriculture is also a major cause of increased sedimentation from the loss of soil from cultivated land. Organic pollution has been, and still is in many countries, one of the great challenges not only to river systems but also to human health and

wellbeing. We understand it well and know how to combat it—at least in its more 'conventional' forms (such as sewage, animal and other oxygen-demanding wastes). Water pollution is a form of habitat degradation for living things, and our almost universal use of rivers for the disposal, dispersal and dilution of wastes is a major cause of diversity loss. However, the nature of pollution is changing rapidly along with technological and industrial developments. 'Emerging contaminants' include a new generation of pesticides and herbicides, pharmaceuticals and 'personal care products', nanomaterials and plastic particles (large and small), plus a plethora of industrial chemicals. We have known for some time that some of these act as endocrine disruptors, leading to the feminisation of fish in rivers (e.g. Tyler & Jobling 2008) but many others pose novel threats (see section 10.3 below).

Various kinds of *flow modification* are again a long-standing threat to river habitats and river life, but a threat that is expanding extremely rapidly—exacerbated by human population growth, an increase in demand for water *per capita*, the rapid development of so-called green energy from river flow (hydropower), a food crisis and the associated expansion of irrigation agriculture, and ongoing climate change altering precipitation patterns.

Overexploitation of river organisms, mainly for food, also has an extremely long history and is widespread. In the UK alone there were apparently increasing numbers of very large, fixed fish traps set in rivers and estuaries during the bronze age (beginning almost 5,000 years ago)—targeting large migratory fish (such as salmon and sturgeon), and the Magna Carta of 1215 stated there were far too many fixed fishing stations in English rivers. A litany of overexploited fisheries is found in large rivers everywhere, including the Yangtze River of China, the Mekong of South-East Asia, the Amazon, and the west-coast rivers of North America (see Dudgeon 2020). The introduction of *alien species* (including pathogens) can also be an important cause of biodiversity decline if they become invasive—an issue that also has a long history but is again accelerating sharply, due to new connections among river catchments (through navigation canals for instance) and increased global trade (see Topic Box 3.2, and Chapter 7 for examples of their impact). Remarkably, the expanding global trade in exotic animals and plants is still largely unregulated. The spread of a fungal disease of frogs and toads (chytridiomycosis) has led to widespread declines and loss of species. Indeed, amphibians may be the most threatened of all vertebrates (see also Chapter 7, sections 7.3.1, p. 233 and 7.4, p. 239).

Newly appreciated threats to freshwater life include increasing *artificial light* at night and environmental *noise*. Night light may affect the behaviour of fish and invertebrates in urban rivers and streams, and it disrupts the dispersal and reproduction of flying semi-aquatic insects. The most pervasive emerging threat is *climate change*—the focus of intense contemporary political and scientific scrutiny. It brings not only higher mean and maximum temperatures but also more extreme and more frequent hydrological events (flood and droughts). Important in its own right, the changing climate often interacts with other stressors (pollution, the spread of invasive species, and human demand for fresh water), binding them together in complexes of 'multistressors' (see section 10.5.4).

10.2.2 Biodiversity losses in rivers

Whereas statistical changes in overall biodiversity are clearly of concern, it is the local or global loss of large and spectacular river species that has the greatest public impact. Among the most depressing real examples is the loss of the endemic Yangtze River dolphin (the 'baiji': *Lipotes vexillifer*), the only cetacean (marine or freshwater) yet to have been driven to extinction by humans. Until recently this was the only extant member of the Lepotidae, but this entire family is now gone, the last sighting probably being in 2002. The loss of this iconic species was driven by human activities, primarily pollution, damming and flow regulation, illegal fishing and injuries caused by shipping (Dudgeon 2020). The Chinese paddlefish (*Psephurus gladius*), one

of the world's largest freshwater fish at up to 7 m in length, went extinct around 2005 (the last sighting was in 2003) and, like the river dolphin, its demise is attributable to deterioration of the Yangtze River (Zhang et al. 2020). The loss of these two species and similar ongoing declines of the Manatee in Florida, are part of the 'decline of the megafauna' (Chapter 5). More generally, Dudgeon (2020) speaks of a 'great shrinking' of the fish fauna of fresh water. Larger (mainly migratory) species are more susceptible to loss than small ones and, in addition, the mean size of many species of fish declines with fishing pressure.

Among the invertebrates, the diverse order Unionida (850 species of bivalve molluscs, mainly in the family Unionidae) are among the most vulnerable. Between 27 and 37 species of Unionida in the USA are extinct or presumed extinct (10% or more of the US fauna), the group being amongst North America's most imperilled animals of any kind (Strayer et al. 2004; see Topic Box 3.2). They have been overexploited for their shells (the shiny internal lining, or nacre, was used for making buttons) and freshwater pearls (genus *Margaritifera*). Freshwater mussels are particularly vulnerable to pollution, clogging of the stream bed by eroded sediment from agricultural land or forestry, and to invasive species. Their other centre of diversity is in the large rivers of China and South-East Asia, where they are also endangered but for which data are poor. Among vertebrates other than fish, frogs and salamanders are among the most threatened globally, with one estimate that 3.1% of all frogs are extinct (~ 200 species), due to an invasive African chytrid fungus in South America, *Batrachochytrium dendrobatidis* (see section 7.4). The loss of their tadpoles from streams is ecologically significant (see section 7.3). Sadly, salamanders are also now at risk from another chytrid fungus, *Batrachochytrium salamandrivorans*, discovered only in 2013 (Dudgeon 2020).

These, and many other examples of the loss of species from streams and rivers across the globe, have become well known. Nevertheless, despite the overall global declines some contrary examples of regional-scale recovery have recently emerged, as indicated by meta-analyses of data now available from numerous long-term surveillance schemes for a number of regions, habitats and taxa—particularly in the developed world. For example, van Klink et al. (2020) analysed data on insect abundance (*not* diversity itself) from 166 long-term (10+ years) surveys of 1,676 sites (mainly but not exclusively in North America and Europe). They found a decline in the abundance of terrestrial insects of ~ 9% per decade, but an increase in that of freshwater insects by ~ 11% per decade.

Running waters were well represented in monitoring schemes and the positive trend was strongest in temperate areas, where progress has been made in improving water quality. Pilotto et al. (2020) also assessed biodiversity trends throughout Europe as shown by 161 long-term (15–91 years) biological time series that included over 6,000 marine, freshwater and terrestrial taxa. The richness and diversity of aquatic invertebrates (including those from running waters) increased, again probably due to recovery from stressors such as water pollution and other factors. Water quality has improved in many large rivers in western Europe (e.g. Hildrew 2022), and there have been enormous long-term gains in the 'health' of numerous formerly grossly polluted urban systems (e.g. Langford et al. 2009; Whelan et al. 2022; and see Chapter 6)—though continuous monitoring and investment in pollution prevention is clearly important and improvements can quickly be reversed, as they are in some developed countries during economic crises. The more positive examples, such as the recolonisation of several large river systems by the European Beaver (*Castor fiber*) and the widespread recovery of the Eurasian otter (*Lutra lutra*) in Europe, show what is possible and are encouraging. However, the overall global situation of biodiversity decline is undoubtedly extremely bleak and, even without the added threat of ongoing climate change, represents a great challenge to science and humanity.

10.3 Alien and invasive species—'the great mixing'

10.3.1 Some basic terms

This field is plagued by terms—including exotic, non-indigenous, alien and invasive—that are often used loosely (we plead guilty too!). Different taxa originated in different places and may have been prevented from mixing by natural barriers to their dispersal. If such barriers are breached by human activities (whether intentional or not), species may disperse across them and become established in the new area, where they can be described as alien, exotic or non-indigenous. There are differing shades of meaning among these three terms (for instance, exotic species tend to originate from some 'distant land') but they are all essentially similar. The term *invasive* is distinct, however, and describes an alien/exotic/non-indigenous species (and occasionally even an indigenous one) that becomes markedly abundant in an area and causes some ecological, economic or societal concern.

It may do so quite soon after arrival or only after a delay. Most alien species probably never become invasive, but many do—certainly in rivers and other fresh waters—and the problem is growing. This constitutes a third 'thread' in the loss of freshwater biodiversity—'a great mixing' to go alongside 'thinning' (reductions in abundance) and 'shrinking' (reductions in body size).

We are currently faced with a trend towards biotic homogenisation (the increasing similarity of faunas and floras in different areas), some calling the modern age the 'homogenocene' (Mann 2011). At its simplest, *ecological homogenisation* involves the increasing dominance of invasive species and a loss of native biota through disease, competition, predation, hybridisation etc so that one place becomes rather like another. There is great and growing concern around this issue (Dudgeon 2020; Cuthbert et al. 2020). A group of papers assembled by Padial et al. (2020) provide strong evidence for the process in a wide variety of organisms and situations. They included the highly diverse crayfish fauna of different regions of North America, the fish fauna of the Great Plains/Rocky Mountains of the USA, macroinvertebrates of US prairie potholes, fish of the eastern USA, fish of the Yellow River (China), phytoplankton of Brazilian reservoirs and other cases.

10.3.2 A real example

Among the better-known, and economically important, examples of an invasive species is the zebra mussel (*Dreissena polymorpha*; see more in Topic Box 3.2, p. 74, and Chapter 7, section 7.3.1, p. 233). Indeed, a few invasive bivalves (including *D. polymorpha*) are among the most troublesome of all freshwater alien invertebrates, along with some highly invasive crayfish. To this we should add the golden mussel, *Limnoperna fortunei*, a native of the Yangtze River basin in China—an extremely damaging invasive bivalve introduced to South America in ballast water (Moutinho 2021). *Dreissena polymorpha* itself is of Ponto-Caspian origin—that is, originating from an area north of the Black Sea and east to the Caspian Sea and beyond, a source of many aquatic invasive species. Since the 1800s this mussel species 'has spread over most of western Europe and eastern North America' (Strayer & Malcolm 2006) causing, among other things, the loss or reduction of many native bivalves (at least locally) and incurring great economic damage by blocking pipes and intakes carrying water supplies (the adults are sessile, attached to the substratum by strong byssus threads). Indeed, Haubrock et al. (2022) estimate an economic cost of invasive biofouling bivalves (dominated by

Dreissenidae and Corbiculidae) at a staggering US $63.7 billion between 1980 and 2020, mainly in North America.

Zebra mussels arrived in the freshwater tidal section of the lower Hudson River in New York State, USA, in around 1991, probably via their planktonic veliger larvae arriving in ballast water pumped from ships from western Europe. There was an initial population boom in 1992 (see Figure 7.8c, p. 235) followed, in the subsequent 20-year period, by cyclic fluctuations of 11-fold with a periodicity of 2–4 years. The cycles were dominated by particular year classes of adults that suppressed larval survival (probably via competition for food—phytoplankton (see Figure 7.8)) until that dominant year-class passed from the population (Strayer & Malcolm 2006). This is just one population of *Dreissena polymorpha*, and a variety of other population trajectories is possible post invasion. Thus, Strayer et al. (2019) assembled data from 67 long-term (> 10 year) data sets for lakes and rivers across Europe and North America. These data included two species of *Dreissena*, the zebra mussel itself (*D. polymorpha*) and its congener the quagga mussel (*D. rostriformis*, sometimes named *D. bugensis*). The quagga is also a Ponto-Caspian invader and is found with the zebra mussel at many sites.

These populations showed a wide range of dynamics, but with two general patterns evident: (a) populations of both species grew rapidly in the first two years after appearance, (b) quagga mussels were slower to colonise (arrived later) and caused the decline of zebra mussels (although they do seem to coexist at some sites). The combined populations of the two species did not show a general decline over time—as is sometimes the case with invasive species (Figure 10.2). Strayer (2020) has pointed out that any single invasive species can have different patterns of abundance in different ecosystems (depending on its environment and complement of resident species), and that the impact of an invader (its effect on resident species and on ecosystem processes) may also differ among systems depending on its abundance and the characteristics of the ecosystem. Consequently, predictions of the impact of an invader in a particular situation remain problematic.

10.3.3 The extent of invasions

The extent of invasion of some large and prominent river systems, and of many connected lakes, is astonishing (reviewed by Dudgeon 2020). Most of these invasive species have been transferred as a side-effect of globalised trade and the building of navigable connections between even quite remotely separate river systems. Others have been transported intentionally, as in attempts to control mosquitoes (using mosquito fish, *Gambusia* spp.), for the aquarium trade or aquaculture (from which they escape), or to improve fishery yields and for sport. Perhaps the most obvious example of the latter is the brown trout, *Salmo trutta*, which is now an extremely widely distributed salmonid outside its essentially European native range (see section 7.5.1, p. 242).

Among the most extensively invaded rivers are the Colorado of western North America and, in Europe, the Thames and the Rhine, though practically all the major rivers of western Europe have been invaded. The Thames is not one of the larger European rivers yet has around 100 non-native species in total, and the arrival rate of new species averages around one per year, similar to the Hudson River that flows from Canada through to New York (Jackson & Grey 2012). A common feature of the Hudson, Rhine and Thames is that they have ports for extensive trans-Atlantic shipping, on which species 'hitch-hike'—although around half the exotic species in the Thames were introduced intentionally.

There are apparently 72 species of non-native fish established in the Colorado, although many more than that that have been introduced at one time or another. Alien species thus outnumber the 49 indigenous fish found previously, 42 of them endemic (native nowhere else) (Blinn & Poff 2005). Invasion has been facilitated by extensive damming throughout the catchment and by major and unsustainable water withdrawals which have created conditions more suitable for invasive generalists than the native species, which tend to prefer the periodically fast-flowing and turbid water characteristic of the unimpounded river.

The Rhine (whose own catchment encompasses part of the land area of nine European countries) is one of the most connected rivers in the world. Leuven et al. (2009) calculated that the total area of river catchments now connected to the Rhine via navigation canals, increased over 20-fold in the last two centuries (Figure 10.3). The Rhine is also highly modified for navigation, and ports in its delta (principally Rotterdam) connect it via shipping to the rest of the world. Through the canals there is a fully aquatic route to eastern European waters including the Black and Caspian Seas, to the Mediterranean and, to the north, the Baltic, North and White Seas. The most important routes for animal dispersal have been via shipping to and from the eastern seaboard of North America (the St Lawrence Seaway and onward to the Great Lakes and beyond), and via navigation canals connecting the Rhine to the Black and Caspian seas (most recently connecting the

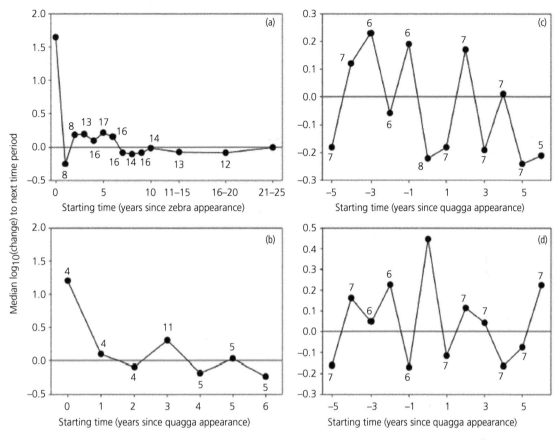

Figure 10.2 Temporal changes in a number of *Dreissena* populations (zebra and quagga mussels) in Europe and North America. Values on the *y*-axis are median $\log_{10}$ population changes between consecutive time periods (e.g. a value of 1 indicates that the population increased ten-fold during that time interval): (a) temporal changes in zebra mussels at sites without quagga mussels (NB non-linear scale on *x*-axis), (b) changes in quagga mussel populations, (c) changes zebra mussel populations before (shown as negative values on the *x*-axis) and after quagga mussels arrived, (d) changes in combined *Dreissena* populations before and after quagga mussels arrived (the number of data points for that time interval is plotted alongside each point).
Source: from Strayer et al. 2019; published under a Creative Commons CC BY licence.

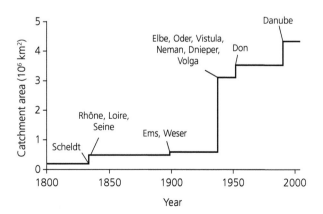

Figure 10.3 The total area of river catchments connected to the River Rhine via the construction of canals from 1800–2000 (names of the various rivers are added to the trajectory at the appropriate date).
Source: from Leuven et al. 2009; published under a Creative Commons Attribution Noncommercial Licence.

River Main—a major tributary of the Rhine—to the Danube, thus completing in 1992 a 'southern invasion corridor': Leuven et al. 2009).

The original benthos of the River Rhine was dominated by insects but was greatly depleted by severe water pollution and, most spectacularly, by the Sandoz disaster of 1986 (the release of 20 tonnes of pesticides and other chemicals from a factory near Basle, Switzerland, killing nearly all fish for 400 km downstream) (see Wantzen et al. 2022 and references therein). Efforts since then have greatly cleaned up the river but have not resulted in a return of the former community but one dominated by invasive species, predominantly crustaceans and molluscs. Alien species now make up > 90% of the total benthic biomass and density. Wantzen et al. (2022) speak of non-native species 'performing a real round-about of ever more competitive species'. The zebra mussel colonised very early (probably in the 1820s) but is now being replaced (since around 2000) by quagga mussels (recall that both are of Ponto-Caspian origin), while two species of Asian clams (*Corbicula*) have colonised

the Rhine from downstream, having arrived in ballast water (Leuven et al. 2009). A Ponto-Caspian polychaete worm (*Hypania invalida*) reaches densities of 12,000 m^{-2} in the lower Rhine and the amphipod crustacean *Chelicorophium curvispinum* is also abundant. Other Ponto-Caspian amphipods include the so-called killer shrimp (*Dikerogammarus villosus*), while Chinese mitten crabs (*Eriocheir sinensis*) invaded the Rhine from downstream and there are also several Asian and American crayfish. Overall, only about a quarter of the benthic species found in the early 20th century are still present. It is thought that almost 45% of the alien species in the Rhine are of Ponto-Caspian origin, close to 30% are from North America, and the rest from elsewhere. The number of alien species in the Rhine has increased exponentially since the 19th century (Figure 10.4) and is apparently related to the increase in cumulative surface area of catchments attached to the river.

Other European rivers have also been widely invaded by alien fish species, though none perhaps quite as extensively as the American Colorado River. Iberian rivers have diverse fish faunas with high

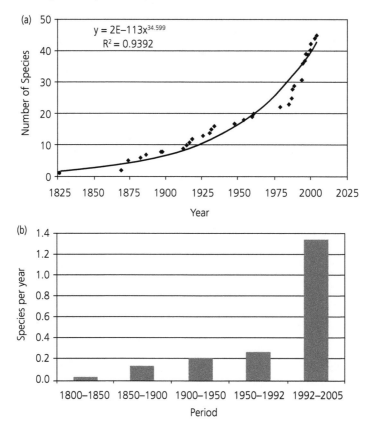

Figure 10.4 (a) The number of non-indigenous species in the freshwater sections of the River Rhine 1825–2000 (points are the data, the line is an exponential curve). (b) The mean number of non-indigenous species colonising per year from the mid-19th to the early 21st centuries.
Source: from Leuven et al. 2009; published under a Creative Commons Attribution Noncommercial Licence.

endemism, with a natural pattern of species of African origin found in the south and of solely European origin to the north. Some are also heavily invaded (Tockner et al. 2022). For instance, the Guadiana (flowing 818 km south-southwest from the dry central plateau of Spain to the Atlantic in southern Portugal) has 13 non-native species out of a total of 42 (including at least 13 endemics). In France, 23 of 59 species are non-native in the Seine, four of 17 in the Adour-Garonne, 15 of 45 in the Rhône, and 31 of 57 in the Loire. In Italy, 25 of 69 fish species are non-native in the Po, a large river in north-central Italy, which flows east from the Alps across a large flat plain to the Adriatic and drains the main agricultural areas of the country. In the Tiber, which flows from the central 'spine' of Italy through Rome, 21 of 41 fish species are non-native, but only 6 of 32 in the morphologically largely intact Tagliamento River, which drains south from the Alps to the Adriatic, are non-native. In contrast to these rivers of western Europe, the diverse rivers of eastern Europe as yet have relatively fewer exotic species (Tockner et al. 2022).

10.3.4 Invasional meltdown?

The rapid and accelerating arrival and establishment of non-indigenous species has stimulated some speculation as to its cause, beyond the simple increase in trade and river traffic described above. One early speculation was that there could be facilitation among invaders—that is, mutually positive population interactions (see Chapter 7, section 7.7, p. 250). Thus, once one was established, others would be more likely to succeed. This might occur, for example, if they came from the same original area and coexisted there. This became known as the *invasional meltdown hypothesis* (Simberloff & Von Holle 1999), a metaphor that has proved at once persuasive but controversial and a hypothesis that needs rigorous testing (Simberloff 2006). Nevertheless, Gallardo & Aldridge (2015) determined that three-quarters of the interactions among 23 Ponto-Caspian species considered most likely to enter Great Britain via mainland Europe appeared to be positive or neutral, and warned that an 'invasional meltdown' from the Ponto-Caspian region could lie ahead.

10.3.5 Ecosystem effects of invasions

Since species invasions are often accompanied by changes in biodiversity, switches in relative abundance, and/or the arrival of species with different traits, we can expect consequent changes in ecosystem processes in the recipient river. For instance, the

invader may be a significant *ecosystem engineer*. We deal with ecosystem engineering in more detail in section 10.6.4 below but, put simply, an ecosystem (or ecological) engineer is a species that alters the habitat substantially, thus changing conditions and the resources available to others. This was a concept introduced by Jones et al. (1994) that has subsequently attracted a great deal of attention. In the particular context of invasive species, for instance, the North American beaver (*Castor canadensis*) was introduced into southern Chile (the Cape Horn Biosphere Reserve) in the 1940s where it is now common (Anderson & Rosemond 2007). It is an obvious and active engineer via its dam-building and tree-felling along stream channels. It reduced the taxonomic diversity, and the variety of functional feeding groups of benthic invertebrates in beaver ponds, compared with unimpacted stretches and sites downstream of dams, On the other hand, it greatly increased the overall density, biomass and productivity of the benthos by retaining fine particulate organic matter and increasing food for deposit feeders. Thus, diversity and productivity were affected in opposite ways at the local habitat scale, though of course overall habitat heterogeneity and diversity at the scale of the whole river system could be increased. A review by Emery-Butcher et al. (2020) suggested that both negative and positive effects of invasive ecosystem engineers on diversity have occurred in fresh waters, though most were negative.

Further examples of ecosystem effects consequent upon species invasions are found among amphipod crustaceans in Europe, where there has been a succession of species from North America and again, in particular, the Ponto-Caspian region. The larger, more predatory Ponto-Caspian species of *Dikerogammarus* (*D. villosus* and *D. haemobaphes*) have widely replaced the more detritivorous/herbivorous native species such as *Gammarus pulex*. Because the latter is a much more effective shredder of leaves than *Dikerogammarus* (see e.g. Constable & Birkby 2016), its loss may influence the dynamics of coarse particulate organic matter breakdown in European streams and rivers. The wider consequences of the loss of such important shredders of leaves, and of the shredding process, are essentially unknown, though in forested streams they can be profound (e.g. Cuffney et al. 1990; see Chapter 8, section 8.5.3). Species invasions in the Rhine food webs also led to marked dietary changes in common fish species such as the eel (*Anguilla anguilla*) and perch (*Perca fluviatilis*) (Kelleher et al. 1998).

10.3.6 Invasions are affected by other stressors

The spread of alien species, and the likelihood of aliens becoming invasive, is influenced by other deleterious factors—we can call them 'stressors'—as environments change. Various stressors seem almost invariably to result in an increase in the spread and success of alien species, and include different kinds of flow regulation, such as damming and water withdrawals, increasing connectivity between previously separate catchments through canals, and climate change. We have highlighted the role of cross-catchment navigational canals, shipping and world trade in general in moving species around—as in the spectacular case of the Ponto-Caspian region, the European Rhine and the east coast of North America into the St Lawrence Seaway and the Hudson River (see section 10.3.3).

Widespread damming of large rivers creates reservoirs with lake-like conditions, replacing the natural fluvial dynamics of free-flowing rivers. This in turn facilitates incursions by widely distributed, often alien, fish species with generalist life histories and diets, and the decline of native rheophilous ('current-loving') species and large-bodied, often migratory, specialists. Some of the world's great rivers have been reduced to what is in effect a series of lakes with cascades between them. An example at the easternmost extent of Europe is the River Volga, Europe's longest river and an icon of Russian culture (it drains 32% of European

Russia), now reduced to a series of enormous reservoirs, and heavily invaded by alien and generalist limnophilous ('lake-loving') species (Mineeva et al. 2022). The manipulation of flows in regulated rivers just might offer an option to managers seeking to control the spread of invaders. For instance, Mathers et al. (2020) showed that invasive signal crayfish became established in English rivers more readily in low-flow years and suggested that the maintenance of occasional flushing flows just might reduce their deleterious effects (including reductions in taxonomic and 'functional' richness) on the benthos.

Rahel & Olden (2008) reviewed and summarised ways in which climate change is likely to affect species invasions (Figure 10.5), not only by changing the thermal regime but also by modifying stream flow with more frequent and extreme floods and droughts. A likely managerial response to drought is to build more reservoirs. Rahel & Olden (2008) perceived that local biotic communities are the result of species being able (or unable) to pass through 'environmental filters', that is, the prevailing environmental conditions (that can include both biotic and abiotic factors). Such filters will change with the climate, enabling the spread and growth of species formerly unable to persist. These latter will include newly arrived non-indigenous species and some native species, which become more competitive or able to increase their ranges or habitat occupancy. Thus, fish species tolerating warm water,

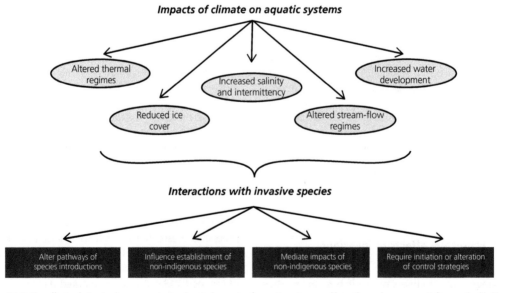

Figure 10.5 The effects of climate change on physicochemical aspects of aquatic systems and how these changes may influence the likelihood and impact of species invasions.
Source: from Rahel & Olden 2008; with permission of John Wiley & Sons.

greater salinity and lake-like conditions will thrive in impoundments at the expense of species needing cool running water of low salinity. Interactions between species invasions, climate change and other stressors are further examples of how 'suites' of factors may affect species—so-called multistressors—a topic to which we will return.

10.4 New and emerging contaminants

10.4.1 What is there?

River pollution has long been a source of the loss of biodiversity (and other deleterious changes) in running waters, while the range of contaminants has expanded greatly over the decades, a process that continues apace. There are apparently over 100 million (and rising) different chemical substances in existence, of which less than one half of one percent are regulated (Gessner & Tlili 2016; Figure 10.6). Around 400 million tonnes were produced in the year 2000, no less than 400,000 times more than in 1930! These are truly astonishing figures: even the most remote areas of the earth's surface are contaminated to some extent by long-distance atmospheric or ocean current transport of both manufactured chemicals and other toxic substances such as mercury. As the vast majority of these new chemicals are unregulated in the environment, and their effects are often essentially unknown, it seems reasonable to label them *emerging contaminants*. Many of these manufactured substances are pesticides, herbicides or pharmaceuticals designed to have biological effects even at low concentrations.

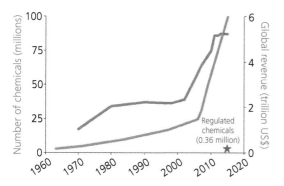

Figure 10.6 The number of chemical substances produced by the global chemical industry over the last 60 years (orange line) and the revenue accruing (blue line). The number subject to environmental regulation (red star) is shown for 2015.
Source: from Gessner & Tlili 2016, with permission of John Wiley & Sons.

They, or their degradation products, are released into the environment and enter rivers and streams via runoff and in industrial or domestic wastewater. The endocrine disrupters are a case in point that arise from a range of such chemicals and pharmaceuticals and cause well-documented fertility and growth problems, particularly with freshwater fish (e.g. Tyler & Jobling 2008; Scholz & Kluver 2009). Other emerging contaminants include pieces of plastic of widely varying sizes which are beginning to cause serious problems particularly in the developing world's rivers, as well as *engineered nanoparticles* (manufactured particles < 100 nm). These often have unique physicochemical properties, such as electric surface charges that may determine the adsorption and release of toxic chemicals. Chemical pollution is much more widespread, more complex and more insidious than ever before, even though its gross effects have been alleviated in some parts of the developed world.

The cocktail of contaminants reaching rivers in industrial areas is extremely complex and imposes real difficulties for managers and regulators. For instance, apparently about 10,000 organic compounds are released into the heavily industrialised Rhine in western Europe, mostly at low concentrations, but only around 150 of them are analysed routinely (Wantzen et al. 2022). Ideally, we would have long-term data on the concentration in rivers of any pollutant as its production begins and (often) declines after regulation, improved water treatment or simply industrial decline. However, such data are usually not available or are incomplete. Nevertheless, if places can be found where deposited riverine sediment has accumulated evenly and regularly over decades without substantial disturbance from floods or river engineering, it is sometimes possible to take sediment cores for the analysis of pollutants of interest. Not surprisingly, such situations are uncommon but there are some records. A good example comes from the French River Seine, which flows through Paris, drains a basin that is home to 16 million people and has a heavily industrial, urbanised and agricultural catchment. Cores of sediment taken from a flood plain downstream of Paris have revealed a fascinating record of 'POPs' (persistent organic pollutants: Figure 10.7). The four POP 'families' assessed each showed a sequential pattern of increase and, in three cases, decline—essentially following their pattern of emission in the catchment.

Globally, agriculture uses ever-larger quantities of pesticides, as we attempt to feed the burgeoning human population. These pesticides enter freshwater ecosystems via surface runoff, spray drift and in

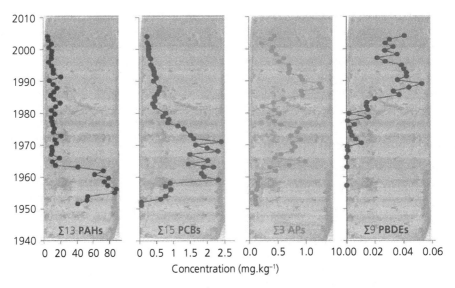

Figure 10.7 Profiles of the concentration (mg. kg^{-1} dry mass) of four families of persistent organic pollutants (POPs) in a sediment core from the Seine floodplain 100 km downstream from Paris, receiving water from 96% of the entire catchment. PAHs (polyaromatic hydrocarbons) are the product of fossil fuel combustion (here mainly coal), already high in 1950 but declining in the 1960s onwards as coal was replaced by natural gas. PCBs (polychlorinated biphenyls) are synthetic compounds used as additives in pesticides, plasticisers and paint and in insulators. Their use in France increased around 1955, about a decade after their first production. They were banned first in paints and pesticides and, finally, as insulators in electrical transformers in 2011 (they were banned altogether in the late 1980s in the USA and Japan). APs (alkyl phenyls) are synthetic surfactants and emulsifying agents, added to plastics and personal care products; large-scale use began in about 1960, peaked around 1990 and has declined thereafter due to regulation and decline in use. PBDEs (polybromodiphenyl ethers) are flame retardants, widely used since the 1970s. *Source:* from Lorgeoux et al. 2016; with permission of Elsevier.

wastewater (see, for instance, Knillmann & Liess 2019). Modelling studies have suggested that around 40% of the global land surface is at risk from pesticide runoff from agriculture, with about 18% at high risk ('risk' here is estimated from the modelled runoff potential from agricultural land; Figure 10.8). The predicted global distribution of risk is very patchy, and evidently depends upon such factors as pesticide use, proportion of land devoted to cultivation of crops, rainfall, slope and the nature of the soil profile.

Pharmaceuticals and 'personal care products' (perfumes, stimulants, analgesics, antibiotics, antihistamines, hormones etc) are also widespread and increasingly enter rivers and streams in wastewater after their use and incorrect disposal, or in insufficiently treated effluent from production plants (e.g. Rosi-Marshall et al. 2013; see Topic Box 10.1 by Emma Rosi on the ecological contaminants of emerging concern). This is a global problem, although predominantly known from North America and Europe, but may be particularly severe in areas with poor provision of wastewater treatment and ungauged, unregulated and often illegal effluent sources (e.g. in Nigeria; Ogunbanwo et al. 2020). Wilkinson et al.

(2022) assessed data on active pharmaceutical ingredients in 258 rivers worldwide (1,052 locations in 104 countries spread across all continents, although the global coverage was still very uneven). The highest concentrations were found in sub-Saharan Africa, south Asia and South America, in low- and middle-income countries with poor wastewater infrastructure. They concluded that 'pharmaceutical pollution poses a global threat to environmental and human health'.

Most of the plastic ever produced is still with us somewhere—in use, in landfill sites, or circulating in the environment. With annual global production of plastic at around 350 billion tonnes and rising, it is accumulating rapidly after its widespread use began around the middle of the 20th century. Plastic pollution in marine and freshwater ecosystems comes from a variety of land- and aquatic-based sources and includes a diverse mixture of shapes, sizes, polymers and chemistries (see e.g, Jones et al. 2020). Particles smaller than 5 mm are termed *microplastics* and are particularly prone to ingestion by aquatic animals and can pass up food chains and become concentrated in top predators. They are either produced as small particles (primary microplastics) or by fragmentation of bigger

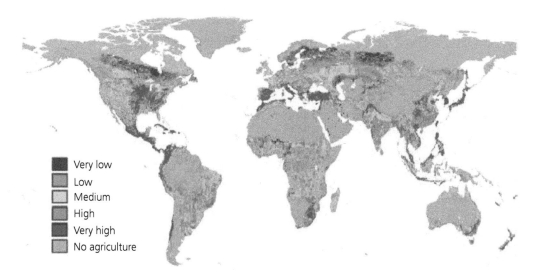

Very low
Low
Medium
High
Very high
No agriculture

Figure 10.8 A world map showing the potential for insecticide runoff to running waters based on agricultural activities, geomorphology and climate.
Source: see Knillmann & Liess 2019; figure from Ippolito et al. 2015, with permission of Elsevier.

pieces (secondary microplastics). Large rivers are the major source of plastics to the oceans, focusing inputs from their various tributaries. Schmidt et al. (2017) estimated that just 10 of the world's major rivers transport around 90% of the global load of plastic into the ocean. They all drain large catchments with a high human population density, eight of them in Asia (in declining order of plastic load: Yangtze, Indus, Yellow, Haihe, Brahmaputra plus Ganges (these two have a common delta), Pearl, Amur, Mekong) and two in Africa (Nile (5th), Niger (9th)). While these major rivers may be particularly important, high loads of plastic are found in practically all urban rivers. For example, Rowley et al. (2020) reported microplastics transported along the tidal Thames through London at an annual mean of 15 and 25 particles m^{-3} at two sites, loads similar to some of the higher values reported globally. These particles were released largely from combined sewer outfalls (carrying a mixture of treated sewage and surface drainage) and were mainly secondary microplastics resulting from the breakdown of packaging (bottles, food wrappers and bags).

10.4.2 What are the effects?

The biological and ecological impact of contaminants is still mainly assessed by 'conventional' ecotoxicological methods, although increasingly it is done by combining ecotoxicological and ecological approaches. Whereas we normally address the direct effects of

toxins on individuals and single species, a new synthesis stresses 'indirect effects'—whereby a toxin that affects a focal 'species A' directly, potentially has indirect 'ramifying' effects on its predators, competitors, symbionts and food resources (see Gessner & Tlili 2016 and citations therein). Interactions between different toxic stressors may also produce outcomes that differ from those of single stressors, and their effect can differ in turn in different environments. As well as influencing the diversity and abundance of the biota, contaminants in natural environments can affect ecosystem processes such as decomposition and primary production. It is well known that many organic contaminants, such as PCBs, can accumulate in aquatic food webs (see e.g. Walters et al. 2016) and that insects emerging from streams can then transfer these contaminants to terrestrial predators. Some river fish are now unsuitable for human consumption because their loads of PCBs are too great.

A number of studies have shown that pharmaceuticals can affect key biofilm communities and their activity in streams. For instance, Rosi-Marshall et al. (2013) used diffusion substrata, similar to those commonly used to study nutrient limitation (see Chapter 9), containing six common pharmaceuticals (alone and as mixtures) in three streams in the USA (Indiana, Maryland and New York) and found that they variously affected algal biomass, gross primary

Topic Box 10.1 Ecological effects of contaminants of emerging concern in fresh waters

Emma Rosi

Pharmaceuticals and personal care products used in our everyday lives have been detected in freshwater ecosystems around the world and are increasingly considered contaminants of emerging concern. The diversity and use of these compounds have increased exponentially over time (Box Figure 10.1; Bernhardt et al. 2017). During the last 20 years, analytical approaches to measure these compounds have been improved, research now having demonstrated that many of them are frequently present in ground waters, in surface waters ranging from headwater streams to rivers to lakes, and in sediments and animal tissues. Concentrations are particularly high in waters receiving wastewater effluents, including most larger river systems as well as coastal waters in populated areas.

Many of these compounds have been detected in relatively pristine environments, including tourist destinations

such as US national parks and even the Antarctic. Because these pharmaceuticals and personal care products are not fully metabolised or degraded during use, wherever human beings live or visit we leave a trace of the compounds that we use in our everyday lives. The number of pharmaceuticals and personal care products released to the environment is difficult to ascertain because of the wide diversity of compounds in use, and the volumes used are proprietary information. In the United States alone, however, there are over 1,400 compounds approved as prescription medications and there are far more in over-the-counter medications, detergents, fragrances, insect repellents and other personal care products. These compounds represent a threat to the health of freshwater ecosystems that is especially challenging because of their chemical diversity and unknown interactions that may occur once they are released into the environment.

Ascertaining the ecological effects of contaminants of emerging concern continues to be a research challenge

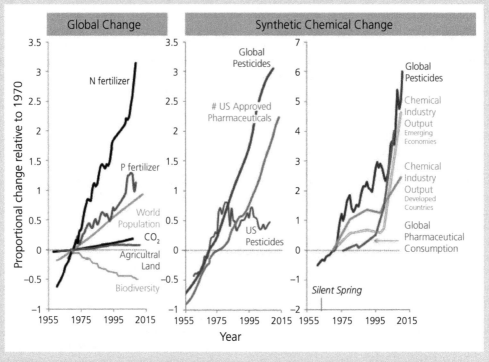

Box Figure 10.1 (a) Trajectories for drivers of global environmental change as defined by the Millennium Ecosystem Assessment (2005); (b) increases in the diversity of US pharmaceuticals and the application of pesticides within the USA and globally; (c) trends for the global trade value of synthetic chemicals and for the pesticide and pharmaceutical chemical sectors individually, used as a proxy for the mass of chemicals produced in the absence of national or international estimates of the amounts of pharmaceuticals produced. All trends are shown relative to values reported in 1970, with the exception of pharmaceutical consumption, where the earliest data reported are from 1975. Expenditures in (c) were adjusted for inflation by the Consumer Price Index reported by the US Department of Labor Bureau of Labor Statistics using 1982–1984 as a base before relating prices to 1970 and 1975 values.
Source: from Bernhardt et al. 2017.

Topic Box 10.1 *Continued*

(Rosi-Marshall & Royer 2012). Many of these compounds are used by humans precisely because of their ability to affect biological systems at low concentrations. For example, antibiotics target bacteria in our bodies, and antidepressants alter brain chemistry in humans very effectively. In many cases, however, the biological mechanisms that pharmaceuticals target in humans exist in a wide variety of other species. An obvious example is the wide variety of bacteria that exist in freshwater ecosystems; research demonstrates that the structure and function of bacterial communities are sensitive to antibiotic exposure (Rosi-Marshall et al. 2013, Rosi-Marshall et al. 2017).

Perhaps less obvious is that drugs used to target human systems may affect even evolutionarily distant species, albeit with different impacts. Research has just begun to discover some of the ways in which these contaminants may disrupt aquatic animal physiology and life cycles. For instance, drugs designed to influence the serotonin system in humans, such as selective serotonin reuptake inhibitors (SSRIs) prescribed to treat depression, have been shown to affect the emergence of aquatic insects (Richmond et al. 2019; Lee et al. 2016). Further examples are histamines, which in invertebrates function as neurotransmitters for sensing light and detecting prey; exposure to antihistamines can significantly affect their life history and behaviour (Hoppe et al. 2012). Insect repellents targeting mosquitoes can result in increased deformities and mortality in salamander larvae (Almeida et al. 2018). Illicit drugs such as cocaine can disrupt the reproduction of freshwater mussels (Parolini et al. 2015). Anti-anxiety medication and antidepressants can affect the behaviour of freshwater animals (Brodin et al. 2013; Reisinger et al. 2021). The list of compounds shown to affect freshwater organisms continues to grow as more research specifically investigates their biological and ecological effects.

Some pharmaceuticals and personal care products also bioaccumulate in organisms and are transferred in food webs. Downstream of wastewater effluents, aquatic insects have high concentrations of many pharmaceuticals in their tissues—a study that investigated the concentrations of 97 pharmaceuticals found over 60 compounds in insect tissues in such a stream (Richmond et al. 2018). The number investigated in that study, however, is only a small fraction of the compounds used today. In addition to their being present in aquatic insect tissues, Richmond et al. (2018) also found high concentrations of pharmaceuticals in terrestrial spiders in trees overhanging the stream, presumably because they rely on aquatic insects as their main food resource. This kind of transfer to terrestrial animals may also occur with birds, bats, frogs and lizards. More research is needed to understand the influence of these novel contaminants on both aquatic and riparian food webs.

The ongoing proliferation of pharmaceuticals and personal care products and their ever-increasing use worldwide—accompanied by their inevitable release into surface waters—presents a formidable challenge for scientists. We need to understand how trace levels of these compounds may disrupt ecological processes through their potent biological activity, and how the complex and variable mixtures found in the environment may exert synergistic or antagonistic effects that may vary among species (Richmond et al. 2017). The transport and ultimate fate of these compounds are not well understood, nor is their propensity to enter food webs, extending to consumers in riparian zones. As with many of our other activities in the Anthropocene, we are effectively engaged in a global experiment with no control and no end in sight!

Dr Emma Rosi is a Senior Scientist at the Cary Institute of Ecosystem Studies, Millbrook, New York, USA.

production and community respiration (Figure 10.9). An antihistamine (diphenhydramine) reduced the relative abundance of the bacterium *Flavobacterium* and increased that of *Pseudomonas*. More recently, Robson et al. (2020) used slightly different methods to expose natural and experimental substrata (some with already established biofilms) to three pharmaceuticals at 'realistic' concentrations (i.e. those occurring in real rivers and streams) in recirculating stream

microcosms. These were ciprofloxacine (an antibiotic), diphenhydramine (an antihistamine) and fluoxetine (an antidepressant). Again, a mix of effects were found, with reductions in primary production and community respiration in developing biofilms (although not in established ones) and a reduction in denitrification in the shade but not in ambient light. These and other studies (e.g. Lee et al. 2016; Sabater et al. 2016; Richmond et al. 2016) suggest extensive and subtle effects of

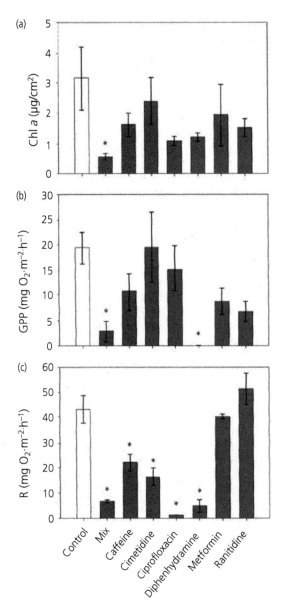

Figure 10.9 The effect of six pharmaceuticals on (a) algal biofilm biomass (as chlorophyll *a*), (b) gross primary production (GPP) and (c) community respiration (R) in autumn experiments in a stream in New York (data are means ± SE). In several cases, the three measures were significantly depressed compared with controls. In humans, caffeine is a stimulant; cimetidine, ranitidine and diphenhydramine are anti-histamines; ciprofloxacin is an anti-biotic; and metformin is an anti-diabetic. These six are all commonly detected in surface waters across the USA.

Source: from Rosi-Marshall et al. 2013, with permission of John Wiley & Sons.

the pharmaceuticals to which streams are increasingly exposed, particularly on biofilms, with likely concomitant impacts on food webs and the many ecosystem processes associated with biofilms.

A study by Richmond et al. (2018) is particularly informative. They surveyed aquatic invertebrates and riparian spiders for traces of 98 different pharmaceutical substances in six streams receiving varying amounts of wastewater near Melbourne (Australia). Sixty-nine of these chemicals were detected, the most abundant being a drug used to treat Alzheimer's disease (memantine), an analgesic (codeine), two antifungals (flucanozol and clotrimazole) and an antidepressant (mianserin). Filter-feeding caddis larvae (Hydropsychidae) had some of the highest total pharmaceutical concentrations of the aquatic invertebrates. Riparian spiders had similar, and sometimes higher, concentrations than the emerging aquatic insects on which they feed, confirming trophic transfer from water to land as well as evidence of biomagnification through the food chain. Two vertebrate predators of aquatic insects, the iconic Australian monotreme mammal the platypus (*Ornithorhynchus anatinus*) and the non-native brown trout (*Salmo trutta*), would thus be expected to consume pharmaceutical compounds via their prey. It was calculated, based on food consumption, that a platypus feeding on aquatic invertebrates from the most contaminated stream would take in about half the daily dose of antidepressants typically prescribed to humans, while brown trout would consume about one quarter of that dose. The ecological effects of such contamination by pharmaceuticals remain unknown.

Despite widespread environmental contamination of rivers by plastics, and it being well known that microplastics are ingested by invertebrates, their biological effects are poorly understood. D'Souza et al. (2020, and references therein) demonstrated the transfer of microplastic particles from invertebrate prey to the riparian insectivorous bird the Eurasian dipper (*Cinclus cinclus*), and to its offspring (as adults carried aquatic prey to nestlings). They found plastic particles in 50% of regurgitates (in common with many predatory birds, dippers produce pellets of hard, indigestible prey parts) and in 45% of the faecal pellets produced by adults and nestlings collected at 15 sites on five rivers in the former South Wales coalfield (UK). Most particles were fibres of polyester, polypropylene, polyvinyl chloride and vinyl chloride copolymers, with about 200 particles being ingested and egested per day by dippers (Figure 10.10). Again,

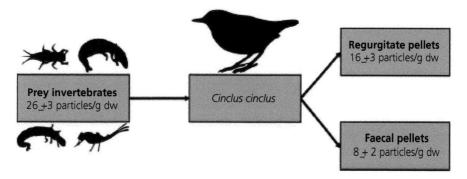

Figure 10.10 The flux of plastic particles (measured as numbers of particles per g dry weight) through various biological compartments relating to dippers in rivers in South Wales. Inputs and outputs were similar and it was estimated that an average of about 200 particles per day were ingested and egested by the birds.
Source: from D'Souza et al. 2020; published under a Creative Commons CCBY licence.

the potential ecotoxicological effects of this ingested plastic itself are unknown. Contaminants are adsorbed onto charged plastic particles and can be released in the guts of animals that have ingested them. Thus, Siri et al. (2021) found that progesterone (an endocrine disruptor present in river water via sewage effluent) is released from plastics in intestinal fluids and could exacerbate the feminisation of fish, for instance.

Studies of the potentially cascading, community-wide effects of contaminants on whole-river ecosystems remain rare, involving as they do an inevitably complex 'knot' of direct and indirect effects of a cocktail of substances. One opportunistic study attempted to characterise, from 'gene to ecosystem', the impacts of a single 'accidental' spill of a pesticide in a British river (Thompson et al. 2015). In 2013 around 15 km of the River Kennett in southern England (a biodiverse and economically important tributary of the Thames) was polluted by the organophosphate insecticide and miticide Chlorpyrifos following a relatively short-lived (3–4 days) 'pulse' of the chemical from a sewage treatment works. The incident, which probably had its origin in an illegal 'down-the-drain' discharge, was first detected by citizen scientists, who reported to local authorities a large-scale kill of macroinvertebrates. Although the peak concentration of insecticide in the river was inevitably missed as the discharge 'slug' travelled downstream from the treatment works outfall, acutely toxic concentrations remained in the vicinity several days later. A sobering range of effects occurred at various 'levels' in the ecosystem, including direct toxic effects on biota and indirect consequences mediated through the food web. Using molecular techniques, difference in the genetic makeup of the microbial community were detected

which indicated a switch in key processes, such that genes conferring the ability to metabolise the pesticide became prevalent, and ammonium-oxidising bacteria increased in response to increased concentrations of ammonia arising from the decomposition of dead metazoans. The biovolume of diatom cells at the impacted site was up to 10 times greater than that at a control site, probably because of a reduction in grazing at the former. Additionally, the abundance of the leaf-shredding amphipod *Gammarus* was greatly reduced at the impacted site and although litter decomposition due to animal feeding declined, it was compensated for by greater microbial decomposition. Not all invertebrates were affected by the chemical spill, however, and the biomass of some, including oligochaete worms and the progeny of those on the wing as adults during the spill, increased—largely compensating for the loss of those that were killed.

10.5 Ecological assessment of running waters: looking back, looking forward

10.5.1 Biological monitoring—a (very) brief introduction

The detection of pollution based on the assemblages of animals and/or plants found at different places and at different times has a very long history going back many decades to the development of the 'saprobic system', mainly in continental Europe (see Sládeček 1965), and to work summarised in the classic *The Biology of Polluted Waters* by H.B.N. Hynes (1960) (see Friberg et al. 2011 for a review). Biological monitoring in general has achieved great prominence, and probably employs more professional freshwater ecologists/biologists than any other single career path related to this field. Biomonitoring is underpinned by pioneering taxonomic research to enable the consistent identification of species. The ecological status of rivers and streams is then based on the known preferences and tolerances of the species that are found in relation to major 'stressors', including organic pollution and over-enrichment (and indirectly to reduced oxygen concentration). This latter information on tolerances and preferences of individual taxa (often at the species level) is again the product of an enormous amount of basic pioneering ecological and ecophysiological research. Bioassessment using species lists (most typically of benthic macroinvertebrates or benthic algae), and various biotic indices of pollution based on those lists, is an important product of applied freshwater biology.

The most obvious form of river pollution is that due to 'oxygen-demanding wastes'. Originally and primarily associated with sewage, these wastes result in the disappearance of organisms requiring clean, oxygen-rich water and enable the proliferation of other taxa that tolerate a low oxygen supply and/or require plentiful plant nutrients. The original impetus and focus for biomonitoring was rightly on public health, and the relationship between sewage pollution, pathogens and human disease (for a long-forgotten example, see 'The case of the *Princess Alice*' below). Great efforts have been made in many countries to clean up and then 'monitor' rivers. As part of this effort, a variety of pollution indices based on the relationships between organic pollution and invertebrates, algae and heterotrophic microorganisms have been devised, described and used (see e.g. Rosenberg & Resh 1993; Bonada et al. 2006; Birk et al. 2012).

Water quality has been vastly improved in many developed countries, and this continues where legislators and water management authorities are vigilant. For instance, Goertzen et al. (2022) found an ongoing improving trend over a recent 12-year monitoring record of 56 urban stream sites in north central Germany. Sewage and other organic wastes remain very important pollutants in many places, however. Indeed, much of the UK at least has seen a fairly recent partial reversal in water quality in many rivers (sadly including the Thames) due to a failure to invest in water treatment (despite a growing human population) and cuts in professional monitoring (e.g. see House of Commons Environmental Audit Committee 2022). Further, and as discussed above, a much wider and growing array of new contaminants also now threatens the ecology of rivers. Added to this are physical modifications to river channels and flow. Further

developments in biomonitoring involve the association of various species traits with stressors, the identification of particular stressors other than organic pollution, and detection of the effects of several stressors at the same time. Many of these approaches still demand the reliable identification of species and here new methods offer further progress.

10.5.2 New methods for assessing diversity and their application

10.5.2.1 Some basics

Why are we interested in diversity and species composition? Firstly, they are of fundamental scientific interest in understanding the dynamics of population and communities. In terms of application, we have discussed declines in river biodiversity. Are rare species being lost from communities, or are they still present, and can we try to detect and conserve them? Further, we might want to detect the early spread of potentially invasive species beyond their native range. Finally, in biomonitoring we use the occurrence of species to tell us things about the state of the environment and its 'health' or 'ecological status'.

The assessment of species diversity in lotic habitats has relied on, and in large part still does, collecting and sorting samples from the environment using appropriate methods for whatever fraction of the biota is of interest. For all but some microbial groups, identification then revolves around morphological features and the use of identification guides to produce a list of species or, in many cases, coarser taxa such as species groups, genera, families etc. There are difficulties, however. In some instances, it is important to be able to detect the presence of rare species or others that are cryptic or difficult to 'catch' or where sampling disturbance is potentially damaging to the habitat or the species in question. Put simply, using conventional methods we often cannot collect or process sufficiently large samples to have a good chance of finding very rare and newly arrived taxa. This is where detecting species using molecular methods can potentially transform the field (e.g. Deiner et al. 2017).

All organisms shed their DNA into the environment, via their faeces, mucus, shed skin cells, gametes and exocellular exudates. *Environmental DNA* (eDNA) is that occurring in an environmental sample (including water, sediment etc collected along with the sample) without isolating any target organisms. *Community DNA* is DNA extracted from the mixture of organisms isolated from the environmental sample

The case of the *Princess Alice*

A single terrible incident captured public concern in the United Kingdom, among the first countries to go through an industrial revolution and to witness the subsequent flood of humans from the countryside to large cities—despite the lack of adequate, or indeed any, hygienic sewage-disposal facilities. The year 1858 saw the 'great stink' in London, where famously Parliament was suspended because of the foul smell from the filthy River Thames. There followed an engineering solution whereby wastes were collected in main sewers and delivered to the Thames well downstream from Parliament, then being pumped into the river on the ebb tide. There it often formed a slick of London's sewage which took some time to make its way downriver, periodically reversing upriver on the flood tide. In 1878 a heavily over-loaded pleasure paddle steamer, the *Princess Alice*, was making its way back upriver after an evening excursion down the Thames estuary. Off Woolwich reach it was hit by a much larger and heavier iron collier ship, the *Bywell Castle*. The steamer was cut almost in half and sank within four minutes, and most of the people on board were thrown into the river. Dead bodies were deposited over several miles of the Thames foreshore over the next few days. Many of them were reported to be blackened and had not drowned but were asphyxiated by hydrogen sulphide, having been pitched into the slick of sewage in the river at the time, while others were poisoned having swallowed the foetid water. It is not known exactly how many people died but it is thought the number could have exceeded 700 people—the greatest death toll from a single maritime incident in peacetime in British history. The ensuing public outcry pre-cipitated steps to clean up the river and reduce the discharge of raw sewage. Many decades later, these measures led to a vast improvement of water quality in the Thames and to its substantial biological recovery. Similar situations of terrible pollution pertained in many industrial countries, setting in train a variety of solutions to enable the monitor-ing of water quality and improve public and environmental health.

'The silent highwayman'—Death rows on the River Thames during 'the great stink' (a cartoon appearing in *Punch* magazine, London, 10 July 1858).

itself. Particular organisms can then be detected from diagnostic portions of their DNA (such as frag-ments of mitochondrial cytochrome oxidase, COI) found in the environment, using 'barcodes' which are being developed rapidly for large numbers of species (akin to species-specific fingerprints). Modern high-throughput sequencers using polymerase chain reac-tion (PCR) now allow for the simultaneous and rapid analysis of millions of sequences (looking for these 'fin-gerprints') and can potentially identify species present in whole communities together—a technique known as *metabarcoding* (see Deiner et al. 2017 and references therein). Although the taxonomic identification of all species using such methods is not possible at present, 'molecular operational taxonomic units' (identified by their genetic dissimilarity from each other) can be used instead, although of lesser value in relation to bioassessment. Of course, preliminary work has to be undertaken to develop reference metabarcode libraries for all or the majority of species (previously identified taxonomically) that are potentially likely to occur in a particular system or region and against which the eDNA samples can be compared.

10.5.2.2 Applications in conservation and species invasions

The analysis of eDNA has been used successfully in freshwater and particularly lentic habitats to detect the presence of specific fish and crayfish species, snails, turtles, snakes, mosquitoes and some water plants (e.g. Kuehne et al. 2020). Difficulties remain, however, particularly in rivers, where studies have had varied results (Jane et al. 2015). For example, detection success of the loach *Misgurnus fossilis* was 100% in ponds but 54% over a 225 km river system known to contain the species (Thomsen et al. 2012). Despite the potential of these techniques, some key questions remain around their usefulness for running waters. For example, how far is eDNA carried in the current and how long does it remain detectable? How does the sampling site relate to the actual (upstream) location of the species itself? Can we yet establish a means of determining relative abundance using eDNA? Are all life stages equally detectable? How does seasonal or flow variation influence detection of species eDNA and are there particular sets of conditions that inhibit the biochemical processes associated with the technique? These and other hurdles are being actively researched, some with greater likelihood of success than others. Of course, there are questions using conventional detection methods—thus, not all life stages are equally well known or distinguishable. Conditions at the time samples are taken can profoundly affect the conventional assessment of diversity and relative abundance. Overall, while arming ourselves with these molecular techniques, this should not be at the cost of the loss of skills in natural history and conventional taxonomy.

As an example of tracing a specific invasion, the extent of the upstream colonisation of the Columbia River basin in the north-west USA (the spawning grounds of endangered native salmonids) by the non-native smallmouth bass (*Micropterus dolomieu*) was investigated using eDNA by Rubenson & Olden (2020). The molecular data were used to locate the range boundary extension of bass, relating this to a 'species distribution model' for the species in relation to various environmental factors (temperature, hydrology and geomorphology). The modelled overlap between the bass and breeding habitat in the various salmonids varied greatly among salmonid species (from 3 to 62%) while the range of the bass was predicted to increase by two-thirds (from 18,000 to 30,000 river km) under moderate climate change.

In a further example, the (originally) Asian bighead carp, *Hypophthalmichthys nobilis*, and the silver carp, *H. molitrix*, were introduced to North America in the 1970s and have become established in the Mississippi, Missouri and Ohio River basins. They also threaten to invade the Great Lakes, mainly via a navigation canal (the Chicago Sanitary and Ship Canal) linking a tributary of the Ohio River and the Mississippi basin to Lake Michigan. These filter-feeding fish can reach high densities, can be hazardous to humans (they grow to over 40 kg and can jump up to 3 m in the air!), and are ecologically competitive with native species. Monitoring the 'invasion front' of bighead carp along rivers can be done using DNA techniques (Fritts et al. 2019) for screening bulk net samples of ichthyoplankton containing fish eggs and larvae. This has advantages over more laborious conventional detection methods (visual sorting and identification).

In the context of the detection of very rare and threatened species, Lor et al. (2020) used eDNA to test for the presence of the federally endangered spectacled mussel (*Margaritifera monodonta*) in two rivers in the Midwest of the USA, two of its remaining strongholds, following a 55% reduction in its range. Conventional surveys in large rivers are challenging as the mussels occur beneath large boulders in fast-flowing water and are difficult for divers to access. Lor et al. (2020) detected mussel eDNA in a greater fraction of water samples on the smaller St Croix River (Wisconsin) than in the much larger upper Mississippi, probably due to dilution related to differences in river discharge, which was 14 times greater in the Mississippi during sampling. They also obtained greater eDNA detection rates at the time when mussels were shedding their larvae into the water column than in post-reproductive periods, and the rate was also higher in water samples taken near the riverbed than at the surface. eDNA detected in water samples taken upstream of any known mussel beds on the St Croix River (selected as a 'negative control') led to the notable discovery of mussels in the upstream stretch, thereby demonstrating further potential benefits of the eDNA method focused on individual species.

In a further example, Atkinson et al. (2019) developed eDNA methods that successfully detected the presence of the widely endangered, and strictly protected, white-clawed crayfish (*Austropotamobius pallipes*) in Irish streams. This is one of only five native crayfish in Europe and has been lost from many sites because of its extreme susceptibility to the non-native 'crayfish plague' (the fungus *Aphanomyces astaci*), among other factors. In this case, conventional surveys and searches are also effective, but time-consuming and relatively expensive in terms of manpower.

Studies on another endangered mussel species in the USA, *Lasmigona decorata*, also demonstrated that eDNA assays could detect the species where it was known to occur but did not identify any new locations. There was also clear evidence that pH was a strong predictor of whether PCR inhibition (and hence ability to detect the 'species signal') occurred or not in samples (Schmidt et al. 2021). Highly controlled tests to detect caged brook trout in fishless streams also highlighted the potential problem of inhibition, and interactions between distance and flow rate may be confounding factors in attempts to infer species abundance based on eDNA (Jane et al. 2015). Such applications of eDNA in focused conservation are now becoming more widespread, although further developments in their precision and application are necessary.

Analysis of the diversity of some relatively poorly known groups, and others for which conventional methods have proved challenging, could be facilitated by analysis of DNA sequences. For instance, the ubiquitous and diverse freshwater meiofauna (very small metazoans—see Chapter 3, section 3.4.1, p. 69) remains largely a specialist business, because of the difficulties in taxonomic identification, despite their evident ecological importance (e.g. Tod & Schmid-Araya 2009; Majdi et al. 2017). The adoption of metabarcoding of the meiofauna is now on the horizon, though there remain many challenges, not least in expanding their coverage in DNA libraries (Schenk & Fontaneto 2020; Weigand & Macher 2018). This diverse group could become much more important in environmental assessment if these practical difficulties can be overcome.

10.5.2.3 Applications in biomonitoring

The greatest potential contribution of molecular assessments of the presence of freshwater organisms is in biological monitoring. Indeed, Friberg (2011) predicted that we are on a threshold of widespread adoption of DNA-based techniques. Where this will lead and how far it will go is still somewhat unclear because of some significant remaining questions (see above). However, there can be little doubt that DNA-based techniques, including further promised and necessary technical developments, will potentially enrich our ability to assess the diversity of a much wider range of organisms, and to relate species composition (where suitable species DNA libraries have been developed) to a wide range of environmental perturbations. Leese et al. (2018) presented a useful overview of different approaches used in biomonitoring: conventional sampling followed by species identification, assessment and interpretation; metabarcoding (in which DNA in the sample is matched to a library of known species and produces a list of taxa, as in a conventional assessment); metagenomics (where a list of 'operational taxonomic units'—OTUs—is obtained). While less rich in information than a list of known species (or genera), their relative abundance and their traits, it is still possible to relate OTUs to environmental conditions.

Hering et al. (2018) assessed the utility of DNA-based identification for the assessment of ecological status currently used in the European Water Framework Directive. Here, sites to be assessed are compared to 'reference conditions' (as far as possible, pristine), using metrics derived from the presence and abundance of different taxa. They considered eDNA techniques to be best suited to the assessment of fish assemblages, as they would eventually replace potentially (and actually) damaging methods such as gillnetting, trawling and electrofishing. Currently, problems arise for benthic invertebrates and algae, for which DNA libraries are incomplete. For these taxa, the indices used successfully at present rely on abundance estimates, which are more challenging for molecular methods than simple presence/absence of taxa, particularly in flowing water. Larger plants are currently usually surveyed in the field to estimate relative cover, rather than samples being taken for further analysis in the laboratory. This makes them less suitable for assessment based on eDNA. Overall, Hering et al. (2018; see also Pawlowski et al. 2018) considered that there is considerable potential for eDNA techniques and that many of the presently perceived problems can be overcome. Nevertheless, they recommended using conventional and molecular techniques side by side in the meantime—which is clearly a sensible and necessary way forward.

Other researchers have assessed the specific advantages of eDNA for particular problems. For instance, Smucker et al. (2020) identified eDNA-derived OTUs of diatoms in streams across a nutrient gradient at 25 stream sites in south-west Ohio (USA). Changes in diatom assemblages (judged by OTUs) began at phosphorus (P) concentrations as low as 20 μg TP L^{-1}, with further sharp changes between 75 and 150 μg TP L^{-1} and with only 'high P' diatoms above 150 μg TP L^{-1}. Diatom OTUs characteristic of low nitrogen (N) declined between 280 and 525 μg TN L^{-1}, whereas 'high N' OTUs dominated above 525 μg TN L^{-1}. They claimed that the use of eDNA methods had advantages over conventional assessments in terms of speed and cost-effectiveness, although the actual species present remain unknown—thus losing potentially valuable ecological information. Diatoms have of course been used very effectively for years in the assessment of stream eutrophication with conventional methods, such as the Trophic Diatom Index adopted in Europe (Kelly & Whitton 1995).

As mentioned above, traditional methods are not without their own problems. Other taxonomic groups have proved more problematic than diatoms for use in biomonitoring—particularly when identification based on morphology is presently impossible or demanding. For instance, while in many regions there are keys for identifying larval Chironomidae, and the exuviae (cast 'skins') of their emerging adults, in various parts of the world they are not routinely used in bioassessment because the group is still perceived as taxonomically 'difficult' (e.g. Anderson et al. 2013). Unfortunately, the chironomids are the most diverse group of insects found in streams and rivers, where they can frequently make up 50% or more of all 'macroinvertebrate' species. Their smaller instars are often considered to be part the meiofauna and the various species fall into all the 'functional feeding groups'. However, identification usually stops at the level of subfamily or even family, meaning that much potentially valuable ecological information, in terms of environmental assessment, is lost.

Nevertheless, there is some potential for DNA metabarcoding identification—if only to OTUs (operational taxonomic units)—of larval chironomids, as shown by Beermann et al. (2018). They manipulated three common stressors (increased salinity, fine sediment and reduced flow) in replicated stream mesocosms that had been 'seeded' with a supply of benthic invertebrates from a neighbouring stream in Germany and were also open to colonisation via drift in stream water pumped into the mesocosms. Each mesocosm provided two substratum types—a covering of leaf litter (alder leaves) and a mix of representative stream sediments (fines < 2 mm, gravel 2–30 mm and stones > 30 mm). At the family level, chironomids responded positively to added sediment and reduced flow on the mixed stream sediments and on the leaf litter, negatively to reduced velocity; salinity had no overall effect. Using barcoding molecular methods, over 100 chironomid OTUs were identified—comparable to the richness found using conventional techniques in specialised studies by taxonomic experts (Cranston 1995). Among the 35 most commonly observed OTUs, 15 different patterns of response to the three stressors were found, including little response to any of them, a specific response by a single OTU to one stressor only, additive effects (see section 10.5.4) of more than one stressor, and more complex multistressor effects. This study and others of its kind show considerable potential for the use of DNA-based identification in environmental assessment of a variety of stressors on highly diverse groups. This can then be enhanced if it can subsequently be linked to taxonomic identification, which then allows consideration of the ecological traits, adaptive history and general biology of the species in efforts to understand the impacts of stressors.

10.5.3 Detecting specific stressors

Environmental managers require methods of assessing that 'holy grail' of monitoring—*ecological status* (or the similar alternatives *environmental quality, ecosystem health* or *ecosystem integrity*). These are somewhat nebulous concepts without precise scientific definition, yet are metaphors that are attractive to non-specialists and politicians and elicit support and resources for the production of appropriate measurement 'tools'. Most early biological monitoring was based on knowledge of the sensitivity of river organisms to oxygen supply, which is profoundly affected by inputs of oxygen-demanding wastes, and over-enrichment with plant nutrients. Essentially, most indices of ecological status rely on comparisons of assemblages at monitored sites with those at 'reference sites'—most simply thought of as places where anthropogenic stress is absent or at least negligible. Feio et al. (2022) provide a comprehensive overview of the biomonitoring approaches and the general status of rivers globally.

Dominant as organic and nutrient enrichment is in environments affected by humans and their activities, there are evidently other important stressors which may have their own independent effects. Thus, efforts have been made to derive separate biological indices that are sensitive to these other stressors, including factors such as excessive fine sediment (operationally defined as deposited particles < 2 mm), reduced flows (due to water withdrawals, for instance), acidification, agricultural and industrial chemicals, mine drainage and others. Each index is generally known by a set of initial letters of the words describing them—producing a 'blizzard' of such acronyms!

Applied ecologists and managers have produced indices appropriate to their own jurisdiction, often the nation state, but others are local modifications of indices initiated elsewhere. Some consortia of nations have sought to cross-calibrate and compare numerous methods among themselves, at best learning from each other and promoting similar standards (see e.g. Poikane et al. 2014; Charles et al. 2021). An early example is the LIFE score (Lotic-invertebrate Index for Flow Evaluation) that links flow variables with the invertebrates present in British rivers. Invertebrates are grouped into 'flow groups' based on their association with differing water velocity (six classes: 'rapid', > 1m s^{-1}; 'moderate-fast', 0.2–1.0 m s^{-1}; 'slow-sluggish', < 0.2 m s^{-1}; 'slow-standing'; 'standing only'; 'drying-drought impacted'), the groups being based on ecological knowledge acquired over many years of natural historical research (Macan 1963; Hynes 1970b; see Extence et al. 1999). The numbers of invertebrates (in different logarithmic categories of abundance) in these flow groups were then assigned 'scores' and a LIFE index calculated as:

$$\frac{\Sigma\, fs}{n}.$$

Here, $\Sigma\, fs$ is the sum of the scores of individual taxa for the whole sample, and n is the number of taxa counted. Higher flows should result in high values of LIFE score. It was found that LIFE scores in rivers draining permeable geology (chalk/limestone) were mainly correlated with summer flows (they were depressed at low flow), whereas scores in rivers draining impermeable geology were linked to short-term fluctuations in flow. Thus, we can see that the LIFE score can be useful for assessing flow conditions in rivers in various natural background settings (i.e. catchment characteristics)—in other words it is necessary to judge scores against what might be expected in an unimpacted river in that setting. This latter requirement is far from simple but illustrates the process by which indices need to be developed. An added complication is that running waters are dynamic and assessment indices can vary in the face of normal environmental variation, as well as to the anthropogenic stressor. It is necessary

to establish what are in effect 'confidence limits' for the index at reference sites. For example, this approach was used to assess the impact of forestry on streams by Johnson et al. (2005).

There are many other examples of tools now used in management to indicate other stressors. The excessive accumulation of fine sediments is recognised as an ecological stress in rivers and can be due to increased erosion from disturbed and agricultural catchments, resulting from deforestation, from eroded banks and exacerbated by low flows insufficient to transport the material. Extence et al. (2013) proposed an index, this time of sediment stress, the PSI (Proportion of Sediment-sensitive Invertebrates) to evaluate its effect. Macroinvertebrates were assigned to one of four 'fine sediment sensitivity ratings' and the abundance of taxa in these categories in standard samples used to calculate a site score (in a similar way to the LIFE score above). Here, an expected score for an unimpacted reference site was calculated from a database of reference sites using the well-known system RIVPACS (River InVertebrate Prediction And Classification System) that was developed, originally at the UK's Freshwater Biological Association, for assessment of organic pollution (see Extence et al. 2013 and references therein). Observed (O) and expected (E) site scores are used to calculate an Environmental Quality Index (EQI) by dividing O by E. This method has successfully distinguished sites and times differentially affected by fine sediments. Similar systems have been developed elsewhere (see e.g. Turley et al. 2016 and references therein).

Assessment tools have also been developed, among others, for pesticides (Liess & Von der Ohe 2005; Knillmann et al. 2018), stream acidity (Murphy et al. 2013) and mine drainage (Gray & Harding 2012), and to detect the ecological effects of drought development (Chadd et al. 2017). Use of such tools, while often producing a relatively straightforward scale of numbers so beloved by decision-makers, does require a nuanced checking and interpretation of the underlying data. For instance, SPEAR$_{pesticides}$ is a tool to indicate chronic (i.e. prolonged, persistent) exposure to pesticides, and is based on a categorisation of species 'at risk' and 'not at risk' (SpEcies At Risk, SPEAR; SpEcies not At Risk, SPEnotAR), based mainly on their supposed vulnerability to toxins. It responds effectively to a consistent replacement of vulnerable by less-vulnerable species. While very effective in routine monitoring of *chronic* pollution, SPEAR proved relatively ineffective as an indicator of recovery in the case of an *acute* (episodic, one-off) and catastrophic insecticide (cypermethrin) spill into a German river, in which nearly all taxa were eradicated, and where recovery depended on factors such as the ease of recolonisation rather than the toxicity itself (Reiber et al. 2021).

Hitherto, we have considered 'tools' relying, at their simplest, on lists of species (i.e. on the taxonomic composition of communities or the genetic OTU equivalent). A much-heralded alternative is found in the replacement of such taxonomic lists with alternative lists representing the various biological traits of communities (Dolédec et al. 1999; see Topic Box 6.1 by Sylvain Dolédec and Nuria Bonada). We introduced species traits in Chapter 4 and this approach has a clear basis in community ecology and promises (a) to yield tools that are applicable in a wider variety of biogeographical areas—since the trait composition of communities under similar conditions is theoretically more stable than their taxonomic composition, and (b) to relate more clearly to ecosystem processes rather than simply to community structure. This latter concept is because the biological traits of species, such as feeding mode and their resulting diet, are intended to capture their activities—that is, they can indicate what species 'do' in ecosystems rather than just identify who they are. A good example of the value of the trait approach to bioassessment has been provided by Murphy et al. (2017), who found that species with protected eggs (i.e. ovoviparous species in which eggs are retained in the body until their development is advanced) were characteristic of UK river sites affected by fine sediment, thus linking a biological trait with the stressor of sediment deposition. An obvious hurdle to the utility of this approach, however, is that databases of biological traits are extremely demanding of basic biological information, and more or less absent or incomplete in many geographical areas (Dolédec et al. 1999; Bonada et al. 2006; Friberg et al. 2011). Nevertheless, knowledge of biological trait characteristics of European and North American invertebrates has accumulated over many years (Tachet et al. 2003; Schmidt-Kloiber & Hering 2015; Twardochleb et al. 2021) and trait databases have also been developed for groups, including diatoms, used in biological assessment (see e.g. Rimet & Bouchez 2012).

Linking the traits of animal species with particular 'functions' in ecosystems is sometimes problematic. The question arises, how faithfully does the 'functional composition' (based on species traits) of an animal assemblage reflect the processes that really occur and thus indicate environmental conditions and stressors? In microbial communities on the other hand, the metabolic functions (via the activity of particular genes) can be measured more directly. Molecular methods also allow the taxonomic assessment of microbial assemblages in different sites. This then offers an exciting route forward in biomonitoring and environmental assessment, and more fundamental understanding of ecosystem processes. This is a field in rapid development. For instance, Fasching et al. (2020) examined the relationship between the occurrence of microbial functional genes and the composition of dissolved organic matter (DOM), nutrients and microbial communities in 11 streams across southern Ontario (Canada), divided across three groups based on land cover (agriculture, forest and wetland). They found a remarkably strong pattern. In the agricultural streams, microbial functions reflected the more labile, proteinaceous DOM and higher nutrient concentrations versus more humic-like DOM in wetlands and forest. Taxonomic diversity per function was also less in the agricultural streams than in the other, more natural, land covers, perhaps indicating reduced functional resilience (to disturbances) under agriculture.

10.5.4 Detecting multiple stressors

It is evident that stressors do not apply singly, but in combinations, which may interact in complex and unexpected ways (see Ormerod et al. 2010 and references therein). In streams and rivers, common combinations might include agrochemicals (including nutrients) occurring with high loads of sediment, morphological changes to river channels, loss of riparian vegetation and excessive water withdrawals. For both research and management purposes we need to know whether the effects of such combinations are simply additive, or more than ('synergistic') or less than ('antagonistic') expected based on their individual effects. These seem like simple ideas. However, different fields (toxicology, ecology) and researchers on different types of ecosystems (freshwater, marine, terrestrial) often use different terms for similar things, inhibiting the exchange of ideas. Further, multiple stressors may have different effects depending on the timescale of interest (e.g. within a generation to over many generations), they may differ with the 'level of organisation' (individual to ecosystem effects), and they may interact within food webs (e.g. be modified via species interactions) (see Orr et al. 2020). Despite having a long history, the field of multistressor research is changing and expanding extraordinarily quickly and adopting new methods. Here, we briefly describe two approaches applicable to stream and river ecology; the first is the diagnosis of multistressor effects using a multivariate analysis of species and/or species trait occurrences in large sets of data, the second is the design of experiments for testing the effects of multiple stressors.

Possibly the first to advocate the potential use of species traits in biomonitoring in running waters were Dolédec et al. (1999). Based on the 'habitat templet' concept of Southwood (1977; see Chapter 4, section 4.6.2 and Text Box 6.1), they had begun work on the relationship between the traits of invertebrate species found at different sites along the French River Rhône and natural environmental conditions. They appreciated that the 'adaptiveness' of different traits could respond to human perturbations of rivers—that resulted in stressors on organisms and systems—as well as to natural gradients of conditions. This insight opened the door to the prospect of relating multiple species traits to the tangle of stressors operating at different places and times. In an early test of the concept, they found that community structure based on biological traits (such as body size, number of descendants per reproductive cycle, parental care and mobility) reliably indicated human impact, and was not greatly confounded by natural environmental gradients.

Subsequently, this approach has been further developed (see Bonada et al. 2006; Statzner & Bêche 2010). In one example, Lange et al. (2014) found that species traits held promise as indicators of stressors in an agricultural catchment in New Zealand, in which 0–95% of sub-catchments had been converted to intensively managed grasslands and stream flow had been reduced through abstraction by 0–92%. The life-history traits of invertebrates, such as those conferring resistance and resilience, were closely related to agricultural intensity, whereas traits such as feeding habit, diet and mode of respiration were more clearly associated with water abstraction. Progressively larger-scale and more inclusive (more than one taxon) studies are beginning to analyse multiple trait-based analyses of river biota. De Castro-Catala et al. (2020) addressed the trait composition of both diatoms and invertebrates in three widely separate European rivers (the Adige in northern Italy; the Sava draining parts of Slovenia, Croatia and Serbia; and the Evrotas in Greece) in relation to human modifications to hydrology and geomorphology, and to toxic pollution by both pesticides and pharmaceuticals. Hydrological alteration was most closely related to community structure (taxonomic diversity) and trait-based composition of both diatoms and invertebrates, while pharmaceutical toxicity was also associated with community traits in diatoms, and pharmaceuticals and a composite index of pesticide toxicity were important for invertebrates.

The European Water Framework Directive relies upon the classification of rivers of EU member states into five categories (high, good, moderate, poor and bad) based on ecological assessments of biotic groups (Biological Quality Elements) including fish, invertebrates, macrophytes and benthic diatoms. Only about 40% of European rivers reach 'good' ecological status', which is the target of the EU. Rather than assessment of the representation of species traits, Lemm et al. (2020) looked at multiple stressors as drivers of variation in ecological status in 12 broad river types across the EU, including > 50,000 sub-catchments incorporating close to 80% of the entire surface area of the European Union. For measures of ecological status, they relied on statutory monitoring by member states from 2010–2015. Environmental data were used to estimate seven stressors on the basis of: (a) remote sensing of pressures such as the extent of urban and agricultural land in the riparian zone (a proxy for morphological and habitat degradation), and (b) well-established models for flow variables, nutrients and a composite measure of toxic substances. The stressors explained 61% of the total variation ('deviance') in ecological status across all 12 river types. Around 23% of this was explained by alterations in river morphology (channelisation, dredging etc), 16% by hydrology, 34% by nutrient enrichment and 26% by toxic substances. More than half of the total deviance in ecological status was best explained statistically by stressor interactions—particularly between nutrient enrichment and toxicity.

A more direct and explicit exploration of multistressor effects involves experiments, which in principle are easier to control than taking real data from the field. If properly designed, experiments are also more straightforward to analyse and interpret, and individual factors can be applied singly and in combination. On the other hand, there are problems of scale (very large-scale and long-term experiments are too demanding or impossible), realism (are the conditions of the experiment really like those in nature?) and replication (replicating several stressors applied singly, or in all possible combinations, quickly becomes intractable).

An excellent example of the experimental approach, illustrating both its advantages and challenges, can be found in Piggott et al. (2015a & b). They used stream-side flow-through 'mesocosms' to manipulate dissolved nutrients, fine sediment and temperature (raised above ambient conditions) in an attempt to

simulate perturbations brought about by both agricultural activities and projected climate change. Water
was pumped from an oligotrophic stream in the South
Island of New Zealand and supplied, suitably modified according to treatment, to the mesocosms, independently. There were (a) two nutrient treatments,
ambient and supplemented with nitrogen and phosphorus, (b) two levels of fine sediment, none and
added, and (c) eight temperature treatments ranging
from 0–6°C above ambient. There were eight experimental blocks (each at a different temperature) of 16
mesocosms—a total of 128—in a fully factorial replicated design, with four replicates of each treatment
combination. The experiment ran for six weeks in the
austral spring/summer. This was evidently a complex setup and approached the limits of what could
be achieved with a 'normal' research budget available
in this field. During the first 21 days, all mesocosms
were held under ambient conditions to allow for natural colonisation by algae and invertebrates (mainly
via incoming drift) and there was some direct addition
of invertebrates. Manipulation of conditions began at
that point. The experiment was intended to elucidate
effects on the algal and bacterial components of the
epilithic biofilm (Piggott et al. 2015a) and invertebrates
(Piggott et al. 2015b) and revealed complex and interactive effects of all three physicochemical factors, some
of which were additive and others antagonistic or
synergistic. One of the more straightforward examples, depicting total algal cell density, is shown in
Figure 10.11.

In conceptually similar experiments in the laboratory, Macaulay et al. (2020) tested the toxicity of three
common neonicotinoid insecticides, individually and
in combination, on the New Zealand mayfly *Deleatidium*. These highly persistent insecticides are still

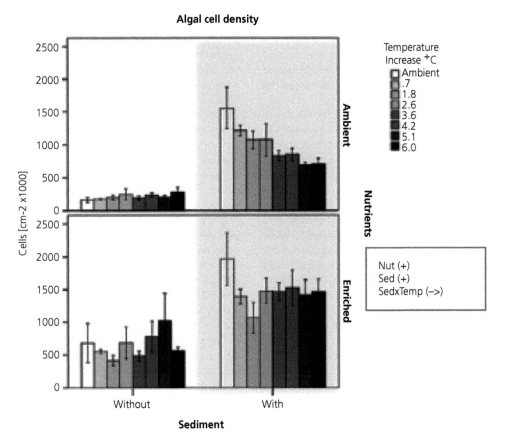

Figure 10.11 The effect on algal cell density of added nutrients and sediment at increasing temperature (0–6°C above ambient) in stream-side
mesocosms in New Zealand comparing 'ambient' (natural) levels with 'enriched' (addition of nutrients and in some treatments also sediment).
There are positive main effects of both nutrients and sediment, but a negative sediment–temperature interaction.
Source: after Piggott et al. 2015a, with permission of John Wiley & Sons.

commonly used against terrestrial insect pests but find their way into waterways through the local hydrological pathways. In the early stages of exposure, the effects of the three insecticides together were strongly negative and synergistic (i.e. greater than expected from their individual toxicities), but after 25 days the effect of imidacloprid (a particularly toxic and commonly used neonicotinoid) dominated and obscured any interaction with the other two.

Combinations of the many different stressors now acting on running waters, and indeed all natural ecosystems, may have profound and somewhat unpredictable effects. In attempting to understand and manage such complex situations, combinations of 'description' (collecting and analysing ecological samples from a large number of sites over many generations) and experimental manipulation may be necessary.

10.6 The uncertain future of rivers and streams in a changing world

10.6.1 Introduction

This book concentrates on the science of life in running waters, building from the small-scale details of the traits and natural history of river organisms to the large-scale interactions of river ecosystems with the rest of the biosphere. However, it is evident that the ecological fate of rivers and streams over the longer term depends on their interactions with people. In addition to all the skills required of anyone who wishes to understand the biology and ecology of rivers, to manage them we must also appreciate how they 'cut across' the activities of the people who live and work in their catchments and beyond. In other words, we must start to understand the socioeconomics of river systems. Humans really do need to appreciate the value of rivers to them and to bring that value into the societal 'balance sheet'. Such an appreciation is still far from the case. For instance, Sabater et al. (2022) write of the rivers of Spain and Portugal:

People in arid and semi-arid regions have the least respect towards rivers; rivers are often dry or produce catastrophic floods, and are viewed more as a danger or nuisance than natural resources to be preserved. Moreover, there is a well-rooted perception in the Iberian mentality that any water that reaches the sea is wasted.

The evidence suggests that such views are by no means restricted to Spain and Portugal!

This section touches upon several contemporary issues emanating from the way humans interact with the natural world, with a particular focus on rivers. Whereas the conventional question is to ask 'what do we humans do to rivers?', its more novel reciprocal is 'what do rivers do for us?'. Much in this book has touched upon the first of these questions (in the context of pollution, habitat and biodiversity loss, etc)—while studies addressing the second are expanding as we begin to appreciate the environmental crisis that is upon us (much of it centred around fresh water and rivers and their feedbacks upon us). Here we seek to make such questions a little more tangible by providing a few, mainly recent, examples. Most of these 'research threads' have a much older basis in ecology and environmental science, but are fields in rapid development, and we can provide only a taster here.

10.6.2 Valuing nature—a few basics

To many people the idea of trying to place an economic value upon nature is distasteful and probably hopeless. Surely the value of the natural world and of living things is incalculable and self-evident? With that perspective, the conservation of nature is intrinsically the right thing to do and, accordingly, must place powerful constraints on the activities of humans. The ongoing hectic overexploitation of natural resources and destruction of natural ecosystems unfortunately shows the inadequacy of this view. We do need to find some way to calculate the value of the natural world to humans—a value that currently remains largely unaccounted for in decisions about resource use and economic development. This value is encapsulated in the concepts of *natural capital* and *ecosystem services*.

Natural capital can be defined as the world's stocks of natural assets which include geology, soil, air, water and all living things (see Helm 2015). It is from this natural capital, and the associated ecological or ecosystem processes, that humans derive a wide range of benefits, nowadays often called *ecosystem services*, which make human life possible. If an economic value can be assigned to these benefits (often based on the estimated cost of their replacement), then the loss of this value can be considered as a direct cost of, for example, habitat and biodiversity loss. Natural capital can be thought of as one of two types, renewable and non-renewable. *Non-renewable resources*, once used, are unavailable in the future. An obvious example is fossil fuels. *Renewable resources* can continue to be harvested in the future, as long as we balance use with the rate of regeneration and do not reduce the stock so far that the resource becomes unsustainable. Fresh water is one such example. *Sustainability* is then defined as a situation where the natural capital available for future generations does not decline. *Overexploitation* can be explained because we do not yet take account of potentially sustainable natural capital that is freely available; that is, which an enterprise does not have to replace and pay for in the long term. The 'bill' is simply left to be picked up by our descendants. River systems and their supplies of fresh water are undeniably part of natural capital and provide many kinds of goods and services that humans have long exploited. The degraded state of many rivers points to their unsustainable overexploitation, although much of their natural capital could be restored if we choose to do so. We can think of this as a task of maintenance and repair, which does itself incur a cost. We now visit a few examples concerning the 'value' of river ecosystems and ecosystem processes.

10.6.3 What do rivers do for us?

The Millenium Ecosystem Assessment (2005), a kind of Doomsday Book for the biosphere, identified, at its simplest, four kinds of ecosystem service supported by natural capital. Firstly, there is *provisioning*—the supply of food, fresh water, fuel/energy, medicines, wood and fibre. River ecosystems are involved in a number of these. Then there are so-called *supporting services*—ways in which natural ecosystems support this provisioning—and include nutrient cycling, soil/sediment formation and primary production. Here rivers play some part in all three. Next come *regulating services*—ways in which our world is made habitable—including climate regulation, flood regulation, water purification and disease regulation. Rivers are involved in at least the first three. Finally, there are *cultural services*, supplying *Homo sapiens* ('thinking man') with aesthetic value, spiritual well-being, educational opportunities and recreation. Rivers here are evidently involved in all four (see also Covich et al. 2004). These are all things 'rivers do for us' (Table 10.1 shows a simplified list). Taking them properly into economic account is not straightforward. For example, Dalal et al. (2018) essentially concluded that, while the economic approach is promising, a great deal of progress is needed before it can be routinely incorporated into management decisions. It will always be difficult to convince people that they need, for the first time, to include the long-term cost of using up natural capital that has always apparently been 'free'.

Multidisciplinary work is now exploring factors affecting the delivery of ecosystem services by rivers (see Yeakley et al. 2016 for more on riverine ecosystem services). Natural features are important. Thus, a flat

Table 10.1 The major goods and services provided by various freshwater ecosystems.

	Running waters	Lakes/ponds	Freshwater wetlands
Water supply			
Drinking, domestic uses	X	X	
Manufacture, industry	X	X	
Irrigation	X	X	
Aquaculture	X	X	X
Goods other than water (e.g. food production, construction materials)			
Fish	X	X	X
Waterfowl	X	X	X
'Shellfish' (mussels, crayfish etc)	X	X	X
Plant products (e.g. timber, reeds, fibres)	X		X
Non-extractive			
Biodiversity	X	X	X
Flood control	X	X	
Transport	X	X	X
Recreation and aesthetic	X	X	X
Pollution dilution/waste disposal	X	X	X
Hydroelectricity	X		
Wildlife habitat	X	X	X
Property values	X	X	X

Source: Modified from Poff et al. (2002) and Postel & Carpenter (1997).

catchment has little potential for hydropower development. Karki et al. (2021) considered the influence of river topology and estimated, based on models, that three broad river network types, 'long-trellis narrow', 'coastal dendritic' and 'inland dendritic' (see Chapter 2, section 2.2.1, p. 22), deliver six distinct types of ecosystem service. These are water supply, flood attenuation, retention of sediments, retention/transformation of nutrients, provision of aquatic habitats, and potential for hydropower generation. Results of such modelling exercises will need careful testing before being used in making management decisions on a particular river system, but the approach represents a valuable start. Assessing the role of particular groups of organisms in delivery of various ecosystem services is also important. For instance, Covich et al. (2004) review how benthic organisms can provide and alter ecosystem services, and point out that the overexploitation by humans of one service can lead to a negative effect on a second or more service(s). This whole area clearly warrants much more work.

Perhaps of most interest are the relationships between the delivery of ecosystem services and human perturbations, as such relationships could then form the basis for decisions whether or not to conserve or to attempt the restoration of vulnerable systems (Gilvear et al. 2013). For instance, Grizzetti et al. (2019) mapped ecosystem service capacity (estimates of what the system could provide) and actual use

('service flow') for European fresh waters (prominently featuring river systems) using the extensive data now available, though important gaps and uncertainties remain. They then related use to the ecological status of appropriate water bodies and compared the two.

The data indicate a very general relationship, as shown in Figure 10.12. 'Regulating' services (including self-purification, prevention of erosion, flood protection) and 'cultural' services (recreation, aesthetics) decline with a decline in status while, somewhat counterintuitively, the provision of water supplies by rivers initially increases with declining water quality then finally declines only when ecological status declines severely. The natural capacity to supply water (Figure 10.13a) is greater in larger rivers, which tend to be in worse condition yet have naturally higher flows available for abstraction. The actual amount of water taken ('service flow') also increases steeply with a decline in status (we take more water from larger, downstream reaches, then have to treat it) (Figure 10.13b). Sustainability of supply, measured as water stress (WEI, the fraction of flows available that is actually used), then declines in more degraded systems (Figure 10.13c). Thus, there are feedbacks of water abstraction on ecological status, such that overexploitation of this single service—water withdrawal—conflicts with regulating and cultural services of river ecosystems.

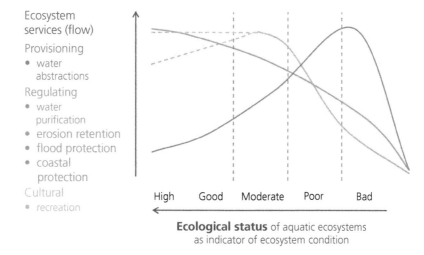

Figure 10.12 Conceptual diagram of the relationship between three different kinds of ecosystem services available in European fresh waters in relation to their ecological status.
Source: from Grizzetti et al 2019; published under a Creative Commons CCBY licence.

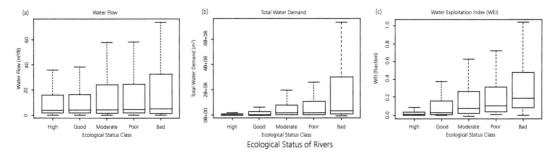

Figure 10.13 The estimated ecosystem services of water supply from European rivers in relation to their ecological status: (a) the natural capacity of rivers to provide water (water flow), (b) the water actually taken ('service flow' or water demand), (c) the sustainability of that demand (as the fraction of natural capacity actually exploited, WEI, the water exploitation index).
Source: from Grizzetti et al. 2019; published under a Creative Commons CCBY licence.

10.6.4 Biotic effects on the lotic environment—ecological engineers

We are most used to thinking of the physical and chemical environment imposing limits on living things—captured in terms like 'physical and chemical drivers' and 'limiting factors'. In this section, we consider the reverse—what do organisms do that 'feeds back' onto the physical and chemical environment? The activities of organisms in rivers and streams obviously affect ecosystem processes and, in some cases, the 'delivery' of ecosystem services from the perspective of humans. Note that strictly we should not think of species having a particular 'function' or 'role' in ecosystems, in the sense of them having a 'purpose' as such. The activities of organisms are governed ultimately by the action of natural selection—the 'blind watchmaker' of Richard Dawkins—on the genes, individuals and traits available. But nevertheless, those activities have implications for the lives of other organisms. Here we discuss the concept of organisms as *ecological engineers*, something we have touched on before in the context of invasive species (see section 10.3.5 above). Recall that an ecosystem, or ecological, engineer is a species that alters habitat structure (most obviously physically) substantially, thus changing the conditions and resources available to other species (Jones et al. 1994). Virtually all ecosystems contain native species that engineer the habitat to some degree or another, though not all of them necessarily affect ecosystem services.

The major ecological engineers of river channels and river flows are undoubtedly the larger water plants—submerged, floating and emergent (despite plants not being what we usually think of as 'engineers')—plus riparian trees, with their extensive root mats and supply of dead wood to the channel (see e.g. Gurnell 2014, 2016). Plants growing within river corridors respond to fluvial processes—that is, river plants are subject to 'large-scale physical forcing' (see Chapter 4). Turning that round, however, it has become clear that river plants also profoundly affect fluvial processes, such as flow and channel form. Gurnell (2014) described 'hotspots of plant engineering' as occurring between central areas of the river corridor (e.g. the main channel)—where flow forces may be sufficient to prevent colonisation by higher plants—and more peripheral areas, where engineering pioneers (capable of colonising bare sediments as flow forces ameliorate) are finally outcompeted by riparian plants less tolerant of disturbance (Figure 10.14). The exact position of such hotspots within the river channel will then differ between systems of differing hydraulic energy. In low-energy rivers the hotspot may be in the central channel, with aquatic plants engineering the river bed. As river energy increases, the hotpot shifts towards the channel margins, where emergent engineers trap sediments to form submerged 'shelves'; then further to exposed ridges at the margins of the channel (at base flow), where seedlings of emergent plants can survive, creating levees; and finally to exposed bars within the channel where snagged trees and large wood can sprout and create islands (e.g. the Italian Tagliamento River, see Chapter 6, section 6.5.2.2, p. 220). The different plants create 'pioneer land forms' with a morphology resulting from an interaction between (a) sediment retention and stabilisation by plants and (b) flow forces tending to transport and erode sediments. Higher plants are indeed true ecosystem engineers of river corridors and help create the biological patchiness that provides habitat for other creatures and contributes to high biodiversity.

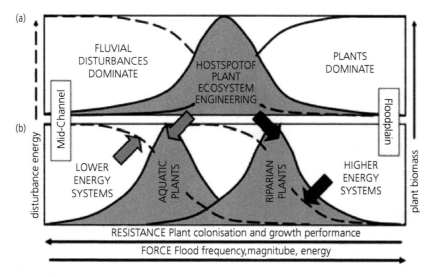

Figure 10.14 Schematic of the distribution of higher plants that are divided into aquatic engineers (resistant to flow but poor competitors) and riparian plants (competitive but susceptible to flow disturbance), laterally from the mid channel of a river out onto the floodplain. Dotted line shows the distribution of flow forces ('disturbance energy', left-hand vertical axis) and solid lines show plant biomass (right-hand axis). Upper panel (a) shows a low or zero plant biomass in mid-channel, where disturbance dominates, and a high biomass of riparian emergent plants, at low disturbance energy, with a 'hotspot of plant ecosystem engineering' of aquatic pioneer plants in between. Lower panel (b) shows the hotspot at different positions in rivers systems of differing energy (see text for further details). Plant resistance to flow forces increases along the horizontal axis from right to left—higher for aquatic plants, lower for riparian plants.
Source: from Gurnell 2014; with permission of John Wiley & Sons.

Animals also play a big part in modifying river habitats (see reviews by Moore 2006 and Bylak & Kukula 2020). The wider natural history of animals as ecological engineers of rivers is fascinating, their effects variable and somewhat unpredictable. Some large terrestrial mammals modify river habitats simply by crossing channels, stirring up sediments and entraining small animals, particulate organic matter and nutrients in the flow, while otters (*Lutra* spp.) build shelters in river banks. The effects of such activities on other organisms are probably usually short lived or on a small spatial scale. However, many of them can alter what we regard as the ecosystem services provided by rivers—either positively or negatively. There is an approximate relationship between body size of the animal concerned and the timescale (and likely spatial scale) of the impact of its 'engineering' activities (Figure 10.15). Thus, hungry insect predators foraging among sediments disturb fine particles and 'winnow' them into the flow, altering sediment transport (e.g. Statzner et al. 1996). Crayfish dig burrows and can destabilise the river bank and bed (Statzner 2012). Net-spinning caddis spin silken shelters that can stabilise the bed, increasing the energy required to entrain particles in the flow (e.g. Statzner et al. 1999; Maguire

et al. 2020). Fish of various kinds (but particularly large migratory Pacific salmon) disturb bed sediments by digging nests, affecting the habitats of benthic algae and invertebrates (e.g. Harding et al. 2014). In essence, these activities often facilitate the activities and success of some species while negatively affecting others (see also Chapter 7, section 7.7, p. 250). On the other hand, the most famous riverine animal engineers are the two species of beavers, North American (*Myocastor canadensis*) and European (*M. fiber*), which build dams, and fell and eat the bark of riparian trees. The longevity and spatial scale of their effects is exceeded only by human river engineers, whose activities (at least in all but small streams) probably now outweigh all others (Figure 10.15).

Beavers have extremely important hydrological and morphological effects on the rivers they inhabit. The creation of ponds behind complete beaver dams gives the channel gradient a 'step profile'. This usually increases the total diversity of microhabitats in the locality and, in turn, the overall biodiversity (reviewed by Stringer & Gaywood 2016). Damming activities also create wetlands and retain organic matter in the streams. Naiman et al. (1986) estimated that beaver dams on some Canadian streams retained about

SCALE of IMPACT:

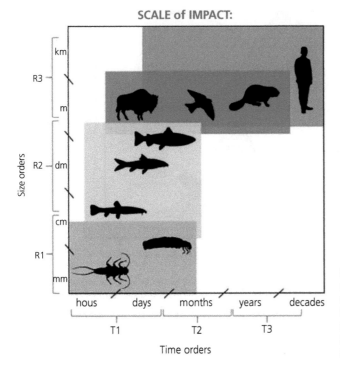

Figure 10.15 Relative patterns of 'engineering' impact of animal groups on mountain stream channels. The spatial scale of impact (R1–R3, small to large) and how long impact lasts (T1–T3, short to long) are shown. 'Animal groups' are aquatic invertebrates, fish, terrestrial and semi-aquatic mammals and birds, plus humans.
Source: from Bylak & Kukula 2020; with permission of Elsevier.

10,000 m³ km⁻¹ of organic matter in the channel. Downstream of dams, the invertebrate benthos responds to changes in stream flow, while the supply and calibre of organic detritus is also modified. The felling of mature trees in the riparian zone opens up the canopy, potentially increasing primary production in beaver streams and plant diversity the adjacent riparian zone.

Having been lost from most of their range in Europe, beavers are now recovering and extending their distribution. This is proving controversial. Some see them as part of 'soft engineering' mitigation of climate change, while many fishermen fear that beaver dams will prevent fish migration, while farmers have concerns about local flooding and waterlogging of their land (Campbell-Palmer et al. 2016). When it comes to river water 'beavers slow it, spread it and store it' (Fairfax & Whittle 2020). They create extensive wetlands, small ponds and channels and generally render the landscape wetter. This can alleviate flooding downstream—normally where more humans live and where land is more expensive and productive—since water falling in intense storms can be temporarily accumulated. Beaver wetlands retain organic carbon and usually raise overall aquatic biodiversity. Unless compensation payments are made,

however, the upland riparian landowner often pays the price.

In an unusual twist to the story, streams with beavers have recently been shown to offer protection to adjacent terrestrial vegetation against fire—increasingly important in dry areas under climate change. Thus, Fairfax & Whittle (2020) compared the 'greenness' of riparian vegetation persisting through five large wildfires in the western United States. Beaver-dammed riparian corridors were relatively unaffected by the fire compared to those without beaver dams—suggesting that beavers increase resistance of the riparian vegetation to fire and provide fire refugia (though they did not affect the rebound of the greenness of vegetation—'resilience') (Figure 10.16). This also increases fire resistance (by acting as fire breaks) in the overall landscape during dry weather—a definite 'ecosystem service'.

10.6.5 Rivers and the climate

An increasing research effort is being devoted to examining and predicting the effects of climate change on running-water ecosystems. The reason is that lotic systems are expected to be sensitive to temperature change (see Woodward et al. 2010) and there is clear evidence of a variable increase in temperature in streams and

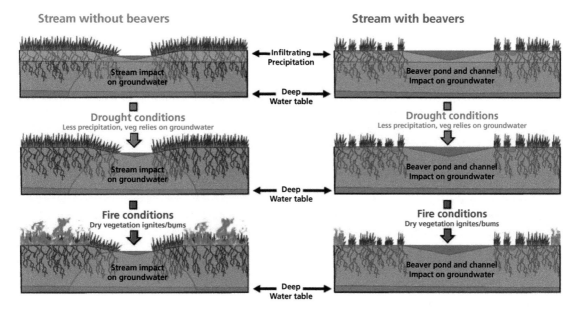

Figure 10.16 A schematic comparing streams with and without beavers illustrates how beavers can positively alter the response of riparian vegetation to both drought and wildfire in dry areas.
Source: after Fairfax & Whittle 2020, with permission of John Wiley & Sons.

rivers when studied over sufficient periods of time (Chapter 2, section 2.6.1, p. 53). For instance, estimates of the rise in mean maximum summer temperature were around 0.1°C and 0.22°C per decade (between 1968 and 2013) for a small, thickly forested headwater in the Welsh uplands and in the main river, respectively (Hildrew et al. 2017). Comparable estimates for three sites in the lowland course of the (very much larger) French River Loire ranged between *c.*0.4 and 0.8°C per decade (1989–2005). Overall, it appears that the temperature increase in small streams (in Europe at least) has normally been < 0.5°C per decade, with somewhat larger increases in lowland rivers (Hildrew et al. 2017). This increase may well accelerate as air temperature warms with progressive climate change.

There have been three broad approaches to investigating the response of running waters to climate change. First, predictions have been made based on projected changes in the climate—that is, essentially assuming that species and communities will respond to expected shifts in temperature according to their past occupation of similar habitats (e.g. Hering et al. 2009; Domisch et al. 2011; Jähnig et al. 2017). Second, if there are 'long-term' data on both biota and conditions, one can ask whether biotic changes have been consistent with a shift in climate (e.g. Daufresne et al. 2007; Jourdan et al. 2018). Third, a novel approach is

to find field situations ('spatially analogous model systems') in which conditions vary in ways consistent with predicted climate change scenarios. An example of this latter approach is the comparison of closely neighbouring streams at a range of temperatures variously heated by geothermal energy (see Topic Box 10.2 by Jon Benstead and Wyatt Cross describing a special study system in Iceland). In addition, at such sites, heat and/or warm water from geothermal springs has been diverted to raise the temperature of cool streams experimentally (e.g. Nelson et al. 2017).

Unsurprisingly, many such studies suggest effects of a changing climate from alterations in communities, to longitudinal distributional change, to local extinction (e.g. communities of invertebrates and fish in the French River Rhône, Daufresne et al. 2003; French stream fish, Comte & Grenouillet 2013; stoneflies in the Great Smoky Mountains of the USA, Sheldon 2012; apparent local extinction of an Arctic-Alpine relict flatworm from Welsh streams, Durance & Ormerod 2010). A common finding, however, is that distributional shifts seem less than would be 'required' to keep pace with environmental change (e.g. Hildrew et al. 2017; Chapter 2, section 2.6.1, p. 55).

As well as climate having a forcing effect on running waters, in this section we want also to stress the possible reciprocal effects of river systems on the

Topic Box 10.2 Studying global change in streams using 'model systems'—the case of the Hengill Valley, Iceland

Jonathan Benstead & Wyatt Cross

How can we understand and predict the responses of lotic ecosystems to key drivers of global change, including climate warming? This is a challenge because the drivers of change take their effect over the long term (i.e. decades or more)—much longer than the life time of most research projects—at spatial scales covering entire river networks or more, and often involve complex indirect interactions across entire ecosystems. A variety of investigative approaches have been used, each with its own pros and cons. Some have used 'landscape gradients', employing observations across sites that differ naturally in one key aspect of global change, such as temperature or nutrient concentrations. Such studies incorporate the complexity of natural communities and provide a window into ecological responses that may only emerge over relatively long periods or at large spatial scales. Yet these studies suffer from a lack of control (it is difficult to 'rule out' the influence of other factors) and little or no replication. Others have used small or short-term experimental studies to test the effect of one or a few treatments that is not possible at the whole-system scale. These latter studies have the benefit of control and replication, but often lack realism and may only incorporate a subset of the diverse species interactions that occur in nature. Clearly, there is no single 'best' approach for studying global change in streams and rivers, but much can be learned from programmes that combine and integrate observations and experiments across multiple scales.

A research programme, based at the Hengill geothermal area in south-west Iceland, is aimed at developing a deeper understanding of how global change, especially warming and nutrient enrichment, is affecting stream ecosystems (O'Gorman et al. 2014). At Hengill, local geothermal heating of soil and bedrock leads to wide differences in the temperature ($6°C$ to $> 60°C$) of streams sometimes as little as a few metres apart, with relatively few differences in other physical or chemical characteristics, making the valley an ideal natural laboratory for studying effects of temperature on stream ecosystems (Box Figure 10.2). Pioneering 'landscape-scale' studies at Hengill focused on observations along the natural thermal gradient, revealing large structural and functional differences among nearby streams (Friberg et al. 2009; Woodward et al. 2010). Temperature seems to be a principal driver of community assembly and structure, while acclimation to warming can influence algal and insect life-history characteristics, including body size and voltinism (Hannesdóttir et al. 2012). Later studies explored variation in food-web structure (O'Gorman et al. 2017) and secondary production (Junker et al. 2020) in response to

temperature, using the Hengill system to confront food-web and metabolic theory with detailed empirical observations.

Hengill's significance as a study site lies in the valley's many streams, each acclimated to its natural temperature regime for decades to centuries, that are otherwise similar and close neighbours. These streams represent a 'natural thermal laboratory', but we can also add the power and control of experimental manipulation. The valley's unusual arrangement of warm and cold streams makes large-scale temperature manipulations a viable experimental approach. Two of the Hengill streams flow within ~ 2 m of each other yet contrast greatly in mean annual temperature ($8°C$ vs $22°C$). This remarkable juxtaposition allowed a unique experimental design in which a simple heat exchanger was placed in the warm stream (Box Figure 10.2b). Water from the upper reaches of the adjacent cold stream was diverted through the heat exchanger and routed back to the cold stream to create an experimentally warmed reach. After a year of pre-warming study of this reach and of a control reach in a second cold stream, the heat exchanger was deployed in October 2011, resulting in a mean warming of ~ $4°C$ over the following two years. One effect of warming was dramatic: extensive blooms of an otherwise uncommon alga (*Ulva* sp.) transformed the benthic environment during the summer growing season (Hood et al. 2018 see Box Figure 10.2c). The abundance of macroinvertebrates was reduced by 60%, while their total biomass remained unchanged because of a relative increase in large-bodied, warm-adapted taxa such as larval black flies and snails (Nelson et al. 2017).

Similar heat exchangers have made possible a third, much smaller spatial scale of experimental approach. By taking advantage of multiple hot springs ($> 50°C$) in the valley, we have heated cold stream water to supply experimental stream channels with a wide range of water temperatures (~ $6°C$ to $24°C$). These channel arrays have allowed detailed and replicated studies of the effects of temperature on the development, structure and activity of algal biofilms (the base of the food web in these upland streams) and how temperature interacts with nitrogen (N) and phosphorus (P) supply (Box Figure 10.2d). These channel experiments show that warming leads to the dominance of N_2-fixing cyanobacteria at naturally low dissolved N:P ratios (Williamson et al. 2016), a competitive effect that is removed by nitrogen enrichment, leading to replacement of N_2-fixers by diatoms and green algae, even at higher temperatures (Collis et al. *unpublished*).

What have we learned by combining these different spatial scales of enquiry? Reassuringly, many findings have corresponded across scales and matched theoretical

Topic Box 10.2 *Continued*

predictions. Shifts towards N_2-fixing primary producers and warm-adapted invertebrate taxa at higher temperatures occurred across all studies, while invertebrate production and respiration of microbial communities scaled with temperature as predicted by metabolic theory (Perkins et al. 2012; Junker et al. 2020). Other results have been more difficult to interpret. The striking response of *Ulva* to the whole-reach warming experiment highlights the potential for even multi-year, whole-system manipulations to reveal surprising and transient species-level responses that differ from those seen in natural, 'acclimated' ecosystems. Another challenge has been to understand the role of interactions in driving responses to warming. For example, the relationship between biomass production by stream consumers and temperature is modified by the thermal responses of their algal resource base (Junker et al. 2020), while the addition of nutrients changes the effects of temperature on algal community structure (Collis et al. *unpublished*). Finally, Hengill is a single, relatively species-poor site (on

a remote and overall cold island), so the results are driven in part by the idiosyncrasies of regional and local community assembly and the limits this may set on trait diversity (Nelson et al. 2017). Thus, although transient responses, interactions among trophic levels and drivers, and context dependence observed at Hengill certainly complicate prediction, future progress on understanding the effects of global change will require acknowledgement of this complexity. Advances in knowledge will also depend on integrated, multi-scale approaches that are guided by ecological theory and that combine the replicated control of experiments with the realism of studying and manipulating natural ecosystems.

Professor Jonathan Benstead is in the Department of Biological Sciences at the University of Alabama, Tuscaloosa, USA.

Professor Wyatt Cross is in the Department of Ecology, Montana State University, Bozeman, USA.

Box Figure 10.2 (a) Panoramic photo showing the Hengill Valley and location of study streams that vary in ambient temperature (mean annual temperatures in Celsius in red circles); (b) a heat exchanger supplied with cold water and immersed in a warm stream; (c) dramatic growth of *Ulva* downstream of point at which warmed water is returned to the experimental stream (black arrow, stream flowing away from the camera); (d) streamside channels supplied with water variably warmed by heat exchangers and enriched with different concentrations of nitrogen and phosphorus.
Source: Photo credits: (a) J.M. Hood; (b), (c) and (d) J.P. Benstead.

climate, which are increasingly being acknowledged. These could be globally substantial and a case can be made to give them some weight in policy for the management of rivers, riparian zones and floodplains, for instance, with a view to maximising the storage of carbon in accumulating sediments and aggrading riverine wetlands and deltas. Certainly, the still widespread drainage of wetlands, usually to provide new agricultural land, results in the emission of vast quantities of CO_2 as stored carbon is mineralised.

Climate change is substantially attributable to increases in 'greenhouse gases' (most importantly CO_2, CH_4 and N_2O) in our atmosphere, with recent increases in CH_4 and N_2O being particularly marked (Figure 10.17) (Dean et al. 2018; Quick et al. 2019). River systems, including tributaries and main channels, are normally net emitters of all three of these gases, and evidence is accumulating that these emissions are sizeable (Battin et al. 2009). Calculation of the global-scale emissions of greenhouse gases from rivers and streams is fraught with difficulty, however, so their relative role in our climate remains uncertain (e.g. Wallin et al. 2020; Blackburn & Stanley 2021). Carbon dioxide is the best-known and most abundant greenhouse gas. Put simply, rivers essentially receive vast amounts of terrestrial carbon, but as we have seen, they are not simply unreactive 'pipes' for carbon transport from the land to the ocean; they store some in sediments, transport some to the ocean, and mineralise a great deal to CO_2.

The distribution and fate of methane in and around river systems is also gaining attention, particularly as methane is a much more powerful greenhouse gas than carbon dioxide. Thus, freshwater wetlands, many essentially the flooded terrestrial extensions of rivers, probably dominate global methane emissions from natural sources, particularly in the tropics (e.g. seasonally flooded areas of the Amazonian forest) (Dean et al. 2018). Moreover, in their review, Stanley et al. (2016) estimated a global annual emission from stream and river channels themselves of around 27 Tg, (a teragram is one million tonnes), a substantial (but still uncertain) fraction of all freshwater sources. Estimates of methane emissions (and other greenhouse gases) from river systems at a variety of scales are accumulating. Thus, Borges et al. (2015) calculated that emissions of CH_4, CO_2 and N_2O from African river channels alone would offset around two thirds of the estimated terrestrial carbon sink for Africa—that is, the overall continental sink is probably much less than previously estimated. Including emissions from the wetlands of the Congo River basin gives estimates around one quarter of the *entire* global ocean and terrestrial carbon sink—and

much of this is as methane. At a somewhat smaller scale, Siezko et al. (2016) addressed river–floodplain interactions in the emission of methane from the Danube floodplain downstream of Vienna. Methane concentrations in isolated waters of the floodplain were high but emissions were greatest from the river during floods, the river acting as an 'exhaust pipe' for methane as these formerly isolated water bodies were reconnected to the mainstem by high and turbulent flows.

Emissions of methane from rivers and streams are strongly related to human land use and management. Sanders et al. (2007) found that increased sedimentation (eroded material of terrestrial origin) on the bed of UK chalk streams resulted in increased methane production and efflux of methane to the atmosphere, particularly via the stems of water plants and ebullition (bubbling). Similarly, Crawford & Stanley (2016) found that sedimentation increased methane production in agricultural streams of Wisconsin (USA). In a recent compilation of data from 236 streams in the UK, Zhu et al. (2022) estimated that mean stream-bed organic matter has increased from about 23 g m^{-2} (a pre-1940 baseline) to 100 g m^{-2} at present, which has increased stream-bed methane production and ultimately tripled methane emissions from 0.2 to 0.7 mmol CH_4 m^{-2} d^1. Surface waters in large conurbations are also sources of methane to the atmosphere, and are likely to grow in importance given global trends towards increased urbanisation. Many urban water bodies are shallow and rich in nutrients and sediment, conditions conducive to methanogenesis (the production of methane by methanogenic bacteria). Estimates of methane emissions have been made for Mexico City (Martinez-Cruz et al. 2017) and Berlin (Herrero et al. 2019), but more are necessary. Water quality (much poorer in Mexico City than in Berlin) and temperature contribute to differences (emissions are much greater in Mexico City). Streams were not particularly important sources of methane in either city, compared with urban ponds for instance, and the rates of methane emission from urban streams were markedly variable in Berlin.

Nitrous oxide (N_2O) is the third important greenhouse gas emitted from streams and rivers (see Quick et al. 2019 for a detailed review) and one whose atmospheric concentration has increased rapidly over the last century (Figure 10.17b). The wide use of nitrogenous fertilisers since the mid 20th century probably underlies this substantial increase. Nitrous oxide is produced by microbes and involves oxidation and reduction of the reactive compounds of inorganic nitrogen, ammonia, nitrate and nitrite and is favoured by

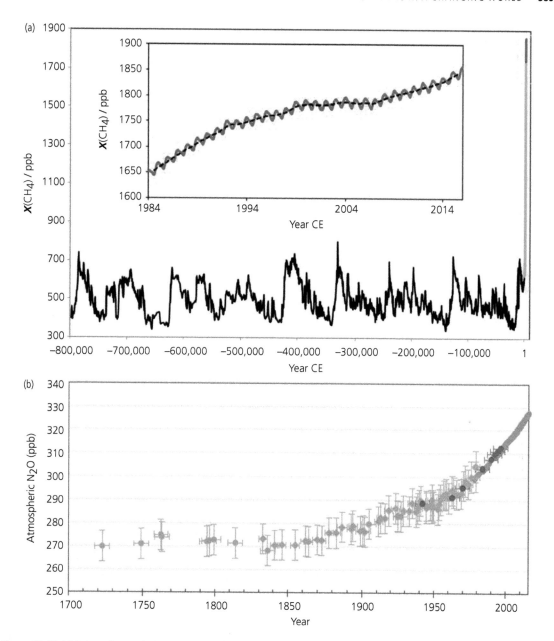

Figure 10.17 (a) Estimated atmospheric concentration of methane over the long term (cᴇ—years of the Common Era) and since 1984 (inset). Long-term data (black) are reconstructed from ice-core measurements, more recent values are a merger of ice-core and atmospheric measurements (blue), and entirely from direct atmospheric measurements (red, inset). (b) Atmospheric concentration of nitrous oxide from the early 1700s to the present. Blue and red values are from ice, while green values are direct measurements.
Source: (a) from Dean et al. 2018; published under a Creative Commons CCBY licence; (b) from Quick et al. 2019; with permission of Elsevier.

high riverine nitrate concentration, suboxic conditions and sufficient organic carbon to promote reduction (see Chapter 9). The main (though not the only) pathway producing N_2O is probably incomplete denitrification (with nitrous oxide emitted rather than gaseous

nitrogen) which occurs if organic carbon runs short, and/or if turbulence exposes suboxic sediments of rivers to the atmosphere via resuspension.

The hyporheic zone of streams is probably the most important in the production of N_2O (Figure 10.18),

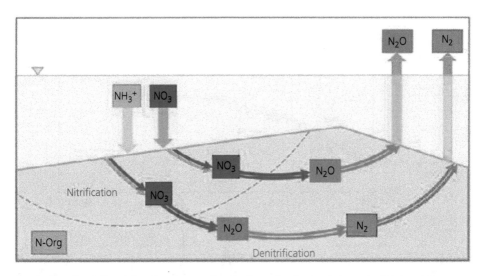

Figure 10.18 A bedform dune (a 'dune' of sandy sediment produced by flow, rather than by wind as on land) and overlying stream water, with potential nitrogen transformations along hyporheic flow paths (left to right). The dashed line indicates an anoxic (denitrification, below the dashed line) and an oxygenated zone (nitrification, above the line). Both nitrous oxide (N_2O) and nitrogen (N_2) can reach the atmosphere. Not all the N_2O produced is emitted. Fixed inorganic nitrogen (as nitrate and ammonium) can be imported to the system or generated internally by decomposition of organic N in the sediment.
Source: from Quick et al. 2019; with permission of Elsevier.

along with the water column of large, turbulent and turbid rivers of low water quality (high inputs of nitrate and ammonia) (Quick et al. 2019). Nevertheless, nitrous oxide can also be emitted from streams in quite unpolluted landscapes, as demonstrated by Audet et al. (2019). They assessed its concentration in a large number of low-order streams draining agricultural and forested land in much of Sweden and found, not surprisingly, that agricultural streams had higher concentrations of total N, as a result of local fertilisation. However, agricultural and forested streams had similar concentrations of N_2O, which Audet et al. (2019) attributed to the lower average pH of forest streams that inhibits complete denitrification to elemental nitrogen. Forest streams do get some fixed nitrogen from the occasional fertilisation of forests (applied to improve timber yield) and from atmospheric deposition (which remains quite high). Due to the great area of forest in Sweden, forest streams emit around 80% of the total N_2O (~ $1.8\ 10^9$ g N_2O-N) from all its streams, while forest streams contribute around 25% of the total N_2O emitted from the entire Swedish agricultural sector.

There is evidence of positive feedbacks in which climate change can exacerbate the emission of greenhouse gases from river systems, potentially contributing to further climate change. One recent example comes from agricultural streams of southern Wisconsin (USA) (Blackburn & Stanley 2021). There, high flows (more frequent with climate change) were associated with increased emissions of CO_2 and CH_4, probably due to flushing of these gases from soils, respiration of organic matter on the stream bed and increased gas exchange between the streams and the atmosphere due to extra turbulence during the floods. Indeed, frequent and prolonged high flows during the crop growing season led to sustained high emissions from the streams. Since agriculture occupies around one third of the global ice-free land area and is increasing, fluvial emissions of greenhouse gases can be expected to gain in significance as well as those from the land itself.

In addition, melting of permafrost in river catchments at high altitudes and latitudes is also a growing problem. Thus, Zhang et al. (2020) found highly significant increases in methane emissions via ebullition (bubbling) in headwaters draining the vast East Qinghai–Tibet plateau. This area has been described as the 'third pole' of the earth, has a mean altitude > 4,000m and is the source of 10 large Asian rivers (including the Yangtze, Yellow and Mekong). There are enormous carbon stocks of Pleistocene age in its surrounding permafrost; this latter is melting rapidly in the face of rising global temperature, and there is also rapid glacial retreat. The labile organic matter (carbon) formerly locked in the permafrost is then exposed and enters headwater streams where there are abundant, cold-tolerant methanogenic bacteria. The low air pressure at high altitude then favours the bubbling of methane from the streams. This bubbling declines in favour of diffusive emissions in higher-order systems, which are typically at lower altitude and higher atmospheric pressure. Releases of methane from these systems are globally significant in terms of greenhouse gas potential; they are undoubtedly 'hotspots' for the delivery of methane to the atmosphere (Zhang et al. 2020).

10.6.6 Methanotrophy—an underappreciated 'ecosystem service'?

Far more methane is produced in nature than is released to the atmosphere. Much of it is assimilated by methanotrophic bacteria to form their own biomass while some of its carbon is respired as CO_2. These methanotrophic bacteria can then be consumed by animals. Although this phenomenon is well known in some lakes and wetlands (e.g. Jones & Grey 2011), methane oxidation and the consumption of methanotrophic bacteria is less well studied in running water, despite the abundance of methane under oxygenated conditions (methanotrophic bacterial are aerobic) (see Chapter 8). In one study, Shelley et al. (2017) assessed methane oxidation in 15 southern English streams (of 'good ecological status' or better). Production by methanotrophic bacteria ranged from 16–650 nmol C cm^{-2} d^{-1} and was much higher in the shade than in reaches in full sun. Unlike the methane itself, the carbon assimilated by methanotrophs is potentially available to animal consumers, in addition to the carbon fixed as net primary (photosynthetic) production. The contribution of methane-derived carbon (in the form of methanotrophic bacteria) to stream and river food webs remains uncertain but may not be negligible. Further, well-oxygenated river-bed gravels with sufficient methane (some of it probably upwelling from the ground water) are active zones of methane oxidation and may provide 'an unnoticed ecosystem service' that prevents methane reaching the atmosphere. We need to know more about the conditions in running waters that influence methane oxidation and its extent.

10.7 Water scarcity, impoundments and related issues

Ultimately, the greatest threat to life in rivers is the loss and degradation of the aquatic habitat itself, increasingly due to 'competition' for water with humans. This involves physical interventions in river systems, most obviously in abstraction of water for 'off-river' purposes, use of water to generate hydroelectricity, and manipulation of flows and channels to prevent overland flooding and to reroute floodwater rapidly downstream or into storage. Manipulation of flows, water storage and diversion of water for irrigation go back thousands of years in human history, to the dawn of agriculture and the growth of towns and cities (see Chapter 1). Obviously, such activities have increased enormously. Cohen (2020) quotes figures of $< 700 \text{ km}^3$ (671 billion m^3) for global water withdrawals in 1901 (covering agricultural, industrial and domestic uses) growing to 3,800 km^3 in 2000. This is a more than a five-fold increase during which the world's human population went up a little less than four-fold. Consumption per person thus accounted for a substantial fraction of the increase. Withdrawals reached 4,000 km^3 in 2014 (see Our World in Data at ourworldindata.org).

Total global annual runoff from land to the ocean through our river networks is just under 40,000 km^3 (Chapter 1 section 1.1, p. 2), more than enough to satisfy human needs. However, Postel (2000) points out that around half runs off rapidly as floods, while another 20% is geographically too remote to be economically usable; that leaves around 30% that is potentially accessible without constructing new dams. The rate of increase in supply of usable water by dam-building has been slower than the rate of increase in demand—particularly for irrigation of agricultural land and to supply growing cities (Postel 2000). If we include in-stream uses, such as dilution of pollutants and generation of hydropower, it is likely that humans already exploit around half of all accessible runoff. Increasing degradation of river systems and the loss of their biodiversity (and attendant ecosystem services) is a consequence of this (over)exploitation. We could say that the 'natural capital' of river systems is being exhausted, to the detriment of future generations.

To set this briefly in a wider 'human' context, the United Nations has set a broad set of development goals in its '2030 Agenda for Sustainable Development' that recognises a basic human right of access to safe drinking water and sanitation. Yet the UNESCO (2019) World Water Development Report finds that one third of all humans do not have access to a safe water supply and two fifths have no access to safe sanitation facilities. Its report asserts that the development goals relating to water are 'entirely achievable'. Remarkably however, it made almost no reference to any tension with environmental sustainability, nor to the freshwater biodiversity crisis, nor any of the other environmental problems applying to rivers—many of which feed back onto humans. Clearly, the potential of ecological knowledge and sciences around rivers has yet to be recognised—surely a development goal of ensuring the environmental sustainability of 'functioning river ecosystems' is required.

Signs of water stress are clearly apparent where the pumping of groundwater for above-surface use exceeds recharge. Such overexploitation is found at local scale around major cities and on a larger scale in many parts of the world including the northern plains of China, the US Great Plains, parts of California (see Topic Box 10.3 by Vince Resh), and much of the Middle East, North Africa and crucial agricultural areas of India—these latter supplying the food needs of hundreds of millions of people, but only by using the groundwater unsustainably. For instance, Oiro et al. (2020) show that exploitation of the Nairobi aquifer system (Kenya) has led to a decline in the groundwater level beneath the city by around 6 m per decade since the mid 1970s, effectively 'mining' 1.5 billion m^3 of groundwater. This increase in water use has been due to the rapid growth in the human population (by about 10-fold since 1970), rather than climate change. Major rivers also increasingly run dry in some areas, including parts of the Ganges, the Nile and the Colorado. Food security is thus endangered (agriculture accounts for around 70% of human water use), the 'health' of river systems is increasingly threatened, while international rivers (those which cross national or even state boundaries) are increasingly likely to become the subject of political conflict.

Technological 'solutions' to the problems of both water stress and climate change are sought in the great expansion in river impoundments, in generating 'green energy' in the form of hydroelectricity and by inter-basin water transfers from river basins in water surplus to dry areas. None is without profound environmental costs, which are not properly taken into account (see Topic Box 2.1, Chapter 2, by Christiane Zarfl, p. 31). Grill et al. (2019) estimated that only 37% of rivers longer than 1,000 km remain free-flowing throughout their length. There are almost none of these remaining in the mainland USA and Mexico, Europe and the Middle East, much of India, southern Africa, southern South America, China, much of South-East Asia and southern Australia. Large free-flowing rivers are now largely restricted to northern parts of North America (Canada and Alaska), northern Eurasia, the Amazon and Orinoco basins in South America, the Congo basin of Africa, and the Irrawaddy and Salween basins of South-East Asia. Construction of dams proceeds apace,

particularly in the Amazon basin (Winemiller et al. 2016).

The small but highly developed continent of Europe provides a stark example (see Tockner et al. 2022; and see Topic Box 2.1). There are > 6,000 dams higher than 15 m, 20% of them in Spain alone, and only one of the 20 largest rivers in Europe is free flowing (the Northern Dvina in Russia). As a consequence, 95% of Europe's floodplains have effectively been lost, along with 88% of its alluvial forests. Many of Europe's rivers, including the Meuse in Belgium and the Netherlands, have been reduced to the status of 'heavily modified or artificial water bodies'. Such a river is essentially either a created water body, where none existed before, or one so substantially altered that it cannot attain a good ecological status. Even Europe's greatest river, the Russian Volga, has 12 very large reservoirs, nine of them on the main river, and much of its course is a cascade of large, shallow water bodies whose flow regime is largely determined by river managers. Similar stories are repeated throughout the developed world. The consequences of dams for river biodiversity, and particularly for large and/or migratory animals, are profound. For example, the Ural River in European Russia is now the only system in the whole of Europe in which the formerly widespread sturgeon (*Acipenser gueldenstaedtii*) spawns naturally in relatively large numbers, and even there, the annual run is down to around 2,500 fish from numbers around 70,000 in the 1970s (Eremkina & Yarushina 2022). In rivers of south-western (including the Upper and Lower Colorado) and south-eastern USA, Kominoski et al. (2018) found that the presence of dams was an overriding factor in the loss of native fish species from river basins.

Many reservoirs are constructed mainly for the generation of hydropower, or so-called green energy. Hydropower contributes around 80% of all renewable energy globally, achieved by constructing > 8,600 dams more than 15 m high. Investment in hydropower in 2010–2012 was around six times greater than a decade earlier (Hermoso 2017). A further 3,700 large hydropower plants were under construction or planned in 2015 (Zarfl et al. 2015), mainly in developing countries, and promoted by international investors and the drive for renewables. As an extreme example, at the beginning of 2021, Norway had 1,681 hydropower plants generating some 154 TWh (trillion watts per hour), representing 90% of the country's total power needs (see https://energifaktanorge.no/).

Many of these large-scale developments are controversial and, as many commentators have indicated, much more serious consideration needs to be given to

---

Topic Box 10.3 Water stress in California: the Sacramento–San Joaquin Delta

Vince Resh

California has 39.5 million people, about 12% of the US population. At least 27 million residents depend on water flowing through the Sacramento–San Joaquin Delta (Box Figure 10.3a) for their domestic, industrial and agricultural water supplies. The rest comes from local supplies or the Colorado River. One of the issues with water in California is that roughly 80% of it is in the northern part of the state but about 80% of the population lives in the south. Coupled with this, much of California has areas with a Mediterranean-type climate, with highly seasonal rainfall from October to April, or areas with an arid climate with minimal rainfall. Moreover, climate change is predicted to increase interannual variability in rainfall, resulting in more extremes of precipitation (and consequently more severe floods and droughts), and to reduce the snowpack in the mountains which is crucial for meeting the summer water demand. Also, the western US is currently (2022) in a long-term (> 20 years) 'megadrought'.

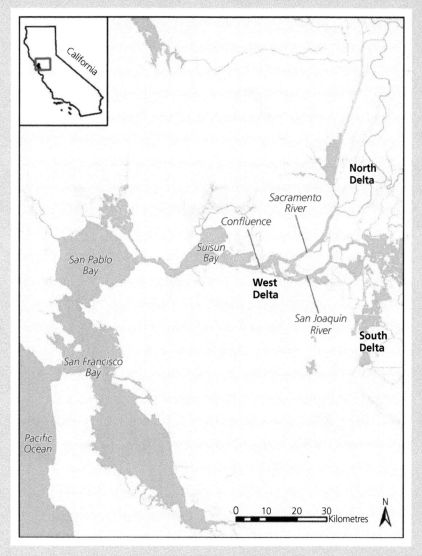

Box Figure 10.3a A sketch map of the upper San Francisco Estuary, receiving water from the Sacramento and San Joaquin Rivers (inset in California).

Source: from Davis et al. 2019 under Creative Commons Licence.

Topic Box 10.3 *Continued*

Agriculture in California uses about 80% of the water, and industrial and domestic use are each about 10%. Although California provides about 45% of fruits, nuts and vegetables grown in the US (around 70% of the global supply of almonds is grown there), it comprises less than 2% of the economy because of the strong contribution of Silicon Valley for the lucrative electronic, and many other, industries throughout the state. In terms of the Delta, major problems include poor drinking water quality because of a high salt content, threats to endangered fish, and concern over whether levee protection is sufficient to survive floods and earthquakes and to prevent salinity intrusion from these events. In addition, 80% of California's commercial fishery species live in or migrate through San Francisco Bay and the Delta and it is the habitat for > 700 species, including five listed under the US Endangered Species Act.

The inflow of water to the Delta and Bay has many sources and variable water quality: 80% comes from the Sacramento River (good water quality); 5% from east-side rivers (good water quality); and 15% from the San Joaquin River (poor water quality). In contrast, the outflow consists of (a) water to San Francisco Bay (69% in an average flow year), (b) in-Delta use (7%), while (c) the remainder is exported south of the Delta and used in the San Francisco Bay Area, the Central Valley, and Southern California (24%).

A common question is why so much water is 'lost to the sea', that is, allowed to flow into the Bay and the Pacific Ocean. Because the twice-daily incoming tides increase salinity, there is a need for 'environmental water' releases from the Bay to maintain freshwater habitats in the Delta.

A major ecological issue that has occurred in the Delta is referred to as the 'pelagic organism decline' or POD. Federal and state government agencies have been conducting a mid-water trawl survey in the Delta and San Francisco Bay each autumn since 1967 (Sommer et al. 2007). They collect data to determine the abundance of the pelagic species in the Delta, such as the Delta smelt (*Hypomesus transpacificus*), longfin smelt (*Spirinchus thaleichthys*), splittail (*Pogonichthys macrolepidotus*), American shad (*Alosa sapidisssima*), threadfin shad (*Dorosoma petenense*) and striped bass (*Morone saxatilis*). The first three of these are native, the last three introduced. Their populations have shown extreme annual fluctuations in the past, given the variable nature of estuaries and, historically, the lowest populations have been in dry years. However, the decline to record or near-record lows of the abundance indices for several of the species (Box Figure 10.3b) raised concerns, especially since flows were moderate during this period. The species abundance indices further deteriorated over the next several years. By 2004, the declines had been

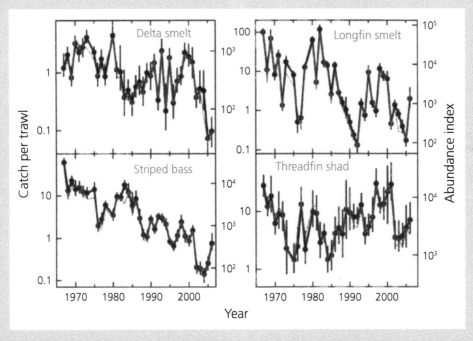

Box Figure 10.3b. Trends in four pelagic fishes (1967–2006) caught by pelagic midwater trawling in the upper San Francisco Estuary (log scales on y-axes; bold line with error bars, catch per trawl ± 95CL, finer line 'Abundance index').
Source: from Sommer et al. 2007, with permission of the American Fisheries Society.

Topic Box 10.3 *Continued*

---

recognised as a major ecological problem requiring research and management decisions.

Much of the concern about these declines has been because of increased diversion from the Delta and its potential environmental consequences. For example, these reductions have resulted from: changes in water movement such as ingress of saline water into previous freshwater habitats and their effects on fish distribution; competition for zooplankton with invasive Asian filter-feeding clams (*Corbula amurensis*); the increased presence of toxic *Microcystis* cyanobacterial blooms; and high concentrations of pesticides and nutrients. Further culprits may be habitat loss from past reductions in wetlands and the channelisation of rivers.

Legislation covering water in the Delta requires both environmental protection and the availability of reliable water for agricultural, domestic and industrial needs south of the Delta. However, climate change and the resulting annual variability means that water supplies are not entirely reliable. All sectors will have to make do with less water. Currently, water managers and regulators favour a portfolio approach involving water conservation, better conveyance of water though the Delta, increased storage through reservoirs and groundwater, desalination, reuse of water, taking land out of agricultural production, voluntary agreements and other broad approaches. One approach under consideration is a proposed water-transfer system. This would consist of an underground tunnel, 56 km long, 46 m below the Delta, to bring fresh water from higher up in the Sacramento River to

the state and federal water-distribution systems located at the base of the Delta. Modified from a two-tunnel plan, this project is still highly controversial.

Agriculture, consuming 80% of the water, is a key issue. Precision and drip irrigation, avoiding runoff, monitoring water use, and using technology so that evapotranspiration loss equals water applications, are necessary approaches to consider. However, the problem is that water savings are presently simply used to grow more crops. Perennial, high-value crops including almonds and walnuts require a great deal of water, with 30% coming from ground water, rising to 60% in drought years. Already, the withdrawal of ground water exceeds the sustainable supply by 13%. A commonly discussed and probable scenario is a future reduction of the footprint of agriculture in California.

Pressure from competing interests for water in the Delta is expected to increase the import of water from the Colorado River, already insufficient to meet the legal demands of other US states and Mexico. California is currently using 20% more Colorado River water than it is entitled to under the 'Law of the River'. Climate change will certainly exacerbate the situation both there and in the Delta. Further information on the Delta and its environment can be found in Bashevkin & Mahardja (2022) and Sommer (2020).

Professor Vincent R. Resh is in the Department of Environmental Science, Policy and Management at the University of California, Berkeley.

---

the environmental, social and economic costs and benefits of hydropower development than has been the case hitherto (see Topic Box 2.1, p. 31). Winemiller et al. (2016) point out that perhaps one third of all freshwater fish species are threatened by large-scale hydropower developments in the Amazon, Congo and Mekong rivers alone, with some structures already built and many more planned. A further tropical example is the Pantanal, a vast floodplain wetland in the Paraguay River hydrological basin which lies mainly within Brazil and with smaller elements in Paraguay and Bolivia. Around 140,000 km² of land are flooded annually and hydro schemes are relatively few and modest in size. However, 149 larger-scale hydroelectric facilities are planned or under construction on tributaries upstream of the Pantanal. There are 23 species of long-distance migratory fish in the Pantanal, 12 species of characins (see Chapter 3) and 11 catfish, some of them

of great economic and cultural importance. Completion of the proposed schemes will block a further 25–32% of the river system to fish migration, that is an additional 11,000–12,000 km of channel. The spawning areas of most of these migratory fish species are not well known and fish passes are either absent or likely to be ineffective (Medinas de Campos et al. 2020).

In addition to the proposed large hydro schemes there has been a rapid proliferation of small hydropower plants (definitions of 'small' vary widely, but these schemes operate on smaller rivers and typically each generate < 10 MW; Couto & Olden 2018). These authors estimate that > 80,000 such plants are operating or under construction and up to three times that number are possible if generating potential is to be realised. The cumulative environmental impact of such developments must be great, especially since they tend to be poorly controlled and indiscriminate in terms of local ecology.

Water shortages, particularly for large-scale irrigation projects, are an increasing problem in arid areas, despite the large-scale damming activity we have just discussed. An apparently attractive prospect is to move water across watersheds ('drainage divides') from a donor basin with a 'water surplus' to a recipient basin with a demand for extra water (e.g. Shumilova et al. 2018; Sinha et al. 2020; Topic Box 10.3). The definitions of surplus and demand seem to be mainly economic and take little account of ecological costs and benefits—which can be profound. Such schemes exist or are planned on all continents except Antarctica and, if all are realised, could transfer almost 2,000 km^3 y^{-1}, equivalent to 26 times the annual flow of the River Rhine. In one scheme of moderate size by international standards, water is transferred from the Tagus basin in central Spain to the dry Segura basin in the south-east of the country, especially to the province of Murcia, the driest area in Europe. The Tagus already supplies water to the Spanish capital Madrid and the Portuguese capital Lisbon, and supports 230,000 ha of irrigated agriculture. The Tagus–Segura transfer scheme diverts 350 Mm^3 (thousands of cubic metres) y^{-1} along a 292 km canal. Further irrigation agriculture in the Segura basin reached 270,000 ha in 2017 and supports > 100,000 jobs. This area supplies north-western Europe with fresh fruit, vegetables and salad in the winter, thus effectively transporting much of the transferred water by heavy goods vehicles hundreds of miles north! Apart from the unsustainability of this activity, the water transfer has contributed to the ecological degradation of the Segura and periodic drying of parts of the Tagus, and has facilitated the transfer of alien and invasive species between them (see Sabater et al. 2022).

10.8 Conservation and restoration

10.8.1 Ecological flows and working at the whole-catchment scale

Dealing with water stress (impinging on food security and sanitation) is an agonisingly difficult problem, and one which will only grow in the future. However, if rivers are to provide all the ecosystem services that are asked of them, and the remaining biodiversity is to be protected, the ecological requirements of river systems must be met. Where they are not, there is a clear need for restoration that can maintain sustainable ecological services and ecological status (Giller 2005). Some fraction of the flow must remain in the

channel and approximate a natural flow regime. Satisfying the requirements of the environment, along with a much greater efficiency of water use by humans, are the cornerstones of future water policy, as outlined by Postel (2000). In helping to define the environmental requirements of rivers systems, river ecologists have played a key role and will do so in the future.

Establishing and implementing 'ecological flows' in river basin plans is a first step in the conservation and restoration of river ecosystems. This is a very active field in running water research, in which we try to reach consensus as to how to manage flows in rivers treated as 'socio-ecological systems' (e.g. Poff, et al. 2010; Poff et al. 2017; Kennan, Stein & Webb 2018). The ELOHA framework (Ecological Limits of Hydrological Alteration; Poff et al. 2010) laid the modern basis for assessing ecological responses to flow alterations, and how to assemble and analyse the hydrological and ecological data necessary and useful for prediction. This field of research now goes far beyond river ecology, however, encroaching into socioeconomics and even the principles of social equity and causes such as 'water for all' (the principle that there is a human right to a clean source of water; UNESCO 2019). This is sometimes used to justify water-transfer schemes. Ecological science should play a crucial role in supporting policy, but other disciplines are also important.

It is much easier to conserve a natural system rather than attempt to restore it once it has become degraded. However, river conservation is particularly problematic (e.g. see Hermoso et al. 2016). Conservation has historically centred upon terrestrial and, more recently, marine protected areas or on particular species. It is often possible to 'draw a line' around a terrestrial area of importance for threatened species and to protect that area. Even for terrestrial systems it is doubtful if that strategy is ethical or effective indefinitely—given what we now know about the importance of the size of reserves and the dispersal of organisms in the landscape, particularly under the shadow of climate change. If we have learned anything of river systems, it is their extreme openness to the movements of water, organisms and materials within the river network and between the river and its upstream catchment and surrounding landscape (encapsulated in Hynes's concept of 'the stream and its valley'; Hynes 1975). At its most obvious, adjacent terrestrial and in-stream disturbances in the catchment upstream have profound influence downstream (and the reverse may be true to some extent). Large rivers have enormous catchments and it is inconceivable that we could ever 'put a fence around' the drainage basins of even moderate-sized rivers in most areas. Existing protected areas may not contain important rivers (they are rarely aimed primarily at running waters) and, even if they do, large, unprotected areas usually lie elsewhere in the river network and drainage basin, a problem exacerbated by the trans-border nature of large river systems. In other words, a large-scale ecosystem approach, involving the human population and its cooperation, is required, rather than simply a species-centric approach, even though the latter is often the starting point for river conservation efforts.

River conservation therefore demands not only strict protection of as much of the catchment as is feasible but, most importantly, modification of human activities over more extensive areas—usually including private land and/or areas devoted to agriculture, industry or urban development. Catchments important for the supply of drinking water to large cities (rather than

for the purposes of river conservation more generally) are often protected to ensure high water quality and reduce treatment costs. A well-known example is that used by New York City to protect the 8,300 km^2 Catskill/Delaware catchments west of the Hudson River (e.g. Ashendorff et al. 1997; Figure 10.19). The overall system supplies 1.2 billion US gallons (4.5 billion litres) of drinking water for > 8–9 million consumers, and is built around 19 reservoirs and a network of aqueducts and tunnels and coordinated catchment land-use management and forest conservation developed through an urban–rural partnership.

An example of the apparent ineffectiveness of protected areas in conserving rivers comes from the Russian Volga. It has a vast catchment of 1.4 million km^2 in which 'protected areas make up an appreciable part' (Mineeva et al. 2022). There are five large biosphere reserves and two national parks incorporating long stretches of the river, while a network of more local

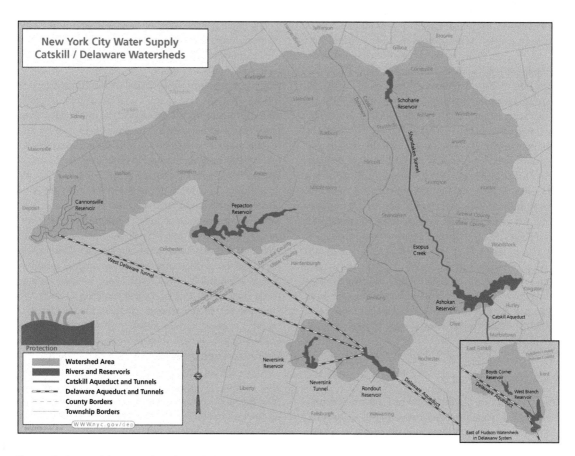

Figure 10.19 Map of the New York City 'watershed' in the Catskill mountains and various supply routes and reservoirs.
Source: from New York City Department of Environmental Protection web page at www.nyc.gov/dep.

nature reserves covers > 6,000 km², designated mainly as terrestrial reserves. As noted above (section 10.6.7), the mainstem of this river is now little more than a series of very large reservoirs. Iconic species such as the beluga sturgeon (*Huso huso*) have lost access to 90% of their spawning area and have virtually disappeared from the Volga. More generally, a native fauna of specialist lotic fish species has been largely replaced by one of generalist, lake-dwelling species, many of them invasive. Protected areas along the Volga have evidently done relatively little to conserve its biodiversity.

There have been attempts to judge the degree of protection of river systems in terms of direct inclusion of the channel in protected areas, and of the degree of protection afforded to upstream catchments (as in the Catskill/Delaware example)—an extremely challenging problem. At a global scale, Abell et al. (2016) attempted this by assessing 'local protection' (the fraction of a river length included within a protected areas) and 'integrated protection' (an assessment of the degree of protection both locally and in the catchment—for technical details see Abell et al. 2016) (Figure 10.20). As the authors stress, the real protection afforded to rivers may be much less than suggested by this approach, particularly for large rivers. This analysis needs to be treated very much as in progress, requiring far more work 'on the ground' to estimate the real level of protection that is effective. Nevertheless, the need to protect rivers at the *whole-basin* scale, short of completely 'fencing off' whole catchments, was confirmed, for instance, by Leal et al. (2018) in a study of fish in 83 low-order streams in three basins in the eastern Amazon, an area subject to agricultural expansion. Such was the degree of species turnover among streams in this hyper-diverse fauna, that local-scale protection of just some streams was insufficient at the basin level. They concluded that 'To safeguard the species-rich freshwater biota of small Amazonian streams, conservation actions must shift towards managing whole basins and drainage networks, as well as agricultural practices in already-cleared land' (Leal et al. 2018). While the message is clear, implementation is a real challenge.

10.8.2 Restoring rivers

Where systems are already damaged, restoration may be attempted. Whereas conservation aims (ideally) at the maintenance of systems and processes in a state that reflects a modest impact by humans (or at least a legacy of impact that has produced a state that people consider 'desirable'), river restoration aims to increase

ecosystem goods and services, and ideally to convert damaged freshwater systems into sustainable ones whilst protecting downstream and coastal ecosystems (Palmer et al. 2005). In rivers, as we have seen, this is a particularly challenging enterprise both in theory and practice. But what is the target of such restoration—and how do we know when we have successfully restored a river system? Six criteria have been suggested (Jansson et al. 2005; Palmer et al 2005): (a) the establishment of a dynamic ecological endpoint to the restoration, a clear view of what a healthier river would look like (i.e. a 'target' or 'guiding image' for restoration), and (b) an improvement in the river's ecological condition, that ideally leads to (c) a more self-sustaining and resilient system; (d) no lasting harm should be inflicted on the system during the restoration, (e) pre- and post-assessment and monitoring must be incorporated into the overall restoration project, and (f) a description or prediction of the ecological mechanisms should be determined through which the intended restoration strategy will achieve its goals.

The chemical recovery of many streams and rivers from organic pollution since the industrial revolution and large-scale urbanisation is an example of the restoration of water quality, although we do not normally use the term 'restoration' in the context of pollution. This has had many benefits to human health and well-being and to biological diversity. However, much work remains to be done on water quality and on the many other stressors that we have discussed. If we really understand how river ecosystems work, it should be possible to restore them using that knowledge, though evidently there is some way to go. Nevertheless, we have learned a lot from our 'failures' and rather more limited successes. Restoration is an enormously active area, in terms of river management and research, and very large sums of money are being spent on it—so far with rather limited ecological monitoring or application of real criteria for success such as those suggested above. As our experience with different kinds of restoration expands, and as long as sufficient repeated monitoring of restoration schemes is carried out, we can look forward to more effective river restoration in future.

As is almost always true of running waters, this area crosses ecology with hydrology and fluvial morphology and socioeconomics. Here we briefly touch on biological and ecological responses to restoration. What have we learned so far? Palmer et al. (2010) reviewed work aimed at increasing physical 'habitat heterogeneity' and at increasing ecological diversity—mainly of benthic invertebrates. They included 78

(a)

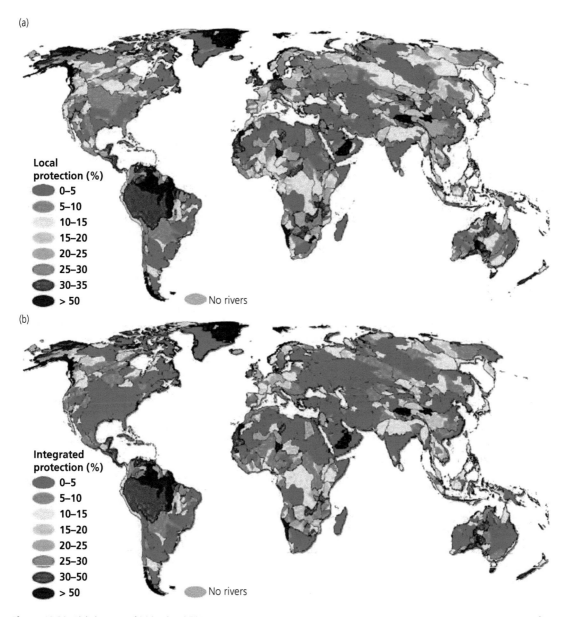

(b)

Figure 10.20 Global pattern of (a) local and (b) integrated protection of rivers. Colours are resolved into sub-basins of around 100,000 km^2, while the black lines identify major river basins. Substantial parts of the Amazon basin and Alaska—hitherto sparsely populated and developed—have 'well-protected rivers' on this evidence, while 'integrated protection' is much less than local protection in the developed world.
Source: from Abell et al. 2016; published under the Creative Commons Attribution Licence.

independent restoration schemes in their review, varying in scale from 'reinstatement' of a few riffles or flow deflectors in a short stretch of channel to more extensive re-meandering of the channel to what was presumed to be closer to an original condition. Many schemes did increase habitat heterogeneity, at least in the short term, although the response of invertebrate diversity was either absent or weak. Given the prevailing paradigm that habitat heterogeneity does increase ecological diversity, this was a disappointing result, but one from which lessons can be learnt. Favoured explanatory hypotheses for the apparent failures of impact (at least for invertebrates) were that: (a) the scale of habitat intervention was important, in that reinstating just a few riffles within an overall degraded agricultural catchment was insufficient, (b) other stressors pertaining locally, such as excess sediment supply or poor water quality, could overwhelm the restoration scheme (a 'multistressor' hypothesis), (c) sources of colonists were too distant for dispersal to be likely, (d) the restoration was inappropriate for the particular site (e.g. coarse substratum particles were added in a catchment of fine alluvium), (e) assessment of the response was insufficient. These are, of course, not mutually exclusive hypotheses.

Attempts to restore in-stream habitat and increase diversity continue (see reviews by Griffith & McManus 2020; dos Reis Oliveira et al. 2020). Results have frequently been equivocal, as found by Palmer et al (2010) above, although there have also been examples of success (see Lehane et al, 2002 and Thompson et al. 2018 for examples of success by adding large woody debris to lowland forested and agricultural rivers at small (former) and large (latter) scales). A particularly successful scheme, at a very large scale, was undertaken on the River Skern, Denmark's biggest river—Denmark being an early pioneer in river restoration (see Feld et al. 2011). The Skern originally had a floodplain 65–100 m wide but this was much reduced after a large-scale regulation scheme in 1968 that 'reclaimed' 3,500 ha of agricultural land from periodically inundated areas. The scheme was reversed from 1999, to improve nutrient retention, reinstate habitat for migratory birds, improve fishing and promote ecotourism. About 19.5 km² of land was purchased from farmers, the pre-regulation channel was reinstated and the overall length of river greatly increased. Water quality had remained good, there were remnant populations of key species in the catchment providing potential colonists, and inputs of fixed nitrogen were reduced. Perhaps not surprisingly, restoration of the River Skern was highly successful ecologically. Of course, restoration at this scale is rarely possible, and in general we probably need to be less ambitious with respect to our targets and as to what we can expect of small-scale projects within large catchments. Measures to reduce the wider catchment-scale impact on rivers of intensive agricultural practice, particularly through reducing erosion and chemical inputs, are all likely to contribute to the success of more local habitat-restoration schemes. Good, wide buffer strips of native vegetation (> 15 m on each bank, and > 1 km in length) were also helpful in schemes surveyed by Feld et al. (2011) and are also acknowledged as successful in mediating impacts of forestry practices on adjacent river systems (e.g. Giller & O'Halloran 2004).

In a novel approach, Levi & McIntyre (2020) assessed the success of channel-restoration projects by examining the response of ecosystem processes (nutrient uptake, whole-stream metabolism, primary production) rather than biodiversity *per se*. The setting was a heavily urbanised area of Milwaukee, Wisconsin (USA), and the study included six pairs of contiguous restored and concrete channel reaches at various points in the network from headwaters to the mainstem (baseflow discharge at the six sites ranged from 10–196 L s⁻¹). Water velocity was greatly reduced in restored reaches and water residence time was 50–5,000% greater than in the concrete reaches. Restored reaches had shorter uptake lengths for ammonium, nitrate and phosphate (a greater rate of uptake by primary producers), and higher whole-stream metabolism. Streams were autotrophic (GPP exceeded ecosystem respiration) and effect sizes were greatest at sites with the lowest discharge (upstream reaches) (Figure 10.21). This example demonstrates that restoration can have benefits in terms of processes and the ecosystem services provided.

As we have mentioned, restoration schemes are often not successful ecologically. At its worst, restoration can effectively involve inappropriate structures being imposed on a resistant system. Rivers 'work against' such schemes, which inevitably require constant maintenance. 'All' a river requires morphologically is a natural flow regime, a sufficient but not excessive supply of sediment, and space in which to move (from side to side—i.e not physically constrained). Given sufficient time, a river will then rework its own dynamic and heterogenous channel and, given suitable water quality, its ecological status should improve. When the value of a 'damaged' river is high, however, or when one or more of these requirements is not met (e.g. water withdrawals have reduced the flow), careful restoration schemes can certainly help.

10.9 Skills for the coming years?

In this final chapter we have tried to give a glimpse of some of the important recent and ongoing applied issues and techniques that occupy freshwater ecologists. Our impression in writing this book, with personal involvements in the field going back over many years, is of enormous shifts in emphasis, expansion and 'internationalisation' of published outputs and major advances, both conceptual and methodological. A student beginning a career in river science and management in its broadest sense evidently needs a much greater range of skills, knowledge and expertise than

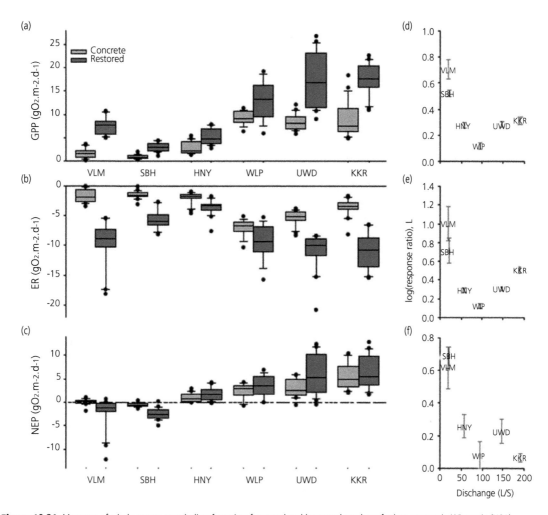

Figure 10.21 Measures of whole-stream metabolism for pairs of restored and 'concrete' reaches of urban streams in Wisconsin (USA) (abbreviations are site names). (a), gross primary production; (b), ecosystem respiration; (c), net ecosystem production—negative values show respiration exceeds primary production. Panels (d), (e) and (f) indicate 'effect sizes' of restoration on the three measurements in relation to baseflow discharge at the sites.

Source: from Levi & McIntyre 2020, with permission of John Wiley & Sons.

was the case 30–40 years ago. In this final section we risk a few opinions about what this background should contain. The physicist Niels Bohr is widely credited with saying 'Prediction is very difficult, particularly when it's about the future'—so we may be wrong, but we hope it is helpful.

We have no doubt that an absolute must is a passion for the natural world, and a good grounding in ecology and natural history. These remain cornerstones for a career, whether in the biological side of freshwater science, the policy sector or management. We stress here biology and ecology, but a sound knowledge of the physical environmental sciences (fluvial morphology

and dynamics, hydrology, river biogeochemistry etc) is also essential. Rivers are dynamic physicochemical systems as well as ecosystems, and the modern river scientist needs knowledge of both. Added to that, a facility to work with 'big data' is invaluable—an ability to deal with databases, martialling and extracting information from online sources.

A knowledge of and familiarity with molecular tools and methods is increasingly relevant in the fields of biodiversity assessment, food web ecology, dispersal of organisms, invasive species biology and conservation biology. It is also apparent that an understanding of the socio-economic underpinning of sustainable river management is highly desirable—including the dispersed 'societal' benefits accruing from the natural world (natural capital and ecosystem services), and the costs that may be imposed on developers, users of the river and riparian owners and managers.

Finally, river scientists need to be good communicators—people who can 'enthuse the young of all ages', who can make the case for rivers to the general public, environmental managers and, increasingly, to politicians. Good, simple communication and the avoidance of jargon is an increasingly valuable skill. In this context, a very strong development over the last decade or more has been the increasing involvement of 'citizen scientists'—members of the public who take an interest in their 'local river' (see e.g. Thornhill et al. 2019 and papers therein). They can supplement and

extend surveillance by regulatory authorities, though not replace it. A further fascinating but distinct movement in some areas has been to include the traditional knowledge of indigenous peoples in conservation of natural resources. Combining 'western science' with these often very different approaches and forms of evidence is challenging but has been particularly prominent among the First Nation peoples of North America and the Maori of New Zealand (e.g. Gadgil et al. 2021).

This list of attributes and skills is challenging—no one course or degree programme is going to supply them all, and no one person is going to acquire them all. Similarly, although no book can provide all of the answers, we do hope this one is a useful introduction to 'the biology and ecology of streams and rivers'.

References

Abell, R., Kottelat, M., Thieme, M. & Bogutskaya, N. (2008). Freshwater ecoregions of the world: a new map of biogeographic units for freshwater biodiversity conservation. *Bioscience*, 58, 403–414.

Abell, R., Lehne, B., Thieme, M. & Linke, S. (2016). Looking beyond the fenceline: assessing protection gaps for the world's rivers. *Conservation Letters*, 10, 384–394.

Abril, M., Bastias, E., von Schiller, D. et al. (2019). Uptake and trophic transfer of nitrogen and carbon in a temperate forested headwater stream. *Aquatic Sciences*, 81, 75, https://doi.org/10.1007/s00027-019-0672-x.

Acuña, V., Vilches, C. & Giorgi, A. (2011). As productive and slow as a stream can be – the metabolism of a Pampean stream. *Journal of the North American Benthological Society*, 30, 71–83.

Adler, P. & Courtney, G. (2019). Ecological and societal services of aquatic diptera. *Insects*, 10, 70; doi:10.3390/insects10030070.

Albertson, L.K., Sklar, L.S., Cooper, S.D. & Cardinale, B.J. (2019). Aquatic macroinvertebrates stabilise gravel bed sediment: A test using silk net-spinning caddisflies in semi-natural river channels. *PLoS ONE* 14(1): e0209087. https://doi.org/10.1371/journal.pone.0209087.

Albrecht, C., Stelbrink, B., Gauffre-Autelin, P., Marwoto, R.M. et al. (2020). Diversification of epizoic freshwater limpets in ancient lakes on Sulawesi, Indonesia: Coincidence or coevolution? *Journal of Great Lakes Research*, 46, 1187–1198.

Ali, M.M., Murphy, K.J. & Abernethy, V.J. (1999). Macrophyte functional variables as predictors of trophic status in flowing waters. *Hydrobiologia*, 415, 131–138.

Allan, J.D. (1982). The effects of reduction in trout density on the invertebrate community of a mountain stream. *Ecology*, 63, 1444–1455.

Allan, J.D. & Castillo, M.M. (2007). Stream ecology: structure and function of running waters. *Springer*, Dodrecht, 436.

Allan, J.D. & Feifarek, B.P. (1989). Distances travelled by drifting mayfly nymphs: factors influencing return to the substrate. *Journal of the North American Benthological Society*, 8, 320–330.

Allen, G.H. & Pavelsky, T.M. (2018). Global extent of rivers. *Science*, DOI: 10.1126/science.aat0636.

Almeida, R.M., Han, B.A., Reisinger, A.J. et al. (2018). High mortality in aquatic predators of mosquito larvae caused by exposure to insect repellent. *Biology Letters*, 14, 20180526.

Alnoee, A.B., Riis, T. & Baattrup-Peddersen, A. (2016). Comparison of metabolic rates among macrophyte and non-macrophyte habitats in streams. *Freshwater Science*, 35, 834–844.

Altermatt, F. (2013). Diversity in riverine metacommunities: a network perspective. *Aquatic Ecology*, 47, 365–377.

AmphibiaWeb 2020. https://amphibiaweb.org.

Anderson, C. & Cabana, G. (2009). Anthropogenic alterations of lotic food web structure: evidence from the use of nitrogen isotopes. *Oikos*, 118, 1929–1939.

Anderson, C.B. & Rosemond, A.D. (2007). Ecosystem engineering by invasive exotic beavers reduces in-stream diversity and enhances ecosystem function in Cape-Horn, Chile. *Oecologia*, 154, 141–153.

Anderson, K.E., Nisbet, R.M., Diehl, S. & Cooper, S.D. (2005). Scaling population responses to spatial environmental variability in advection-dominated systems. *Ecology Letters*, 8, 933–943.

Anderson, T., Cranston, P.S. & Epler, J.H. (eds) (2013). *Chironomidae of the Holarctic Region: keys and diagnoses, Part 1: larvae. Insect Systematics and Evolution Supplements*, 66, 573.

Anholt, B.R. (1995). Density-dependence resolves the stream drift paradox. *Ecology*, 76, 2235–2239.

Armitage, P.D. (1995). Behaviour and ecology of adults. In P. Armitage, P.S. Cranston and L.C.V. Pinder (eds), *The Chironomidae. The Biology and Ecology of Non-biting Midges*, 194–224. Chapman and Hall, London.

Arscott, D.B., Larned, S., Scarsbrook, M.R. & Lambert, P. (2010). Aquatic invertebrate community structure along an intermittence gradient: Selwyn River, New Zealand. *Freshwater Science*, 29. doi.org/10.1899/08-124.1.

Ashendorff, A., Principe, M.A., Seely, A. et al. (1997). Watershed protection for New York City's supply. *Journal of the American Water Works Association*, 89, 75–88.

Aspin, T.W.H., Khamis, K., Matthews, T.J. et al. (2018) Extreme drought pushes stream invertebrate communities over functional thresholds. *Global Change Biology*, doi: 10.1111/gcb. 14495.

Atkinson, C.L., Encalada, A.C., Rugenski, A.T. et al. (2018). Determinants of food resource assimilation by stream insects along a tropical elevation gradient. *Oecologia*, 187, 731–744.

Atkinson, C.L. & Forshay, K.J. (2022). Community patch dynamics governs direct and indirect nutrient recycling by aggregated animals across spatial scales. *Functional Ecology*. DOI: 10.1111/1365-2435.13982.

Atkinson, S., Carlsson, J.E.L., Ball, B. Kelly-Quinn, M. et al. (2019). Field application of eDNA asset for the threatened white-claw crayfish *Austropotamobius pallipes*. *Freshwater Biology*, 38, 503–509.

Atkinson, W.D. & Shorrocks, B. (1981). Competition on a divided ephemeral resource: a simulation model. *Journal of Animal Ecology*, 50, 461–471.

Atkore, V., Keikar, N., Badiger, S., Shanker, K, et al. (2020). Multiscale investigation of water chemistry effects on fish guild species richness in regulated and nonregulated rivers of India's Western Ghats: implications for restoration. *Transactions of the American Fisheries Society*, 149, 298–319.

Audet, J., Bastviken, D., Bundschu, M. et al. (2019). Forest streams are important sources for nitrous oxide emissions. *Global Change Biology*, 26, 629–641.

Audet, J., Olsen, T.M., Elsborg, T. et al. (2020). Influence of plant habitats on denitrification in lowland agricultural streams. *Journal of Environmental Management*, 286, 112193.

Baattrup-Pedersen, A., Gothe, E., Larsen, S., O'Hare, M. et al. (2015). Plant trait characteristics in European lowland streams. *Journal of Applied Ecology*, 52, 1617–1628.

Baattrup-Pedersen, A., Göthe, E., Riis, T. & O'Hare, M. (2016). Functional trait composition of aquatic plants can serve to disentangle multiple interacting stressors in lowland streams. *Science of the Total Environment*, 543, 230–238.

Baattrup-Pedersen, A., Larsen, S.E. & Riis, T. (2002). Long term effects of stream management on plant communities in two Danish lowland streams. *Hydrobiologia*, 481, 33–45.

Bagnoud, A., Pramataftaki, P., Bogard, M.J. et al. (2020). Microbial ecology of methanotrophy in streams along a gradient of CH_4 availability. *Frontiers in Microbiology*, 11, 771. Doi 10.3389/fmicb.2020.00771.

Baker, K., Chadwick, M.A., Wahab, R.A. & Kahar, R. (2017). Benthic community structure and ecosystem functions in above-and below-waterfall pools in Borneo. *Hydrobiologia*, 787, 307–322.

Baker, M.A. & Webster, J.R. (2017). Conservative and reactive solute dynamics. In G.A. Lamberti & F.R. Hauer, (eds) *Methods in Stream Ecology. Volume 2: Ecosystem Function*, 3rd edn, Academic Press, London, 129–145.

Baker, V.R., Hamilton, C.W., Burr, D.M., Gulik, V.C. et al. (2015). Fluvial geomorphology on earth-like planetary surfaces: a review. *Geomorphology*, 245, 149–182.

Balian, E.V., Lévêque, C. Segers, H. & Martens, K. (2008). *Freshwater Animal Bioiversity Assessment*, Springer, Berlin.

Ballinger, A. & Lake, P.S. (2006). Energy and nutrient fluxes from rivers and streams into terrestrial food webs. *Marine and Freshwater Research*, 57, 15–28.

Baptista, D.F., Buss, D.F., Dias, L.G. et al. (2006). Functional feeding groups of Brazilian Ephemeroptera nymphs: ultrastructure of mouthparts. *Annales de Limnologie—International Journal of Limnology*, 42, 87–96.

Baranov, V., Jourdan, J., Pilotto, F., Wagner, R. et al. (2020). Complex and non-linear climate-driven changes in freshwater insect communities over 42 years. *Conservation Biology*, 34, 1241–1251. DOI: 10.1111/cobi.13477.

Bärlocher F. (ed) (1992). *The Ecology of Aquatic Hyphomycetes*. Springer, Berlin.

Bärlocher F. (ed) (2016). Aquatic fungi. *Fungal Ecology*, 19, 1–218.

Bärlocher, F., Gessner, M.O. & Graça, M.A.S. (eds) (2020). *Methods to Study Litter Decomposition: A Practical Guide* (2nd edn). Springer Nature, Cham, Switzerland.

Bärlocher F. & Sridhar, K.R. (2014). Associations of animals and fungi in leaf decomposition. In E.B.G. Jones, K.D. Hyde & K.-L. Pang (eds) *Freshwater Fungi and Fungal-like Organisms*, De Gruyter, Berlin, 423–442.

Barr, W.B. (1984). Prolegs and attachment of *Simulium vittatum* (sibling 1S-7) (Diptera: *Simuliidae*) larvae. *Canadian Journal of Zoology*, 62, 1355–1362.

Barrat-Segretain, M.H. (1996). Strategies of reproduction, dispersion and competition in river plants: A review. *Vegetation*, 123, 13–37.

Baschien C., Tsui, C.K.M., Gulis, V., Szewzyk U. et al. (2013). The molecular phylogeny of aquatic hyphomycetes with affinity to the Leotiomycetes. *Fungal Biology*, 117, 660–672.

Bashevkin, S.M. & Mahardja, B. (2022). Seasonally variable relationships between surface water temperature and inflow in the Upper San Francisco Estuary. *Limnology & Oceanography*, doi 10.1002/lno.12027.

Bastos, D.A., Zuanon, J., Rapp Py-Daniel, L. & de Santana, C.D. (2021). Social predation in electric eels. *Ecology and Evolution*, 11, 1088–1092.

Battin, T.J., Luysaart, S., Kaplan, L.A. et al. (2009). The boundless carbon cycle. *Nature Geosciences*, 2, 598–600.

Battin, T.J., Besemer, K., Bengtsson, M.M., Romani, A.M. et al. (2016). The ecology and biogeochemistry of stream biofilms. *Nature Reviews Microbiology*, 14, 251–263 (doi 10.1038/nmicro. 2016.15).

Battin, T.J., Lauerwald, R., Bernhardt, E.S., Bertuzzo, E. et al. (2023). River ecosystem metabolism and carbon biogeochemistry in a changing world. *Nature*. DOI: 10.1038/s41586-022-05500-8.

Bauer, R.T. (2013). Amphidromy in shrimps: a life cycle between rivers and the sea. *Latin American Journal of Aquatic Research*, 41, 633–650.

Baxter, C.V., Kausch, K.D. & Saunders, C.W. (2005). Tangled webs: reciprocal flows of invertebrate prey link streams and riparian zones. *Freshwater Biology*, 50, 201–220.

Bayley, P.B. & Li, H.W. (1996). Riverine Fishes. In G. Petts and P. Galow (eds) *River Biota Diversity and Dynamics*, Blackwell, Oxford, 92–122.

Beauchamp, R.S. & Ullyot, P. (1932). Competitive relationships between certain species of freshwater triclads. *Journal of Ecology*, 20, 200–208.

Beauchesne, D., Cazelles, K., Archambault, P., Dee, L.E. Gravel, D. (2021). On the sensitivity of food webs to multiple stressors. *Ecology letters*, 24, 2219–2237.

Beaulieu, J.K., Arango, C.P., Balz, D.A & Shuster, W.D. (2013). Continuous monitoring reveals multiple controls on ecosystem metabolism in a suburban stream. *Freshwater Biology*, 58, 918–937.

Beck, W.S., Rugenski, A.T. & Poff, N.L. (2017). Influence of experimental, environmental and geographic factors on nutrient-diffusing substrate experiments in running waters. *Freshwater Biology*, 62, 1667–1680.

Beer, T. & Young, P.C. (1983). Longitudinal dispersion in natural streams. *Journal of Environmental Engineering*, 109, 1047–1067.

Beermann, A.J., Zizka, V.M.A., Elbrecht, V. et al. (2018). DNA metabarcodong reveals the complex and hidden responses of chironomids to multiple stressors. *Environmental Sciences Europe*, 30, 26 doi.org/101186/s12303-018-0157-x.

Begon, M., Townsend, C.R. & Harper, J.L. (2006). *Ecology: From Individuals to Ecosystems* (4th edn) Blackwell Publishing, Oxford.

Bellamy, A.R., Bauer, J.E. & Grottoli, A.G. (2019). Contributions of autochthonous, allochthonous and aged carbon and organic matter to macroinvertebrate nutrition in the Susquehanna River Basin. *Freshwater Science*, doi 10.1086/705017.

Belletti, B., Garcia de Leaniz, C., Jones, J. et al. (2020). More than one million barriers fragment Europe's rivers. *Nature*, 588, 436–441.

Benbow, M.E., Pechal, J.L. & Ward, A.K. (2017). Heterotrophic bacterial production and microbial community assessment. In F.R. Hauer & G.A. Lamberti (eds) *Methods in Stream Ecology. Volume 1: Ecosystem Structure* (3rd edn), Academic Press, London, 161–176.

Bencala, K.E. & Walters, R.A. (1983). Simulation of solute transport in a mountain pool-and riffle stream: a transient storage model. *Water Resources Research*, 19, 718–724.

Benfield, E.F., Fritz, K.M. & Tiegs, S.D. (2017). Leaf litter breakdown. In G.A. Lamberti & F.R. Hauer (eds) *Methods in Stream Ecology. Volume 2: Ecosystem Function* (3rd edn), Academic Press, London, 71–82.

Benke, A. (2018). River food webs: an integrative approach to bottom-up flow webs and top-down impact webs, and trophic position. *Ecology*, 99, 1370–1381.

Benke, A.C. & Cushing, C.E. (eds) (2005). *Rivers of North America*. Elsevier.

Benke, A.C. & Huryn, A.D. (2017). Secondary production and quantitiative food webs. In G.A. Lamberti & F.R. Hauer (eds) *Methods in Stream Ecology Volume 2: Ecosystem Function* (3rd edn), Academic Press, London, 235–254.

Benke, A.C. & Jacobi, D.I. (1994). Production dynamics and resource utilization of snag-dwelling mayflies in a blackwater river. *Ecology*, 75, 1219–1232.

Benke, A.C. & Wallace, J.B. (1980). Trophic basis of production among net-spinning caddisflies in a southern Appalachian stream. *Ecology*, 61, 108–118.

Benstead J.P., Barnes, K. & Pringle, C.M. (2001). Diet, activity patterns, foraging movement and responses to deforestation of the aquatic tenrec *Limnogale mergulus* (Lipotyphla:

Tenrecidae) in eastern Madagascar. *Journal of Zoology*, London, 254, 119–129.

Benstead, J.P., Cross, W.F., Gulis, V. & Rosemond, A.D. (2021). Combined carbon flows through detritus, microbes, and animals in reference and experimentally enriched stream ecosystems. *Ecology*, 102, e03279.

Benstead, J.P., Evans-White, M.A. Gibson, C.A. & Hood, J.H. (2017). Elemental content of stream biota. In F.R. Hauer & G.A. Lamberti (eds) *Methods in Stream Ecology Volume 2: Ecosystem Function* (3rd edn), Academic Press, London, 255–273.

Benstead, J.P., March, J.G., Pringle, C.M., Ewel, K.C. et al. (2009). Biodiversity and ecosystem function in species-poor communities: community structure and leaf litter breakdown in a Pacific island stream. *Journal of the North American Benthological Society*, 28, 454–465.

Bere, T. & Tundisi, J.D. (2011). The effects of substrate type on diatom-based multivariate quality assessment in a tropical river (Monjolinho), Sao Carlos, SP, Brazil. *Water, Air and Substrate Pollution*, 216, 391–409.

Berg, M.B. (1995). Larval food and feeding behaviour. In P. Armitage, P.S. Cranston, & L.C.V. Pinder (eds) *The Chironomidae*. Chapman and Hall, London, 136–168.

Bergey, E.A. (2005). How protective are refuges? Quantifying algal protection in rock crevices. *Freshwater Biology*, 50, 1163–1177.

Bergfeld, T., Scherwass, A., Ackerman, B., Arndt, H. et al. (2009). Comparison of the components of the planktonic food web in three large rivers (Rhine, Moselle and Saar). *River Research and Applications*, 25, 1232–1250.

Berggren, M., Bengtson, P, Soares, A.R.A. & Karlsson, J. (2018). Terrestrial support of zooplankton biomass in northern rivers. *Limnology & Oceanography*, 63, 2479–2492.

Bernal, S., von Schiller, D., Sabater, F. & Marti, E. (2013). Hydrological extremes modulate nutrient dynamics in mediterranean climate streams across different spatial scales. *Hydrobiologia*, 719, 31–42.

Berner, E.K. & Berner, R.A. (1987). *The Global Water Cycle*. Prentice-Hall, Englewood Cliffs, NJ.

Bernhardt, E.S., Heffernan, J.B., Grimm, N.B. et al. (2018). The metabolic regime of flowing waters. *Limnology & Oceanography*, 63, S99–S118.

Bernhardt, E.S., Rosi, E.J. & Gessner, M.O. (2017). Synthetic chemicals: a neglected driver of global change. *Frontiers in Ecology and the Environment*, 15, 84–90.

Berthon, V. Bouchez, A. and Rimet, F. (2011). Using diatom life-forms and ecological guilds to assess organic pollution and trophic level in rivers: a case study of rivers in southeastern France. *Hydrobiologia*, 673, 259–271.

Biffi, M., Gillet, F., Laffaille, P., Colas, F. et al. (2017). Novel insights into the diet of the Pyrenean desman (*Galemys pyrenaicus*) using next-generation sequencing molecular analyses. *Journal of Mammology*, 98, 1497–1507.

Biggs, B.J.F. (2000). Eutrophication of streams and rivers: dissolved nutrient chlorophyll relationships for benthic algae. *Journal of the North American Benthololgical Society*, 19, 17–31.

Billen, G., Decamps, H., Garnier, J., Boet, P., Meybeck, M., & Servais, P. (1995). River and stream ecosystems: Atlantic river systems of Western Europe (France, Belgium, The Netherlands). In: C.E. Cushing et al. (eds) *River and stream ecosystems, ecosystems of the world*, Vol. 22, Chapter 12. Elsevier, Amsterdam, 389–418.

Bilton, D., Ribera, I. & Short, A.E. (2019). Water beetles as models in ecology and evolution. *Annual Review of Entomology*, 64, 359–377.

Bino, G., Kingsford, R., Archer, M., Connolly, J. et al. (2019). The platypus: evolutionary history, biology and an uncertain future. *Journal of Mammology*, 100, 308–327.

Birk, S., Bonne, W. & Borja, A. et al. (2012). Three hundred ways to assess Europe's surface waters: an almost complete overview of biological methods to implement the Water Framework Directive. *Ecological Indicators*, 18, 31–41.

Bishop, K., Buffam, I., Erlandsson, M. et al. (2008). *Aqua Incognita*: the unknown headwaters. *Hydrological Processes*, 22, 1239–1242.

Blackburn, S.R. & Stanley, E.H. (2021). Floods increase carbon dioxide and methane fluxes in agricultural streams. *Freshwater Biology*, 66, 62–77.

Blinn, D.M. & Poff, N.L. (2005). Colorado River Basin. In A.C. Benke & C.E. Cushing (eds) *Rivers of North America* Elsevier Academic Press, Amsterdam, 483–538.

Böck, K., Polt, R. & Schülting, L. (2018). Ecosystem services in river landscapes. In S. Schmutz and J. Sendzimir (eds) *Riverine Ecosystem Management*. Aquatic Ecology Series, 8. Springer Open.

Bogan, M.T. (2017). Hurry up and wait: life cycle and distribution of an intermittent stream specialist (*Mesocapnia arizonensis*). *Freshwater Science*, 36, 805–815.

Bogan, M.T., Boersma, K.S. & Lytle, D.A. (2013). Flow intermittency alters longitudinal patterns of invertebrate diversity and assemblage composition in an arid-land stream network. *Freshwater Biology*, 58, 1016–1028.

Bogan, M.T., Chester, E.T., Datry, T., Murphy, A.L. et al. (2017). Resistance, resilience and community recovery in intermittent rivers and ephemeral streams. In N. Bonada, A. Boulton & T. Datry (eds) *Intermittent Rivers and Ephemeral Streams: Ecology and Management*. Academic Press, Burlington, 624.

Bogan, M.T. & Lytle, D.A. (2011). Severe drought drives novel community trajectories in desert stream pools. *Freshwater Biology*, 56, 2070–2081.

Bonada, N. & Dolédec S. (2018). Does the Tachet trait database report voltinism variability of aquatic insects between Mediterranean and Scandinavian regions? *Aquatic Sciences*, 80, 1–11.

Bonada, N., Dolédec, S. & Statzner, B. (2007). Taxonomic and biological trait differences of stream macroinvertebrate communities between Mediterranean and temperate regions: implication for future climatic scenarios. *Global Change Biology*, 13, 1658–1671.

Bonada, N., Prat, N., Resh, V.H. & Statzner, B. (2006). Developments in aquatic insect biomonitoring: a comparative analysis of recent approaches. *Annual Reviews of Entomology*, 51, 485–523.

Borchardt, D. (1993). Effects of flow and refugia on drift loss of benthic macroinvertebrates: implications for habitat restoration in lowland streams. *Freshwater Biology*, 29, 221–227.

Borges, A.V., Darchambeau, F., Teodoru, C.R. et al. (2015). Globally significant greenhouse-gas emissions from African inland waters. *Nature Geoscience*, 8, 637–642.

Bothwell, M.L. & Taylor, B.W. (2017). Blooms of benthic diatoms in phosphorus-poor streams. *Frontiers in Ecology and Environment*, 15, 110–111.

Bott, T.L. & Kaplan, L.A. (1985). Bacterial biomass, metabolic state, and activity in stream sediments: relation to environmental variables and multiple assay comparisons. *Applied & Environmental Microbiology*, 50, 508–522.

Boulton, A.J. & Lake, P.S. (2007). Effects of drought on stream insects and its ecological consequences. In J. Lancaster & R.A. Briers (eds) *Aquatic Insects: Challenges to Populations, 24th Symposium of the Royal Entomological Society*, 81–102.

Boulton, A.J., Boyero, L., Covich, A.P. et al. (2008). Are tropical streams ecologically different from temperate streams? In D. Dudgeon (ed) *Tropical Stream Ecology*, Aquatic Ecology, 257–284. https://doi.org/10.1016/B978-012088449-0.50011-X.

Bour, R. (2008). Global diversity of turtles (Chelonii: Reptilia) in freshwater. *Hydrobiologia*, 595, 593–598.

Bournaud, M. (1975). Eléments d'observation sur la cinématique, la dynamique et l'énergétique de la locomotion dans le courant chez une larve de Trichoptère à fourreau. *Hydrobiologia*, 46, 489–513.

Bowden, W.B., Glime, J.M. & Riis, T. (2017). Macrophytes and Bryophytes. In F.R. Hauer & G.A. Lamberti (eds) *Methods in Stream Ecology* (3rd edn). Academic Press, San Diego, London, 243–271.

Bowman, J.J. & Bracken, J.J. (1993). Effect of run-off from afforested and nonafforested catchments on the survival of brown trout *Salmo trutta* L. in two acid sensitive rivers in Wicklow, Ireland. *Biology and Environment. Proceedings of the Royal Irish Academy*, 93B, 143–152.

Boyero, L., López-Rojo, N., Tonin, A.M. et al. (2021). Impacts of detritivore diversity loss on instream decomposition are greatest in the tropics. *Nature Communications*, 12 (1), 3700 doi 10.1038/s41467-021-23930-2.

Boyero, L., Pearson, R.G., Dudgeon, D. et al. (2012). Global patterns of stream detritivore distribution: implications for biodiversity loss in changing climates. *Global Ecology & Biogeography*, 21, 134–141.

Boyero, L. Pearson, R.G., Gessner, M.O. et al. (2015). Leaf-litter breakdown in tropical streams: is variability the norm? *Freshwater Science*, 34, 759–769.

Boyero, L., Pearson, R.G., Hui, C. et al. (2016). Biotic and abiotic variables influencing plant litter breakdown in streams: a global study. *Proceedings of the Royal Society of London B: Biological Sciences*, 283, 2015.2664.

Boyero, L., Perez, J. et al. (2021). Latitude dictates plant diversity effects on instream decomposition. *Science Advances*, 7, eabe7860.

Boyero, L., Ramirez, A. Dudgeon. D. & Pearson, R.G. (2009). Are tropical streams really different? *Journal of the North American Benthological Society*, 28, 397–403.

Bradley, D.C. & Ormerod, S.J. (2001). Community persistence among stream invertebrates tracks the North Atlantic Oscillation. *Journal of Animal Ecology*, 70, 987–996.

Brand, C. & Miserendino, M. (2011). Life history strategies and production of caddisflies in a perennial headwater stream in Patagonia. *Hydrobiologia*, 673(1),137–151.

Brauns, M., Kneis, D., Brabender, M. & Weitere, M. (2022). Habitat availability determines food chain length and interaction strength in food webs of a large lowland river. *River Research and Applications*, 38, 323–333.

Brendelburger, H. & Klauke, C. (2009). Pedal feeding in freshwater unionid mussels: particle-size selectivity. *Verhandlungen der Internationale Vereinigung für Theoretische und Angewandte Limnologie*, 30, 1082–1084.

Brett, M.T., Bunn, S.E., Chandra, S. et al. (2017). How important are terrestrial organic carbon inputs for secondary production in freshwater ecosystems? *Freshwater Biology*, 62, 833–853.

Briand, F. & Cohen, J.E. (1987). Environmental correlates of food chain length. *Science*, 238, 956–960.

Bridcut, E. & Giller, P.S. (1993). Movement and site fidelity in young brown trout (*Salmo trutta* L.) populations in a southern Irish stream. *Journal of Fish Biology*, 43, 889–899.

Bridcut, E. & Giller, P.S. (1995). Diet variability and foraging strategies in brown trout (Salmo trutta L.): an analysis from sub-populations to individuals. *Canadian Journal of Fisheries and Aquatic Science*, 52, 2543–2552.

Briers, R.A., Gee, J.H.R., Carris, H.M. & Geoghan, R. (2004). Inter-population dispersal by adult stoneflies detected by stable isotope enrichment. *Freshwater Biology*, 49, 425–431.

Brighenti, S., Hotaling, S., Finn, D.S. et al. (2021). Rock glaciers and related cold rocky landforms: Overlooked climate refugia for mountain biodiversity. *Global Change Biology*, 27, 1504–1517.

Brinkhurst, R.O. & Gelder, S.R. (1991). Annelids. In *Ecology and Classification of North American Freshwater Invertebrates*. First edition. (ed by J.H. Thorp and A.P. Covich). 401–436, Academic Press, San Diego.

Brittain, J.E. (1982). Biology of mayflies. *Annual Review of Entomology*, 27, 119–147.

Brittain, J.E. & Ekeland, T.J. (1988). Invertebrate drift—a review. *Hydrobiologia*, 166, 77–93.

Brodin, T., Fick, J., Jonsson, M. et al. (2013). Dilute Concentrations of a Psychiatric Drug Alter Behavior of Fish from Natural Populations. *Science*, 339, 814–815.

Brönmark, C. & Hansson, L.-A. (2017). *The Biology of Lakes and Ponds* (3rd edn) Biology of Habitats, Oxford University Press, 338 pages.

Brönmark,C., Malmqvist, B. & Otto, C. (1985). Dynamics and structure of a *Velia caprai* (Heteroptera) population in a south-Swedish stream. *Holarctic Ecology*, 8, 253–258.

Brown, B.L., Creed, R.P., Skelton, J., Rollins, M.A. et al. (2012). The fine line between mutualism and parasitism: Complex effects in a cleaning symbiosis demonstrated by multiple field experiments. *Oecologia*, 170, 199–207.

Brown, B.L., Swan, C.M., Auerbach, D.A., Campbell Grant, E.H. et al. (2011). Metacommunity theory as a multispecies framework for studying the influence of river network structure on riverine communities and ecosystems. *Journal of the North American Benthological Society*, 30, 310–327.

Brown, J., Gillooly, J., Allen, A., Savage, V. et al. (2004). Toward a metabolic theory of ecology. *Ecology*, 85, 1771–1789.

Brown, K.M. (1991). Mollusca: Gastropoda. In J.H. Thorp & A.P. Covich (eds) *Ecology and Classification of North American Freshwater Invertebrates*, Academic Press, San Diego, 285–314.

Brown, L.E. & Milner, A.M. (2012). Rapid loss of glacial ice reveals stream community assembly processes. *Global Change Biology*, 18, 2195–2204.

Bu, H., Tan, X., Li, S. & Zhang, Q. (2010). Temporal and spatial variations of water quality in the Jinshui River of the South Qinling Mts., China. *Ecotoxicology and Environmental Safety*, 73, 907–913.

Buchwalter, D.B., Lamberti, G.A., Resh, V.F.H. & Verberk, W.C. (2019). Aquatic insect respiration. In R.W. Merritt, K.W. Cummins & M.B. Berg (eds) *An Introduction to the Aquatic Insects of North America* (5th edn). Kendall/Hunt, Dubuque, Iowa, 39–54.

Buckton, S.T. & Ormerod S.J. (2002). Global patterns of diversity among the specialist birds of riverine landscapes. *Freshwater Biology*, 47, 695–709. https://doi.org/10.1046/j.1365-2427.2002.00891.x.

Buckton, S.T. & Ormerod, S.J. (2008). Niche segregation of Himalayan river birds. *Journal of Field Ornithology*, 79: 176–185. https://doi.org/10.1111/j.1557-9263.2008.00160.x.

Buckwalter, J.D., Frimpong, E.A., Angermeier, P.L. & Barney, J.N. (2018). Seventy years of stream fish collections reveal invasions and native range contractions in an Appalachian (USA) watershed. *Diversity and Distributions*, 24, 219–232.

Bumpers, P.M., Rosemond, A., Maerz, J.C. & Benstead, J.P. (2017). Experimental nutrient enrichment of forest streams increases energy flow to predators along greener food-web pathways. *Freshwater Biology*, 62, 1794–1805.

Bunn, S.E., Davies, P.M. & Winning, M. (2003). Sources of organic carbon supporting the food web of an arid-zone flood plain river. *Freshwater Biology*, 48, 619–635.

Burgin, A.J. & Hamilton, S.K. (2007). Have we overemphasised the role of denitrification in aquatic ecosystems? A review of nitrate removal pathways. *Frontiers in Ecology and the Environment*, 5, 89–96.

Burgis, M.J. & Morris, P. (1987). *The Natural History of Lakes*. Cambridge University Press.

Burt, T.P. (1996). The hydrology of headwater catchments. In G.Petts & P. Calow (eds) *River Flows and Channel Forms*. Blackwell Science, Oxford, 2–31.

Buss, D.F., Baptista, D.F., Silveira, M.P. et al. (2002). Influence of water chemistry and environmental degradation on macroinvertebrate assenblages in a river basin in southeast Brazil. *Hydrobiologia*, 481, 125–136.

Butler, M.G. (1984). Life histories of aquatic insects. In V.H. Resh & D.M. Rosenberg (eds) *The Ecology of Aquatic Insects*. Praeger, New York, 24–55.

Byers, G.W. (1996). Tipulidae. In R.W. Merritt & K.W. Cummins (eds) *An Introduction to the Aquatic insects of North America* (3rd edn). Kendall/Hunt, Dubuque, IA, 549–570.

Bylack, A. & Kukula, K. (2020). Geomorphological effects of animals in mountain streams: impact and role. *Science of the Total Environment*, 749, doi.org/10.1016/j.scitotenv.2020.141283.

Caissie, D. (2006). The thermal regime of rivers: a review. *Freshwater Biology*, 5, 1389–1406.

Callisto, M. & Graça, M.A.S. (2013). The quality and quantity of fine particulate organic matter for collector species in streams. *International Review of Hydrobiology*, 98, 132–140.

Campbell Grant., E.H., Nichols, J.D., Lowe, W.H. & Fagan, W.F. (2010). Use of multiple dispersal pathways facilitates amphibian persistence in stream networks. *Proceedings of the National Academy of Sciences*, 107, 6936–6940.

Campbell, R.E., Winterbourn, M.J., Cochrane, T.A. et al. (2015). Flow-related disturbance creates a gradient of metacommunity types within stream networks. *Landscape Ecology*, 30, 667–680.

Campbell-Palmer, R., Jones, S., Parker, H. et al. (2016). *The Eurasian Beaver Handbook: Ecology and Management of Castor fiber*. Pelagic Publishing, Exeter, UK.

Caraco, N.F. & Cole, J.J. (2002). Contrasting impacts of a native and alien macrophyte on dissolved oxygen in a large river. *Ecological Applications*, 12, 1496–1509.

Caraco, N.F., Cole, J.J., Raymond, P.A., Strayer, D.L. et al. (1997). Zebra mussel invasion in a large turbid river: phytoplankton response to increased grazing. *Ecology*, 78, 588–602.

Carey, N., Chester, E.T. & Robson, B.J. (2021). Life-history traits are poor predictors of species responses to flow regime changes in headwater streams. *Global Change Biology*, 27, 3547–3564.

Carlson, A.K., Fincel, A.J., Longhenry, C.M. & Graeb, B.D. (2016). Effects of historic flooding on fishes and aquatic habitats in a Missouri River delta. *Journal of Freshwater Ecology*, 31, 271–288.

Carlson, R.L & Lauder, G.V. (2011). Escaping the flow: boundary layer use by the darter *Etheostoma tetrazonum* (Percidae) during benthic station holding. *The Journal of Experimental Biology*, 214, 1181–1193.

Carrizo, S.F., Jähnig, S.C., Bermerich, V., Freyhof, J. et al. (2017). Freshwater megafauna: flagships for freshwater biodiversity under threat. *BioScience*, 10, 919–927.

Carroll, T.M., Thorp, J.H. & Roach, K.A. (2016). Authochthony in karst spring food webs. *Hydrobiologia*, 776, 173–191.

Ceneviva-Bastos, M., Casatti, L. & Uieda, V.S. (2012). Can seasonal differences influence food web structure on preserved habitats? Responses from two Brazilian streams. *Community Ecology*, 13, 243–252.

Ceron, G. (2015). Torrent ducks (*Merganetta armata*) diving and feeding in hot springs. *Waterbirds*, 38, 214–216.

Chadd, R.P., England, J.A., Constable, D., Dunbar, M.J. et al. (2017). An index to track the ecological effects of drought development and recovery on reversion invertebrate communities. *Ecological Indicators*, 82, 344–356.

Chambers, P.A., Macoul, P., Murphy, K.J. & Thomaz, S.M. (2008). Global diversity of aquatic macrophytes in freshwater. *Hydrobiologia*, 595, 9–26.

Chance, J.M. & Craig, D.A. (1986). Hydrodynamics and behaviour of Simuliidae larvae (Diptera). *Canadian Journal of Zoology*, 64, 1295–1309.

Chang H. (2008). Spatial analysis of water quality trends in the Han River basin, South Korea. *Water Research*, 42, 3285–3304.

Charbonnel, A., D'Amico, F., Besnard, A., Blanc, F. et al. (2014). Spatial replciates as an alternative to temporal replicates for occupancy modelling when surveys are based on linear features of the landscape. *Journal of Applied Ecology*, 51, 1425–1433.

Charles, D.F., Kelly, M.G., Stevenson, R.J. et al. (2021). Benthic algae assessments in the EU and the US: striving for consistency in the face of great ecological diversity. *Ecological Indicators*, 121, doi.org/10.1016/j.ecolind.2020.107082.

Charlson, R.J. & Rodhe, H. (1982). Factors controlling the acidity of natural rainwater. *Nature*, 295, 683–685.

Chasse, P., Blatrix, C. & Frascaria-Lacoste, N. (2020). What is wrong between ecological science and policy? *Ecology Letters*, 23, 1736–1738.

Chauvet, E., Cornut, J., Sridhar, K.R., Selosse, M.A. et al. (2016). Beyond the water column: Aquatic hyphomycetes outside their preferred habitat. *Fungal Ecology*, 19, 112–127.

Chauvet, E., Ferreira, V., Giller, P.S. et al. (2016). Litter decomposition as an indicator of stream ecosystem functioning at local-to-continental scales: insights from the European RivFunction project. *Advances in Ecological Research*, 55, 99–182.

Chellaiah, D. & Yule, C.M. (2018). Litter decomposition is driven by microbes and is more influenced by litter quality than by environmental conditions in oil palm streams with different riparian types. *Aquatic Sciences*, 80, 43. doi.org/10.1007/s00027-018-0595-y.

Cheney, K.N., Roy, A.H., Smith, R.F & Dewalt, R.E. (2019). Effects of temperature and substrate type on emergence patterns of Plecoptera and Trichoptera from Northeastern United States headwater streams. *Environmental Entomology*, 48, 1349–1359.

Chester, E.T., Miller, A.D., Valenzuela, I., Wickson, S.J. et al. (2015). Drought survival strategies, dispersal potential and persistence of invertebrate species in an intermittent stream landscape. *Freshwater Biology*, 60, 2066–2083.

Chester, E.T. & Robson, B.J. (2013). Anthropogenic refuges for freshwater biodiversity: their ecological characteristics and management. *Biological Conservation*, 166, 64–75.

Chester E.T., Robson, B.J., Chambers, J.M. et al. (2013). *Novel Methods for Managing Freshwater Refuges against Climate Change in Southern Australia: Anthropogenic Refuges for Freshwater Biodiversity*. National Climate Change Adaptation Research Facility, Gold Coast, Australia.

Childress, E.S., Allan, J.D. & McIntyre, P.B. (2014). Nutrient subsidies from iteroparous fish migrations can enhance stream productivity. *Ecosystems*, 17, 522–534.

Christie, K.S., Hocking, M.D. & Reimchen, T.E. (2008). Tracing salmon nutrients in riparian food webs: isotopic evidence in a ground-foraging passerine. *Canadian Journal of Zoology*, 86, 1317–1323.

Christie, K.S. & Reimchen, T.E. (2008). Presence of salmon increase passerine density on Pacific Northwest streams. *The Auk*, 125, 51–59.

Church, M. (2006). Bed material transport and the morphology of alluvial river channels. *Annual Review of Earth and Planetary Science*, 34, 325–354.

Cilleros, K., Allard, L., Vigouroux, R. & Brosse, S. (2017). Disentangling spatial and environmental determinants of fish species richness and assemblage structure in Neotropical rainforest streams. *Freshwater Biology*, 62, 1707–1720.

Clare, E.L., Symondsen, W.O.C., Broders, H. et al. (2014). The diet of *Myotis lucifugus* across Canada: assessing foraging quality and diet variability. *Molecular Ecology*, 23, 3618–3632.

Clarke, A., MacNally, R., Bond, N. & Lake, P.S. (2008). Macroinvertebrate diversity in headwater streams: a review. *Freshwater Biology*, 53, 1707–1721.

Clements, F.E. (1916) Plant Succession. Carnegie Institute Washington Publication 242.

Clenaghan, C., O'Halloran, J., Giller, P. & Roche, N. (1998). Longitudinal and temporal variation in the hydrochemistry of streams in an Irish conifer afforested catchment. *Hydrobiologia*, 389, 63–71.

Cohen, J.E. (2020). Population, population, population. *The Bulletin of the Ecological Society of America*, 10.1002/bes2.1694.

Colas, F., Baudoin, J.M., Danger, M., Usseglio-Polatera, P. et al. (2013). Synergistic impacts of sediment contamination and dam presence on river functioning. *Freshwater Biology*, 58, 320–336.

Collins, S.L. Kohler, T.J., Thomas, S.A., Fetzer, W.W. et al. (2016). The importance of terrestrial subsidies in stream food webs varies along a stream size gradient. *Oikos*, 125, 674–685.

Comer-Warner, S.A., Gooddy, D.C., Ullah, S. et al. (2020). Seasonal variability of sediment controls of nitrogen cycling in an agricultural stream. *Biogeochemistry*, 148, 31–48. https://doi.org/10.1007/s10533-020-00644-z.

Comte, L., Cucherousset, J., Bouletreau, S & Olden, J.D. (2016). Resource partitioning and fiunctional diversity of worldwide freshwater fish communities. *Ecosphere*, 7 (6), Article e01356, 10.1002/ecs2.1356.

Comte, L. & Grenouillet, G. (2013). Do stream fish track climate change? Assessing distribution shifts in recent decades. *Ecography*, 36, 1236–1246.

Connolly, N.M. & Pearson, R.G. (2018). Colonisation, emigration and equilibrium of stream invertebrates in patchy habitats. *Freshwater Biology*, 63, 1446–1456.

Constable, D. & Birkby, N.J. (2016). The impact of the invasive amphipod *Dikerogammarus haemobaphes* on litter processing in UK rivers. *Aquatic Ecology*, 50, 273–281.

Cooper, S.D., Diehl, S., Kratz, K. & Sarnelle, O. (1998). Implications of scale for patterns and processes in stream ecology. *Australian Journal of Ecology*, 23, 27–40.

Cooper, S.D., Walde, S.J. & Peckarsky, B.L. (1990). Prey exchange rates and the impact of predators on prey populations in streams. *Ecology*, 71, 1503–1514.

Corbet, P.S., (1999). *Dragonflies; Behaviour and Ecology of Odonata*. Harley Books, Essex, England.

Corkum, L.D. (1989). Patterns of benthic invertebrate assemblages in rivers of northwestern North America. *Freshwater Biology*, 21, 191–205.

Corkum, L.D. (1991). Spatial patterns of macroinvertebrate distribution along rivers in eastern decidious forest and grassland biomes. *Journal of the North American Benthological Society*, 10, 358–371.

Corkum, L.D. & Currie, D. (1987). Distributional patterns of immature Simuliidae (Diptera) in Northwestern North America. *Freshwater Biology*, 17, 201–221.

Corliss, J.O. (1994). An interim utilitarian ('user-friendly') hierarchical classification and characterisation of the protists. *Acta Protozoologica*, 33, 1–51.

Corman, J.R., Moody, E.K. & Elser, J.J. (2016). Calcium carbonate deposition drives nutrient cycling in a calcareous headwater stream. *Ecological Monographs*, 86, 448–461.

Cornut J., De Respinis, S., Tonolla, M., Petrini, O. et al. (2019). Rapid characterization of aquatic hyphomycetes by matrix-assisted laser desorption/ionization time-of-flight mass spectrometry. *Mycologia*, 111, 177–189.

Cousins, S.H. (1987). The decline of the trophic level concept. *Trends in Ecology & Evolution*, 2, 312–316.

Couto, T.B.A. & Olden, J.D. (2018). Global proliferation of small hydropower plants—science and policy. *Frontiers in Ecology & Management*, 16, 91–100.

Cover, M.R. & Resh, V.H. (2008). Global diversity of dobsonflies, fishflies, and alderflies (Megaloptera; Insecta) and spongillaflies, nevrorthids, and osmylids (Neuroptera; Insecta) in freshwater. *Hydrobiologia*, 595, 409–417.

Cover, M.R., Seo, J.H. and Resh, V.H. (2015). Life history, burrowing behaviour and distribution of *Neohermes filicornis* (Megaloptera: Corydalidae), a long-lived aquatic insect in intermittent streams. *Western North American Naturalist*, 75, 474–490.

Covich, A., Ewel, K., Giller, P.S. et al. (2004). Ecosystem services provided by freshwater benthos, In D.H. Wall (ed) *Sustaining Biodiversity and Functioning in Salts and Sediments*. SCOPE Series, Vol. 64, Island Press, Washington, USA, 45–73.

Covich, A.P. (1988). Geographical and historical comparisons of neotropical streams: biotic diversity and detrital processing in highly variable habitats. *Journal of the North American Benthological Society*, 7, 361–386.

Covich, A.P., Crowl, T.A. & Heartsill-Scalley, T. (2006). Effects of drought and hurricane disturbances on headwater distributions of palaemonid river shrimp (*Macrobrachium* spp.) in the Luquillo Mountains, Puerto Rico. *Journal of the North American Benthological Society*, 25, 99–107.

Covich, A.P., Crowl, T.A., Hein, C.L., Townsend, M.J. et al. (2009). Importance of geomorphic barriers to predator–prey interactions in river networks. *Freshwater Biology*, 54, 450–465.

Covich, A.P. & Thorp, J.H. (2010). Introduction to the Crustacea. In J.H. Thorp & A.P. Covich (eds) *Ecology and Classification of North American Freshwater Invertebrates* (4th edn). Elsevier.

Crabot, J., Heino, J., Launay, B. & Datry, T. (2020). Drying determines the temporal dynamics of stream invertebrate structural and functional beta diversity. *Ecography*, 43, 620–635.

Craig, D.A. (2003). Geomorphology, development of running water habitats, and evolution of black flies on Polynesian islands. *BioScience*, 53, 1079–1093.

Crandall, K.A. & Buhay, J. (2008). Global diversity of crayfish (Astacidae, Cambaridae, and Parastacidae—Decapoda) in freshwater. *Hydrobiologia*, 595, 295–301.

Cranston, P.S. (1995). Introduction. In P.D. Armitage, P.S. Cranston & L.C.V. Pinder (eds) *The Chironomidae*. Chapman & Hall, London, 1–7.

Creed, R.P. (1994). Direct and indirect effects of crayfish grazing in a stream community. *Ecology*, 75, 2091–2103.

Creed, R.P. & Brown, B.L. (2018). Multiple mechanisms can stabilize a freshwater mutualism. *Freshwater Science*, 37, 760–769.

Crenier, C., Arce-Funck, J., Bec, A. et al. (2017). Minor food sources can play a major role in secondary production in detritus-based ecosystems. *Freshwater Biology*, 62, 1155–1167.

Cross, W.F., Benstead, J.P., Frost, P.C. & Thomas, S.A. (2005). Ecological stoichiometry in freshwater benthic systems: recent progress and perspectives. *Freshwater Biology*, 50, 1895–1912.

Cross, W.F., Covich, A.P., Crowl, T.A., Benstead, J.P. et al. (2008). Secondary production, longevity and resource consumption rates of freshwater shrimps in two tropical streams with contrasting geomorphology and food web structure. *Freshwater Biology*, 53, 2504–2519.

Cross, W.F., Wallace, J.B., Rosemond, A.D. & Eggert, S.L. (2006). Whole-system nutrient enrichment increases secondary production in a detritus-based stream. *Ecology*, 87, 1556–1565.

Crosskey, R.W. (1990). *The Natural History of Blackflies*. Wiley, Chichester.

Crowl, T.A. & Covich, A.P. (1994). Responses of a freshwater shrimp to chemical and tactile stimuli from a large decapod predator. *Journal of the North American Benthological Society*, 13, 291–298.

Crowl, T.A., McDowell, W.H., Covich, A.P. and Johnson, S.L. (2001). Species-specific responses in leaf litter processing in a tropical headwater stream (Puerto Rico). *Ecology*, 82, 775–783.

Cuffney, T.F., Wallace, J.B. & Lugthart, G.J. (1990). Experimental evidence quantifying the role of benthic invertebrates in organic matter dynamics of headwater streams. *Freshwater Biology*, 23, 281–299.

Culp, J.M., Armanini, D.G., Dunbar, M.J., Orlofske, J.M. et al. (2011). Incorporating traits in aquatic biomonitoring to enhance causal diagnosis and prediction. *Integrated Environmental Assessment and Management*, 7, 187–197.

Cumberlidge, N., Ng, P., Yeo, D. et al. (2009). Freshwater crabs and the biodiversity crisis: Importance, threats, status, and conservation challenges. *Biological Conservation*, 142, 1665–1673.

Cummins, K.W. (1973). Trophic relations of aquatic insects. *Annual Review of Entomology*, 18, 183–206.

Cummins, K.W. (1974). Structure and function of stream ecosystems. *BioScience*, 24, 631–641.

Cummins, K.W. (2016). Combining taxonomy and function in the study of stream macroinvertebrates. *Journal of Limnology*, 75, 235–241.

Cummins, K.W. & Merritt, R.W. (1996). General morphology of aquatic insects. Chapter 2 In R.W. Merritt & K.W. Cummins (eds) *An Introduction to the Aquatic Insects of North America*, Kendall/Hunt Publishing Company, Dubuque, Iowa, 5–11.

Currier, C.M., Chaloner, D.T., Rüegg, J. et al. (2020). Beyond nitrogen and phosphorus subsidies: Pacific salmon (*Onchorhynchus* spp.) as potential vectors of micronutrients. *Aquatic Sciences*, 82, 50, https://doi.org/10.1007/s00027-020-00725-z.

Cuthbert, R.N., Bacher, S., Blackburn, T.M. et al. (2020). Invasion costs, impacts, and human agency: response to Sagoff 2020. *Conservation Biology*, doi: 10.1111/cobi.13592.

D'Souza, J.M., Windsor, F.M., Santillo, D. & Ormerod, S.J. (2020). Food web transfer of plastics to an apex riverine predator. *Global Change Biology*, 26, 3846–3857, https://doi.org/10.1111/gcb.15139.

Dahl, J. & Peckarsky, B.L. 2002. Induced morphological defences in the wild: predator effects on a mayfly, Drunella coloradensis. *Ecology*, 83, 1620–1634.

Dai, A. & Trenberth, K.E. (2002). Estimates of freshwater discharge from continents: latitudinal and seasonal variations. *Journal of Hydrometeorology*, 6, 660–687.

Dalal, E.L.H., Tomscha, S.A., Dallaire, C.O. & Bennett, E.M. (2018). A review of riverine ecosystem services quantification: research gaps and recommendations. *Journal of Applied Ecology*, 55, 1299–1311.

Dalu, T. Wassermann, R.J., Tonkin, J.D., Alexander, M.E. et al. (2017). Assessing drivers of benthic macroinvertebrate community structure in African highland streams: An exploration using multivariate analysis. *Science of the Total Enviroment*, 601–602, 1340–1348.

Dang, C.K., Gessner, M.O. & Chauvet, E. (2007). Influence of conidial traits and leaf structure on attachment success of aquatic hyphomycetes on leaf litter. *Mycologia*, 99, 24–32.

Dangles, O. & Malmqvist, B. (2004). Species richness-decomposition relationships depend on species dominance. *Ecology Letters*, 7, 395–402.

Darwel, W.R. & Freyhof, J. (2016). Lost fishes, who is counting? The extent of the threat to freshwater fish biodiversity. Chapter 1. In G.P Closs, M. Krkosek, & J.D. Olden (eds) (2016). *Conservation of Freshwater Fishes*. Cambridge University Press, Cambridge, UK, 1–36.

Das, N., Das, R., Chaudhury, G.R. & Das, S.N., (2010). Chemical composition of precipitation at background level. *Atmospheric Research*, 95, 108–111.

Datry, T., Bonada, N. & Heino, J. (2015). Towards understanding the organisation of metacommunities in highly dynamic ecological systems. *Oikos*, 125, 149–159.

Datry, T., Corti, R., Foulquier, A., von Schiller, D. et al. (2016a). One for all, all for one: A global river research network, *Eos*, *97*, https://doi.org/10.1029/2016EO053587. Published 7 June 2016.

Datry, T., Fritz, K. & Leigh, C. (2016b). Challenges, developments and perspectives in intermittent river ecology. *Freshwater Biology*, 61, 1171–1180.

Datry, T., Bonada, N. & Boulton, A. (eds) (2017a). *Intermittent Rivers and Ephemeral Streams: Ecology and Management*. Academic Press, London, 584p.

Datry, T., Vander Vorste, R., Goitia, E., Moya, N. et al. (2017b). Context-dependent resistance of freshwater invertebrate communities to drying. *Ecology & Evolution*, 7, 3201–3211.

Datry, T., Foulquier, A., Corti, R. et al. (2018). A global analysis of terrestrial plant litter dynamics in non-perennial waterways. *Nature Geoscience*, doi 10.1038/s41561-018-0134-4.

Daufresne, M., Bady, P. & Fruget, J.-F. (2007). Impacts of global change and extreme hydroclimatic events on macroinvertebrate community structures in the French Rhône River. *Oecologia*, 151, 544–559. https://doi.org/10.1007/s00442-006-0655-1.

Daufresne, M., Roger, M.C., Capra, H. & Lamouroux, N. (2003). Long-term changes within the invertebrate and fish communities of the Upper Rhône River: effects of climatic factors. *Global Change Biology*, 10, 124–140.

Davis, B.E., Cocherell, D.E., Sommer, T., Baxter, R.D. et al. (2019). Sensitivities of an endemic, endangered California smelt and two non-native fishes to serial increases in temperature and salinity: implications for shifting community structure with climate change. *Conservation Physiology*, 7, https://doi.org/10.1093/conphys/coy076.

Davis, J.A., Pavlova, A., Thompson, R. & Sunnucks, P. (2013). Evolutionary refugia and ecological refuges: key concepts for conserving Australian arid zone freshwater biodiversity under climate change. *Global Change Biology*, 19, 1970–1984.

Davis, J.M., Rosemond, A.D., Eggert, S.L., Cross, W.F. et al. (2010). Long-term nutrient enrichment decouples predator and prey production. *Proceedings of the National Academy of Sciences*, 107, 121–126.

Day, N.K. & Hall, R.O. (2017). Ammonium uptake kinetics and nitrification in mountain streams. *Freshwater Science*, 36, 41–54.

De Castro-Catala, N., Dolédec, S., Kalogianni, E., Skoulikidis, N. Th. et al. (2020). Unravelling the effects of multiple stressors on diatom and macroinvertebrate communities in European river basins using structural and functional approaches. *Science of the Total Environment* 742: 14053. doi 10.1016/j.scitotenv.2020.140543.

de Eyto, E, Dalton, C., Dillane, M. et al. (2016). The response of North Atlantic diadromous fish to multiple stressors, including land use change: a multidecadal study. *Canadian Journal of Fisheries and Aquatic Sciences*, 73, 1759–1769.

de Jong, H., Oosterbroek, P., Gelhaus, J., Reusch, H. et al. (2008). Global diversity of craneflies (Insecta, Diptera: Tipulidea or Tipulidae sensu lato) in freshwater. *Hydrobiologia*, 595, 457–467.

De Santana, C.D., Crampton, W.G.R., Dillman, C.B., Frederico, R.G. et al. (2019). Unexpected species diversity in electric eels with a description of the strongest living bioelectricity generator. *Nature Communications*, 10, 4000. https://doi.org/10.1038/s41467-019-11690-z

Dean, J.F., Middelberg, J.J., Röckmann, T. et al. (2018). Methane feedbacks to the global climate system in a warmer world. *Reviews of Geophysics*, 56, 207–250.

Death, R.G. (2007). The effect of floods on aquatic invertebrate communities. In J. Lancaster & R.A. Briers (eds) *Aquatic Insects: Challenges to Populations*. Proceedings of the Royal Entomological Society's 24th Symposium, CAB International, Wallingford, UK, 103–121.

Death, R.G., Death, F. Stubbington, R., Joy, M.K. et al. (2015). How good are Bayesian belief networks for environmental management? A test with data from an agricultural river catchment. *Freshwater Biology*, 60, 2297–2309.

Debata, S. & Kar, T. (2021). Factors affecting nesting success of threatened riverine birds: a case from Odisha, Eastern India. Proceedings of the Zoological Society, Kolkata. https://doi.org/10.1007/s12595-021-00376-4.

Dehorter, O. & Guillemain, M. (2008). Global diversity of freshwater birds. *Hydrobiologia*, 595, 619–626.

Deiner, K., Bik, H.M., Machier, E. et al. (2017). Environmenal DNA metabarcoding: transforming how we survey animal and plant communities. *Molecular Ecology*, 26, 5872–5895.

DelVecchia, A.G., Stanford, J.A. & Xu, X. (2016). Ancient and methane-derived carbon subsidizes contemporary food webs. *Nature Communications*, 7, 1–9.

Demars, B.O.L. & Edwards, A.C. (2009) Distribution of aquatic macrophytes in contrasting river systems: a cricritique of compositional-based assessment of water quality. *Science of the Total Environment*, 407, 975–990.

Demars, B.O.L., Manson, J.R., Olafsson, J.S. et al. (2011). Temperature and the metabolic balance of streams. *Freshwater Biology*, 56, 1106–1121.

Demars, B.O.L., Friberg, N. & Thornton, B. (2020). Pulse of organic matter alters reciprocal carbon subsidies between autotrophs and bacteria in stream food webs. *Ecological Monographs*, doi 10.1002/ecm.1399.

Depetris, P.J. (2021). The importance of monitoring river water discharge. *Policy Brief. Frontiers in Water*, https://doi.org/10.3389/frwa.2021.745912.

Descy, J-P., Latli, A., Roland, F. et al. (2021). The Meuse River basin. In K. Tockner, C. Zarfl, C. Robinson & A.C. Benke (eds) *Rivers of Europe* (2nd edn), Elsevier, 230–244.

DeWalt, R.E. & Stewart, K.W. (1995). Life histories of stoneflies (Plecoptera) in the Rio Conejos of southern Colorado. *The Great Basin Naturalist*, 55, 1–18.

Di Sabatino, A., Gerecke, R. & Martin, P. (2000) The biology and ecology of lotic water mites. *Freshwater Biology*, 44, 47–62.

Di Sabatino, A., Smit, H., Gerecke, R., et al. Global diversity of water mites (Acari; Hydrachnidia; Arachnida) in freshwater (2008). *Hydrobiologia*, 595, 303–315.

Dietrich, M. & Anderson, N.H. (1995). Life cycles and food habits of mayflies and stoneflies from temporary streams in western Oregon. *Freshwater Biology*, 34, 47–60.

Dineen, G., Harrison, S.S.C., Giller, P.S. (2007). Diet partitioning in sympatric Atlantic salmon and brown trout in streams with contrasting riparian vegetation. *Journal of Fish Biology*, 71, 17–38.

Ditsche-Kuru, P. & Koop, J.H. (2009). New insights into a life in current: do the gill lamellae of *Epeorus assimilis* and *Iron alpicola* larvae (Ephemeropteera: Heptageniidae) function as a sucker or as friction pads? *Aquatic Insects*, 31, 495–506.

Ditsche-Kuru, P., Koop, J.H. & Gorb, S.N. (2010). Underwater attachment in current: the role of setose attachment structures on the gills of the mayfly larvae *Epeorus assimilis* (Ephemeroptera, Heptageniidae). *Journal of Experimental Biology*, 213, 1950–1959.

Dobson, M. & Hildrew, A.G. (1992). A test of resource limitation among shredding detritivores in low order streams in southern England. *Journal of Animal Ecology*, 61, 69–78.

Dobson M., Hildrew, A.G., Orton, S. & Ormerod, S.J (1995). Increasing litter retention in moorland streams: ecological and management aspects of a field experiment. *Freshwater Biology*, 33, 325–337.

Dodd, J.A., Dick, J.T., Alexander, M., Macneil, C. et al. (2014). Predicting the ecological impacts of a new freshwater invader: functional responses and prey selectivity of the 'killer shrimp' *Dikerogammarus villosus*, compared to the native *Gammarus pulex*. *Freshwater Biology*, 59, 337–352.

Dodds, W.K., Bruckerhoff, L., Batzer, D., Schechner, A. et al. (2019). The freshwater gradient framework: predicting macroscale properties based on latitude, altitude and precipitation. Ecosphere, 10(7), e02786.10.1002/ecs2.2786.

Dodds, W.K. & Smith, V.H. (2016). Nitrogen, phosphorus, and eutrophication in streams. *Inland Waters*, 6, 155–164.

Dodds, W.K. & Whiles, M.R. (2019). *Freshwater Ecology: Concepts and Applications of Limnology.* Elsevier, Oxford, 981, https://doi.org/10.1016/C2016-0-03667-7.

Dolédec, S., Statzner, B. & Bournaud, M. (1999). Species traits for future biomonitoring across regions: patterns along a human-impacted river. *Freshwater Biology*, 42, 737–758.

Dole-Olivier, M., Galassi, D., Marmonier, P. & Creuzé des Châtelliers, M. (2000). The biology and ecology of lotic microcrustaceans. *Freshwater Biology*, 44, 63–91.

Dole-Olivier, M.-J. (2011). The hyporheic refuge hypothesis reconsidered: a review of hydrological aspects. *Marine and Freshwater Research*, 62, 1281–1304.

Dole-Olivier, M.-J., Marmonier, P. & Beffy, J.L. (1997). Response of invertebrates to lotic disturbance: is the hyporheic zone a patchy refugium? *Freshwater Biology*, 37, 257–276.

Dole-Olivier, M.-J., Marmonier, P., Creuzé Des Châtelliers, M. & Martin, D. (1994). Interstitial fauna associated with the alluvial floodplains of the Rhône River (France). In J. Gibert, D.L. Danielopol & J.A. Stanford (eds) *Groundwater Ecology*. Academic Press, San Diego, California, 313–346.

Domisch, S., Jähnig, S.C. & Haase, P. (2011). Climate change winners and losers: stream macroinvertebrates of a submontane region in central Europe. *Freshwater Biology*, 56, 2009–2020.

Dong, X., Ruhl, A. & Grimm, N.B. (2017). Evidence for self-organization in determining spatial patterns of stream nutrients, despite primacy of the geomorphic template. *Proceedings of the National Academy of Sciences*, 114, E4744–E4752.

Dopheide, A., Lear, G. Stott, R & Lewis, G. (2009). Relative Diversity and Community Structure of Ciliates in Stream Biofilms According to Molecular and Microscopy Methods. *Applied Environmental Microbiology*, 75, 5261–5272.

Doretto, A., Piano, E. & Larsen, C.E. (2020). The River Continuum Concept: lessons from the past and perspectives for the future. *Canadian Journal of Fisheries and Aquatic Sciences*, 77, 1853–1864.

dos Reis Oliveira, P., van der Geest, H.G., Kraak, M.H.S. et al. (2020). Over forty years of lowland stream restoration; lessons learned? *Journal of Environmental Management*, 264, 110417.

Dosskey, M.G., Vidon, P., Gurwick, N.P., Allan, C.J. et al. (2010). The Role of Riparian Vegetation in Protecting and Improving Chemical Water Quality in Streams. *Journal of the American Water Resources Association*, 46(2), 261–277.

Douglas, M.E. & Matthews, W.J. (1992). Does morphology predict ecology? Hypothesis testing within a freshwater stream fish assemblage. *Oikos*, 65, 213–224.

Downes, B.J. (1990). Patch dynamics and mobility of fauna in streams and other habitats. *Oikos*, 59, 411–413.

Downes, B.J., Lake, P.S. & Schreiber, E.S. (1995). Habitat structure and invertebrate assemblages on stream stones: A multivariate view from the riffle. *Australian Journal of Ecology*, 20, 503–514.

Downing, J.A., Cole, J.J., Duarte, C.M., Middelburg, J.J. et al. (2012). Global abundance and size distribution of streams and rivers. *Inland Waters*, 2, 229–236.

Dudgeon, D. (1995). The ecology of rivers and streams in tropical Asia. In G.E. Gushing, K.W. Cummins & G.W. Minshall (eds) *Ecosystems of the World 22: River and Stream Ecosystems*. Elsevier, Amsterdam, 615–657.

Dudgeon, D. (2000). The ecology of Asian rivers and streams in relation to biodiversity conservation. *Annual Review of Ecology and Systematics*, 31, 239–263.

Dudgeon, D. (2020). *Freshwater Biodiversity: Status, Threats and Conservation*. Cambridge University Press, Cambridge.

Dudgeon, D., Arthington, A.H., Gessner, M.O., Kawabata, Z.-I. et al. (2006). Freshwater biodiversity: importance, threats, status and conservation challenges. *Biological Reviews*, 81, 163–182.

Dudley, T. & D'Antonio, C. (1991). The effects of substrate texture, grazing, and disturbance on macroalgal establishment in streams. *Ecology*, 72, 297–309.

Dudley, T.L., D'Antonio, C.M. & Cooper, S.D. (1990). Mechanisms and consequences of interspecific competition between two stream insects. *Journal of Animal Ecology*, 59, 849–866.

Dunkle, M.R., Lampman, R.T., Jackson, A.D. & Caudill, C.C. (2020). Factors affecting the fate of Pacific lamprey carcasses and resource transport to riparian and stream macrohabitats. *Freshwater Biology*, 65, 1429–1439.

Dunn, F.E., Nicholls, R.J. Darby, S.E. et al. (2018). Projections of historical and 21st century fluvial sediment delivery to the Ganges-Brahmaputra-Meghna, Mahandi, and Volta deltas. *Science of the Total Environment*, 642, 105–116.

Dunne, T. & Leopold, L.B. (1978). *Water in Environmental Planning*. W.H. Freeman and Co. San Francisco.

Durance, I. & Ormerod, S.J. (2007). Climate change effects on upland stream macroinvertebrates over a 25-year period. *Global Change Biology*, 13, 942–957, DOI: 10.1111/j.1365-2486.2007.01340.x.

Durance, I. & Ormerod, S.J. (2010). Evidence for the role of climate in the local extinction of a cool-water triclad. *Journal of the North American Benthological Society*, 29, 1367–1378.

Ebersole, J.L., Wigington, P.J., Leibowitz, S.G., Comeleo, R.L. et al. (2015). Predicting the occurrence of cold-water patches at intermittent and ephemeral tributary confluences with warm rivers. *Freshwater Science*, 34, 111–124.

Eddy, A.M., Mark, B.G., Baraer, M., McKenzie, J. et al. (2017). Exploring Patterns and Controls on the Hydrochemistry of Proglacial Streams in the Upper Santa River, *Peru Revista de Glaciares y Ecosistemas de Montaña* 3, 41–57.

Edington, J.M. (1968). Habitat preferences in net-spinning caddis larvae with special reference to the influence of water velocity. *Journal of Animal Ecology*, 37, 675–692.

Edington J.M. & Hildrew A.G. (1995). A revised key to the case-less caddis larvae of the British Isles with notes on their ecology. *Scientific Publications of the Freshwater Biological Association*, 53, 1–119.

Edington, J.M, Edington, M.A. & Dorman, J.A. (1984). Habitat partitioning amongst hydropsychid larvae of a Malasian stream. *Entomologica*, 30, 123–129.

Edwards, P.J., Kollman, J., Gurnell, A.M., Petts, G.E. et al. (1999). A conceptual model of vegetation dynamics on gravel bars of a large, Alpine river. *Wetlands Ecology and Management*, 7, 141–153.

Einarsson, A., Gardarsson, A., Gislason, G. & Gudbergsson, G. (2006). Populations of ducks and trout on the River Laxa, Iceland, in relation to variation in food resources. *Hydrobiologia*, 567, 183–194.

Elliott, J.M. (1971). The distances travelled by drifting invertebrates in a Lake District stream. *Oecologia*, 6, 350–379.

Elliott, J.M. (1982). The life cycle and spatial distribution of the aquatic parasitoid *Agriotypus armatus* (Hymenoptera: Agriotypidae) and its caddis host *Silo pallipes* (Trichoptera: Goeridae). *Journal of Animal Ecology*, 51, 923–941.

Elliott, J.M. (1983). The response of the aquatic parasitoid *Agriotypus armatus* (Hymenoptera: Agriotypidae) to the spatial distribution and density of its caddis host *Silo pallipes* (Trichoptera: Goeridae). *Journal of Animal Ecology*, 52, 315–330.

Elliott, J.M. (1994). *Quantitative Ecology and the Brown Trout*. Oxford Series in Ecology & Evolution, Oxford University Press, New York.

Elliott, J.M. (1995). Egg hatching and ecological partitioning in carnivorous stoneflies (Plecoptera). *Comptes Rendus de l'Académie des Sciences, Paris, Sciences de la Vie, Biologie et Pathologie Animale*, 318, 237–243.

Elliott, J.M. (1996). *British Freshwater Megaloptera and Neuroptera: A Key with Ecological Notes*. Scientific Publications of the Freshwater Biological Association, Ambleside, UK.

Elliott, J. M. (2000). Contrasting diel activity and feeding patterns of four species of carnivorous stoneflies. *Ecological Entomology*, 25, 26–34.

Elliott, J.M. (2003). A comparative study of the dispersal of 10 species of stream invertebrates. *Freshwater Biology*, 48, 1652–1668.

Elliott, J.M. (2005). Ontogenetic shifts in the functional response and interference interactions of *Rhyacophila dorsalis* larvae (Trichoptera). *Freshwater Biology*, 50, 2021–2033.

Elliott, J.M. (2008). The ecology of riffle beetles. *Freshwater Reviews* 18, 189–203.

Elliott, J.M. (2013). Contrasting dynamics from egg to adult in the life cycle of summer and overwintering generations of *Baetis rhodani* in a small stream. *Freshwater Biology*, 58, 866–879.

Elliott, J.M. & Hurley, M.A. (1998). Population regulation in adult, but not juvenile, resident trout (*Salmo trutta*) in a Lake District stream. *Journal of Animal Ecology*, 67, 280–286.

Emery-Butcher, H.E., Beatty, S.J. & Robson, B.J. (2020). The impacts of invasive ecosystem engineers in freshwaters: a review. *Freshwater Biology*, 65, 999–1015.

Encalada, A. & Peckarsky, B.L. (2011). The influence of recruitment on within generation population dynamics of a mayfly. *Ecosphere*, 2(10), 107.

Encalada, A. & Peckarsky, B.L. (2012). Large-scale manipulation of mayfly recruitment affects population size. *Oecologia*, 168, 967–976.

Englund, G. (1991). Effects of disturbance on stream moss and invertebrate community structure. *Journal of the North American Benthological Society*, 10, 143–153.

Englund, G. & Cooper, S. (2003). Scale effects and extrapolation in ecological experiments. *Advances in Ecological Research*, 33, 161–213.

Englund, G., Johansson, F. & Olsson, T. (1992). Asymmetric competition between distant taxa: poecilid fishes and water striders. *Oecologia* 92, 1432–1939.

Epele, B., Miserendino, M.L. & Brand, C. (2012). Does nature and persistence of substrate at a mesohabitat scale matter for Chironomidae assemblages? A study of two perennial mountain streams in Patagonia, Argentina. *Journal of Insect Science*, 12:68, available online: insectscience.org/12.68.

Erdozain, M., Kidd, K., Kreutzweiser, D. & Sibley, P. (2019). Increased reliance of stream macroinvertebrates on terrestrial food sources linked to forest management intensity. *Ecological Applications*, 29, e01889.

Eremkina, T.V. & Yarushina, M.I (2022). Ural River Basin. In K. Tockner, C. Zarfl, C. Robinson & A.C. Benke (eds) *Rivers of Europe* (2nd edn) Elsevier, Amsterdam, 883–900.

Eriksen, T.E., Brittain, J.E., Soli, G., Jaconsen, D. et al. (2021). A global perspective on the application of riverine macroinvertebrates as biological indicators in Africa, South-Central America, Mexico and Southern Asia. *Ecological Indicators*, 126, https://doi.org/10.1016/j.ecolind.2021.107609.

Esnaola, A., Arrizabalage, A., Estaban, J., Elosegi, A. et al. (2018). Determining diet from faeces: selection of metabarcoding primers for the insectivore Pyranean Desman (Galemys pyrenaicus). *Plos One*, 13(12), e0208986. https://doi.org/10.1371/journal.pone.0208986.

Estes, J.A., Brashares, J.S. & Power, M.E. (2013). Predicting and detecting reciprocity between indirect ecological interactions and evolution. *American Naturalist*, 181, S76–899.

Estes, J.A., Steneck, R.S. & Lindberg, D.R. (2013). Exploring the consequences of species interactions through the assembly and disassembly of food webs: a Pacific-Atlantic comparison. *Bulletin of Marine Science*, 89, 11–29.

Extence, C.A., Balbi, D.M. & Chadd, R.P. (1999). River flow indexing using British benthic macroinvertebrates: a framework for setting hydrological objectives. *Regulated Rivers: Research and Management*, 15, 543–574.

Extence, C.A., Chadd, R.P., England, J. et al. (2013). The assessment of fine sediment accumulation in rivers using macro-invertebrate community responses. *River Research and Applications*, 29, 17–55.

Fairchild, G.W., Lowe, R.L. & Richardson, W.B. (1985). Algal periphyton growth on nutrient-diffusing substrates: an *in situ* bioassay. *Ecology*, 66, 465–472.

Fairfax, E. & Whittle, SA. (2020). Smokey the Beaver: beaver-dammed riparian corridors stay green during wildfire throughout the western United States. *Ecological Applications*, 30, e02225.

Fasching, C., Akotioye, C., Bizic, M. et al. (2020). Linking stream microbial community functional genes to dissolved organic matter and inorganic nutrients. *Limnology and Oceanography*, 65, S71–S87.

Fausch, K. B. & White, A. J. (1981). Competition between brook trout (*Salvelinus fontinalis*) and brown trout (*Salmo trutta*) for positions in a Michigan stream. *Canadian Journal of Fisheries and Aquatic Science*, 38, 1220–1227.

Fausch, K.D., Nakano, S., Kitano, S., Kanno, Y. et al. (2021). Interspecific social dominance networks reveal mechanisms promoting coexistence in sympatric char in Hokkaido, Japan. *Journal of Animal Ecology*, 90, 515–527.

Feeley, H., Giller, P.S., Baars, J-R. & Kelly Quinn, M, (2020). Benthic macroinvertebrates: the lifeblood of a river. Chapter 7 in "Irish Rivers". M. Kelly-Quinn and J. Reynolds (eds). UCD Press, Dublin, 95–139.

Feio, M.J., Hughes, R.M., Serra, S.R.Q., Nichols, S.J. et al., (2022). Fish and macroinvertebrate assemblages reveal extensive degradation of the world's rivers. *Global Change Biology*, 29, 355–374.

Feld, C.K., Birk, S., Bradley, D.C. et al. (2011). From natural to degraded rivers and back again: a test of restoration ecology theory and practice. *Advances in Ecological Research*, 44, 119–209.

Feminella, J.W. & Resh, V.H. (1991). Herbivorous caddisflies, macroalgae and epilithic microalgae: dynamic interactions in a stream grazing system. *Oecologia*, 87, 247–256.

Fenchel, T. & Finlay, B.J. (1989). Kentrophoros: a mouthless ciliate with a symbiotic kitchen garden. *Ophelia* 30, 75±93.

Ferreira V., Gulis, V., Pascoal, C. & Graça, M.A.S. (2014). Stream pollution and fungi. In E.B.G. Jones, K.D. Hyde, K.L. Pang (eds) *Freshwater Fungi and Fungal-like Organisms*. De Gruyter, Berlin, 389–412.

Ferrington, L.C. (2008). Global diversity of non-biting midges (Chironomidae; Insecta-Diptera) in freshwater. *Hydrobiologia* 595, 447–455.

Ficetola, G.F., Rondinini, C., Bonardi, A., Baisero, D. et al. (2015). Habitat availability for amphibians and extinction threat: a global analysis. *Diversity and Distributions*, 21, 302–311.

Ficsór, M. & Csabai, Z. (2021). Longitudinal zonation of larval *Hydropsyche* (Trichoptera: Hydropsychidae): abiotic environmental factors and biotic intercations behind the downstream sequence of central European species. *Hydrobiologia*, 848, 3371–3388.

Fiebig, D.M. & Lock, M.A. (1991). Immobilization of dissolved organic matter from groundwater discharging through the stream bed. *Freshwater Biology*, 26, 45–55.

Findlay, S. (2010). Stream microbial ecology. *Journal North American Benthological Society*, 29, 170–181.

Findlay, S., Strayer, D., Bain, M. & Nieder, W.C. (2006). Ecology of Hudson River submerged aquatic vegetation. Final Report to the New York State Department of Environmental Conservation. http://www.css.cornell.edu/iris/pdf%20files/HRSAVEcologyReport20063.pdf.

Findlay, S.E.G. & Parr, T.B. (2017). Dissolved organic matter. In G.A. Lamberti & F.R. Hauer (eds), *Methods in Stream Ecology Volume 2: Ecosystem Function* (3rd edn), Academic Press, London, 21–36.

Finlay, B.J. & Esteban G.F. (1998). Freshwater protozoa: biodiversity and ecological function. *Biodiversity and Conservation*, 7, 1163–1186.

Finlay, J.C. (2001). Stable carbon isotope ratios of river biota: implications for energy flow in lotic food webs. *Ecology*, 82, 1052–1064.

Finn, D.S., Blouin, M.S. & Lytle, D.A. (2007). Population genetic structure reveals terrestrial affinities for a headwater stream insect. *Freshwater Biology*, 52, 1881–1897.

Finn, D.S., Bogan, M.T. & Lytle, D.A. (2009). Demographic Stability Metrics for Conservation Prioritization of Isolated Populations. *Conservation Biology*, 23, 1185–1194.

Finn, D.S., Bonada, N., Múrrie, C. & Hughes, J.M. (2011). Small but mighty: headwaters are vital to stream network biodiversity at two levels of organization. *Journal of the North American Benthological Society*, 30, 963–980.

Finn, D.S., Encalada, A.C. & Hampel, H.A. (2016). Genetic isolation among mountains but not between stream types in a tropical high-altitude mayfly. *Freshwater Biology*, 61, 702–714.

Finn D.S., Theobald D.M., Black W.C. IV & Poff N.L. (2006). Spatial population genetic structure and limited dispersal in a Rocky Mountain alpine stream insect. *Molecular Ecology*, 15, 3553–3566.

Finney, B.P., Gregory-Eaves, I., Douglas, M.S.V. & Smol, J.P. (2002). Fisheries productivity in the northeastern Pacific Ocean over the past 2,200 years. *Nature*, 416, 729–733.

Fischer, H. & Pusch, M. (2001). Comparison of bacterial production in sediments, epiphyton and the pelagic zone of a lowland river. *Freshwater Biology*, 46, 1335–1348.

Fisher, S.G. (1994). Pattern, process and scale in freshwater systems: some unifying thoughts. In P.S. Giller, A.G. Hildrew & D.G. Raffaelli (eds) 'Aquatic ecology: scale, pattern and process', 34th Symposium of the British Ecological Society, Blackwell Scientific Publications, Oxford, 575–591.

Fisher, S.G., Gray, L.G., Grimm, N.B. & Busch, D.E. (1982). Temporal succession in a desert stream ecosystem. *Ecological Monographs*, 52, 92–110.

Fisher, S.G. & Likens, G.L. (1973). Energy flow in Bear Brook, New Hampshire: an integrative approach to stream ecosystem metabolism. *Ecological Monographs*, 43, 421–439.

Flenner, I., Richter, O. & Suhling, F. 2010. Rising temperature and development in dragonfly populations at different latitudes. *Freshwater Biology*, 55, 397–410.

Floury, M., Souchon, Y. & Van Looy, K. (2018). Climatic and trophic processes drive long-term changes in functional diversity of freshwater invertebrate communities. *Ecography*, 41, 209–218.

Fochetti, R. & Tierno de Figueroa, J.M. (2008). Global diversity of stoneflies (Plecoptera; Insecta) in freshwater. *Hydrobiologia*, 595, 365–377.

Foote, K.J., Joy, M.K. & Death, R.G. (2015). New Zealand dairy farming: milking our environment for all its worth. *Environmental Management*, 56, 709–720.

Forrester, G.E. (1994). Influences of predatory fish on the drift dispersal and local density of stream insects. *Ecology*, 75, 1208–1218.

Forseth, T., Barlaup, B.T. Finstad, B. et al. (2017). The major threats to Atlantic salmon in Norway. *ICES Journal of Marine Science*, doi:10.1093/icesjms/fsx020.

Frainer, A., Jabiol, J., Gessner, M. et al. (2015). Stoichiometric imbalances between detritus and detritivores are related to shifts in ecosystem functioning. *Oikos*, 125, 861–871.

Francoeur, S.N., Biggs, B.J.F. & Lowe, R.L. (1998). Microform bed clusters as refugia for periphyton in a flood-prone headwater stream. *New Zealand Journal of Marine and Freshwater Research*, 32, 363–374.

Franken, R.J.M., Storey, R.G. & Dudley Williams, D. (2001). Biological, chemical and physical characteristics of downwelling and upwelling zones in a hyporheic zone in a north temperate stream. *Hydrobiologia*, 444, 183.

Franklin, P., Dunbar, M. & Whitehead, P. (2008). Flow controls on lowland river macrophytes: a review. *Science of the Total Environment*, 400, 369–378.

Frasson, R.P. de M., Pavelsky, T.M., Fonstad, M.A., Durand, M.T. et al. (2019). Global relationships between river width, slope, catchment area, meander wavelength, sinuosity, and discharge. *Geophysical Research Letters*, 46, 3252–3262. https://doi.org/10.1029/2019GL082027.

Freilich, J.E. (1991). Movement patterns and ecology of *Pteronarcys* nymphs (Plecoptera): observations of marked individuals in a Rocky Mountain stream. *Freshwater Biology*, 25, 379–394.

Friberg, N., Bonada, N., Bradley, D.C. et al. (2011). Biomonitoring of human impacts in freshwater ecosystems: the good the bad and the ugly. *Advances in Ecological Research*, 44, 1–68.

Friberg, N., Dybkjær, J.B., Ólafsson J.S., Gíslason, G.M., Larsen, S.E. & Lauridsen, T.L. (2009). Relationships between structure and function in streams contrasting in temperature. *Freshwater Biology*, 54, 2051–2068.

Frimpong, E.A. (2018). A case for conserving common species. *PLoS Biology*, 16 (2), e2004261.

Frissell, C.A., Liss, W.J., Warren, C.E. & Hurley, M.D. (1986). A hierarchical framework for stream habitat classification: viewing streams in a watershed context. *Environmental Management* 10, 199–214.

Fritts, A.K., Knights, B.C., Larson, J.H. Amberg, J.J. et al. (2019). Development of a quantitative PCR method for screening ichthyoplankton samples for bigheaded carps. *Biological Invasions*, 21, 1143–1153.

Fritz, K.M., Evans, M.A. & Feminella, J.W. (2004). Factors affecting biomass allocation in the riverine macrophyte *Justicia americana*. *Aquatic Botany*, 78, 279–288.

Frost, P.C., Benstead, J.P., Cross, W.F. et al. (2006). Threshold elemental ratios of carbon and phosphorus in aquatic consumers. *Ecology Letters*, 9, 774–779.

Frost, T.M. (1991). Porifera. In J.H. Thorp & A.P. Covich (eds) *Ecology and Classification of North American Freshwater Invertebrates*. Academic Press, San Diego, 95–124.

Frutiger, A. (2002). The function of the suckers of larval net-winged midges (Diptera: Blephariceridae). *Freshwater Biology*, 47, 293–302.

Fryer, G. (1993). The Freshwater Crustacea of Yorkshire; a faunistic and ecological survey. Yorkshire Naturalists' Union & Leeds Philosophical and Literary Society, 312.

Fuller, M.R., Doyle, M.W. & Strayer, D.L. (2015). Causes and consequences of habitat fragmentation in river networks. *Annals of the New York Academy of Sciences (The Year in Ecology and Conservation Biology)*, 1355, 31–51.

Fuquing, W., Yang, Z., Xi, C. & Heng, Z. (2019). Hydrochemistry and its controlling factors of rivers in the source region of the Nujiang River on the Tibetan Plateau. *Water*, 11, 2166, DOI:10.3390/w11102166.

Gadgil, M., Barkes, F. & Folke, C. (2021). Indigenous knowledge; from local to global. *Ambio*, doi.org/10.1007/s13280-020-01478-7.

Gallardo, B. & Aldridge, D.C. (2015) Is Great Britain heading for a Ponto-Caspian meltdown? *Journal of Applied Ecology*, 52, 41–49.

Garnier, J., Meybeck, M., Ayrault, S. et al. (2022). Continental Atlantic Rivers: the Seine Basin. In K. Tockner, C. Zarfl, C. Robinson & A.C. Benke (eds) *Rivers of Europe* (2nd edn), Elsevier, 292–330.

Gatley, G.K. (1988). Competition and the structure of hydropsychid guilds in southern Sweden. *Hydrobiologia* 164, 23–32.

Gauthier, M., Launay, B., Le Goff, B., Pella, H. et al. (2020). Fragmentation promotes the role of dispersal in determining ten intermittent headwater stream metacommunities. *Freshwater Biology*, 65, 2169–2185.

Gee, A.S. & Stoner, J.H. (1989). A review of the causes and effects of acidification of surface waters in Wales and potential mitigation techniques. *Archives of Environmental Contamination and Toxicology*, 18, 121–130.

Geesey, G.G., Mutch, R., Costerton, J.W. & Green, R.B. (1978). Sessile bacteria: an important component of the microbial population in small mountain streams. *Limnology and Oceanography*, 23, 1214–1223.

Gelwick, F.P. & Matthews, W.J. (1992). Effects of an algivorous minnow on temperate stream ecosystem properties. *Ecology*, 73, 1630–1645.

Gende, S.M., Edwards, R.T., Wilson, M.F. & Wipfli, M.S. (2002). Pacific salmon in aquatic and terrestrial ecosystems: Pacific salmon subsidize freshwater and terrestrial ecosystems through several pathways, which generates unique management and conservation issues but also provides valuable research opportunities. *BioScience*, 52, 917–928.

Georgian, T. & Wallace, J.B. (1983). Seasonal production dynamics in a guild of periphyton-grazing insects in a Southern Appalachian stream. *Ecology*, 64, 1236–1248.

Gessner, M.O. & Chauvet, E. (1994). Growth and production of stream microfungi in controlling breakdown rates of leaf litter. *Ecology*, 1807–1817.

Gessner, M.O. & Newell, S.Y. (2002). Biomass, growth rate, and production of filamentous fungi in plant litter. In C.J. Hurst, R.L. Crawford, G. Knudsen et al. (eds) *Manual of Environmental Microbiology*. American Society for Microbiology, Washington, DC, 390–408.

Gessner, M.O. & C.T. Robinson (2003). Aquatic hyphomycetes in alpine streams. In: J.V. Ward & U. Uehlinger (eds), *Ecology of a Glacial Flood Plain*. Kluwer, Dordrecht, The Netherlands, 123–136.

Gessner, M.O. & Tlili, A. (2016). Fostering integration of freshwater ecology with ecotoxicology. *Freshwater Biology*, 61, 1991–2001.

Gessner, M.O. Inchausti, P., Persson, et al. (2004) Biodiversity effects on ecosystem functioning: insights from aquatic systems. *Oikos*, 104, 419–422.

Gessner M.O., Gulis, V., Kuehn, K.A., Chauvet, E. et al. (2007). Fungal decomposers of plant litter in aquatic ecosystems. In: C.P. Kubicek & I.S. Druzhinina (eds) *The Mycota Vol. 4: Environmental and Microbial Relationships*. Springer, Berlin, 301–324.

Gessner, M.O., Swan, C.M., Dang, C.K. et al. (2010). Diversity meets decomposition. *Trends in Ecology & Evolution*, 25(6), 372–380.

Giller, P.S. (1984). Community structure and the niche. Chapman and Hall, London.

Giller, P.S. (1986). The natural diet of the Notonectidae: Field trials using electrophoresis. *Ecological Entomology*, 11, 163–172.

Giller, P.S. (1996). The biodiversity of soils: the poor man's rainforest. *Biodiversty and Conservation*, 5, 135–168.

Giller, P.S. (2005). River restoration: seeking ecological standards. Editor's introduction. *Journal of Applied Ecology*, 42, 201–207.

Giller P.S. (2020). The Araglin Valley: a Catchment Research Study Area. In M. Kelly-Quinn & J. Reynolds (eds) *Irish Rivers*. UCD Press, Dublin, 383–409.

Giller, P.S. & Malmquist, B. (1998). *The Biology of Streams and Rivers*. Oxford University Press, Oxford.

Giller, P.S. & O'Halloran, J. (2004). Forestry and the aquatic environment: studies in an Irish context. *Hydroloy & Earth System Science*, 8, 314–326.

Giller, P.S. & Sangpradub, N. (1993). Predatory foraging behaviour and activity patterns of larvae of two species of limnephilid cased caddis. *Oikos*, 67, 351–357.

Giller, P.S. & Greenberg, L. (2015). The relationship between individual habitat use and diet in brown trout. *Freshwater Biology*, 60, 256–266.

Giller, P.S., Sangpradup, N. & Twomey, H. (1991). Catastrophic flooding and macroinvertebrate community structure. *Vernhandlungen der Internationale Vereinigung für Theoretische und Angewandte Limnologie*, 24, 1724–1729.

Giller, P.S., Hildrew, A.G. & Raffaelli, D.G. (eds) (1994). *Aquatic Ecology: Scale, Pattern and Process*. Symposia of the British Ecological Society, Blackwell Scientific Publications.

Gilvear, D.J., Spray, C. & Casa-Mulet, R. (2013). River rehabilitation for the delivery of multiple ecosystem services at the river network scale. *Journal of Environmental Management*, 126, 30–43.

Gionchetta, G., Oliva, F., Menéndez, M., Laseras, P.L. et al. (2019). Key role of streambed moisture and flash storms for microbial resistance and resilience to long-term drought. *Freshwater Biology*, 64, 306–322.

Gleason, H.A. (1926). The individualistic concept of plant association. *Bulletin of the Torrey Botanical Club*, 53, 7–26.

Glime, J.M. (2020). Streams: Structural Modifications— Leaves and Stems. Chapters 2 and 3 in: J. M. Glime, *Bryophyte Ecology. 2-3-1 Volume 1. Habitat and Role*. E-book sponsored by Michigan Technological University and the International Association of Bryologists. Last updated 21 July 2020 and available at http://digitalcommons.mtu.edu/bryophyte-ecology/.

Goedkoop, W., Demandt, M. & Ahlgren, G. (2007). Interactions between food quantity and quality (long-chain polyunsaturated fatty acid concentrations) effects on growth and development of *Chironomus riparius*. *Canadian Journal of Fisheries and Aquatic Sciences*, 64, 425–436.

Goertzen, D., Schneider, A.-K., Eggers, T.O. et al. (2022). Temporal changes of biodiversity in urban running waters – results of a twelve-year monitoring study. *Basic and Applied Ecology*, 58, 74–87.

Goldschmidt, T. (2016). Water mites (Acari, Hydrachnidia): powerful but widely neglected bioindicators—a review. *Neotropical Biodiversity*, 2, 12–25.

Gomard, Y., Cornuault, J., Licciardi, S., Lagadec, E., Belqat, B., Dsouli, N., Mavingui, P. & Tortosa, P. (2018). Evidence of multiple colonizations as a driver of black fly diversification in an oceanic island. *Plos One*, 13, e0202015.

Gomez, R., Arce, M.I., Baldwin, D.S. & Dahm, C.N. (2017). Water physicochemistry in intermittent rivers and ephemeral streams. In T. Datry, N. Bonada & A. Boulton (eds) *Intermittent Rivers and Ephemeral Streams*. Elsevier, 109–134.

Goodman, K.M. & Hay, M.E. (2013). Activated chemical defences suppress herbivory on freshwater red algae. *Oecologia*, 171, 921–933.

Gordon, N.D., McMahon, T.A. & Finlayson, B.L. (2006). *Stream Hydrology: An Introduction for Ecologists* (2nd edn). John Wiley & Sons, Chichester.

Goudie, R. (2008). Aspects of distribution and ecology of Harlequin Ducks on the Torrent River, Newfoundland. *Waterbirds*, 31, 92–103.

Grabner, D.S. (2017). Hidden diversity: Parasites of stream arthropods. *Freshwater Biology*, 62, 52–64.

Graça, M.A.S., Pozo, J., Canhoto, C. & Elosegi, A. (2002). Effects of *Eucalyptus* plantations on detritus, decomposers, and detritivores in streams. *The Scientific World Journal*, 2, 1173–1185.

Graf, D.L. & Cummings, K.S. (2007). Review of the systematics and global diversity of freshwater mussels (Bivalvia: Unionoida). *Journal of Molluscan Studies*, 73, 291–314.

Graf, D.L. & Cummings, K.S. (2021). A 'big data' approach to global freshwater mussel diversity (Bivalvia: Unionoida), with an updated checklist of genera and species. *Journal of Molluscan Studies*, 87, eyaa034.

Gray, L.A. (1981). Species composition and life histories of aquatic insects in a lowland Sonoran desert stream. *American Midland Naturalist*, 106 (2), 229–242.

Gray, D.P. & Harding, J.S. (2012). Acid Mine Drainage Index (AMDI): a benthic invertebrate biotic index for assessing coal mining impacts in New Zealand streams. *New Zealand Journal of Marine & Freshwater Research*, 46, 335–352.

Greenberg, L.A. & Giller, P.S. (2001). Individual variation in habitat use and growth of male and female brown trout. *Ecography*, 24, 212–224.

Greig, H.S. & McIntosh, A.R. (2006). Indirect effects of predatory trout on organic matter processing in detritus-base stream food webs. *Oikos*, 112, 31040.

Griffith, M. & McManus, M. (2020). Consideration of spatial and temporal scales in stream restoration and biotic monitoring to assess restoration outcomes: a literature review, part 2. *River Research and Application*, 36, 1398–1415.

Grill G., Lehner B., Thieme M. et al. (2019). Mapping the world's free-flowing rivers. *Nature*, 569, 215–221.

Grime, J.P. (1977). Evidence for the existence of three primary strategies in plants and its relevance to ecological and evolutionary theory. *American Naturalist*, 111, 1169–1194.

Grime, J.P. (2006). Trait convergence and trait divergence in herbaceous plant communities: mechanisms and consequences. *Journal of Vegetation Science*, 17, 255–260.

Grimm, N.B. (1988). Role of macroinvertebrates in nitrogen dynamics of a desert stream. *Ecology*, 69, 1884–1893.

Grimm, N.B. (1994). Disturbance, succession and ecosystem processes in streams: a case study from the desert. In P.S. Giller, A.G. Hildrew & D.G. Rafaelli (eds) *Aquatic Ecology: Scale Pattern and Process*, 34th Symposium of the British Ecological Society. Blackwell, Oxford, 93–112.

Grimm, N. B. & Fisher, S.G. (1984). Exchange between interstitial and surface water: implications for stream metabolism and nutrient cycling. *Hydrobiologia*, 111, 219–228.

Grizzetti, B., Liquetti, C., Pistocchi, A., Vigiak, O. et al. (2019). Relationship between ecological condition and ecosystem services in European rivers, lakes and coastal waters. *Science of the Total Environment*, 671, 452–465.

Gross, E. (2003). Allelopathy of aquatic autotrophs. *Critical Reviews in Plant Science*, 22, 313–339.

Grossman, G.D. & Sabo, J.L. (2010). Incorporating environmental variation into models of community stability: examples of stream fish. *American Fisheries Society Symposium*, 73, 407–426.

Grossman, G.D., Moyle, P.B. & Whitaker, J.R. (1982). Stochasticity in structural and functional characteristics of an Indiana stream fish assemblage: a test of community theory. *American Naturalist*, 120, 423–454.

Grossman, G., Ratajczak, R.E., Petty, J.T., Hunter, M.D. et al. (2006). Population dynamics of mottled sculpin (Pisces) in a variable environment: information theoretic approaches. *Ecological Monograph*, 76, 217–234.

Grossman, G., Sundin, G. & Ratajczak, R.E. (2016). Long-term persistence, density-dependence and effects of climate on rosyside dace (Cyprinidae). *Freshwater Biology*, 61, 832–847.

Grossman, G., Carline, R. & Wagner, T. (2017). Population dynamics of brown trout (*Salmo trutta*) in Spruce Creek, Pennsylvania: a quarter-century perspective. *Freshwater Biology*, 55, 1–12.

GS Australia (2006). GEODATA TOPO 250K Series 3. Shape Files Format. Commonwealth of Australia (GeoSciences Australia), Canberra, ACT, Australia.

Gulis, V. & Bärlocher, F. (2017). Fungi: biomass, production and community structure. In F.R. Hauer & G.A. Lamberti (eds) *Methods in Stream Ecology* (3rd edn, Volume 1—Ecosystem Structure). Academic Press, London, 177–192.

Gulis V., Su, R. & Kuehn, K.A. (2019). Fungal decomposers in freshwater environments. In: C.J. Hurst (ed) *The Structure and Function of Aquatic Microbial Communities*, Springer Nature, Cham, Switzerland, 121–155.

Gulis, V. & Suberkropp, K. (2003). Leaf litter decomposition and microbial activity in nutrient-enriched and unaltered reaches of a headwater stream. *Freshwater Biology*, 48, 123–134.

Guo, F., Ebm, N., Bunn, S.E., Brett, M.T. et al. (2021). Longitudinal variation in the nutritional quality of basal food sources and its effect on invertebrates and fish in sub-alpine rivers. *Journal of Animal Ecology*, 90, 2678–2691.

Guo, F., Kainz, M.J., Sheldon, F. & Bunn, S.E. (2016a). The importance of high-quality algal food sources in stream food webs—current status and future perspectives. *Freshwater Biology*, 61, 815–831.

Guo, F., Kainz, M., Valdez, D. et al. (2016b). High quality algae attached to leaf litter boost invertebrate shredder growth. *Freshwater Science*, 35, 1213–1221.

Gurnell, A.M. (2014). Plants as river system engineers. *Earth Surface Processes and Landforms*, 39, 4–25.

Gurnell, A.M. (2016). Trees, wood and river morphodynamics: results from 15 years research on the Tagliamento River, Italy. In D.J. Gilvear, M.T. Greenwood, M.C. Thoms & P.J. Wood (eds) *River Science: Research and Management for the 21st Century*. Wiley-Blackwell, Oxford, 132–155.

Gutiérrez-Fonseca, P.E., Ramírez, A., Pringle, C.M., Torres, P.J. et al. (2020). When the rainforest dries: Drought effects on a montane tropical stream ecosystem in Puerto Rico. *Freshwater Science* 39, 197–212.

Haag, W.H. (2012). *North American Freshwater Mussels: Natural History, Ecology, and Conservation*. Cambridge University Press, New York.

Haines, A.T., Findlayson, B.L. & McMahon, T.A. (1988). A global classification of river regimes. *Applied Geography*, 8, 255–272.

Hall, R.O., Kock, B.J. Marshall, M.C. Taylor, B. et al. (2007). How body size mediates the role of animals in nutrient cycles in aquatic systems. In A.G. Hildrew, D. Raffaelli & R. Edmonds-Brown (eds) *Body Size: The Structure and Function of Aquatic Systems. Ecological Reviews*, British Ecological Society, Cambridge University Press, 286–305.

Hall, R.O., Wallace, J.B. & Eggert, S.L. (2000). Organic matter flow in stream food webs with a reduced detrital resource base. *Ecology*, 81, 3445–3463.

Hall, R.O. Jr, Yackulic, C.B., Kennedy, T.A. et al. (2015). Turbidity, light, temperature, and hydropeaking control primary productivity in the Colorado River, Grand Canyon. *Limnology and Oceanography*, 60, 512–526.

Hall., R.O. Jr, Tank, J., Sobota, D.J. et al. (2009). Nitrate removal in stream ecosystems measured by 15N addition experiments: total uptake. *Limnology and Oceanography*, 54, 653–665.

Halvorson, H.M., Fuller, C.L., Entrekin, S.A. Scott, J.T et al. (2018). Detrital nutrient content and leaf species differentially affect growth and nutritional regulation of detritivores. *Oikos*, 127, 1471–1481.

Handa, I.T., Aerts, R., Berendse, F. et al. (2014). Consequences of biodiversity loss for litter decomposition across biomes. *Nature*, 509, 218–222.

Hannesdóttir, E. R., G. M. Gíslason, & J. S. Ólafsson. (2012). Life cycles of *Eukiefferiella claripennis* (Lundbeck 1898) and *Eukiefferiella minor* (Edwards 1929) (Diptera: Chironomidae) in spring-fed streams of different temperatures with reference to climate change. *Fauna norvegica*, 31, 35–35.

Hansen, R.A., Hart, D.D. & Merz, R.A. (1991). Flow mediates predator-prey interactions between triclad flatworms and larval blackflies. *Oikos*, 60, 187–196.

Harding, J. (1997). Strategies for coexistence in two species of New Zealand Hydropsychidae (Trichoptera). *Hydrobiologia*, 350, 25–33.

Harding, J.N., Harding, J.M.S. & Reynolds, J.D. (2014). Movers and shakers: nutrient subsidies and benthic disturbance predict biofilm biomass and stable isotope signatures in coastal streams. *Freshwater Biology*, 59, 1361–1377.

Harries, J., Sheahan, D. Jobling, S. Matthiessen, P. et al. (2009). Estrogenic activity in five United Kingdom rivers detected by measurement of vitellogenesis in caged male trout. *Environmental Toxicology and Chemistry*, 16, 534–542.

Hart, D.D. (1978). Diversity in stream insects: regulation by rock size and microspatial complexity. *Verhandlungen der Internationale Vereinigung für Theoretische und Angewandte Limnologie*, 20, 1376–1381.

Hart, D.D. (1983). The importance of competitive interactions within stream populations and communities. In J.R. Barnes and G.W. Minshall (eds) *Stream Ecology*. Plenum Press, New York, 99–136.

Hart, D.D. (1986). The adaptive significance of territoriality in filter-feeding larval blackflies (Diptera: Simuliidae). *Oikos*, 46, 88–92.

Hart, D.D. (1992). Community organization in streams: the importance of species interactions, physical factors, and chance. *Oecologia*, 91, 220–228.

Hart, D.D., Clark, B.D. & Jasentuliyana, A. (1996). Fine-scale field measurement of benthic flow environments inhabited by stream invertebrates. *Limnology & Oceanography*, 41, 297–308.

Hart, D.D., Merz, R.A., Genovese, S.J. & Clark, B.D. (1991). Feeding posture of suspension-feeding larval blackflies: the conflicting demands of drag and food acquisition. *Oecologia*, 85, 457–463.

Harvey, B.C. & Marti, C.D. (1993). The impact of dipper, *Cinclus mexicanus*, predation on stream benthos. *Oikos*, 68, 431–436.

Hassan, F.A. (2011). *Water History for Our Times*. UNESCO International Hydrological Programme, IHP essays on water history, 2.

Haubrock, P.J., Cuthbert, R.N., Ricciardi, A. et al. (2022). Economic costs of invasive bivalves in freshwater ecosystems. *Diversity & Distributions*, doi: 10.1111/ddi.13501.

Hauer, F.R. Locke, H., Dreitz, V.J., Hebblewhite, M., Lowe, W.H., Muhlfeld, C.C., Nelson, C.R., Proctor, M.F. & Rood, S.B. (2016). Gravel-bed river floodplains are the ecological nexus of glaciated mountain landscapes. *Science Advances*, 2, e1600026.

Hawkins, C.P. & Furnish, J.K. (1987). Are snails important competitors in stream ecosystems? *Oikos*, 49, 209–220.

Hay, S.E., Jenkins, K.M. & Kingsford, R.T. (2018). Diverse invertebrate fauna using dry sediment as a refuge in semi-arid and temperate Australian Rivers. *Hydrobiologia*, 806, 95–109.

Hayden, B., McWilliam-Hughes, S.M. & Cunjak, R.A. (2016). Evidence for limited trophic transfer of allochthonous energy in temperate food webs. *Freshwater Science*, 35, 544–558.

Heffernan, J.B. & Cohen, M.J. (2010). Direct and indirect coupling of primary production and diel nitrate dynamics in a subtropical spring-fed river. *Limnology and Oceanography*, 55, 677–688.

Heggenes, J., Bagliniere, J.L. & Cunjak, R.A. (1999). Spatial niche variability for young Atlantic salmon (Salmo salar) and brown trout (S. trutta) in heterogeneous streams. *Ecology of Freshwater Fish*, 8, 1–21

Hein, C.L., Pike, A.S., Blanco, J.F., Covich, A.P. et al. (2011). Effects of coupled natural and anthropogenic factors on the community structure of diadromous fish and shrimp species in tropical island streams. *Freshwater Biology*, 56, 1002–1015.

Heino, J., Melo, A.S., Siqueira, T., Soininen, J. et al. (2015). Metacommunity organisation, spatial extent and dispersal in aquatic systems: patterns, processes and prospects. *Freshwater Biology*, 60, 845–869.

Heino, J., Schmera, D. & Eros, T. (2013). A macroecological perspective on trait patterns in stream communities. *Freshwater Biology*, 58, 1539–1555.

Helfield, J.M. & Naiman, R.J. (2006). Keystone interactions: salmon and bear in riparian forests of Alaska. *Ecosystems*, 9, 167–180.

Helm, D. (2015). *Natural Capital: Valuing the Planet*. Yale University Press, New Haven, Connecticut.

Hemphill, N. Competition between two stream dwelling filter-feeders, *Hydropsyche oslari* and *Simulium virgatum*. *Oecologia*, 77, 73–80.

Henriques-Silva, R., Logez, M., Reynaud, R., Tedesco, P.A. et al. (2018). A comprehensive examination of the network position hypothesis across multiple river matacommunities. *Ecography*, 41, 1–11.

Hering, D., Borja, A., Jones, J.I. et al. (2018). Implementation options for DNA-based identification into ecological status assessment under the European Water Framework Directive. *Water Research*, 138, 195–205.

Hering, D., Schmidt-Kloiber, A., Murphy, J. et al. (2009). Potential impacts of climate change on aquatic insects: a sensitivity analysis for European caddisflies (Trichoptera) based on distribution patterns and ecological preferences. *Aquatic Sciences*, 71, 3–14.

Hermoso, V. (2017). Freshwater ecosystems could become the biggest losers of the Paris Agreement. *Global Change Biology*, 23, 3433–3436.

Hermoso, V., Abell, R., Linke, S. & Boon, P. (2016). The role of protected areas for freshwater biodiversity conservation: challenges, and opportunities in a rapidly changing world. *Aquatic Conservation: Marine and Freshwater Ecosystems*, 26 (Suppl. 1), 3–11.

Herrero Ortega, S., Gonzalez-Quijano, C.R., Casper, P. et al. (2019). Methane emissions from contrasting urban freshwaters: rates, drivers, and a whole city footprint. *Global Change Biology*, 25, 4234–4243.

Hershey, A.E., Northington, R.M., Finlay, J. C. & Peterson, B.J. (2017). Stable isotopes in stream food webs. In G.A. Lamberti & F.R. Hauer (eds) *Methods in Stream Ecology Volume 2: Ecosystem Function* (3rd edn). Academic Press, London, 3–20.

Hershey, A.E., Pastor, J., Peterson, B.J. & Kling, G. W. (1993). Stable isotopes resolve the drift paradox for *Baetis* mayflies in an Arctic river. *Ecology*, 74, 2315–2325.

Hieber, M. & Gessner, M.O. (2002). Contribution of stream detritivores, fungi, and bacteria to leaf breakdown based on biomass estimates. *Ecology*, 83, 1026–1038.

Hildrew, A.G. (1986). Aquatic insects: patterns in life history, environment and community structure. *Proceedings of the IIIrd European Congress of Entomology*, Amsterdam, Part. 1, 23–34.

Hildrew, A.G. (1992). Food webs and species interactions. In P. Calow and G.E. Petts (eds) *The Rivers Handbook, 1*, 309–330.

Hildrew, A.G. (2009). Sustained research on stream communities: a model system and the comparative approach. *Advances in Ecological Research*, 41, 175–312.

Hildrew, A.G. (2018). Freshwater acidification: natural history, ecology and environmental policy. *Excellence in Ecology* (Vol 27), International Ecology Institute, Oldendorf/Luhe.

Hildrew, A.G. (2022). European Rivers: a personal perspective. In K. Tockner, C. Zarfl & C. Robinson (eds) *Rivers of Europe* (2nd edn) Elsevier, Amsterdam, 899–909.

Hildrew, A.G. & Townsend, C.R. (1977). The influence of substrate on the functional response of *Plectrocnemia conspersa* (Curtis) larvae (Trichoptera: Polycentropodiae). *Oecologia*, 31, 21–26.

Hildrew, A.G. & Edington, J.M. (1979). Factors facilitating the coexistence of hydropsychid caddis larvae (Trichoptera) in the same river system. *Journal of Animal Ecology*, 48, 557–576.

Hildrew, A.G. & Townsend, C.R. (1980). Aggregation, interference and foraging by larvae of *Plectrocnemia conspersa* (Trichoptera: Polycentropodidae). *Animal Behaviour*, 28, 553–560.

Hildrew, A.G. & Townsend, C.R. (1987) Organization in freshwater benthic communities. In J.H.R. Gee & P.S. Giller (eds) *Community organization: past and present*. Symposia of the British Ecological Society, Blackwell Scientific Publications, Oxford, 347–371.

Hildrew, A.G. &Wagner, R. (1992). The briefly colonial life of hatchlings of the net-spinning caddisfly *Plectrocnemia conspersa*. *Journal of the North American Benthological Society*, 11, 60–68.

Hildrew, A.G & Giller, P.S. (1994). Patchiness, species interactions and disturbance in the stream benthos. In P.S. Giller, A.G. Hildrew and D.G. Raffaelli (eds) *Aquatic Ecology: Scale, Pattern and Process*, 34th Symposium of the British Ecological Society, Blackwell Scientific Publications, Oxford, 31–62.

Hildrew, A.G., Townsend, C.R. Francis, J. & Finch, K. (1984). Cellulolytic decomposition in streams of contrasting pH and its relationship with invertebrate community structure. *Freshwater Biology*, 14, 323–328.

Hildrew, A.G., Townsend, C.R. & Hasham, A. (1985). The predatory Chironomidae of an iron-rich stream: feeding ecology and food web structure. *Ecological Entomology*, 10, 403–413.

Hildrew, A.G., Woodward, G., Winterbotton, J.H. & Orton, S. (2004). Strong density-dependence in a predatory insect: large-scale experiments in a stream. *Journal of Animal Ecology*, 73, 448–458.

Hildrew, A.G., Raffaelli, D. & Edmonds-Brown, V. (2007). (eds) Body size: the structure and function of aquatic ecosystems. Symposia of the British Ecological Society, *Ecological Reviews*, Cambridge University Press, 1–343.

Hildrew, A.G., Durance, I. & Statzner, B. (2017). Persistence in the longitudinal distribution of lotic insects in a changing climate: a tale of two rivers. *Science of the Total Environment*, 574, 1294–1304.

Hill, W.R. & Griffiths, N.A. (2017). Nitrogen processing by grazers in a headwater stream: riparian connections. *Freshwater Biology*, 62, 17–29.

Hill, W.R., Fanta, S.R. & Roberts, B.J. (2009). Quantifying phosphorus and light effects in stream algae. *Limnology and Oceanography*, 54, 368–380.

Hladyz, S., Abjornsson, K., Giller P.S. & Woodward, G. (2011). Impacts of aggressive riparian invader on community structure and ecosystem functioning in stream food webs. *Journal of Applied Ecology*, 48, 432–452.

Hladyz, S., Gessner, M.O. Giller, P.S. et al. (2009). Resource quality and stoichiometric constraints on stream ecosystem functioning. *Freshwater Biology*, 54, 957–970.

Hocking, M.D. & Reimchen, T.E. (2002). Salmon-derived nitrogen in terrestrial invertebrates from coniferous forests of the Pacific Northwest. *BMC Ecology*, 2, https://doi.org/10.1186/1472-6785-2-4.

Hocking, M.D. & Reynolds, J.D. (2011). Impacts of salmon on riparian plant diversity. *Science*, 331, 1609–1612.

Hocking, M.D. & Reynolds, J.D. (2012). Nitrogen uptake by plants subsidized by Pacific salmon carcasses: a hierarchical experiment. *Canadian Journal of Forest Research*, 42, 908–917.

Hof, C., Brändle, M. & Brandl, R. (2008). Latitudinal variation of diversity in European freshwater animals is not concordant across habitat types. *Global Ecology and Biogeography*, 17, 539–546.

Hogsden, K.L. & Harding, J.S. (2012). Anthropogenic and natural sources of acidity and metals and their influence on the structure of stream food webs. *Environmental Pollution*, 162, 466–474.

Holdich, D.M. & Reeve, I.D. (1991). Distribution of freshwater crayfish in the British Isles, with particular reference to crayfish plague, alien introductions and water quality. *Aquatic Conservation: Marine and Freshwater Ecosystems*, 1, 139–158.

Holomuzki, J.R. & Biggs, B.J. (2006). Food limitation affects algivory and grazer performance for New Zealand stream macroinvertebrates. *Hydrobiologia*, 561, 83–94.

Holomuzki, J.R., Feminella, J.W. & Power, M.E. (2010). Biotic interactions in freshwater benthic habitats.

Journal of the North American Benthological Society, 29, 220–244.

Holt, G. & Chesson, P. (2018). The role of branching in the maintenance of diversity in watersheds. *Freshwater Science*, DOI 10.1086/700680.

Holt, R.D. (1977). Predation, apparent competition, and the structure of prey communities. *Theoretical Population Biology*, 12, 197–229.

Hood, J.M., Benstead, J.P. Cross, W.F. Huryn, A.D. et al. (2018). Increased resource use efficiency amplifies positive response of aquatic primary production to experimental warming. *Global Change Biology*, 24, 1069–1084.

Hopkins, S. R., Wyderko, J.A., Sheehy, R.R., Belden, L.K. & Wojdak, J.M. (2013). Parasite predators exhibit a rapid numerical response to increased parasite abundance and reduce transmission to hosts. *Ecology and Evolution*, 3, 4427–4438.

Hoppe, P.D., Rosi-Marshall, E.J. & Bechtold, H.A. (2012). The antihistamine cimetidine alters invertebrate growth and population dynamics in artificial streams. *Freshwater Science*, 31, 379–388.

Horn, M.H., Correa, S.B., Parolin, P., Pollux, B.J.A. et al. (2011). Seed dispersal by fishes in tropical and temperate fresh waters: the growing evidence. *Acta Oecologica*, 37, 561–577.

Hornung, M., Le-Grice, S., Brown, N. & Norris, D. (1990). The role of geology and soils in controlling surface water acidity in Wales. In R.W. Edwards et al. (eds) *Acid waters in Wales*. Kluwer, Dordrecht, 55–66.

Hornung, M. & Reynolds, B. (1995). The effects of natural and anthropogenic environmental changes on ecosystem processes at the catchment scale. *Trends in Ecology and Evolution*, 10, 443–449.

Hosen, J.D., Aho, K.S., Appling, A.P. et al. (2019). Enhancement of primary production during drought in a temperate watershed is greater in larger rivers than headwater streams. *Limnology and Oceanography*, 64, 1458–1472.

Houde, A.L., Smith, A.D., Wilson, C.C., Peres-Neto, P.R. et al. (2016). Competitive effects between rainbow trout and Atlantic salmon in natural and artificial streams. *Ecology of Freshwater Fish*, 25, 248–260.

House of Commons Environmental Audit Committee (2022). Water quality in rivers. HC 74. https://committees.parliament.uk/publications/8460/documents/88412/default/.

Howard, J.K. & Coffey, K.M. (2006). The functional role of native freshwater mussels in the fluvial benthic environment. *Freshwater Biology*, 51, 460–474.

Howden, N.J.K. & Burt, T.P. (2009). Statistical analysis of nitrate concentrations from the Rivers Frome and Piddle (Dorset, UK) for the period 1965–2007. *Ecohydrology*, 2, 55–65.

Hrbek, T., da Silva, V.M.F., Dutra, N. et al. (2014). A new species of River Dolphin from Brazil or: How little do we know our biodiversity. *PLoS ONE*, 9(1), e83623.

Huang C.F. & Sih A. (1991). An experimental-study on the effects of salamander larvae on isopods in stream pools. *Freshwater Biology*, 25, 451–459.

Hubbell, S.P. (2001). *The Unified Neutral Theory of Biodiversity and Biogeography*. Princeton University Press, Princeton, NJ.

Huggins, X., Gleeson, T., Kummu, M. et al. (2022). Hotspots for social and ecological impacts from freshwater stress and storage loss. *Nature Communications*, 13, 439. https://doi.org/10.1038/s41467-022-28029-w.

Hughes, J.M., Schmidt, D.J & Finn, D.S. (2009). Genes in streams: using DNA to understand the movement of freshwater fauna and their riverine habitat. *BioScience*, 59, 573–583.

Hui, Z., Jaric, I., Roberts, D.L., Yongfeng, H. et al. (2020). Extinction of one of the world's largest freshwater fishes: Lessons for conserving the endangered Yangtze fauna. *Science of the Total Environment*, 710, https://doi.org/10.1016/j.scitotenv.2019.136242.

Humphries, S. & Ruxton, G.D. (2002). Is there really a drift paradox? *Journal of Animal Ecology*, 71, 151–154.

Humphries, S. & Ruxton, G.D. (2003). Estimation of intergenerational drift dispersal distances and mortality risk for aquatic macroinvertebrates. *Limnology and Oceanography*, 48, 2117–2124.

Huryn, A.D. (1996). An appraisal of the Allen Paradox in a New Zealand trout stream. *Limnology and Oceanography*, 41, 243–252.

Huryn, A.D. (1998). Ecosystem-level evidence for top-down and bottom-up control of production in a grassland stream ecosystem. *Oecologia*, 115, 173–183.

Huryn, A.D. & Wallace, J.B (2000). Life history and production of stream insects. *Annual Review of Entomology*, 45, 83–110.

Hutchens, J.J., Wallace, J.B. & Grubaugh, J.W. (2017). Transport and storage of fine particulate organic matter. In G.A. Lamberti & F.R. Hauer (eds) *Methods in Stream Ecology Volume 2: Ecosystem Function* (3rd edn). Academic Press, London, 37–53.

Hynes, H.B.N. (1960). *The Biology of Polluted Waters*. Liverpool University Press.

Hynes, H.B.N. (1970a). The ecology of stream insects. *Annual Review of Entomology*, 15, 25–42.

Hynes, H.B.N. (1970b). *The Ecology of Running Waters*. Liverpool University Press.

Hynes, H.B.N. (1975). The stream and its valley. *Verhandlungen der Internationale Vereinigung für Theoretische und Angewandte Limnologie*, 19, 1–15.

Hynes, H.B.N. (1983). Groundwater and stream ecology. *Hydrobiologia*, 100, 93–99.

Illies, J. & Botosaneanu, L. (1963). Problemes et méthodes de la classification et de la zonation des eaux courantes, considerées surtout du point de vue faunistique. *Mitteilungen der internationale Vereinigung für theoretische und angewandte Limnologie*, 12, 1–57.

Ingold C.T. (1942). Aquatic hyphomycetes on decaying alder leaves. *Transactions of the British Mycological Society*, 25, 339–417.

Ings, N.L., Hildrew, A.G. & Grey, J. (2010). Gardening by the psychomyiid caddisfly *Tinodes waeneri*: Evidence from stable isotopes. *Oecologia*, 163, 127–139.

IPCC. (2022). *Climate Change 2022: Impacts, Adaptation, and Vulnerability*. Contribution of Working Group II to the Sixth Assessment Report of the Intergovernmental Panel on Climate Change (H.-O. Pörtner, D.C. Roberts, M. Tignor, E.S. Poloczanska et al. (eds)) Cambridge University Press, Cambridge, UK and New York, NY, USA, 3056. doi:10.1017/9781009325844.

Ippolito, A., Kattwinkel, M., Rasmussen, J.J. et al. (2015). Modelling global distribution of agricultural insecticides in surface waters. *Environmental Pollution*, 198, 54–60.

Irons, J.G. III, Oswood, M.W., Stout, R.J. & Pringle, C.M. (1994). Latitudinal patterns in leaf litter breakdown: is temperature really important? *Freshwater Biology*, 32, 401–411.

Ittner L.D., Junghans, M. & Werner, I. (2018). Aquatic fungi: a disregarded trophic level in ecological risk assessment of organic fungicides. *Frontiers in Environmental Science*, 6, 105.

Ives, A.R. (1988). Aggregation and coexistence of competitors. *Annales Zoologici Fennici*, 25, 329–335.

Jackson D.A., Peres-Neto P.R. & Olden J.D. (2001). What controls who is where in freshwater fish communities—the roles of biotic, abiotic and spatial factors? *Canadian Journal of Fisheries and Aquatic Sciences*, 58, 157–170.

Jackson, M.C. & Grey, J. (2012). Accelerating rates of freshwater invasions in the catchment of the River Thames. *Biological Invasions*, 15, 945–951.

Jacobi, G.Z. & Gary, S.J. (1996). Winter stoneflies (Plecoptera) in seasonal habitats in New Mexico, USA. *Journal of the North American Benthological Society*, 15, 690–699.

Jacobsen, D. (2020). The dilemma of altitudinal shifts: caught between high temperature and low oxygen. *Frontiers in Ecology & Environment*, 18, 211–218.

Jacobsen, D. Schultz. R. & Encalada, A. (1997). Structure and diversity of stream invertebrate assemblages: the influence of temperature with altitude and latitude. *Freshwater Biology*, 38, 247, 263.

Jacobus, L., Macadum, C. & Sartori, M. (2019). Mayflies (Ephemerotera) and their contributions to ecosystem services. *Insects*, 10, 1–26.

Jahfer, S., Vinayachandran, P.N. & Nanjundiah, R.S. (2017). Long-term impact of Amazon river runoff on northern hemispheric climate. *Scientific Reports*, 7, 10989. https://doi.org/10.1038/s41598-017-10750-y.

Jähnig, S.C., Tonkin, J.D., Gles, M. et al. (2017). Severity multipliers as a methodology to explore potential effects of climate change on stream bioassessment programs. *Water*, 9, 188, https://doi.org/10.3390/w9040188.

Jane, S.J., Wilcox, T.M., McKelvey, K.S., Young, M.K. et al. (2015). Distance, flow and PCR inhibition: eDNA dynamics in two headwater streams. *Molecular Ecology Resources*, 15, 216–227.

Jansson, R., Backs, H., Boulton, A.J. et al. (2005). Stating mechanisms and refining criteria for ecologically successful river restoration: a comment on Palmer et al. (2005). *Journal of Applied Ecology*, 42, 218–222.

Jardine, T.D., Rayner, T.S., Pettit, N.E. et al. (2017). Body-size drives allochthony in food webs of tropical rivers. *Oecologia*, 183, 505–517.

Jarvie, H.P., Smith, D.R., Norton, L.R. et al. (2018). Phosphorus and nitrogen limitation and impairment of headwater streams relative to rivers in Great Britain: a national perspective on eutrophication. *Science of the Total Environment*, 621, 849–862.

Jenkins, A.P. & Jupiter, S.D. (2011). Spatial and seasonal patterns in freshwater ichthyofaunal communities of a tropical high island in Fiji. *Environmental Biology of Fishes*, 91, 261–274.

Jenkins, G.B., Woodward, G. & Hildrew, A.G. (2013). Long-term amelioration of acidity accelerates decomposition in headwater streams. *Global Change Biology*, 19, 1100–1106.

Jennings, S. (2005). Size-based analysis of aquatic food webs. In A. Belgrano, U.M. Scharler, J. Dunne & R.E Ulanowicj (eds) *Aquatic Food Webs*. Oxford University Press, New York, 86–97.

Jensen, K.W. & Snekvik, E. (1972). Low pH levels wipe out salmon and trout populations in southernmost Norway. *Ambio*, 1, 223–225.

Johnson, B.R. & Wallace, J.B. (2005). Bottom-up limitation of a stream salamander in a detritus-based food web. *Canadian Journal of Fisheries & Aquatic Sciences*, 62, 301–311.

Johnson, M. J., Giller, P. S., O'Halloran, J., O'Gorman, K. & Gallagher, M. (2005). A novel approach to assess the impact of land use activity on chemical and biological parameters in river catchments. *Freshwater Biology*, 50, 1273–1289.

Jones, C.G., Lawton, J.H. & Shachak, M. (1994). Organsims as ecological engineers. *Oikos*, 69, 373–386.

Jones E.B.G., Hyde, K.D. & Pang, K.-L. (eds) (2014). *Freshwater Fungi and Fungal-like Organisms*. De Gruyter, Berlin.

Jones, J.I., Vdovchenko, A., Cooling, D. et al. (2020). Systematic analysis of the relative abundance of polymers, occurring as microplastics in freshwater and estuaries. *International Journal of Environmental Research and Public Health*, 17, 9304; doi:10.3390/ijerph17249304.

Jones, J.R.E. (1950). A further ecological study of the River Rheidol: the food of the common insects of the main stream. *Journal of Animal Ecology*, 19, 159–174.

Jones, N.E. & Schmidt, B.J. (2018). Influence of tributaries on the longitudinal patterns of benthic invertebrate communties. *River Research & Applications*, 34, 165–173.

Jones, R.I., Carter, C.E., Kelly, A. et al. (2008). Widespread contribution of methane-cycle bacteria to the diets of lake profundal chironomid larvae. *Ecology*, 89, 857–864.

Jones, R.I. & Grey, J. (2011). Biogenic methane in freshwater food webs. *Freshwater Biology*, 56, 213–229.

Jonsson, B. & Jonsson, N. (2004). Factors affecting marine production of Atlantic salmon (*Salmo salar*). *Canadian Journal of Fisheries and Aquatic Sciences*, 61, 2369–2383.

Jonsson, M. & Malmqvist, B. (2000). Ecosystem process rate increases with animal specis richness: evidence from leaf-eating, aquatic insects. *Oikos*, 89, 519–523.

Jonsson, N., Jonsson, B. & Hansen, L.P. (1998). The relative role of density-dependent and density-independent survival in the life-cycle of the Atlantic salmon (*Salmo salar*). *Journal of Animal Ecology*, 67, 751–762.

Jourdan, J., O'Hara. R.B, Bottarin, R. et al. (2018). Effects of changing climate on European stream invertebrate communities: a long-term data analysis. *Science of the Total Environment*, 621, 588–599.

Joy, M. (2015). *Polluted Inheritance: New Zealand's Freshwater Crisis*. BWB texts, Wellington.

Joy, M. & Canning, A.D. (2020). Shifting baselines and political expediency in New Zealand's freshwater management. *Marine & Freshwater Research*, 72, 456–461.

Julian, J.P., De Beurs, K.M., Owsley, B. et al. (2017). River water quality changes in New Zealand over 26 years: reponse to land use intensity. *Hydrology & Earth System Science*, 21, 1149–1171.

Junk, W.J., Bayley, P.B. & Sparks, R.E. (1989). The flood pulse concept in river-flood-plain systems. In D.P. Dodge (ed) *Proceedings of the Internationl Large River Symposium. Special Publications of the Canadian Journal of Fisheries and Aquatic Sciences*, 106, 110–127.

Junker, J.R., Cross, W.F., Benstead, J.P., Huryn, A.D. et al. (2020). Resource supply governs the apparent temperature dependence of animal production in stream ecosystems. *Ecology Letters*, 23, 1809–1819.

Kagami, M., Miki, T. & Takimoto, G. (2014). Mycoloop: chytrids in aquatic food webs. *Frontiers in Microbiology*, 5, 166, doi: 10.3389/fmicb.2014.00166.

Kalkman, V., Calusnitzer, V., Dijkstra, K., Orr, A., Paulson, D. & Tol, J. (2008). Global diversity of dragonflies (Odonata) in freshwater. *Hydrobiologia*, 595, 351–363.

Kareiva, P., Marvier, M. & McClure, M. (2000). Recovery and management options for spring/summer Chinook salmon in the Columbia River basin. *Science*, 290, 977–979.

Karki, S., Stewardson, M.J., Webb, J.A. et al. (2021). Does the topology of the river network influence the delivery of river ecosystem services? *River Research and Applications*, 37, 256–269.

Kaushik, N.K. & Hynes, H.B.N. (1968). Experimental study on the role of autumn-shed leaves in aquatic environments. *Journal of Ecology*, 56, 229–243.

Kaushik, N.K. & Hynes, H.B.N. (1971). The fate of dead leaves that fall into streams. *Archiv für Hydrobiologie*, 68, 465–515.

Kautza, A. & Sullivan, S.M.P. (2016). The energetic contributions of aquatic primary producers to terrestrial food webs in a mid-size river system. *Ecology*, 97, 694–705.

Keddy, P. & Weiher, E. (1999). Introduction: the scope and goals of research on assembly rules. In E. Weker & P. Keddy (eds) *Ecological Assembly Rules*. Cambridge University Press, Cambridge, 1–20.

Keiper, J.B. & Foote, B.A. (2000). Biology and larval feeding habits of coexisting Hydroptilidae (Trichoptera) from a small woodland stream in Northeastern Ohio. *Annals of the Entomological Society of America*, 93, 225–234.

Keith, D.A., Ferrer-Paris, J.R., Nicholson, E. & Kingsford, R.T. (eds) (2020). IUCN Global ecosystem typology 2.0. Descriptive profiles for biomes and ecosystem functional groups. IUCN, Gland, Switzerland.

Keith, D.A., Ferrer-Paris, J.R., Nicholson, E., Giller, P.S. et al. (2022). A function-based typology for Earth's ecosystems. *Nature*, 610, 513–518.

Keitzer, S.C. & Goforth, R.R. (2013). Salamander diversity alters stream macroinvertebrate community structure. *Freshwater Biology*, 58, 2114–2125.

Kelleher, B., Bergers, P.J.M., Van den Brink, F.W.B. et al. (1998). Effects of exotic amphipod invasions on fish diet in the Lower Rhine. *Archiv für Hydrobiologie*, 143, 363–382.

Keller, R. (1984). The world's fresh water: yesterday, today, tomorrow. *Applied Geography and Development*, 24, 7–23.

Kelly, F. & King, J.J. (2001). A review of the ecology and distribution of three lamprey species, *Lampetra fluviatilis* (L.), *Lampetra planeri* (Bloch) and *Petromyzon marinus* (L.): a context for conservation and biodiversity considerations in Ireland. *Biology & Environment Proceedings of the Royal Irish Academy* 101, 165–185.

Kelly, M., Juggins, S. Guthrie, R., Pritchard, S. et al. (2008). Assessment of ecological status in U.K. rivers using diatoms. *Freshwater Biology*, 53, 403–422.

Kelly, M.G. & Whitton, B.A. (1995). The Trophic Diaton Index: a new index for monitoring eutrophication in rivers. *Journal of Applied Phycology*, 7, 433–444.

Kelso, J.E., Rosi, E.J. & Baker, M.A. (2020). Towards more realistic estimates of DOM decay in streams: incubation methods, light, and non-aditive effects. *Freshwater Science*, 39, 559–575.

Kennan, J.G., Stein, E.D. & Webb, J.A. (2018). Evaluating and managing environmental water regimes in a water-scarce and uncertain future. *Freshwater Biology*, 63, 733–737.

Kennard, M.J., Pusey, B.J., Olden, J.D., Mackay, S.J. et al. (2010). Classification of natural flow regimes in Australia to support environmental flow management. *Freshwater Biology*, 55, 171–193.

Kennedy, G.J.A. & Strange, C.D. (1982). The distribution of salmonids in upland streams in relation to depth and gradient. *Journal of Fish Biology*, 20, 579–591.

Kennedy, K.T.M. & El-Sabaawi, R.W. (2017). A global meta-analysis of exotic versus native leaf decay in stream ecosystems. *Freshwater Biology*, 62, 977–989.

Kennedy, T.A. & Hobbie, S.E. (2004). Saltcedar (*Tamarix ramisissima*) invasion alters organic matter dynamics in a desrt stream. *Freshwater Biology*, 49, 65–76.

Kerezsy, A., Gido, K., Magalhaes, M. & Skelton, P.H. (2017). The biota of intermittent rivers and ephemeral streams. In T. Datry, N. Bonada & A.J. Boulton (eds) *Intermittent Rivers and Ephemeral Streams: Ecology and Management*. Academic Press, Burlington, 273–298.

Kim, S., Peoples, B.K. & Kanno, Y. (2020). Diverse reproductive patterns of Bluehead Chub (*Nocomis leptocephalus*) and their relationships with nest size and interactions with an associate, Yellowfin Shiner (*Notropis lutipennis*). *Environmental Biology of Fish*, 103, 783–794.

Kim, S.-K. (2015). Morphology and Ecological Notes on the Larvae and Pupae of Simulium (Simulium) from Korea. *Animal Systematics, Evolution and Diversity*, 31(4), 209–246. https://doi.org/10.5635/ASED.2015.31.4.209.

Kinzie III, R.A. (1988). Habitat utilization by Hawaiian stream fishes with reference to community structure in oceanic island streams. *Environmental Biology of Fishes*, 22, 179–192.

Klaus, S., Selvandran, S., Goh, J.W., Wowor, D. et al. (2013). Out of Borneo: Neogene diversification of Sundaic freshwater crabs (Crustacea: Brachyura: Gecarcinucidae: Parathelphusa). *Journal of Biogeography*, 40, 63–74.

Klose, K. & Cooper, S. (2011). Contrasting effects of an invasive crayfish on two temperate stream communities. *Freshwater Biology*, 57, 526–540.

Knillmann, S. & Liess, M. (2019). Pesticide Effects on Stream Ecosystems. In M. Schröter, A. Bonn, S. Klotz, R. Seppelt et al. (eds) *Atlas of Ecosystem Services*. Springer, Cham, Switzerland, 211–214. https://doi.org/10.1007/978-3-319-96229-0_33.

Knillmann, S., Orlinskiy, P., Kaska, O., Foit, K. et al. (2018). Indication of pesticide effects and recolonization in streams. *Science of the Total Environment*, 630, 1619–1627.

Kohler, S.L. (1992). Competition and the structure of a benthic stream community. *Ecological Monographs*, 62, 165–188.

Kohler S.L. & Wiley M.J. (1997). Pathogen outbreaks reveal large-scale effects of competition in stream communities. *Ecology*, 78, 2164–2176.

Kolasa, J. (2000). The biology and ecology of lotic microturbellarians. *Freshwater Biology*, 44, 5–14.

Kolmakova, A.A., Gladyshev, M.I., Kalachova, G.S et al. (2013). Amino acid composition of epilithic biofilm and benthic animals in a large Siberian river. *Freshwater Biology*, 58, 2180–2195.

Kominoski, J.S., Rosemond, A.D., Benstead, J.P., Gulis, V. et al. (2018). Nitrogen and phosphorus additions increase rates of stream ecosystem respiration and carbon loss. *Limnology and Oceanography*, 63, 22–36.

Kominoski, J.S., Ruhi, A., Hagler, M.M. et al. (2018). Patterns and drivers of fish extirpations in rivers of the American Southwest and Southeast. *Global Change Biology*, 24, 1175–1185.

Kondrashov, D., Feliks, Y. & Ghil, M. Oscillatory modes of extended Nile River records (A.D. 622–1922) (2005). *Geophysical Research Letters*, 322005, L10702, doi:10.1029/2004GL022156.

Koperski, P. (2017). Taxonomic, phylogenetic and functional diversity of leeches (Hirudinea) and their suitability in biological assessment of environmental quality. *Knowledge and Management of Aquatic Systems*, 418. DOI: https://doi.org/10.1051/kmae/2017040.

Krause, S., Hannah, D.M., Fleckenstein, J.H., Heppell, C.M. et al. (2011). Interdisciplinary perspectives on processes in the hyporheic zone. *Ecohydrology*, 4, 481–499.

Krauss, G-J., Sole, M., Krauss, G., Schlosser, D. et al. (2011). Fungi in freshwaters: ecology, physiology and biochemical potential. *FEMS Microbiology Review*, 35, 620–651.

Kreuzinger-Janik, B., Bruchner-Huttemann, H. & Traunspurger, W. (2019). Effect of prey size and structural complexity on the functional response in a nematode-nematode system. *Scientific Reports*, 9, 5696. doi: 10.1038/s41598-019-42213-x.

Kuehne, L.M., Ostberg, C.O., Chase, D. et al. (2020). The use of environmental DNA to detect the invasive aquatic plants *Myriophyllum spicatum* and *Egeria densa* in lakes. *Freshwater Science*, 39, 521–533.

Kühmayer, T., Guo, F., Ebm, N. et al. (2019). Preferential retention of algal carbon in benthic invertebrates: stable isotope and fatty acid evidence from an outdoor flume experiment. *Freshwater Biology*, 10.1111/fwb.13492.

Kullberg, A. (1988). The case, mouthparts, silk and silk formation of *Rheotanytarsus musciola* Kieffer (Chironomidae: Tanytarsini). *Aquatic Insects*, 10, 249–255.

Kuzmanovic, M., Dolédec, S., de Castro-Catala, N., Ginebreda, A. et al. (2017). Environmental stressors as a driver of the trait composition of benthic macroinvertebrate assemblages in polluted Iberian rivers. *Environmental Research*, 156, 485–493.

Labed-Veydert, T., Danger, M., Felton, V. et al. (2022). Microalgal food sources greatly improve macroinvertebrate growth in detritus-based headwater streams: evidence from an instream experiment. *Freshwater Biology*, 67, 1380–1394.

Ladle, M., Cooling, D.A., Welton, J.S. & Bass, J.A.B. (1985). Studies on Chironomidae in experimental recirculating stream systems, II. The growth, development and production of a spring generation of *Orthocladius* (*Euorthocladius*) *calvus* Pinder. *Freshwater Biology*, 15, 243–255.

Lagarde, R., Teichert, N., Valade, P. & Ponton, D. (2021). Structure of small tropical island freshwater fish and crustacean communities: a niche-or dispersal-based process? *Biotropica* 53, 243–254.

Lake, P.S. (2000). Disturbance, patchiness, and diversity in streams. *Journal of the North American Benthological Society*, 19, 573–592.

Lake, P.S. (2003). Ecological effects of perturbation by drought in flowing water. *Freshwater Biology*, 48, 1161–1172.

Lake, P.S. (2011). *Drought and Aquatic Ecosystems: Effects ands Responses*. Wiley-Blackwell, Oxford.

Lamberti, G.A. & Resh, V.H. (1987). Seasonal patterns of suspended bacteria and algae in two northern California streams. *Archiv fur Hydrobiologie*, 110, 45–57.

Lamberti, G.A. & Hauer, F.R. (eds) (2017). *Methods in Stream Ecology Volume 2: Ecosystem Function* (3rd edn). Academic Press, London.

Lancaster, J. (1996). Scaling the effects of predation and disturbance in a patchy environment. *Oecologia*, 107, 321–331.

Lancaster, J. (2020). Coexistence of predatory caddisfly species may be facilitated by variations in the morphology of feeding apparatus and diet. *Freshwater Biology*, 66, 745–752.

Lancaster, J. & Hildrew, A.G. (1993a). Characterizing in-stream flow refugia. *Canadian Journal of Fisheries and Aquatic Sciences*, 50, 1663–1675.

Lancaster, J. & Hildrew, A.G. (1993b). Flow refugia and the microdistribution of lotic macroinvertebrates. *Journal of the North American Benthological Society*, 12, 385–393.

Lancaster, J. & Robertson, A.L. (1995). Microcrustacean prey and macroinvertebrate predators in a stream food web. *Freshwater Biology*, 34(1), 123–134.

Lancaster, J. & Belyea, L. (1997). Nested hierarchies and scale-dependence of mechanisms of flow refugium use. *Journal of the North American Benthological Society*, 16, 221–238.

Lancaster, J. & Ledger, M.E. (2015). Population-level reponses of stream macroinvertebrates to drying can be density-independent or density-dependent. *Freshwater Biology*, 60, 2559–2570.

Lancaster, J. & Downes, B. (2021). Multiyear resource enrichment creates persistently higher species diversity in a landscale-scale field experiment. *Ecology*, 102, e03451.

Lancaster, J., Hildrew, A.G. & Townsend, C.R. (1988). Competition for space by predators in streams: field experiments on a net-spinning caddisfly. *Freshwater Biology*, 20, 185–193.

Lancaster, J., Hildrew, A.G. & Townsend, C.R. (1991). Invertebrate predation on patchy and mobile prey in streams. *Journal of Animal Ecology*, 60, 625–641.

Lancaster, J., Hildrew, A.G. & Gjerlov, C. (1996). Invertebrate drift and longitudinal transport processes in streams. *Canadian Journal of Fisheries and Aquatic Sciences*, 53, 572–582.

Lancaster, J., Downes, B.J. & Arnold, A. (2010). Environmentsl constraints on oviposition limit egg supply of a stream insect at multiple scales. *Oecologia*, 163, 373–384.

Lange, K., Townsend, C.R. & Matthaei, C.D. (2014). Can biological traits of stream invertebrates help disentangle the effects of multiple stressors in an agricultural catchment? *Freshwater Biology*, 59, 2431–2446.

Langford, T.E.L., Shaw, P.J., Ferguson, A.J.D. & Howard, S.R. (2009). Long-term recovery of macroinvertebrate biota in grossly polluted streams: recolonisation as a constraint to ecological quality. *Ecological Indicators*, 9, 1064–1077.

Langhans, S.D., Tiegs, S.D., Gessner, M.O. & Tockner, K. (2008). Leaf-decomposition hetereogeneity across a flood-plain mosaic. *Aquatic Sciences*, 70, 337–346.

Lansdown, K., McKew, B.A., Whitby, C. et al. (2016). Importance and controls of anaerobic ammonium oxidation influenced by riverbed geology. *Nature Geoscience*, 9, 357–360. https://doi.org/10.1038/ngeo2684.

Larned, S.T., Kinzie III, R.A., Covich, A.P. & Chong, C.T. (2003). Detritus processing by endemic and non-native Hawaiian stream invertebrates: a microcosm study of species-specific effects. *Archiv für Hydrobiologie*, 156, 241–254.

Larned, S.T., Snelder, T., Unwin, M.J. & McBride, G.B. (2016). Water quality changes in New Zealand rivers: current state and trends. *New Zealand Journal of Marine & Freshwater Research*, 50, 389–417.

Larsen, T.E.L & Ormerod, S.J. (2010). Combined effects of habitat modification on trait composition and species nesteness in river invertebrates. *Biological Conservation*, 143, 2638–2646.

Larsen, S., Chase, J.M., Durance, I. & Ormerod, S.J. (2018). Lifting the veil: richness measurements fail to detect systematic biodiversity change over three decades. *Ecology*, 99, 1316–1326.

Lau, D.C.P., Leung, K.M.Y. & Dudgeon, D. (2009). What does stable isotope analysis reveal about trophic relationships and the relative importance of allochthonous and autochthonous resources in tropical streams? A synthetic study from Hong Kong. *Freshwater Biology*, 54, 127–141.

Lauridsen, R.B., Edwards, F.K., Cross, W.F., Woodward, G.W. et al. (2014). Consequences of inferring diet from feeding guilds when estimating and interpreting consumer-resource stoichiometry. *Freshwater Biology*, 59, 1497–1508.

Layer, K., Riede, J.O, Hildrew, A.G. & Woodward, G. (2010). Food web structure and stability in 20 streams across a wide pH gradient. *Advances in Ecological Research*, 42, 265–299.

Layer, K., Hildrew, A.G., Jenkins, G.B., Riede, J.O. et al. (2011). Long-term dynamics of a well-characterised food web: four decades of acidification and recovery in the Broadstone Stream model system. *Advances in Ecological Research*, 44, 69–116.

Layer. K., Hildrew, A.G. & Woodward, G. (2012). Grazing and detritivory in 20 stream food webs across a broad pH gradient. *Oecologia*, 171, 459–471.

Leal, C.G., Barlow, J. & Gardner, T.A. (2018). Is environemtnal legislation conserving tropical stream faunas? A large-scale assessment of local, riparian and catchment-scale influences on Amazonian fish. *Journal of Applied Ecology*, 55, 1312–1326.

Leberfinger, K., Bohman, I. & Herrmann, J. (2011). The importance of terrestrial resource subsidies for shredders in open-canopy streams revealed by stable isotope analysis. *Freshwater Biology*, 56, 470–480.

LeBourdais, S.V., Ydenburg, R.C. & Esler, D. (2009). Fish and Harlequin duck compete on breeding streams. *Canadian Journal of Zoology*, 87, 31–40.

Lechner, A., Keckeis. H. & Humphries, P. (2016). Patterns and processes in the drift of early developmental stages of fish in rivers: a review. *Reviews in Fish Biology and Fisheries,* 26, 471–489.

Ledger, M.E. & Hildrew, A.G. (1998) Temporal and spatial variation in the epilithic biofilm of an acid stream. *Freshwater Biology* 40, 655–670.

Ledger, M.E. & Hildrew, A.G. (2000). Herbivory in an acid stream. *Freshwater Biology,* 43, 545–556.

Ledger, M.E. & Hildrew, A.G. (2001). Recolonisation by the benthos of an acid stream following a drought. *Archiv für Hydrobiologie,* 152-1-17.

Ledger, M.E. & Hildrew, A.G. (2005). The ecology of acidification and recovery: changes in herbivore-algal food web linkages across a stream pH gradient. *Environmental Pollution,* 137, 103–118.

Ledger, M.E. & Milner, A.M. (2015). Extreme events in running waters. *Freshwater Biology,* 60, 2455–2460.

Lee, J.H., Kim, T.W., & Choe, J.C. (2009). Commensalism or mutualism: Conditional outcomes in a branchiobdellid–crayfish symbiosis. *Oecologia,* 159, 217–224.

Lee, S.S., Paspalof, D.D., Snow, E.K. et al. (2016). Occurrence and potential biological effects of amphetamine on stream communities. *Environmental Science & Technology,* 50, 9727–9735.

Leese, F., Bouchez, A., Abarenkov, K. et al. (2018). Why we need sustainable netoworks bridging countries, disciplines, cultures and generations for Aquatic Monitoring 2.0: a perspective derived from the DNAqua-Net COST Action. *Advances in Ecological Research,* 58, 63–99. Doi 10.1016/bs.aecr.2018.01.001.

Leflaive, J. & Ten-Hage, L. (2007). Algal and cyanobacterial secondary metabolites in freshwaters: a comparison of allelopathic compounds and toxins. *Freshwater Biology,* 52, 199–214.

Lehane, B.M., Giller, P.S. O'Halloran, J. et al. (2002). Experimental provision of large woody debris in streams as a trout management technique. *Aquatic Conservation: Marine and Freshwater Systems,* 12, 289–311.

Lehner, B., Verdin, K. & Jarvis, A. (2008). New global hydrography derived from spaceborne elevation data, *Eos Transactions of the American Geophysical Union,* 89(10), 93–104. https://doi.org/10.1029/2008EO100001.

Leibold, M.A., Holyoak, M., Mouquet, N., Amarasekare, P. et al. (2004). The metacommunity concept: a framework for multi-scale community ecology. *Ecology Letters,* 7, 601–613.

Leivestad, H., Henry, G., Muniz, I.P. & Snekvik, E. (1976). Effects of acid precipitation on freshwater organisms. In G. Abrahamsen, H. Dovland, E.T. Gjessing, H. Leivestad (eds) *Impacts of acid precipitation on forest and freshwater ecosystems in Norway.* Oslo, SNSF Project, 86–111.

Lemm, J.U., Venohr, M., Globevnik, L. et al. (2020). Multiple stressors determine river ecological status at the European scale: towards an integrated understanding of river status deterioration. *Global Change Biology,* 10.1111/gcb.15504.

Leopold, L.B. (1962). Rivers. *American Scientist,* 50, 511–537.

Leopold, L.B. & Maddock, T. (1953). The hydraulic geometry of stream channels and some physiographic implications. Geological Survey Professional Paper 252, https://doi.org/10.3133/pp252.

Leopold, L.B., Woolman, M.G. & Miller, J.F. (1964). *Fluvial Processes in Geomorphology.* Freeman, San Francisco.

Leuven, R.S.E.W., van der Velde, G., Baijens, I. et al. (2009). The River Rhine: a global highway for dispersal of aquatic invasive species. *Biological Invasions,* 11, 1989–2008.

Levi, P.S. & McIntyre, P.B. (2020). Ecosystem responses to channel restoration decline with stream size in urban river networks. *Ecological Applications,* 30, e02107.

Levi, P.S., Riis, T., Baisner, A.J., Peipoch, M et al. (2015). Macrophyte complexity controls nutrient uptake in lowland streams. *Ecosystems,* 18, 914–931.

Levine, S. (1980). Several measures of trophic structure applicable to complex food webs. *Journal of Theoretical Biology,* 83, 195–207.

Lewis W.M. Jr., Hamilton S.K. & Saunders J.F. (1995). Rivers of northern South America. In C.E. Cushing, K.W. Cummins & G.W. Minshall (eds) *River and Stream Ecosystems.* Elsevier, Amsterdam, 219–256.

Leyer, I. (2006). Dispersal, diversity and distribution patterns in pioneer vegetation; the role of river–floodplain connectivity. *Journal of Vegetation Science,* 17, 407–416.

Leys, M., Keller, I., Räsänen, K., Gattolliat, J.-L. et al. (2016). Distribution and population genetic variation of cryptic species of the Alpine mayfly *Baetis alpinus* (Ephmeroptera: Baetidae) in the central Alps. *BMC Ecology & Evolution,* 77, doi.org/10.1186/s12862-016-0643-y.

Liess, M. & Von der Ohe, P.C. (2005). Analyzing effects of pesticides on invertebrate comunities in streams. *Environmental Toxicity and Chemistry,* 24, 954–965.

Likens, G.E. & Bormann, F.H. (1974). Linkages between terrestrial and aquatic ecosystems. *Bioscience,* 24, 447–456.

Likens, G.E. & Bormann, F.H. (1995). *Biogeochemistry of a Forested Ecosystem* (2nd edn). Springer, New York.

Lillehammer, A., Brittain, J.E., Saltveit, S.J. & Nielsen, P.S. (1989). Egg development, nymphal growth and life cycle, strategies in Plecoptera, *Holarctic Ecology,* 12, 173–186.

Lintern, A., Webb, J.A., Ryu, D., Liu, S. et al. (2017). Key factors influencing differences in stream water quality across space. *WIREs Water,* 5, e1260. Doi 10.1002/wat2.1260.

LINX collaborators (2014). The Lotic Intersite Nitrogen Experiments: an example of successful ecological research collaboration. *Freshwater Science,* 33, 700–710.

Lobon-Cervia, J. (2012). Density-dependent mortality of adult, but not of juveniles, of stream-resident brown trout (*Salmo trutta*). *Freshwater Biology,* 57, 2181–2189.

Lock, M.A., Wallace, R.R., Costerton, J.W., Ventullo, R.M. et al. (1984). River eplithon: towards a structural-functional model. *Oikos,* 42, 10–22.

Lodge, D.M. (1991). Herbivory on freshwater macrophytes. *Aquatic Botany*, 41, 195–224.

Lods-Crozet, B., Lencioni, V., Ólafsson, J.S., Snook, D.L. et al. (2001). Chironomid (Diptera: Chironomidae) communities in six European glacier-fed streams. *Freshwater Biology*, 46, 1791–1809.

Logie, J.W. (1995). Effects of stream acidity on non-breeding dippers *Cinclus cinclus* in the south-central highlands of Scotland. *Aquatic Conservation*, 5, 23–35.

Longo, M. & Blanco, J.F. (2014). Patterns at Multi-Spatial Scales on Tropical Island Stream Insect Assemblages: Gorgona Island Natural National Park, Colombia, Tropical Eastern Pacific. *Revista de Biología Tropical*, 62, 65–83.

Lopes-Lima, M., Burlakova, L., Karatayev, A., Meyler, K. et al. (2018). Conservation of freshwater bivalves at the global scale: diversity, threats and research needs. *Hydrobiologia*, 810, 1–14.

Lopéz-Delgado, E.O., Winemiller, K.O. & Villa-Navarro, F.A. (2019). Do metacommunity theories explain spatial variation in fish assemblage structure in a pristine tropical river? *Freshwater Biology*, 64, 367–379.

Lopez-Rodrıguez M.J., Tierno de Figueroa J.M. & Alba-Tercedor J. (2008). Life history and larval feeding of some species of Ephemeroptera and Plecoptera (Insecta) in the Sierra Nevada (Southern Iberian Peninsula). *Hydrobiologia*, 610, 277–295.

Lor, Y., Schreier, T., Waller, D.L. & Merkes, C.M. (2020). Using environmental DNA (eDNA) to detect the endangered Spectaclecase Mussel (*Margartifera monodonta*). *Freshwater Science*, 39. https://doi.org/10.1086/711673.

Lorgeoux, C., Moilleron, R., Gasperi, J. et al. (2016). Temporal trends of persistent organic pollutants in dated sediment cores: chemical fingerprinting of the anthropogenic impacts in the Seine River basin, Paris. *Science of the Total Environment*, 541, 1355–1363.

Loudon, C. & Alstad, D.N. (1990). Theoretical mechanisms of particle capture: predictions for hydropsychid caddisfly distributional ecology. *American Naturalist*, 135, 360–381.

Lucy, F.E., Burlakova, L.E., Karatayev, A.Y., Mastitsky, S.E. et al. (2014). Zebra mussel impacts on unionids: a synthesis of trends in North America and Europe. In T.F. Nalepa & D.W. Schloesser (eds) *Quagga and zebra mussels: biology, impacts, and control* (2nd edn). CRC Press, Boca Raton, FL, 623–646.

Lutscher, F., Nisbet, R.M. & Pachepsky, E. (2010). Population persistence in the face of advection. *Theoretical Ecology*, 3, 271–284.

Lyons, W.B., Carey, A.E., Gardner, C.B., Welch, S.A. et al. (2021). The geochemistry of Irish rivers. *Journal of Hydrology Regional Studies*, 37, 100881.

Lytle, D.A. (2008). Life-history and behavioural adaptations to flow regime in aquatic insects. Chapter 7 In J. Lancaster & R.A. Briers (eds) *Aquatic Insects: Challenges to Populations*. CAB International, Wallingford, UK, 122–138.

Lytle, D.A. & Poff, L. (2004). Adaptation to natural flow regimes. *Trends in Ecology and Evolution*, 19, 94–100

Lytle, D.A., Bogan, M.T. & Finn, D.S. (2008). Evolution of aquatic insect behaviours across a gradient of disturbance predictability. *Proceedings of the Royal Society B—Biological Sciences*, 275, 453–462.

Maasri, A., Jahnig, S.C. et al. (2022). A global agenda for advancing freshwater biodiversity research. *Ecology Letters*, 25, 255–263.

Maavara, T., Chen, Q., Van Meter, K. et al. (2020). River dam impacts on biogeochemical cycling. *Nature Reviews Earth & Environment*, 1, 103–116.

Macan, T.T. (1961). A review of running water studies. *Verhandlungen der Internationale Vereinigung für theoretische und angewandte Limnologie*, 14, 587–602.

Macan, T.T. (1963). *Freshwater Ecology*, Longman, London.

Macaulay, S.J., Hageman, K.J., Piggott, J.J. & Matthaei, C.D. (2020). Imidacloprid dominates toxicities of neonicotenoid mixtures to stream mayfly nymphs. *Science of the Total Environment*, 761. https://doi.org/10.1016/j.scitotenv.2020.143263.

Macías, N.A., Colón-Gaud, C., Duggins, J.W. & Ramírez, A. (2014). Do omnivorous shrimp influence mayfly nymph life history traits in a tropical island stream? *Revista de Biología Tropical*, 62, 41–51.

Mackay, R.J. (1995). River and stream ecosystems of Canada. In C.E. Gushing, K.W. Cummins & G.W. Minshall (eds) *Ecosystems of the World: 22. River and Stream Ecosystems*. Elsevier, Amsterdam, 33–60.

Macklin, M.G. & Lewin, J. 2015. The rivers of civilisation. *Quaternary Science Reviews*, 114, 228–244.

Macneale, K.H., Peckarsky, B.L. & Likens, G.E. (2005). Stable isotopes identify dispersal patterns of stonefly populations living along stream corridors. *Freshwater Biology*, 50, 1117–1130.

Madsen, B.L. (1972). Detritus on stones in small streams. *Memorie dell'Istituto Italiano di Idrobiologia*, 29 (Supplement), 385–403.

Maguire. Z., Tumulo, B.B. & Albertson, L.K. (2020). Retreat but no surrender: net-spinning caddisfly (Hydropsychidae) silk has enduring effects on stream channel hydraulics. *Hydrobiologia*, 847, 1539–1551.

Maier, M. & Peterson, T.D. (2017). Prevalence of chytrid parasitism among diatom populations in the lower Columbia River (2009–2013). *Freshwater Biology*, 62, 414–428.

Maitland, P.S. (2003). *Ecology of the River, Brook and Sea lamprey. Conserving Natura 2000 Rivers. Ecology Series 5. English Nature*, Peterborough, UK, 52.

Majdi, N., Traunsperger, W., Richardson, J.S. & Lecerf, A. (2015). Small stonefly predators affect microbenthic and meiobenthic communities in stream leaf packs. *Freshwater Biology*, 60, 1930–1943. doi:10.1111/fwb.12622.

Majdi, N., Threis, I. & Traunspurger, W. (2017). It's the little things that count: meiofaunal density and production in the sediment of two headwater streams. *Limnology and Oceanography*, 62, 151–163.

Majdi, N., Schmid-Araya, J.M. & Traunspurger, W. (2020). Preface: Patterns and processes of meiofauna in freshwater ecosystems. *Hydrobiologia*, 847, 2587–2595.

Malas, D. & Wallace, J.B. (1977). Strategies for coexistence in three species of net-spinning caddisflies (Trichoptera) in second-order southern Appalachian streams. *Canadian Journal of Zoology*, 55, 1829–1840.

Malmqvist, B. (1991). Stonefly functional responses: Influence of substrate heterogeneity and predator interaction. *Verhandlungen der Internationale Vereinigung für Theoretische und Angewandte Limrnologie*, 24, 2895–900.

Malmqvist, B. (1993). Interactions in stream leaf packs: effects of a stonefly predator on detritivores and organic matter processing. *Oikos*, 66, 454–462.

Malmqvist, B. & Maki, M. (1994). Benthic macroinvertebrate assemblages in north Swedish streams: environmental relationships. *Ecography*, 17, 9–16.

Malmqvist, B. & Sackman, G. (1996). Changing risk of predation for a filter feeding insect along a current velocity gradient. *Oecologia*, 108, 450–458.

Malmqvist, B. & Hoffsten, P. (2000). Macroinvertebrate taxonomic richness, community structure and nestedness in Swedish streams. *Archiv für Hydrobiologie*, 150, 29–54.

Malmqvist, B., Sjöström, P. & Frick, K. (1991). The diet of two species of *Isoperla* (Plecoptera: Perlodidae) in relation to season, site and sympatry. *Hydrobiologia*, 213, 191–203.

Malmqvist, B., Wotton, R.S. & Zhang, Y. (2001). Suspension feeders transform massive amounts of seston in large northern rivers. *Oikos*, 92, 35–43.

Mani, T., Hauk, A., Walter, U. & Burkhardt-Holm, P. (2016). Microplastics profile along the River Rhine. *Scientific Reports*, 5, 17988. DOI: 10.1038/srep17988.

Mann C.C. (2011). *1493: Uncovering the New World Columbus Created*. Knopf publishers, New York.

Mann, R.H.K. & Penczak, T. (1986). Fish production in rivers: a review. *Polskie Archiwum Hydrobiologii*, 33, 233–247.

Manning, D.W.P., Rosemond, A.D., Gulis, V. et al. (2016). Convergence of detrital stoichiometry predicts thresholds of nutrient-stimulated breakdown in streams. *Ecological Applications*, 26, 1745–1757.

Manning, D.W.P., Rosemond, A.D., Benstead, J.P. et al. (2020). Transport of N and P in US streams and rivers differs with land use and between dissolved and particulate forms. *Ecological Applications*, 30, e02130. https://doi.org/10.1002/eap.2130.

Manolaki, P., Mouridsen, M.B., Nielsen, E., Olesen, A. et al. (2020). A comparison of nutrient uptake efficiency and growth rate between different macrophyte growth forms. *Journal of Environmental Management*, 274, DOI: 10.1016/j.jenvman.2020.111181.

Marcarelli, A.M., Baxter, C.V., Mineau, M.M. & Hall, R.O. (2011). Quantity and quality: unifying food web and ecosystem perspectives on the role of resource subsidies in freshwaters. *Ecology*, 92, 1215–1225.

Marcarelli, A., Baxter, C.V., Benjamin, J.R., Miyake, Y. et al. (2020). Magnitude and direction of stream-forest community interactions change with timescale. *Ecology*, 101(8), e03064.

March, J.G., Benstead, J.P., Pringle, C.M. & Scatena, F.N. (1998). Migratory drift of larval freshwater shrimps in two tropical streams, Puerto Rico. *Freshwater Biology*, 40, 261–273.

Marchant, R. (2021). Long-term fluctuations in density of two species of caddisfly from south-east Australia and the importance of density-dependent mortality. *Freshwater Biology*, doi 10.1111/fwb.13821.

Marchant, R. & Yule, C.M. (1996). A method for estimating larval life spans of aseasonal aquatic insects from streams on Bougainville Island, Papua New Guinea. *Freshwater Biology*, 35, 101–107.

Marden, J.H. & Kramer, M.G. (1995). Locomotor performance of insects with rudimentary wings. *Nature*, 337, 332–334.

Marks, J.C. (2019). Revisiting the fates of dead leaves that fall into streams. *Annual Review of Ecology, Evolution, and Systematics*, 50, 547–568.

Martin, I.D. & Mackay, R.J. (1982). Interpreting the diet of *Rhyacophila* larvae (Trichoptera) from gut analysis: An evaluation of techniques. *Canadian Journal of Zoology*, 60, 783–789.

Martin, P., Martinez-Ansemil, E., Pinder, A., Timm, T. et al. (2008). Global diversity of oligochaetous clitellates ('Oligochaeta': Clitellata) in freshwater. *Hydrobiologia*, 595, 117–127.

Martin, S. (2008). Global diversity of crocodiles (Crocodilia: Reptilia) in freshwater. *Hydrobiologia*, 595, 587–591.

Martinez-Cruz, K., Gonzalez-Valencia, R., Sepulveda-Jaurugui, A. et al. (2017). Methane emission from aquatic systems of Mexico City. *Aquatic Sciences*, 79, 159–169.

Marvanova, L. & Muller-Haeckel, A. (1980). Waterborne spores in foam in a subarctic stream system in Sweden. Sydowia, *Annales Mycologica Series*, 11, 33, 210–220.

Marxsen, J. (1996). Measurement of bacterial production in stream-bed sediments via leucine incorporation. *FEMS Microbiology Ecology*, 21, 313–325.

Marzolf, N.S. & Ardón, M. (2021). Ecosystem metabolism in tropical streams and rivers: a review and synthesis. *Limnology and Oceanography*, 66, 1627–1638.

Masters, Z., Petersen, I., Hildrew, A.G. & Ormerod, S.J. (2007). Insect dispersal does not limit the biological recovery of streams from acidification. *Aquatic Conservation: Marine and Freshwater Ecosystems*, 17, 375–383.

Matczak, T. Z. & R. J. Mackay, (1990). Territoriality in filter-feeding caddisfly larvae: laboratory experiments. *Journal of the North American Benthological Society*, 9, 26–34.

Mathers, K.L., White, J.C., Fornarolli, R. & Chadd, R. (2020). Flow regimes control the establishment of invasive crayfish and alter their effects on lotic macroinvertebrate communities. *Journal of Applied Ecology*, 57, 886–902.

Matthaei C.D. & Huber H. (2002). Microform bed clusters: are they preferred habitats for invertebrates in a flood-prone stream? *Freshwater Biology*, 47, 2174–2190.

Matthaei, C.D., Arbuckle, C.J. & Townsend, C.R. (2000). Stable surface stones as refugia for invertebrates during disturbance in a New Zealand stream. *Journal of the North American Benthological Society*, 19, 82–93.

Matthews, W.J. & Marsh-Matthews, E. (2016). Dynamics of an upland stream community over 40 years: trajectories and support for the loose-equilibrium hypothesis. *Ecology*, 97, 706–719.

Maude, S.H. & Williams, D.D. (1983). Behavior of crayfish in water currents: hydrodynamics of eight species with reference to their distribution patterns in southern Ontario. *Canadian Journal of Fisheries and Aquatic Science*, 40, 68–77.

McAuliffe, J. R. (1984). Competition for space, disturbance and the structure of a benthic stream community. *Ecology* 65, 894–908.

McBride, J. & Cohen, M.J (2020). Controls on productivity of submerged aquatic vegetation in 2 spring-fed rivers. *Freshwater Science*, 39, 1–7.

McCaffery, M. & Eby, L. (2016). Beaver activity increases aquatic subsidies to terrestrial consumers. *Freshwater Biology*, 61, 518–532.

McCarthy, M.D., Rinella, D.J. & Finney, B.P. (2018). Sockeye salmon population dynamics over the past 4000 years in Upper Russian Lake, south-central Alaska. *Journal of Palaeolimnology*, 60, 67–75.

McCreadie, J.W. & Adler, P.H. (1999). Parasites of larval blackflies (Diptera: Simuliidae) and environmental factors associated with their distributions. *Invertebrate Biology*, 118, 310–318.

McDowall, R.M. (2010). Why be amphidromous: expatrial dispersal and the place of source and sink dynamics. *Review of Fish Biology and Fisheries*, 20, 87–100.

McHugh, P.A., McIntosh, A.R. & Jellyman, P.G. (2010). Dual influences of ecosystem size and disturbance on food chain length in streams. *Ecology Letters*, 13, 881–890.

McIntosh, A.R. & Townsend, C.R. (1995). Impacts of an introduced predatory fish on mayfly grazing in New Zealand streams. *Limnology and Oceanography*, 40, 1508–1512.

McIntosh, A.R., Death, R.G., Greenwood, M.J. & Paterson, R.A. (2016). Food webs of streams and rivers. In P. Jellyman, C. Davie, C. Pearson & J. Harding, (eds) *Advances in New Zealand Freshwater Science*. New Zealand Freshwater Sciences and Hydrological Societies, Christchurch, 261–282.

McIntosh, A.R., Greig, H.S. & Howard, S. (2022). Regulation of open populations of a stream insect through larval density dependence. *Journal of Animal Ecology*, doi: 10.1111/1365-2656.13696.

McKenna, J.E., Slattery, M.T. & Clifford, K.M. (2013). Broad-scale patterns of Brook trout responses to introduced Brown trout in New York. *North American Journal of Fisheries management*, 33, 1221–1235, DOI: 10.1080/02755947.2013.830998.

McKie, B.G., Woodward, G., Hladyz, S. et al. (2008). Ecosystem functioning in stream assemblages from different regions: contrasting responses to variation in detritivore richness, evenness and density. *Journal of Animal Ecology*, 77, 495–504.

McLay, C. (1970). A theory concerning the distance travelled by animals entering the drift of a stream. *Journal of the Fisheries Research Board of Canada*, 27, 359–370.

McLennan, D., Auer, S.K., Anderson, G.J. et al. (2019). Simulating nutrient release from parental carcasses increases the growth, biomass, and genetic diversity of juvenile Atlantic salmon. *Journal of Applied Ecology*, 56, 1937–1947.

McMahon, R.F. (1991) Mollusca: Bivalvia. In J.H. Thorpe & A.P. Covich (eds) *Ecology and classification of North American freshwater invertebrates*. Academic Press, New York, 321–405.

McManamay, R., DeRolph, C. (2019). A stream classification system for the conterminous United States. *Scientific Data* 6, 190017. https://doi.org/10.1038/sdata.2019.17.

McMullen, L.E. & Lytle, D.A. (2012). Quantifying invertebrate resistance to floods: a global-scale meta-analysis. *Ecological Applications*, 22, 2164–2175.

McNair, J.N. & Newbold, J.D. (2012). Turbulent particle transport in streams: can exponential settling be reconciled with fluid dynamics? *Journal of Theoretical Biology*, 300, 62–80.

McShaffrey, D. & McCafferty, W.P. (1987). The behaviour and form of *Psephenus herricki* (DeKay) (Coleoptera: Psephenidae) in relation to water flow. *Freshwater Biology*, 18, 319–324.

McShaffrey, D. & McCafferty, W.P. (1988). Feeding behaviour of *Rhithrogena pellucida* (Ephemeroptera: Heptageniidae). *Journal of the North American Benthological Society*, 7, 87–99.

Medinas de Campos, M., Tritico, H.M., Girard, P. et al. (2020). Predicted impacts of proposed hydroelectric facilities on fish, migration routes upstream from the Pantanal wetland (Brazil). *River Research and Applications*, doi:10.1002/rra.3588.

Meffe, G. & Vrijenhoek, R. (1988). Conservation genetics in the management of desert fishes. *Conservation Biology*, 2, 157–169.

Melo-Santos, G., Figueiredo Rodrigues, A.L., Tardin, R.H., de Sá Maciel, I. et al. (2019). The newly described Araguaian river dolphins, *Inia araguaiaensis* (Cetartiodactyla, Iniidae), produce a diverse repertoire of acoustic signals. *PeerJ*, 7, e6670. https://doi.org/10.7717/peerj.6670.

Mendoza-Lera, C., Ribot, M., Foulquier, A. et al. (2019). Exploring the role of hydraulic conductivity on the contribution of the hyporheic zone to in-stream nitrogen uptake. *Ecohydrology*, 12, e2139. https://doi.org/10.1002/eco.2139.

Menezes, S., Baird, D.J., & Soares, A.M.V.M. (2010). Beyond taxonomy: a review of macroinvertebrate trait-based community descriptors as tools for freshwater biomonitoring. *Journal of Applied Ecology*, 711–719.

Merbt, S.N., Proia, L., Prosser, J. et al. (2016). Stream drying drives microbial ammonia oxidation and first-flush nitrate export. *Ecology*, 97, 2192–2198.

Mérigoux, S. & Dolédec, S. (2004). Hydraulic requirements of stream communities: a case study on invertebrates. *Freshwater Biology*, 49, 600–613.

Merritt, R.W., Cummins, K.W & Berg, M.B. (2017). Trophic relationships of macroinvertebrates. In R. Hauer & G.A. Lamberti (eds) *Methods in Stream Ecology. Volume 1* (3rd edn). Elsevier, Burlington, MA, 413–433.

Merritt, R.W., Cummins, K.W. & Berg, M.B. (2019). *An Introduction to the Aquatic Insects of North America* (5th edn). Kendall Hunt, Dubuque, IA.

Methvin, B.R. & Suberkropp, K. (2003). Annual production of leaf-decaying fungi in 2 streams. *Journal of the North American Benthological Society*, 22, 554–564.

Meyer, J.L., & J.B. Wallace. (2001). Lost linkages and lotic ecology: rediscovering small streams. In M.C. Press, N.J. Huntly & S. Levin (eds) Proceedings of the 41st Symposium of the British Ecological Society, jointly sponsored by the Ecological Society of America, 10–13 April 2000, Orlando, FL. Blackwell Science, Oxford, 295–317.

Meyer, J.L., Strayer, D., Wallace, J.B. et al. (2007). The contribution of headwater streams to biodiversity in river networks. *Journal of the American Water Resources Association*, 43, 86–103.

Michel, R.L. (1992). Residence times in river basins as determined by analysis of long-term tritium records. *Journal of Hydrology*, 130, 367–378.

Mihuc, T.B. (1997). The functional trophic role of lotic primary consumers: generalists versus specialist strategies. *Freshwater Biology*, 37, 455–462.

Millenium Ecosystem Assessment (2005). *Ecosystems and Human Well-being: Biodiversity Synthesis*. Island Press, Washington, DC.

Milner, A.M. & Petts, G.E. (1994). Glacial rivers: physical habitat and ecology. *Freshwater Biology*, 32, 295–307.

Milner, A.M., Knudsen, E.E., Soiseth, C., Robertson, A.L. et al. (2000). Colonisation and development of stream communities across a 200-year gradient in Glacier Bay National Park, Alaska, USA. *Canadian Journal of Fisheries and Aquatic Sciences*, 57, 2319–2335.

Milner, A.M., Roberston, A.L. Monaghan, K.A. et al. (2008). Colonization and development of an Alaskan stream community over 28 years. *Frontiers in Ecology & Environment*, 6, 413–419.

Milner, A.M., Picken, J.L., Klaar, M.J., Robertson, A.L. et al. (2018). River ecosystem resilience to extreme flood events. *Ecology & Evolution*, 8, 8354–8363. Doi.org/10.1002/ece3.4300.

Mims, M.C., Olden, J.D., Shattuck, Z.R. & Poff, N.L. (2010). Life history trait diversity of native freshwater fishes in North America. *Ecology of Freshwater Fish*, 19, 390–400.

Min, J.K. & Kong, D-S. (2020) Distribution patterns of benthic macroinvertaebrate communities based on multi-scale environmental variables in the river systems of Republic of Korea. *Journal of Freshwater Ecology*, 35, 323–347.

Minaudo, C., Meybeck, M., Moatar, F. et al. (2015). Eutrophication mitigation in rivers: 30 years of trends in spatial and seasonal patterns of biogeochemistry of the Loire River (1980-2012). *Biogeosciences*, 12, 2549–2563.

Mineeva, N., Lazareva, V., Litvinov, A. et al. (2022). The Volga River. In K. Tockner, C. Zarfl & C. Robinson (eds) *Rivers of Europe* (2nd edn). Elsevier, Amsterdam, 27–80.

Miniat, C.F., Oishi, A.C., Bolstad, P.V., Jackson, C.R. et al. (2021). The Coweeta Hydrologic laboratory and the Coweeta Long-Term Ecological Research project. *Hydrological Processes*, doi.org/10.1002/hyp.14302.

New Zealand Ministry for the Environment (2019). *Action for healthy waterways—a discussion document on national direction for our essential freshwater*. Ministry for the Environment, Wellington, New Zealand.

Minshall, G.W. (1967). Role of allochthonous detritus in the trophic structure of a woodland springbrook community. *Ecology*, 48, 139–149.

Minshall, G.W. (1988). Stream Ecosystem Theory: A Global Perspective. *Journal of the North American Benthological Society*, 7, 263–288 (Community Structure and Function in Temperate and Tropical Streams: Proceedings of a Symposium).

Moatar, F. et al. (2021). The Loire River Basin. In K. Tockner, C. Zarfl & C. Robinson (eds) *Rivers of Europe* (2nd edn). Elsevier, Amsterdam, 245–271.

Moerke, A.H., Ruetz, C.R., Simon, T.N & Pringle, C.M. (2017). Macroconsumer–resource interactions. In R. Hauer & G.A. Lamberti (eds) *Methods in Stream Ecology*. Elsevier, 399–412.

Molloy, D.P. (1981). Mermithid parasitism of blackflies (Diptera: Simuliidae). *Journal of Nematology*, 13, 250–256.

Molloy, J.M. (1992). Diatom communities along stream longitudinal gradients. *Freshwater Biology*, 28, 59–69.

Moore, J.W. (2006). Animal ecosystem engineers in streams. *BioScience*, 56, 237–246.

Morrissey, C.A., Bendell-Young, L.I. & Elliott, J.E. (2004). Linking contaminant profiles to the diet and breeding location of American dippers using stable isotopes. *Journal of Applied Ecology*, 41, 502–512. https://doi.org/10.1111/j.0021-8901.2004.00907.x.

Morita, K., Sahashi, G. & Tsuboi, J. (2016). Altitudinal niche partitioning between white-spotted charr (*Salvelinus leucomaenis*) and masu salmon (*Oncorhynchus masou*) in a Japanese river. *Hydrobiologia*, 783, 93–103.

Morrison, W.E. & Hay, M.E. (2011). Induced chemical defenses in a freshwater macrophyte suppress herbivore fitness and the growth of associated microbes. *Oecologia*, 165, 427–436.

Morse, J.C., Frandsen, P.B., Graf, W. & Thomas, J.A. (2019). Diversity and ecosystem services of Trichoptera. *Insects*, 10, 125. doi:10.3390/insects10050125.

Moss, B. (1988). *The Ecology of Freshwaters*. Blackwell Science, Oxford.

Moss, B. (2015). Mammals, freshwater reference states, and the mitigation of climate change. *Freshwater Biology*, 60, 1964–1976.

Moss, D., Furse, M.T., Wright, J.F. & Armitage, P.D. (1987). The prediction of the macroinvertebrate fauna of unpolluted running-water sites in Great-Britain using environmental data. *Freshwater Biology*, 17, 41–52.

Mouthon, J. & Daufresne, M. (2015). Resilience of mollusc communities of the River Saone (eastern France) and its two main tributaries after the 2003 heatwave. *Freshwater Biology*, doi 10.1111/fwb.12540.

Moutinho, S. (2021). A golden menace: an invasive mussel is devastating ecosystems as it speeds through South American rivers, threatening the Amazon basin. *Science*, doi: 10.1126/science.acx9398.

Muehlbauer, J.D., Collins, S. F., Doyle, M.W. & Tockner, K. (2014). How wide is a stream? Spatial extent of the potential 'stream signature' in terrestrial food webs using meta-analysis. *Ecology*, 95, 44–55.

Mulholland, P.J., Fellows, C.S., Tank, J.L. et al. (2001). Inter-biome comparison of factors controlling stream metabolism. *Freshwater Biology*, 46, 1503–1517.

Mulholland, P.J., Tank, J.L., Webster, J.R. et al. (2002). Can uptake length in streams be determined by nutrient addition experiments? Results from an interbiome comparison study. *Journal of the North American Benthological Society*, 21, 544–560.

Mulholland, R.S., Helton, A.M., Poole, G.C. et al. (2008). Stream denitrification across biomes and its response to anthropogenic nitrate loading. *Nature*, 7184, 202–205.

Müller, K. (1954). Investigations on the organic drift in north Swedish streams. *Report of the Institute of Freshwater Research, Drottningholm*, 35, 133–148.

Müller, K. (1982). The colonization cycle of freshwater insects. *Oecologia*, 52, 202–207.

Mulvihill, R.S., Newell, F.L. & Latta, S.C. (2008). Effects of acidity on the breeding ecology of a stream-dependent songbird, the Lousiana waterthrush (*Seiurus motacilla*). *Freshwater Biology*, 53, 2158–2169.

Muotka, T. & Penttinen, A. (1994). Detecting small-scale spatial patterns in lotic predator-prey relationships: statistical methods and a case study. *Canadian Journal of Fisheries and Aquatic Science*, 51, 2210–2218.

Murphy, J., Giller, P.S. & Horan, M.A. (1998). Spatial scale and the aggregation of stream macroinvertebrates associated with leaf packs. *Freshwater Biology*, 39, 325–339.

Murphy, J.F., Davy-Bowker, J., McFarland, B. & Ormerod, S. J. (2013). A diagnostic biotic index for assessing acidity in sensitive streams in Britain. *Ecological Indicators*, 24, 562–572.

Murphy, J.F., Jones, J.I., Arnold, A., Duerdoth, C.P. et al. (2017). Can Macroinvertebrates biological traits indicate fine-grained sediment conditions in streams. *River Research and Applications*, 33, 1606–1617.

Murphy, M.O., Jones, K.S., Price, S.J. & Weisrock, D.W. (2018). A genomic assessment of population structure and gene flow in an aquatic salamander identifies the roles of spatial scale, barriers, and river architecture. *Freshwater Biology*, 63(5). 407–419.

Myers, M.J., Meyer, C.P. & Resh, V.H. (2000). Neritid and thiarid gastropods from French Polynesian streams: how reproduction (sexual, parthenogenetic) and dispersal (active, passive) affect population structure. *Freshwater Biology*, 44, 535–545.

Myrstener, M., Rocher-Ros, G., Burrows, R.M. et al. (2018). Persistent nitrogen limitation of stream biofilam communities along climate gradients in the Arctic. *Global Change Biology*, 24, 3680–3691.

Naiman, R.J., Melillo, J.M. & Hobbie, J.E. (1986). Ecosystem alteration of boreal forest streams by beaver. *Ecology*, 67, 1254–1269.

Naiman, R.J., Melillo, J.M., Lock, M.A & Ford, T.E. (1987). Longitudinal patterns of ecosystem processes and community structure in a subarctic river continuum. *Ecology*, 68, 1139–1156.

Nakano, S. & Murakami, M. (2001). Reciprocal subsidies: dynamic interdependence between terrestrial and aquatic food webs. *Proceedings of the National Academy of Sciences*, 98, 166–170.

Nakano, S., Fausch, K.D. & Kitano, S. (1999). Flexible niche partitioning via a foraging mode shift: a proposed mechanism for coexistence in stream-dwelling charrs. *Journal of Animal Ecology*, 68, 1079–1092.

Naman, S.M., Rosenfeld, J.S. & Richardson, J.S. (2016). Causes and consequences of invertebrate drift in running waters: from individuals to populations and trophic fluxes. *Canadian Journal of Fisheries and Aquatic Sciences*, 73, 1–14.

NCEI (2021). National Center for Environmental Information, Annual 2021 Global Climate report; https://www.ncei.noaa.gov/access/monitoring/monthly-report/global/202113#precip.

Nelson, D., Benstead, J.P., Huryn, A.D. et al. (2017). Shifts in community size structure drive temperature invariance of secondary production in a stream-warming experiment. *Ecology*, 98, 1797–1806.

Nelson, D.R. & Marley, N.J. (2000). The biology and ecology of lotic Tardigrada. *Freshwater Biology*, 44, 93–108.

Neres-Lima, V., Machado-Silva, F., Baptista, D.F., Oilveira, R.B.S. et al. (2017). Allochthonous and autochthonous carbon flows in food webs of tropical forest streams. *Freshwater Biology*, 62, 1012–1023.

Newbold, J.D. (1992). Cycles and spirals of nutrients. In P.Calow & G.E. Petts (eds) *The Rivers Handbook*. Blackwell Scientific Publications, Oxford, 370–408.

Newbold, J.D., Mulholland, P.J., Elwood, J.W. & N'Neill, R.V. (1982). Organic matter spiralling in stream ecosystems. *Oikos*, 38, 266–272.

Newbold, J.D., Elwood, J.W. O'Neil, R.V. & Sheldon, A.L. (1983). Phosphorus dynamics in a woodland stream ecosystem: a study of nutrient spiralling. *Ecology*, 64, 1249–1265.

Nikolaidis, N.P., Phillips, G., Poikane, S., Vábiró, G. et al. (2022). River and lake nutrient targets that support ecological status: European scale gap analysis and strategies for the implementation of the Water Framework Directive. *Science of the Total Environment*, 813, 151898.

Nilsson, C. & Jansson, R. (1995). Floristic differences between riparian corridors of regulated and free-flowing boreal rivers. *Regulated Rivers: Research and Management*, 11, 55–66.

Nilsson, A.L.K., L'Abée-Lund, J.H., Vøllestad, L.A., Jerstad, K. et al. (2018). The potential influence of Atlantic salmon *Salmo salar* and brown trout *Salmo trutta* on density and breeding of the white-throated dipper *Cinclus cinclus*. *Ecology and Evolution*, 8, 4065–4073. https://doi.org/10.1002/ece3.3958.

Norman, B.C., Whiles, M.R., Collins, S.M. et al. (2017). Drivers of nitrogen transfer in stream food webs across continents. *Ecology*, 98, 3044–3055.

O'Gorman, E.J., Benstead, J.P., Cross, W.F., Friberg, N. et al. (2014). Climate change and geothermal ecosystems: natural laboratories, sentinel systems, and future refugia. *Global Change Biology*, 20, 3291–3299.

O'Gorman, E.J., Zhao, L., Pichler, D.E., Adams, G. et al. (2017). Unexpected changes in community size structure in a natural warming experiment. *Nature Climate Change*, 7, 659–663.

Ocasio-Torres, M.E., Crowl, T.A. & Sabat, A.M. (2021). Effect of multimodal cues from a predatory fish on refuge use and foraging on an amphidromous shrimp. *PeerJ*, 9, e11011.

Ochs, C.A., Pongruktham, O. & Zimba, P.V. (2013). Darkness at the break of noon: phytoplankton production in the Lower Mississipi River. *Limnology and Oceanography*, 58, 555–568.

Ogunbanwo, O., Kay, P., Boxall, A. et al. (2020). High Concentrations of Pharmaceuticals in a Nigeria River Catchment. *Environmental Toxicology & Chemistry*, 10.1002/etc.4879.

Öhlund, G., Nordwall, F., Degerman, E. & Eriksson, T. (2008). Life history and large-scale habitat use of Brown Trout *(Salmo trutta)* and Brook Trout *(Salvelinus fontinalis)*—implications for species replacement patterns. *Canadian Journal of Fisheries and Aquatic Sciences*, 65, 633–644.

Ohta, T., Matsunaga, S., Niwa, S. et al. (2016). Detritivore stoichiometric diversity alters litter processing efficiency in a freshwater ecosystem. *Oikos*, 125, 1162–1172.

Oiro, S., Comte, J.-C., Soulsby, C. et al. (2020). Depletion of groundater resources under rapid urbanisation in Africa: recent and future trends in the Nairobi Aquifer System, Kenya. *Hydrogeology Journal*, 28, 2635–2656.

Okafor N. (2011). Taxonomy, Physiology, and Ecology of Aquatic Microorganisms. In N. Okafor *Environmental Microbiology of Aquatic and Waste Systems*. Springer, Dordrecht, 47–107. https://doi.org/10.1007/978-94-007-1460-1_4.

Okamura, B., Hartikainen, H., Schmidt-Posthaus et al. (2011). Life cycle complexity, environmental change and the emerging status of salmonid proliferative kidney disease. *Freshwater Biology*, 56, 735–753.

Olden, J.D., Kennard, M.J. & Pusey, B.J. (2012). A framework for hydrologic classification with a review of methodologies and applications in ecohydrology. *Ecohydrology*, 5, 503–518.

Olden, J.D. & Poff, N.L. (2003). Redundancy and the choice of hydrologicindices for characterising streamflow regimes. *River Research and Applications*, 19, 101–121.

Oldmeadow, D.F., Lancaster, J. & Rice, S.P. (2010). Drift and settlement of stream insects in a complex hydraulic environment. *Freshwater Biology*, 55, 1020–1035.

Olesen, A., Mundbjerg, S., Alnoee, A.B., Baattrup-Pedersen, A. et al. (2018). Nutrient kinetics in submerged plant beds: a mesocosm study simulating constructed drainage wetlands. *Ecological Engineering*, 122, 263–270.

Oliver, R.L. & Merrick, C.J (2006). Partitioning of river metabolism identifies phytoplankton as a major contributor in the regulated Murray River (Australia). *Freshwater Biology*, 51, 1131–1148.

Olsson, K., Stenroth, P., Nystrom, P. & Graneli, W. (2009). Invasions and niche width: does niche width of an introduced crayfish differ from a native crayfish? *Freshwater Biology*, 54, 1731–1740.

Ormerod, S.J. & Tyler, S.J. (1991). Exploitation of prey by a river bird, the dipper *Cinclus cinclus* (L.), along acidic and circumneutral streams in upland Wales. *Freshwater Biology*, 25, 105–116.

Ormerod S.J. & Tyler S.J. (1993). Birds as indicators of changes in water quality. In R.W Furness & J.J.D. Greenwood (eds) *Birds as Monitors of Environmental Change*. Springer, Dordrecht, 179–216. https://doi.org/10.1007/978-94-015-1322-7_5.

Ormerod, S.J., Allinson, N., Hudson, D. & Tyler S.J. (1986). The distribution of breeding dippers (*Cinclus cinclus* L.; Aves) in relation to stream acidity in upland Wales. *Freshwater Biology*, 16, 501–507.

Ormerod, S.J., Boole, P., McCahon, C.P., Weatherley, N.S. et al. (1987). Short-term experimental acidification of a Welsh stream: comparing the biological effects of hydrogen ions and aluminium. *Freshwater Biology*, 17, 341–356.

Ormerod, S.J., O'Halloran, J., Gribbin, S.D. & Tyler, S.J. (1991). The Ecology of Dippers *Cinclus cinclus* in Relation to Stream Acidity in Upland Wales: Breeding Performance, Calcium Physiology and Nestling Growth. *Journal of Applied Ecology*, 28, 419–433. https://doi.org/10.2307/2404559.

Ormerod, S.J., Dobson, M., Hildrew, A.G. & Townsend, C.R. (2010). Multiple stressors in freshwater ecosystems. *Freshwater Biology*, 55(Suppl. 1), 1–4.

Orr, J.A., Vinebrooke, R.D., Jackson, M.C. et al. (2020). Towards a unified study of multiple stressors: division and common goals across research disciplines. *Proceeding of the Royal Society B*, 287, 20200421.

Osborn, G. & du Toit, C. 1991. Lateral planation of rivers as a geomorphic agent. *Geomorphology*, 4, 249–260.

Osorio, E.D., Tanchuling, M.A. & Diola, M.B. (2021). Microplastics occurrence in surface waters and sediments in five river mouths of Manila Bay. *Frontiers in Environmental Science*, https://doi.org/10.3389/fenvs.2021. 719274.

Oswood, M.W. (1976). Comparative life histories of the *Hydropsyche* in a Montana lake outlet. *American Midland Naturalis*, 96, 493–497.

Oswood, M.W., Irons, J.G. & Milner, A.M. (1995). River and steam ecosystems of Alaska. In Cushing, C., Cummins, K. & Minshall, G. (eds) *Ecosystems of the World: 22. River and Stream Ecosystems*. Elsevier, Amsterdam, 9–32.

Ottino, P. & Giller, P.S. (2004). Distribution, density, diet and habitat use of the otter in relation to land use in the Araglin valley, Southern Ireland. *Biology and Environment: Proceedings of the Royal Irish Academy*, 104, 1–17.

Padial, A.A., Vituel, J.R.S. & Olden, J.D. (2020). Preface: aquatic homogenocene—understanding the era of biological reshuffling in aquatic ecosystems. *Hydrobiologia*, 847, 3705–3709.

Page, L.M. & Schemske, D.W. (1978). The effect of interspecific competition on the distribution and size of darters of the subgenus *Catonotus* (Percidae: *Etheostoma*). *Copeia*, 406–412.

Pagotto, J.P.A., Goulart, E., Oliveira, E.F., & Yamamura, C.B. (2011). Trophic ecomorphology of Siluriformes (Pisces, Osteichthyes) from a tropical stream. *Brazilian Journal of Biology*, 71, 469–479.

Palmer, M.A., Bely, A.E. & Berg, K.E. (1992). Response of invertebrates to lotic disturbance: a test of the hyporheic refuge hypothesis. *Oecologia*, 89, 182–194.

Palmer, M.A., Bernhardt, E.S., Allan, J.D. et al. (2005). Standards for ecologically successful river restoration. *Journal of Applied Ecology*, 42, 208–217.

Palmer, M.A., Menninger, H.L. & Bernhardt, E. (2010). River restoration, habitat heterogeneity and biodiversity: a failure of theory or practice? *Freshwater Biology*, 55, 205–222.

Parker J.D., Burkepile, D.E., Collins, D.O., Kubanek, J. et al. (2007). Stream mosses as chemically defended refugia for freshwater macroinvertebrates. *Oikos*, 116, 302–312.

Parker, S.M & Huryn, A.D. (2006). Food web structure and function in two arctic streams with contrasting disturbance regimes. *Freshwater Biology*, 51, 1249–1263.

Parolini, M., Magni, S., Castiglioni, S. et al. (2015). Realistic mixture of illicit drugs impaired the oxidative status of the zebra mussel (*Dreissena polymorpha*). *Chemosphere*, 128, 96–102.

Pasquini, A. I. & Depetris, P. J. (2007). Discharge trends and flow dynamics of South American rivers draining the southern Atlantic seaboard: an overview. *Journal of Hydrology*, 333, 385–399. doi: 10.1016/j.jhydrol. 2006.09.005.

Passy, S.I. (2007). Diatom ecological guilds display distinct and predictable behaviour along nutrient and disturbance gradients in running waters. *Aquatic Botany*, 86, 171–178.

Patterson, M.A., Mair, R.A., Eckert, N.L., Gatenby, C.M et al. (eds) (2018). *Freshwater Mussel Propagation for Restoration*. Cambridge University Press, Cambridge.

Pawlowski, J., Kelly-Quinn, M., Altermatt, F. et al. (2018). The future of biotic indices in the ecogenomic era: integrating (e)DNA metabarcoding in biological assessment of aquatic ecosystems. *Science of the Total Environment*, 637/638, 1295–1310.

Pearson, R.G. & Boyero, L. (2009). Gradients in regional diversity of freshwater taxa. *Journal of the North American Benthological Society*, 28, 504–514.

Peckarsky, B.L. (1987). Mayfly cerci as defense against stonefly predation: deflection and detection. *Oikos* 48, 161–170.

Peckarsky, B.L. & Penton, M.A. (1988). Why do *Ephemerella* nymphs scorpion posture: a 'ghost of predation past'? *Oikos*, 53, 185–193.

Peckarsky, B.L. & Wilcox, R.S. (1989). Stonefly nymphs use hydrodynamic cues to discriminate between prey. *Oecologia*, 79, 265–270.

Peckarsky, B.L. & Cowan, C.A. (1991). Consequences of larval intraspecific competitition to stonefly growth and fecundity. *Oecologia*, 88, 277–288.

Peckarsky, B. L. & Lamberti, G. A. (2017). Invertebrate consumer—resource interactions. Chapter 18 In: F. R. Hauer and G. A. Lamberti (eds) *Methods in Stream Ecology* (3rd edn), Volume 1: Ecosystem Structure. Elsevier, Academic Press 379–398.

Peckarsky, B.L., Kerans, B.L., Taylor, B.W. & McIntosh, A.R. (2008). Predator effects on prey population dynamics in open systems. *Oecologia*, 156, 431–440.

Peckarsky, B.L., McIntosh, A.R., Taylor, B.W. & Dahl, J. (2002). Predator chemicals induce changes in mayfly life history traits: a whole-stream manipulation. *Ecology* 83, 612–618.

Peckarsky, B.L., Taylor, B.W. & Caudill, C.C. (2000). Hydrological and behavioral constraints on oviposition of stream insects: implications for adult dispersal. *Oecologia*, 125, 186–200.

Pennak, R.W. (1989). *Fresh-water Invertebrates of the United States. Protozoa to Mollusca* (3rd edn). Wiley, New York.

Peoples. B.K. & Frimpong, E.A. (2016). Biotic interactions and habitat drive positive co-occurrence between facilitating and beneficiary stream fishes. *Journal of Biogeography*, 43, 923–931.

Perkins, D.M., Yvon-Durocher, G., Demars, O, Reiss, J. et al. (2012). Consistent temperature dependence of respiration across ecosystems contrasting in thermal history. *Global Change Biology*, 18, 1300–1311.

Perkins, D.M., Bailey, R.A. Dossena, M. I. et al. (2015). Higher biodiversity is required to sustain multiple ecosystem processes across temperature regimes. *Global Change Biology*, 21, 396–406.

Perkins, D.M., Durance, I., Edwards, F.K et al. (2018). Bending the rules: exploitation of allochthonous resources by a top-predator modifies size-abundance scaling in stream food webs. *Ecology Letters*, 21, 1771–17870.

Peters R., Berlekamp J., Lucía A. et al. (2021). Integrated impact assessment for sustainable hydropower planning in the Vjosa Catchment (Greece, Albania). *Sustainability*, 13, 1514.

Petersen, I., Winterbottom, J.H., Orton, J. & Hildrew, A.G. (1999). Does the colonization cycle exist? In N. Friberg & J.D. Carl (eds) *Proceedings of the Nordic Benthological Meeting in Silkeborg, 13–14th November 1997*. National Environmental Research Institute, Silkeborg, Denmark, 123–126.

Petersen, I., Masters, Z., Hildrew, A.G. & Ormerod, S.J. (2004). Dispersal of adult aquatic insects in catchments of differing land use. *Journal of Applied Ecology*, 41, 934–950.

Petersen, N.R., & Jensen, K. (1997). Nitrification and denitrification in the rhizosphere of the aquatic macrophyte *Lobelia dortmanna* L. *Limnology and Oceanography*, 42, 529–537. doi:10.4319/lo.1997.42.3.0529.

Peterson, B.J., Hobbie, J.E., Hershey, A.E. et al. (1985). Transformation of a tundra river from heterotrophy to autotrophy by addition of phosphorus. *Science*, 229, 1383–1386.

Peterson, B.J., Wollheim, W.M., Mulholland, P.J. et al. (2001). Control of nitrogen export from watersheds by headwater streams. *Science*, 292(5514), 86–90.

Peterson, C.G., Dudley, T.L., Hoagland, K.D. & Johnson, L.M. (1993). Infection, growth, and community-level consequences of a diatom paythogen in a Sonoran desert stream. *Journal of Phycology*, 29, 442–452.

Phillipsen, I.C. & Lytle, D.A. (2012). Aquatic insects in a sea of desert: population genetic structure is shaped by limited dispersal in a naturally fragmented landscape. *Ecography*, 36, 731–743.

Phillipsen, I.C., Kirk, E.H., Bogan, M.T., Mims, M.C. et al. (2015). Dispersal ability and habitat requirements determine landscape-level genetic patterns in desert aquatic insects. *Molecular Ecology*, 24, 54–69.

Piano, E., Doretto, A., Mammola, S., Falasco, W.E. et al. (2020). Taxonomic and functional homogenisation of macroinvertebrate communities in recently intermittent Alpine watercourses. *Freshwater Biology*, 65, 2096–2107.

Pickett, S.T.A. & White, P.S. (eds) (1985). *The Ecology of Natural Disturbance and Patch Dynamics*. Academic Press, Orlando, FL.

Piggott, J.J., Salis, R.K., Lear, G. et al. (2015a). Climate warming and agricultural stressors interact to determine stream periphyton community composition. *Global Change Biology*, 21, 206–222.

Piggott, J.J., Townsend, C.R. & Matthaei, C.D. (2015b). Climate warming and agricultural stressors interact to determine stream macroinvertebrate community dynamics. *Global Change Biology*, 21, 1887–1906.

Pikitch, E.K., Doukakis, P., Lauck, L., Chakrarty, P. et al. (2005). Status, trends and management of sturgeon and paddle fisheries. *Fish & Fisheries*, 6, 233–265.

Pilecky, M., Kämmer, S.K., Mathieu-Resuge, M. et al. (2022). Hydrogen isotopes (d2H) of polyunsaturated fatty acids track bioconversion by zooplankton. *Functional Ecology*, 36, 538–549.

Pilotto, F., Kühn, I., Adrian. R., Alber, R. et al. (2020). Meta-analysis of multidecadal biodiversity trends in Europe. *Nature Communications*, 11, 3486. doi 10.1038/s41467-020-17171-y.

Pincheira-Donoso, D., Bauer, A.M., Meiri, S. & Uetz, P. (2013). Global taxonomic diversity of living reptiles. *PLOS ONE*, https://doi.org/10.1371/journal.pone.0059741.

Pinkert, S., Dijkstra, K.-B. B., Zeuss, D., Reudenbach, C. et al. (2018). Evolutionary processes, dispersal limitation and climatic history shape current diversity patterns of European dragonflies. *Ecography*, 41, 795–804.

Planas, D. (1996). Acidification effects: In Stevenson, R.J., Bothwell, M.I. & Lowe, R.L. (eds) *Algal Ecology: Freshwater Benthic Systems*, Academic Press, San Diego, 497–530.

Platts, W.S., Megahan, W.F. & Minshall, G.W. (1983). *Methods for evaluating atream, riparian and biotic conditions*. USDA Forest Service General Technical Report INT-221.

Pletterbauer F., Melcher A. & Graf W. (2018). Climate Change Impacts in Riverine Ecosystems. In S. Schmutz & J. Sendzimir (eds) *Riverine Ecosystem Management*. Aquatic Ecology Series, vol. 8. Springer, Cham. https://doi.org/10.1007/978-3-319-73250-3_11.

Plont, S., O'Donnell, B.M., Gallagher, M.T. & Hotchkiss, E.R. (2020). Linking carbon and nitrogen spiralling in streams. *Freshwater Science*, 39, 126–136.

Poff, N.L. (1992) Why disturbances can be predictable: a perspective on the definition of disturbance in streams. *Journal of the North American Benthological Society*, 11, 86–92.

Poff, N.L. (1996). A hydrogeography of unregulated streams in the United States and an examination of scale-dependence in some hydrological descriptors. *Freshwater Biology*, 36, 71–91.

Poff, N.L. (1997). Landscape filters and species traits: towards mechanistic understanding and prediction in stream ecology. *Journal of the North American Benthological Society*, 16, 391–409.

Poff, N.L. & Ward, J.V. (1989). Implications of streamflow variability and predictability for lotic community structure: a regional analysis of streamflow patterns. *Canadian Journal of Fisheries and Aquatic Sciences*, 46, 1805–1818.

Poff N.L. & Ward, J.V. (1990). The physical habitat template of lotic systems: recovery in the context of historical pattern of spatio-temporal heterogeneity. *Environmental Management*, 14, 629–646.

Poff, N.L., Allan, J.D., Bain, M.B., Karr, J.R. et al. (1997). The natural flow regime. A paradigm for river conservation and restoration. *Bioscience*, 47, 769–784.

Poff, N.L, M. Brinson, & J.B. Day. (2002). *Freshwater and Coastal Ecosystems and Global Climate Change: A Review of Projected Impacts for the United States*. Pew Center on Global Climate Change, Arlington, VA.

Poff, N.L., Bledsoe, B.P. & Cuhaciyan, C.O. (2006). Hydrologic variation with land use across the contiguous United States: geomorphic and ecological consequences for stream ecosystems. *Geomorphology*, 79, 264–285.

Poff, N.L., Olden, J.D., Vieira, N.K., Finn, D.S., Simmons, M.P. & Kondratieff, B.C. (2006). Functional trait niches of North American lotic insects: traits-based ecological applications in light of phylogenetic relationships. *Journal of the North American Benthological Society*, 25, 730–755.

Poff, N.L., Richter, B.D., Arthington, A.H., Bunn, S.E. et al. (2010). The ecological limits of hydrologic alteration (ELOHA): a new framework for developing regional environmental flow standards. *Freshwater Biology*, 55, 147–170.

Poff, N.L., Tharne, R.E. & Arthington, A.H. (2017). Evolution of environmental flows assessment science, principles, and methodologies. In A.C. Horne, J.A. Webb, M.J. Stewardson, B. Richter & M. Acreman (eds) *Water for the Environment: From Policy and Science to Implementation and Management* Academic Press, San Diego, 203–236.

Poikane, S., Zampoukas, N., Borja, A. et al. (2014). Intercalibration of aquatic ecological assessment methods in the European Union: lessons learned and way forward. *Environmental Science & Policy*, 44, 237–246.

Poikane, S., Kelly, M.G., Herrero, F.S. et al. (2019). Nutrient criteria for surface waters under the European Water Framework Directive: current state of the art, challenges and future outlook. *Science of the Total Environment*, 695, 133888.

Polhemus J.T. & Polhemus D.A. (2007). Global diversity of true bugs (Heteroptera; Insecta) in freshwater. In E.V. Balian, C. Lévêque, H. Segers & K. Martens (eds) *Freshwater Animal Diversity Assessment. Developments in Hydrobiology*, vol 198. Springer, Dordrecht, 379–391.

Politi, N., Martinuzzi, S., Aragón, P., Miranda, V. et al. (2020). Conservation status of the threatened and endemic Rufous-throated Dipper *Cinclus schulzi* in Argentina. *Bird Conservation International*, 30, 396–405. doi:10.1017/S0959270919000467.

Pomeranz, J.P.F., Wesner, J.S. & Harding, J.S. (2020). Changes in stream food web structure across a gradient of acid mine drainage increase local community stability. *Ecology*, 101, e03102.

Ponniah, M. & Hughes, J.M. (2004). The evolution of Queensland spiny mountain crayfish of the genus *Euastacus* I Testing vicariance and dispersal with interspecific mitochondrial DNA. *Evolution*, 58, 1073–1085.

Ponniah, M. & Hughes, J.M. (2006). The evolution of Queensland spiny mountain crayfish of the genus *Euastacus* II Investigating simultaneous vicariance with intraspecific genetic data. *Marine & Freshwater Research*, 57, 349–362.

Portella, T., Lobon-Cervia, J., Manna, L.R., Bergallo, H.G. et al. (2017). Ecomorphological attributes and feeding habits in coexisting characins. *Journal of Fish Biology*, 90, 129–146.

Postel, S.L. (2000). Entering an era of water scarcity: the challenges ahead. *Ecological Applications*, 10, 941–948.

Postel, S. & Carpenter, S. (1997). Freshwater ecosystem services. In G.C. Daily (ed) *Nature's Services: Societal Dependence on Natural Ecosystems*. Island Press, Washington, DC, 195–214.

Power, M.E. (1990). Effects of fish in river food webs. *Science*, 250, 811–814

Power, M.E. (1992). Top down and bottom-up forces in food webs: do plants have primacy? *Ecology*, 73, 733–746.

Power, M.E. & Matthews, W.J. (1983). Algae-grazing minnows *(Compostoma anomalum)*, piscivorous bass *(Micropterus* spp.) and the distribution of attached algae in a small prairie-margin stream. *Oecologia*, 60, 328–332.

Power, M.E., Matthews, W.J. & Stewart, A.J. (1985). Grazing minnows, piscivorous bass and stream algae: dynamicsof a strong interaction. *Ecology*, 66, 1448–1456.

Power, M.E., Stewart, A.J. & Matthews, W.J. (1988). Grazer control of algae in an Ozark mountain stream: effects of short term exclusion. *Ecology*, 69, 1894–1898.

Power, M.E., Stout, R.J., Cushing, C.E., Harper, P.P. et al. (1988). Biotic and abiotic controls in river and stream communities. *Journal of the North American Benthological Society*, 7, 456–479.

Power, M.E., Dudley, T.L. & Cooper, S.D. (1989). Grazing catfish, fishing birds, and attached algae in a Panamanian stream. *Environmental Biology of Fishes*, 26, 285–294.

Power, M.E., Holomuzki, J.R. & Lowe, R.L. (2013). Food webs in Mediterranean rivers. *Hydrobiologia*, 719, 119–136.

Prairie, J., Nowak, K., Rajagopalan, B., Lall, U. & Fulp, T. (2008). A stochastic nonparametric approach for streamflow generation combining observational and paleoreconstructed data. *Water Resources Research*, 44, W06423, doi:10.1029/2007WR006684.

Preston, D.L., Layden, T.J., Segui, L.M., Falke, L.P. et al. (2021). Trematode parasites exceed aquatic biomass in Oregon stream food webs. *Journal of Animal Ecology*, 90, 766–775.

Pretty, J.L., Hildrew, A.G. & Trimmer, M. (2006). Nutrient dynamics in relation to surface-subsurface hydrological exchange in a groundwater-fed chalk stream. *Journal of Hydrology*, 330, 84–100, https://doi.org/10.1016/j.jhydrol.2006.04.013.

Price, A., Weadlock, C., Shim, J. & Rodd, F.H. (2009). Pigments, patterns and fish behaviour. *Zebrafish* 5, 297–307.

Price, T.L., Harper, J., Francoeur, S.N. et al. (2021). Brown meets green: light and nutrients alter detritivore assimilation of microbial nutrients from leaf litter. *Ecology*, 102, e03358.

Pritchard, G. (1983). Biology of Tipulidae. *Annual Review of Entomology*, 28, 1–22.

Ptatscheck, C., Brüchner-Hüttemann, H., Kreuzinger-Janik, B., et al. (2020). Are meiofauna a standard meal for macroinvertebrates and juvenile fish? *Hydrobiologia* 847, 2755–2778. https://doi.org/10.1007/s10750-020-04189-y.

Pulliam, H.R. (1988). Sources, sinks, and population regulation. *American Naturalist*, 132, 652–661.

Pusch, M., Anderson, H.E., Bäthe, J. et al. (2021). Rivers of the central European highlands and plains. In K. Tockner, C. Zarfl & C. Robinson (eds) *Rivers of Europe* (2nd edn), Elsevier, Amsterdam, 717–773.

Quick, A.M., Reeder, W.J., Farrell, T.B. et al. (2019). Nitrous oxide from streams and rivers: a review of primary biogeochemical pathways and environmental variables. *Earth-Science Reviews*, 191, 224–262.

Radke, R.L. & Kinzie, R.A. (1996). Evidence of a marine larval stage in endemic Hawaiian stream gobies from isolated high-elevation locations. *Transactions of the American Fisheries Society*, 125, 613–621.

Rahel, F.J. & Olden, J.D. (2008). Assessing the effects of climate change on aquatic invasive species. *Conservation Biology*, 22, 521–533.

Ramirez, F., Davenport, T.L. & Mojica, J.I. (2015). Dietary–morphological relationships of nineteen fish species from an Amazonian terra firma blackwater stream in Colombia. *Limnologica*, 52, 89–102.

Ranvestel, A.W., Lips, K.R., Pringle, C.M., Whiles, M.R. et al. (2004). Neotropical tadpoles influence stream benthos: evidence for the ecological consequences of decline in amphibian populations. *Freshwater Biology*, 49, 274–285.

Raymond, P. A., Hartmann, J., Lauerwald, R., Sobek, S. et al. (2013). Global carbon dioxide emissions from inland waters. *Nature*, 503, 355–359.

Razeng, E., Smith, A.E., Harrisson, K.A. et al. (2017). Evolutionary divergence in freshwater insects with contrasting dispersal capacity across a sea of desert. *Freshwater Biology*, 62, 1443–1459.

Recalde, F.C., Postali, T.C. & Romero, G.Q. (2016). Unravelling the trole of allochthonous aquatic resources to food web struture in a tropical riparian forest. *Journal of Animal Ecology*, 85, 525–536.

Reiber, L., Knillmann, S., Kaske, O. et al. (2021). Long-term effects of a catastrophic insecticide spill on stream invertebrates. *Science of the Total Environment*, 768, 144456.

Reid, A.J., Carlson, A. K., Creed, I.F., Eliason, E.J. et al. (2019). Emerging threats and persistant conservation challenges for freshwater biodiversity. *Biological Reviews*, 94, 849–873.

Reisinger, A.J., Reisinger, L.S., Richmond, E.K. et al. (2021). Exposure to a common antidepressant alters crayfish behavior and has potential subsequent ecosystem impacts. *Ecosphere*, 12(6), e03527. 10.1002/ecs2.3527.

Reiss, J. (2018). Microorganisms 2. Viruses, prokaryotes, fungi, protozoans, and microscopic metazoans. Chapter 8 In J. Hughes (ed) *Freshwater Ecology and Conservation: Approaches and Techniques*. Oxford University Press, Oxford, 157–172.

Reiss, J. & Schmid-Araya, J. (2008). Existing in plenty: abundance, biomass and diversity of ciliates and meiofauna in small streams. *Freshwater Biology*, 53, 652–668.

Reiss, J., Bridle, J., Montoya, J.M. & Woodward, G. (2009). Emeging horizons in biodiversity and ecosystem functioning research. *Trends in Ecology & Evolution*, 24(9), 505–514.

Reiss, J., Bailey, R.A., Perkins, D.M., Pluchinotta, A. et al. (2011). Testing effects of consumer richness, evenness and body size on ecosystem functioning. *Journal of Animal Ecology*, 80, 1145–1154.

Rempel, L.L., Richardson, J.S. & Healy, M.C. (1999). Flow refugia for benthic macroinvertebrates during flooding of a large river. *Journal of North American Benthological Society*, 18, 34–48.

Resetarits, W.J., Jr. (1997). Interspecific competition and qualitative competitive asymmetry between two benthic stream fishes. *Oikos*, 78, 428–439.

Resh, V.H. & De Szalay, F.A. (1995). Streams and rivers of Oceania: River and stream ecosystems. In C.E. Cushing, K.W. Cummins & G.W. Minshall (eds) *Ecosystems of the World*. Elsevier, Amsterdam, 717–737.

Resh, V.H. & Rosenberg, D.M. (2010). Recent trends in life history research on benthic macroinvertebrates. *Freshwater Science*, 29, 207–219.

Resh, V.H., Brown, A.V., Covich, A.P. et al. (1988). The role of disturbance in stream ecology. *Journal of the North American Benthological Society*, 7, 433–455.

Resh, V.H., Hildrew, A.G., Statzner, B. & Townsend, C.R. (1994). Theoretical habitat templets, species traits, and species richness—a synthesis of long-term ecological research on the Upper Rhône River in the context of currently developed ecological theory. *Freshwater Biology*, 31, 539–554.

Rey-Boissezon, A. & Auderset Joye, D. (2015). Habitat requirements of charophytes—evidence of species discrimination through distribution analysis. *Aquatic Botany*, 120, 84–91.

Reynolds, C.S. (1984). Phytoplankton periodicity: the interactions of form, function and environmental variability. *Freshwater Biology*, 14, 111–142.

Reynolds, C.S. (1994). The role of fluid motion in the dynamics of phytoplankton in lakes and rivers. In P.S. Giller, A.G. Hildrew & D.G. Raffaelli (eds) *Aquatic Ecology: Scale, Pattern and Process*, 34th Symposium of the British Ecological Society. Blackwell, Oxford, 141–187.

Reynolds, C.S. (2000). Hydroecology of river plankton: the role of variability in channel flow. *Hydrological Processes*, 14, 3119–3132.

Reynolds, C.S., Carling, P.A. & Beven, K.J. (1991). Flow in river channels. *Archiv für Hydrobiologie*, 121, 171–179.

Ricci, C. & Balsamo, M. (2000). The biology and ecology of logic rotifers and gastrotrichs. *Freshwater Biology*, 44, 15–28.

Richard, J.C., Leis, E., Dunn, C.D., Agbalog, R. et al. (2020). Mass mortality in freshwater mussels (*Actinonaias pectorosa*) in the Clinch River, USA, linked to a novel densovirus. *Scientific Reports*, 10(1), 1–10.

Richardson, J. (2019). Biological diversity in headwater streams. *Water*, 11, 366.

Richardson, J.S. (1991). Seasonal food limitation of detritivores in a montane stream: an experimental test. *Ecology*, 72, 873–887.

Richmond, E.K., Rosi-Marshall, E.J., Lee, S.S. et al. (2016). Antidepressants in stream ecosystems: influence of selective serotonin reuptake inhibitors (SSRIs) on algal production and insect emergence. *Freshwater Science*, 35, 845–855.

Richmond, E.K., Grace, M.R., Kelly, J.J. et al. (2017). Pharmaceuticals and personal care products (PPCPs) are ecological disrupting compounds (EcoDC). *Elementa*, 5, 66. https://doi.org/10.1525/elementa.252.

Richmond, E.K., Rosi, E.J., Walters, D.M. et al. (2018). A diverse suite of pharmaceuticals contaminate stream and riparian food webs. *Nature Communications*, 9, 4491.

Richmond, E.K., Rosi, E.J. & Reisinger, A.J. (2019). Influences of the antidepressant fluoxetine on stream ecosystem function and aquatic insect emergence at environmental. *Journal of Freshwater Ecology*, 34, 513–531. DOI: 10.1080/02705060.2019.1629546.

Ricklefs, R.E. & Miles, D.B. (1994). Ecological and evolutionary inferences from morphology: an ecological perspective. In P.G. Wainwright & S.M. Reilly (eds) *Ecological Morphology: Integrative Organismal Biology*. University of Chicago Press, Chicago, 13–41.

Rico-Sanchez, A.E., Rodriguez-Romero, A.J., Sedeno-Diaz, J.E., Lopez-Lopez, E. et al. (2022). Aquatic macroinvertebrate assemblages in rivers influenced by mining activities. *Scientific Reports*, 12, 3209. https://doi.org/10.1038/s41598-022-06869-2.

Riis, T., Sand-Jensen, K. & Larsen, S.E. (2001). Plant distribution and abundance in relation to physical conditions and location within Danish stream systems. *Hydrobiologia*, 448, 217e228.

Riis, T., Dodds, W.K., Kristensen, P.B. & Baisner, A.J. (2012). Nitrogen cycling and dynamics in a macrophyte-rich stream as determined by a release. *Freshwater Biology*, 57, 1579–1591.

Riis, T, Olesen, A., Mundbjerg, S., Alnoee, A.B. et al. (2018). Submerged plant communities do not show species complementarity effect in freshwater wetland mesocosms. *Biology Letters*, 14(12), 20180635.

Riis, T., Tank, J.L., Reisinger, A.J. et al. (2019). Riverine macrophytes control seasonal nutrient uptake via both physical and biological pathways. *Freshwater Biology*, 65, 178–192.

Riley, W.D., Potter, C.E. Biggs, J. et al. (2018). Small water bodies in Great Britian and Ireland: ecosystem function, human-generated degradation, and options for restorative action. *Science of the Total Environment*, 645, 1598–1616.

Rimet, F. & Bouchez, A. (2012). Life-forms, cell-sizes and ecological guilds of diatoms in European rivers. *Knowledge and Management of Aquatic Ecosystems*, 406, 01, doi: 10.1051/kmae/2012018.

Rinaldi, M., Gurnell, A.M., González del Tánago, M. et al. (2016). Classification of river morphology and hydrology to support management and restoration. *Aquatic Sciences: Research across Boundaries*, 78, 17–33.

Rinne, A. & Wiberg-Larsen, P. (2017) *Trichoptera Larvae of Finland*. Trificon, Tampere.

Ritz, S. & Fischer, H. (2019). A mass balance of nitrogen in a large lowland river (Elbe, German). *Water*, 11, 2383, https://doi.org/10.3390/w11112383.

Ritz, S., Dähnke, K. & Fischer, H. (2017a). Open-channel measurement of denitrification in a large lowland river. *Aquatic Sciences*, 80, 11. doi.org.10.1007/s00027-017-0560-1.

Ritz, S., Esser, M., Arndt, H. & Weitere, M. (2017b). Large scale patterns of biofilm-dwelling ciliate communities in a river network: only small effects of stream order. *Hydrobiologia*, 102, 114–124.

Roberts, B.J., Mulholland, P.J. & Hill, W.R. (2007). Multiple scales of temporal variability in ecosystem metabolism rates: results from 2 years of continuous monitoring in a forested headwater stream. *Ecosystems*, 10, 588–606.

Robertson, A.L., Rundle, S.D. & Scmid-Araya, J.M. (2000). An introduction to a special issue on lotic meiofauna. *Freshwater Biology*, 44, 1–3.

Robson, B.J., Matthews, T.G., Lind, P.R. & Thomas, N.A. (2008). Pathways for algal recolonization in seasonally-flowing streams. *Freshwater Biology*, 53, 2385–2401.

Robson, S.V., Rosi, E.J., Richmond, E.K. & Grace, M.R. (2020). Environmental concentrations of pharmaceuticals alter metabolism, denitrification, and diatom assemblages in articial streams. *Freshwater Science*, 39, 256–267.

Rockwood, L.R. (2015). *Introduction to Population Ecology* (2nd edn). John Wiley & Sons, Chichester, W. Sussex.

Rohde, K. (1992). Latitudinal gradients in species diversity: the search for the primary cause. *Oikos*, 65, 514–527.

Rome, N.E., Connoer, S.L. & Bauer, R.T. (2009). Delivery of hatching larvae to estuaries by an amphidromous river shrimp: tests of hypotheses based on larval moulting and distribution. *Freshwater Biology*, 54, 1924–1932.

Rosas, K.G., Colón-Gaud, C. & Ramirez, A. (2020). Trophic basis of production in tropical headwater streams, Puerto Rico: an assessment of the importance of allochthonous resources in fueling food webs. *Hydrobiologia* 847, 1961–1975.

Rosemond, A.D., Benstead, J.P., Bumpers, PM. et al. (2015). Experimental nutrient additions accelerate terrestrial carbon loss from stream ecosystems. *Science*, 347, 1142–1145.

Rosemond, A.D., Bumpers, P.M., Eggert, S.L. & Paul, M.J. (2021). Ecoregion 8.4.4 Blue Ridge: Coweeta Hydrologic Laboratory, North Carolina. In D.F. Ryan (ed) *Biological Responses to Stream Nutrients: A Synthesis of Science From Experimental Forests and Ranges*. Forest Service, USDA, Pacific NW Res. Sta. General Tech. Rept. PNW-GTR-981, Dec. 2021, 349–387.

Rosenberg, D.M. & Resh, V.H. (Eds) (1993). *Freshwater Monitoring and Benthic Macroinvertebrates*. Chapman & Hall, New York, 488.

Rosi, E. J., Bechtold, H.A., Snow, D. et al. (2017). Urban stream microbial communities show resistance to pharmaceutical exposure. *Ecosphere*, 8(12), e02041. 10.1002/ecs2.2041.

Rosi-Marshall, E.J. & Wallace, J.B. (2002). Invertebrate food webs along a stream resource gradient. *Freshwater Biology*, 47, 129–141.

Rosi-Marshall, E.J. & Royer, T.V. (2012). Pharmaceutical compounds and ecosystem function: an emerging research challenge for aquatic ecologists. *Ecosystems*, 15, 867–880. doi:10.1007/s10021-012-9553-z.

Rosi-Marshall, E.J., Kincaid, D.W., Bechtold, H.A. et al. (2013). Pharmaceuticals suppress algal growth and microbial respiration and alter bacterial communities in stream biofilms. *Ecological Applications*, 23, 583–593.

Rosi-Marshall, E.J., Vallis, K.L., Baxter, C.V. & Davis, J.M (2016). Retesting a prediction of the River Continuum Concept: autochthonous versus allochthonous resources in the diets of invertebrates. *Freshwater Science*, 35, 534–543.

Roslin. T., Traugott, M., Jonsson, M., Stone, G.N. et al. (2019). Introduction: Special Issue on species interactions, ecological networks and community dynamics—untangling the tangled bank using molecular techniques. *Molecular Ecology*, 28, 157–164.

RoTAP. (2012). Review of Transboundary Air Pollution: acidification, eutrophication, ground level ozone and heavy metals in the UK. Report to the Dept of Environment, Food and Rural Affairs. Centre for Ecology & Hydrology. www.rotap.ceh.ac.uk.

Rott, E., Cantonati, M., Fureder, L. & Pfister, P. (2006). Benthic algae in high altitude streams of the Alps: a neglected component of the aquatic biota. *Hydrobiologia*, 562, 195–216.

Rounick, J.S. & Winterbourn, M.J. (1983). The formation, structure and utilization of stone surface organic layers in two New Zealand streams. *Freshwater Biology*, 13, 57–72.

Rounick, J.S., Winterbourn, M.J. & Lyon, G.L. (1982). Differential utilization of allochthonous and autochthonous inputs by aquatic invertebrates in some New Zealand streams: a stable carbon isotope study. *Oikos*, 39, 191–198.

Rowley, K.H., Cucknell, A.C., Smith, B.D. et al. (2020). London's river of plastic: high levels of microplastics in the Thames water column. *Science of the Total Environment*, 740, 10.1016/j.scitotenv.2020.140018.

Roy, A., Chatterjee, A., Tiwari, S., Sarkar C. et al. (2016). Precipitation chemistry over urban, rural and high altitude Himalayan stations in eastern India. *Atmospheric Research*, 181, 44–53.

Roy, K., Karim, M.R., Akter, F., Islam, S. et al. (2018). Hydrochemistry, water quality and land use signatures in an ephemeral tidal river: implications in water management in the southwestern coastal region of Bangladesh. *Applied Water Science* 8, 78. https://doi.org/10.1007/s13201-018-0706-x.

Rubneson, E.S. & Olden, J.D. (2020). An invader in salmonid rearing habitat: current and future distributions of smallmouth bass (*Micropterus dolomite*) in the Columbia River basin. *Canadian Journal of Fisheries & Aquatic Sciences*, 77. https://doi.org/10.1139/cjfas-2018-0357.

Rüegg, J., Chaloner, D., Ballantyne, F. et al. (2020). Understanding the relative roles of salmon spawner enrichment and disturbance: a high-frequency, multi-habitat field and modelling approach. *Frontiers in Ecology & Evolution*, 8, https://doi.org/10.3389/fevo.2020.00019.

Rundle, S.D. & Hildrew, A.G. (1990). The distribution of micro-arthropods in some southern English chalkstreams: the influence of physiochemistry. *Freshwater Biology*, 23, 411–432.

Rundle, S.D., Robertson, A.L. & Schmid-Araya, J.M. (2002). *Freshwater Meiofauna: Biology and Ecology*. Backhuys Publishers, Netherlands.

Sabater, S.D., Barceló, N., De-Castro-Catala, A. et al. (2016). Shared effects of organic microcontaminants and environmental stressors on biofilms and invertebrates in impaired rivers. *Environmental Pollution*, 210, 303–314.

Sabater, S., Timoner, X., Bornette, G., de Wilde, M. et al. (2017). Algae and vascular plants. In T. Datry, N. Bonada & A.J. Boulton (eds) *Intermittant Rivers and Ephemeral Streams: Ecology and Management*. Academic Press, 189–216.

Sabater, S., Elosegi, A., Feio, M.J. et al. (2022). The Iberian Rivers. In K. Tockner, C. Zarfl, C. Robinson & A.C. Benke (eds) *Rivers of Europe* (2nd edn). Elsevier, Amsterdam, 181–224.

Sabo, J.L., Finlay, J.C., Kennedy, T. & Post, D.M. (2010). Scaling of drainage area and food chain length in rivers. *Science*, 330, 965–967.

Saigo, M., Zilli, F.L., Marchese, M.R. & Demonte, D. (2015). Trophic level, food chain length and omnivory in the Parana River: a food web model approach in a floodplain river system. *Ecological Research*, 30(5), DOI 10.1007/s11284-015-1283-1.

Salo, J., Kalliola, R., Häkkinen, I., Mäkinen, Y. et al. 1986). River dynamics and the diversity of Amazonian floodplain forest. *Nature*, 322, 254–258.

Sampson, A., Ings, N., Shelley, F. Tuffin, S. et al. (2018). Geographically widespread 13C -depletion of grazing caddis larvae: a third way of fuelling stream food webs? *Freshwater Biology*, 64, 787–798.

Sanchez-Bayo, F. & Wyckhuys, K. (2019). World-wide decline of the entomofauna: a review of its drivers. *Biological Conservation*, 232, 8–27.

Sánchez-Hernández, J., Gabler, H.-M., & Amundsen, P.-A. (2016). Food resource partitioning between stream-dwelling Arctic charr *Salvelinus alpinus* (L.), Atlantic salmon *Salmo salar* L. and alpine bullhead *Cottus poecilopus* Heckel, 1836: an example of water column segregation. *Hydrobiologia*, 783, 105–115.

Sanchez-Hernandez, J., Gabler, H-M. & Amundsen, P.-A. (2017). Prey diversity as a driver of resource partitioning between river-dwelling fish species. *Ecology and Evolution*, 7, 2058–2068.

Sanchez-Hernandez, J., Nunn, A.D., Adams, C.E. & Amundsen, P.-A. (2019). Causes and consequences of ontogenetic dietary shifts: a global synthesis using fish models. *Biological Reviews*, 94, 539–554.

Sand, K. & Brittain, J. E. (2009). Life cycle shifts in *Baetis rhodani* (Ephemeroptera) in the Norwegian mountains, *Aquatic Insects*, 31(sup. 1), 283–291.

Sanders, I.A., Heppell, C.M., Cotton, J.A. et al. (2007). Emissions of methane from chalk streams has potential implications for agricultural practices. *Freshwater Biology*, 52, 1176–1186.

Sandin, L. & Johnson, R.K. (2004) Local, landscape and regional factors structuring benthic macroinvertebrate assemblages in Swedish streams. *Landscape Ecology*, 19, 501–514.

Sand-Jensen K. (1998). Influence of submerged macrophytes on sediment composition and near-bed flow in lowland streams. *Freshwater Biology*, 39, 663–679. doi:10.1046/j.1365-2427.1998.00316.x.

Sangpradub, N., Giller, P.S. & O'Connor, J. (1999). Life history patterns of stream-dwelling caddis. *Archiv für Hydrobiologie*, 44, 471–493.

Santos, F.J.M., Protazio, A.S., Moura, C.W.N & Junca, F.A. (2016). Diet and food resource partition among benthic tadpoles of three anuran species in Atlantic forest tropical streams. *Journal of Freshwater Ecology*, 31, 53–60.

Sanzone, D.M, Meyer, J.L., Marti, E., Gardiner, E.P. et al. (2003). Carbon and nitrogen transfer from a desert stream to riparian predators. *Oecologia*, 134, 238–250.

Sato, T., Iritani, R., & Sakura, M. (2019). Host manipulation by parasites as a cryptic driver of energy flow through food webs. *Current Opinion in Insect Science*, 33, 69–76.

Schenk, J. & Fontaneto, D. (2019). Biodiversity analyses in freshwater meiofauna through DNA sequence data. *Hydrobiologia*, 847, 2597–2611.

Schlosser, L.J. (1991). Stream fish ecology: a landscape perspective. *BioScience*, 41, 704–712.

Schmera, D., Arva, D., Boda, P., Bodis, E. et al. (2018). Does isolation influence the relative role of environmental and dispersal-related processes in stream networks? An empirical test of the network position hypotheses using multiple taxa. *Freshwater Biology*, 1, 74–85.

Schmera, D., Podani, J., Heino, J., Eros, T. et al. (2015). A proposed unified terminology of species traits in stream ecology. *Freshwater Science*, 34(3), 823–830.

Schmidt, B.C., Spear, S.F., Tomi, A. & Bodinot Jachowski, C.M. (2021). Evaluating the efficacy of environmental DNA (eDNA) to detect an endangered freshwater mussel *Lasmigona decorate* (Bivalvia: Unionidae). *Freshwater Science*, 40, 354–367.

Schmidt, C., Krauth, T. & Wagner, S. (2017). Export of plastic debris into the sea. *Environmental Science & Toxicology*, 51, 12246–12253.

Schmid-Araya, J.M., Hildrew, A.G., Robertson, A., Schmid, P.E. et al. (2002a). The importance of meiofauna in food webs: evidence from an acid stream. *Ecology*, 83, 1271–1285.

Schmid-Araya, J., Schmid, P, Robertson, A., Winterbottom, J. et al. (2002b). Connectence in stream food webs. *Journal of Animal Ecology*, 71, 1056–1062.

Schmidt-Kloiber, A. & Hering, D. (2015). www.freshwaterecology.info—An online tool that unifies, standardises and codifies more than 20,000 European freshwater organisms and their ecological preferences. *Ecological Indicators*, 53, 271–282.

Schneider, C., Laize, C.L.R., Acreman, M.C. and Florke, M. (2013). How will climate change modify river flow regimes in Europe? *Hydrology and Earth System Sciences*, 17, 325–339.

Scholz, S. & Kluver, N. (2009). Effects of endocrine disruptors on sexual, gonadal development in fish. *Sexual Development*, 3, 136–151.

Schubart, C.D. & Santl, T. (2014). Differentiation within a river system: ecology or geography driven? Evolutionary significant units and new species in Jamaican freshwater crabs. In D.C.J. Yeo, N. Cumberlidge & S. Klaus (eds) *Advances in Freshwater Decapod Systematics and Biology*. Brill, Leiden/Boston, 173–193.

Schultheis, A.S. & Hughes, J.M. (2005). Spatial patterns among populations of a stone-cased caddis (Trichoptera: Tasimiidae) in south-east Queensland, Australia. *Freshwater Biology*, 50, 2002–2010.

Sedell, J.R., Everest, F.H. & Swanson, F.J. (1982). Fish habitat and streamside management: past and present. — In: *Proceedings of the Society of American Foresters, Annual Meeting*: 244-255. (September 27–30, 1981). Society of American Foresters, Bethesda, MD.

Seebacher, F., White, C.R. & Franklin, C.E. (2015). Physiological plasticity increases resilience of ectothermic animals to climate change. *Nature Climate Change*, 5, 61–66.

Seehausen, O. & Wagner, C. (2014). Speciation in freshwater fishes. *Annual Review of Ecology and Systematics*, 45, 621–665.

Seena S., Bärlocher, F., Sobral, O., Gessner, M.O. et al. (2019). Biodiversity of leaf litter fungi in streams along a latitudinal gradient. *Science of the Total Environment*, 661, 306–315.

Sevenster, J.G. (1996). Aggregation and coexistence. I. Theory and analysis. *Journal of Animal Ecology*, 65, 297–307.

Seymour, R.S., Jones, K.K. & Hetz S.K. (2015). Respiratory function of the plastron in the aquatic bug *Aphelocheirus aestivalis* (Hemiptera, Aphelocheiridae). *Journal of Experimental Biology*, 218, 2840–2846.

Shearer, C.A., Descals, E., Kohlmeyer, B., Kohlmeyer, J. et al. (2007). Fungal biodiversity in aquatic habitats. *Biodiversity and Conservation*, 16, 49–67.

Sheldon, A.J. (2012). Possible climate-induced shift of stoneflies in a southern Appalachian catchment. *Freshwater Science*, 31, 765–774.

Shelley, F., Ings, N., Hildrew, A.G. et al. (2017). Bringing methanotrophy in rivers out of the shadows. *Limnology and Oceanography*, 62, 2345–2359.

Shibabaw, T., Betene, A., Awoke, A., Tirfie, M. et al. (2021). Diatom community structure in relation to environmental factors in human-influenced rivers and streams in tropical Africa. *PLoS ONE*, 16(2), e0246043. https://doi.org/10.1371/journal.pone.0246043.

Shockaert, E., Hooge, M., Sluys, R., Schilling, S. et al. (2008). Global diversity of free living flatworms (Platy-helminthes, "Turbellaria") in freshwater. *Hydrobiologia*, 595, 41–48.

Shumilova, O., Tockner, K., Thieme, M., Koska, A. et al (2018). Global water transfer megaprojects: a potential solution for the water-food-energy nexus? *Frontiers in Environmental Science*. https://doi.org/10.3389/fenvs.2018.00150.

Siddell, M.E., Budinoff, R.B. & Borda, E. (2005). Phylogenetic evaluation of systematics and biogeography of the leech family Glossiphoniidae. *Invertebrate Systematics*, 19, 105–112.

Siezko, A.K., Demeter, K., Singer, G.A. et al. (2016). Aquatic methane dynamics in a human-impacted river floodplain of the Danube. *Limnology and Oceanography*, 10.1002/lno.10346.

Sih, A. & Wooster, D.E. (1994). Prey behavior, prey dispersal, and predator impacts on stream prey. *Ecology*, 75, 1199–1207.

Silknetter, S., Creed, R.P., Brown, B.L., Frimpong, E.A. et al. (2020). Positive biotic interactions in freshwaters: A review and research directive. *Freshwater Biology*, 65, 811–832.

Silva, T.S.F., Costa, M.P.F. & Melack, J.M. (2009). Annual net primary production of macrophytes in the Eastern Amazon floodplain. *Wetlands*, 29, 747–758.

Simberloff, D. (2006). Invasional meltdown 6 years later: important phenomenon, unfortunate metaphor, or both? *Ecology Letters*, 9, 912–919.

Simberloff, D. & Von Holle, B. (1999). Positive interactions of nonindigenous species: invasion meltdown? *Biological Invasions*, 1, 21–32.

Sinha, A., Chatterjee, N, Ormerod, S.J., Bhupendra, S.A. et al. (2019). River birds as potential indicators of local- and catchment-scale influences on Himalayan river ecosystems. *Ecosystems and People*, 15, 90–101.

Sinha, A., Chatterjee, N., Ramesh, K. & Ormerod, S.J. (2022). Community assembly, functional traits and phylogeny in Himalayan river birds. *Ecology and Evolution*, 12, DOI: 10.1002/ece3.9012.

Sinha, P., Rollason, E. Bracken, L.J., Wainwrighy, J. et al. (2020). A new framework for integrated, holistic, and transparent evaluation of inter-basin water transfer schemes. *Science of the Total Environment*, 721. https://doi.org/10.1016/j.scitotenv.2020.137646.

Sioli, H. (1975). Tropical River: the Amazon. In B.A. Whitton (ed) *River Ecology*. University of California Press, Berkeley, 461–488.

Siri, C., Liu, Y., Masset, T. et al. (2021). Adsorption of progesterone onto microplastics and its desorption in simulated gastric and intestinal fluids. *Environmental Science: Processes & Impacts*, 23, 1566–1577.

Sjöström, P. (1985). Territoriality in nymphs of *Dinocras cephalotes* (Plecoptera). *Oikos*, 45, 353–357.

Skoulikidis, N.T. (1993). Significance evaluation of factors controlling river water composition. *Environmental Geology*, 22, 178–185.

Sládeček, V. (1965). The future of the saprobity system. *Hydrobiologia*, 25, 518–537.

Slavic, K., Peterson, B.J. Deegan, L.A. et al. (2004). Long-term responses of the Kuparuk River ecosystem to phosphorus fertilization. *Ecology*, 85, 939–954.

Sleith, R.S., Wehr, J.D. & Karol, K.G. (2018) Untangling climate and water chemistry to predict changes in freshwater macrophyte distribiutions. *Ecology & Evolution*, 8, 2802–2811.

Small, G.E., Ardón, M. Duff, J.H. et al. (2016). Phosphorus retention in a lowland Neotropical stream following an eight-year enrichment experiment. *Freshwater Science*, 35, 1–11.

Small, G.E., Duff, J.H., Pedro, J.T. & Pringle, C.M. (2013). Insect emergence as a nitrogen flux in Neotropical streams: comparisons with microbial denitrification across a stream phosphorus gradient. *Freshwater Science*, 32, 1178–1187.

Smircich, M.G., Strayer, D.L. & Schultz, E.T. (2017). Zebra mussel (*Dreissena polymorpha*) affects the feeding ecology of early stage striped bass (*Morone saxitilis*) in the Hudson River estuary. *Environmental Biology of Fish*, 100, 395–406.

Smith, B.D. & Reeves, R.R. (2012). River cetaceans and habitat change: generalist resilience or specialist vulnerability? *Journal of Marine Biology*, Article ID 718935, doi:10.1155/2012/718935.

Smith, B.P. (1988). Host-parasite interaction and impact of larval water mites on insects. *Annual Review of Entomology*, 33, 53–64.

Smith, I. & Lyle, A. (1978). *Distribution of Freshwaters in Great Britain*. Institute of Terrestrial Ecology, Edinburgh.

Smith, I.M, Cook, D.R., & Smith, B.P. (2010). Water mites (Hydrachnidiae) and other arachnids. Chapter 15 In Thorp, J.H. and Covich, A.P. (eds) *Ecology and Classification of North American Freshwater Invertebrates* (3rd edn). Academic Press, San Diego.

Smith, N.A., Koeller, K.L., Clarke, J.A., Ksepka, D.T. et al. (2021). Convergent evolution in dippers (Aves, Cinclidae): The only wing-propelled diving songbirds. *The Anatomical Record*, 1–29. https://doi.org/10.1002/ar.24820.

Smith, R.L. (1976). Male brooding behavior of the giant water bug *Abedus herberti* (Hemiptera: Belostomatidae). *Annals of the Entomological Society of America*, 69, 740–747.

Smolders, E., Baetens, E., Verbeeck, M. et al. (2017). Internal loading and redox cycling of sediment iron explain reactive phosphorus concentrations in lowland rivers. *Environmental Science and Technology*, 51, 2584–2592.

Smucker, N., Pilgrim, E.M., Nietch, C.T. et al. (2020). DNA metabarcoding effectively quantifies diatom responses to nutrients in streams. *Ecological Applications*, 30, e02205, doi.org/10.1002/eap.2205.

Solheim, A.L., Globevnik, L., Austnes, K., Kristensen, P. et al. (2019). A new broad typology for rivers and lakes in Europe: development and application for large-scale environmental assessments. *Science of the Total Environment*, 697, 134043.

Soluk, D.A. (1993). Multiple predator effects: predicting combined functional response of stream fish and invertebrate predators. *Ecology*, 74, 219–225.

Soluk, D.A. & Richardson, J.S. (1997). The role of stoneflies in enhancing growth of trout: a test of the importance of predator–predator facilitation within a stream community. *Oikos*, 80, 214–219.

Sommer, T. (2020). How to respond? An introduction to current Bay–Delta natural resource management options. *San Francisco Estuary Watershed Science*, 18, 18iss3art1. Doi 10.15447//sfews.2020v18iss3art1.

Sommer, T., Armor, C., Baxter, R. et al. (2007). The collapse of pelagic fishes in the Upper San Francisco Estuary. *Fisheries*, 32(6), 270–277.

Song, C., Dodds, W.K., Rüegg, J. et al. (2018). Continental scale decrease in net primary productivity in streams due to climate warming. *Nature Geoscience*, 11. 415–420.

Soons, M.B., de Groot, G.A., Ramirez, M.T.C., Fraije, R.G.A. et al. (2017). Directed dispersal by an abiotic vector: wetland plants disperse their seeds selectively to suitable sites along the hydrological gradient via water. *Functional Ecology*, 31, 499–508.

Soria, M., Leigh, C., Datry, T. Bini, L.M. et al. (2017). Biodiversity in perennial and intermittent rivers: a metanalysis. *Oikos*, 126, 1078–1089.

Soria, M., Gutiérrez-Cánovas, C., Bonada, N., Acosta, R. et al. (2020). Natural disturbances can produce misleading bioassessment results: identifying metrics to detect anthropogenic impacts in intermittent rivers. *Journal of Applied Ecology*, 57, 283–295.

Sorrell, B.K. & Dromgoole, F.L. (1989). Oxygen diffusion and dark respiration in aquatic macrophytes. *Plant, Cell and Environment*, 12, 293–299.

Soto, B. (2016). Assessment of trends in stream temperatures in the north of the Iberian Peninsula using a nonlinear regression model for the period 1950–2013. *River Research and Applications*, 32, 1355–1364.

Sousa, W.P. (1984). The role of disturbance in natural communities. *Annual Review of Ecology & Systematics*, 15, 353–391.

Southwood, T.R.E. (1977). Habitat, the templet for ecological strategies? *Journal of Animal Ecology*, 46, 337–365.

Southwood T.R.E. (1988). Tactics, Strategies and Templets. *Oikos*, 52, 3–18.

Souza, M. L. & Moulton, T. P. (2005). The effects of shrimps on benthic material in a Brazilian island stream. *Freshwater Biology*, 50, 592–602.

Speirs, D.C. & Gurney, W.S.C. (2001). Population persistence in rivers and estuaries. *Ecology*, 82, 1219–1237.

Stanford, J., & Ward, J. 1988. The hyporheic habitat of river ecosystems. *Nature*, 335, 64–66. https://doi.org/10.1038/335064a0.

Stanford, J.A. & Ward, J.V. (1993). An ecosystem perspective of alluvial rivers: connectivity and the hyporheic corridor. *Journal of the North American Benthological Society*, 12, 391–413.

Stanley, E.H., Casson, N.J., Christel, S.T et al. (2016). The ecology of methane in streams and rivers: patterns, controls, and global significance. *Ecological Monographs*, 86, 146–171.

Stanley, E.H., Powers, S.M. & Lottig, N.R. (2010). The evolving legacy of disturbance in stream ecology: concepts, contributions, and coming challenges. *Journal of the North American Benthological Society*, 29, 67–83.

Statzner, B. (2008). How views about flow adaptations of benthic stream invertebrates changed over the last century. *International Review of Hydrobiology*, 93, 593–605.

Statzner, B. (2012). Geomorphic implications of engineering bed sediments by lotic animals. *Geomorphology*, 157/158, 49–65.

Statzner, B. & Higler, B. (1986). Stream hydraulics as a major determinant of benthic invertebrate zonation patters. *Freshwater Biology*, 16, 127–139.

Statzner, B. & Holm, T.F. (1982). Morphological adaptations of benthic invertebrates to stream flow—an old question studied by means of a new technique (laser doppler anemometry). *Oecologia*, 53, 290–292.

Statzner, B. & Holm, T.F. (1989). Morphological adaptations of shape to flow: microcurrents around lotic macroinvertebrates with known Reynolds numbers at quasi-natural flow conditions. *Oecologia*, 78, 145–157.

Statzner, B. & Bêche, L.A. (2010). Can biological invertebrate traits resolve effects of multiple stressors on running water ecosystems? *Freshwater Biology*, 55, 80–119.

Statzner, B. & Dolédec, S. (2011). Phylogenetic, spatial and species-trait patterns across environmental gradients: the case of *Hydropsyche* (Trichoptera) along the Loire River. *International Review of Hydrobiology*, 96, 121–140.

Statzner, B., Gore J. A. & Resh, V. H. (1988). Hydraulic stream ecology: observed patterns and potential applications. *Journal of the North American Benthological Society*, 7, 307–360.

Statzner, B., Fuchs, U. & Higler, L.W.G. (1996). Sand erosion by mobile predaceous stream insects: implications for ecology and hydrology. *Water Resources Research*, 32, 2279–2287.

Statzner, B., Arens, M.F., Champagne, J.Y. et al. (1999). Silk-producing insects and gravel erosion: significant biological effects on critical shear stress. *Water Resources Research*, 35, 3495–3506.

Statzner, B., Bis, B., Dolédec, S. & Usseglio-Polatera, P. (2001). Perspectives for biomonitoring at large spatial scales: a unified measure for the functional composition of invertebrate communities in European running waters. *Basic and Applied Ecology*, 2, 73–85.

Statzner, B., Hildrew, A.G. & Resh, V.H. (2001). Species traits and environmental constraints: entomological research and the history of ecological theory. *Annual Review of Entomology*, 46, 291–316.

Stead, T.K., Schmid-Araya, J.M. & Hildrew, A.G. (2005). Secondary production of a stream metazoan community: does the meiofauna make a difference? *Limnology and Oceanography*, 50, 398–403.

Sterner, R.W. & Elser, J.J. (2002). *Ecological Stoichiometry: the Biology of Elements from Macromolecules to the Biosphere*. Princeton University Press, Princeton & Oxford.

Strahler, A.N. (1952). Dynamic basis of geomorphology. *Geological Society of America Bulletin*, 63, 1117–1142.

Strayer, D.L. (2008). *Freshwater Mussel Ecology: A Multifactor Approach to Distribution and Abundance*. University of California Press, Oakland, CA.

Strayer, D.L. (2017). What are freshwater mussels worth? *Freshwater Mollusk Biology and Conservation*, 20, 103–113.

Strayer, D.L. (2020). Non-native species have multiple abundance-impact curves. *Ecology & Evolution*, 10, 6833–6843.

Strayer, D.L. & Malcolm, H.M. (2006). Long-term demography of a zebra mussel (*Dreissena polymorpha*) population. *Freshwater Biology*, 51, 117–130.

Strayer, D.L. & Malcolm, H.M. (2018). Long term responses of native bivalves (Unionidae) and Sphaeriidae) to a *Dreissena* invasion. *Freshwater Science*, 37, 697–711.

Strayer, D.L. Hunter, D.C., Smith, L.C., & Borg, C. (1994). Distribution, abundance, and role of freshwater clams (Bivalvia: Unionidae) in the freshwater tidal Hudson River. *Freshwater Biology*, 31, 239–248.

Strayer, D.L., J.A. Downing, W.R. Haag, T.L. King, J.B. Layzer, T.J. Newton, & Nichols, S.J. (2004). Changing perspectives on pearly mussels, North America's most imperiled animals. *BioScience*, 54, 429–439.

Strayer, D.L., Cid, N. & Malcolm, H.M. (2011). Long term changes in a population of an invasive bivalve and its effects. *Oecologia*, 165, 1063–1072.

Strayer D.L., Cole, J.J., Findlay, S.E., Fischer, D.T. et al. (2014). Decadal-scale change in a large-river ecosystem, *Bioscience*, 64, 496–510.

Strayer, D, Adamovich, B., Adrian R. et al. (2019). Long-term population dynamics of dreissenid mussels (*Dreissena polymorpha* and *D. rostriformis*): a cross-system analysis. *Ecosphere*, 10, e02701.

Strayer, D.L., Fischer, D.T., Hamilton, S.K., Malcolm, H.M. et al. (2019). Long-term variability and density dependence in Hudson River *Dreissena* populations. *Freshwater Biology*, 65, 474–489.

Strayer, D.L., Solomon, C., Findlay, S.E. & Rosi, E.J. (2019). Long-term research reveals multiple relatioships between the abundance and impacts of a non-native species. *Limnology and Oceanography*, 64, 5105–5117. doi: 10.1002/lno.11029.

Stream Bryophyte Group. (1999). Roles of Bryophytes in stream ecosystems. *Journal of North American Benthological Society*, 18, 151–184.

Stream Solute Network. (1990). Concepts and methods for assessing solute dynamics in stream ecosystems. *Journal of the North American Benthological Society*, 9, 95–119.

Sternberg, D. & Kennard, M.J. (2014). Phylogentic effects on functional traits and life history strategies of Australian freshwater fish. *Ecography*, 37, 54–64.

Stringer, A.P. & Gaywood, M.J. (2016). The impacts of beavers *Castor* spp on biodiversity and the ecological basis for their reintroduction to Scotland, UK. *Mammal Review*, 46, 270–283.

Strong, E., Gargominy, O., Ponder, W. & Bouchet, P. (2008). Global diversity of gastropods (Gastropoda: Mollusca). *Hydrobiologia*, 595, 149–166.

Stubbington, R. (2012) The hypotheic zone as an invertebrate refuge: a review of variability in space, time, taxa and behaviour. *Marine & Freshwater Research*, 63, 293–311.

Stubbington R & Datry T. (2013). The macroinvertebrate seedbank promotes community persistence in temporary rivers across climate zones. *Freshwater Biology*, 58, 1202–1220.

Stubbington, R., Gunn, J., Little, S., Worrall, T.P. et al. (2016). Macroinvertebrate seedbank composition in relation to antecedent duration of drying and multiple wet-dry cycles in a temporary stream. *Freshwater Biology*, 61, 1293–1307.

Stubbington, R., Bogan, M., Bonada, N., Boulton, A.J. et al. (2017). The biota of intermittent rivers and ephemeral streams: aquatic invertebrates. In T. Datry, N. Bonada & A.J. Boulton (eds) *Intermittent Rivers and Ephemeral Streams: Ecology and Management*. Academic Press, Cambridge, MA, USA, 217–243.

Subalusky, A.L., Dutton, C.L., Rosi-Marshal, E.J. & Post, D. (2015). The hippopotamus conveyor belt: vectors of carbon and nutrients from terrestrial grasslands to aquatic systems in sub-Saharan Africa. *Freshwater Biology*, 60, 512–525.

Subalusky, A.L., Dutton, C.L., Njoroge, L., Rosi, E.J. et al. (2018). Organic matter and nutrient inputs from large wildlife influence ecosystem function in the Mara River, Africa. *Ecology*, 99, 2558–2574.

Suberkropp, K. (1992a). Aquatic hyphomycete communities. In D.T Wicklow & G.C. Carroll (eds) *The Fungal Community: Its Organization and Role in the Ecosystem* (2nd edn). Marcel Dekker, Inc., New York, 729–748.

Suberkropp K. (1992b). Interactions with invertebrates. In: F. Bärlocher (ed) *The Ecology of Aquatic Hyphomycetes*. Springer, Berlin, 118–134.

Suberkropp, K. (1997). Annual production of leaf-decaying fungi in a woodland stream. *Freshwater Biology*, 38, 169–178.

Suberkropp K. & J.B. Wallace (1992). Aquatic hyphomycetes in insecticide-treated and untreated streams. *Journal of the North American Benthological Society*, 11, 165–171.

Suberkropp K., V. Gulis, A.D. Rosemond & J.P. Benstead (2010). Ecosystem and physiological scales of microbial responses to nutrients in a detritus-based stream: Results of a 5-year continuous enrichment. *Limnology and Oceanography*, 55, 149–160.

Suli, A., Watson, G., Rubel. E. & Raibke, D. (2012). Rheotaxis in larval zebrafish by lateral line mechansensory hair cells. *PLoS One* 7, e29727.

Swan, C.M. & Palmer, M.A. (2004). Leaf diversity alters litter breakdown in a Piedmont stream. *Journal of the North American Benthological Society*, 23, 15–28.

Swan C.M., Boyero, L. & Canhoto, C. (eds) (2021). *The Ecology of Plant Litter Decomposition in Stream Ecosystems*, Springer Nature, Cham, Switzerland.

Swank, W.T. & Crossley, D.A., Jr. (1988). *Forest Hydrology and Ecology at Coweeta*. Ecological Studies, vol 66. Springer, New York.

Swank, W.T. & Webster, J.R. (eds). (2014). *Long-term Response of a Forest Watershed Ecosystem: Clearcutting in the Southern Appalachians*. Oxford University Press, New York.

Sweeney, B.W. & Vannote, R.L. (1982). Population synchrony in mayflies: a predator satiation hypothesis. *Evolution*, 36, 810–821.

Tachet, H., Bournaud, M., Richoux, P. & Usseglio-Polatera, P. (2003). Invertébrés d'eau douce: systématique, biologie, écologie. CNRS editions, Paris.

Tadaki, M., Brierley, G. & Cullum, C. (2014). River classification: theory, practice, politics. *WIREs Water*, 1, 349–367. doi: 10.1002/wat2.1026.

Takimoto, G. & Post, D.M. (2012). Environmental determinants of food-chain length: a meta-analysis. *Ecological Research*, 28(5). DOI: 10.1007/s11284-012-0943-7.

Tanentzap, A.J., Szokan-Emilson, E.J. Kielstra, B.W. et al. (2014). Forests fuel fish growth in freshwater deltas. *Nature Communications*, 5, 4077.

Tank, J.L., Rosi-Marshall, E.J., Griffiths, N.A. et al. (2010). A review of allochthonous organic dynamics and metabolism in streams. *Journal of the North American Benthological Society*, 29, 118–146.

Tank, J.L. Reisinger, A.J. & Rosi, E.J. (2017). Nutrient limitation and uptake. In R. Hauer & G.A. Lamberti (eds) *Methods in Stream Ecology. Volume 2: Ecosystem Function* (3rd edn). Academic Press, London, 147–171.

Tank, J.L., Marti, E., Riis, T. et al. (2018). Partitioning assimilatory nitrogen uptake in streams: an analysis of stable isotope tracer additions across continents. *Ecological Monographs*, 88, 120–138.

Tansley, A.G. (1935). The use and abuse of vegetational concepts and terms. *Ecology*, 16, 284–307.

Taylor, A. & O'Halloran, J. (2001). The diet of the Dipper *Cinclus cinclus* during an early winter spate and possible implications for Dipper populations subjected to climate change. *Bird Study*, 48, 173–179.

Taylor, W.D. & Sanders, R.W. (2009). Protozoa. In J.H. Thorp & A.P. Covich (eds) *Ecology and Classification of North American Freshwater Invertebrates*. Elsevier, London, Burlington and San Diego, 49–90.

Tedersoo, L., Bahram, M., Polme, S. et al. (2014). Fungal biogeography, global diversity and geography of soil fungi. *Science*, 346(6213), 1256688.

Teittinen, A., Taka, M., Ruth, O. & Soininen, J. (2015) Variation in stream diatom communities in relation to water quality and catchment variables in a boreal, urbanized region. *Science of the Total Environment*, 530-531, 279–289.

Thompson, M.S.A., Bankier, C., Bell, T. et al. (2015). Gene-to-ecosystem impacts of a catastrophic pesticide spill: testing a multilevel bioassessment approach in a river ecosystem. *Freshwater Biology*, 10.1111/fwb.12676.

Thompson, M.S.A., Brooks, S.J., Sayer, C.D. et al. (2018). Large woody debris 'rewilding' rapidly restores biodiversity in riverine food webs. *Journal of Applied Ecology*, 55, 895–904.

Thompson, R.M & Townsend, C.R. (2005). Energy availability, spatial heterogeneity and ecosystem size predict food-web structure in streams. *Oikos*, 108, 137–148.

Thompson, R. & Townsend, C.R. (2006). A truce with neutral theory: local deterministic factors, species traits and dispersal limitation together determine patterns of diversity in stream invertebrates. *Journal of Animal Ecology*, 75, 476–484.

Thompson, R. M., Brose, U., Dunne, J. A., Jr, R. O. H., Hladyz, S., Kitching, R. L., et al. (2012). Food webs: reconciling the structure and function of biodiversity. *Trends in Ecology & Evolution*, 27(12), 689–697. doi:10.1016/j.tree.2012.08.005.

Thomsen, P.F. & Willersley, E. (2015). Environmentl DNA—an emerging tool in conservation for monitoring past and present biodiversity. *Biological Conservation*, 183, 4–18.

Thorat, L. & Nath, B. (2018) Aquatic silk proteins in Chironomus: A review. *Journal of Limnology*, doi: 10.4081/jlimnol.2018.1797.

Thornhill, I., Loiselle, S., Clymans, W. et al. (2019). How citizen scientists can enrich freshwater science as contributors, collaborators, and co-creators. *Freshwater Science*, 38, 231–235.

Thorp, J.H. & Delong, M.D. (1994). The Riverine Productivity Model: an heuristic view of carbon sources and organic processing in large river ecosystems. *Oikos*, 70, 305–308.

Thorp, J.H. & Delong, M.D. (2002). Dominance of autochthonous autotrophic carbon in food webs of heterotrophic rivers. *Oikos*, 96, 543–550.

Thorp, J.H., & Rogers, D.C. (2011). *Field Guide to Freshwater Invertebrates of North America*. Burlington, MA: Academic Press, 304.

Thorp, J.H & Rogers, D.C. (2014). *Freshwater Invertebrates* (4th edn). *Volume I: Ecology and General Biology*. Academic Press, San Diego, London.

Thorp, J.H & Rogers, D.C. (2016). *Freshwater Invertebrates, Volume II: Keys to Nearctic Fauna*. Elsevier, San Diego, London.

Thorpe, W.H. (1950). Plastron respiration in aquatic insects. *Biological Reviews*, 25, 334–390.

Tiegs, S.D., Costello, D.M., Isken, M.W. et al. (2019). Global drivers of ecosystem functioning of rivers and riparian zones. *Science Advances*, 5, eeav0486.

Tierno del Figueroa, J. & Lopez-Rodriguez, M. (2019). Trophic ecology of Plecoptera: a review. *The European Zoological Journal*, 86, 79–102.

Tison-Rosebery, J., Leboucher, T., Archaimbault, V. et al. (2022). Decadal biodiversity trends in rivers reveal recent community rearrangements. *Science of the Total Environment*, 823, 153431.

Tockner, K., Robinson, C.T. & Uehlinger, U. (eds) (2009). *Rivers of Europe*. Academic Press, Amsterdam.

Tockner, K., Zarfl, C. & Robinson, C. (eds) (2021). *Rivers of Europe* (2nd edn). Elsevier, Amsterdam.

Tod, S.P. & Schmid-Araya, J.M. (2009). Meiofauna versus macrofauna: secondary production of invertebrates in a lowland chalk stream. *Limnology and Oceanography*, 54, 450–456.

Tokeshi, M. (1994). Community ecology and patchy freshwater habitats. In P.S. Giller, A.G. Hildrew & D.G. Raffaelli (eds) *Aquatic Ecology: Scale, Pattern and Process*. 34th Symposium of the British Ecological Society, Blackwell Scientific Publications, Oxford, 63–91.

Tokeshi, M., & Townsend, C. R. (1987). Random patch formation and weak competition: coexistence in an epiphytic chironomid community. *Journal of Animal Ecology*, 56, 833–845.

Tomonova, S., Goitia, E. & Helesic, J. (2006). Trophic levels and functional feeding groups of macroinvertebrates in neotropical streams. *Hydrobiologia*, 556, 251–264.

Tonkin, J.D., Altermatt, F., Finn, D.S., Heino, J. et al. (2018a). The role of dispersal in river network communities: patterns, processes and pathways. *Freshwater Biology*, 63, 141–163.

Tonkin, J.D., Heino, J. & Altermatt, F. (2018b). Metacommunities in river networks: the importance of network structure and connectivity on patterns and processes. *Freshwater Biology*, 63, 1–5.

Toussant, A., Charpin, N., Brosse, S. & Villeger, S. (2016). Global functional diversity of freshwater fish is concentrated in the Neotropics while functional vulnerability is widespread. *Nature Scientific Reports*, 6, 22125. DOI: 10.1038/srep22125.

Townsend, C.R. (1989). The patch dynamics concept of stream community ecology. *Journal of the North American Benthological Society*, 8, 36–50.

Townsend, C.R. (1996). Invasion biology and ecological impacts of brown trout (*Salmo trutta*) in New Zealand. *Biological Conservation*, 78, 13–22.

Townsend, C. R. & Hildrew, A.G. (1976). Field experiments on the drifting, colonization and continuous redistribution of stream benthos. *Journal of Animal Ecology*, 45, 759–772.

Townsend, C.R. & Hildrew, A.G. (1979). Form and function of the prey catching net net of *Plectrocnemia conspersa* larvae (Trichoptera). *Oikos*, 33, 412–418.

Townsend, C.R. & Hildrew, A.G. (1979). Foraging strategies and coexistence in a seasonal environment. *Oecologia*, 38, 231–234.

Townsend, C.R. & Hildrew A.G. (1994). Species traits in relation to a habitat templet for river systems. *Freshwater Biology*, 31, 265–275.

Townsend, C.R., Hildrew, A.G. & Francis, J. (1983). Community structure in some southern English streams: the influence of physicochemical factors. *Freshwater Biology*, 13, 521–544.

Townsend, C.R., Hildrew, A.G. & Schofield, K. (1987). Persistence of stream invertebrate communities in relation to environmental variability. *Journal of Animal Ecology*, 56, 597–613.

Townsend, C.R., Thompson, R.M., McIntosh, A.R., Kilroy, C. et al. (1998). Disturbance, resource supply, and food-web architecture in streams. *Ecology Letters*, 1, 200–209.

Traunspurger, W. & Majdi, N. (2017). Meiofauna. In R. Hauer & G.A. Lamberti (eds) *Methods in Stream Ecology Volume 1: Ecosystem Structure* (3rd edn). Academic Press, London, 273–295.

Traunspurger, W. (2000). The biology and ecology of lotic nematodes. *Freshwater Biology*, 44, 29–45.

Trevelline, B.K., Nuttle, T, Hoenig, B.D. et al. (2018). DNA metabarcoding of nestling feces reveals provisioning of aquatic prey and resource partitioning among neotropical migrant songbirds in a riparian habitat. *Oecologia*, 187, 85–98.

Trimmer, M., Grey, J., Heppell, C.M. et al. (2012). River bed carbon and nitrogen cycling: state of play and some new directions. *Science of the Total Environment*, 434, 143–158.

Trimmer, M., Hildrew, A.G., Jackson, M. et al. (2009). Evidence for the role of methane-derived carbon in a free-flowing, lowland river food web. *Limnology and Oceanography*, 54, 1451–1547.

Trimmer, M., Shelley, F.C., Purdy, K.J. et al. (2015). Riverbed methanotrophy sustained by high carbon conversion efficiency. *ISME*, 9, 2304–2314.

Tsoi, W.Y., Hadwen, W.L. & Sheldon, F. (2017). How do abiotic environmental variables shape benthic diatom assemblages in subtropical streams? *Marine and Freshwater Research*, 68, 863–877.

Turchin, P. (2003). *Complex Population Dynamics: A Theoretical/Empirical Synthesis*. Princeton University Press, Princeton, New Jersey.

Turley, M.D., Bilotta, G.S., Chadd, R.P. Extence, C.A et al. (2016). A sediment-specific family-level biomonitoring tool to identify the impacts of fine sediment in temperate rivers and streams. *Ecological Indicators*, 70, 151–165.

Turner, T.F. & Edwards, M.S. (2012). Aquatic food web structure of the Rio Grande assessed with stable isotopes. *Freshwater Science*, 31, 825–834.

Twardochleb, L. et al. (2021). Freshwater insects CONUS: a database of freshwater insect occurrences and traits for the contiguous United States. *Global Ecology & Biogeography*, 00, 1–16, doi: 10.1111/geb.13527.

Tweedley, J.R., Bird, D.J., Potter, I.C., Gill, H.S. et al. (2013). Species compositions and ecology of the riverine ichthyofaunas in two Sulawesian islands in the biodiversity hotspot of Wallacea. *Journal of Fish Biology* 82, 1916–1950.

Twidale, C.R. (2004). River patterns and their meaning. *Earth-Science Reviews*, 67, 159–218.

Twining, C.W., Josephson, D.C., Kraft, C.E. et al. (2017). Limited seasonal variation in food quality and foodweb structure in an Adirondack stream: insights from fatty acids. *Freshwater Science*, 36, 877–892.

Twining, C.W., Shipley, J.R. & Winkler, D.W. (2018). Aquatic insects rich in omega-3 fatty acids drive breeding success in a widespread bird. *Ecology Letters*, 21, 1812–1820.

Twomey, H. & Giller, P. S. (1991). The effects of catastrophic flooding on the benthos and fish of a tributary of the River Araglin, Co. Cork. In *Irish Rivers: Biology and Management* (ed. by M. Steer). Proceedings of Royal Irish Academy Conference (Dublin 1989), 59–81.

Tyler, C.R. & Jobling, S. (2008). Roach, sex, and gender-bending chemicals: the feminization of wild fish in English rivers. *BioScience*, 58, 1051–1059.

Tyler, S. & Ormerod, S.J (1992). A review of the likely causal pathways relating the reduced density of breeding dippers *Cinclus cinclus* to the acidification of upland streams. *Environmental Pollution*, 78, 49–55.

Tyler, S. J. & Ormerod, S.J. (1994). *The Dippers*. T & A D Poyser, London.

Uehlinger, U. (2006). Annual cycle and inter-annual variability of gross primary production and ecosystem respiration in a flood prone river during a 15-year period. *Freshwater Biology*, 51, 938–950.

UNESCO World Water Assessment Programme. (2019). The World Water Development Report: Leaving No One Behind. UNESCO, Paris.

UNESCO World Water Development Report. (2021). *Valuing Water*. United Nations Educational, Scientific and Cultural Organisation, Paris, France.

Unwin, M.J. & Larned, S.T. (2013). Statistical models, indicators and trend analysis for reporting national-scale river water quality (NEMAR Phase 3). In: *For the Ministry for the Environment*, Vol. NIWAS Client Report NO: CHC2013-033. NIWA, Christchurch, New Zealand.

Uriarte, M., Yackulic, C.B., Lim, Y. & Arce-Nazario, J.A. (2011). Influence of land use on water quality in a tropical landscape: A multi-scale analysis. *Landscape Ecology*, 26, 1151–1164.

Usseglio-Polatera, P. & Beisel, J.N. (2002). Longitudinal changes in macroinvertebrate assemblages in the Meuse River: Anthropogenic effects versus natural change. *River Research and Applications*, 18, 197–211.

Usseglio-Polatera, P., Bournaud, M., Philippe, R. & Tachet, H. (2000). Biological and ecological traits of benthic freshwater macroinvertebrates: relationships and definition of groups with similar traits. *Freshwater Biology*, 43, 175–205.

Vadeboncoeur, Y. & Power, M.E. (2017). Attached algae; the cryptic base of inverted trophic pyramids in freshwaters. *Annual Review of Ecology and Systematics*, 48, 255–279.

Vadher, A.N., Leigh, C., Millett, J. & Stubbington, R. (2017). Vertical movements through subsurface stream sediments by benthic macroinvertebrates during experimental drying are influenced by sediment characteristics and species traits. *Freshwater Biology*, 62, 1730–1740.

Vainola, R., Witt, J.D., Grabowski, M., Bradbury, J.H. et al. (2008). Global diversity of amphipods (Amphipoda: Crustacea) in freshwater. *Hydrobiologia*, 595, 241–255

Valencia-Aguilar, A., Cortés-Gómez, A.M. & Ruiz-Agudelo, C. (2013). Ecosystem services provided by amphibians and reptiles in Neotropical ecosystems, *International Journal of Biodiversity Science, Ecosystem Services & Management*, 9(3), 257–272. DOI: 10.1080/21513732.2013.821168.

van Klink, R., Bowler, D.E. Gongalsky, K.B. et al.(2020). Meta-analysis reveals declines in terrestrial but increases in freshwater insect abundances. *Science*, 368, 417–420. doi 10.1126/science.aax9931.

Van Looy, K., Floury, M., Ferréol, M., Prieto-Montes, M. et al. (2016). Long-term changes in temperate stream communities reveal a synchronous trophic amplification at the turn of the millennium. *Science of the Total Environment*, 565, 481–488.

Van Rijn, L.C. (2019). Critical movement of large rocks in currents and waves. *International Journal of Sediment Research*, 34, 387–398.

Van Rooij, P., Martel, A., Haesebrouck, F. & Pasmans, F. (2015). Amphibian chytridiomycosis: a review with focus on fungus-host interactions. *Veterinary Research*, 46, 137. https://doi.org/10.1186/s13567-015-0266-0.

Vander Vorste, R., Corti, R., Sagouis, A. & Datry, T. (2016). Invertebrate communities in gravel-bed, braided rivers are highly resilient to flow intermittence. *Freshwater Science*, 35, 164–177.

Vander Vorste, R., Obedzinski, M., Nossaman Pierce, S.N., Carlson, S.N. et al. (2020). Refuges and ecological traps: Extreme drought threatens persistence of an endangered fish in intermittent streams. *Global Change Biology*, 26, 3834–3845.

Vander Zanden, J.M. & Fetzer, W.W. (2007). Global patterns of aquatic food chain length. *Oikos*, 116, 1378–1388.

Vanni, M.J. (2002). Nutrient cycling by animals in freshwater ecosystems. *Annual Review of Ecology & Systematics*, 33, 341–370.

Vanni, M.J., McIntyre, P.B., Allen, D. et al. (2017). A global database of nitrogen and phosphorus excretion rates of aquatic animals. *Ecology*, 98, 1475. Doi: 10.1002/ecy.1792.

Vannote, R.L., Minshall, G.W., Cummins, K.W., Sedell, J.R. et al. (1980). The river continuum concept. *Canadian Journal of Fisheries and Aquatic Sciences*, 37, 130–137.

Vári, Á., Podschun, S.A., Erős, T. et al. (2022). Freshwater systems and ecosystem services: Challenges and chances for cross-fertilization of disciplines. *Ambio*, 51, 135–151.

Vaughan, I.P. & Ormerod, S.J. (2014). Large-scale, long-term trends in British river macroinvertebrates. *Global Change Biology*, 18, 2184–2194.

Vaughn, C.C. (2018). Ecosystem services provided by freshwater mussels. *Hydrobiologia*, 810, 15–27.

Vaughn, C.C & Hoellein, T.J. (2018). Bivalve impacts on freshwater and marine ecosystems. *Annual Review of Ecology, Evolution, and Systematics*, 49, 183–208.

Vences, M. & Kohler, J. (2008). Global diversity of amphibians (Amphibia). *Hydrobiologia*, 595, 569–580.

Verberk, W.C.E.P., Siepel, H. & Esselink, H. (2008). Life history strategies in freshwater macroinvertebrates. *Freshwater Biology*, 53, 1722–1738.

Verberk, W.C.E.P., Bilton, D.T., Calosi, P., et al. (2011). Oxygen supply in aquatic ectotherms: partial pressure and solubility together explain biodiversity and size patterns. *Ecology*, 92, 1565–1572.

Verberk, W.C.E.P., van Noordwijk, C.G.E. & Hildrew, A.G. (2013). Delivering on a promise: integrating species traits to transform descriptive community ecology into a predictive science. *Freshwater Science*, 32, 531–547.

Verberk, W.C.E.P., Durance, I., Vaughan, I.P., et al. (2016). Field and laboratory studies reveal interacting effects of stream oxygenation and warming on aquatic ectotherms. *Global Change Biology*, 22, 1769–1778.

Veron, G., Patterson, B. & Reeves, R. (2008). Global diversity of mammals (Mammalia) in freshwater. *Hydrobiologia*, 595, 607–617.

Verpoorter, C., Kutser, T., Seekell, D.A. & Tranvik, L.J. (2014). A global inventory of lakes based on high resolution satellite imagery. *Geophysical Research Letters*, 41, 6396–6402.

Vet, R., Artz, R.S., Carou, S., Shaw, M. et al. (2014). A global assessment of precipitation chemistry and deposition of sulfur, nitrogen, sea salt, base cations, organic acids, acidity and pH and phosphorus. *Atmospheric Environment*, 3, 3–100.

Vogel, S. (1994). *Life in Moving Fluids: The Physical Biology of Flow*. Princeton University press, Princeton, NJ.

Volder, A., Bonnis, A. & Grillas, P. (1997). Effects of drought and flooding on the reproduction of an amphibious plant, *Ranunculus peltatus*. *Aquatic Botany*, 58, 113–120.

von Rintelen, K., Glaubrecht, M., Schubart, C.D., Wessel, A. et al. (2010). Adaptive radiation and ecological diversification of Sulawesi's ancient lake shrimps. *Evolution: International Journal of Organic Evolution*, 64, 3287–3299.

Vorosmarty, C.J., McIntyre, P.B., Gessner, M.O., Dudgeon, D. et al. (2010). Global threats to human water security and river biodiversity. *Nature*, 467, 555–561.

Wagenhoff, A., Clapcott, J.E., Lau, K.E. et al. (2017). Identifying congruence in stream assemblage thresholds in response to nutrient and sediment gradients for limit setting. *Ecological Applications*, 27, 469–484.

Wagner, M.A. & Reynolds, J.D. (2019). Salmon increase forest bird abundance and diversity. *PLoS One*, 14, e0210031.

Wagner, R., Marxsen, J., Zwick, P. & Cox, E.J. (eds). (2011). *Central European Stream Ecosystems: The Long-Term Study of the Breitenbach*. Wiley, Weinheim, Germany.

Wallace, J.B. & Anderson, N.H. (1996). Habitat, life history and behavioural adaptations of aquatic insects. In R.W. Merritt & K.W. Cummins (eds) *An Introduction to the Aquatic Insects of North America*. Kendall/Hunt, Dubuque, IA, 41–73.

Wallace, J.B., Webster, J.R. & Lowe, R.L. (1992). High-gradient streams of the Appalachians. In C.T. Hackney, S.M. Adams & W.A. Martin (eds) *Biodiversity of Southeastern United States Aquatic Communities*. Wiley, New York, 133–191.

Wallace, J.B., Eggert, S.L., Meyer, J.L. & Webster, J.R. (2015). Stream invertebrate productivity linked to forest subsidies: 37 stream-years of reference and experimental data. *Ecology*, 96, 1213–1228.

Wallace, R.L. & Snell, T.W. (2009). Rotifera. In J.H. Thorp & A.P. Covich (eds) *Ecology and Classification of North American Freshwater Invertebrates* (3rd edn). Elsevier, San Diego, 173–188.

Wallin, M.B., Audet, J. & Peacock, M. et al. (2020). Carbon dioxide dynamics in an agricultural headwater stream driven by hydrology and primary production. *Biogeosciences*, 17, 2487–2498.

Wallis, S.G., Young, P.C. & Beven, K.J. (1989). Experimental investigation of the aggregated dead zone model for longitudinal solute transport in stream channels. *Proceedings of the Institute of Civil Engineers*, 87, 1–22.

Walters, D.M., Jardine, T.D., Cade, B.S. et al. (2016). Trophic magnification of organic chemicals: a global synthesis. *Environmental Science & Technology*, 50, 4650–4658.

Wantzen, K. M., Yule, C.M., Mathooko, J.M. & Pringle, C.M. (2008). Organic matter processing in tropical streams. In D. Dudgeon (ed) *Tropical Stream Ecology*. Elsevier, London, 43–64.

Wantzen, K.M., Ballouche, A., Longuet, I., Bao, I. et al. (2016). River Culture: an eco-social approach to mitigate the biological and cultural diversity crisis in riverscapes. *Ecohydrology & Hydrobiology*, 16, 7–18.

Wantzen, K.M., Uehlinger, U., Van der Velde, G., Leuven, R., et al. (2022). The Rhine River Basin. In K. Tockner, C. Zarfl, C. Robinson & A.C. Benke (eds) *Rivers of Europe* (2nd edn). Elsevier, Amsterdam, 181–224.

Ward, J.V. (1985). Thermal characteristics of running waters. *Hydrobiologia*, 125, 31–46.

Ward, J.V. (1989) The four-dimensional nature of lotic ecosystems. *Journal of the North American Benthological Society*, 8, 2–8. https://doi.org/10.2307/1467397.

Ward, J.V. (1992a). River ecosystems. *Encyclopedia of Earth Systems Science*, 4. Academic Press, London.

Ward, J.V. (1992b). *Aquatic Insect Ecology. 1: Biology and Habitat*. Wiley & Sons, New York.

Ward, J.V. & Stanford, J.A. (1983). The serial discontinuity concept of lotic ecosystems. In T.D. Fontaine & S.M. Bartell (eds) *Dynamics of Lotic Ecosystems*. Ann Arbor Scientific Publishers, Ann Arbor, MI, 29–42.

Ward, J.V., Bretschko, G., Brunke, M., Danielopol, D. et al. (1998). The boundaries of river systems: the metazoan perspective. *Freshwater Biology*, 40, 531–569.

Ward, P.R.B., Anders, P.J., Minshall, G.W., Holderman, C. et al. (2018). Nutrient uptake during low-level fertilization of a large, seventh-order oligotrophic river. *Canadian Journal of Fisheries and Aquatic Sciences*, 75, 569–579.

Warfe, D.M., Jardine, T.D., Pettit, N.E., Hamilton, S.K et al. (2013). Productivity, disturbance and ecosystem size have no influence on food chain length in seasonally connected rivers. *PLoS ONE*, 8(6), e66240. doi:10.1371/journal.pone.0066240.

Waringer, J., Vitecek, S., Martini, J., Zittra, C. et al. (2020). Hydraulic stress parameters of a cased caddis larva (*Drusus biguttatus*) using spatio-temporally filtered velocity measurements. *Hydrobiologia*, 847, 3437–3451. doi: 10.1007/s10750-020-04349-0.

Waters, T.F (1966). Production rate, population density, and drift of a stream invertebrate. *Ecology*, 47, 595–604.

Waters, T.F. (1988). Fish production-benthos production relationships in trout streams. *Polskie Archiwum Hydrobiologii*, 35, 545–561.

Watson, D.J. & Balon, E.K. (1984). Ecomorphological analysis of fish taxocenes in rain-forest streams of northern Borneo, *Journal of Fish Biology*, 25, 371–384.

Webb, B. & Nobilis, F. (2007). Long-term changes in river temperature and the influence of climatic and hydrological factors. *Hydrological Sciences Journal*, 52, 74–85.

Webster, J. & Descals, E. (1981). Morphology, distribution and ecology of conidial fungi in freshwater habitats. In: G.T. Cole & B. Kendrick (eds) *The Biology of Conidial Fungi, Vol. 1*. Academic Press, New York, 295–355.

Webster, J.R. & Benfield, E.F. (1986). Vascular plant breakdown in freshwater ecosystems. *Annual Review of Ecology & Systematics*, 17, 567–594.

Webster, J.R. & Meyer, J.L (1997). Organic matter budgets for streams: a synthesis. *Journal of the North American Benthological Society*, 16, 141–161.

Webster, J.R., Wallace, J. & Benfield, E.F. (1995). Organic processes in streams of the eastern United States. In C.E. Cushing, G.W. Minshall & K.W. Cummins (eds) *River and Stream Ecosystems*. Elsevier, Amsterdam, 117–187.

Webster, J.R. Benfield, E.F., Ehrman, T.P., Schaeffer, M.A. et al. (1999). What happens to allochthonous material that falls into streams? A synthesis of new and published information from Coweeta. *Freshwater Biology*, 41, 687–705.

Webster J.R., Newbold, J.D & Thomas, S.A. (2022). The nutrient spiraling concept. In K. Tockner (ed) *Encyclopedia of Inland Waters* (2nd edn). Elsevier, Oxford, 244–248.

Weigand, A.M. & Macher, J.N. (2018). A DNA metabarcoding protocol for hyporheic freshwater meiofauna: evaluating highly degenerate COI primers and replication strategy. *Metabarcoding & Metagenomics* 2, e26869.

Weissenberger, J., Spatz, H.-Ch., Emanns, A. & Schwoerbel, J. (1991). Measurment of lift and drag forces in the mN range experienced by benthic arthropods at flow velocities below 1.2m s-1. *Freshwater Biology*, 25, 21–31.

Weitere, M., Erken, M., Majdi, N. et al. (2018). The food web perspective on aquatic biofilms. *Ecological Monographs*, doi 10.1002/ecm.1315.

Welcomme, R.L. (1979). *Fisheries Ecology of Floodplain Rivers*. Longman, London and New York.

Welcomme, R.L. (1985). *River Fisheries*. FAO Fisheries Technical Paper No. 262. Food and Agriculture Organization, Rome.

Wetzel, R.G. (1983). *Limnology* (2nd edn). Saunders, New York.

Whelan, M.J., Linstead, C. Worrall, F. et al. (2022). Is water quality in British rivers 'better than at any time since the end of the industrial revolution'? *Science of the Total Environment*, 843, 157014.

Whiles, M.R., Lips, K.R., Pringle, C.M., Kilham, S.S. et al. (2006). The effects of amphibian population declines on the structure and function of Neotropical stream ecosystems. *Frontiers in Ecology and the Environment*, 4, 27–34.

Whiles, M.R., Hall, R.O., Dodds, W.K., Verburg, P. et al. (2013). Disease-driven amphibian declines alter ecosystem processes in a tropical stream. *Ecosystems*, 16, 146–157.

White, D.S. & Brigham, W.U. (1996). Aquatic Coleoptera. In R.W. Merritt & K.W. Cummins (eds) *An Introduction to the Aquatic Insects of North America* (3rd edn). Kendall/Hunt, Dubuque, IA, 399–473.

Wiggins, G.B. (1973). A contribution to the biology of caddisflies (Trichoptera) in temporary pools. *Life Sciences Contributions of the Royal Ontario Museum*, 88, 1–28.

Wilco, C., Verberk, E.P., Siepel, H. & Esselink, H. (1998). Life-history strategies in freshwater macroinvertebrates. *Freshwater Biology*, 53, 1722–1738.

Wilcock, H.R., Nichols, R.A. & Hildrew, A.G. (2003). Genetic population structure, and neighbourhood population size estimates of the caddisfly *Plectrocnemia conspersa*. *Freshwater Biology*, 48, 1813–1824.

Wiley, M.J. & Kohler, S.L. (1980). Positioning changes of mayfly nymphs due to behavioral regulation of oxygen consumption. *Canadian Journal of Zoology*, 58, 618–622.

Wilkinson, J.L., Boxall, A.B.A., Kolpin, D.W. et al. (2022). Pharmaceutical pollution of the world's rivers. *Proceedings of the National Academy of Sciences*, 119, No. 8, e2113947119.

Williams, D.D., Williams, N. & Hogg, I. (1995). Life history plasticity of *Nemoura trisinosa* (Plecoptera: Nemouridae) along a permanent-temporary water habitat gradient. *Freshwater Biology*, **34**, 155–163.

Williams, H.C., Ormerod, S.J. & Bruford, M.W. (2006). Molecular systematics and phylogeography of the cryptic species complex *Baetis rhodani* (Ephemeroptera, Baetidae). *Molecular Phylogenetics and Evolution*, 40, 370–382.

Williams, R.J., White, C., Harrow, M.L. & Neal, C. (2000). Temporal and small-scale spatial variations of dissolved oxygen in the Rivers Thames, Pang and Kennett, UK. *Science of the Total Environment*, 251/252, 497–510.

Williams, W.D. (1976). Freshwater isopods (Asellidae) of North America. Water Pollution Control Series, U.S. Environmental Protection Agency, Cincinnati, Ohio, 45.

Williamson, T.J., Cross, W.F., Benstead, J.P., Gíslason, G.M., Hood, J.M., Huryn, A.D., Johnson, P.W. & Welter, J.R. (2016). Warming alters coupled carbon and nutrient cycles in experimental streams. *Global Change Biology*, 22, 2152–2164.

Winemiller, K. (2005). Life history strategies, population regulation and implications for fisheries management. *Canadian Journal of Fisheries and Aquatic Science*, 62, 872–885

Winemiller, K.O. (1989). Ontogenetic diet shifts and resource partitioning among piscivorous fishes in the Venezuelan llanos. *Environmental Biology of Fishes*, 26, 177–199.

Winemiller, K.O., Flecker, A.S. & Hoeinghaus, D.J. (2010). Patch dynamics and environmental heterogeneity in lotic ecosystems. *Journal of the North American Benthological Society*, 29, 84–99.

Winemiller, K.O., McIntyre, P., Castello, L. et al. (2016). Balancing hydropower and biodiversity in the Amazon, Congo and Mekong. *Science*, 6269, 128–129.

Winterbottom, J.H., Orton, S.E., Hildrew, A.G. & Lancaster, J. (1997a). Field experiments on flow refugia in streams. *Freshwater Biology*, 37, 569–580.

Winterbottom. J.H., Orton, S.E. & Hildrew, A.G. (1997b). Field experiments on the mobility of benthic invertebrates in a southern English stream. *Freshwater Biology*, 38, 37–47.

Winterbourn, M.J., Rounick, J.S. & Cowie, B. (1981). Are New Zealand stream ecosystems really different? *New Zealand Journal of Marine and Freshwater Research*, 10, 399–416.

Winterbourn, M.J., Cowie, B. & Rounick, J.S. (1984). Food resources and ingestion patterns of insects along a west coast, South Island, river system. *New Zealand Journal of Marine and Freshwater Research*, 18, 379–388.

Wohl, E. & Merritt, D. (2021). Bedrock channel morphology. *Geological Society of America Bulletin*, 113, 1205–1212.

Wohl, E., Bledsoe, B.P., Jacobson, R.B., Poff, N.L. et al. (2015). The Natural Sediment Regime in Rivers: Broadening the Foundation for Ecosystem Management *BioScience*, 65, 358–371.

Wolda, H. (1987). Seasonality and the community. In J. Gee & P. Giller (eds) *Organisation of Communities Past and Present*. Blackwell, Oxford, 69–98.

Wolters, J-W., Verdonschot, R.C., Schoelynck, J., Brion, N et al. (2018). Stable isotope measurements confirm consumption of submerged macrophytes by macroinvertebrates and fish taxa. *Aquatic Ecology*, 52(4), 269–280.

Woodiwiss, F.S. (1964). The biological system of stream classification used by the Trent River Board. *Chemistry & Industry*, 14, 443–447.

Woodward, G. & Hildrew, A.G. (2001). Invasion of a stream food web by a new top predator. *Journal of Animal Ecology*, 70, 273–288.

Woodward G. & Hildrew, A.G. (2002). Food web structure in riverine landscapes. *Freshwater Biology*, 47, 777–798.

Woodward, G. & Warren, P. (2007). Body size and predatory interactions in freshwaters: scaling from individuals to communities. In A.G. Hildrew, D. Raffaelli & V. Edmonds-Brown (eds) *Body Size: The Structure and Function of Aquatic Ecosystems*. Cambridge University Press, Cambridge, 98–117.

Woodward, G., Blanchard, J., Lauridsen, R.B., Edwards, F. et al. (2010). Individual-based food webs: species identity, body size and sampling effects. *Advances in Ecological Research*, 43, 211–266.

Woodward, G., Bonada, N., Brown, L., Death, R.G. et al. (2016). The effects of climatic fluctuations and extreme events on running water ecosystems. *Philosophical Transactions of the Royal Society B*, 371, 20150274.

Woodward, G., Bonada, N., Feeley, H.B. & Giller, P.S. (2015). Resilience of a stream community to extreme climatic events and long term recovery from a catastrophic flood. *Freshwater Biology*, 60, 2497–2510.

Woodward, G., Gessner, M.O., Giller, P.S. et al. (2012). Continental scale effects of nutrient pollution on stream ecosystem functioning. *Science*, 336, 1438–1440.

Woodward, G., Perkins, D.M. & Brown, L. (2010). Climate change and freshwater ecosystems: impacts across multiple levels of organization. *Philosophical Transactions of the Royal Society of London B, Biological Sciences*, 365, 2093–2106.

Woodward, G. Thompson, R, Townsend, C.R. & Hildrew, A.G. (2005). Pattern and process in food webs: evidence from running waters. In A. Belgrano, U.M. Scharler, J. Dunne & R.E. Ulanowicz (eds) *Aquatic Food Webs: An Ecosystem Approach*. Oxford University Press, Oxford, 51–56.

Woodward, G., Papantoniou, G., Edwards, F. & Lauridsen, R.B. (2008). Trophic trickles and cascades in a complex food web: impacts of a keystone predator on stream community structure and ecosystem processes. *Oikos*, 117, 683–692.

Woodward, G., Speirs, D.C. & Hildrew, A.G. (2005). Quantification and resolution of a complex, size-structured food web. *Advances in Ecological Research*, 36, 85–135.

Wooster, D. (1994). Predator impacts on stream benthic prey. *Oecologia*, 99, 7–15.

Wooton, J.T., Parker, M & Power, M. (1996). Effects of disturbance on river food webs. *Science*, 273, 1558–1561.

Worrall, F., Howden, N.J.K. & Burt, T.P. (2014). A method of estimating in-stream residence time of water in rivers. *Journal of Hydrology*, 512, 274–284.

Wotton, R.S. (1988). Very high secondary production in a lake outlet. *Freshwater Biology*, 20, 341–346.

Wotton, R.S. (ed) (1994). *The Biology of Particles in Aquatic Systems* (2nd edn). Taylor & Francis Inc., Boca Raton, FL.

Wotton, R.S. (1996). Colloids, bubbles, and aggregates – a perspective on their role in suspension feeding. *Journal of the North American Benthological Society*, 15, 127–135.

Wotton, R. S., Malmqvist, B., Muotka, T. & Larsson, K. (1998). Fecal pellets from a dense aggregation of suspension feeders: an example of ecosystem engineering in a stream. *Limnology and Oceanography* 43, 719–725.

Wright, J.F., Cameron, A.C., Hiley, P.D. & Berrie, A.D. (1982). Seasonal changes in biomass of macrophytes on shaded and unshaded sections of the River Lamboum, England. *Freshwater Biology* 12, 271–283.

WWF (2020). Living Planet Report 2020: Bending the curve of biodiversity loss. Almond, R.E.A., Grooten, M. & Petersen, T (eds). WWF, Gland, Switzerland.

Wymore, A.S., Coble, A.A., Rodriguez-Cardona, B. & McDowell, W.H. (2016). Nitrate uptake across biomes and the influence of elemental stoichiometry: a new look at LINX II. *Global Biogeochemical Cycles*, 30, 1183–1191.

Yao, J., Colas, F., Solimini, A.G. et al. (2017). Macroinvertebrate community traits and nitrate removal in stream sediments. *Freshwater Biology*, 62, 929–944.

Yeakley, J.A., Ervin, D., Chang, H. et al. (2016). Ecosystem services of streams and rivers. In D.J. Gilvear, M.T. Greenwood, M.C. Thoms & P.J. Wood (eds) *River Science: Research and Management for the 21st Century*. John Wiley & Sons, Chichester, 335–352.

Yoerg, S.I. (1994). Development of foraging behaviour in the Eurasian dipper, *Cinclus cinclus*, from fledging until dispersal. *Animal Behaviour*, 47(3), 577–588. https://doi.org/10.1006/anbe.1994.1081.

Young, R.G., Quarterman, A.J., Eyles, R.F., Smith, R.A. et al. (2005). Water quality and thermal regime of the Motueka River: influences of land cover, geology and position in the catchment. *New Zealand Journal of Marine and Freshwater Research*, 39, 803–825.

Yule, C.M., Leong, M.Y., Liew, K.C. et al. (2009) Shredders in Malaysia: abundance and richness are higher in cool upland tropical streams. *Journal of the North American Benthological Society*, 28, 404–415.

Yvon-Durocher, G., Jones, J. I., Trimmer, M. et al. (2010). Warming alters the metabolic balance of ecosystems. *Philosophical Transaction of the Royal Society (B)*, 365, 2117–2126. doi: 10.1098/rstb.2010.0038.

Zarat, T.M. & Rand, A.S. (1971). Competition in tropical stream fishes: support for the competitive exclusion principle. *Ecology*, 52, 336–342.

Zarfl, C., Lumsden, A.E., Berlekamp, J. et al. (2015). A global boom in hydropower dam construction. *Aquatic Sciences*, 77, 161–170.

Zeglin, L.H. (2015). Stream microbial diversity in response to environmental changes: review and synthesis of existing research. *Frontiers in Microbiology*, 6, 454 doi:10.3389/fmicb.2015.00454.

Zeug, S.C. & Winemiller, K.O. (2008). Evidence supporting the importance of terrestrial carbon in a large-river food web. *Ecology*, 89, 1733–1743.

Zhang, L., Xia, X., Liu, S. et al. (2020). Significant methane ebullition from alpine permafrost rivers on the East Qinghai–Tibet Plateau. *Nature Geoscience*, 13, 1–6.

Zhang, Y. & Malmqvist, B. (1996). Relationships between labral fan morphology and habitat in North Swedish blackfly larvae (Diptera: Simulidae). *Biological Journal of the Linnean Society*, 59, 261–280.

Zhu, Y., Jones, J.I., Collins, A. et al. (2022). Separating natural from human-enhanced methane emissions in headwater streams. *Nature Communications*, doi.org/10.1038/s41467-022-31559-y.

Index

Note: Tables, figures, and boxes are indicated by an italic *t*, *f*, and *b* following the page number. Rivers and streams are indexed under their country only if unnamed. For countries, see also their named rivers and streams.